S*pringer* **M***onographs* in **M***athematics*

For further volumes:
http://www.springer.com/series/3733

Katsuro Sakai

Geometric Aspects of General Topology

Springer

Katsuro Sakai
Department of Mathematics
Faculty of Engineering
Kanagawa University
Yokohama, Kanagawa
Japan

ISSN 1439-7382
ISBN 978-4-431-54396-1 ISBN 978-4-431-54397-8 (eBook)
DOI 10.1007/978-4-431-54397-8
Springer Tokyo Heidelberg New York Dordrecht London

Library of Congress Control Number: 2013942182

Mathematics Subject Classification: 54-01; 54E35; 54F45; 54C55; 57-01; 57N17; 57Q05; 55M10; 55M15; 55U05

Printed on acid-free paper

Springer is part of Springer Science+Business Media (www.springer.com)

We can imagine and consider many mathematical concepts, such as numbers, spaces, maps, dimensions, etc., that can be indefinitely extended beyond infinity in our minds. Contemplating our mathematical ability in such a manner, I can recall this phrase from the Scriptures:

Everything he has made pretty in its time. Even time indefinite he has put in their heart, that mankind may never find out the work that the true God has made from the start to the finish.—Ecclesiastes 3:11

May our Maker be glorified! Our brain is the work of his hands, as in Psalms 100:3, *Know that Jehovah is God. It is he that has made us, and not we ourselves.* There are many reasons to give thanks to God. Our mathematical ability is one of them.

Preface

This book is designed for graduates studying Dimension Theory, ANR Theory (Theory of Retracts), and related topics. As is widely known, these two theories are connected with various fields in Geometric Topology as well as General Topology. So, for graduate students who wish to research subjects in General and Geometric Topology, understanding these theories will be valuable. Some excellent texts on these theories are the following:

- W. Hurewicz and H. Wallman, *Dimension Theory* (Princeton Univ. Press, Princeton, 1941)
- K. Borsuk, *Theory of Retracts*, MM **44** (Polish Sci. Publ., Warsaw, 1966)
- S.-T. Hu, *Theory of Retracts* (Wayne State Univ. Press, Detroit, 1965)

However, these classical texts must be updated. This is the purpose of the present book.

A comprehensive study of Dimension Theory may refer to the following book:

- R. Engelking, *Theory of Dimensions, Finite and Infinite*, SSPM **10** (Heldermann Verlag, Lembo, 1995)

Engelking's book, however, lacks results relevant to Geometric Topology. In this or any other textbook, no proof is given that $\dim X \times \mathbf{I} = \dim X + 1$ for a metrizable space X,[1] and no example illustrates the difference between the small and large inductive dimensions or a hereditarily infinite-dimensional space (i.e., an infinite-dimensional space that has no finite-dimensional subspaces except for 0-dimensional subspaces).[2]

In the 1980s and 1990s, famous longstanding problems from Dimension Theory and ANR Theory were finally resolved. In the process, it became clear that

[1] This proof can be found in Kodama's appendix of the following book:

- K. Nagami, *Dimension Theory* (Academic Press, Inc., New York, 1970)

[2] As will be mentioned later, a hereditarily infinite-dimensional space is treated in the book of J. van Mill: *Infinite-Dimensional Topology*.

these theories are linked with others. In Dimension Theory, the Alexandroff Problem had long remained unsolved. This problem queried the existence of an infinite-dimensional space whose cohomological dimension is finite. On the other hand, the CE Problem arose as a fascinating question in Shape Theory that asked whether there exists a cell-like map of a finite-dimensional space onto an infinite-dimensional space. In the 1980s, it was shown that these two problems are equivalent. Finally, in 1988, by constructing an infinite-dimensional compact metrizable space whose cohomological dimension is finite, A.N. Dranishnikov solved the Alexandroff Problem.

On the other hand, in ANR Theory, for many years it was unknown whether a metrizable topological linear space is an AR (or more generally, whether a locally equi-connected metrizable space is an ANR). In 1994, using a cell-like map of a finite-dimensional compact manifold onto an infinite-dimensional space, R. Cauty constructed a separable metrizable topological linear space that is not an AR. These results are discussed in the latter half of the final chapter and provide an understanding of how deeply these theories are related to each other. This is also the purpose of this book.

The notion of simplicial complexes is useful tool in Topology, and indispensable for studying both Theories of Dimension and Retracts. There are many textbooks from which we can gain some knowledge of them. Occasionally, we meet non-locally finite simplicial complexes. However, to the best of the author's knowledge, no textbook discusses these in detail, and so we must refer to the original papers. For example, J.H.C. Whitehead's theorem on small subdivisions is very important, but its proof cannot be found in any textbook. This book therefore properly treats non-locally finite simplicial complexes. The homotopy type of simplicial complexes is usually discussed in textbooks on Algebraic Topology using CW complexes, but we adopt a geometrical argument using simplicial complexes, which is easily understandable.

As prerequisites for studying infinite-dimensional manifolds, Jan van Mill provides three chapters on simplicial complexes, dimensions, and ANRs in the following book:

- J. van Mill, *Infinite-Dimensional Topology, Prerequisites and Introduction*, North-Holland Math. Library **43** (Elsevier Sci. Publ. B.V., Amsterdam, 1989)

These chapters are similar to the present book in content, but they are introductory courses and restricted to separable metrizable spaces. The important results mentioned above are not treated except for an example of a hereditarily infinite-dimensional space. Moreover, one can find an explanation of the Alexandroff Problem and the CE Problem in Chap. 3 of the following book:

- A. Chigogidze, *Inverse Spectra*, North-Holland Math. Library **53** (Elsevier Sci. B.V., Amsterdam, 1996)

Unfortunately, this book is, however, inaccessible for graduate students.

The present text has been in use by the author for his graduate class at the University of Tsukuba. Every year, a lecture has been given based on some topic selected from this book except the final chapter, and the same material has been used for an undergraduate seminar. Readers are required to finish the initial courses of Set Theory and General Topology. Basic knowledge of Linear Algebra is also a prerequisite. Except for the latter half of the final chapter, this book is self-contained.

Chapter 2 develops the general material relating to topological spaces appropriate for graduate students. It provides a supplementary course for students who finished an undergraduate course in Topology. We discuss paracompact spaces and some metrization theorems for non-separable spaces that are not treated in a typical undergraduate course.[3] This chapter also contains Michael's theorem on local properties, which can be applied in many situations. We further discuss the direct limits of towers (increasing sequences) of spaces, which are appear in Geometric and Algebraic Topology.[4] A non-Hausdorff direct limit of a closed tower of Hausdorff spaces is included. The author has not found any literature representing such an example. The limitation topology on the function spaces is also discussed.

Chapter 3 is devoted to topological linear spaces and convex sets. There are many good textbooks on these subjects. This chapter represents a short course on fundamental results on them. First, we establish the existing relations between these objects and to General and Geometric Topology. Convex sets are then discussed in detail. This chapter also contains Michael's selection theorem. Moreover, we show the existence of free topological spaces.

In Chap. 4, simplicial complexes are treated without assuming local finiteness. As mentioned above, we provide proof of J.H.C. Whitehead's theorem on small subdivisions. The simplicial mapping cylinder is introduced and applied to prove the Whitehead–Milnor theorem on the homotopy type of simplicial complexes. It is also applied to prove that every weak homotopy equivalence between simplicial complexes is a homotopy equivalence. The inverse limits of inverse sequences are also discussed, and it is shown that every completely metrizable space is homeomorphic to locally finite-dimensional simplicial complexes with the metric topology. These results cannot be found in any other book dealing with simplicial complexes but are buried in old journals. D.W. Henderson established the metric topology version of the Whitehead theorem on small subdivisions, but his proof is valid only for locally finite-dimensional simplicial complexes. Here we offer a complete proof without the assumption of local finite-dimensionality. Knowledge of homotopy groups is not required, even when weak homotopy equivalences are

---

[3]These subjects are discussed in Munkres' book, now a very popular textbook at the *senior* or the *first-year graduate* level:

- J.R. Munkres, *Topology, 2nd ed.* (Prentice Hall, Inc., Upper Saddle River, 2000)

[4]The direct limits are discussed in Appendix of Dugundji's book:

- J. Dugundji, *Topology* (Allyn and Bacon, Inc., Boston, 1966)

But, they are not discussed even in Engelking's book, a comprehensive reference book for General Topology:

- R. Engelking, *General Topology, Revised and completed edition*, SSPM **6** (Heldermann Verlag, Berlin, 1989) discussed. However, we do review homotopy groups in Appendix 4.14 because they are helpful in the second half of Chap. 7.

Chapters 5 and 6 are devoted to Dimension Theory and ANR Theory, respectively. We prove basic results and fundamental theorems on these theories. The contents are very similar to Chaps. 5 and 6 of van Mill's "Infinite-Dimensional Topology". However, as mentioned previously, we do not restrict ourselves to separable metrizable spaces and instead go on to prove further results.

In Chap. 5, we describe a non-separable metrizable space such that the large inductive dimension does not coincide with the small inductive dimension. As mentioned above, such an example is not treated in any other textbook on Dimension Theory (not even Engelking's book). Here, we present Kulesza's example with Levin's proof. The transfinite inductive dimension is also discussed, which is not treated in van Mill's book. Further, we prove that every completely metrizable space with dimension $\leq n$ is homeomorphic to the inverse limit of an inverse sequence of metric simplicial complexes with dimension $\leq n$. Finally, hereditarily infinite-dimensional spaces are discussed based on van Mill's book.

In Chap. 6, we discuss several topics that are not treated in van Mill's book or in the two classical books by Hu and Borsuk mentioned above. Following are examples of such topics: uniform ANRs in the sense of Michael and its completion; Kozlowski's theorem that the metrizable range of a fine homotopy equivalence of an ANR is also an ANR; Cauty's characterization, with Sakai's proof, that a metrizable space is an ANR if and only if every open set has the homotopy type of an ANR; Haver's theorem that every countable-dimensional locally contractible metrizable space is an ANR; and Bothe's theorem, with Kodama's proof, that every n-dimensional metrizable space can be embedded in an $(n+1)$-dimensional AR as a closed set.

In Chap. 7, cell-like maps and related topics are discussed. The first half is self-contained, but the second half is not because some algebraic results are necessary. In the first half, we examine the existing relations between cell-like maps, soft maps, fine homotopy equivalences, etc. The second half is devoted to related topics. In particular, the CE Problem is explained and Cauty's example is presented. Note that Chigogidze's "Inverse Spectra" is the only book dealing with soft maps and provides an explanation of the Alexandroff Problem and the CE Problem.

In the second half of Chap. 7, using the K-theory result of Adams, we present the Taylor example. Eilenberg–MacLane spaces are usually constructed as CW complexes, but here they are constructed as simplicial complexes. To avoid using cohomology, we define the cohomological dimension geometrically. By applying the cohomological dimension, we can prove the equality $\dim X \times \mathbf{I} = \dim X + 1$ for every metrizable space X. We also discuss the Alexandroff Problem and the CE Problem as mentioned above. The equivalence of these problems is proved. Next, we describe the Dydak–Walsh example that gives an affirmative answer to the Alexandroff Problem. However, this part of the text is not self-contained. As a corollary, we can answer the CE Problem, i.e., we can obtain a cell-like mapping of a finite-dimensional compact manifold onto an infinite-dimensional compactum. We also present Cauty's example, i.e., a metrizable topological linear space that is not an absolute extensor. In the proof, we need the above cell-like mapping to be open, and we therefore use Walsh's open mapping approximation theorem. A proof of Walsh's theorem is beyond the scope of this book.

The author would like to express his sincere appreciation to his teacher, Professor Yukihiro Kodama, who introduced him to Shape Theory and Infinite-Dimensional Topology and warmly encouraged him to persevere. He owes his gratitude to Ross Geoghegan for improving the written English text. He is also grateful to Haruto Ohta, Taras Banakh and Zhongqiang Yang for their valuable comments and suggestions. Finally, he also warmly thanks his graduate students, Yutaka Iwamoto, Yuji Akaike, Shigenori Uehara, Masayuki Kurihara, Masato Yaguchi, Kotaro Mine, Atsushi Yamashita, Minoru Nakamura, Atsushi Kogasaka, Katsuhisa Koshino, and Hanbiao Yang for their careful reading and helpful comments.

Katsuro Sakai
Tsukuba, Japan
December 2012

This work was supported by JSPS KAKENHI Grant Number 22540063.

Contents

Chapter 1
Preliminaries

The reader should have finished a first course in Set Theory and General Topology; basic knowledge of Linear Algebra is also a prerequisite. In this chapter, we introduce some terminology and notation. Additionally, we explain the concept of Banach spaces contained in the product of real lines.

1.1 Terminology and Notation

For the standard sets, we use the following notation:

- $\mathbb{N}$ — the set of natural numbers (i.e., positive integers);
- $\omega = \mathbb{N} \cup \{0\}$ — the set of non-negative integers;
- $\mathbb{Z}$ — the set of integers;
- $\mathbb{Q}$ — the set of rationals;
- $\mathbb{R} = (-\infty, \infty)$ — the real line with the usual topology;
- $\mathbb{C}$ — the complex plane;
- $\mathbb{R}_+ = [0, \infty)$;
- $\mathbf{I} = [0, 1]$ — the unit closed interval.

A (topological) **space** is assumed to be **Hausdorff** and a **map** is a **continuous** function. A **singleton** is a space consisting of one point, which is also said to be **degenerate**. A space is said to be **non-degenerate** if it is not a singleton. Let X be a space and $A \subset X$. We denote

- $\mathrm{cl}_X A$ (or $\mathrm{cl}\, A$) — the closure of A in X;
- $\mathrm{int}_X A$ (or $\mathrm{int}\, A$) — the interior of A in X;
- $\mathrm{bd}_X A$ (or $\mathrm{bd}\, A$) — the boundary of A in X;
- id_X (or id) — the identity map of X.

For spaces X and Y,

- $X \approx Y$ means that X and Y are homeomorphic.

K. Sakai, *Geometric Aspects of General Topology*, Springer Monographs in Mathematics,
DOI 10.1007/978-4-431-54397-8_1, © Springer Japan 2013

Given subspaces $X_1, \ldots, X_n \subset X$ and $Y_1, \ldots, Y_n \subset Y$,

- $(X, X_1, \ldots, X_n) \approx (Y, Y_1, \ldots, Y_n)$ means that there exists a homeomorphism $h : X \to Y$ such that $h(X_1) = Y_1, \ldots, h(X_n) = Y_n$;
- $(X, x_0) \approx (Y, y_0)$ means $(X, \{x_0\}) \approx (Y, \{y_0\})$.

We call (X, x_0) a **pointed space** and x_0 its **base point**.

For a set Γ, the cardinality of Γ is denoted by $\operatorname{card}\Gamma$. The **weight** $w(X)$, the **density** $\operatorname{dens} X$, and the **cellurality** $c(X)$ of a space X are defined as follows:

- $w(X) = \min\{\operatorname{card}\mathcal{B} \mid \mathcal{B} \text{ is an open basis for } X\}$;
- $\operatorname{dens} X = \min\{\operatorname{card} D \mid D \text{ is a dense set in } X\}$;
- $c(X) = \sup\{\operatorname{card}\mathcal{G} \mid \mathcal{G} \text{ is a pair-wise disjoint open collection}\}$.

As is easily observed, $c(X) \leq \operatorname{dens} X \leq w(X)$ in general. If X is metrizable, all these cardinalities coincide.

Indeed, let D be a dense set in X with $\operatorname{card} D = \operatorname{dens} X$, and $\mathcal{G}$ be a pairwise disjoint collection of non-empty open sets in X. Since each $G \in \mathcal{G}$ meets D, we have an injection $g : \mathcal{G} \to D$, hence $\operatorname{card}\mathcal{G} \leq \operatorname{card} D = \operatorname{dens} X$. It follows that $c(X) \leq \operatorname{dens} X$. Now, let $\mathcal{B}$ be an open basis for X with $\operatorname{card}\mathcal{B} = w(X)$. By taking any point $x_B \in B$ from each $B \in \mathcal{B}$, we have a dense set $\{x_B \mid B \in \mathcal{B}\}$ in X, which implies $\operatorname{dens} X \leq w(X)$.

When X is metrizable, we show the converse inequality. The case $\operatorname{card} X < \aleph_0$ is trivial. We may assume that $X = (X, d)$ is a metric space with $\operatorname{diam} X \geq 1$ and $\operatorname{card} X \geq \aleph_0$. Let D be a dense set in X with $\operatorname{card} D = \operatorname{dens} X$. Then, $\{\mathrm{B}(x, 1/n) \mid x \in D,\ n \in \mathbb{N}\}$ is an open basis for X, which implies $w(X) \leq \operatorname{dens} X$. For each $n \in \mathbb{N}$, using Zorn's Lemma, we can find a maximal 2^{-n}-discrete subset $X_n \subset X$, i.e., $d(x, y) \geq 2^{-n}$ for every pair of distinct points $x, y \in X_n$. Then, $\mathcal{G}_n = \{\mathrm{B}(x, 2^{-n-1}) \mid x \in X_n\}$ is a pairwise disjoint open collection, and hence we have $\operatorname{card} X_n = \operatorname{card}\mathcal{G}_n \leq c(X)$. Observe that $X_* = \bigcup_{n\in\mathbb{N}} X_n$ is dense in X, which implies $\sup_{n\in\mathbb{N}} \operatorname{card} X_n = \operatorname{card} X_* \geq \operatorname{dens} X$. Therefore, $c(X) \geq \operatorname{dens} X$.

For the product space $\prod_{\gamma\in\Gamma} X_\gamma$, the γ-coordinate of each point $x \in \prod_{\gamma\in\Gamma} X_\gamma$ is denoted by $x(\gamma)$, i.e., $x = (x(\gamma))_{\gamma\in\Gamma}$. For each $\gamma \in \Gamma$, the projection $\mathrm{pr}_\gamma : \prod_{\gamma\in\Gamma} X_\gamma \to X_\gamma$ is defined by $\mathrm{pr}_\gamma(x) = x(\gamma)$. For $\Lambda \subset \Gamma$, the projection $\mathrm{pr}_\Lambda : \prod_{\gamma\in\Gamma} X_\gamma \to \prod_{\lambda\in\Lambda} X_\lambda$ is defined by $\mathrm{pr}_\Lambda(x) = x|\Lambda\ (= (x(\lambda))_{\lambda\in\Lambda})$. In the case that $X_\gamma = X$ for every $\gamma \in \Gamma$, we write $\prod_{\gamma\in\Gamma} X_\gamma = X^\Gamma$. In particular, $X^{\mathbb{N}}$ is the product space of countable infinite copies of X. When $\Gamma = \{1, \ldots, n\}$, $X^\Gamma = X^n$ is the product space of n copies of X. For the product space $X \times Y$, we denote the projections by $\mathrm{pr}_X : X \times Y \to X$ and $\mathrm{pr}_Y : X \times Y \to Y$.

A compact metrizable space is called a **compactum** and a connected compactum is called a **continuum**.[1] For a metrizable space X, we denote

- $\mathrm{Metr}(X)$ — the set of all admissible metrics of X.

Now, let $X = (X, d)$ be a metric space, $x \in X$, $\varepsilon > 0$, and $A, B \subset X$. We use the following notation:

[1]Their plurals are **compacta** and **continua**, respectively.

- $\mathrm{B}_d(x,\varepsilon) = \{y \in X \mid d(x,y) < \varepsilon\}$ — the ε-neighborhood of x in X (or the open ball with center x and radius ε);
- $\overline{\mathrm{B}}_d(x,\varepsilon) = \{y \in X \mid d(x,y) \leq \varepsilon\}$ — the closed ε-neighborhood of x in X (or the closed ball with center x and radius ε);
- $\mathrm{N}_d(A,\varepsilon) = \bigcup_{x\in A} \mathrm{B}_d(x,\varepsilon)$ — the ε-neighborhood of A in X;
- $\operatorname{diam}_d A = \sup\{d(x,y) \mid x,y \in A\}$ — the diameter of A;
- $d(x,A) = \inf\{d(x,y) \mid y \in A\}$ — the distance of x from A;
- $\operatorname{dist}_d(A,B) = \inf\{d(x,y) \mid x \in A,\ y \in B\}$ — the distance of A and B.

It should be noted that $\mathrm{N}_d(\{x\},\varepsilon) = \mathrm{B}_d(x,\varepsilon)$ and $d(x,A) = \operatorname{dist}_d(\{x\},A)$. For a collection $\mathcal{A}$ of subsets of X, let

- $\operatorname{mesh}_d \mathcal{A} = \sup\{\operatorname{diam}_d A \mid A \in \mathcal{A}\}$ — the mesh of $\mathcal{A}$.

If there is no possibility of confusion, we can drop the subscript d and write $\mathrm{B}(x,\varepsilon)$, $\overline{\mathrm{B}}(x,\varepsilon)$, $\mathrm{N}(A,\varepsilon)$, $\operatorname{diam} A$, $\operatorname{dist}(A,B)$, and $\operatorname{mesh}\mathcal{A}$.

The standard spaces are listed below:

- $\mathbb{R}^n$ — the n-dimensional Euclidean space with the norm
$$\|x\| = \sqrt{x(1)^2 + \cdots + x(n)^2},$$
$\mathbf{0} = (0,\ldots,0) \in \mathbb{R}^n$ — the origin, the zero vector or the zero element,
$\mathbf{e}_i \in \mathbb{R}^n$ — the unit vector defined by $\mathbf{e}_i(i) = 1$ and $\mathbf{e}_i(j) = 0$ for $j \neq i$;
- $\mathbf{S}^{n-1} = \{x \in \mathbb{R}^n \mid \|x\| = 1\}$ — the unit $(n-1)$-sphere;
- $\mathbf{B}^n = \{x \in \mathbb{R}^n \mid \|x\| \leq 1\}$ — the unit closed n-ball;
- $\Delta^n = \{x \in (\mathbb{R}_+)^{n+1} \mid \sum_{i=1}^{n+1} x(i) = 1\}$ — the standard n-simplex;
- $\boldsymbol{Q} = [-1,1]^{\mathbb{N}}$ — the Hilbert cube;
- $\boldsymbol{s} = \mathbb{R}^{\mathbb{N}}$ — the space of sequences;
- $\mu^0 = \{\sum_{i=1}^{\infty} 2x_i/3^i \mid x_i \in \{0,1\}\}$ — the Cantor (ternary) set;
- $\nu^0 = \mathbb{R} \setminus \mathbb{Q}$ — the space of irrationals;
- $\mathbf{2} = \{0,1\}$ — the discrete space of two points.

Note that $\mathbf{S}^{n-1}$, $\mathbf{B}^n$, and Δ^n are not product spaces, even though the same notations are used for product spaces. The indexes $n-1$ and n represent their dimensions (the indexes of μ^0 and ν^0 are identical).

As is well-known, the countable product $\mathbf{2}^{\mathbb{N}}$ of the discrete space $\mathbf{2} = \{0,1\}$ is homeomorphic to the Cantor set μ^0 by the correspondence:

$$x \mapsto \sum_{i\in\mathbb{N}} \frac{2x(i)}{3^i}.$$

On the other hand, the countable product $\mathbb{N}^{\mathbb{N}}$ of the discrete space $\mathbb{N}$ of natural numbers is homeomorphic to the space ν^0 of irrationals. In fact, $\mathbb{N}^{\mathbb{N}} \approx (0,1) \setminus \mathbb{Q} \approx (-1,1) \setminus \mathbb{Q} \approx \nu^0$. These three homeomorphisms are given as follows:

$$x \mapsto \cfrac{1}{x(1) + \cfrac{1}{x(2) + \cfrac{1}{x(3) + \cfrac{1}{\ddots}}}}; \quad t \mapsto 2t - 1; \quad s \mapsto \frac{s}{1 - |s|}.$$

That the first correspondence is a homeomorphism can be verified as follows: for each $n \in \mathbb{N}$, let $a_n : \mathbb{N}^{\mathbb{N}} \to \mathbf{I}$ be a map defined by

$$a_n(x) = \cfrac{1}{x(1) + \cfrac{1}{x(2) + \cfrac{1}{\ddots + \cfrac{1}{x(n)}}}}.$$

Then, $0 < a_2(x) < a_4(x) < \cdots < a_3(x) < a_1(x) \leq 1$. Using the fact shown below, we can conclude that the first correspondence $\mathbb{N}^{\mathbb{N}} \ni x \mapsto \alpha(x) = \lim_{n\to\infty} a_n(x) \in (0, 1)$ is well-defined and continuous.

Fact. *For every $m > n$, $|a_n(x) - a_m(x)| < (n + 1)^{-1}$.*

This fact can be shown by induction on $n \in \mathbb{N}$. First, observe that

$$|a_1(x) - a_2(x)| = \frac{1}{x(1)(x(1)x(2) + 1)} < 1/2,$$

which implies the case $n = 1$. When $n > 1$, for each $x \in \mathbb{N}^{\mathbb{N}}$, define $x^* \in \mathbb{N}^{\mathbb{N}}$ by $x^*(i) = x(i + 1)$. By the inductive assumption, $|a_{n-1}(x^*) - a_{m-1}(x^*)| < n^{-1}$ for $m > n$, which gives us

$$\begin{aligned} |a_n(x) - a_m(x)| &= \frac{|a_{n-1}(x^*) - a_{m-1}(x^*)|}{(x(1) + a_{n-1}(x^*))(x(1) + a_{m-1}(x^*))} \\ &\leq \frac{|a_{n-1}(x^*) - a_{m-1}(x^*)|}{(1 + a_{n-1}(x^*))(1 + a_{m-1}(x^*))} \\ &< \frac{|a_{n-1}(x^*) - a_{m-1}(x^*)|}{1 + |a_{n-1}(x^*) - a_{m-1}(x^*)|} \\ &\leq \frac{n^{-1}}{1 + n^{-1}} = \frac{1}{n+1}. \end{aligned}$$

Let $t = q_1/q_0 \in (0, 1) \cap \mathbb{Q}$, where $q_1 < q_0 \in \mathbb{N}$. Since $q_0/q_1 = t^{-1} > 1$, we can choose $x_1 \in \mathbb{N}$ so that $x_1 \leq q_0/q_1 < x_1 + 1$. Then, $1/(x_1 + 1) < t \leq 1/x_1$. If $t \neq 1/x_1$, then $x_1 < q_0/q_1$, and hence $t^{-1} = q_0/q_1 = x_1 + q_2/q_1$ for some $q_2 \in \mathbb{N}$ with $q_2 < q_1$. Now, we choose $x_2 \in \mathbb{N}$ so that $x_2 \leq q_1/q_2 < x_2 + 1$. Thus, $x_1 + 1/(x_2 + 1) < x_1 + q_2/q_1 \leq x_1 + 1/x_2$, so $1/(x_1 + 1/x_2) \leq t < 1/(x_1 + 1/(x_2 + 1))$. If $t \neq 1/(x_1 + 1/x_2)$, then $x_2 < q_1/q_2$. Similarly, we write $q_1/q_2 = x_2 + q_3/q_2$, where $q_3 \in \mathbb{N}$ with $q_3 < q_2$ $(< q_1)$, and choose $x_3 \in \mathbb{N}$ so that $x_3 \leq q_2/q_3 < x_3 + 1$. Then, $1/(x_1 + 1/(x_2 + 1/x_3)) \leq t < 1/(x_1 + 1/(x_2 + 1/(x_3 + 1)))$. This process has only a finite number of steps (at most q_1 steps). Thus, we have the following unique representation:

$$t = \cfrac{1}{x_1 + \cfrac{1}{x_2 + \cfrac{1}{\ddots + \cfrac{1}{x_n}}}}, \quad x_1, \ldots, x_n \in \mathbb{N}.$$

It follows that $\alpha(\mathbb{N}^{\mathbb{N}}) \subset (0,1) \setminus \mathbb{Q}$.

For each $t \in (0,1) \setminus \mathbb{Q}$, choose $x_1 \in \mathbb{N}$ so that $x_1 < t^{-1} < x_1 + 1$. Then, $1/(x_1+1) < t < 1/x_1$ and $t^{-1} = x_1 + t_1$ for some $t_1 \in (0,1) \setminus \mathbb{Q}$. Next, choose $x_2 \in \mathbb{N}$ so that $x_2 < t_1^{-1} < x_2+1$. Thus, $x_1+1/(x_2+1) < x_1+t_1 < x_1+1/x_2$, and so $1/(x_1+1/x_2) < t < 1/(x_1+1/(x_2+1))$. Again, write $t_1^{-1} = x_2+t_2, t_2 \in (0,1)\setminus\mathbb{Q}$, and choose $x_3 \in \mathbb{N}$ so that $x_3 < t_2^{-1} < x_3+1$. Then, $1/(x_1+1/(x_2+1/(x_3+1))) < t < 1/(x_1+1/(x_2+1/x_3))$. We can iterate this process infinitely many times. Thus, there is the unique $x = (x_n)_{n\in\mathbb{N}} \in \mathbb{N}^{\mathbb{N}}$ such that $a_{2n}(x) < t < a_{2n+1}(x)$ for each $n \in \mathbb{N}$, where $\alpha(x) = \lim_{n\to\infty} a_n(x) = t$. Therefore, $\alpha : \mathbb{N}^{\mathbb{N}} \to (0,1) \setminus \mathbb{Q}$ is a bijection.

In the above, let $a_{2n}(x) < s < a_{2n-1}(x)$ and define $y = (y_i)_{i\in\mathbb{N}} \in \mathbb{N}^{\mathbb{N}}$ for this s similar to x for t. Then, $\alpha(y) = s$ and $x_i = y_i$ for each $i \leq 2n-1$, i.e., the first $2n-1$ coordinates of x and y are all the same. This means that α^{-1} is continuous.

Let $f : A \to Y$ be a map from a closed set A in a space X to another space Y. The **adjunction space** $Y \cup_f X$ is the quotient space $(X \oplus Y)/\sim$, where $X \oplus Y$ is the topological sum and $\sim$ is the equivalence relation corresponding to the decomposition of $X \oplus Y$ into singletons $\{x\}$, $x \in X \setminus A$, and sets $\{y\} \cup f^{-1}(y)$, $y \in Y$ (the latter is a singleton $\{y\}$ if $y \in Y \setminus f(A)$). In the case that Y is a singleton, $Y \cup_f X \approx X/A$. One should note that, in general, the adjunction spaces are *not Hausdorff*. Some further conditions are necessary for the adjunction space to be Hausdorff.

Let $\mathcal{A}$ and $\mathcal{B}$ be collections of subsets of X and $Y \subset X$. We define

- $\mathcal{A} \wedge \mathcal{B} = \{A \cap B \mid A \in \mathcal{A},\ B \in \mathcal{B}\}$;
- $\mathcal{A}|Y = \{A \cap Y \mid A \in \mathcal{A}\}$;
- $\mathcal{A}[Y] = \{A \in \mathcal{A} \mid A \cap Y \neq \emptyset\}$.

When each $A \in \mathcal{A}$ is contained in some $B \in \mathcal{B}$, it is said that $\mathcal{A}$ **refines** $\mathcal{B}$ and denoted by:

$$\mathcal{A} \prec \mathcal{B} \text{ or } \mathcal{B} \succ \mathcal{A}.$$

It is said that $\mathcal{A}$ **covers** Y (or $\mathcal{A}$ is a **cover** of Y in X) if $Y \subset \bigcup \mathcal{A}$ $(= \bigcup_{A\in\mathcal{A}} A)$. When $Y = X$, a cover of Y in X is simply called a cover of X. A cover of Y in X is said to be **open** or **closed** in X depending on whether its members are open or closed in X. If $\mathcal{A}$ is an open cover of X then $\mathcal{A}|Y$ is an open cover of Y and $\mathcal{A}[Y]$ is an open cover of Y in X. When $\mathcal{A}$ and $\mathcal{B}$ are open covers of X, $\mathcal{A} \wedge \mathcal{B}$ is also an open cover of X. For covers $\mathcal{A}$ and $\mathcal{B}$ of X, it is said that $\mathcal{A}$ is a **refinement** of $\mathcal{B}$ if $\mathcal{A} \prec \mathcal{B}$, where $\mathcal{A}$ is an **open** (or **closed**) **refinement** if $\mathcal{A}$ is an open (or closed) cover. For a space X, we denote

- $\operatorname{cov}(X)$ — the collection of all open covers of X.

Let $(X_\gamma)_{\gamma\in\Gamma}$ be a family of (topological) spaces and $X = \bigcup_{\gamma\in\Gamma} X_\gamma$. The **weak topology** on X with respect to $(X_\gamma)_{\gamma\in\Gamma}$ is defined as follows:

$$U \subset X \text{ is open in } X \;\Leftrightarrow\; \forall\gamma \in \Gamma, U \cap X_\gamma \text{ is open in } X_\gamma$$

$$\Big(A \subset X \text{ is closed in } X \;\Leftrightarrow\; \forall\gamma \in \Gamma, A \cap X_\gamma \text{ is closed in } X_\gamma\Big).$$

Suppose that X has the weak topology with respect to $(X_\gamma)_{\gamma\in\Gamma}$, and that the topologies of X_γ and $X_{\gamma'}$ agree on $X_\gamma \cap X_{\gamma'}$ for any $\gamma, \gamma' \in \Gamma$. If $X_\gamma \cap X_{\gamma'}$ is closed (resp. open) in X_γ for any $\gamma, \gamma' \in \Gamma$ then each X_γ is closed (resp. open) in X and the original topology of each X_γ is a subspace topology inherited from X. In the case that $X_\gamma \cap X_{\gamma'} = \emptyset$ for $\gamma \neq \gamma'$, X is the **topological sum** of $(X_\gamma)_{\gamma\in\Gamma}$, denoted by $X = \bigoplus_{\gamma\in\Gamma} X_\gamma$.

Let $f : X \to Y$ be a map. For $A \subset X$ and $B \subset Y$, we denote

$$f(A) = \big\{f(x) \bigm| x \in A\big\} \text{ and } f^{-1}(B) = \big\{x \in X \bigm| f(x) \in B\big\}.$$

For collections $\mathcal{A}$ and $\mathcal{B}$ of subsets of X and Y, respectively, we denote

$$f(\mathcal{A}) = \big\{f(A) \bigm| A \in \mathcal{A}\big\} \text{ and } f^{-1}(\mathcal{B}) = \big\{f^{-1}(B) \bigm| B \in \mathcal{B}\big\}.$$

The restriction of f to $A \subset X$ is denoted by $f|A$. It is said that a map $g : A \to Y$ **extends over** X if there is a map $f : X \to Y$ such that $f|A = g$. Such a map f is called an **extension** of g.

Let $[a, b]$ be a closed interval, where $a < b$. A map $f : [a, b] \to X$ is called a **path** (from $f(a)$ to $f(b)$) in X, and we say that two points $f(a)$ and $f(b)$ are connected by the path f in X. An embedding $f : [a, b] \to X$ is called an **arc** (from $f(a)$ to $f(b)$) in X, and the image $f([a, b])$ is also called an **arc**. Namely, a space is called an **arc** if it is homeomorphic to $\mathbf{I}$. It is known that each pair of distinct points $x, y \in X$ are connected by an arc if and only if they are connected by a path.[2]

For spaces X and Y, we denote

- $\mathrm{C}(X, Y)$ — the set of (continuous) maps from X to Y.

For maps $f, g : X \to Y$ (i.e., $f, g \in \mathrm{C}(X, Y)$),

- $f \simeq g$ means that f and g are **homotopic** (or f is **homotopic** to g),

that is, there is a map $h : X \times \mathbf{I} \to Y$ such that $h_0 = f$ and $h_1 = g$, where $h_t : X \to Y$, $t \in \mathbf{I}$, are defined by $h_t(x) = h(x, t)$, and h is called a **homotopy** from f to g (between f and g). When g is a constant map, it is said that f is **null-homotopic**, which we denote by $f \simeq 0$. The relation $\simeq$ is an equivalence relation on $\mathrm{C}(X, Y)$. The equivalence class $[f] = \{g \in \mathrm{C}(X, Y) \mid g \simeq f\}$ is called the **homotopy class** of f. We denote

---

[2]This will be shown in Corollary 5.14.6.

- $[X, Y] = \{[f] \mid f \in \mathrm{C}(X, Y)\} = \mathrm{C}(X, Y)/\simeq$
  — the set of the homotopy classes of maps from X to Y.

For each $f, f' \in \mathrm{C}(X, Y)$ and $g, g' \in \mathrm{C}(Y, Z)$, we have the following:

$$f \simeq f',\ g \simeq g' \Rightarrow gf \simeq g'f'.$$

Thus, we have the composition $[X, Y] \times [Y, Z] \to [X, Z]$ defined by $([f], [g]) \mapsto [g][f] = [gf]$. Moreover,

- $X \simeq Y$ means that X and Y are **homotopy equivalent** (or X is **homotopy equivalent** to Y),[3]

that is, there are maps $f : X \to Y$ and $g : Y \to X$ such that $gf \simeq \mathrm{id}_X$ and $fg \simeq \mathrm{id}_Y$, where f is called a **homotopy equivalence** and g is a **homotopy inverse** of f.

Given subspaces $X_1, \ldots, X_n \subset X$ and $Y_1, \ldots, Y_n \subset Y$, a map $f : X \to Y$ is said to be a map from $(X, X_1, \ldots, X_n)$ to $(Y, Y_1, \ldots, Y_n)$, written

$$f : (X, X_1, \ldots, X_n) \to (Y, Y_1, \ldots, Y_n),$$

if $f(X_1) \subset Y_1, \ldots, f(X_n) \subset Y_n$. We denote

- $\mathrm{C}((X, X_1, \ldots, X_n), (Y, Y_1, \ldots, Y_n))$
  — the set of maps from $(X, X_1, \ldots, X_n)$ to $(Y, Y_1, \ldots, Y_n)$.

A homotopy h between maps $f, g \in \mathrm{C}((X, X_1, \ldots, X_n), (Y, Y_1, \ldots, Y_n))$ requires the condition that $h_t \in \mathrm{C}((X, X_1, \ldots, X_n), (Y, Y_1, \ldots, Y_n))$ for every $t \in \mathbf{I}$, i.e., h is regarded as the map

$$h : (X \times \mathbf{I}, X_1 \times \mathbf{I}, \ldots, X_n \times \mathbf{I}) \to (Y, Y_1, \ldots, Y_n).$$

Thus, $\simeq$ is an equivalence relation on $\mathrm{C}((X, X_1, \ldots, X_n), (Y, Y_1, \ldots, Y_n))$. We denote

- $[(X, X_1, \ldots, X_n), (Y, Y_1, \ldots, Y_n)] = \mathrm{C}((X, X_1, \ldots, X_n), (Y, Y_1, \ldots, Y_n))/\simeq$.

When there exist maps

$$f : (X, X_1, \ldots, X_n) \to (Y, Y_1, \ldots, Y_n),$$
$$g : (Y, Y_1, \ldots, Y_n) \to (X, X_1, \ldots, X_n)$$

such that $gf \simeq \mathrm{id}_X$ and $fg \simeq \mathrm{id}_Y$, we denote

- $(X, X_1, \ldots, X_n) \simeq (Y, Y_1, \ldots, Y_n)$.

[3] It is also said that X and Y have the **same homotopy type** or X has the **homotopy type** of Y.

Similarly, for each pair of pointed spaces (X, x_0) and (Y, y_0),

- $\mathrm{C}((X, x_0), (Y, y_0)) = \mathrm{C}((X, \{x_0\}), (Y, \{y_0\}))$;
- $[(X, x_0), (Y, y_0)] = \mathrm{C}((X, x_0), (Y, y_0))/\simeq$;
- $(X, x_0) \simeq (Y, y_0)$ means $(X, \{x_0\}) \simeq (Y, \{y_0\})$.

For $A \subset X$, a homotopy $h : X \times \mathbf{I} \to Y$ is called a **homotopy relative to** A if $h(\{x\} \times \mathbf{I})$ is degenerate (i.e., a singleton) for every $x \in A$. When a homotopy from f to g is a homotopy relative to A (where $f|A = g|A$), we denote

- $f \simeq g$ rel. A.

Let $f, g : X \to Y$ be maps and $\mathcal{U}$ a collection of subsets of Y (in usual, $\mathcal{U} \in \mathrm{cov}(Y)$). It is said that f and g are **$\mathcal{U}$-close** (or f is **$\mathcal{U}$-close** to g) if

$$\big\{\{f(x), g(x)\} \bigm| x \in X\big\} \prec \mathcal{U} \cup \big\{\{y\} \bigm| y \in Y\big\},$$

which implies that $\mathcal{U}$ covers the set $\{f(x), g(x) \mid f(x) \neq g(x)\}$. A homotopy h is called a **$\mathcal{U}$-homotopy** if

$$\big\{h(\{x\} \times \mathbf{I}) \bigm| x \in X\big\} \prec \mathcal{U} \cup \big\{\{y\} \bigm| y \in Y\big\},$$

which implies that $\mathcal{U}$ covers the set

$$\bigcup \big\{h(\{x\} \times \mathbf{I}) \bigm| h(\{x\} \times \mathbf{I}) \text{ is non-degenerate}\big\}.$$

We say that f and g are **$\mathcal{U}$-homotopic** (or f is **$\mathcal{U}$-homotopic** to g) and denoted by $f \simeq_{\mathcal{U}} g$ if there is a $\mathcal{U}$-homotopy $h : X \times \mathbf{I} \to Y$ such that $h_0 = f$ and $h_1 = g$.

When $Y = (Y, d)$ is a metric space, we define the distance between $f, g \in \mathrm{C}(X, Y)$ as follows:

$$d(f, g) = \sup \big\{d(f(x), g(x)) \bigm| x \in X\big\}.$$

In general, it may be possible that $d(f, g) = \infty$, in which case d is not a metric on the set $\mathrm{C}(X, Y)$. If Y is bounded or X is compact, then this d is a metric on the set $\mathrm{C}(X, Y)$, called the **sup-metric**. For $\varepsilon > 0$, we say that f and g are **ε-close** or f is **ε-close** to g if $d(f, g) < \varepsilon$. A homotopy h is called an **ε-homotopy** if $\mathrm{mesh}\{h(\{x\} \times \mathbf{I}) \mid x \in X\} < \varepsilon$, where $f = h_0$ and $g = h_1$ are said to be **ε-homotopic** and denoted by $f \simeq_\varepsilon g$.

In the above, even if d is not a metric on $\mathrm{C}(X, Y)$ (i.e., $d(f, g) = \infty$ for some $f, g \in \mathrm{C}(X, Y)$), it induces a topology on $\mathrm{C}(X, Y)$ such that each f has a neighborhood basis consisting of

$$\mathrm{B}_d(f, \varepsilon) = \big\{g \in \mathrm{C}(X, Y) \bigm| d(f, g) < \varepsilon\big\}, \ \varepsilon > 0.$$

This topology is called the **uniform convergence topology**.

The **compact-open topology** on $C(X, Y)$ is generated by the sets

$$\langle K; U\rangle = \{f \in C(X, Y) \mid f(K) \subset U\},$$

where K is any compact set in X and U is any open set in Y. With respect to this topology, we have the following:

Proposition 1.1.1. *Every map $f : Z \times X \to Y$ (or $f : X \times Z \to Y$) induces the map $\bar{f} : Z \to C(X, Y)$ defined by $\bar{f}(z)(x) = f(z, x)$ (or $\bar{f}(z)(x) = f(x, z)$).*

Proof. For each $z \in Z$, it is easy to see that $\bar{f}(z) : X \to Y$ is continuous, i.e., $\bar{f}(z) \in C(X, Y)$. Thus, $\bar{f}$ is well-defined.

To verify the continuity of $\bar{f} : Z \to C(X, Y)$, it suffices to show that $\bar{f}^{-1}(\langle K; U\rangle)$ is open in Z for each compact set K in X and each open set U in Y. Let $z \in \bar{f}^{-1}(\langle K; U\rangle)$, i.e., $f(\{z\} \times K) \subset U$. Using the compactness of K, we can easily find an open neighborhood V of z in Z such that $f(V \times K) \subset U$, which means that $V \subset \bar{f}^{-1}(\langle K; U\rangle)$. □

With regards to the relation $\simeq$ on $C(X, Y)$, we have the following:

Proposition 1.1.2. *Each $f, g \in C(X, Y)$ are connected by a path in $C(X, Y)$. When X is metrizable or locally compact, the converse is also true, that is, $f \simeq g$ if and only if f and g are connected by a path in $C(X, Y)$ if $f \simeq g$.*[4]

Proof. By Proposition 1.1.1, a homotopy $h : X \times \mathbf{I} \to Y$ from f to g induces the path $\bar{h} : \mathbf{I} \to C(X, Y)$ defined as $\bar{h}(t)(x) = h(x, t)$ for each $t \in \mathbf{I}$ and $x \in X$, where $\bar{h}(0) = f$ and $\bar{h}(1) = g$.

For a path $\varphi : \mathbf{I} \to C(X, Y)$ from f to g, we define the homotopy $\tilde{\varphi} : X \times \mathbf{I} \to Y$ as $\tilde{\varphi}(x, t) = \varphi(t)(x)$ for each $(x, t) \in X \times \mathbf{I}$. Then, $\tilde{\varphi}_0 = \varphi(0) = f$ and $\tilde{\varphi}_1 = \varphi(1) = g$. It remains to show that $\tilde{\varphi}$ is continuous if X is metrizable or locally compact.

In the case that X is locally compact, for each $(x, t) \in X \times \mathbf{I}$ and for each open neighborhood U of $\tilde{\varphi}(x, t) = \varphi(t)(x)$ in Y, x has a compact neighborhood K in X such that $\varphi(t)(K) \subset U$, i.e., $\varphi(t) \in \langle K; U\rangle$. By the continuity of φ, t has a neighborhood V in $\mathbf{I}$ such that $\varphi(V) \subset \langle K; U\rangle$. Thus, $K \times V$ is a neighborhood of $(x, t) \in X \times \mathbf{I}$ and $\tilde{\varphi}(K \times V) \subset U$. Hence, $\tilde{\varphi}$ is continuous.

In the case that X is metrizable, let us assume that $\tilde{\varphi}$ is not continuous at $(x, t) \in X \times \mathbf{I}$. Then, $\tilde{\varphi}(x, t)$ has some open neighborhood U in Y such that $\tilde{\varphi}(V) \not\subset U$ for any neighborhood V of (x, t) in $X \times \mathbf{I}$. Let $d \in \mathrm{Metr}(X)$. For each $n \in \mathbb{N}$, we have $x_n \in X$ and $t_n \in \mathbf{I}$ such that $d(x_n, x) < 1/n$, $|t_n - t| < 1/n$ and $\tilde{\varphi}(x_n, t_n) \notin U$. Because $x_n \to x$ $(n \to \infty)$ and $\varphi(t)$ is continuous, we have $n_0 \in \mathbb{N}$ such that $\varphi(t)(x_n) \in U$ for all $n \geq n_0$. Note that $K = \{x_n, x \mid n \geq n_0\}$ is compact and $\varphi(t)(K) \subset U$. Because $t_n \to t$ $(n \to \infty)$ and φ is continuous at t, $\varphi(t_{n_1})(K) \subset U$ for some $n_1 \geq n_0$. Thus, $\tilde{\varphi}(x_{n_1}, t_{n_1}) \in U$, which is a contradiction. Consequently, $\tilde{\varphi}$ is continuous. □

Remark 1. It is easily observed that Proposition 1.1.2 is also valid for

---

[4]More generally, this is valid for every k-space X, where X is a ***k*-space** provided U is open in X if $U \cap K$ is open in K for every compact set $K \subset X$. A k-space is also called a **compactly generated space**.

$$C((X, X_1, \dots, X_n), (Y, Y_1, \dots, Y_n)).$$

Some Properties of the Compact-Open Topology 1.1.3.

The following hold with respect to the compact-open topology:

(1) For $f \in C(Z, X)$ and $g \in C(Y, Z)$, the following are continuous:

$$f^* : C(X, Y) \to C(Z, Y),\ f^*(h) = h \circ f;$$
$$g_* : C(X, Y) \to C(X, Z),\ g_*(h) = g \circ h.$$

(2) When Y is locally compact, the following (composition) is continuous:

$$C(X, Y) \times C(Y, Z) \ni (f, g) \mapsto g \circ f \in C(X, Z).$$

Sketch of Proof. Let K be a compact set in X and U an open set in Z with $f \in C(X, Y)$ and $g \in C(Y, Z)$ such that $g \circ f(K) \subset U$. Since Y is locally compact, we have an open set V in Y such that $\mathrm{cl}\, V$ is compact, $f(K) \subset V$ and $g(\mathrm{cl}\, V) \subset U$. Then, $f'(K) \subset V$ and $g'(\mathrm{cl}\, V) \subset U$ imply $g' \circ f'(K) \subset U$.

(3) For each $x_0 \in X$, the following (evaluation) is continuous:

$$C(X, Y) \ni f \mapsto f(x_0) \in Y.$$

(4) When X is locally compact, the following (evaluation) is continuous:

$$C(X, Y) \times X \ni (f, x) \mapsto f(x) \in Y.$$

In this case, for every map $f : Z \to C(X, Y)$, the following is continuous:

$$Z \times X \ni (z, x) \mapsto f(z)(x) \in Y.$$

(5) In the case that X is locally compact, we have the following inequalities:

$$w(Y) \le w(C(X, Y)) \le \aleph_0 w(X) w(Y).$$

Sketch of Proof. By embedding Y into $C(X, Y)$, we obtain the first inequality. For the second, we take open bases $\mathcal{B}_X$ and $\mathcal{B}_Y$ for X and Y, respectively, such that $\mathrm{card}\, \mathcal{B}_X = w(X)$, $\mathrm{card}\, \mathcal{B}_Y = w(Y)$, and $\mathrm{cl}\, A$ is compact for every $A \in \mathcal{B}_X$. The following is an open sub-basis for $C(X, Y)$:

$$\mathcal{B} = \{\langle \mathrm{cl}\, A, B\rangle \mid (A, B) \in \mathcal{B}_X \times \mathcal{B}_Y\}.$$

Indeed, let K be a compact set in X, U be an open set in Y, and $f \in C(X, Y)$ with $f(K) \subset U$, i.e., $f \in \langle K, U\rangle$. First, find $B_1, \dots, B_n \in \mathcal{B}_Y$ so that $f(K) \subset B_1 \cup \dots \cup B_n \subset U$. Next, find $A_1, \dots, A_m \in \mathcal{B}_X$ so that $K \subset A_1 \cup \dots \cup A_m$ and each $\mathrm{cl}\, A_i$ is contained in some $f^{-1}(B_{j(i)})$. Then, $f \in \bigcap_{i=1}^m \langle \mathrm{cl}\, A_i, B_{j(i)}\rangle \subset \langle K, U\rangle$.

(6) If X is compact and $Y = (Y, d)$ is a metric space, then the sup-metric on $\mathrm{C}(X, Y)$ is admissible for the compact-open topology on $\mathrm{C}(X, Y)$.

Sketch of Proof. Let K be a compact set in X and U be an open set in Y with $f \in \mathrm{C}(X, Y)$ such that $f(K) \subset U$. Then, $\delta = \mathrm{dist}(f(K), Y \setminus U) > 0$, and $d(f, f') < \delta$ implies $f'(K) \subset U$. Conversely, for each $\varepsilon > 0$ and $f \in \mathrm{C}(X, Y)$, we have $x_1, \ldots, x_n \in X$ such that $X = \bigcup_{i=1}^n f^{-1}(\mathrm{B}(f(x_i), \varepsilon/4))$. Observe that

$$f'(f^{-1}(\overline{\mathrm{B}}(f(x_i), \varepsilon/4))) \subset \mathrm{B}(f(x_i), \varepsilon/2) \ (\forall i = 1, \ldots, n)$$
$$\Rightarrow d(f, f') < \varepsilon.$$

(7) Let $X = \bigcup_{n \in \mathbb{N}} X_n$, where X_n is compact and $X_n \subset \operatorname{int} X_{n+1}$. If $Y = (Y, d)$ is a metric space, then $\mathrm{C}(X, Y)$ with the compact-open topology is metrizable.

Sketch of Proof. We define a metric ρ on $\mathrm{C}(X, Y)$ as follows:

$$\rho(f, g) = \sup_{n \in \mathbb{N}} \min \left\{ n^{-1}, \sup_{x \in X_n} d(f(x), g(x)) \right\}.$$

Then, ρ is admissible for the compact-open topology on $\mathrm{C}(X, Y)$. To see this, refer to the proof of (6).

1.2 Banach Spaces in the Product of Real Lines

Throughout this section, let Γ be an infinite set. We denote

- $\mathrm{Fin}(\Gamma)$ — the set of all non-empty finite subsets of Γ.

Note that $\operatorname{card} \mathrm{Fin}(\Gamma) = \operatorname{card} \Gamma$. The product space $\mathbb{R}^\Gamma$ is a linear space with the following scalar multiplication and addition:

$$\mathbb{R}^\Gamma \times \mathbb{R} \ni (x, t) \mapsto tx = (tx(\gamma))_{\gamma \in \Gamma} \in \mathbb{R}^\Gamma;$$
$$\mathbb{R}^\Gamma \times \mathbb{R}^\Gamma \ni (x, y) \mapsto x + y = (x(\gamma) + y(\gamma))_{\gamma \in \Gamma} \in \mathbb{R}^\Gamma.$$

In this section, we consider various (complete) norms defined on linear subspaces of $\mathbb{R}^\Gamma$. In general, the unit closed ball and the unit sphere of a normed linear space $X = (X, \|\cdot\|)$ are denoted by $\mathbf{B}_X$ and $\mathbf{S}_X$, respectively. Namely, let

$$\mathbf{B}_X = \{x \in X \mid \|x\| \leq 1\} \text{ and } \mathbf{S}_X = \{x \in X \mid \|x\| = 1\}.$$

The zero vector (the zero element) of X is denoted by $\mathbf{0}_X$, or simply $\mathbf{0}$ if there is no possibility of confusion.

Before considering norms, we first discuss the product topology of $\mathbb{R}^\Gamma$. The scalar multiplication and addition are continuous with respect to the product topology. Namely, $\mathbb{R}^\Gamma$ with the product topology is a topological linear space.[5] Note that $w(\mathbb{R}^\Gamma) = \operatorname{card} \Gamma$.

Let $\mathcal{B}_0$ be a countable open basis for $\mathbb{R}$. Then, $\mathbb{R}^\Gamma$ has the following open basis:

$$\left\{\bigcap_{\gamma\in F} \operatorname{pr}_\gamma^{-1}(B_\gamma) \;\middle|\; F \in \operatorname{Fin}(\Gamma),\ B_\gamma \in \mathcal{B}_0\ (\gamma \in F)\right\}.$$

Thus, we have $w(\mathbb{R}^\Gamma) \le \aleph_0 \operatorname{card} \operatorname{Fin}(\Gamma) = \operatorname{card} \Gamma$. Let $\mathcal{B}$ be an open basis for $\mathbb{R}^\Gamma$. For each $B \in \mathcal{B}$, we can find $F_B \in \operatorname{Fin}(\Gamma)$ such that $\operatorname{pr}_\gamma(B) = \mathbb{R}$ for every $\gamma \in \Gamma \setminus F_B$. Then, $\operatorname{card} \bigcup_{B\in\mathcal{B}} F_B \le \aleph_0 \operatorname{card} \mathcal{B}$. If $\operatorname{card} \mathcal{B} < \operatorname{card} \Gamma$ then $\operatorname{card} \bigcup_{B\in\mathcal{B}} F_B < \operatorname{card} \Gamma$, so we have $\gamma_0 \in \Gamma \setminus \bigcup_{B\in\mathcal{B}} F_B$. The open set $\operatorname{pr}_{\gamma_0}^{-1}((0,\infty)) \subset \mathbb{R}^\Gamma$ contains some $B \in \mathcal{B}$. Then, $\operatorname{pr}_{\gamma_0}(B) \subset (0,\infty)$, which means that $\gamma_0 \in F_B$. This is a contradiction. Therefore, $\operatorname{card} \mathcal{B} \ge \operatorname{card} \Gamma$, and thus we have $w(\mathbb{R}^\Gamma) \ge \operatorname{card} \Gamma$.

For each $\gamma \in \Gamma$, we define the unit vector $\mathbf{e}_\gamma \in \mathbb{R}^\Gamma$ by $\mathbf{e}_\gamma(\gamma) = 1$ and $\mathbf{e}_\gamma(\gamma') = 0$ for $\gamma' \ne \gamma$. It should be noted that $\{\mathbf{e}_\gamma \mid \gamma \in \Gamma\}$ is not a Hamel basis for $\mathbb{R}^\Gamma$, and the linear span of $\{\mathbf{e}_\gamma \mid \gamma \in \Gamma\}$ is the following:[6]

$$\mathbb{R}^\Gamma_f = \left\{x \in \mathbb{R}^\Gamma \;\middle|\; x(\gamma) = 0 \text{ except for finitely many } \gamma \in \Gamma\right\},$$

which is a dense linear subspace of $\mathbb{R}^\Gamma$. The subspace $\mathbb{R}^{\mathbb{N}}_f$ of $\boldsymbol{s} = \mathbb{R}^{\mathbb{N}}$ is also denoted by $\boldsymbol{s}_f$, which is the space of finite sequences (with the product topology). When $\operatorname{card} \Gamma = \aleph_0$, the space $\mathbb{R}^\Gamma$ is linearly homeomorphic to the space of sequences $\boldsymbol{s} = \mathbb{R}^{\mathbb{N}}$, i.e., there exists a linear homeomorphism between $\mathbb{R}^\Gamma$ and $\boldsymbol{s}$, where the linear subspace $\mathbb{R}^\Gamma_f$ is linearly homeomorphic to $\boldsymbol{s}_f$ by the same homeomorphism. The following fact can easily be observed:

Fact. *The following are equivalent:*

(a) $\mathbb{R}^\Gamma$ *is metrizable;*
(b) $\mathbb{R}^\Gamma_f$ *is metrizable;*
(c) $\mathbb{R}^\Gamma_f$ *is first countable;*
(d) $\operatorname{card} \Gamma \le \aleph_0$.

The implication (c) $\Rightarrow$ (d) is shown as follows: Let $\{U_i \mid i \in \mathbb{N}\}$ be a neighborhood basis of $\mathbf{0}$ in $\mathbb{R}^\Gamma_f$. Then, each $\Gamma_i = \{\gamma \in \Gamma \mid \mathbb{R}\mathbf{e}_\gamma \not\subset U_i\}$ is finite. If Γ is uncountable, then $\Gamma \setminus \bigcup_{i\in\mathbb{N}} \Gamma_i \ne \emptyset$, i.e., $\mathbb{R}\mathbf{e}_\gamma \subset \bigcap_{i\in\mathbb{N}} U_i$ for some $\gamma \in \Gamma$. In this case, $U_i \not\subset \operatorname{pr}_\gamma^{-1}((-1,1))$ for every $i \in \mathbb{N}$, which is a contradiction.

Thus, every linear subspace L of $\mathbb{R}^\Gamma$ containing $\mathbb{R}^\Gamma_f$ is non-metrizable if Γ is uncountable, and it is metrizable if Γ is countable. On the other hand, due to the following proposition, every linear subspaces L of $\mathbb{R}^\Gamma$ containing $\mathbb{R}^\Gamma_f$ is non-normable if Γ is infinite.

Proposition 1.2.1. *Let Γ be an infinite set. Any norm on $\mathbb{R}^\Gamma_f$ does not induce the topology inherited from the product topology of $\mathbb{R}^\Gamma$.*

---

[5]For topological linear spaces, refer Sect. 3.4.

[6]The linear subspace generated by a set B is called the **linear span** of B.

Proof. Assume that the topology of $\mathbb{R}^\Gamma_f$ is induced by a norm $\|\cdot\|$. Because $U = \{x \in \mathbb{R}^\Gamma_f \mid \|x\| < 1\}$ is an open neighborhood of $\mathbf{0}$ in $\mathbb{R}^\Gamma_f$, we have a finite set $F \subset \Gamma$ and neighborhoods V_γ of $0 \in \mathbb{R}$, $\gamma \in F$, such that $\mathbb{R}^\Gamma_f \cap \bigcap_{\gamma\in F} \mathrm{pr}_\gamma^{-1}(V_\gamma) \subset U$. Take $\gamma_0 \in \Gamma \setminus F$. As $\mathbb{R}\mathbf{e}_{\gamma_0} \subset U$, we have $\|\mathbf{e}_{\gamma_0}\|^{-1}\mathbf{e}_{\gamma_0} \in U$ but $\big\| \|\mathbf{e}_{\gamma_0}\|^{-1}\mathbf{e}_{\gamma_0}\big\| = \|\mathbf{e}_{\gamma_0}\|^{-1}\|\mathbf{e}_{\gamma_0}\| = 1$, which is a contradiction. □

The Banach space $\ell_\infty(\Gamma)$ and its closed linear subspaces $\boldsymbol{c}(\Gamma) \supset \boldsymbol{c}_0(\Gamma)$ are defined as follows:

- $\ell_\infty(\Gamma) = \big\{x \in \mathbb{R}^\Gamma \;\big|\; \sup_{\gamma\in\Gamma} |x(\gamma)| < \infty\big\}$ with the sup-norm

$$\|x\|_\infty = \sup_{\gamma\in\Gamma} |x(\gamma)|;$$

- $\boldsymbol{c}(\Gamma) = \big\{x \in \mathbb{R}^\Gamma \;\big|\; \exists t \in \mathbb{R} \text{ such that } \forall \varepsilon > 0,\ |x(\gamma) - t| < \varepsilon \text{ except for finitely many } \gamma \in \Gamma\big\}$;
- $\boldsymbol{c}_0(\Gamma) = \big\{x \in \mathbb{R}^\Gamma \;\big|\; \forall \varepsilon > 0,\ |x(\gamma)| < \varepsilon \text{ except for finitely many } \gamma \in \Gamma\big\}$.

These are linear subspaces of $\mathbb{R}^\Gamma$, but are not topological subspace according to Proposition 1.2.1. The space $\boldsymbol{c}(\Gamma)$ is linearly homeomorphic to $\boldsymbol{c}_0(\Gamma) \times \mathbb{R}$ by the correspondence

$$\boldsymbol{c}_0(\Gamma) \times \mathbb{R} \ni (x,t) \mapsto (x(\gamma) + t)_{\gamma\in\Gamma} \in \boldsymbol{c}(\Gamma).$$

This correspondence and its inverse are Lipschitz with respect to the norm $\|(x,t)\| = \max\{\|x\|_\infty, |t|\}$. Indeed, let $y = (x(\gamma)+t)_{\gamma\in\Gamma}$. Then, $\|y\|_\infty \leq \|x\|_\infty + |t| \leq 2\|(x,t)\|$. Because $x \in \boldsymbol{c}_0(\Gamma)$ and $|t| \leq |y(\gamma)| + |x(\gamma)| \leq \|y\|_\infty + |x(\gamma)|$ for every $\gamma \in \Gamma$, it follows that $|t| \leq \|y\|_\infty$. Moreover, $|x(\gamma)| \leq |y(\gamma)| + |t| \leq 2\|y\|_\infty$ for every $\gamma \in \Gamma$. Hence, $\|x\|_\infty \leq 2\|y\|_\infty$, and thus we have $\|(x,t)\| \leq 2\|y\|_\infty$.

Furthermore, we denote $\mathbb{R}^\Gamma_f$ with this norm as $\ell^f_\infty(\Gamma)$. We then have the inclusions:

$$\ell^f_\infty(\Gamma) \subset \boldsymbol{c}_0(\Gamma) \subset \boldsymbol{c}(\Gamma) \subset \ell_\infty(\Gamma).$$

The topology of $\ell^f_\infty(\Gamma)$ is different from the topology inherited from the product topology. Indeed, $\{\mathbf{e}_\gamma \mid \gamma \in \Gamma\}$ is discrete in $\ell^f_\infty(\Gamma)$, but $\mathbf{0}$ is a cluster point of this set with respect to the product topology.

We must pay attention to the following fact:

Proposition 1.2.2. *For an arbitrary infinite set Γ,*

$$w(\ell_\infty(\Gamma)) = 2^{\operatorname{card}\Gamma} \ \ \textit{but} \ \ w(\boldsymbol{c}(\Gamma)) = w(\boldsymbol{c}_0(\Gamma)) = w(\ell^f_\infty(\Gamma)) = \operatorname{card}\Gamma.$$

Proof. The characteristic map $\chi_\Lambda : \Gamma \to \{0,1\} \subset \mathbb{R}$ of $\Lambda \subset \Gamma$ belongs to $\ell_\infty(\Gamma)$ ($\chi_\emptyset = \mathbf{0} \in \ell_\infty(\Gamma)$), where $\|\chi_\Lambda - \chi_{\Lambda'}\|_\infty = 1$ if $\Lambda \neq \Lambda' \subset \Gamma$. It follows that $w(\ell_\infty(\Gamma)) = c(\ell_\infty(\Gamma)) \geq 2^{\operatorname{card}\Gamma}$. Moreover, $\mathbb{Q}^\Gamma \cap \ell_\infty(\Gamma)$ is dense in $\ell_\infty(\Gamma)$, and hence we have

$$w(\ell_\infty(\Gamma)) = \operatorname{dens} \ell_\infty(\Gamma) \le \operatorname{card} \mathbb{Q}^\Gamma = \aleph_0^{\operatorname{card}\Gamma} = 2^{\operatorname{card}\Gamma}.$$

On the other hand, $\mathbf{e}_\gamma \in \ell_\infty^f(\Gamma)$ for each $\gamma \in \Gamma$ and $\|\mathbf{e}_\gamma - \mathbf{e}_{\gamma'}\|_\infty = 1$ if $\gamma \neq \gamma'$. Since $\ell_\infty^f(\Gamma) \subset \boldsymbol{c}_0(\Gamma)$, it follows that

$$w(\boldsymbol{c}_0(\Gamma)) \ge w(\ell_\infty^f(\Gamma)) = c(\ell_\infty^f(\Gamma)) \ge \operatorname{card} \Gamma.$$

Moreover, $\boldsymbol{c}_0(\Gamma)$ has the following dense subset:

$$\mathbb{Q}_f^\Gamma = \{x \in \mathbb{Q}^\Gamma \mid x(\gamma) = 0 \text{ except for finitely many } \gamma \in \Gamma\},$$

and so it follows that

$$w(\boldsymbol{c}_0(\Gamma)) = \operatorname{dens} \boldsymbol{c}_0(\Gamma) \le \operatorname{card} \mathbb{Q}_f^\Gamma \le \aleph_0 \operatorname{card} \operatorname{Fin}(\Gamma) = \operatorname{card} \Gamma.$$

Thus, we have $w(\boldsymbol{c}_0(\Gamma)) = w(\ell_\infty^f(\Gamma)) = \operatorname{card} \Gamma$. As already observed, $\boldsymbol{c}(\Gamma) \approx \boldsymbol{c}_0(\Gamma) \times \mathbb{R}$, hence $w(\boldsymbol{c}(\Gamma)) = w(\boldsymbol{c}_0(\Gamma))$. □

When $\Gamma = \mathbb{N}$, we write

- $\ell_\infty(\mathbb{N}) = \ell_\infty$ — the space of bounded sequences,
- $\boldsymbol{c}(\mathbb{N}) = \boldsymbol{c}$ — the space of convergent sequences,
- $\boldsymbol{c}_0(\mathbb{N}) = \boldsymbol{c}_0$ — the space of sequences convergent to 0, and
- $\ell_\infty^f(\mathbb{N}) = \ell_\infty^f$ — the space of finite sequences with the sup-norm,

where $\ell_\infty^f \not\approx \boldsymbol{s}_f$ as (topological) spaces. According to Proposition 1.2.2, $\boldsymbol{c}$ and $\boldsymbol{c}_0$ are separable, but ℓ_∞ is non-separable. When $\operatorname{card} \Gamma = \aleph_0$, the spaces $\ell_\infty(\Gamma)$, $\boldsymbol{c}(\Gamma)$, and $\boldsymbol{c}_0(\Gamma)$ are linearly isometric to these spaces ℓ_∞, $\boldsymbol{c}$ and $\boldsymbol{c}_0$, respectively.

Here, we regard $\operatorname{Fin}(\Gamma)$ as a directed set by $\subset$. For $x \in \mathbb{R}^\Gamma$, we say that $\sum_{\gamma\in\Gamma} x(\gamma)$ is **convergent** if $\left(\sum_{\gamma\in F} x(\gamma)\right)_{F\in\operatorname{Fin}(\Gamma)}$ is convergent, and define

$$\sum_{\gamma\in\Gamma} x(\gamma) = \lim_{F\in\operatorname{Fin}(\Gamma)} \sum_{\gamma\in F} x(\gamma).$$

In the case that $x(\gamma) \ge 0$ for all $\gamma \in \Gamma$, $\sum_{\gamma\in\Gamma} x(\gamma)$ is convergent if and only if $\left(\sum_{\gamma\in F} x(\gamma)\right)_{F\in\operatorname{Fin}(\Gamma)}$ is upper bounded, and then

$$\sum_{\gamma\in\Gamma} x(\gamma) = \sup_{F\in\operatorname{Fin}(\Gamma)} \sum_{\gamma\in F} x(\gamma).$$

By this reason, $\sum_{\gamma\in\Gamma} x(\gamma) < \infty$ means that $\sum_{\gamma\in\Gamma} x(\gamma)$ is convergent.

For $x \in \mathbb{R}^\mathbb{N}$, we should distinguish $\sum_{i\in\mathbb{N}} x(i)$ from $\sum_{i=1}^\infty x(i)$. When the sequence $\left(\sum_{i=1}^n x(i)\right)_{n\in\mathbb{N}}$ is convergent, we say that $\sum_{i=1}^\infty x(i)$ is **convergent**, and define

$$\sum_{i=1}^{\infty} x(i) = \lim_{n\to\infty} \sum_{i=1}^{n} x(i).$$

Evidently, if $\sum_{i\in\mathbb{N}} x(i)$ is convergent, then $\sum_{i=1}^{\infty} x(i)$ is also convergent and $\sum_{i=1}^{\infty} x(i) = \sum_{i\in\mathbb{N}} x(i)$. However, $\sum_{i\in\mathbb{N}} x(i)$ is not necessary convergent even if $\sum_{i=1}^{\infty} x(i)$ is convergent. In fact, due to Proposition 1.2.3 below, we have the following:

$$\sum_{i\in\mathbb{N}} x(i) \text{ is convergent } \Leftrightarrow \sum_{i=1}^{\infty} |x(i)| \text{ is convergent.}$$

Proposition 1.2.3. *For an infinite set Γ and $x \in \mathbb{R}^{\Gamma}$, $\sum_{\gamma\in\Gamma} x(\gamma)$ is convergent if and only if $\sum_{\gamma\in\Gamma} |x(\gamma)| < \infty$. In this case, $\Gamma_x = \{\gamma \in \Gamma \mid x(\gamma) \neq 0\}$ is countable, and $\sum_{\gamma\in\Gamma} x(\gamma) = \sum_{i=1}^{\infty} x(\gamma_i)$ for any sequence $(\gamma_i)_{i\in\mathbb{N}}$ in Γ such that $\Gamma_x \subset \{\gamma_i \mid i \in \mathbb{N}\}$ and $\gamma_i \neq \gamma_j$ if $i \neq j$.*

Proof. Let us denote $\Gamma_+ = \{\gamma \in \Gamma \mid x(\gamma) > 0\}$ and $\Gamma_- = \{\gamma \in \Gamma \mid x(\gamma) < 0\}$. Then, $\Gamma_x = \Gamma_+ \cup \Gamma_-$.

If $\sum_{\gamma\in\Gamma} x(\gamma)$ is convergent, we have $F_0 \in \operatorname{Fin}(\Gamma)$ such that

$$F_0 \subset F \in \operatorname{Fin}(\Gamma) \Rightarrow \left| \sum_{\gamma\in\Gamma} x(\gamma) - \sum_{\gamma\in F} x(\gamma) \right| < 1.$$

Then, for each $E \in \operatorname{Fin}(\Gamma_+) \cup \operatorname{Fin}(\Gamma_-)$ (i.e., $E \in \operatorname{Fin}(\Gamma_+)$ or $E \in \operatorname{Fin}(\Gamma_-)$),

$$\sum_{\gamma\in E\setminus F_0} |x(\gamma)| = \left| \sum_{\gamma\in E\setminus F_0} x(\gamma) \right| = \left| \sum_{\gamma\in E\cup F_0} x(\gamma) - \sum_{\gamma\in F_0} x(\gamma) \right| < 2.$$

Hence, $\sum_{\gamma\in F} |x(\gamma)| < \sum_{\gamma\in F_0} |x(\gamma)| + 4$ for every $F \in \operatorname{Fin}(\Gamma)$, which means that $\left(\sum_{\gamma\in F} |x(\gamma)|\right)_{F\in\operatorname{Fin}(\Gamma)}$ is upper bounded, i.e., $\sum_{\gamma\in\Gamma} |x(\gamma)| < \infty$.

Conversely, we assume that $\sum_{\gamma\in\Gamma} |x(\gamma)| < \infty$. Then, for each $n \in \mathbb{N}$, $\Gamma_n = \{\gamma \in \Gamma \mid |x(\gamma)| > 1/n\}$ is finite, and hence $\Gamma_x = \bigcup_{n\in\mathbb{N}} \Gamma_n$ is countable. Note that $\sum_{\gamma\in\Gamma_+} |x(\gamma)| < \infty$ and $\sum_{\gamma\in\Gamma_-} |x(\gamma)| < \infty$. We show that

$$\sum_{\gamma\in\Gamma} x(\gamma) = \sum_{\gamma\in\Gamma_+} |x(\gamma)| - \sum_{\gamma\in\Gamma_-} |x(\gamma)|.$$

For each $\varepsilon > 0$, we can find $F_+ \in \operatorname{Fin}(\Gamma_+)$ and $F_- \in \operatorname{Fin}(\Gamma_-)$ such that

$$F_\pm \subset E \in \operatorname{Fin}(\Gamma_\pm) \Rightarrow \sum_{\gamma\in\Gamma_\pm} |x(\gamma)| - \varepsilon/2 < \sum_{\gamma\in E} |x(\gamma)| \leq \sum_{\gamma\in\Gamma_\pm} |x(\gamma)|.$$

Then, it follows that, for each $F \in \mathrm{Fin}(\Gamma)$ with $F \supset F_+ \cup F_-$,

$$\begin{aligned}
&\left| \sum_{\gamma \in F} x(\gamma) - \left(\sum_{\gamma \in \Gamma_+} |x(\gamma)| - \sum_{\gamma \in \Gamma_-} |x(\gamma)| \right) \right| \\
&\quad \leq \left| \sum_{\gamma \in F \cap \Gamma_+} |x(\gamma)| - \sum_{\gamma \in \Gamma_+} |x(\gamma)| \right| + \left| \sum_{\gamma \in F \cap \Gamma_-} |x(\gamma)| - \sum_{\gamma \in \Gamma_-} |x(\gamma)| \right| \\
&\quad < \varepsilon/2 + \varepsilon/2 = \varepsilon.
\end{aligned}$$

Now, let $(\gamma_i)_{i \in \mathbb{N}}$ be a sequence in Γ such that $\Gamma_x \subset \{\gamma_i \mid i \in \mathbb{N}\}$ and $\gamma_i \neq \gamma_j$ if $i \neq j$. We define

$$n_0 = \max\{i \in \mathbb{N} \mid \gamma_i \in F_+ \cup F_-\}.$$

For each $n \geq n_0$, it follows from $F_+ \cup F_- \subset \{\gamma_1, \ldots, \gamma_n\}$ that

$$\left| \sum_{i=1}^{n} x(\gamma_i) - \left(\sum_{\gamma \in \Gamma_+} |x(\gamma)| - \sum_{\gamma \in \Gamma_-} |x(\gamma)| \right) \right| < \varepsilon.$$

Thus, we also have $\sum_{\gamma \in \Gamma} x(\gamma) = \sum_{i=1}^{\infty} x(\gamma_i)$. □

For each $p \geq 1$, the Banach space $\ell_p(\Gamma)$ is defined as follows:

- $\ell_p(\Gamma) = \left\{x \in \mathbb{R}^\Gamma \mid \sum_{\gamma \in \Gamma} |x(\gamma)|^p < \infty\right\}$ with the norm

$$\|x\|_p = \left(\sum_{\gamma \in \Gamma} |x(\gamma)|^p \right)^{1/p}.$$

Similar to $\ell_\infty^f(\Gamma)$, we denote the space $\mathbb{R}_f^\Gamma$ with this norm by $\ell_p^f(\Gamma)$.

The triangle inequality for the norm $\|x\|_p$ is known as the Minkowski inequality, which is derived from the following Hölder inequality:

$$\sum_{\gamma \in \Gamma} a_\gamma b_\gamma \leq \left(\sum_{\gamma \in \Gamma} a_\gamma^p \right)^{1/p} \left(\sum_{\gamma \in \Gamma} b_\gamma^{\frac{1}{1-1/p}} \right)^{1-1/p} \quad \text{for every } a_\gamma, b_\gamma \geq 0.$$

Indeed, for every $x, y \in \ell_p(\Gamma)$,

$$\begin{aligned}
\|x + y\|_p^p &= \sum_{\gamma \in \Gamma} |x(\gamma) + y(\gamma)|^p \\
&\leq \sum_{\gamma \in \Gamma} \big(|x(\gamma)| + |y(\gamma)|\big)|x(\gamma) + y(\gamma)|^{p-1} \\
&= \sum_{\gamma \in \Gamma} |x(\gamma)| \cdot |x(\gamma) + y(\gamma)|^{p-1} + \sum_{\gamma \in \Gamma} |y(\gamma)| \cdot |x(\gamma) + y(\gamma)|^{p-1}
\end{aligned}$$

$$\le \Big(\sum_{\gamma\in\Gamma}|x(\gamma)|^p\Big)^{1/p}\Big(\sum_{\gamma\in\Gamma}|x(\gamma)+y(\gamma)|^{(p-1)\frac{1}{1-1/p}}\Big)^{1-1/p}$$
$$+\Big(\sum_{\gamma\in\Gamma}|y(\gamma)|^p\Big)^{1/p}\Big(\sum_{\gamma\in\Gamma}|x(\gamma)+y(\gamma)|^{(p-1)\frac{1}{1-1/p}}\Big)^{1-1/p}$$
$$= (\|x\|_p+\|y\|_p)\big(\|x+y\|_p^p\big)^{1-1/p} = (\|x\|_p+\|y\|_p)\frac{\|x+y\|_p^p}{\|x+y\|_p},$$

so it follows that $\|x+y\|_p \le \|x\|_p + \|y\|_p$.

As for $\boldsymbol{c}_0(\Gamma)$, we can show $w(\ell_p(\Gamma)) = \operatorname{card}\Gamma$. When $\operatorname{card}\Gamma = \aleph_0$, the Banach space $\ell_p(\Gamma)$ is linearly isometric to $\ell_p = \ell_p(\mathbb{N})$, which is separable. The space $\ell_2(\Gamma)$ is the Hilbert space with the inner product

$$\langle x, y\rangle = \sum_{\gamma\in\Gamma} x(\gamma)y(\gamma),$$

which is well-defined because

$$\sum_{\gamma\in\Gamma}|x(\gamma)y(\gamma)| \le \tfrac{1}{2}(\|x\|_2^2+\|y\|_2^2) < \infty.$$

For $1 \le p < q$, we have $\ell_p(\Gamma) \subsetneqq \ell_q(\Gamma) \subsetneqq \boldsymbol{c}_0(\Gamma)$ as sets (or linear spaces). These inclusions are continuous because $\|x\|_\infty \le \|x\|_q \le \|x\|_p$ for every $x \in \ell_p(\Gamma)$. When Γ is infinite, the topology of $\ell_p(\Gamma)$ is distinct from that induced by the norm $\|\cdot\|_q$ or $\|\cdot\|_\infty$ (i.e., the topology inherited from $\ell_q(\Gamma)$ or $\boldsymbol{c}_0(\Gamma)$). In fact, the unit sphere $\mathbf{S}_{\ell_p(\Gamma)}$ is closed in $\ell_p(\Gamma)$ but not closed in $\ell_q(\Gamma)$ for any $q > p$, nor in $\boldsymbol{c}_0(\Gamma)$. To see this, take distinct $\gamma_i \in \Gamma$, $i \in \mathbb{N}$, and let $(x_n)_{n\in\mathbb{N}}$ be the sequence in $\mathbf{S}_{\ell_p(\Gamma)}$ defined by $x_n(\gamma_i) = n^{-1/p}$ for $i \le n$ and $x_n(\gamma) = 0$ for $\gamma \ne \gamma_1, \dots, \gamma_n$. It follows that $\|x_n\|_\infty = n^{-1/p} \to 0$ $(n\to\infty)$ and

$$\|x_n\|_q = \big(n\cdot n^{-q/p}\big)^{1/q} = n^{(p-q)/pq} \to 0 \quad (n\to\infty)$$

because $(p-q)/pq < 0$.

For $1 \le p \le \infty$, we have $\mathbb{R}^\Gamma_f \subset \ell_p(\Gamma)$ as sets (or linear spaces). We denote by $\ell_p^f(\Gamma)$ this $\mathbb{R}^\Gamma_f$ with the topology inherited from $\ell_p(\Gamma)$, and we write $\ell_p^f(\mathbb{N}) = \ell_p^f$ (when $\Gamma = \mathbb{N}$). From Proposition 1.2.1, we know $\ell_p^f(\Gamma) \ne \mathbb{R}^\Gamma_f$ as spaces for any infinite set Γ. In the above, the sequence $(x_n)_{n\in\mathbb{N}}$ is contained in the unit sphere $\mathbf{S}_{\ell_p^f(\Gamma)}$ of $\ell_p^f(\Gamma)$, which means that $\mathbf{S}_{\ell_p^f(\Gamma)}$ is not closed in ℓ_q^f, hence $\ell_p^f \ne \ell_q^f$ as spaces for $1 \le p < q \le \infty$. Note that $\mathbf{S}_{\ell_p^f(\Gamma)}$ is a closed subset of ℓ_q^f for $1 \le q < p$.

Concerning the convergence of sequences in $\ell_p(\Gamma)$, we have the following:

Proposition 1.2.4. *For each $p \in \mathbb{N}$ and $x \in \ell_p(\Gamma)$, a sequence $(x_n)_{n\in\mathbb{N}}$ converges to x in $\ell_p(\Gamma)$ if and only if*

$$\|x\|_p = \lim_{n\to\infty} \|x_n\|_p \quad \text{and} \quad x(\gamma) = \lim_{n\to\infty} x_n(\gamma) \ \text{for every}\ \gamma \in \Gamma.$$

Proof. The "only if" part is trivial, so we concern ourselves with proving the "if" part for $\ell_p(\Gamma)$. For each $\varepsilon > 0$, we have $\gamma_1, \dots, \gamma_k \in \Gamma$ such that

$$\sum_{\gamma \neq \gamma_i} |x(\gamma)|^p = \|x\|_p^p - \sum_{i=1}^k |x(\gamma_i)|^p < 2^{-p}\varepsilon^p/4.$$

Choose $n_0 \in \mathbb{N}$ so that if $n \geq n_0$ then $\big|\|x_n\|_p^p - \|x\|_p^p\big| < 2^{-p}\varepsilon^p/8$,

$$\big||x_n(\gamma_i)|^p - |x(\gamma_i)|^p\big| < 2^{-p}\varepsilon^p/8k \ \text{ and } \ |x_n(\gamma_i) - x(\gamma_i)|^p < \varepsilon^p/4k$$

for each $i = 1, \dots, k$. Then, it follows that

$$\begin{aligned}
\sum_{\gamma \neq \gamma_i} |x_n(\gamma)|^p =& \|x_n\|_p^p - \sum_{i=1}^k |x_n(\gamma_i)|^p \\
=& \|x_n\|_p^p - \|x\|_p^p + \|x\|_p^p - \sum_{i=1}^k |x(\gamma_i)|^p \\
&+ \sum_{i=1}^k \big(|x(\gamma_i)|^p - |x_n(\gamma_i)|^p\big) \\
<& 2^{-p}\varepsilon^p/8 + 2^{-p}\varepsilon^p/4 + 2^{-p}\varepsilon^p/8 = 2^{-p}\varepsilon^p/2,
\end{aligned}$$

and hence we have

$$\begin{aligned}
\|x_n - x\|_p^p &\leq \sum_{i=1}^k |x_n(\gamma_i) - x(\gamma_i)|^p + \sum_{\gamma \neq \gamma_i} 2^p \max\big\{|x_n(\gamma)|,\ |x(\gamma)|\big\}^p \\
&< \varepsilon^p/4 + \sum_{\gamma \neq \gamma_i} 2^p |x_n(\gamma)|^p + \sum_{\gamma \neq \gamma_i} 2^p |x(\gamma)|^p \\
&< \varepsilon^p/4 + \varepsilon^p/2 + \varepsilon^p/4 = \varepsilon^p,
\end{aligned}$$

that is, $\|x_n - x\|_p < \varepsilon$. □

Remark 2. It should be noted that Proposition 1.2.4 is valid not only for sequences, but also for nets, which means that the unit spheres $\mathbf{S}_{\ell_p(\Gamma)}$, $p \in \mathbb{N}$, are subspaces of the product space $\mathbb{R}^\Gamma$, whereas $\mathbb{R}^\Gamma$ and $\mathbb{R}^\Gamma_f$ are not metrizable if Γ is uncountable. Therefore, if $1 \leq p < q \leq \infty$, then $\mathbf{S}_{\ell_p(\Gamma)}$ is also a subspace of $\ell_q(\Gamma)$, although, as we have seen, $\mathbf{S}_{\ell_p(\Gamma)}$ of $\ell_p(\Gamma)$ is not closed in the space $\ell_q(\Gamma)$. The unit sphere $\mathbf{S}_{\ell_p^f(\Gamma)}$ of $\ell_p^f(\Gamma)$ is a subspace of $\mathbb{R}_f^\Gamma$ ($\subset \mathbb{R}^\Gamma$), and also a subspace of $\ell_q(\Gamma)$ for $1 \leq q \leq \infty$.

Remark 3. The "if" part of Proposition 1.2.4 does not hold for the space $\boldsymbol{c}_0(\Gamma)$ (although the "only if" part obviously does hold), where Γ is infinite. For instance, take distinct $\gamma_n \in \Gamma$, $n \in \omega$, and let $(x_n)_{n\in\mathbb{N}}$ be the sequence in $\boldsymbol{c}_0(\Gamma)$ defined by $x_n = \mathbf{e}_{\gamma_n} + \mathbf{e}_{\gamma_0}$. Then, $\|x_n\|_\infty = 1$ for each $n \in \mathbb{N}$,

$$\lim_{n\to\infty} x_n(\gamma_0) = 1 = \mathbf{e}_{\gamma_0}(\gamma_0) \ \text{ and } \ \lim_{n\to\infty} x_n(\gamma) = 0 = \mathbf{e}_{\gamma_0}(\gamma) \ \text{ for } \gamma \neq \gamma_0,$$

but $\|x_n - \mathbf{e}_{\gamma_0}\|_\infty = 1$ for every $n \in \Gamma$. In addition, the unit sphere $\mathbf{S}_{\boldsymbol{c}_0(\Gamma)}$ of $\boldsymbol{c}_0(\Gamma)$ is not a subspace of $\mathbb{R}^\Gamma$, because $\mathbf{e}_{\gamma_n} \in \mathbf{S}_{\boldsymbol{c}_0(\Gamma)}$ but $(\mathbf{e}_{\gamma_n})_{n\in\mathbb{N}}$ converges to $\mathbf{0}$ in $\mathbb{R}^\Gamma$.

Concerning the topological classification of $\ell_p(\Gamma)$, we have the following:

Theorem 1.2.5 (Mazur). *For each $1 < p < \infty$, $\ell_p(\Gamma)$ is homeomorphic to $\ell_1(\Gamma)$. By the same homeomorphism, $\ell_p^f(\Gamma)$ is also homeomorphic to $\ell_1^f(\Gamma)$.*

Proof. We define $\varphi : \ell_1(\Gamma) \to \ell_p(\Gamma)$ and $\psi : \ell_p(\Gamma) \to \ell_1(\Gamma)$ as follows:

$$\varphi(x)(\gamma) = \operatorname{sign} x(\gamma) \cdot |x(\gamma)|^{1/p} \ \text{ for } x \in \ell_1(\Gamma),$$
$$\psi(x)(\gamma) = \operatorname{sign} x(\gamma) \cdot |x(\gamma)|^p \ \text{ for } x \in \ell_p(\Gamma),$$

where $\operatorname{sign} 0 = 0$ and $\operatorname{sign} a = a/|a|$ for $a \neq 0$. We can apply Proposition 1.2.4 to verify the continuity of φ and ψ. In fact, the following functions are continuous:

$$\ell_1(\Gamma) \ni x \mapsto \|\varphi(x)\|_p = \left(\|x\|_1\right)^{1/p} \in \mathbb{R}, \ \ell_1(\Gamma) \ni x \mapsto \varphi(x)(\gamma) \in \mathbb{R}, \ \gamma \in \Gamma;$$
$$\ell_p(\Gamma) \ni x \mapsto \|\psi(x)\|_1 = \left(\|x\|_p\right)^p \in \mathbb{R}, \ \ell_p(\Gamma) \ni x \mapsto \psi(x)(\gamma) \in \mathbb{R}, \ \gamma \in \Gamma.$$

Observe that $\psi\varphi = \mathrm{id}$ and $\varphi\psi = \mathrm{id}$. Thus, φ is a homeomorphism with $\varphi^{-1} = \psi$, where $\varphi(\ell_p^f(\Gamma)) \subset \ell_1^f(\Gamma)$ and $\psi(\ell_1^f(\Gamma)) \subset \ell_p^f(\Gamma)$. □

For each space X, we denote $\mathrm{C}(X) = \mathrm{C}(X, \mathbb{R})$. The Banach space $\mathrm{C}^B(X)$ is defined as follows:

- $\mathrm{C}^B(X) = \left\{ f \in \mathrm{C}(X) \mid \sup_{x\in X} |f(x)| < \infty \right\}$ with the sup-norm

$$\|f\| = \sup_{x\in X} |f(x)|.$$

This sup-norm of $\mathrm{C}^B(X)$ induces the uniform convergence topology. If X is discrete and infinite, then we have $\mathrm{C}^B(X) = \ell_\infty(X)$, and so, in particular, $\mathrm{C}^B(\mathbb{N}) = \ell_\infty$. When X is compact, $\mathrm{C}^B(X) = \mathrm{C}(X)$ and the topology induced by the norm coincides with the compact-open topology.

The **uniform convergence topology** of $\mathrm{C}(X)$ is induced by the following metric:

$$d(f,g) = \sup_{x \in X} \min\{|f(x) - g(x)|,\ 1\}.$$

As can be easily observed, $\mathrm{C}^B(X)$ is closed and open in $\mathrm{C}(X)$ under the uniform convergence topology. Note that $\mathrm{C}^B(X)$ is a component of the space $\mathrm{C}(X)$ because $\mathrm{C}^B(X)$ is path-connected as a normed linear space.

Regarding $\mathrm{C}(X)$ as a subspace of the product space $\mathbb{R}^X$, we can introduce a topology on $\mathrm{C}(X)$, which is called the **pointwise convergence topology**. With respect to this topology,

$$\lim_{n\to\infty} f_n = f \quad \Leftrightarrow \quad \lim_{n\to\infty} f_n(x) = f(x) \ \text{ for every } x \in X.$$

The space $\mathrm{C}(X)$ with the pointwise convergence topology is usually denoted by $\mathrm{C}_p(X)$. The space $\mathrm{C}_p(\mathbb{N})$ is simply the space of sequences $\boldsymbol{s} = \mathbb{R}^{\mathbb{N}}$.

In this chapter, three topologies on $\mathrm{C}(X)$ have been considered — the compact-open topology, the uniform convergence topology, and the pointwise convergence topology. Among them, the uniform convergence topology is the finest and the pointwise convergence topology is the coarsest.

Notes for Chap. 1

Theorem 1.2.5 is due to Mazur [3]. Zhongqiang Yang pointed out that Proposition 1.2.4 can be applied to show the continuity of φ and ψ in the proof of Theorem 1.2.5. Related to Mazur's result, Anderson [1] proved that $\boldsymbol{s} = \mathbb{R}^{\mathbb{N}}$ is homeomorphic to the Hilbert space ℓ_2. For an elementary proof, refer to [2].

References

1. R.D. Anderson, Hilbert space is homeomorphic to the countable infinite product of lines. Bull. Am. Math. Soc. **72**, 515–519 (1966)
2. R.D. Anderson, R.H. Bing, A complete elementary proof that Hilbert space is homeomorphic to the countable infinite product of lines. Bull. Am. Math. Soc. **74**, 771–792 (1968)
3. S. Mazur, Une remarque sur l'homéomorphie des champs fonctionnels. Stud. Math. **1**, 83–85 (1929)

Chapter 2
Metrization and Paracompact Spaces

In this chapter, we are mainly concerned with metrization and paracompact spaces. We also derive some properties of the products of compact spaces and perfect maps. Several metrization theorems are proved, and we characterize completely metrizable spaces. We will study several different characteristics of paracompact spaces that indicate, in many situations, the advantages of paracompactness. In particular, there exists a useful theorem showing that, if a paracompact space has a certain property *locally*, then it has the same property *globally*. Furthermore, paracompact spaces have partitions of unity, which is also a very useful property.

2.1 Products of Compact Spaces and Perfect Maps

In this section, we present some theorems regarding the products of compact spaces and compactifications. In addition, we introduce perfect maps. First, we present a proof of the TYCHONOFF THEOREM.

Theorem 2.1.1 (TYCHONOFF). *The product space* $\prod_{\lambda \in \Lambda} X_\lambda$ *of compact spaces* X_λ, $\lambda \in \Lambda$, *is compact.*

Proof. Let $X = \prod_{\lambda \in \Lambda} X_\lambda$. We may assume that $\Lambda = (\Lambda, \leq)$ is a well-ordered set. For each $\mu \in \Lambda$, let $p_\mu : X \to \prod_{\lambda \leq \mu} X_\lambda$ and $q_\mu : X \to \prod_{\lambda < \mu} X_\lambda$ be the projections.

Let $\mathcal{A}$ be a collection of subsets of X with the finite intersection property (f.i.p.). Using transfinite induction, we can find $x_\lambda \in X_\lambda$ such that $\mathcal{A}|p_\lambda^{-1}(U)$ has the f.i.p. for every neighborhood U of $(x_\nu)_{\nu \leq \lambda}$ in $\prod_{\nu \leq \lambda} X_\nu$. Indeed, suppose that $x_\lambda \in X_\lambda$, $\lambda < \mu$, have been found, but there exists no $x_\mu \in X_\mu$ with the above property, i.e., any $y \in X_\mu$ has an open neighborhood V_y with an open neighborhood U_y of $(x_\lambda)_{\lambda < \mu}$ in $\prod_{\lambda < \mu} X_\lambda$ such that $\mathcal{A}|q_\mu^{-1}(U_y) \cap \mathrm{pr}_\mu^{-1}(V_y)$ does not have the f.i.p. Because X_μ is compact, we have $y_1, \ldots, y_n \in X_\mu$ such that $X_\mu = \bigcup_{i=1}^n V_{y_i}$. Since $\bigcap_{i=1}^n U_{y_i}$ is a neighborhood of $(x_\lambda)_{\lambda < \mu}$ in

K. Sakai, *Geometric Aspects of General Topology*, Springer Monographs in Mathematics, DOI 10.1007/978-4-431-54397-8_2, © Springer Japan 2013

$\prod_{\lambda<\mu} X_\lambda$, we have $\nu_1, \ldots, \nu_m < \mu$ and neighborhoods W_i of x_{ν_i} in X_{ν_i} such that $\bigcap_{i=1}^m \mathrm{pr}_{\nu_i}^{-1}(W_i) \subset q_\mu^{-1}(\bigcap_{i=1}^n U_{y_i})$. Let $\nu = \max\{\nu_1, \ldots, \nu_m\} < \mu$. Then, we can write $\bigcap_{i=1}^m \mathrm{pr}_{\nu_i}^{-1}(W_i) = p_\nu^{-1}(W)$ for some neighborhood W of $(x_\lambda)_{\lambda\le\nu}$ in $\prod_{\lambda\le\nu} X_\lambda$. Because $p_\nu^{-1}(W) \subset \bigcap_{i=1}^n q_\mu^{-1}(U_{y_i})$, no $\mathcal{A}|p_\nu^{-1}(W) \cap \mathrm{pr}_\mu^{-1}(V_{y_i})$ have the f.i.p. Since $X = \bigcup_{i=1}^n \mathrm{pr}_\mu^{-1}(V_{y_i})$, it follows that $\mathcal{A}|p_\nu^{-1}(W)$ does not have the f.i.p., which contradicts the inductive assumption.

Now, we have obtained the point $x = (x_\lambda)_{\lambda\in\Lambda} \in X$. For each neighborhood U of x in X, we have $\lambda_1, \ldots, \lambda_n \in \Lambda$ and neighborhoods U_i of x_{λ_i} in X_{λ_i} such that $\bigcap_{i=1}^n \mathrm{pr}_{\lambda_i}^{-1}(U_i) \subset U$. Let $\lambda_0 = \max\{\lambda_1, \ldots, \lambda_n\} \in \Lambda$. Then, we can write $\bigcap_{i=1}^n \mathrm{pr}_{\lambda_i}^{-1}(U_i) = p_{\lambda_0}^{-1}(U_0)$ for some neighborhood U_0 of $(x_\nu)_{\nu\le\lambda_0}$ in $\prod_{\nu\le\lambda_0} X_\nu$. Since $p_{\lambda_0}^{-1}(U_0) \subset U$, $\mathcal{A}|U$ has the f.i.p. Consequently, every neighborhood U of x in X meets every member of $\mathcal{A}$. This means that $x \in \bigcap_{A\in\mathcal{A}} \mathrm{cl}\, A$, and so $\bigcap_{A\in\mathcal{A}} \mathrm{cl}\, A \neq \emptyset$. □

Note. There are various proofs of the Tychonoff Theorem. In one familiar proof, Zorn's Lemma is applied instead of the transfinite induction. Let $\mathcal{A}$ be a collection of subsets of X with the f.i.p. and Φ be all of collections $\mathcal{A}'$ of subsets of X such that $\mathcal{A}'$ has the f.i.p. and $\mathcal{A} \subset \mathcal{A}'$. Applying Zorn's Lemma to the ordered set $\Phi = (\Phi, \subset)$, we can obtain a maximal element $\mathcal{A}^* \in \Phi$. Because of the maximality, $\mathcal{A}^*$ has the following properties:

(1) The intersection of any finite members of $\mathcal{A}^*$ belongs to $\mathcal{A}^*$;
(2) If $B \subset X$ meets every member of $\mathcal{A}^*$, then $B \in \mathcal{A}^*$.

For each $\lambda \in \Lambda$, $\mathrm{pr}_\lambda(\mathcal{A}^*)$ has the f.i.p. Since X_λ is compact, we have $x_\lambda \in \bigcap_{A\in\mathcal{A}^*} \mathrm{cl}\, \mathrm{pr}_\lambda(A)$. It follows from (2) that $\mathrm{pr}_\lambda^{-1}(V) \in \mathcal{A}^*$ for every neighborhood V of x_λ in X_λ. Now, it is easy to see that

$$x = (x_\lambda)_{\lambda\in\Lambda} \in \bigcap_{A\in\mathcal{A}^*} \mathrm{cl}\, A \subset \bigcap_{A\in\mathcal{A}} \mathrm{cl}\, A.$$

Next, we prove Wallace's Theorem:

Theorem 2.1.2 (Wallace). *Let $A = \prod_{\lambda\in\Lambda} A_\lambda \subset X = \prod_{\lambda\in\Lambda} X_\lambda$, where each A_λ is compact. Then, for each open set W in X with $A \subset W$, there exists a finite subset $\Lambda_0 \subset \Lambda$ and open sets V_λ in X_λ, $\lambda \in \Lambda_0$, such that $A \subset \bigcap_{\lambda\in\Lambda_0} \mathrm{pr}_\lambda^{-1}(V_\lambda) \subset W$.*

Proof. When Λ is finite, we may take $\Lambda_0 = \Lambda$. Then, $\bigcap_{\lambda\in\Lambda_0} \mathrm{pr}_\lambda^{-1}(V_\lambda)$ coincides with $\prod_{\lambda\in\Lambda} V_\lambda$. This case can be proved by induction on card Λ, which is reduced to the case card $\Lambda = 2$. Proving the case card $\Lambda = 2$ is an excellent exercise.[1]

We will show that the general case is derived from the finite case. For each $x \in A$, we have a finite subset $\Lambda(x) \subset \Lambda$ and an open set $U(x)$ in $\prod_{\lambda\in\Lambda(x)} X_\lambda$ such that $x \in \mathrm{pr}_{\Lambda(x)}^{-1}(U(x)) \subset W$. Because of the compactness of A, there exist finite $x_1, \ldots, x_n \in A$ such that $A \subset \bigcup_{i=1}^n \mathrm{pr}_{\Lambda(x_i)}^{-1}(U(x_i))$. Thus, we have a finite subset

[1] Use the same strategy used in the proof of normality of a compact Hausdorff space.

$\Lambda_0 = \bigcup_{i=1}^n \Lambda(x_i) \subset \Lambda$. For each $i = 1, \dots, n$, let $p_i : \prod_{\lambda \in \Lambda_0} X_\lambda \to \prod_{\lambda \in \Lambda(x_i)} X_\lambda$ be the projection. Then, $W_0 = \bigcup_{i=1}^n p_i^{-1}(U(x_i))$ is an open set in $\prod_{\lambda \in \Lambda_0} X_\lambda$.

Note that $\bigcup_{i=1}^n \mathrm{pr}_{\Lambda(x_i)}^{-1}(U(x_i)) = \mathrm{pr}_{\Lambda_0}^{-1}(W_0)$. From the finite case, we obtain open sets V_λ, $\lambda \in \Lambda_0$, such that $\prod_{\lambda \in \Lambda_0} A_\lambda \subset \prod_{\lambda \in \Lambda_0} V_\lambda \subset W_0$. Hence,

$$A \subset \bigcap_{\lambda \in \Lambda_0} \mathrm{pr}_\lambda^{-1}(V_\lambda) \subset \mathrm{pr}_{\Lambda_0}^{-1}(W_0) \subset W. \qquad \square$$

For any space X, we define the evaluation map $e_X : X \to \mathbf{I}^{\mathrm{C}(X,\mathbf{I})}$ by $e_X(x) = (f(x))_{f \in \mathrm{C}(X,\mathbf{I})}$ for each $x \in X$. The continuity of e_X follows from the fact that $\mathrm{pr}_f \circ e_X = f$ is continuous for each $f \in \mathrm{C}(X, \mathbf{I})$, where $\mathrm{pr}_f : \mathbf{I}^{\mathrm{C}(X,\mathbf{I})} \to \mathbf{I}$ is the projection (i.e., $\mathrm{pr}_f(\xi) = \xi(f)$).

Proposition 2.1.3. *For every Tychonoff space X, the map $e_X : X \to \mathbf{I}^{\mathrm{C}(X,\mathbf{I})}$ is an embedding.*

Proof. Let U be an open set in X and $x \in U$. Since X is a Tychonoff space, we have some $f \in \mathrm{C}(X, \mathbf{I})$ such that $f(x) = 0$ and $f(X \backslash U) \subset \{1\}$. Then, $V = \mathrm{pr}_f^{-1}([0, 1))$ is an open set in $\mathbf{I}^{\mathrm{C}(X,\mathbf{I})}$. Since $\mathrm{pr}_f(e_X(x)) = f(x) = 0$, it follows that $e_X(x) \in V$. Since $\mathrm{pr}_f \circ e_X(X \setminus U) = f(X \setminus U) \subset \{1\}$, we have $e_X(X \setminus U) \cap V = \emptyset$. Therefore, $e_X(x) \in V \cap e_X(X) \subset e_X(U)$. This implies that $e_X : X \to e_X(X)$ is an open map.

For $x \neq y \in X$, applying the above argument to $U = X \setminus \{y\}$, we can see that $e_X(x)(f) = 0 \neq 1 = e_X(y)(f)$. Thus, e_X is an embedding. $\square$

From Tychonoff's Theorem, it follows that the product space $\mathbf{I}^{\mathrm{C}(X,\mathbf{I})}$ is compact. Then, identifying X with $e_X(X)$, we define a compactification βX of X as follows:

$$\beta X = \mathrm{cl}_{\mathbf{I}^{\mathrm{C}(X,\mathbf{I})}}\, e_X(X),$$

which is called the **Stone–Čech compactification**.

Now, let $f : X \to Y$ be a map between Tychonoff spaces. The map $f_* : \mathbf{I}^{\mathrm{C}(X,\mathbf{I})} \to \mathbf{I}^{\mathrm{C}(Y,\mathbf{I})}$ is defined as $f_*(\xi) = (\xi(kf))_{k \in \mathrm{C}(Y,\mathbf{I})}$ for each $\xi \in \mathbf{I}^{\mathrm{C}(X,\mathbf{I})}$, where the continuity of f_* follows from the continuity of $\mathrm{pr}_k \circ f_* = \mathrm{pr}_{kf}$, $k \in \mathrm{C}(Y, \mathbf{I})$. Then, we have $f_* \circ e_X = e_Y \circ f$.

$$\begin{array}{ccc} X & \xrightarrow{\;f\;} & Y \\ {\scriptstyle e_X}\downarrow & & \downarrow{\scriptstyle e_Y} \\ \mathbf{I}^{\mathrm{C}(X,\mathbf{I})} & \xrightarrow[f_*]{} & \mathbf{I}^{\mathrm{C}(Y,\mathbf{I})} \end{array}$$

Indeed, for each $x \in X$ and $k \in \mathrm{C}(Y, \mathbf{I})$,

$$f_*(e_X(x))(k) = e_X(x)(kf) = k(f(x)) = e_Y(f(x))(k).$$

Since f_* is continuous, it follows that $f_*(\beta X) \subset \beta Y$. Thus, f extends to the map $\beta f = f_*|\beta X : \beta X \to \beta Y$.

Further, let $g : Y \to Z$ be another map, where Z is Tychonoff. Then, for each $\xi \in \mathbf{I}^{\mathrm{C}(X,\mathbf{I})}$ and $k \in \mathrm{C}(Z,\mathbf{I})$,

$$g_*(f_*(\xi))(k) = f_*(\xi)(kg) = \xi(kgf) = (gf)_*(\xi)(k),$$

that is, $g_* f_* = (gf)_*$. Therefore, $\beta(gf) = \beta g \beta f$.

The Stone–Čech compactification βX can be characterized as follows:

Theorem 2.1.4 (Stone; Čech). *Let X be a Tychonoff space. For any compactification γX of X, there exists the (unique) map $f : \beta X \to \gamma X$ such that $f|X = \mathrm{id}_X$. If a compactification $\beta' X$ has the same property as above, then there exists a homeomorphism $h : \beta X \to \beta' X$ such that $h|X = \mathrm{id}_X$.*

Proof. Note that $\beta(\gamma X) = \gamma X$ because γX is compact. Let $i : X \hookrightarrow \gamma X$ be the inclusion and let $f = \beta i : \beta X \to \beta(\gamma X) = \gamma X$. Then, $f|X = \mathrm{id}_X$ and f is unique because X is dense in βX.

If a compactification $\beta' X$ of X has the same property, then we have two maps $h : \beta X \to \beta' X$ and $h' : \beta' X \to \beta X$ such that $h|X = h'|X = \mathrm{id}_X$. It follows that $h'h = \mathrm{id}_{\beta X}$ and $hh' = \mathrm{id}_{\beta' X}$, which means that h is a homeomorphism. □

A **perfect map** $f : X \to Y$ is a closed map such that $f^{-1}(y)$ is compact for each $y \in Y$. A map $f : X \to Y$ is said to be **proper** if $f^{-1}(K)$ is compact for every compact set $K \subset Y$.

Proposition 2.1.5. *Every perfect map $f : X \to Y$ is proper. If Y is locally compact, then every proper map $f : X \to Y$ is perfect.*

Proof. To prove the first assertion, let $K \subset Y$ be compact and $\mathcal{U}$ an open cover of $f^{-1}(K)$ in X. For each $y \in K$, choose a finite subcollection $\mathcal{U}_y \subset \mathcal{U}$ so that $f^{-1}(y) \subset \bigcup \mathcal{U}_y$. Since f is closed, each $V_y = Y \setminus f(X \setminus \bigcup \mathcal{U}_y)$ is an open neighborhood of y in Y, where $f^{-1}(V_y) \subset \bigcup \mathcal{U}_y$. We can choose $y_1, \ldots, y_n \in K$ so that $K \subset \bigcup_{i=1}^n V_{y_i}$. Thus, we have a finite subcollection $\mathcal{U}_0 = \bigcup_{i=1}^n \mathcal{U}_{y_i} \subset \mathcal{U}$ such that $f^{-1}(K) \subset \bigcup \mathcal{U}_0$. Hence, $f^{-1}(K)$ is compact.

To show the second assertion, it suffices to prove that a proper map f is closed. Let $A \subset X$ be closed and $y \in \mathrm{cl}\, f(A)$. Since Y is locally compact, y has a compact neighborhood N in Y. Note that $N \cap f(A) \neq \emptyset$, which implies $f^{-1}(N) \cap A \neq \emptyset$. Since f is proper, $f^{-1}(N)$ is compact, and hence $f^{-1}(N) \cap A$ is also compact. Thus, $f(f^{-1}(N) \cap A)$ is compact, so it is closed in Y. If $y \notin f(f^{-1}(N) \cap A)$, y has a compact neighborhood $M \subset N$ with $M \cap f(f^{-1}(N) \cap A) = \emptyset$. Then, observe that

$$f(f^{-1}(M) \cap A) \subset M \cap f(f^{-1}(N) \cap A) = \emptyset,$$

which means that $f^{-1}(M) \cap A = \emptyset$. However, using the same argument as for $f^{-1}(N) \cap A \neq \emptyset$, we can see that $f^{-1}(M) \cap A \neq \emptyset$, which is a contradiction. Thus, $y \in f(f^{-1}(N) \cap A) \subset f(A)$. Therefore, $f(A)$ is closed in Y. □

It follows from the first assertion of Proposition 2.1.5 that the composition of any two perfect maps is also perfect. In the second assertion, the local compactness of Y is not necessary if X and Y are metrizable, which allows the following proposition:

Proposition 2.1.6. *For a map $f : X \to Y$ between metrizable spaces, the following are equivalent:*

(a) *$f : X \to Y$ is perfect;*
(b) *$f : X \to Y$ is proper;*
(c) *Any sequence $(x_n)_{n\in\mathbb{N}}$ in X has a convergent subsequence if $(f(x_n))_{n\in\mathbb{N}}$ is convergent in Y.*

Proof. The implication (a) ⇒ (b) has been shown in Proposition 2.1.5.

(b) ⇒ (c): Let $y = \lim_{n\to\infty} f(x_n) \in Y$ and $K = \{f(x_n) \mid n \in \mathbb{N}\} \cup \{y\}$. Since K is compact, (b) implies the compactness of $f^{-1}(K)$, whose sequence $(x_n)_{n\in\mathbb{N}}$ has a convergent subsequence.

(c) ⇒ (a): For each $y \in Y$, every sequence $(x_n)_{n\in\mathbb{N}}$ in $f^{-1}(y)$ has a convergent subsequence due to (c), which means that $f^{-1}(y)$ is compact because $f^{-1}(y)$ is metrizable.

To see that f is a closed map, let $A \subset X$ be a closed set and $y \in \mathrm{cl}_Y f(A)$. Then, we have a sequence $(x_n)_{n\in\mathbb{N}}$ in A such that $y = \lim_{n\to\infty} f(x_n)$. Due to (c), $(x_n)_{n\in\mathbb{N}}$ has a convergent subsequence $(x_{n_i})_{i\in\mathbb{N}}$, and since A is closed in X, we have $\lim_{i\to\infty} x_{n_i} = x \in A$. Then, $y = f(x) \in f(A)$, and therefore $f(A)$ is closed in Y. This completes the proof. □

Lemma 2.1.7. *Let D be a dense subset of X such that $D \neq X$. Any perfect map $f : D \to Y$ cannot extend over X.*

Proof. Assume that f extends to a map $\tilde{f} : X \to Y$. Let $x_0 \in X \setminus D$, $y_0 = \tilde{f}(x_0)$, $\widetilde{D} = D \cup \{x_0\}$, and $g = \tilde{f}|\widetilde{D} : \widetilde{D} \to Y$. Since $f^{-1}(y_0)$ is compact and $x_0 \notin f^{-1}(y_0)$, $\widetilde{D}$ has disjoint open sets U and V such that $x_0 \in U$ and $f^{-1}(y_0) \subset V$. Since f is a closed map, $f(D \setminus V)$ is closed in Y, hence $g^{-1}(f(D \setminus V))$ is closed in $\widetilde{D}$. Because $g^{-1}(y) = f^{-1}(y)$ for any $y \in Y \setminus \{y_0\}$, we have

$$D \setminus V \subset g^{-1}(f(D \setminus V)) = f^{-1}(f(D \setminus V)) \subset D.$$

On the other hand, $x_0 \notin \mathrm{cl}_{\widetilde{D}} V$. Therefore, $D = \mathrm{cl}_{\widetilde{D}} V \cup g^{-1}(f(D \setminus V))$ is closed in $\widetilde{D}$, which contradicts the fact that D is dense in $\widetilde{D}$. □

Theorem 2.1.8. *For a map $f : X \to Y$ between Tychonoff spaces, the following are equivalent:*

(a) *f is perfect;*
(b) *For any compactification γY of Y, f extends to a map $\tilde{f} : \beta X \to \gamma Y$ so that $\tilde{f}(\beta X \setminus X) \subset \gamma Y \setminus Y$;*
(c) *$\beta f(\beta X \setminus X) \subset \beta Y \setminus Y$.*

Proof. The implication (b) $\Rightarrow$ (c) is obvious.

(a) $\Rightarrow$ (b): Applying Theorem 2.1.4, we can obtain a map $g : \beta Y \to \gamma Y$ with $g|Y = \mathrm{id}$. Then, $\tilde{f} = g(\beta f)$ is an extension of f. Moreover, we can apply Lemma 2.1.7 to see that $\tilde{f}(\beta X \setminus X) \subset \gamma Y \setminus Y$.

(c) $\Rightarrow$ (a): For each $y \in Y$, $f^{-1}(y) = (\beta f)^{-1}(y)$ is compact. For each closed set A in X,

$$(\beta f)(\mathrm{cl}_{\beta X} A) \cap Y = f(\mathrm{cl}_{\beta X} A \cap X) = f(A),$$

which implies that $f(A)$ is closed in Y. Therefore, f is perfect. □

Remark 1. In Theorem 2.1.4, the map $f : \beta X \to \gamma X$ with $f|X = \mathrm{id}_X$ satisfies the condition $f(\beta X \setminus X) \subset \gamma X \setminus X$ that follows from Theorem 2.1.8.

Using Tychonoff's Theorem 2.1.1 and Wallace's Theorem 2.1.2, we can prove the following:

Theorem 2.1.9. *For each $\lambda \in \Lambda$, let $f_\lambda : X_\lambda \to Y_\lambda$ be a perfect map. Then, the map $f = \prod_{\lambda \in \Lambda} f_\lambda : X = \prod_{\lambda \in \Lambda} X_\lambda \to Y = \prod_{\lambda \in \Lambda} Y_\lambda$ is also perfect.*

Proof. Owing to Tychonoff's Theorem 2.1.1, $f^{-1}(y) = \prod_{\lambda \in \Lambda} f_\lambda^{-1}(y(\lambda))$ is compact for each $y \in Y$. To show that f is a closed map, let A be a closed set in X and $y \in Y \setminus f(A)$. Since $f^{-1}(y) \subset X \setminus A$, we can apply Wallace's Theorem 2.1.2 to obtain $\lambda_1, \ldots, \lambda_n \in \Lambda$ and open sets U_i in X_{λ_i}, $i = 1, \ldots, n$, such that

$$f^{-1}(y) = \prod_{\lambda \in \Lambda} f_\lambda^{-1}(y(\lambda)) \subset \bigcap_{i=1}^{n} \mathrm{pr}_{\lambda_i}^{-1}(U_i) \subset X \setminus A.$$

Since f_{λ_i} is a closed map, $V_i = Y_{\lambda_i} \setminus f_{\lambda_i}(X_{\lambda_i} \setminus U_i)$ is an open neighborhood of $y(\lambda_i)$ in Y_{λ_i} and $f_{\lambda_i}^{-1}(V_i) \subset U_i$. Then, $V = \bigcap_{i=1}^{n} \mathrm{pr}_{\lambda_i}^{-1}(V_i)$ is a neighborhood of y in Y and $f^{-1}(V) \subset X \setminus A$, i.e., $V \cap f(A) = \emptyset$. Therefore, f is a closed map. □

2.2 The Tietze Extension Theorem and Normalities

In this section, we prove the Tietze Extension Theorem and present a few concepts that strengthen normality. For $A, B \subset X$, it is said that A and B are **separated** in X if $A \cap \mathrm{cl}\, B = \emptyset$ and $B \cap \mathrm{cl}\, A = \emptyset$.

Lemma 2.2.1. *Let A and B be separated F_σ sets in a normal space X. Then, X has disjoint open sets U and V such that $A \subset U$ and $B \subset V$.*

Proof. Let $A = \bigcup_{n \in \mathbb{N}} A_n$ and $B = \bigcup_{n \in \mathbb{N}} B_n$, where $A_1 \subset A_2 \subset \cdots$ and $B_1 \subset B_2 \subset \cdots$ are closed in X. Set $U_0 = V_0 = \emptyset$. Using normality, we can inductively choose open sets $U_n, V_n \subset X$, $n \in \mathbb{N}$, so that

$$A_n \cup \mathrm{cl}\, U_{n-1} \subset U_n \subset \mathrm{cl}\, U_n \subset X \setminus (\mathrm{cl}\, B \cup \mathrm{cl}\, V_{n-1}) \quad \text{and}$$

$$B_n \cup \mathrm{cl}\, V_{n-1} \subset V_n \subset \mathrm{cl}\, V_n \subset X \setminus (\mathrm{cl}\, A \cup \mathrm{cl}\, U_n).$$

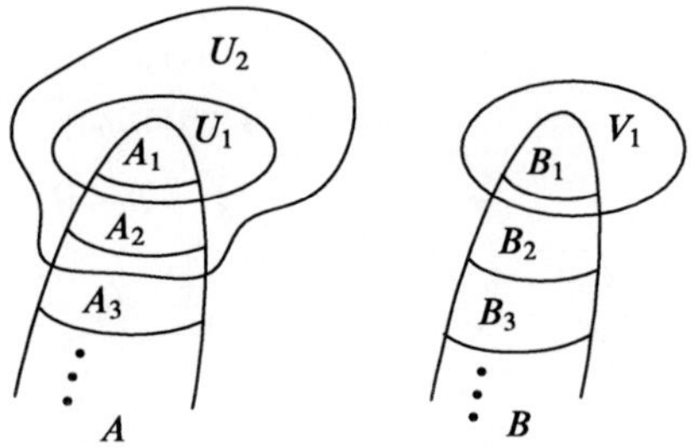

Fig. 2.1 Construction of U_n and V_n

Then, $U = \bigcup_{n\in\mathbb{N}} U_n$ and $V = \bigcup_{n\in\mathbb{N}} V_n$ are disjoint open sets in X such that $A \subset U$ and $B \subset V$ — Fig. 2.1. □

We can now prove the following extension theorem:

Theorem 2.2.2 (TIETZE EXTENSION THEOREM). *Let A be a closed set in a normal space X. Then, every map $f : A \to \mathbf{I}$ extends over X.*

Proof. We first construct the open sets $W(q)$ in X, $q \in \mathbf{I} \cap \mathbb{Q}$, so that

(1) $q < q' \Rightarrow \operatorname{cl} W(q) \subset W(q')$,
(2) $A \cap W(q) = f^{-1}([0, q))$.

To this end, let $\{q_n \mid n \in \mathbb{N}\} = \mathbf{I} \cap \mathbb{Q}$, where $q_1 = 0$, $q_2 = 1$ and $q_i \neq q_j$ if $i \neq j$. We define $W(q_1) = W(0) = \emptyset$ and $W(q_2) = W(1) = X \setminus f^{-1}(1)$. Assume that $W(q_1), W(q_2), \cdots, W(q_n)$ have been defined so as to satisfy (1) and (2). Let

$$q_l = \min\{q_i \mid q_i > q_{n+1}, i = 1, \cdots, n\} \quad \text{and}$$
$$q_m = \max\{q_i \mid q_i < q_{n+1}, i = 1, \cdots, n\}.$$

Note that $f^{-1}([0, q_{n+1}))$ and $f^{-1}((q_{n+1}, 1])$ are separated F_σ sets in X. Using Lemma 2.2.1, we can find an open set U in X such that $f^{-1}([0, q_{n+1})) \subset U$ and $f^{-1}((q_{n+1}, 1]) \cap \operatorname{cl} U = \emptyset$. Then, $V = U \setminus f^{-1}(q_{n+1})$ is open in X and $A \cap V = f^{-1}([0, q_{n+1}))$. Again, using normality, we can obtain an open set G in X such that

$$\operatorname{cl} W(q_m) \cup f^{-1}([0, q_{n+1}]) \subset G \subset \operatorname{cl} G \subset W(q_l).$$

Then, $A \cap (V \cap G) = f^{-1}([0, q_{n+1}))$ and $\operatorname{cl}(V \cap G) \subset W(q_l)$. Yet again, using normality, we can take an open set H in X such that

$$\operatorname{cl} W(q_m) \subset H \subset \operatorname{cl} H \subset G \setminus f^{-1}([q_{n+1}, 1]) \ (\subset W(q_l)).$$

Then, $W(q_{n+1}) = (V \cap G) \cup H$ is the desired open set in X (Fig. 2.2).

Now, we define $\tilde{f} : X \to \mathbf{I}$ as follows:

$$\tilde{f}(x) = \begin{cases} 1 & \text{if } x \notin W(1), \\ \inf\{q \in \mathbf{I} \cap \mathbb{Q} \mid x \in W(q)\} & \text{if } x \in W(1). \end{cases}$$

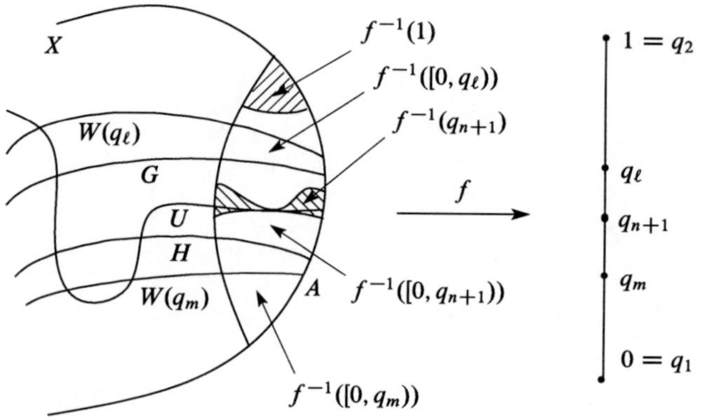

Fig. 2.2 $W(q_{n+1}) = ((U \setminus f^{-1}(q_{n+1})) \cap G) \cup H$

Then, $\tilde{f}|A = f$ because, for each $x \in A \cap W(1) = A \setminus f^{-1}(1)$,

$$\tilde{f}(x) = \inf\{q \in \mathbf{I} \cap \mathbb{Q} \mid x \in f^{-1}([0,q))\} = f(x).$$

To see the continuity of $\tilde{f}$, let $0 < a \leq 1$ and $0 \leq b < 1$. Since $\tilde{f}(x) < a$ if and only if $x \in W(q)$ for some $q < a$, it follows that $\tilde{f}^{-1}([0,a)) = \bigcup_{q<a} W(q)$ is open in X. Moreover, from (1), it follows that $\tilde{f}(x) > b$ if and only if $x \notin \operatorname{cl} W(q)$ for some $q > b$. Then, $\tilde{f}^{-1}((b,1]) = X \setminus \bigcap_{q>b} \operatorname{cl} W(q)$ is also open in X. Therefore, $\tilde{f}$ is continuous. □

As a corollary, we have Urysohn's Lemma:

Corollary 2.2.3 (Urysohn's Lemma). *For each disjoint pair of closed sets A and B in a normal space X, there exists a map $f : X \to \mathbf{I}$ such that $A \subset f^{-1}(0)$ and $B \subset f^{-1}(1)$.* □

Such a map f as in the above is called a **Urysohn map**.

Note. In the standard proof of the Tietze Extension Theorem 2.2.2, the desired extension is obtained as the uniform limit of a sequence of approximate extensions that are sums of Urysohn maps. On the other hand, Urysohn's Lemma is directly proved as follows:

Using the normality property yields the open sets $W(q)$ in X corresponding to all $q \in \mathbf{I} \cap \mathbb{Q}$ satisfying condition (1) in our proof of the Tietze Extension Theorem and

$$A \subset W(0) \subset \operatorname{cl} W(0) \subset W(1) = X \setminus B.$$

A Urysohn map $f : X \to \mathbf{I}$ can be defined as follows:

$$f(x) = \begin{cases} 1 & \text{if } x \notin W(1), \\ \inf\{q \in \mathbf{I} \cap \mathbb{Q} \mid x \in W(q)\} & \text{if } x \in W(1). \end{cases}$$

In general, a subspace of a normal space is not normal (cf. Sect. 2.10). However, we have the following proposition:

Proposition 2.2.4. *Every F_σ set in a normal space is also normal.*

Proof. Let Y be an F_σ set in a normal space X. Every pair of disjoint closed sets in Y are F_σ sets in X that are separated in X. Then, the normality of Y follows from Lemma 2.2.1. □

A space X is **hereditarily normal** if every subspace of X is normal. Evidently, every metrizable space is hereditarily normal. It is said that X is **completely normal** provided that, for each pair of separated subsets $A, B \subset X$, there exist disjoint open sets U and V in X such that $A \subset U$ and $B \subset V$. These concepts meet in the following theorem:

Theorem 2.2.5. *For a space X, the following are equivalent:*

(a) *X is hereditarily normal;*
(b) *Every open set in X is normal;*
(c) *X is completely normal.*

Proof. The implication (a) $\Rightarrow$ (b) is obvious.

(c) $\Rightarrow$ (a): For an arbitrary subspace $Y \subset X$, each pair of disjoint closed sets A and B in Y are separated in X. Then, (a) follows from (c).

(b) $\Rightarrow$ (c): Let $A, B \subset X$ be separated, i.e., $A \cap \operatorname{cl} B = \emptyset$ and $B \cap \operatorname{cl} A = \emptyset$. Then, $W = X \setminus (\operatorname{cl} A \cap \operatorname{cl} B)$ is open in X and $A, B \subset W$. Moreover,

$$\operatorname{cl}_W A \cap \operatorname{cl}_W B = W \cap \operatorname{cl} A \cap \operatorname{cl} B = \emptyset.$$

From the normality of W, we have disjoint open sets U and V in W such that $A \subset U$ and $B \subset V$. Then, U and V are open in X, and hence we have (c). □

A normal space X is **perfectly normal** if every closed set in X is G_δ in X (equivalently, every open set in X is F_σ in X). Clearly, every metrizable space is perfectly normal. A closed set $A \subset X$ is called a **zero set** in X if $A = f^{-1}(0)$ for some map $f : X \to \mathbb{R}$, where $\mathbb{R}$ can be replaced by $\mathbf{I}$. The complement of a zero set in X is called a **cozero set**.

Theorem 2.2.6. *For a space X, the following conditions are equivalent:*

(a) *X is perfectly normal;*
(b) *Every closed set in X is a zero set (equivalently, every open set in X is a cozero set);*
(c) *For every pair of disjoint closed sets A and B in X, there exists a map $f : X \to \mathbf{I}$ such that $A = f^{-1}(0)$ and $B = f^{-1}(1)$.*

Proof. The implication (c) $\Rightarrow$ (a) is trivial.

(a) $\Rightarrow$ (b): Let A be a closed set in X. Then, we can write $A = \bigcap_{n \in \mathbb{N}} G_n$, where each G_n is open in X. Using Urysohn's Lemma, we take maps $f_n : X \to \mathbf{I}$, $n \in \mathbb{N}$, such that $f_n(A) \subset \{0\}$ and $f_n(X \setminus G_n) \subset \{1\}$. We can define a map $f : X \to \mathbf{I}$ as $f(x) = \sum_{n\in\mathbb{N}} 2^{-n} f_n(x)$. Then, it is easy to see that $A = f^{-1}(0)$.

(b) $\Rightarrow$ (c): Let A and B be disjoint closed sets in X. Condition (b) provides two maps $g, h : X \to \mathbf{I}$ such that $g^{-1}(0) = A$ and $h^{-1}(0) = B$. Then, the desired map $f : X \to \mathbf{I}$ can be defined as follows:

$$f(x) = \frac{g(x)}{g(x) + h(x)}.$$ □

Theorem 2.2.7. *Every perfectly normal space is hereditarily normal (= completely normal).*

Proof. Let X be perfectly normal. Then, each open set in X is an F_σ set, which is normal as a consequence of Proposition 2.2.4. Hence, it follows from Theorem 2.2.5 that X is hereditarily normal. □

Remark 2. Let $A_0, A_1, \dots, A_n$ be pairwise disjoint closed sets in a normal space X. We can apply the Tietze Extension Theorem 2.2.2 to obtain a map $f : X \to \mathbf{I}$ such that $A_i \subset f^{-1}(i/n)$ (i.e., $f(A_i) \subset \{i/n\}$) for each $i = 0, 1, \dots, n$. When X is perfectly normal and $n > 2$, the condition $A_i \subset f^{-1}(i/n)$ cannot be replaced by $A_i = f^{-1}(i/n)$. For example, let $X = \mathbf{S}^1$ be the unit circle (the unit 1-sphere of $\mathbb{R}^2$), $A_0 = \{\mathbf{e}_1\}$, $A_1 = \{\mathbf{e}_2\}$, and $A_2 = \{-\mathbf{e}_1\}$, where $\mathbf{e}_1 = (1, 0), \mathbf{e}_2 = (0, 1) \in \mathbb{R}^2$. Since $X \setminus A_1$ is (path-)connected, there does not exist a map $f : X \to \mathbf{I}$ such that $A_0 = f^{-1}(0)$, $A_1 = f^{-1}(1/2)$ and $A_2 = f^{-1}(1)$.

2.3 Stone's Theorem and Metrization

In this section, we prove Stone's Theorem and characterize the metrizability using open bases. Let $\mathcal{A}$ be a collection of subsets of a space X and $B \subset X$. Recall that

$$\mathcal{A}[B] = \{A \in \mathcal{A} \mid A \cap B \neq \emptyset\}.$$

When $B = \{x\}$, we write $\mathcal{A}[\{x\}] = \mathcal{A}[x]$. It is said that $\mathcal{A}$ is **locally finite** (resp. **discrete**) in X if each $x \in X$ has a neighborhood U that meets only finite members (resp. at most one member) of $\mathcal{A}$, i.e., $\operatorname{card} \mathcal{A}[U] < \aleph_0$ (resp. $\operatorname{card} \mathcal{A}[U] \leq 1$). When $w(X) \geq \aleph_0$, if $\mathcal{A}$ is locally finite in X, then $\operatorname{card} \mathcal{A} \leq w(X)$. For the sake of convenience, we introduce the notation $\mathcal{A}^{\mathrm{cl}} = \{\operatorname{cl} A \mid A \in \mathcal{A}\}$. The following is easily proved and will be used frequently:

Fact. *If $\mathcal{A}$ is locally finite (or discrete) in X, then so is $\mathcal{A}^{\mathrm{cl}}$ and also $\operatorname{cl} \bigcup \mathcal{A} = \bigcup \mathcal{A}^{\mathrm{cl}}$ $(= \bigcup_{A\in\mathcal{A}} \operatorname{cl} A)$.*

A collection of subsets of X is said to be **σ-locally finite** (resp. **σ-discrete**) in X if it can be represented as a countable union of locally finite (resp. discrete) collections.

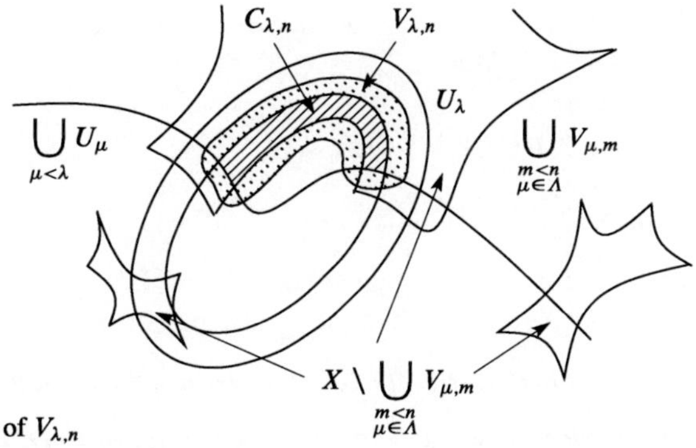

Fig. 2.3 Definition of $V_{\lambda,n}$

Theorem 2.3.1 (A.H. STONE). *Every open cover of a metrizable space has a locally finite and σ-discrete open refinement.*

Proof. Let $X = (X, d)$ be a metric space and $\mathcal{U} \in \mathrm{cov}(X)$. We may index all members of $\mathcal{U}$ by a well-ordered set $\Lambda = (\Lambda, \leq)$, that is, $\mathcal{U} = \{U_\lambda \mid \lambda \in \Lambda\}$. By induction on $n \in \mathbb{N}$, we define open collections $\mathcal{V}_n = \{V_{\lambda,n} \mid \lambda \in \Lambda\}$ as follows:

$$V_{\lambda,n} = \mathrm{N}(C_{\lambda,n}, 2^{-n}) = \{x \in X \mid d(x, C_{\lambda,n}) < 2^{-n}\},$$

where

$$C_{\lambda,n} = \{x \in X \mid d(x, X \setminus U_\lambda) > 2^{-n}3\} \setminus \left(\bigcup_{\mu<\lambda} U_\mu \cup \bigcup_{\substack{m<n \\ \mu\in\Lambda}} V_{\mu,m} \right).$$

For each $x \in X$, let $\lambda(x) = \min\{\lambda \in \Lambda \mid x \in U_\lambda\}$ and choose $n \in \mathbb{N}$ so that $2^{-n}3 < d(x, X \setminus U_{\lambda(x)})$. Then, $x \in C_{\lambda(x),n} \subset V_{\lambda(x),n}$ or $x \in V_{\mu,m}$ for some $\mu \in \Lambda$ and $m < n$. Hence, we have $\mathcal{V} = \bigcup_{n\in\mathbb{N}} \mathcal{V}_n \in \mathrm{cov}(X)$. Since each $V_{\lambda,n}$ is contained in U_λ, it follows that $\mathcal{V} \prec \mathcal{U}$. See Fig. 2.3.

The discreteness of each $\mathcal{V}_n$ follows from the claim:

Claim (1). If $\lambda \neq \mu$ then $\mathrm{dist}_d(V_{\lambda,n}, V_{\mu,n}) \geq 2^{-n}$.

To prove this claim, we may assume $\mu < \lambda$. For each $x \in V_{\lambda,n}$ and $y \in V_{\mu,n}$, choose $x' \in C_{\lambda,n}$ and $y' \in C_{\mu,n}$ so that $d(x, x') < 2^{-n}$ and $d(y, y') < 2^{-n}$, respectively. Then, $x' \notin U_\mu$ and $d(y', X \setminus U_\mu) > 2^{-n}3$, hence $d(x', y') > 2^{-n}3$. Therefore,

$$d(x, y) \geq d(x', y') - d(x, x') - d(y, y') > 2^{-n}.$$

The local finiteness of $\mathcal{V}$ follows from the discreteness of each $\mathcal{V}_n$ and the claim:

Claim (2). If $\mathrm{B}(x, 2^{-k}) \subset V_{\mu,m}$, then $\mathrm{B}(x, 2^{-k-1}) \cap V_{\lambda,n} = \emptyset$ for all $\lambda \in \Lambda$ and $n > \max\{k, m\}$.

For each $y \in V_{\lambda,n}$, choose $y' \in C_{\lambda,n}$ so that $d(y, y') < 2^{-n}$. Since $y' \notin V_{\mu,m}$, it follows that $d(x, y') \geq 2^{-k}$. Hence,

$$d(x, y) \geq d(x, y') - d(y, y') > 2^{-k} - 2^{-n} \geq 2^{-k-1}.$$

The proof is complete. □

Applying Theorem 2.3.1 to the open covers $\mathcal{B}_n = \{\mathrm{B}(x, 2^{-n}) \mid x \in X\}$, $n \in \mathbb{N}$, of a metric space $X = (X, d)$, we have the following corollary:

Corollary 2.3.2. *Every metrizable space has a σ-discrete open basis.* □

Lemma 2.3.3. *A regular space X with a σ-locally finite open basis is perfectly normal.*

Proof. Let $\mathcal{B} = \bigcup_{n\in\mathbb{N}} \mathcal{B}_n$ be an open basis for X where each $\mathcal{B}_n$ is locally finite in X. Instead of proving that every closed set in X is a G_δ set, we show that every open set $W \subset X$ is F_σ. For each $x \in W$, choose $k(x) \in \mathbb{N}$ and $B(x) \in \mathcal{B}_{k(x)}$ so that $x \in B(x) \subset \operatorname{cl} B(x) \subset W$. For each $n \in \mathbb{N}$, let

$$W_n = \bigcup \bigl\{B(x) \bigm| x \in W,\ k(x) = n\bigr\}.$$

Because of the local finiteness of $\mathcal{B}_n$, we have

$$\operatorname{cl} W_n = \bigcup \bigl\{\operatorname{cl} B(x) \bigm| x \in W,\ k(x) = n\bigr\} \subset W.$$

Since $W = \bigcup_{n\in\mathbb{N}} W_n$, it follows that $W = \bigcup_{n\in\mathbb{N}} \operatorname{cl} W_n$, which is F_σ in X.

To prove normality, let A and B be disjoint closed sets in X. As seen above, we have open sets $V_n, W_n \subset X$, $n \in \mathbb{N}$, such that $X \setminus A = \bigcup_{n\in\mathbb{N}} V_n = \bigcup_{n\in\mathbb{N}} \operatorname{cl} V_n$ and $X \setminus B = \bigcup_{n\in\mathbb{N}} W_n = \bigcup_{n\in\mathbb{N}} \operatorname{cl} W_n$. For each $n \in \mathbb{N}$, let

$$G_n = W_n \setminus \bigcup_{m\leq n} \operatorname{cl} V_m \quad\text{and}\quad H_n = V_n \setminus \bigcup_{m\leq n} \operatorname{cl} W_m.$$

Then, $G = \bigcup_{n\in\mathbb{N}} G_n$ and $H = \bigcup_{n\in\mathbb{N}} H_n$ are disjoint open sets in X such that $A \subset G$ and $B \subset H$. □

Theorem 2.3.4 (Bing; Nagata–Smirnov). *For a regular space X, the following conditions are equivalent:*

(a) *X is metrizable;*
(b) *X has a σ-discrete open basis;*
(c) *X has a σ-locally finite open basis.*

Proof. The implication (a) ⇒ (b) is Corollary 2.3.2 and (b) ⇒ (c) is obvious. It remains to show the implication (c) ⇒ (a).

(c) ⇒ (a): Let $\mathcal{B} = \bigcup_{n\in\mathbb{N}} \mathcal{B}_n$ be an open basis for X where each $\mathcal{B}_n$ is locally finite in X. Since X is perfectly normal by Lemma 2.3.3, we have maps $f_B : X \to \mathbf{I}$,

$B \in \mathcal{B}$, such that $f_B^{-1}(0) = X \setminus B$ (Theorem 2.2.6). For each $n \in \mathbb{N}$, since $\mathcal{B}_n$ is locally finite, we can define a map $f_n : X \to \ell_1(\mathcal{B}_n)$ by $f_n(x) = (f_B(x))_{B \in \mathcal{B}_n} \in \ell_1(\mathcal{B}_n)$. Let $f : X \to \prod_{n \in \mathbb{N}} \ell_1(\mathcal{B}_n)$ be the map defined by $f(x) = (f_n(x))_{n \in \mathbb{N}}$. Since $\prod_{n \in \mathbb{N}} \ell_1(\mathcal{B}_n)$ is metrizable, it suffices to show that f is an embedding.

For each $x \neq y \in X$, choose $B \in \mathcal{B}_n \subset \mathcal{B}$ so that $x \in B$ and $y \notin B$. Then, $f_B(x) > 0 = f_B(y)$, so $f_n(x) \neq f_n(y)$. Hence, f is an injection.

For each $n \in \mathbb{N}$ and $B \in \mathcal{B}_n$, $V_B = \{y \in \ell_1(\mathcal{B}_n) \mid y(B) > 0\}$ is open in $\ell_1(\mathcal{B}_n)$. Observe that for $x \in X$,

$$x \in B \Leftrightarrow f_n(x)(B) = f_B(x) > 0 \Leftrightarrow f_n(x) \in V_B.$$

Then, it follows that $f(B) = \mathrm{pr}_n^{-1}(V_B) \cap f(X)$ is open in $f(X)$, where $\mathrm{pr}_n : \prod_{n \in \mathbb{N}} \ell_1(\mathcal{B}_n) \to \ell_1(\mathcal{B}_n)$ is the projection. Thus, f is an embedding. □

The equivalence of (a) and (b) in Theorem 2.3.4 is called the BING METRIZATION THEOREM, and the equivalence of (a) and (c) is called the NAGATA–SMIRNOV METRIZATION THEOREM. As a corollary, we have the URYSOHN METRIZATION THEOREM:

Corollary 2.3.5. *A space is separable and metrizable if and only if it is regular and second countable.* □

For a metrizable space X, let Γ be an infinite set with $w(X) \leq \operatorname{card} \Gamma$. In the proof of Theorem 2.3.4, note that $\operatorname{card} \mathcal{B}_n \leq \operatorname{card} \Gamma$ because of the local finiteness of $\mathcal{B}_n$ in X. Then, every $\ell_1(\mathcal{B}_n)$ can be embedded into $\ell_1(\Gamma)$. Therefore, we can state the following corollary:

Corollary 2.3.6. *Let X be a metrizable space and Γ an infinite set such that $w(X) \leq \operatorname{card} \Gamma$. Then, X can be embedded in the completely metrizable topological linear space*[2] $\ell_1(\Gamma)^{\mathbb{N}}$. □

Here, $w(\ell_1(\Gamma)^{\mathbb{N}}) = w(\ell_1(\Gamma)) = \operatorname{card} \Gamma$. In fact, $w(\ell_1(\Gamma)) \geq \operatorname{card} \Gamma$ because $\ell_1(\Gamma)$ has a discrete open collection with the same cardinality as Γ. Let

$$D = \big\{x \in \ell_1(\Gamma) \,\big|\, x(\gamma) \in \mathbb{Q} \text{ for all } \gamma \in \Gamma \text{ and } x(\gamma) = 0 \text{ except for finitely many } \gamma \in \Gamma\big\}.$$

Then, $\{\mathrm{B}(x, n^{-1}) \mid x \in D, n \in \mathbb{N}\}$ is an open basis for $\ell_1(\Gamma)$ with the same cardinality as Γ, hence $w(\ell_1(\Gamma)) \leq \operatorname{card} \Gamma$.

The **hedgehog** $J(\Gamma)$ is the closed subspace of $\ell_1(\Gamma)$ defined as follows:

$$J(\Gamma) = \bigcup_{\gamma \in \Gamma} \mathbf{I}\mathbf{e}_\gamma = \big\{x \in \ell_1(\Gamma) \,\big|\, x(\gamma) \in \mathbf{I} \text{ for all } \gamma \in \Gamma \text{ and } x(\gamma) \neq 0 \text{ at most one } \gamma \in \Gamma\big\},$$

[2] For topological linear spaces, refer to Sect. 3.4.

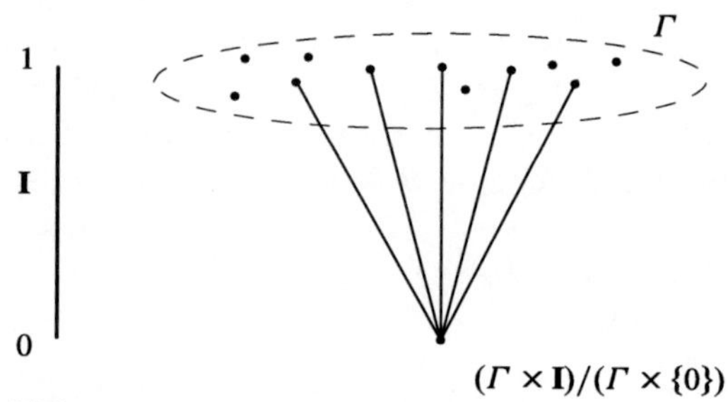

Fig. 2.4 The hedgehog $J(\Gamma)$

where $\mathbf{e}_\gamma \in \ell_1(\Gamma)$ is the unit vector defined by $\mathbf{e}_\gamma(\gamma) = 1$ and $\mathbf{e}_\gamma(\gamma') = 0$ for $\gamma' \neq \gamma$ (Fig. 2.4). The hedgehog $J(\Gamma)$ can also be defined as the space $(\Gamma \times \mathbf{I})/(\Gamma \times \{0\})$ with the metric induced from the pseudo-metric ρ on $\Gamma \times \mathbf{I}$ defined as follows:

$$\rho((\gamma, t), (\gamma', s)) = \begin{cases} |t - s| & \text{if } \gamma = \gamma', \\ t + s & \text{if } \gamma \neq \gamma'. \end{cases}$$

Note that $w(J(\Gamma)^{\mathbb{N}}) = \operatorname{card} \Gamma$. In the proof of Theorem 2.3.4, if each $\mathcal{B}_n$ is discrete in X, then $f_n(X) \subset J(\mathcal{B}_n)$. Similar to Corollary 2.3.6, we have the following:

Corollary 2.3.7. *Let X be a metrizable space and Γ an infinite set such that $w(X) \leq \operatorname{card} \Gamma$. Then, X can be embedded in $J(\Gamma)^{\mathbb{N}}$.* □

In the second countable case, X can be embedded in $\mathbf{I}^{\mathbb{N}}$, since we can take $\mathcal{B} = \bigcup_{n \in \mathbb{N}} \mathcal{B}_n$ in the proof of Theorem 2.3.4 so that each $\mathcal{B}_n$ contains only one open set. Thus, we have the following embedding theorem for separable metrizable spaces:

Corollary 2.3.8. *Every separable metrizable space can be embedded in the Hilbert cube $\mathbf{I}^{\mathbb{N}}$, and hence in $\mathbb{R}^{\mathbb{N}}$.* □

In association with Corollary 2.3.6, we state the following theorem:

Theorem 2.3.9. *Every metric space $X = (X, d)$ can be isometrically embedded into the Banach space $C^B(X)$.*

Sketch of Proof. Fix $x_0 \in X$ and define $\varphi : X \to C^B(X)$ as follows:

$$\varphi(x)(z) = d(x, z) - d(x_0, z), \ z \in X.$$

It is easy to see that $\|\varphi(x)\| = d(x, x_0)$ and $\|\varphi(x) - \varphi(y)\| = d(x, y)$.

The **(metric) completion** of a metric space $X = (X, d)$ is a complete metric space $\widetilde{X} = (\widetilde{X}, \tilde{d})$ containing X as a dense set and as a metric subspace, that is, d is the restriction of $\tilde{d}$. Since a closed set in a complete metric space is also complete, Theorem 2.3.9 implies the following:

Corollary 2.3.10. *Every metric space has a completion.* □

2.4 Sequences of Open Covers and Metrization

In this section, we characterize metrizable spaces via sequences of open covers. Given a cover $\mathcal{V}$ of a space X and $A \subset X$, we define

$$\mathrm{st}(A, \mathcal{V}) = \bigcup \mathcal{V}[A],$$

which is called the **star** of A with respect to $\mathcal{V}$. When $A = \{x\}$, we write $\mathrm{st}(\{x\}, \mathcal{V}) = \mathrm{st}(x, \mathcal{V})$.

Theorem 2.4.1 (ALEXANDROFF–URYSOHN; FRINK). *For a space X, the following conditions are equivalent:*

(a) *X is metrizable;*
(b) *X has open covers $\mathcal{U}_1, \mathcal{U}_2, \ldots$ such that $\{\mathrm{st}(x, \mathcal{U}_n) \mid n \in \mathbb{N}\}$ is a neighborhood basis of each $x \in X$ and*

$$U, U' \in \mathcal{U}_{n+1},\ U \cap U' \neq \emptyset \Rightarrow \exists U'' \in \mathcal{U}_n \text{ such that } U \cup U' \subset U'';$$

(c) *Each $x \in X$ has an open neighborhood basis $\{V_n(x) \mid n \in \mathbb{N}\}$ satisfying the condition that, for each $x \in X$ and $i \in \mathbb{N}$, there exists a $j(x,i) \geq i$ such that*

$$V_{j(x,i)}(x) \cap V_{j(x,i)}(y) \neq \emptyset \Rightarrow V_{j(x,i)}(y) \subset V_i(x).$$

Proof. (a) $\Rightarrow$ (c): A metric space $X = (X, d)$ satisfies (c) because

$$\mathrm{B}(x, 3^{-n}) \cap \mathrm{B}(y, 3^{-n}) \neq \emptyset \Rightarrow \mathrm{B}(y, 3^{-n}) \subset \mathrm{B}(x, 3^{-n+1}).$$

(c) $\Rightarrow$ (b): For each $x \in X$, let $k(x,1) = 1$ and inductively define

$$k(x,n) = \max\{n, j(x,i) \mid i = 1, \ldots, k(x,n-1)\} \geq n.$$

For each $n \in \mathbb{N}$, let $U_n(x) = \bigcap_{i=1}^{k(x,n)} V_i(x)$. Then, $\{U_n(x) \mid n \in \mathbb{N}\}$ is an open neighborhood basis of x and

$$\begin{aligned} U_n(x) \cap U_n(y) \neq \emptyset \ \Rightarrow U_n(x) \cup U_n(y) &\subset U_{n-1}(x) \text{ or} \\ U_n(x) \cup U_n(y) &\subset U_{n-1}(y). \end{aligned}$$

In fact, assume that $U_n(x) \cap U_n(y) \neq \emptyset$. In the case $k(x,n) \leq k(y,n)$, $V_{j(x,i)}(y) \subset V_i(x)$ for each $i = 1, \ldots, k(x,n-1)$ because $V_{j(x,i)}(x) \cap V_{j(x,i)}(y) \neq \emptyset$. Then, it follows that

$$U_n(y) \subset \bigcap_{i=1}^{k(x,n)} V_i(y) \subset \bigcap_{i=1}^{k(x,n-1)} V_{j(x,i)}(y) \subset \bigcap_{i=1}^{k(x,n-1)} V_i(x) = U_{n-1}(x).$$

Since $U_n(x) \subset U_{n-1}(x)$ by definition, we have $U_n(x) \cup U_n(y) \subset U_{n-1}(x)$. As above, $k(y,n) \leq k(x,n)$ implies $U_n(x) \cup U_n(y) \subset U_{n-1}(y)$.

For each $n \in \mathbb{N}$, we have $\mathcal{U}_n = \{U_n(x) \mid x \in X\} \in \operatorname{cov}(X)$. It remains to be prove that $\{\operatorname{st}(x,\mathcal{U}_n) \mid n \in \mathbb{N}\}$ is a neighborhood basis of $x \in X$. Evidently, each $\operatorname{st}(x,\mathcal{U}_n)$ is a neighborhood of $x \in X$. Then, it suffices to show that $\operatorname{st}(x,\mathcal{U}_{j(x,n)}) \subset V_n(x)$. If $x \in U_{j(x,n)}(y)$, then

$$V_{j(x,n)}(x) \cap V_{j(x,n)}(y) \supset V_{j(x,n)}(x) \cap U_{j(x,n)}(y) \neq \emptyset,$$

and hence $U_{j(x,n)}(y) \subset V_{j(x,n)}(y) \subset V_n(x)$.

(b) $\Rightarrow$ (a): First, note that $\mathcal{U}_{i+1} \prec \mathcal{U}_i$ for each $i \in \mathbb{N}$. Let $\mathcal{U}_0 = \{X\} \in \operatorname{cov}(X)$. For each $x, y \in X$, define

$$\delta(x, y) = \inf\{2^{-i} \mid \exists U \in \mathcal{U}_i \text{ such that } x, y \in U\}.$$

Note that if $\delta(x, y) > 0$, then $\delta(x, y) = 2^{-n}$ for some $n \geq 0$. As can easily be shown, the following hold for each $x, y, z \in X$:

(1) $\delta(x, y) = 0$ if and only if $x = y$;
(2) $\delta(x, y) = \delta(y, x)$;
(3) $\delta(x, y) \leq 2\max\{\delta(x, z), \delta(z, y)\}$.

Furthermore, we claim that

(4) for every $n \geq 3$ and each $x_1, \ldots, x_n \in X$,

$$\delta(x_1, x_n) \leq 2(\delta(x_1, x_2) + \delta(x_{n-1}, x_n)) + 4\sum_{i=2}^{n-2} \delta(x_i, x_{i+1}).$$

In fact, when $n = 3$, the inequality follows from (3). Assuming claim (4) holds for any $n < k$, we show (4) for $n = k$. Then, we may assume that $x_k \neq x_1$. For each $x_1, \ldots, x_k \in X$, let

$$m = \min\{i \mid \delta(x_1, x_k) \leq 2\delta(x_1, x_i)\} \geq 2.$$

Then, $\delta(x_1, x_k) \leq 2\delta(x_1, x_m)$. From (3) and the minimality of m, we have $\delta(x_1, x_k) \leq 2\delta(x_{m-1}, x_k)$. If $m = 2$ or $m = k$, then the inequality in (4) holds for $n = k$. In the case $2 < m < k$,

$$\delta(x_1, x_k) = \frac{1}{2}\delta(x_1, x_k) + \frac{1}{2}\delta(x_1, x_k) \leq \delta(x_1, x_m) + \delta(x_{m-1}, x_k).$$

By the inductive assumption, we have

$$\delta(x_1, x_m) \leq 2(\delta(x_1, x_2) + \delta(x_{m-1}, x_m)) + 4\sum_{i=2}^{m-2} \delta(x_i, x_{i+1}) \quad \text{and}$$

$$\delta(x_{m-1}, x_k) \leq 2(\delta(x_{m-1}, x_m) + \delta(x_{k-1}, x_k)) + 4\sum_{i=m}^{k-2} \delta(x_i, x_{i+1}),$$

so the desired inequality is obtained. By induction, (4) holds for all $n \in \mathbb{N}$.

Now, we can define $d \in \operatorname{Metr}(X)$ as follows:

$$d(x, y) = \inf\left\{ \sum_{i=1}^{n-1} \delta(x_i, x_{i+1}) \;\middle|\; n \in \mathbb{N},\ x_i \in X,\ x_1 = x,\ x_n = y \right\}.$$

In fact, $d(x, y) = d(y, x)$ by (2) and the above definition. The triangle inequality follows from the definition of d. Since $\delta(x, y) \leq 4d(x, y)$ by (4), it follows from (1) that $d(x, y) = 0$ implies $x = y$. Obviously, $x = y$ implies $d(x, y) = 0$. Moreover, it follows that

$$d(x, y) \leq 2^{-n-2} \Rightarrow \exists U \in \mathcal{U}_n \text{ such that } x, y \in U,$$

which means that $\overline{\mathrm{B}}_d(x, 2^{-n-2}) \subset \operatorname{st}(x, \mathcal{U}_n)$ for each $x \in X$ and $n \in \mathbb{N}$. Since $d(x, y) \leq \delta(x, y)$, we have $\operatorname{mesh}_d \mathcal{U}_n \leq 2^{-n}$, so $\operatorname{st}(x, \mathcal{U}_n) \subset \overline{\mathrm{B}}_d(x, 2^{-n})$. Therefore, $\{\mathrm{B}_d(x, 2^{-n}) \mid n \in \mathbb{N}\}$ is a neighborhood basis of $x \in X$. □

Remark 3. In the above proof of (b) ⇒ (a), the obtained metric $d \in \operatorname{Metr}(X)$ has the following property:

$$\operatorname{st}(x, \mathcal{U}_{n+2}) \subset \overline{\mathrm{B}}_d(x, 2^{-n-2}) \subset \operatorname{st}(x, \mathcal{U}_n).$$

Moreover, $d(x, y) \leq 1$ for every $x, y \in X$.

In Theorem 2.4.1, the equivalence between (a) and (b) is called the ALEXANDROFF–URYSOHN METRIZATION THEOREM and the equivalence between (a) and (c) is called the FRINK METRIZATION THEOREM.

Let $\mathcal{U}$ and $\mathcal{V}$ be covers of X. When $\{\operatorname{st}(x, \mathcal{V}) \mid x \in X\} \prec \mathcal{U}$, we call $\mathcal{V}$ a **Δ-refinement** (or **barycentric refinement**) of $\mathcal{U}$ and denote

$$\mathcal{V} \overset{\Delta}{\prec} \mathcal{U} \quad (\text{or } \mathcal{U} \overset{\Delta}{\succ} \mathcal{V}).$$

The following corollary follows from the Alexandroff–Urysohn Metrization Theorem:

Corollary 2.4.2. *A space X is metrizable if and only if X has a sequence of open covers*

$$\mathcal{U}_1 \overset{\Delta}{\succ} \mathcal{U}_2 \overset{\Delta}{\succ} \mathcal{U}_3 \overset{\Delta}{\succ} \cdots$$

such that $\{\operatorname{st}(x, \mathcal{U}_n) \mid n \in \mathbb{N}\}$ is a neighborhood basis of each $x \in X$. □

For covers $\mathcal{U}$ and $\mathcal{V}$ of X, we define

$$\operatorname{st}(\mathcal{V}, \mathcal{U}) = \left\{ \operatorname{st}(V, \mathcal{U}) \;\middle|\; V \in \mathcal{V} \right\},$$

which is called the **star** of $\mathcal{V}$ with respect to $\mathcal{U}$. We denote $\operatorname{st}(\mathcal{V}, \mathcal{V}) = \operatorname{st} \mathcal{V}$, which is called the **star** of $\mathcal{V}$. When $\operatorname{st} \mathcal{V} \prec \mathcal{U}$, we call $\mathcal{V}$ a **star-refinement** of $\mathcal{U}$ and denote

$$\mathcal{V} \overset{*}{\prec} \mathcal{U} \quad (\text{or } \mathcal{U} \overset{*}{\succ} \mathcal{V}).$$

For each $n \in \mathbb{N}$, the ***n*-th star** of $\mathcal{V}$ is inductively defined as follows:

$$\mathrm{st}^n\,\mathcal{V} = \mathrm{st}(\mathrm{st}^{n-1}\,\mathcal{V}, \mathcal{V}),$$

where $\mathrm{st}^0\,\mathcal{V} = \mathcal{V}$. Observe that $\mathrm{st}(\mathcal{V}, \mathrm{st}\,\mathcal{V}) = \mathrm{st}^3\,\mathcal{V}$ and $\mathrm{st}(\mathrm{st}\,\mathcal{V}) = \mathrm{st}^4\,\mathcal{V}$. When $\mathrm{st}^n\,\mathcal{V} \prec \mathcal{U}$, $\mathcal{V}$ is called an ***n*-th star-refinement** of $\mathcal{U}$. There is the following relation between Δ-refinements and star-refinements:

Proposition 2.4.3. *For every three open covers $\mathcal{U}, \mathcal{V}, \mathcal{W}$ of a space X,*

$$\mathcal{W} \overset{\Delta}{\prec} \mathcal{V} \overset{\Delta}{\prec} \mathcal{U} \Rightarrow \mathcal{W} \overset{*}{\prec} \mathcal{U}.$$

Sketch of Proof. For each $W \in \mathcal{W}$, take any $x \in W$ and choose $U \in \mathcal{U}$ so that $\mathrm{st}(x, \mathcal{V}) \subset U$. Then, we see that $\mathrm{st}(W, \mathcal{W}) \subset U$.

By virtue of this proposition, Δ-refinements in Corollary 2.4.2 can be replaced by star-refinements, which allows us to sate the following corollary:

Corollary 2.4.4. *A space X is metrizable if and only if X has a sequence of open covers*

$$\mathcal{U}_1 \overset{*}{\succ} \mathcal{U}_2 \overset{*}{\succ} \mathcal{U}_3 \overset{*}{\succ} \cdots$$

such that $\{\mathrm{st}(x, \mathcal{U}_n) \mid n \in \mathbb{N}\}$ is a neighborhood basis of each $x \in X$. □

Remark 4. By tracing the proof of Theorem 2.4.1, we can directly prove Corollary 2.4.4. This direct proof is simpler than that of Theorem 2.4.1, and the obtained metric $d \in \mathrm{Metr}(X)$ has the following, more acceptable, property than the previous remark:

$$\mathrm{st}(x, \mathcal{U}_{n+1}) \subset \overline{\mathrm{B}}_d(x, 2^{-n}) \subset \mathrm{st}(x, \mathcal{U}_n).$$

Similar to the previous metric, $d(x, y) \leq 1$ for every $x, y \in X$.

Sketch of the direct proof of Corollary 2.4.4. To see the "if" part, replicate the proof of (b) ⇒ (a) in Theorem 2.4.1 to construct $d \in \mathrm{Metr}(X)$. Let $\mathcal{U}_0 = \{X\}$. For each $x, y \in X$, we define

$$\delta(x, y) = \inf\{2^{-i+1} \mid \exists U \in \mathcal{U}_i \text{ such that } x, y \in U\} \text{ and}$$
$$d(x, y) = \inf\{\textstyle\sum_{i=1}^n \delta(x_{i-1}, x_i) \mid n \in \mathbb{N},\ x_0 = x,\ x_n = y\}.$$

The admissibility and additional property of d are derived from the inequality $d(x, y) \leq \delta(x, y) \leq 2d(x, y)$. To prove the right-hand inequality, it suffices to show the following:

$$\delta(x_0, x_n) \leq 2\sum_{i=1}^{n} \delta(x_{i-1}, x_i) \text{ for each } x_0, x_1, \ldots, x_n \in X.$$

This is proved by induction on $n \in \mathbb{N}$. Set $\sum_{i=1}^n \delta(x_{i-1}, x_i) = \alpha$ and let k be the largest number such that $\sum_{i=1}^k \delta(x_{i-1}, x_i) \leq \alpha/2$. Then, $\sum_{i=k+2}^n \delta(x_{i-1}, x_i) < \alpha/2$. By the inductive assumption, $\delta(x_0, x_k) \leq \alpha$ and $\delta(x_{k+1}, x_n) < \alpha$. Note that $\delta(x_k, x_{k+1}) \leq \alpha$. Let $m = \min\{i \in \mathbb{N} \mid 2^{-i+1} \leq \alpha\}$. Since $\mathrm{st}\,\mathcal{U}_m \prec \mathcal{U}_{m-1}$, we can find $U \in \mathcal{U}_{m-1}$ such that $x_0, x_n \in U$, and hence $\delta(x_0, x_n) \leq 2^{-m+2} \leq 2\alpha$.

Additional Results on Metrizability 2.4.5.

(1) The perfect image of a metrizable space is metrizable, that is, if $f : X \to Y$ is a surjective perfect map of a metrizable space X, then Y is also metrizable.

Sketch of Proof. For each $y \in Y$ and $n \in \mathbb{N}$, let

$$U_n(y) = \mathrm{N}_d(f^{-1}(y), 2^{-n}) \text{ and } V_n(y) = Y \setminus f(X \setminus U_n(y)),$$

where d is an admissible metric for X. Show that $\{V_n(y) \mid n \in \mathbb{N}\}$ is a neighborhood basis of $y \in Y$ that satisfies condition 2.4.1(c). For each $y \in Y$ and $i \in \mathbb{N}$, since $f^{-1}(y)$ is compact, we can choose $j \geq i$ so that $U_j(y) \subset f^{-1}(V_{i+1}(y))$. Then, the following holds:

$$V_{j+1}(y) \cap V_{j+1}(z) \neq \emptyset \Rightarrow V_{j+1}(z) \subset V_i(y).$$

To see this, observe that

$$\begin{aligned} V_{j+1}(y) \cap V_{j+1}(z) \neq \emptyset &\Rightarrow U_j(y) \cap f^{-1}(z) \neq \emptyset \\ &\Rightarrow f^{-1}(z) \subset f^{-1}(V_{i+1}(y)) \subset U_{i+1}(y) \\ &\Rightarrow f^{-1}(V_{j+1}(z)) \subset U_{j+1}(z) \subset U_i(y). \end{aligned}$$

(2) A space X is metrizable if it is a locally finite union of metrizable closed subspaces.

Sketch of Proof. To apply (1) above, construct a surjective perfect map $f : \bigoplus_{\lambda \in \Lambda} X_\lambda \to X$ such that each X_λ is metrizable and $f|X_\lambda$ is a closed embedding. The metrizability of $\bigoplus_{\lambda \in \Lambda} X_\lambda$ easily follows from Theorem 2.3.4. (The metrizability of $\bigoplus_{\lambda \in \Lambda} X_\lambda$ can also be seen by embedding $\bigoplus_{\lambda \in \Lambda} X_\lambda$ into the product space $\Lambda \times \ell_1(\Gamma)^{\mathbb{N}}$ for some Γ, where we give Λ the discrete topology.)

2.5 Complete Metrizability

In this section, we consider complete metrizability. A space X has the **Baire property** or is a **Baire space** if the intersection of countably many dense open sets in X is also dense; equivalently, every countable intersection of dense G_δ sets in X is also dense. This property is very valuable. In particular, it can be used to prove various existence theorems. Observe that the Baire property can also be expressed as follows: if a countable union of closed sets has an interior point, then at least one of the closed sets has an interior point. The following statement is easily proved:

- Every open subspace and every dense G_δ subspace of a Baire space is also Baire.

Complete metrizability is preferable because it implies the Baire property.

Theorem 2.5.1 (Baire Category Theorem). *Every completely metrizable space X is a Baire space. Consequently, X cannot be written as a union of countably many closed sets without interior points.*

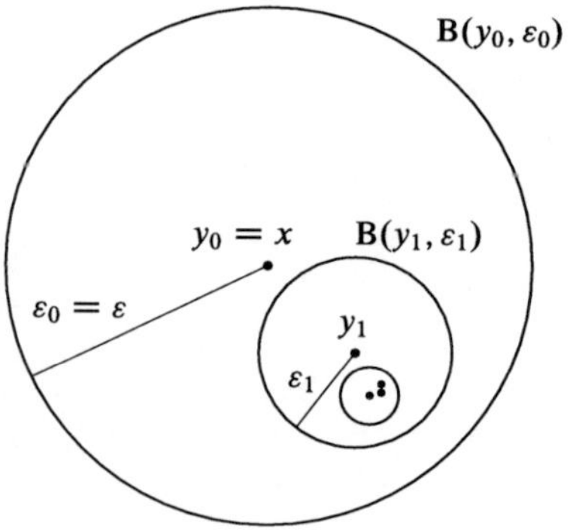

Fig. 2.5 Definition of $y_n \in X$ and $\varepsilon_n > 0$

Proof. For each $i \in \mathbb{N}$, let G_i be a dense open set in X and $d \in \mathrm{Metr}(X)$ be a complete metric. For each $x \in X$ and $\varepsilon > 0$, we inductively choose $y_i \in X$ and $\varepsilon_i > 0$, $i \in \mathbb{N}$, so that

$$y_i \in \mathrm{B}(y_{i-1}, \tfrac{1}{2}\varepsilon_{i-1}) \cap G_i,\ \ \mathrm{B}(y_i, \varepsilon_i) \subset G_i \text{ and } \varepsilon_i \leq \tfrac{1}{2}\varepsilon_{i-1},$$

where $y_0 = x$ and $\varepsilon_0 = \varepsilon$ (Fig. 2.5). Then, $(y_i)_{i\in\mathbb{N}}$ is d-Cauchy, hence it converges to some $y \in X$. For each $n \in \omega$,

$$d(y_n, y) \leq \sum_{i=n}^{\infty} d(y_i, y_{i+1}) < \sum_{i=n}^{\infty} \tfrac{1}{2}\varepsilon_i \leq \sum_{i=1}^{\infty} 2^{-i}\varepsilon_n = \varepsilon_n.$$

Thus, $y \in \mathrm{B}(x, \varepsilon)$ and $y \in \mathrm{B}(y_i, \varepsilon_i) \subset G_i$ for each $i \in \mathbb{N}$, that is, $y \in \mathrm{B}(x, \varepsilon) \cap \bigcap_{i\in\mathbb{N}} G_i$. Therefore, $\bigcap_{i\in\mathbb{N}} G_i$ is dense in X. □

A metrizable space X is said to be **absolutely G_δ** if X is G_δ in an arbitrary metrizable space that contains X as a subspace. This concept characterizes complete metrizability, which leads us to the following:

Theorem 2.5.2. *A metrizable space is completely metrizable if and only if it is absolutely G_δ.*

This follows from Corollary 2.3.6 (or 2.3.10) and the following theorem:

Theorem 2.5.3. *Let $X = (X, d)$ be a metric space and $A \subset X$.*

(1) *If A is completely metrizable, then A is G_δ in X.*
(2) *If X is complete and A is G_δ in X, then A is completely metrizable.*

Proof. (1): Since $\mathrm{cl}\, A$ is G_δ in X, it suffices to show that A is G_δ in $\mathrm{cl}\, A$. Let $\rho \in \mathrm{Metr}(A)$ be a complete metric. For each $n \in \mathbb{N}$, let

$$G_n = \{x \in \operatorname{cl} A \mid x \text{ has a neighborhood } U \text{ in } X \text{ with } \operatorname{diam}_d U < 2^{-n} \text{ and } \operatorname{diam}_\rho U \cap A < 2^{-n}\}.$$

Then, each G_n is clearly open in $\operatorname{cl} A$ and $A \subset \bigcap_{n\in\mathbb{N}} G_n$. Each $x \in \bigcap_{n\in\mathbb{N}} G_n$ has neighborhoods $U_1 \supset U_2 \supset \cdots$ in X such that $\operatorname{diam}_d U_n < 2^{-n}$ and $\operatorname{diam}_\rho U_n \cap A < 2^{-n}$. Since $x \in \operatorname{cl} A$, we have points $x_n \in U_n \cap A$, $n \in \mathbb{N}$. Then, $(x_n)_{n\in\mathbb{N}}$ converges to x. Since $(x_n)_{n\in\mathbb{N}}$ is ρ-Cauchy, it is convergent in A. Thus, we can conclude that $x \in A$. Therefore, $A = \bigcap_{n\in\mathbb{N}} G_n$, which is G_δ in $\operatorname{cl} A$.

(2): First, we show that any open set U in X is completely metrizable. We can define an admissible metric ρ for U as follows:

$$\rho(x, y) = d(x, y) + \left|d(x, X \setminus U)^{-1} - d(y, X \setminus U)^{-1}\right|.$$

Every ρ-Cauchy sequence $(x_n)_{n\in\mathbb{N}}$ in U is d-Cauchy, so it converges to some $x \in X$. Since $(d(x_n, X \setminus U)^{-1})_{n\in\mathbb{N}}$ is a Cauchy sequence in $\mathbb{R}$, it is bounded. Then,

$$d(x, X \setminus U) = \lim_{n\to\infty} d(x_n, X \setminus U) > 0.$$

This means that $x \in U$, and hence $(x_n)_{n\in\mathbb{N}}$ is convergent in U. Thus, ρ is complete.

Next, we show that an arbitrary G_δ set A in X is completely metrizable. Write $A = \bigcap_{n\in\mathbb{N}} U_n$, where $U_1, U_2, \ldots$ are open in X. As we saw above, each U_n admits a complete metric $d_n \in \operatorname{Metr}(U_n)$. Now, we can define a metric $\rho \in \operatorname{Metr}(A)$ as follows:

$$\rho(x, y) = \sum_{n\in\mathbb{N}} \min\{2^{-n},\ d_n(x, y)\}.$$

Every ρ-Cauchy sequence in A is d_n-Cauchy, which is convergent in U_n. Hence, it is convergent in $A = \bigcap_{n\in\mathbb{N}} U_n$. Therefore, ρ is complete. □

Analogous to compactness, the completeness of metric spaces can be characterized by the finite intersection property (f.i.p.).

Theorem 2.5.4. *In order for a metric space $X = (X, d)$ to be complete, it is necessary and sufficient that, if a family $\mathcal{F}$ of subsets of X has the finite intersection property and contains sets with arbitrarily small diameter, then $\mathcal{F}^{\mathrm{cl}}$ has a non-empty intersection, which is a singleton.*

Proof. (*Necessity*) Let $\mathcal{F}$ be a family of subsets of X with the f.i.p. such that $\mathcal{F}$ contains sets with arbitrarily small diameter. For each $n \in \mathbb{N}$, choose $F_n \in \mathcal{F}$ so that $\operatorname{diam} F_n < 2^{-n}$, and take $x_n \in F_n$. For any $n < m$, $F_n \cap F_m \neq \emptyset$, hence

$$d(x_n, x_m) \le \operatorname{diam} F_n + \operatorname{diam} F_m < 2^{-n} + 2^{-m} < 2^{-n+1}.$$

Thus, $(x_n)_{n\in\mathbb{N}}$ is a Cauchy sequence, therefore it converges to a point $x \in X$. Then, $x \in \bigcap \mathcal{F}^{\mathrm{cl}}$. Otherwise, $x \notin \operatorname{cl} F$ for some $F \in \mathcal{F}$. Choose $n \in \mathbb{N}$ so that $d(x, x_n),\ 2^{-n} < \frac{1}{2} d(x, F)$. Since $F \cap F_n \neq \emptyset$, it follows that

$$d(x, F) \le d(x, x_n) + \operatorname{diam} F_n < d(x, x_n) + 2^{-n} < d(x, F),$$

which is a contradiction.

(*Sufficiency*) Let $(x_n)_{n \in \mathbb{N}}$ be a Cauchy sequence in X. For each $n \in \mathbb{N}$, let $F_n = \{x_i \mid i \ge n\}$. Then, $F_1 \supset F_2 \supset \cdots$ and $\operatorname{diam} F_n \to 0$ $(n \to \infty)$. From this condition, we have $x \in \bigcap_{n \in \mathbb{N}} \operatorname{cl} F_n$. For each $\varepsilon > 0$, choose $n \in \mathbb{N}$ so that $\operatorname{diam} \operatorname{cl} F_n = \operatorname{diam} F_n < \varepsilon$. Then, $d(x_i, x) < \varepsilon$ for $i \ge n$, that is, $\lim_{n \to \infty} x_n = x$. Therefore, X is complete. □

Using compactifications, we can characterize complete metrizability as follows:

Theorem 2.5.5. *For a metrizable space X, the following are equivalent:*

(a) *X is completely metrizable;*
(b) *X is G_δ in an arbitrary compactification of X;*
(c) *X is G_δ in the Stone–Čech compactification βX;*
(d) *X has a compactification in which X is G_δ.*

Proof. The implications (b) ⇒ (c) ⇒ (d) are obvious. We show the converse (d) ⇒ (c) ⇒ (b) and the equivalence (a) ⇔ (b).

(d) ⇒ (c): Let γX be a compactification of X and $X = \bigcap_{n \in \mathbb{N}} G_n$, where each G_n is open in γX. Then, by Theorem 2.1.4, we have a map $f : \beta X \to \gamma X$ such that $f|X = \mathrm{id}$, where $X = f^{-1}(X)$ by Theorem 2.1.8. Consequently, $X = \bigcap_{n \in \mathbb{N}} f^{-1}(G_n)$ is G_δ in βX.

(c) ⇒ (b): By condition (c), we can write $\beta X \setminus X = \bigcup_{n \in \mathbb{N}} F_n$, where each F_n is closed in βX. For any compactification γX of X, we have a map $f : \beta X \to \gamma X$ such that $f|X = \mathrm{id}$ (Theorem 2.1.4). From Theorem 2.1.8, $\gamma X \setminus X = f(\beta X \setminus X) = \bigcup_{n \in \mathbb{N}} f(F_n)$ is F_σ in γX, hence X is G_δ in γX.

(b) ⇒ (a): To prove the complete metrizability of X, we show that X is absolutely G_δ (Theorem 2.5.2). Let X be contained in a metrizable space Y. Since $\operatorname{cl}_{\beta Y} X$ is a compactification of X, it follows from (b) that X is G_δ in $\operatorname{cl}_{\beta Y} X$, and hence it is G_δ in $Y \cap \operatorname{cl}_{\beta Y} X = \operatorname{cl}_Y X$, where $\operatorname{cl}_Y X$ is also G_δ in Y. Therefore, X is G_δ in Y.

(a) ⇒ (b): Let γX be a compactification of X and d an admissible complete metric for X. For each $n \in \mathbb{N}$ and $x \in X$, let $G_n(x)$ be an open set in γX such that $G_n(x) \cap X = \mathrm{B}_d(x, 2^{-n})$. Then, $G_n = \bigcup_{x \in X} G_n(x)$ is open in γX and $X \subset G_n$. We will show that each $y \in \bigcap_{n \in \mathbb{N}} G_n$ is contained in X. This implies that $X = \bigcap_{n \in \mathbb{N}} G_n$ is G_δ in γX.

For each $n \in \mathbb{N}$, choose $x_n \in X$ so that $y \in G_n(x_n)$. Since $y \in \operatorname{cl}_{\gamma X} X$ and $G_n(x_n) \cap X = \mathrm{B}_d(x_n, 2^{-n})$, it follows that $\{\mathrm{B}_d(x_n, 2^{-n}) \mid n \in \mathbb{N}\}$ has the f.i.p. By Theorem 2.5.4, we have $x \in \bigcap_{n \in \mathbb{N}} \operatorname{cl}_X \mathrm{B}_d(x_n, 2^{-n})$, where $\lim_{n \to \infty} x_n = x$ because $d(x_n, x) \le 2^{-n}$. Thus, we have $y = x \in X$. Otherwise, there would be disjoint open sets U and V in γX such that $x \in U$ and $y \in V$. Since $y \in \bigcap_{n\in\mathbb{N}} G_n \cap V$, $\{\mathrm{B}_d(x_n, 2^{-n}) \cap V \mid n \in \mathbb{N}\}$ has the f.i.p. Again, by Theorem 2.5.4, we have

$$x' \in \bigcap_{n\in\mathbb{N}} \mathrm{cl}_X(\mathrm{B}_d(x_n, 2^{-n}) \cap V) \subset \mathrm{cl}_X V.$$

Since $\lim_{n\to\infty} x_n = x'$ is the same as x, it follows that $x' = x \in U$, which is a contradiction. □

Note that conditions (b)–(d) in Theorem 2.5.5 are equivalent without the metrizability of X, but X should be assumed to be Tychonoff in order that X has a compactification. A Tychonoff space X is said to be **Čech-complete** if X satisfies one of these conditions.

Every compact metric space is complete. Since a non-compact locally compact metrizable space X is open in the one-point compactification $\alpha X = X \cup \{\infty\}$, X is completely metrizable because of Theorem 2.5.5. Thus, we have the following corollary:

Corollary 2.5.6. *Every locally compact metrizable space is completely metrizable.* □

We now state and prove the LAVRENTIEFF G_δ-EXTENSION THEOREM:

Theorem 2.5.7 (LAVRENTIEFF). *Let $f : A \to Y$ be a map from a subset A of a space X to a completely metrizable space Y. Then, f extends over a G_δ set G in X such that $A \subset G \subset \mathrm{cl}\, A$.*

Proof. We may assume that Y is a complete metric space. The oscillation of f at $x \in \mathrm{cl}\, A$ is defined as follows:

$$\mathrm{osc}_f(x) = \inf\big\{\mathrm{diam}\, f(A \cap U) \;\big|\; U \text{ is an open neighborhood of } x\big\}.$$

Let $G = \{x \in \mathrm{cl}\, A \mid \mathrm{osc}_f(x) = 0\}$. Then, $A \subset G$ because f is continuous. Since each $\{x \in \mathrm{cl}\, A \mid \mathrm{osc}_f(x) < 1/n\}$ is open in $\mathrm{cl}\, A$, it follows that G is G_δ in X. For each $x \in G$,

$$\mathcal{F}_x = \big\{f(A \cap U) \;\big|\; U \text{ is an open neighborhood of } x\big\},$$

has the f.i.p. and contains sets with arbitrarily small diameter. By Theorem 2.5.4, we have $\bigcap \mathcal{F}_x^{\mathrm{cl}} \neq \emptyset$, which is a singleton because $\mathrm{diam} \bigcap \mathcal{F}_x^{\mathrm{cl}} = 0$. The desired extension $\tilde{f} : G \to Y$ of f can be defined by $\tilde{f}(x) \in \bigcap \mathcal{F}_x^{\mathrm{cl}}$. □

If A is a subspace of a metric space X and Y is a complete metric space, then every uniformly continuous map $f : A \to Y$ extends over $\mathrm{cl}\, A$. This result can be obtained by showing that $G = \mathrm{cl}\, A$ in the above proof. However, a direct proof is easier.

We will modify Theorem 2.5.7 into the following, known as the LAVRENTIEFF HOMEOMORPHISM EXTENSION THEOREM:

Theorem 2.5.8 (LAVRENTIEFF). *Let X and Y be completely metrizable spaces and let $f : A \to B$ be a homeomorphism between $A \subset X$ and $B \subset Y$. Then, f extends to a homeomorphism $\tilde{f} : G \to H$ between G_δ sets in X and Y such that $A \subset G \subset \operatorname{cl} A$ and $B \subset H \subset \operatorname{cl} B$.*

Proof. By Theorem 2.5.7, f and f^{-1} extend to maps $g : G' \to Y$ and $h : H' \to X$, where $A \subset G' \subset \operatorname{cl} A$, $B \subset H' \subset \operatorname{cl} B$ and G', H' are G_δ in X and Y, respectively. Then, we have G_δ sets $G = g^{-1}(H')$ and $H = h^{-1}(G')$ that contain A and B as dense subsets, respectively. Consider the maps $h(g|G) : G \to X$ and $g(h|H) : H \to Y$. Since $h(g|G)|A = \mathrm{id}_A$ and $g(h|H)|B = \mathrm{id}_B$, it follows that $h(g|G) = \mathrm{id}_G$ and $g(h|H) = \mathrm{id}_H$. Then, as is easily observed, we have $g(G) \subset H$ and $h(H) \subset G$. Hence, $\tilde{f} = g|G : G \to H$ is a homeomorphism extending f. □

In the above, when $X = Y$ and $A = B$, we can take $G = H$, that is, we can show the following:

Corollary 2.5.9. *Let X be a completely metrizable space and $A \subset X$. Then, every homeomorphism $f : A \to A$ extends to a homeomorphism $\tilde{f} : G \to G$ over a G_δ set G in X with $A \subset G \subset \operatorname{cl} A$.*

Proof. Using Theorem 2.5.8, we extend f to a homeomorphism $g : G' \to G''$ between G_δ sets $G', G'' \subset X$ with $A \subset G' \cap G''$ and $G', G'' \subset \operatorname{cl} A$. We inductively define a sequence of G_δ sets $G' = G_1 \supset G_2 \supset \cdots$ in X as follows:

$$G_{n+1} = G_n \cap g(G_n) \cap g^{-1}(G_n).$$

Then, $G = \bigcap_{n \in \mathbb{N}} G_n$ is G_δ in X and $g(x), g^{-1}(x) \in G$ for each $x \in G$. Indeed, for each $n \in \mathbb{N}$, since $x \in G_{n+1}$, it follows that $g(x) \in G_n$ and $g^{-1}(x) \in G_n$. Thus, $\tilde{f} = g|G : G \to G$ is the desired extension of f. □

Additional Results on Complete Metrizability 2.5.10.

(1) Let $f : X \to Y$ be a surjective perfect map between Tychonoff spaces. Then, X is Čech-complete if and only if Y is Čech-complete. When X is metrizable, X is completely metrizable if and only if Y is completely metrizable.

Sketch of Proof. See Theorem 2.1.8.

(2) A space X is completely metrizable if it is a locally finite union of completely metrizable closed subspaces.

Sketch of Proof. Emulate 2.4.5(2). To prove the complete metrizability of the topological sum $\bigoplus_{\lambda \in \Lambda} X_\lambda$ of completely metrizable spaces, embed $\bigoplus_{\lambda \in \Lambda} X_\lambda$ into the product space $\Lambda \times \ell_1(\Gamma)^{\mathbb{N}}$ for some Γ.

2.6 Paracompactness and Local Properties

A space X is **paracompact** if each open cover of X has a locally finite open refinement.[3] According to Stone's Theorem 2.3.1, every metrizable space is paracompact. A space X is **collectionwise normal** if, for each discrete collection $\mathcal{F}$ of closed sets in X, there is a pairwise disjoint collection $\{U_F \mid F \in \mathcal{F}\}$ of open sets in X such that $F \subset U_F$ for each $F \in \mathcal{F}$. Obviously, every collectionwise normal space is normal. In the definition of collectionwise normality, $\{U_F \mid F \in \mathcal{F}\}$ can be discrete in X. Indeed, choose an open set V in X so that $\bigcup \mathcal{F} \subset V \subset \operatorname{cl} V \subset \bigcup_{F\in\mathcal{F}} U_F$. Then, $F \subset V \cap U_F$ for each $F \in \mathcal{F}$, and $\{V \cap U_F \mid F \in \mathcal{F}\}$ is discrete in X.

Theorem 2.6.1. *Every paracompact space X is collectionwise normal.*

Proof. To see the regularity of X, let A be a closed set in X and $x \in X \setminus A$. Each $a \in A$ has an open neighborhood U_a in X so that $x \notin \operatorname{cl} U_a$. Let $\mathcal{U}$ be a locally finite open refinement of

$$\{U_a \mid a \in A\} \cup \{X \setminus A\} \in \operatorname{cov}(X).$$

Then, $V = \operatorname{st}(A, \mathcal{U}) = \bigcup \mathcal{U}[A]$ is an open neighborhood of A. Since $\mathcal{U}$ is locally finite, it follows that $\operatorname{cl} V = \bigcup \mathcal{U}[A]^{\mathrm{cl}}$. Since each $U \in \mathcal{U}[A]$ is contained in some U_a, it follows that $x \notin \operatorname{cl} U$, and hence $x \notin \operatorname{cl} V$.

We now show that X is collectionwise normal. Let $\mathcal{F}$ be a discrete collection of closed sets in X. Since X is regular, each $x \in X$ has an open neighborhood V_x in X such that $\operatorname{card} \mathcal{F}[\operatorname{cl} V_x] \leq 1$. Let $\mathcal{U}$ be a locally finite open refinement of $\{V_x \mid x \in X\} \in \operatorname{cov}(X)$. For each $F \in \mathcal{F}$, we define

$$W_F = X \setminus \bigcup \{\operatorname{cl} U \mid U \in \mathcal{U},\ F \cap \operatorname{cl} U = \emptyset\}.$$

Then, W_F is open in X and $F \subset W_F \subset \operatorname{st}(F, \mathcal{U}^{\mathrm{cl}})$ (Fig. 2.6). Since $\operatorname{card} \mathcal{F}[\operatorname{cl} U] \leq 1$ for each $U \in \mathcal{U}$, it follows that $\operatorname{st}(F, \mathcal{U}^{\mathrm{cl}}) \cap W_{F'} = \emptyset$ if $F' \neq F \in \mathcal{F}$. Therefore, $\{W_F \mid F \in \mathcal{F}\}$ is pairwise disjoint. □

Lemma 2.6.2. *If X is regular and each open cover of X has a locally finite refinement (consisting of arbitrary sets), then for any open cover $\mathcal{U}$ of X there is a locally finite closed cover $\{F_U \mid U \in \mathcal{U}\}$ of X such that $F_U \subset U$ for each $U \in \mathcal{U}$.*

Proof. Since X is regular, we have $\mathcal{V} \in \operatorname{cov}(X)$ such that $\mathcal{V}^{\mathrm{cl}} \prec \mathcal{U}$. Let $\mathcal{A}$ be a locally finite refinement of $\mathcal{V}$. There exists a function $\varphi : \mathcal{A} \to \mathcal{U}$ such that $\operatorname{cl} A \subset \varphi(A)$ for each $A \in \mathcal{A}$. For each $U \in \mathcal{U}$, define

$$F_U = \bigcup \{\operatorname{cl} A \mid A \in \varphi^{-1}(U)\} \subset U.$$

[3] Recall that spaces are assumed to be **Hausdorff**.

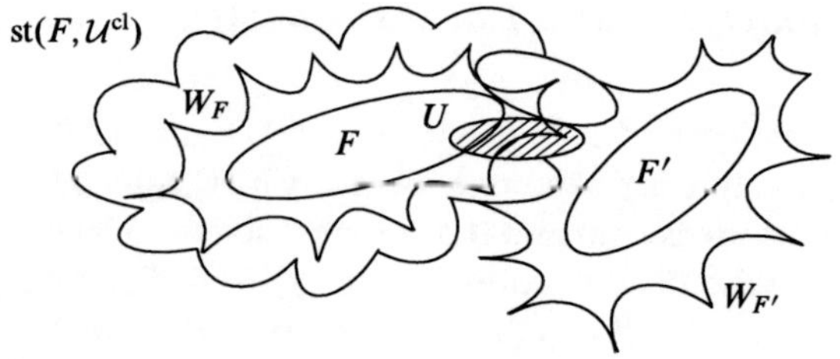

Fig. 2.6 The pairwise disjoint collection $\{W_F \mid F \in \mathcal{F}\}$

Since each $x \in X$ is contained in some $A \in \mathcal{A}$ and $A \subset F_{\varphi(A)}$, $\{F_U \mid U \in \mathcal{U}\}$ is a cover of X. Since $\mathcal{A}$ is locally finite, each F_U is closed in X and $\{F_U \mid U \in \mathcal{U}\}$ is locally finite. □

We have the following characterizations of paracompactness:

Theorem 2.6.3. *For a space X, the following conditions are equivalent:*

(a) *X is paracompact;*
(b) *Each open cover of X has an open Δ-refinement;*
(c) *Each open cover of X has an open star-refinement;*
(d) *X is regular and each open cover of X has a σ-discrete open refinement;*
(e) *X is regular and each open cover of X has a locally finite refinement.*

Proof. (a) ⇒ (b): Let $\mathcal{U} \in \operatorname{cov}(X)$. From Lemma 2.6.2, it follows that X has a locally finite closed cover $\{F_U \mid U \in \mathcal{U}\}$ such that $F_U \subset U$ for each $U \in \mathcal{U}$. For each $x \in X$, define

$$W_x = \bigcap\{U \in \mathcal{U} \mid x \in F_U\} \setminus \bigcup\{F_U \mid U \in \mathcal{U},\ x \notin F_U\}.$$

Then, W_x is an open neighborhood of x in X, hence $\mathcal{W} = \{W_x \mid x \in X\} \in \operatorname{cov}(X)$. For each $x \in X$, choose $U \in \mathcal{U}$ so that $x \in F_U$. If $x \in W_y$ then $y \in F_U$, which implies that $W_y \subset U$. Therefore, $\operatorname{st}(x, \mathcal{W}) \subset U$ for each $x \in X$, which means that $\mathcal{W}$ is a Δ-refinement of $\mathcal{U}$.

(b) ⇒ (c): Due to Proposition 2.4.3, for $\mathcal{U}, \mathcal{V}, \mathcal{W} \in \operatorname{cov}(X)$,

$$\mathcal{W} \overset{\Delta}{\prec} \mathcal{V} \overset{\Delta}{\prec} \mathcal{U} \Rightarrow \mathcal{W} \overset{*}{\prec} \mathcal{U}.$$

This gives (b) ⇒ (c).

(c) ⇒ (d): To prove the regularity of X, let A $\subset X$ be closed and $x \in X \setminus A$. Then, $\{X \setminus A, X \setminus \{x\}\} \in \operatorname{cov}(X)$ has an open star-refinement $\mathcal{W}$. Choose $W \in \mathcal{W}$ so that $x \in W$. Then, $\operatorname{st}(W, \mathcal{W}) \subset X \setminus A$, i.e., $W \cap \operatorname{st}(A, \mathcal{W}) = \emptyset$. Hence, X is regular.

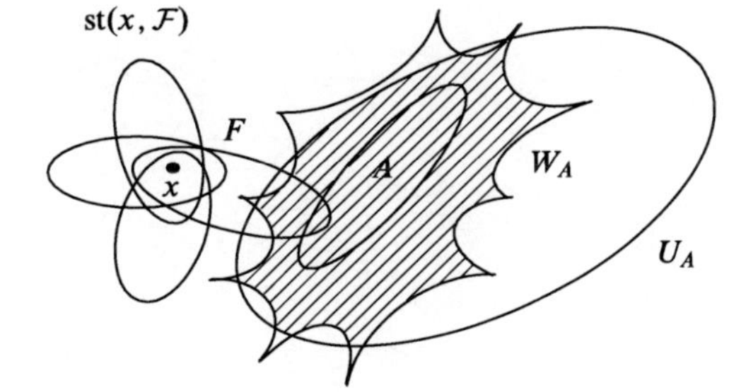

Fig. 2.7 Definition of W_A

Next, we show that each $\mathcal{U} \in \operatorname{cov}(X)$ has a σ-discrete open refinement. We may assume that $\mathcal{U} = \{U_\lambda \mid \lambda \in \Lambda\}$, where $\Lambda = (\Lambda, \leq)$ is a well-ordered set. By condition (c), we have a sequence of open star-refinements:

$$\mathcal{U} \overset{*}{\succ} \mathcal{U}_1 \overset{*}{\succ} \mathcal{U}_2 \overset{*}{\succ} \cdots .$$

For each $(\lambda, n) \in \Lambda \times \mathbb{N}$, let

$$U_{\lambda,n} = \bigcup \{U \in \mathcal{U}_n \mid \operatorname{st}(U, \mathcal{U}_n) \subset U_\lambda\} \subset U_\lambda .$$

Then, we have

$(*)$ $\operatorname{st}(U_{\lambda,n}, \mathcal{U}_{n+1}) \subset U_{\lambda,n+1}$ for each $(\lambda, n) \in \Lambda \times \mathbb{N}$.

Indeed, each $U \in \mathcal{U}_{n+1}[U_{\lambda,n}]$ meets some $U' \in \mathcal{U}_n$ such that $\operatorname{st}(U', \mathcal{U}_n) \subset U_\lambda$. Since $U \subset \operatorname{st}(U', \mathcal{U}_{n+1})$, it follows that

$$\operatorname{st}(U, \mathcal{U}_{n+1}) \subset \operatorname{st}^2(U', \mathcal{U}_{n+1}) \subset \operatorname{st}(U', \operatorname{st}\mathcal{U}_{n+1}) \subset \operatorname{st}(U', \mathcal{U}_n) \subset U_\lambda ,$$

which implies that $U \subset U_{\lambda,n+1}$. Thus, we have $(*)$.

Now, for each $(\lambda, n) \in \Lambda \times \mathbb{N}$, let

$$V_{\lambda,n} = U_{\lambda,n} \setminus \operatorname{cl} \bigcup_{\mu<\lambda} U_{\mu,n+1} \subset U_\lambda .$$

Then, each $\mathcal{V}_n = \{V_{\lambda,n} \mid \lambda \in \Lambda\}$ is discrete in X. Indeed, each $x \in X$ is contained in some $U \in \mathcal{U}_{n+1}$. If $U \cap V_{\mu,n} \neq \emptyset$, then $U \subset \operatorname{st}(U_{\mu,n}, \mathcal{U}_{n+1}) \subset U_{\mu,n+1}$ by $(*)$. Hence, $U \cap V_{\lambda,n} = \emptyset$ for all $\lambda > \mu$. This implies that U meets at most one member of $\mathcal{V}_n$ — Fig. 2.8.

It remains to be proved that $\mathcal{V} = \bigcup_{n\in\mathbb{N}} \mathcal{V}_n \in \operatorname{cov}(X)$. Each $x \in X$ is contained in some $U \in \mathcal{U}_1$. Since $\operatorname{st}(U, \mathcal{U}_1) \subset U_\lambda$ for some $\lambda \in \Lambda$, it follows that $x \in U_{\lambda,1}$. Thus, we can define

$$\lambda(x) = \min \{\lambda \in \Lambda \mid x \in U_{\lambda,n} \text{ for some } n \in \mathbb{N}\}.$$

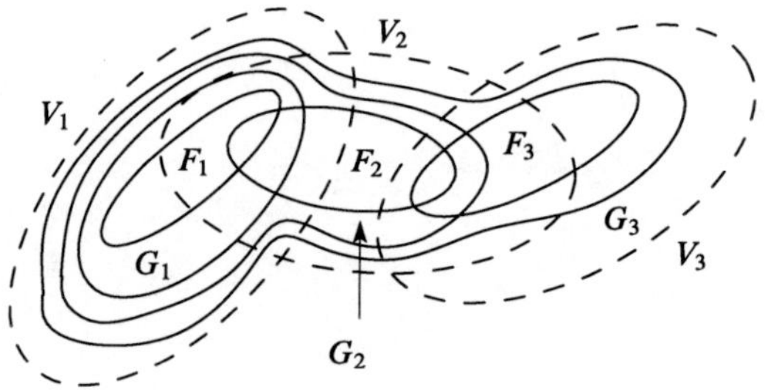

Fig. 2.8 Construction of G_n

Then, $x \in U_{\lambda(x),n}$ for some $n \in \mathbb{N}$. It follows from $(*)$ that

$$\operatorname{cl}\bigcup_{\mu<\lambda(x)} U_{\mu,n+1} \subset \operatorname{st}\big(\bigcup_{\mu<\lambda(x)} U_{\mu,n+1}, \mathcal{U}_{n+2}\big)$$
$$= \bigcup_{\mu<\lambda(x)} \operatorname{st}(U_{\mu,n+1}, \mathcal{U}_{n+2}) \subset \bigcup_{\mu<\lambda(x)} U_{\mu,n+2},$$

hence $x \notin \operatorname{cl}\bigcup_{\mu<\lambda(x)} U_{\mu,n+1}$. Therefore, $x \in V_{\lambda(x),n}$, and hence $\mathcal{V} \in \operatorname{cov}(X)$. Consequently, $\mathcal{V}$ is a σ-discrete open refinement of $\mathcal{U}$.

(d) $\Rightarrow$ (e): It suffices to show that every σ-discrete open cover $\mathcal{U}$ of X has a locally finite refinement. Let $\mathcal{U} = \bigcup_{n\in\mathbb{N}} \mathcal{U}_n$, where each $\mathcal{U}_n$ is discrete in X and $\mathcal{U}_n \cap \mathcal{U}_m = \emptyset$ if $n \neq m$. For each $U \in \mathcal{U}_n$, let $A_U = U \setminus \bigcup_{m<n}(\bigcup \mathcal{U}_m)$. Then, $\mathcal{A} = \{A_U \mid U \in \mathcal{U}\}$ is a cover of X that refines $\mathcal{U}$. For each $x \in X$, choose the smallest $n \in \mathbb{N}$ such that $x \in \bigcup \mathcal{U}_n$ and let $x \in U_0 \in \mathcal{U}_n$. Then, U_0 misses A_U for all $U \in \bigcup_{m>n} \mathcal{U}_m$. For each $m \leq n$, since $\mathcal{U}_m$ is discrete, x has a neighborhood V_m in X such that $\operatorname{card}\mathcal{U}_m[V_m] \leq 1$. Then, $V = U_0 \cap V_1 \cap \cdots \cap V_n$ is a neighborhood of x in X such that $\operatorname{card}\mathcal{A}[V] \leq n$. Hence, $\mathcal{A}$ is locally finite in X — Fig. 2.9.

(e) $\Rightarrow$ (a): Let $\mathcal{U} \in \operatorname{cov}(X)$. Then $\mathcal{U}$ has a locally finite refinement $\mathcal{A}$. For each $x \in X$, choose an open neighborhood V_x of x in X so that $\operatorname{card}\mathcal{A}[V_x] < \aleph_0$. According to Lemma 2.6.2, $\{V_x \mid x \in X\} \in \operatorname{cov}(X)$ has a locally finite closed refinement $\mathcal{F}$. Then, $\operatorname{card}\mathcal{A}[F] < \aleph_0$ for each $F \in \mathcal{F}$. For each $A \in \mathcal{A}$, choose $U_A \in \mathcal{U}$ so that $A \subset U_A$ and define

$$W_A = U_A \setminus \bigcup\{F \in \mathcal{F} \mid A \cap F = \emptyset\}.$$

Then, $A \subset W_A \subset U_A$ and W_A is open in X, hence $\mathcal{W} = \{W_A \mid A \in \mathcal{A}\}$ is an open refinement of $\mathcal{U}$. Since $\mathcal{F}$ is a locally finite closed cover of X, $\operatorname{st}(x, \mathcal{F})$ is a neighborhood of $x \in X$. For each $F \in \mathcal{F}$ and $A \in \mathcal{A}$, $F \cap W_A \neq \emptyset$ implies $F \cap A \neq \emptyset$. Then, $\operatorname{card}\mathcal{W}[F] \leq \operatorname{card}\mathcal{A}[F] < \aleph_0$ for each $F \in \mathcal{F}$. Since $\operatorname{card}\mathcal{F}[x] < \aleph_0$, $\operatorname{st}(x, \mathcal{F})$ meets only finitely many members of $\mathcal{W}$. Hence, $\mathcal{W}$ is locally finite in X. □

A space X is **Lindelöf** if every open cover of X has a countable open refinement. By verifying condition (d) above, we have the following:

Corollary 2.6.4. *Every regular Lindelöf space is paracompact.* □

Let $\mathcal{P}$ be a property of subsets of a space X. It is said that X has property $\mathcal{P}$ **locally** if each $x \in X$ has a neighborhood U in X that has property $\mathcal{P}$. Occasionally, we need to determine whether X has some property $\mathcal{P}$ if X has property $\mathcal{P}$ locally. Let us consider this problem now. A property $\mathcal{P}$ of *open* sets in X is said to be ***G*-hereditary** if the following conditions are satisfied:

(G-1) If U has property $\mathcal{P}$, then every open subset of U has $\mathcal{P}$;
(G-2) If U and V have property $\mathcal{P}$, then $U \cup V$ has property $\mathcal{P}$;
(G-3) If $\{U_\lambda \mid \lambda \in \Lambda\}$ is discrete in X and each U_λ has property $\mathcal{P}$, then $\bigcup_{\lambda \in \Lambda} U_\lambda$ has property $\mathcal{P}$.

The following theorem is very useful to show that a space has a certain property:

Theorem 2.6.5 (E. Michael). *Let $\mathcal{P}$ be a G-hereditary property of open sets in a paracompact space X. If X has property $\mathcal{P}$ locally, then X itself has property $\mathcal{P}$.*

Proof. Since X has property $\mathcal{P}$ locally, there exists $\mathcal{U} \in \operatorname{cov}(X)$ such that each $U \in \mathcal{U}$ has property $\mathcal{P}$. According to Theorem 2.6.3, $\mathcal{U}$ has an open refinement $\mathcal{V} = \bigcup_{n \in \mathbb{N}} \mathcal{V}_n$ such that each $\mathcal{V}_n$ is discrete in X. Each $V \in \mathcal{V}$ has property $\mathcal{P}$ by (G-1). For each $n \in \mathbb{N}$, let $V_n = \bigcup \mathcal{V}_n$. Then, each V_n has property $\mathcal{P}$ by (G-3), hence $V_1 \cup \cdots \cup V_n$ has property $\mathcal{P}$ by (G-2). From Lemma 2.6.2, it follows that X has a closed cover $\{F_n \mid n \in \mathbb{N}\}$ such that $F_n \subset V_n$ for each $n \in \mathbb{N}$.[4] Inductively choose open sets G_n ($n \in \mathbb{N}$) so that

$$F_n \cup \operatorname{cl} G_{n-1} \subset G_n \subset \operatorname{cl} G_n \subset V_1 \cup \cdots \cup V_n,$$

where $G_0 = \emptyset$ (Fig. 2.7). For each $n \in \mathbb{N}$, let $W_n = G_n \setminus \operatorname{cl} G_{n-2}$, where $G_{-1} = \emptyset$. Then, each W_n also has property $\mathcal{P}$ by (G-1). Let $X_i = \bigcup_{n \in \omega} W_{3n+i}$, where $i = 1, 2, 3$. Since $\{W_{3n+i} \mid n \in \omega\}$ is discrete in X, each X_i has property $\mathcal{P}$ by (G-3). Hence, $X = X_1 \cup X_2 \cup X_3$ also has property $\mathcal{P}$ by (G-2). □

There are many cases where we consider properties of closed sets rather than open sets. In such cases, Theorem 2.6.5 can also be applied. In fact, let $\mathcal{P}$ be a property of *closed* sets of X. We define the property $\mathcal{P}^\circ$ of *open* sets in X as follows:

$$U \text{ has property } \mathcal{P}^\circ \underset{\text{def}}{\iff} \operatorname{cl} U \text{ has property } \mathcal{P}.$$

It is said that $\mathcal{P}$ is ***F*-hereditary** if it satisfies the following conditions:

(F-1) If A has property $\mathcal{P}$, then every closed subset of A has property $\mathcal{P}$;

[4] Closed sets $F_n \subset X$, $n \in \mathbb{N}$ can be inductively obtained so that $X = \bigcup_{i \leq n} \operatorname{int} F_i \cup \bigcup_{i > n} V_i$.

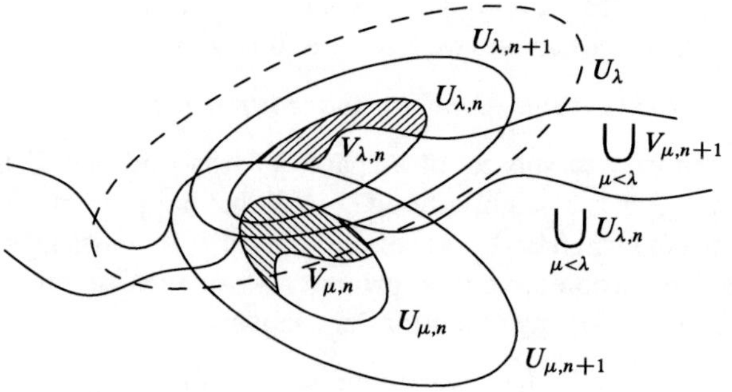

Fig. 2.9 Definition of $V_{\lambda,n}$

(F-2) If A and B have property $\mathcal{P}$, then $A \cup B$ has property $\mathcal{P}$;
(F-3) If $\{A_\lambda \mid \lambda \in \Lambda\}$ is discrete in X and each A_λ has property $\mathcal{P}$, then $\bigcup_{\lambda\in\Lambda} A_\lambda$ has property $\mathcal{P}$.

Evidently, if property $\mathcal{P}$ is F-hereditary, then $\mathcal{P}^\circ$ is G-hereditary. Therefore, Theorem 2.6.5 yields the following corollary:

Corollary 2.6.6 (E. MICHAEL). *Let $\mathcal{P}$ be an F-hereditary property of closed sets in a paracompact space X. If X has property $\mathcal{P}$ locally, then X itself has property $\mathcal{P}$.* □

Additional Results on Paracompact Spaces 2.6.7.

(1) A space is paracompact if it is a locally finite union of paracompact closed subspaces.

Sketch of Proof. Let $\mathcal{F}$ be a locally finite closed cover of a space X such that each $F \in \mathcal{F}$ is paracompact. To prove regularity, let $x \in X$ and U an open neighborhood of x in X. Since each $F \in \mathcal{F}[x]$ is regular, we have an open neighborhood U_F of x in X such that $\mathrm{cl}(F \cap U_F) \subset U$. The following U_0 is an open neighborhood of x in X:

$$U_0 = \bigcap_{F\in\mathcal{F}[x]} U_F \setminus \bigcup(\mathcal{F} \setminus \mathcal{F}[x]) \left(\subset \bigcup \mathcal{F}[x] = \mathrm{st}(x, \mathcal{F})\right).$$

Observe that $\mathrm{cl}_X U_0 = \mathrm{cl} \bigcup_{F\in\mathcal{F}[x]}(U_0 \cap F) = \bigcup_{F\in\mathcal{F}[x]} \mathrm{cl}(U_0 \cap F) \subset U$. Thus, it suffices to show that X satisfies condition 2.6.3(e).

(2) Every F_σ subspace A of a paracompact space X is paracompact.

Sketch of Proof. It suffices to show that A satisfies condition 2.6.3(d). Let $A = \bigcup_{n\in\mathbb{N}} A_n$, where each A_n is closed in X. For each $\mathcal{V} \in \mathrm{cov}(A)$ and $n \in \mathbb{N}$, let

$$\mathcal{U}_n = \{X \setminus A_n\} \cup \{\widetilde{V} \mid V \in \mathcal{V}\} \in \mathrm{cov}(X),$$

where each $\widetilde{V}$ is open in X with $\widetilde{V} \cap A = V$. Note that $\mathcal{V}_n \prec \mathcal{U}_n$ implies that $\mathcal{V}_n[A_n]|A \prec \mathcal{V}$.

(3) Let X be a paracompact space. If every open subspace of X is paracompact, then every subspace of X is also paracompact.

Sketch of Proof. To find a locally finite open refinement of $\mathcal{U} \in \mathrm{cov}(A)$, take an open collection $\widetilde{\mathcal{U}}$ in X such that $\widetilde{\mathcal{U}}|A = \mathcal{U}$ and use the paracompactness of $\bigcup \widetilde{\mathcal{U}}$.

(4) A paracompact space X is (completely) metrizable if it is locally (completely) metrizable.

Sketch of Proof. To apply 2.4.5(2) (2.5.10(2)), construct a locally finite cover of X consisting of (completely) metrizable closed sets.

A space X is **hereditarily paracompact** if every subspace of X is paracompact. The following theorem comes from (2) and (3).

Theorem 2.6.8. *Every perfectly normal paracompact space is hereditarily paracompact.* □

2.7 Partitions of Unity

A collection $\mathcal{A}$ of subsets of X is said to be **point-finite** if each point $x \in X$ is contained in only finitely many members of $\mathcal{A}$, that is, $\mathrm{card}\,\mathcal{A}[x] < \aleph_0$. Obviously, every locally finite collection is point-finite. We prove the following, which is called the OPEN COVER SHRINKING LEMMA.

Lemma 2.7.1. *Each point-finite open cover $\mathcal{U}$ of a normal space X has an open refinement $\{V_U \mid U \in \mathcal{U}\}$ such that $\mathrm{cl}\, V_U \subset U$ for each $U \in \mathcal{U}$.*

Proof. Let $\mathcal{T}$ be the topology of X (i.e., the collection of all open sets in X) and define an ordered set $\Phi = (\Phi, \leq)$ as follows:

$$\Phi = \{\varphi : \mathcal{U} \to \mathcal{T} \mid \textstyle\bigcup_{U \in \mathcal{U}} \varphi(U) = X;\ \mathrm{cl}\,\varphi(U) \subset U \ \text{ if } \ \varphi(U) \neq U\},$$

$$\varphi_1 \leq \varphi_2 \underset{\mathrm{def}}{\iff} \varphi_1(U) \neq U \ \text{ implies } \ \varphi_1(U) = \varphi_2(U).$$

Observe that if Φ has a maximal element φ_0 then $\mathrm{cl}\,\varphi_0(U) \subset U$ for each $U \in \mathcal{U}$. Then, the desired open refinement $\{V_U \mid U \in \mathcal{U}\}$ can be defined by $V_U = \varphi_0(U)$.

We apply Zorn's Lemma to show that Φ has a maximal element. It suffices to show that every totally ordered subset $\Psi \subset \Phi$ is upper bounded in Φ. For each $U \in \mathcal{U}$, let $\varphi(U) = \bigcap_{\psi \in \Psi} \psi(U)$. Then, $\varphi(U) \neq U$ implies $\psi_U(U) \neq U$ for some $\psi_U \in \Psi$, which means that $\varphi(U) = \psi_U(U)$ because $\psi(U) = \psi_U(U)$ or $\psi(U) = U$ for every $\psi \in \Psi$. Thus, we have $\varphi : \mathcal{U} \to \mathcal{T}$ such that $\mathrm{cl}\,\varphi(U) \subset U$ if $\varphi(U) \neq U$. To verify $X = \bigcup_{U \in \mathcal{U}} \varphi(U)$, let $x \in X$. If $\varphi(U) = U$ for some $U \in \mathcal{U}[x]$ then $x \in U = \varphi(U)$. When $\varphi(U) \neq U$ for every $U \in \mathcal{U}[x]$, by the same argument as above, we can see that $\varphi(U) = \psi_U(U)$ for each $U \in \mathcal{U}[x]$. Since $\mathcal{U}[x]$ is finite, we have $\psi_0 = \max\{\psi_U \mid U \in \mathcal{U}[x]\} \in \Psi$. Then, $\varphi(U) = \psi_U(U) = \psi_0(U)$ for each $U \in \mathcal{U}[x]$. Since $X = \bigcup_{U \in \mathcal{U}} \psi_0(U)$, it follows that $x \in \psi_0(U)$

$(\subset U)$ for some $U \in \mathcal{U}$, which implies $x \in \varphi(U)$ because $U \in \mathcal{U}[x]$. Consequently, $\varphi \in \Phi$. It follows from the definition that $\psi \le \varphi$ for any $\psi \in \Psi$. □

Remark 5. The above lemma can be proved using the transfinite induction instead of Zorn's Lemma.

For a map $f : X \to \mathbb{R}$, let

$$\operatorname{supp} f = \operatorname{cl}\{x \in X \mid f(x) \neq 0\} \subset X,$$

which is called the **support** of f. A **partition of unity** on X is an indexed family $(f_\lambda)_{\lambda\in\Lambda}$ of maps $f_\lambda : X \to \mathbf{I}$ such that $\sum_{\lambda\in\Lambda} f_\lambda(x) = 1$ for each $x \in X$. It is said that $(f_\lambda)_{\lambda\in\Lambda}$ is **locally finite** if each $x \in X$ has a neighborhood U such that

$$\operatorname{card}\{\lambda \in \Lambda \mid U \cap \operatorname{supp} f_\lambda \neq \emptyset\} < \aleph_0.$$

A partition of unity $(f_\lambda)_{\lambda\in\Lambda}$ on X is said to be (**weakly**) **subordinated** to $\mathcal{U} \in \operatorname{cov}(X)$ if $\{\operatorname{supp} f_\lambda \mid \lambda \in \Lambda\} \prec \mathcal{U}$ $(\{f_\lambda^{-1}((0,1]) \mid \lambda \in \Lambda\} \prec \mathcal{U})$.

Theorem 2.7.2. *Let $\mathcal{U}$ be a locally finite open cover of a normal space X. Then, there is a partition of unity $(f_U)_{U\in\mathcal{U}}$ on X such that* $\operatorname{supp} f_U \subset U$ *for each $U \in \mathcal{U}$.*

Proof. By Lemma 2.7.1, we have $\{V_U \mid U \in \mathcal{U}\}, \{W_U \mid U \in \mathcal{U}\} \in \operatorname{cov}(X)$ such that $\operatorname{cl} W_U \subset V_U \subset \operatorname{cl} V_U \subset U$ for each $U \in \mathcal{U}$. For each $U \in \mathcal{U}$, let $g_U : X \to \mathbf{I}$ be a Urysohn map with $g_U(\operatorname{cl} W_U) = 1$ and $g_U(X \setminus V_U) = 0$. Since $\mathcal{U}$ is locally finite and $\operatorname{supp} g_U \subset \operatorname{cl} V_U \subset U$ for each $U \in \mathcal{U}$, we can define a map $\varphi : X \to [1,\infty)$ by $\varphi(x) = \sum_{U\in\mathcal{U}} g_U(x)$. For each $U \in \mathcal{U}$, let $f_U : X \to \mathbf{I}$ be the map defined by $f_U(x) = g_U(x)/\varphi(x)$. Then, $(f_U)_{U\in\mathcal{U}}$ is the desired partition of unity. □

Since every open cover of a paracompact space has a locally finite open refinement, we have the following corollary:

Corollary 2.7.3. *A paracompact space X has a locally finite partition of unity subordinated to each open cover of X.* □

There exists a partition of unity which is not locally finite. For example, the hedgehog $J(\mathbb{N})$ has a non-locally finite partition of unity $(f_n)_{n\in\omega}$ defined as follows: $f_0(x) = 1 - \|x\|_1$ and $f_n(x) = x(n)$ for each $n \in \mathbb{N}$, where

$$J(\mathbb{N}) = \{x \in \ell_1 \mid x(n) \in \mathbf{I} \text{ for all } n \in \mathbb{N} \text{ and } x(n) \neq 0 \text{ at most one } n \in \mathbb{N}\} \subset \ell_1.$$

However, the existence of a partition of unity implies the existence of a locally finite one.

Proposition 2.7.4. *If X has a partition of unity $(f_\lambda)_{\lambda\in\Lambda}$ then X has a locally finite partition of unity $(g_\lambda)_{\lambda\in\Lambda}$ such that* $\operatorname{supp} g_\lambda \subset f_\lambda^{-1}((0,1])$ *for each $\lambda \in \Lambda$.*

Proof. We define $h : X \to \mathbf{I}$ by $h(x) = \sup_{\lambda\in\Lambda} f_\lambda(x) > 0$. To see the continuity of h, for each $x \in X$, choose $\Lambda(x) \in \mathrm{Fin}(\Lambda)$ so that $\sum_{\lambda\in\Lambda(x)} f_\lambda(x) > 1 - \frac{1}{2}h(x)$. Then, $f_\lambda(x) < \frac{1}{2}h(x)$ for every $\lambda \in \Lambda \setminus \Lambda(x)$, so $h(x) = f_{\lambda(x)}(x)$ for some $\lambda(x) \in \Lambda(x)$. Since $\sum_{\lambda\in\Lambda(x)} f_\lambda$ and $f_{\lambda(x)}$ are continuous, x has a neighborhood U_x in X such that

$$\sum_{\lambda\in\Lambda(x)} f_\lambda(y) > 1 - \frac{1}{2}h(x) \text{ and } f_{\lambda(x)}(y) > \frac{1}{2}h(x) \text{ for all } y \in U_x.$$

Thus, $f_\lambda(y) < \frac{1}{2}h(x) < f_{\lambda(x)}(y)$ for $\lambda \in \Lambda \setminus \Lambda(x)$ and $y \in U_x$. Therefore,

$$h(y) = \max\left\{f_\lambda(y) \mid \lambda \in \Lambda(x)\right\} \text{ for each } y \in U_x.$$

Hence, h is continuous.

For each $\lambda \in \Lambda$, let $k_\lambda : X \to \mathbf{I}$ be a map defined by

$$k_\lambda(x) = \max\left\{0,\ f_\lambda(x) - \frac{2}{3}h(x)\right\}.$$

Then, $\operatorname{supp} k_\lambda \subset f_\lambda^{-1}((0,1])$. Indeed, if $f_\lambda(x) = 0$ then x has a neighborhood U such that $f_\lambda(y) < \frac{2}{3}h(y)$ for every $y \in U$, which implies $x \notin \operatorname{supp} k_\lambda$. For each $x \in X$, take U_x and $\Lambda(x)$ as in the proof of the continuity of h. Choose an open neighborhood V_x of x in X so that $V_x \subset U_x$ and $h(y) > \frac{3}{4}h(x)$ for all $y \in V_x$. If $\lambda \in \Lambda \setminus \Lambda(x)$ and $y \in V_x$, then

$$f_\lambda(y) - \frac{2}{3}h(y) < f_\lambda(y) - \frac{1}{2}h(x) < 0,$$

which implies that $V_x \cap \operatorname{supp} k_\lambda = \emptyset$ for any $\lambda \in \Lambda \setminus \Lambda(x)$. Thus, $(k_\lambda)_{\lambda\in\Lambda}$ is locally finite. As in the proof of Theorem 2.7.2, for each $\lambda \in \Lambda$, let $g_\lambda : X \to \mathbf{I}$ be the map defined by $g_\lambda(x) = k_\lambda(x)/\varphi(x)$, where $\varphi(x) = \sum_{\lambda\in\Lambda} k_\lambda(x)$. Then, $(g_\lambda)_{\lambda\in\Lambda}$ is the desired partition of unity on X. □

The paracompactness can be characterized by the existence of a partition of unity as follows:

Theorem 2.7.5. *A space X is paracompact if and only if X has a partition of unity (weakly) subordinated to each open cover of X.*

Proof. The "only if" part is Corollary 2.7.3. The "if" part easily follows from Proposition 2.7.4. □

It is said that a real-valued function $f : X \to \mathbb{R}$ is **lower semi-continuous**, abbreviated as **l.s.c.** (or **upper semi-continuous, u.s.c.**) if $f^{-1}((t,\infty))$ (or $f^{-1}((-\infty,t))$) is open in X for each $t \in \mathbb{R}$. Then, $f : X \to \mathbb{R}$ is continuous if and only if f is l.s.c. and u.s.c.

Theorem 2.7.6. *Let $g, h : X \to \mathbb{R}$ be real-valued functions on a paracompact space X such that g is u.s.c., h is l.s.c. and $g(x) < h(x)$ for each $x \in X$. Then, there* *exists a map* $f : X \to \mathbb{R}$ *such that* $g(x) < f(x) < h(x)$ *for each* $x \in X$*. Moreover, given a map* $f_0 : A \to \mathbb{R}$ *of a closed set* A *in* X *such that* $g(x) < f_0(x) < h(x)$ *for each* $x \in A$*, the map* f *can be an extension of* f_0*.*

Proof. For each $q \in \mathbb{Q}$, let

$$U_q = g^{-1}((-\infty, q)) \cap h^{-1}((q, \infty)).$$

For each $x \in X$, we have $q \in \mathbb{Q}$ such that $g(x) < q < h(x)$, hence $\mathcal{U} = \{U_q \mid q \in \mathbb{Q}\} \in \operatorname{cov}(X)$. By Corollary 2.7.3, X has a locally finite partition of unity $(f_\lambda)_{\lambda \in \Lambda}$ subordinated to $\mathcal{U}$. For each $\lambda \in \Lambda$, choose $q(\lambda) \in \mathbb{Q}$ so that $\operatorname{supp} f_\lambda \subset U_{q(\lambda)}$. Then, we define a map $f : X \to \mathbb{R}$ as follows:

$$f(x) = \sum_{\lambda \in \Lambda} q(\lambda) f_\lambda(x).$$

For each $x \in X$, let $\{\lambda \in \Lambda \mid x \in \operatorname{supp} f_\lambda\} = \{\lambda_1, \cdots, \lambda_n\}$. Since $x \in \bigcap_{i=1}^n U_{q(\lambda_i)}$, we have $g(x) < q(\lambda_i) < h(x)$ for each $i = 1, \cdots, n$, hence it follows that

$$\begin{aligned} g(x) = \sum_{i=1}^n g(x) f_{\lambda_i}(x) < f(x) &= \sum_{i=1}^n q(\lambda_i) f_{\lambda_i}(x) \\ < h(x) &= \sum_{i=1}^n h(x) f_{\lambda_i}(x). \end{aligned}$$

To prove the additional statement, apply the Tietze Extension Theorem 2.2.2 to extend f_0 to a map $f' : X \to \mathbb{R}$. Then, we have an open neighborhood U of A in X such that $g(x) < f'(x) < h(x)$ for each $x \in U$. Let $k : X \to \mathbf{I}$ be a Urysohn map with $k(A) = 1$ and $k(X \setminus U) = 0$. We can define $\tilde{f} : X \to \mathbb{R}$ as follows:

$$\tilde{f}(x) = (1 - k(x)) f(x) + k(x) f'(x).$$

Therefore, $\tilde{f}|A = f_0$ and $g(x) < \tilde{f}(x) < h(x)$ for each $x \in X$. □

Refinements by Open Balls 2.7.7.

(1) Let X be a metrizable space and $\mathcal{U}$ an open cover of X. Then, X has an admissible metric ρ such that

$$\{\mathrm{B}_\rho(x, 1) \mid x \in X\} \prec \mathcal{U}.$$

Moreover, for a given $d \in \operatorname{Metr}(X)$, ρ can be chosen so that $\rho \geq d$ (hence, if d is complete then ρ is) and if d is bounded then ρ is also bounded.

Sketch of Proof. Take an open Δ-refinement $\mathcal{V}$ of $\mathcal{U}$ and a locally finite partition of unity $(f_\lambda)_{\lambda \in \Lambda}$ on X subordinated to $\mathcal{V}$. For a given $d \in \operatorname{Metr}(X)$, the desired metric $\rho \in \operatorname{Metr}(X)$ can be defined as follows:

$$\rho(x, y) = d(x, y) + \sum_{\lambda \in \Lambda} |f_\lambda(x) - f_\lambda(y)| \geq d(x, y).$$

If $\rho(x, y) \le 1$ then $x, y \in f_\lambda^{-1}((0, 1]) \subset \operatorname{supp} f_\lambda$ for some $\lambda \in \Lambda$, otherwise we have

$$\sum_{\lambda\in\Lambda} |f_\lambda(x) - f_\lambda(y)| = \sum_{\lambda\in\Lambda} f_\lambda(x) + \sum_{\lambda\in\Lambda} f_\lambda(y) = 2 > 1.$$

Then, it follows that $\overline{\mathrm{B}}_\rho(x, 1) \subset \operatorname{st}(x, \mathcal{V})$.

Sketch of another Proof. The above can be obtained as a corollary of 2.6.3 and 2.4.2 (or 2.4.4) as follows: By 2.4.2 (or 2.4.4), X has a sequence of open covers

$$\mathcal{U}_1 \overset{\Delta}{\succ} \mathcal{U}_2 \overset{\Delta}{\succ} \mathcal{U}_3 \overset{\Delta}{\succ} \cdots \quad \left(\text{or } \mathcal{U}_1 \overset{*}{\succ} \mathcal{U}_2 \overset{*}{\succ} \mathcal{U}_3 \overset{*}{\succ} \cdots\right)$$

such that $\{\operatorname{st}(x, \mathcal{U}_n) \mid n \in \mathbb{N}\}$ is a neighborhood basis of each $x \in X$. By 2.6.3, we can inductively define $\mathcal{V}_n \in \operatorname{cov}(X)$, $n \in \mathbb{N}$, such that

$$\mathcal{V}_n \prec \mathcal{U}_n \text{ and } \mathcal{V}_n \overset{\Delta}{\prec} \mathcal{V}_{n-1} \quad \left(\mathcal{V}_n \overset{*}{\prec} \mathcal{V}_{n-1}\right),$$

where $\mathcal{V}_0 = \mathcal{U}$. Let $d' \in \operatorname{Metr}(X)$ be the bounded metric obtained by applying Corollary 2.4.2 (or 2.4.4) with Remark 3 (or 4). For a given $d \in \operatorname{Metr}(X)$, the desired $\rho \in \operatorname{Metr}(X)$ can be defined by $\rho = 8d' + d$ (or $\rho = 2d' + d$).

(2) Let $X = (X, d)$ be a metric space. For each open cover $\mathcal{U}$ of X, there is a map $\gamma : X \to (0, 1)$ such that

$$\{\mathrm{B}(x, \gamma(x)) \mid x \in X\} \prec \mathcal{U}.$$

Sketch of Proof. For each $x \in X$, let

$$r(x) = \sup_{U\in\mathcal{U}} \min\{1,\ d(x, X \setminus U)\} = \sup_{U\in\mathcal{U}} \bar{d}(x, X \setminus U),$$

where $\bar{d} = \min\{1, d\}$. Show that $r : X \to (0, \infty)$ is l.s.c. Then, we can apply Theorem 2.7.6 to obtain a map $\gamma : X \to (0, 1)$ such that $\gamma(x) < r(x)$ for each $x \in X$.

Remark. If $\mathcal{U}$ is locally finite, r is continuous (in fact, r is 1-Lipschitz), so we can define $\gamma = \frac{1}{2}r$.

2.8 The Direct Limits of Towers of Spaces

In this section, we consider the direct limit of a tower $X_1 \subset X_2 \subset \cdots$ of spaces, where each X_n is a subspace of X_{n+1}. The **direct limit** $\varinjlim X_n$ is the space $\bigcup_{n\in\mathbb{N}} X_n$ endowed with the weak topology with respect to the tower $(X_n)_{n\in\mathbb{N}}$, that is,

$$U \subset \varinjlim X_n \text{ is open in } \varinjlim X_n \Leftrightarrow \forall n \in \mathbb{N}, U \cap X_n \text{ is open in } X_n$$

$$\left(\text{equiv. } A \subset \varinjlim X_n \text{ is closed in } \varinjlim X_n \Leftrightarrow \forall n \in \mathbb{N}, A \cap X_n \text{ is closed in } X_n\right).$$

In other words, the topology of $\varinjlim X_n$ is the finest topology such that every inclusion $X_n \subset \varinjlim X_n$ is continuous; equivalently, every X_n is a subspace of $\varinjlim X_n$. For an arbitrary space Y,

$$f : \varinjlim X_n \to Y \text{ is continuous } \Leftrightarrow \forall n \in \mathbb{N},\ f|X_n \text{ is continuous.}$$

Remark 6. Each point $x \in \varinjlim X_n$ belongs to some $X_{n(x)}$. If V is a neighborhood of x in $\varinjlim X_n$, then $V \cap X_n$ is a neighborhood x in X_n for every $n \geq n(x)$. However, it should be noted that the converse does not hold. For example, consider the direct limit $\mathbb{R}^\infty = \varinjlim \mathbb{R}^n$ of the tower $\mathbb{R} \subset \mathbb{R}^2 \subset \mathbb{R}^3 \subset \cdots$, where each $\mathbb{R}^n$ is identified with $\mathbb{R}^n \times \{0\} \subset \mathbb{R}^{n+1}$. Let $W = \bigcup_{n\in\mathbb{N}}(-2^{-n}, 2^{-n})^n \subset \mathbb{R}^\infty$. Then, every $W \cap \mathbb{R}^n$ is a neighborhood of $0 \in \mathbb{R}^n$ because it contains $(-2^{-n}, 2^{-n})^n$. Nevertheless, W is not a neighborhood of 0 in $\mathbb{R}^\infty$. Indeed,

$$(\operatorname{int}_{\mathbb{R}^\infty} W) \cap \mathbb{R}^n \subset \operatorname{int}_{\mathbb{R}^n}(W \cap \mathbb{R}^n) = (-2^{-n}, 2^{-n})^n \quad \text{for each } n \in \mathbb{N}.$$

Then, it follows that $(\operatorname{int}_{\mathbb{R}^\infty} W) \cap \mathbb{R} \subset \bigcap_{n\in\mathbb{N}}(-2^{-n}, 2^{-n}) = \{0\}$, which means that $(\operatorname{int}_{\mathbb{R}^\infty} W) \cap \mathbb{R} = \emptyset$, and hence $0 \notin \operatorname{int}_{\mathbb{R}^\infty} W$.

It should also be noted that the direct limit $\varinjlim X_n$ is T_1 but, in general, non-Hausdorff. Such an example is shown in 2.10.3.

As is easily observed, $\varinjlim X_{n(i)} = \varinjlim X_n$ for any $n(1) < n(2) < \cdots \in \mathbb{N}$. It is also easy to prove the following proposition:

Proposition 2.8.1. *Let $X_1 \subset X_2 \subset \cdots$ and $Y_1 \subset Y_2 \subset \cdots$ be towers of spaces. Suppose that there exist $n(1) < n(2) < \cdots$, $m(1) < m(2) < \cdots \in \mathbb{N}$ and maps $f_i : X_{n(i)} \to Y_{m(i)}$ and $g_i : Y_{m(i)} \to X_{n(i+1)}$ such that $g_i f_i = \mathrm{id}_{X_{n(i)}}$ and $f_{i+1} g_i = \mathrm{id}_{Y_{m(i)}}$, that is, the following diagram is commutative:*

$$\begin{array}{ccccccc} X_{n(1)} & \subset & X_{n(2)} & \subset & X_{n(3)} & \subset & \cdots \\ f_1 \downarrow & \nearrow g_1 & f_2 \downarrow & \nearrow g_2 & f_3 \downarrow & & \\ Y_{m(1)} & \subset & Y_{m(2)} & \subset & Y_{m(3)} & \subset & \cdots \end{array}$$

Then, $\varinjlim X_n$ is homeomorphic to $\varinjlim Y_n$. □

Remark 7. It should be noted that $\varinjlim X_n$ is not a subspace of $\varinjlim Y_n$ even if each X_n is a closed subspace of Y_n. For example, let $Y_n = \mathbb{R}$ be the real line and

$$X_n = \{0\} \cup [n^{-1}, 1] \subset Y_n = \mathbb{R}.$$

Then, $\mathbf{I} = \bigcup_{n\in\mathbb{N}} X_n$, $\mathbb{R} = \varinjlim Y_n$, and 0 is an isolated point of $\varinjlim X_n$ but is not in the subspace $\mathbf{I} \subset \mathbb{R}$.

On the other hand, as is easily observed, if each X_n is an open subspace of Y_n then $\varinjlim X_n$ is an open subspace of $\varinjlim Y_n$.

The following proposition is also rather obvious:

Proposition 2.8.2. *Let $Y_1 \subset Y_2 \subset \cdots$ be a tower of spaces. If X is a closed (resp. open) subspace of $Y = \varinjlim Y_n$, then $X = \varinjlim(X \cap Y_n)$. Equivalently, if each $X \cap Y_n$ is closed (resp. open) in Y_n, then $\varinjlim(X \cap Y_n)$ is a closed (resp. open) subspace of Y.* □

Remark 8. In general, $X \neq \varinjlim(X \cap Y_n)$ for a subspace $X \subset \varinjlim Y_n$. For example, let Y_n be a subspace of the Euclidean plane $\mathbb{R}^2$ defined by

$$Y_n = \{(0,0),\ (i^{-1},0),\ (j^{-1},k^{-1}) \mid i,k \in \mathbb{N},\ j = 1,\ldots,n\}.$$

Observe that $A = \{(j^{-1},k^{-1}) \mid j,k \in \mathbb{N}\}$ is dense in $\varinjlim Y_n$, hence it is not closed in the following subspace X of $\varinjlim Y_n$:

$$X = \{(0,0)\} \cup \{(j^{-1},k^{-1}) \mid j,k \in \mathbb{N}\},$$

whereas A is closed in $\varinjlim(X \cap Y_n)$.

With regard to products of direct limits, we have:

Proposition 2.8.3. *Let $X_1 \subset X_2 \subset \cdots$ be a tower of spaces. If Y is locally compact then $(\varinjlim X_n) \times Y = \varinjlim(X_n \times Y)$ as spaces.*

Proof. First of all, note that

$$(\varinjlim X_n) \times Y = \varinjlim(X_n \times Y) = \bigcup_{n\in\mathbb{N}}(X_n \times Y) \quad \text{as sets.}$$

It is easy to see that $\mathrm{id} : \varinjlim(X_n \times Y) \to (\varinjlim X_n) \times Y$ is continuous. To see this is an open map, let W be an open set in $\varinjlim(X_n \times Y)$. For each $(x,y) \in W$, choose $m \in \mathbb{N}$ so that $x \in X_m$. Since Y is locally compact, there exist open sets $U_m \subset X_m$ and $V \subset Y$ such that $x \in U_m$, $y \in V$, $U_m \times \mathrm{cl}_Y V \subset W$ and $\mathrm{cl}_Y V$ is compact. Then, by the compactness of $\mathrm{cl}_Y V$, we can find an open set $U_{m+1} \subset X_{m+1}$ such that $U_m \subset U_{m+1}$ and $U_{m+1} \times \mathrm{cl}_Y V \subset W$. Inductively, we can obtain $U_m \subset U_{m+1} \subset U_{m+2} \subset \cdots$ such that each U_n is open in X_n and $U_n \times \mathrm{cl}_Y V \subset W$. Then, $U = \bigcup_{n\geq m} U_n$ is open in $\varinjlim X_n$, and hence $U \times V$ is an open neighborhood of (x,y) in $(\varinjlim X_n) \times Y$ with $U \times V \subset W$. Thus, W is open in $(\varinjlim X_n) \times Y$. □

Proposition 2.8.4. *Let $X_1 \subset X_2 \subset \cdots$ and $Y_1 \subset Y_2 \subset \cdots$ be towers of spaces. If each X_n and Y_n are locally compact, then*

$$\varinjlim X_n \times \varinjlim Y_n = \varinjlim(X_n \times Y_n) \quad \textit{as spaces.}$$

Proof. First of all, note that

$$\varinjlim X_n \times \varinjlim Y_n = \varinjlim (X_n \times Y_n) = \bigcup_{n\in\mathbb{N}} (X_n \times Y_n) \text{ as sets.}$$

It is easy to see that $\mathrm{id} : \varinjlim(X_n \times Y_n) \to \varinjlim X_n \times \varinjlim Y_n$ is continuous. To see that this is open, let W be an open set in $\varinjlim(X_n \times Y_n)$. For each $(x, y) \in W$, choose $m \in \mathbb{N}$ so that $(x, y) \in X_m \times Y_m$. Since X_m and Y_m are locally compact, we have open sets $U_m \subset X_m$ and $V_m \subset Y_m$ such that

$$x \in U_m,\ y \in V_m,\ \mathrm{cl}_{X_m} U_m \times \mathrm{cl}_{Y_m} V_m \subset W$$

and both $\mathrm{cl}_{X_m} U_m$ and $\mathrm{cl}_{Y_m} V_m$ are compact. Then, by the compactness of $\mathrm{cl}_{X_m} U_m$ and $\mathrm{cl}_{Y_m} V_m$, we can easily find open sets $U_{m+1} \subset X_{m+1}$ and $V_{m+1} \subset Y_{m+1}$ such that

$$\mathrm{cl}_{X_m} U_m \subset U_{m+1},\ \mathrm{cl}_{Y_m} V_m \subset V_{m+1},\ \mathrm{cl}_{X_{m+1}} U_{m+1} \times \mathrm{cl}_{Y_{m+1}} V_{m+1} \subset W$$

and both $\mathrm{cl}_{X_{m+1}} U_{m+1}$ and $\mathrm{cl}_{Y_{m+1}} V_{m+1}$ are compact. Inductively, we can obtain $U_m \subset U_{m+1} \subset U_{m+2} \subset \cdots$ and $V_m \subset V_{m+1} \subset V_{m+2} \subset \cdots$ such that U_n and V_n are open in X_n and Y_n, respectively, $\mathrm{cl}_{X_n} U_n$ and $\mathrm{cl}_{Y_n} V_n$ are compact, and $\mathrm{cl}_{X_n} U_n \times \mathrm{cl}_{Y_n} V_n \subset W$. Then, $U = \bigcup_{n\geq m} U_n$ and $V = \bigcup_{n\geq m} V_n$ are open in $\varinjlim X_n$ and $\varinjlim Y_n$, respectively, and $(x, y) \in U \times V \subset W$. Therefore, W is open in $\varinjlim X_n \times \varinjlim Y_n$. □

A tower $X_1 \subset X_2 \subset \cdots$ of spaces is said to be **closed** if each X_n is closed in X_{n+1}; equivalently, each X_n is closed in the direct limit $\varinjlim X_n$. For a pointed space $X = (X, *)$, let

$$X_f^{\mathbb{N}} = \{x \in X^{\mathbb{N}} \mid x(n) = * \text{ except for finitely many } n \in \mathbb{N}\} \subset X^{\mathbb{N}}.$$

Identifying each X^n with $X^n \times \{(*, *, \dots)\} \subset X_f^{\mathbb{N}}$, we have a closed tower $X \subset X^2 \subset X^3 \subset \cdots$ with $X_f^{\mathbb{N}} = \bigcup_{n\in\mathbb{N}} X^n$. We write $X^\infty = \varinjlim X^n$, which is the space $X_f^{\mathbb{N}}$ with the weak topology with respect to the tower $(X^n)_{n\in\mathbb{N}}$. A typical example is $\mathbb{R}^\infty$, which appeared in Remark 6.

Proposition 2.8.5. *Let $X = (X, *)$ be a pointed locally compact space. Then, each $x \in X^\infty = \varinjlim X^n$ has a neighborhood basis consisting of $X^\infty \cap \prod_{n\in\mathbb{N}} V_n$, where each V_n is a neighborhood of $x(n)$ in X^n.*[5]

Sketch of Proof. Let U be an open neighborhood of x in X^∞. Choose $n_0 \in \mathbb{N}$ so that $x \in X^{n_0}$. For each $i = 1, \dots, n_0$, each $x(i)$ has a neighborhood V_i in X such that $\mathrm{cl}\, V_i$ is

---

[5] In other words, the topology of $\varinjlim X^n$ is a relative (subspace) topology inherited from the box topology of $X^{\mathbb{N}}$.

compact and $\prod_{i=1}^{n_0} \operatorname{cl} V_i \subset U \cap X^{n_0}$. Recall that we identify $X^{n-1} = X^{n-1} \times \{*\} \subset X^n$. For $n > n_0$, we can inductively choose a neighborhood V_n of $x(n) = *$ in X so that $\operatorname{cl} V_n$ is compact and $\prod_{i=1}^{n} \operatorname{cl} V_i \subset U \cap X^n$, where we use the compactness of $\prod_{i=1}^{n-1} \operatorname{cl} V_i$ $(= \prod_{i=1}^{n-1} \operatorname{cl} V_i \times \{*\})$. This is an excellent exercise as the first part of the proof of Wallace's Theorem 2.1.2.

Remark 9. Proposition 2.8.3 does not hold without the local compactness of Y even if each X_n is locally compact. For example, $(\varinjlim \mathbb{R}^n) \times \ell_2 \neq \varinjlim(\mathbb{R}^n \times \ell_2)$. Indeed, each $\mathbb{R}^n$ is identified with $\mathbb{R}^n \times \{\mathbf{0}\} \subset \mathbb{R}^{\mathbb{N}}_f \subset \ell_2$. Then, we regard

$$(\varinjlim \mathbb{R}^n) \times \ell_2 = \varinjlim(\mathbb{R}^n \times \ell_2) = \mathbb{R}^{\mathbb{N}}_f \times \ell_2 \text{ as sets.}$$

Consider the following set:

$$D = \left\{(k^{-1}\mathbf{e}_n, n^{-1}\mathbf{e}_k) \in \mathbb{R}^{\mathbb{N}}_f \times \ell_2 \;\middle|\; k, n \in \mathbb{N}\right\},$$

where each $\mathbf{e}_i \in \mathbb{R}^{\mathbb{N}}_f \subset \ell_2$ is the unit vector defined by $\mathbf{e}_i(i) = 1$ and $\mathbf{e}_i(j) = 0$ for $j \neq i$. For each $n \in \mathbb{N}$, let

$$D_n = \left\{(k^{-1}\mathbf{e}_n, n^{-1}\mathbf{e}_k) \;\middle|\; k \in \mathbb{N}\right\}.$$

Since $\{n^{-1}\mathbf{e}_k \mid k \in \mathbb{N}\}$ is discrete in ℓ_2, it follows that D_n is discrete (so closed) in $\mathbb{R}^n \times \ell_2$, hence it is also closed in $\mathbb{R}^m \times \ell_2$ for every $m \geq n$. Observe that $D \cap (\mathbb{R}^n \times \ell_2) = \bigcup_{i=1}^{n} D_i$. Then, D is closed in $\varinjlim(\mathbb{R}^n \times \ell_2)$. On the other hand, for each neighborhood U of $(0, 0)$ in $(\varinjlim \mathbb{R}^n) \times \ell_2$, we can apply Proposition 2.8.5 to take $\delta_i > 0$ $(i \in \mathbb{N})$ and $n \in \mathbb{N}$ so that

$$\left(\mathbb{R}^{\mathbb{N}}_f \cap \prod_{i \in \mathbb{N}} [-\delta_i, \delta_i]\right) \times n^{-1}\mathbf{B}_{\ell_2} \subset U,$$

where $\mathbf{B}_{\ell_2}$ is the unit closed ball of ℓ_2. Choose $k \in \mathbb{N}$ so that $k^{-1} < \delta_n$. Then, $(k^{-1}\mathbf{e}_n, n^{-1}\mathbf{e}_k) \in U$, which implies $U \cap D \neq \emptyset$. Thus, D is not closed in $(\varinjlim \mathbb{R}^n) \times \ell_2$.

Remark 10. In Proposition 2.8.4, it is necessary to assume that both X_n and Y_n are locally compact. Indeed, let $X_n = \mathbb{R}^n$ and $Y_n = \ell_2$ for every $n \in \mathbb{N}$. Then, $\varinjlim X_n \times \varinjlim Y_n \neq \varinjlim(X_n \times Y_n)$, as we saw in the above remark. Furthermore, this equality does not hold even if $X_n = Y_n$. For example, $\varinjlim(\ell_2)^n \times \varinjlim(\ell_2)^n \neq \varinjlim((\ell_2)^n \times (\ell_2)^n)$. Indeed, consider

$$\varinjlim(\ell_2)^n \times \varinjlim(\ell_2)^n = \varinjlim((\ell_2)^n \times (\ell_2)^n) = (\ell_2)^{\mathbb{N}}_f \times (\ell_2)^{\mathbb{N}}_f \text{ as sets.}$$

Identifying $\mathbb{R}^n = (\mathbb{R}\mathbf{e}_1)^n \subset (\ell_2)^n$ and $\ell_2 = \ell_2 \times \{\mathbf{0}\} \subset (\ell_2)^{\mathbb{N}}_f$, we can also consider

$$(\varinjlim \mathbb{R}^n) \times \ell_2 = \varinjlim(\mathbb{R}^n \times \ell_2) = \mathbb{R}^{\mathbb{N}}_f \times \ell_2 \subset (\ell_2)^{\mathbb{N}}_f \times (\ell_2)^{\mathbb{N}}_f \text{ as sets.}$$

By Proposition 2.8.2, $(\varinjlim \mathbb{R}^n) \times \ell_2$ and $\varinjlim(\mathbb{R}^n \times \ell_2)$ are closed subspaces of $\varinjlim(\ell_2)^n \times \varinjlim(\ell_2)^n$ and $\varinjlim((\ell_2)^n \times (\ell_2)^n)$, respectively. As we saw above, $(\varinjlim \mathbb{R}^n) \times \ell_2 \neq \varinjlim(\mathbb{R}^n \times \ell_2)$. Thus, $\varinjlim(\ell_2)^n \times \varinjlim(\ell_2)^n \neq \varinjlim((\ell_2)^n \times (\ell_2)^n)$.

Theorem 2.8.6. *For the direct limit $X = \varinjlim X_n$ of a tower $X_1 \subset X_2 \subset \cdots$ of spaces, the following hold:*

(1) *Every compact set $A \subset X$ is contained in some X_n.*
(2) *For each map $f : Y \to X$ from a first countable space Y to X, each point $y \in Y$ has a neighborhood V in Y such that the image $f(V)$ is contained in some X_n. In particular, if $A \subset X$ is a metrizable subspace then each point of A has a neighborhood in A that is contained in some X_n.*

Proof. (1): Assume that A is not contained in any X_n. For each $n \in \mathbb{N}$, take $x_n \in A \setminus X_n$ and let $D = \{x_n \mid n \in \mathbb{N}\} \subset A$. Then, D is infinite and discrete in $\varinjlim X_n$. Indeed, every $C \subset D$ is closed in $\varinjlim X_n$ because $C \cap X_n$ is finite for each $n \in \mathbb{N}$. This contradicts the compactness of A.

(2): Let $\{V_n \mid n \in \mathbb{N}\}$ be a neighborhood basis of y_0 in Y such that $V_n \subset V_{n-1}$. Assume that $f(V_n) \not\subset X_n$ for every $n \in \mathbb{N}$. Then, taking $y_n \in V_n \setminus f^{-1}(X_n)$, we have a compact set $A = \{y_n \mid n \in \omega\}$ in Y. Due to (1), $f(A)$ is contained in some X_m, and hence $f(y_m) \in X_m$. This is a contradiction. Therefore, $f(V_n) \subset X_n$ for some $n \in \mathbb{N}$. □

By Theorem 2.8.6(2), the direct limit of metrizable spaces is non-metrizable in general (e.g., $\varinjlim \mathbb{R}^n$ is non-metrizable). However, it has some favorable properties, which we now discuss.

Theorem 2.8.7. *For the direct limit $X = \varinjlim X_n$ of a closed tower $X_1 \subset X_2 \subset \cdots$ of spaces, the following properties hold:*

(1) *If each X_n is normal, then X is also normal;*
(2) *If each X_n is perfectly normal, then X is also perfectly normal;*
(3) *If each X_n is collectionwise normal, then X is also collectionwise normal;*
(4) *If each X_n is paracompact, then X is also paracompact.*

Proof. (1): Obviously, every singleton of X is closed, so X is T_1. Let A and B be disjoint closed sets in X. Then, we have a map $f_1 : X_1 \to \mathbf{I}$ such that $f_1(A \cap X_1) = 0$ and $f_1(B \cap X_1) = 1$. Using the Tietze Extension Theorem 2.2.2, we can extend f_1 to a map $f_2 : X_2 \to \mathbf{I}$ such that $f_2(A \cap X_2) = 0$ and $f_2(B \cap X_2) = 1$. Thus, we inductively obtain maps $f_n : X_n \to \mathbf{I}$, $n \in \mathbb{N}$, such that

$$f_n|X_{n-1} = f_{n-1},\ f_n(A \cap X_n) = 0 \text{ and } f_n(B \cap X_n) = 1.$$

Let $f : X \to \mathbf{I}$ be the map defined by $f|X_n = f_n$ for $n \in \mathbb{N}$. Evidently, $f(A) = 0$ and $f(B) = 1$. Therefore, X is normal.

(2): From (1), it suffices to show that every closed set A in X is a G_δ set. Each X_n has open sets $G_{n,m}$, $m \in \mathbb{N}$, such that $A \cap X_n = \bigcap_{m\in\mathbb{N}} G_{n,m}$. For each $n, m \in \mathbb{N}$, let

$G^*_{n,m} = G_{n,m} \cup (X \setminus X_n)$. Since X_n is closed in X, each $G^*_{n,m}$ is open in X. Observe that $A = \bigcap_{n,m\in\mathbb{N}} G^*_{n,m}$. Hence, A is G_δ in X.

(3): Let $\mathcal{F}$ be a discrete collection of closed sets in X. By induction on $n \in \mathbb{N}$, we have discrete collections $\{U_n^F \mid F \in \mathcal{F}\}$ of open sets in X_n such that $(F \cap X_n) \cup \operatorname{cl} U_{n-1}^F \subset U_n^F$ for each $F \in \mathcal{F}$, where $U_0^F = \emptyset$. For each $F \in \mathcal{F}$, let $U_F = \bigcup_{n\in\mathbb{N}} U_n^F$. Then, $F \subset U_F$ and U_F is open in X because $U_F \cap X_n = \bigcup_{i\geq n} U_i^F \cap X_n$ is open in X_n for each $n \in \mathbb{N}$. If $F \neq F'$, then $U_F \cap U_{F'} = \emptyset$ because

$$U_i^F \cap U_j^{F'} \subset U^F_{\max\{i,j\}} \cap U^{F'}_{\max\{i,j\}} = \emptyset \text{ for each } i, j \in \mathbb{N}.$$

Therefore, X is collectionwise normal.

(4): Since every paracompact space is collectionwise normal (Theorem 2.6.1), X is also collectionwise normal by (3), so it is regular. Then, due to Theorem 2.6.3, it suffices to show that each $\mathcal{U} \in \operatorname{cov}(X)$ has a σ-discrete open refinement. By Theorem 2.6.3, we have $\bigcup_{m\in\mathbb{N}} \mathcal{V}_{n,m} \in \operatorname{cov}(X_n)$, $n \in \mathbb{N}$, such that each $\mathcal{V}_{n,m}$ is discrete in X_n and $\mathcal{V}_{n,m} \prec \mathcal{U}$. For each $V \in \mathcal{V}_{n,m}$, choose $U_V \in \mathcal{U}$ so that $V \subset U_V$. Note that each $\mathcal{V}^{\mathrm{cl}}_{n,m}$ is discrete in X, and recall that X is collectionwise normal. So, X has a discrete open collection $\{W_V \mid V \in \mathcal{V}_{n,m}\}$ such that $\operatorname{cl} V \subset W_V$. Let $\mathcal{W}_{n,m} = \{W_V \cap U_V \mid V \in \mathcal{V}_{n,m}\}$. Then, $\mathcal{W} = \bigcup_{n,m\in\mathbb{N}} \mathcal{W}_{n,m} \in \operatorname{cov}(X)$ is a σ-discrete open cover refinement of $\mathcal{U}$. □

From Theorems 2.8.7 and 2.6.8, we conclude the following:

Corollary 2.8.8. *The direct limit of a closed tower of metrizable spaces is perfectly normal and paracompact, and so it is hereditarily paracompact.* □

2.9 The Limitation Topology for Spaces of Maps

Let X and Y be spaces. Recall that $\mathrm{C}(X, Y)$ denotes the set of all maps from X to Y. For each $f \in \mathrm{C}(X, Y)$ and $\mathcal{U} \in \operatorname{cov}(Y)$, we define

$$\mathcal{U}(f) = \{g \in \mathrm{C}(X, Y) \mid g \text{ is } \mathcal{U}\text{-close to } f\}.$$

Observe that if $\mathcal{V} \in \operatorname{cov}(Y)$ is a Δ-refinement (or a star-refinement) of $\mathcal{U}$ then $\mathcal{V}(g) \subset \mathcal{U}(f)$ for each $g \in \mathcal{V}(f)$. Then, in the case that Y is paracompact, $\mathrm{C}(X, Y)$ has a topology such that $\{\mathcal{U}(f) \mid \mathcal{U} \in \operatorname{cov}(Y)\}$ is a neighborhood basis of f. Such a topology is called the **limitation topology**.

The limitation topology is Hausdorff. Indeed, let $f \neq g \in \mathrm{C}(X, Y)$. Then $f(x_0) \neq g(x_0)$ for some $x_0 \in X$. Take disjoint open sets $U, V \subset Y$ with $f(x_0) \in U$ and $g(x_0) \in V$, and define

$$\mathcal{U} = \{U, Y \setminus \{f(x_0)\}\}, \ \mathcal{V} = \{V, Y \setminus \{g(x_0)\}\} \in \operatorname{cov}(Y).$$

Then, $\mathcal{U}(f) \cap \mathcal{V}(g) = \emptyset$.

Remark 11. In the above, $\mathcal{U}(f)$ is not open in general. For example, consider the hedgehog $J(\mathbb{N}) = \bigcup_{n\in\mathbb{N}} \mathbf{I}\mathbf{e}_n$ (see Sect. 2.3) and the map $f : \mathbb{N} \to J(\mathbb{N})$ defined by $f(n) = \mathbf{e}_n$ for each $n \in \mathbb{N}$, where $\mathbf{e}_n(n) = 1$ and $\mathbf{e}_n(i) = 0$ if $i \neq n$. For each $n \in \mathbb{N}$, let

$$U_n = \mathbf{I}\mathbf{e}_n \cup \mathrm{B}(\mathbf{0}, n^{-1}) \subset J(\mathbb{N}).$$

Then, $\mathcal{U} = \{U_n \mid n \in \mathbb{N}\} \in \mathrm{cov}(J(\mathbb{N}))$. We show that $\mathcal{U}(f)$ is not open in $\mathrm{C}(\mathbb{N}, J(\mathbb{N}))$ with respect to the limitation topology. Indeed, $\mathcal{U}(f)$ contains the constant map f_0 with $f_0(\mathbb{N}) = \{\mathbf{0}\}$. For each $\mathcal{V} \in \mathrm{cov}(J(\mathbb{N}))$, choose $k \in \mathbb{N}$ so that $\mathrm{B}(\mathbf{0}, k^{-1}) \subset V_0$ for some $V_0 \in \mathcal{V}$. Then, $\mathcal{V}(f_0)$ contains the map $g : \mathbb{N} \to J(\mathbb{N})$ defined by $g(n) = (k+1)^{-1}\mathbf{e}_{n+1}$ for each $n \in \mathbb{N}$. Observe that $g(k+1) = (k+1)^{-1}\mathbf{e}_{k+2} \notin U_{k+1}$ but $f(k+1) = \mathbf{e}_{k+1} \notin U_n$ if $n \neq k+1$, which means that $g \notin \mathcal{U}(f)$. Thus, $\mathcal{V}(f_0) \not\subset \mathcal{U}(f)$. Hence, $\mathcal{U}(f)$ is not open.

The set of all admissible bounded metrics of a metrizable space Y is denoted by $\mathrm{Metr}^B(Y)$. If Y is completely metrizable, let $\mathrm{Metr}^c(Y)$ denote the set of all admissible bounded complete metrics of Y. The sup-metric on $\mathrm{C}(X, Y)$ defined by $d \in \mathrm{Metr}^B(Y)$ is denoted by the same notation d. For each $f \in \mathrm{C}(X, Y)$ and $d \in \mathrm{Metr}^B(Y)$, let

$$U_d(f) = \mathrm{B}_d(f, 1) = \{g \in \mathrm{C}(X, Y) \mid d(f, g) < 1\}.$$

Then, $U_{n\cdot d}(f) = \mathrm{B}_d(f, n^{-1})$ for each $n \in \mathbb{N}$.

Proposition 2.9.1. *When Y is metrizable, $\{U_d(f) \mid d \in \mathrm{Metr}^B(Y)\}$ is a neighborhood basis of $f \in \mathrm{C}(X, Y)$ in the space $\mathrm{C}(X, Y)$ with the limitation topology. If Y is completely metrizable, then $\{U_d(f) \mid d \in \mathrm{Metr}^c(Y)\}$ is also a neighborhood basis of $f \in \mathrm{C}(X, Y)$.*

Proof. For each $d \in \mathrm{Metr}^B(Y)$, let

$$\mathcal{U} = \{\mathrm{B}_d(y, \tfrac{1}{3}) \mid y \in Y\} \in \mathrm{cov}(Y).$$

Then, clearly $\mathcal{U}(f) \subset U_d(f)$ for each $f \in \mathrm{C}(X, Y)$. Conversely, for each $\mathcal{U} \in \mathrm{cov}(Y)$, choose $d \in \mathrm{Metr}^B(Y)$ (or $d \in \mathrm{Metr}^c(Y)$) so that $\{\mathrm{B}_d(y, 1) \mid y \in Y\} \prec \mathcal{U}$ (cf. 2.7.7(1)). Thus, $U_d(f) \subset \mathcal{U}(f)$ for each $f \in \mathrm{C}(X, Y)$. □

For a space X, let $\mathrm{Homeo}(X)$ be the set of all homeomorphisms of X onto itself. The **limitation topology** on $\mathrm{Homeo}(X)$ is the subspace topology inherited from the space $\mathrm{C}(X, X)$ with the limitation topology. If X is metrizable, for each $f \in \mathrm{Homeo}(X)$ and $d \in \mathrm{Metr}^B(X)$, let

$$U_{d^*}(f) = \mathrm{B}_{d^*}(f, 1) = \{g \in \mathrm{Homeo}(X) \mid d^*(f, g) < 1\},$$

where d^* is the metric on $\mathrm{Homeo}(X)$ defined as follows:

$$d^*(f, g) = d(f, g) + d(f^{-1}, g^{-1}).$$

The following is the homeomorphism space version of Proposition 2.9.1:

Proposition 2.9.2. *When X is metrizable, $\{U_{d^*}(f) \mid d \in \mathrm{Metr}^B(X)\}$ is a neighborhood basis of $f \in \mathrm{Homeo}(X)$ in the space $\mathrm{Homeo}(X)$ with the limitation topology. If X is completely metrizable, then $\{U_{d^*}(f) \mid d \in \mathrm{Metr}^c(X)\}$ is also a neighborhood basis of $f \in \mathrm{Homeo}(X)$.*

Proof. For each $f \in \mathrm{Homeo}(X)$ and $d \in \mathrm{Metr}^B(Y)$, let

$$\mathcal{U} = \left\{\mathrm{B}_d(x, 1/5) \cap f\left(\mathrm{B}_d(f^{-1}(x), 1/5)\right) \,\middle|\, x \in X\right\} \in \mathrm{cov}(X).$$

Then, $\mathcal{U}(f) \cap \mathrm{Homeo}(X) \subset U_{d^*}(f)$. Indeed, for each $g \in \mathcal{U}(f) \cap \mathrm{Homeo}(X)$ and $x \in X$, we can find $y \in X$ such that

$$f(g^{-1}(x)), x = g(g^{-1}(x)) \in f\left(\mathrm{B}_d\left(f^{-1}(y), 1/5\right)\right),$$

which means that $d(g^{-1}(x), f^{-1}(y)) < 1/5$ and $d(f^{-1}(x), f^{-1}(y)) < 1/5$, hence $d(f^{-1}(x), g^{-1}(x)) < 2/5$. Therefore, $d(f^{-1}, g^{-1}) \leq 2/5$. On the other hand, it is easy to see that $d(f, g) \leq 2/5$. Thus, we have $d^*(f, g) < 1$, that is, $g \in U_{d^*}(f)$.

Conversely, for each $f \in \mathrm{Homeo}(X)$ and $\mathcal{U} \in \mathrm{cov}(X)$, choose $d \in \mathrm{Metr}^B(X)$ (or $d \in \mathrm{Metr}^c(X)$) so that $\{\mathrm{B}_d(y, 1) \mid y \in Y\} \prec \mathcal{U}$ (cf. 2.7.7(1)). Then, $U_{d^*}(f) \subset \mathcal{U}(f)$. Indeed, for each $g \in U_{d^*}(f)$ and $x \in X$, $d(f(x), g(x)) < 1$ and $\mathrm{B}_d(f(x), 1)$ is contained in some $U \in \mathcal{U}$, hence $f(x), g(x) \in U$. Therefore, $g \in \mathcal{U}(f)$. □

If $Y = (Y, d)$ is a metric space, for each $f \in \mathrm{C}(X, Y)$ and $\alpha \in \mathrm{C}(Y, (0, \infty))$, let

$$N_\alpha(f) = \left\{g \in \mathrm{C}(X, Y) \,\middle|\, \forall x \in X,\ d(f(x), g(x)) < \alpha(f(x))\right\}.$$

Proposition 2.9.3. *When $Y = (Y, d)$ is a metric space, $\{N_\alpha(f) \mid \alpha \in \mathrm{C}(Y, (0, \infty))\}$ is a neighborhood basis of $f \in \mathrm{C}(X, Y)$ in the space $\mathrm{C}(X, Y)$ with the limitation topology.*

Proof. Let $\alpha \in \mathrm{C}(Y, (0, \infty))$. For each $y \in Y$, choose an open neighborhood U_y so that $\operatorname{diam} U_y \leq \frac{1}{2}\alpha(y)$ and $\alpha(y') > \frac{1}{2}\alpha(y)$ for all $y' \in U_y$. Thus, we have $\mathcal{U} = \{U_y \mid y \in Y\} \in \mathrm{cov}(Y)$. Let $f \in \mathrm{C}(X, Y)$ and $g \in \mathcal{U}(f)$. Then, for each $x \in X$, we have some $y \in Y$ such that $f(x), g(x) \in U_y$, which implies $d(f(x), g(x)) \leq \frac{1}{2}\alpha(y) < \alpha(f(x))$. Therefore, $\mathcal{U}(f) \subset N_\alpha(f)$.

Conversely, let $\mathcal{U} \in \mathrm{cov}(Y)$. For each $y \in Y$, let

$$\gamma(y) = \sup\left\{r > 0 \,\middle|\, \exists U \in \mathcal{U} \text{ such that } \mathrm{B}(y, r) \subset U\right\}.$$

Then, $\gamma : Y \to (0, \infty)$ is lower semi-continuous. Hence, by Theorem 2.7.6, we have $\alpha \in \mathrm{C}(Y, (0, \infty))$ such that $\alpha < \gamma$, which implies that $N_\alpha(f) \subset \mathcal{U}(f)$ for any $f \in \mathrm{C}(X, Y)$. □

The following two theorems are very useful to show the existence of some types of maps or homeomorphisms:

Theorem 2.9.4. *For a completely metrizable space Y, the space* $\mathrm{C}(X, Y)$ *with the limitation topology is a Baire space.*

Proof. Let G_n, $n \in \mathbb{N}$, be dense open sets in $\mathrm{C}(X, Y)$. To see that $\bigcap_{n\in\mathbb{N}} G_n$ is dense in $\mathrm{C}(X, Y)$, let $f \in \mathrm{C}(X, Y)$ and $d \in \mathrm{Metr}^c(Y)$. We can inductively choose $g_n \in \mathrm{C}(X, Y)$ and $d_n \in \mathrm{Metr}(Y)$, $n \in \mathbb{N}$, so that

$$g_n \in U_{2d_{n-1}}(g_{n-1}) \cap G_n, \; U_{d_n}(g_n) \subset G_n \text{ and } d_n \geq 2d_{n-1},$$

where $g_0 = f$ and $d_0 = d$. Observe $d_m \leq 2^{-n} d_{m+n}$ for each $m, n \in \omega$. Since $d(g_{n-1}, g_n) \leq 2^{-n+1} d_{n-1}(g_{n-1}, g_n) < 2^{-n}$ for each $n \in \mathbb{N}$, $(g_n)_{n\in\mathbb{N}}$ is d-Cauchy. From the completeness of d, $(g_n)_{n\in\mathbb{N}}$ converges uniformly to $g \in \mathrm{C}(X, Y)$ with respect to d. Since

$$d(f, g) \leq \sum_{n\in\mathbb{N}} d(g_{n-1}, g_n) < \sum_{n\in\mathbb{N}} 2^{-n} = 1,$$

we have $g \in U_d(f)$ and, for each $n \in \mathbb{N}$,

$$\begin{aligned} d_n(g_n, g) &\leq \sum_{i\in\mathbb{N}} d_n(g_{n+i-1}, g_{n+i}) \\ &\leq \sum_{i\in\mathbb{N}} 2^{-i+1} d_{n+i-1}(g_{n+i-1}, g_{n+i}) < \sum_{i\in\mathbb{N}} 2^{-i} = 1, \end{aligned}$$

hence $g \in U_{d_n}(g_n) \subset G_n$. Thus, $U_d(f) \cap \bigcap_{n\in\mathbb{N}} G_n \neq \emptyset$, hence $\bigcap_{n\in\mathbb{N}} G_n$ is dense in $\mathrm{C}(X, Y)$. □

In the above proof, replace $\mathrm{C}(X, Y)$ and U_{d_n} with $\mathrm{Homeo}(X)$ and $U_{d_n^*}$, respectively. Then, we can see that $(g_n)_{n\in\mathbb{N}}$ is d^*-Cauchy. From the completeness of d^*, we have $g \in \mathrm{Homeo}(X)$ with $\lim_{n\to\infty} d^*(g_n, g) = 0$. By the same calculation, we can see $d_n^*(g_n, g) < 1$, that is, $g \in U_{d_n^*}(g) \subset G_n$ for every $n \in \mathbb{N}$. Then, $U_{d^*}(f) \cap \bigcap_{n\in\mathbb{N}} G_n \neq \emptyset$. Therefore, we have:

Theorem 2.9.5. *For a completely metrizable space X, the space* $\mathrm{Homeo}(X)$ *with the limitation topology is a Baire space.* □

Now, we consider the space of proper maps.

Proposition 2.9.6. *Let $\mathcal{U}$ be a locally finite open cover of Y such that* $\mathrm{cl}\, U$ *is compact for every $U \in \mathcal{U}$ (so Y is locally compact). If a map $f : X \to Y$ is $\mathcal{U}$-close to a proper map g then f is also proper.*

Proof. For each compact set A in Y, $f^{-1}(A) \subset g^{-1}(\mathrm{st}(A, \mathcal{U}^{\mathrm{cl}}))$. Since $\mathcal{U}^{\mathrm{cl}}$ is locally finite, it follows that $\mathcal{U}^{\mathrm{cl}}[A]$ is finite, and hence $\mathrm{st}(A, \mathcal{U}^{\mathrm{cl}})$ is compact. Then, $g^{-1}(\mathrm{st}(A, \mathcal{U}^{\mathrm{cl}}))$ is compact because g is proper. Thus, its closed subset $f^{-1}(A)$ is also compact. □

Let $C^P(X, Y)$ be the subspace of $C(X, Y)$ consisting of all proper maps.[6] Then, Proposition 2.9.6 yields the following corollary:

Corollary 2.9.7. *If Y is locally compact and paracompact, then $C^P(X, Y)$ is clopen (i.e., closed and open) in the space $C(X, Y)$ with the limitation topology, where X is also locally compact if $C^P(X, Y) \neq \emptyset$.* □

From Theorem 2.9.4 and Corollary 2.9.7, we have:

Theorem 2.9.8. *For every pair of locally compact metrizable spaces X and Y, the space $C^P(X, Y)$ with the limitation topology is a Baire space.* □

Some Properties of the Limitation Topology 2.9.9.

(1) For each paracompact space Y, the evaluation map

$$\mathrm{ev} : X \times C(X, Y) \ni (x, f) \mapsto f(x) \in Y$$

is continuous with respect to the limitation topology.

Sketch of Proof. For each $(x, f) \in X \times C(X, Y)$ and each open neighborhood V of $f(x)$ in Y, take an open neighborhood W of $f(x)$ in Y so that $\mathrm{cl}\, W \subset V$ and let $\mathcal{V} = \{V, X \setminus \mathrm{cl}\, W\} \in \mathrm{cov}(Y)$. Show that $(x', f') \in f^{-1}(W) \times \mathcal{V}(f)$ implies $f'(x') \in V$.

(2) If both Y and Z are paracompact, the composition

$$C(X, Y) \times C(Y, Z) \ni (f, g) \mapsto g \circ f \in C(X, Z)$$

is continuous with respect to the limitation topology.

Sketch of Proof. For each $(f, g) \in C(X, Y) \times C(Y, Z)$ and $\mathcal{U} \in \mathrm{cov}(Z)$, let $\mathcal{V} \in \mathrm{cov}(Z)$ be a star-refinement of $\mathcal{U}$. Show that $f' \in g^{-1}(\mathcal{V})(f)$ and $g' \in \mathcal{V}(g)$ implies $g' \circ f' \in \mathcal{U}(g \circ f)$.

(3) For every paracompact space X, the inverse operation

$$\mathrm{Homeo}(X) \ni h \mapsto h^{-1} \in \mathrm{Homeo}(X)$$

is continuous with respect to the limitation topology. Combining this with (1), the group $\mathrm{Homeo}(X)$ with the limitation topology is a topological group.

Sketch of Proof. Let $h \in \mathrm{Homeo}(X)$ and $\mathcal{U} \in \mathrm{cov}(X)$. Show that $g \in h(\mathcal{U})(h)$ implies $g^{-1} \in \mathcal{U}(h^{-1})$.

Remark 12. If $Y = (Y, d)$ is a metric space, for each $f \in C(X, Y)$ and $\gamma \in C(X, (0, \infty))$, let

$$V_\gamma(f) = \{g \in C(X, Y) \mid \forall x \in X,\ d(f(x), g(x)) < \gamma(x)\}.$$

---

[6] If Y is locally compact, $C^P(X, Y)$ is the subspace of $C(X, Y)$ consisting of all perfect maps (Proposition 2.1.5).

We have the topology of $C(X, Y)$ such that $\{V_\gamma(f) \mid \gamma \in C(X, (0, \infty))\}$ is a neighborhood basis of f. This is finer than the limitation topology. In general, these topologies are not equal.

For example, let $\gamma \in C(\mathbb{N}, (0, \infty))$ be the map defined by $\gamma(n) = 2^{-n}$ for $n \in \mathbb{N}$. Then, $V_\gamma(\mathbf{0})$ is not a neighborhood of $\mathbf{0} \in C(\mathbb{N}, \mathbb{R})$ in the limitation topology. Indeed, for any $\alpha \in C(\mathbb{R}, (0, \infty))$, we define $g \in C(\mathbb{N}, \mathbb{R})$ by $g(n) = \frac{1}{2}\alpha(0)$ for every $n \in \mathbb{N}$. Then, $g \in N_\alpha(\mathbf{0})$ but $g \notin V_\gamma(\mathbf{0})$. Thus, $N_\alpha(\mathbf{0}) \not\subset V_\gamma(\mathbf{0})$. Moreover, the composition

$$C(\mathbb{N}, \mathbb{R}) \times C(\mathbb{R}, \mathbb{R}) \ni (f, g) \mapsto g \circ f \in C(\mathbb{N}, \mathbb{R})$$

is not continuous with respect to this topology.

Indeed, let γ be the above map. For any $\alpha \in C(\mathbb{R}, (0, \infty))$, we have $n \in \mathbb{N}$ such that $2^{-n} < \frac{1}{2}\alpha(0)$. Let $h = \mathrm{id} + \frac{1}{2}\alpha \in C(\mathbb{R}, \mathbb{R})$. Then, $h \in V_\alpha(\mathrm{id})$ but $h \circ \mathbf{0} \notin V_\gamma(\mathrm{id} \circ \mathbf{0})$ because $h \circ \mathbf{0}(n) = h(0) = \frac{1}{2}\alpha(0) > 2^{-n} = \gamma(n)$. (Here, id can be replaced by any $g \in C(\mathbb{R}, \mathbb{R})$.)

2.10 Counter-Examples

In this section, we show that the concepts of normality, collectionwise normality, and paracompactness are neither hereditary nor productive, and that the concepts of perfect normality and hereditary normality are not productive either. Moreover, we show that the direct limit of a closed tower of Hausdorff spaces need not be Hausdorff.

The following example shows that *the concepts of normality, collectionwise normality and paracompactness are not hereditary.*

The Tychonoff plank 2.10.1. *Let $[0, \omega_1)$ be the space of all countable ordinals with the order topology. The space $[0, \omega_1]$ is the one-point compactification of the space $[0, \omega_1)$. Let $[0, \omega]$ be the one-point compactification of the space $\omega = [0, \omega)$ of non-negative integers. The product space $[0, \omega_1] \times [0, \omega]$ is a compact Hausdorff space, hence it is paracompact. The following dense subspace of $[0, \omega_1] \times [0, \omega]$ is called the* ***Tychonoff plank****:*

$$T = [0, \omega_1] \times [0, \omega] \setminus \{(\omega_1, \omega)\}.$$

We now prove that

– *The Tychonoff plank T is not normal.*

Proof. We have disjoint closed sets $\{\omega_1\} \times [0, \omega)$ and $[0, \omega_1) \times \{\omega\}$ in T. Assume that T has disjoint open sets U, V such that $\{\omega_1\} \times [0, \omega) \subset U$ and $[0, \omega_1) \times \{\omega\} \subset V$. For each $n \in \omega$, choose $\alpha_n < \omega_1$ so that $[\alpha_n, \omega_1] \times \{n\} \subset U$. Let $\alpha = \sup_{n \in \mathbb{N}} \alpha_n <$

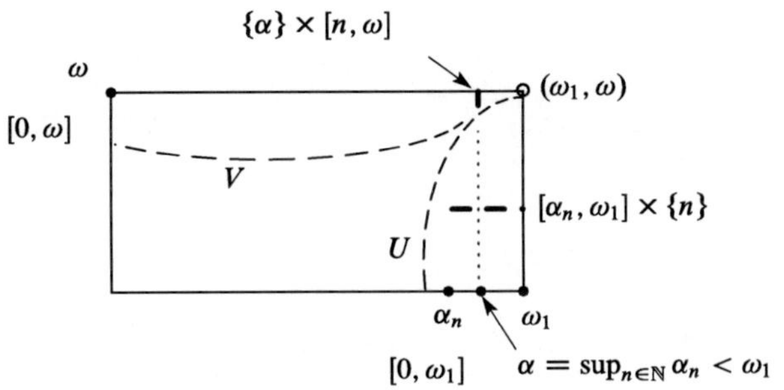

Fig. 2.10 Tychonoff plank

ω_1. Then, $[\alpha, \omega_1] \times \mathbb{N} \subset U$. On the other hand, we can choose $n \in \mathbb{N}$ so that $\{\alpha\} \times [n, \omega] \subset V$. Then, $U \cap V \neq \emptyset$, which is a contradiction (Fig. 2.10). □

The next example shows that *the concepts of normality, perfect normality, hereditary normality, collectionwise normality, and paracompactness are not productive.*

The Sorgenfrey Line 2.10.2. *The* **Sorgenfrey line** *S is the space $\mathbb{R}$ with the topology generated by $[a, b)$, $a < b$. The product S^2 is called the* **Sorgenfrey plane**. *These spaces have the following properties:*

(1) *S is a separable regular Lindelöf space, hence it is paracompact, and so is collectionwise normal;*
(2) *S is perfectly normal, and so is hereditarily normal;*
(3) *S^2 is not normal.*

Proof. (1): It is obvious that S is Hausdorff. Since each basic open set $[a, b)$ is also closed in S, it follows that S is regular. Clearly, $\mathbb{Q}$ is dense in S, hence S is separable. To see that S is Lindelöf, let $\mathcal{U} \in \operatorname{cov}(S)$. We have a function $\gamma : S \to \mathbb{Q}$ so that $\gamma(x) > x$ and $[x, \gamma(x)) \subset U$ for some $U \in \mathcal{U}$. Then, $\{[x, \gamma(x)) \mid x \in S\} \in \operatorname{cov}(S)$ is an open refinement of $\mathcal{U}$. For each $q \in \gamma(S)$, if there exists $\min \gamma^{-1}(q)$, let $R(q) = \{\min \gamma^{-1}(q)\}$. Otherwise, choose a countable subset $R(q) \subset \gamma^{-1}(q)$ so that $\inf R(q) = \inf \gamma^{-1}(q)$, where we mean $\gamma^{-1}(q) = -\infty$ if $\gamma^{-1}(q)$ is unbounded below. Then, the following is a subcover of $\{[x, \gamma(x)) \mid x \in S\} \in \operatorname{cov}(S)$:

$$\big\{[z, q) \bigm| q \in \gamma(S),\ z \in R(q)\big\} \in \operatorname{cov}(S),$$

which is a countable open refinement of $\mathcal{U}$.

(2): Let U be an open set in S. We have a function $\gamma : U \to \mathbb{Q}$ so that $\gamma(x) > x$ and $[x, \gamma(x)) \subset U$. Then, $U = \bigcup_{x \in U} [x, \gamma(x))$. By the same argument as the proof of (1), we can find a countable subcollection

$$\big\{[a_i, b_i) \bigm| i \in \mathbb{N}\big\} \subset \big\{[x, \gamma(x)) \bigm| x \in U\big\}$$

such that $U = \bigcup_{i \in \mathbb{N}} [a_i, b_i)$, hence U is F_σ in S. Thus, S is perfectly normal.

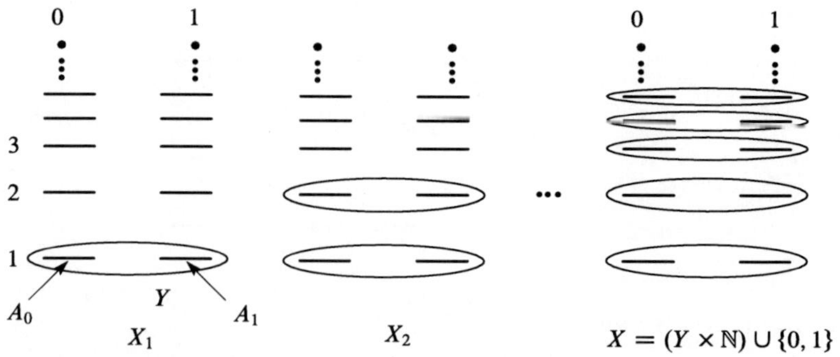

Fig. 2.11 Non-Hausdorff direct limit

(3): As we saw in the proof of (1), $\mathbb{Q}$ is dense in S, hence $\mathbb{Q}^2$ is dense in S^2. It follows that the restriction $\mathrm{C}(S^2, \mathbb{R}) \ni f \mapsto f|\mathbb{Q}^2 \in \mathbb{R}^{\mathbb{Q}^2}$ is injective. Therefore,

$$\operatorname{card} \mathrm{C}(S^2, \mathbb{R}) \leq \operatorname{card} \mathbb{R}^{\mathbb{Q}^2} = 2^{\aleph_0} = \mathfrak{c}.$$

On the other hand, $D = \{(x, y) \in S^2 \mid x + y = 0\}$ is a discrete set in S^2. Then, we have

$$\operatorname{card} \mathrm{C}(D, \mathbb{R}) = \operatorname{card} \mathbb{R}^D = 2^{\mathfrak{c}} > \mathfrak{c} \geq \operatorname{card} \mathrm{C}(S^2, \mathbb{R}).$$

If S^2 is normal, it would follow from the Tietze Extension Theorem 2.2.2 that the restriction $\mathrm{C}(S^2, \mathbb{R}) \ni f \mapsto f|D \in \mathrm{C}(D, \mathbb{R})$ is surjective, which is a contradiction. Consequently, S^2 is not normal. □

Finally, we will construct a closed tower such that the direct limit is *not Hausdorff.*

A Non-Hausdorff Direct Limit 2.10.3. *Let Y be a space which is Hausdorff but non-normal, such as the Tychonoff plank. Let A_0, A_1 be disjoint closed sets in Y that have no disjoint neighborhoods. We define $X = (Y \times \mathbb{N}) \cup \{0, 1\}$ with the topology generated by open sets in the product space $Y \times \mathbb{N}$ and sets of the form*

$$\bigcup_{k>n} (U_k \times \{k\}) \cup \{i\},$$

where $i = 0, 1$ and each U_k is an open neighborhood of A_i. Then, X is not Hausdorff because 0 and 1 have no disjoint neighborhoods in X. For each $n \in \mathbb{N}$, let

$$X_n = Y \times \{1, \dots, n\} \cup (A_0 \cup A_1) \times \{k \mid k > n\} \cup \{0, 1\}.$$

Then, $X_1 \subset X_2 \subset \cdots$ are closed in X and $X = \bigcup_{n\in\mathbb{N}} X_n$ (Fig. 2.11). As is easily observed, every X_n is Hausdorff. We will prove that $X = \varinjlim X_n$, that is,

– *X has the weak topology with respect to the tower $(X_n)_{n\in\mathbb{N}}$.*

Proof. Since id : $\varinjlim X_n \to X$ is obviously continuous, it suffices to show that every open set V in $\varinjlim X_n$ is open in X. To this end, assume that $V \cap X_n$ is open in X_n for each $n \in \mathbb{N}$. Each $x \in V \setminus \{0, 1\}$ is contained in some $Y \times \{n\} \subset X_n$. Then, $V \cap (Y \times \{n\})$ is an open neighborhood of x in $Y \times \{n\}$, and so is an open neighborhood in X. When $0 \in V$, $A_0 \times \{k \mid k > n\} \subset V$ for some $n \in \mathbb{N}$ because $V \cap X_1$ is open in X_1. For each $k > n$, since $V \cap (Y \times \{k\})$ is open in $Y \times \{k\}$, there is an open set U_k in Y such that $V \cap (Y \times \{k\}) = U_k \times \{k\}$. Note that $A_0 \subset U_k$. Then, $\bigcup_{k>n}(U_k \times \{k\}) \cup \{0\} \subset V$, hence V is a neighborhood of 0 in X. Similarly, V is a neighborhood of 1 in X if $1 \in V$. Thus, V is open in X. □

Notes for Chap. 2

For more comprehensive studies on General Topology, see Engelking's book, which contains excellent historical and bibliographic notes at the end of each section.

- R. Engelking, *General Topology, Revised and complete edition*, Sigma Ser. in Pure Math. **6** (Heldermann Verlag, Berlin, 1989)

The following classical books are still good sources.

- J. Dugundji, *Topology*, (Allyn and Bacon, Inc., Boston, 1966)
- J.L. Kelly, *General Topology*, GTM **27** (Springer-Verlag, Berlin, 1975); Reprint of the 1955 ed. published by Van Nostrand

For counter-examples, the following is a good reference:

- L.A. Steen and J.A. Seebach, Jr., *Counterexamples in Topology, 2nd edition* (Springer-Verlag, New York, 1978)

Of the more recent publications, the following textbook is readable and seems to be popular:

- J.R. Munkres, *Topology, 2nd edition* (Prentice Hall, Inc., Upper Saddle River, 2000)

Most of the contents discussed in the present chapter are found in Chaps. 5–8 of this text, although it does not discuss the Frink Metrization Theorem (cf. 2.4.1) and Michael's Theorem 2.6.5 on local properties.

Among various proofs of the Tychonoff Theorem 2.1.1, our proof is a modification of the proof due to Wright [19]. Our proof of the Tietze Extension Theorem 2.2.2 is due to Scott [14]. Theorem 2.3.1 was established by Stone [16], but the proof presented here is due to Rudin [13]. The Nagata–Smirnov Metrization Theorem (cf. 2.3.4) was independently proved by Nagata [12] and Smirnov [15]. The Bing Metrization Theorem (cf. metrization) was proved in [2]. The Urysohn Metrization Theorem 2.3.5 and the Alexandroff–Urysohn Metrization Theorem (cf. 2.4.1) were established in [18] and [1], respectively. The Frink Metrization Theorem (cf. 2nd-metrization) was proved by Frink [5]. The Baire Category Theorem 2.5.1 was first proved by Hausdorff [6] (Baire proved the theorem for the real line in 1889). The equivalence of (a) and (b) in Theorem 2.5.5 was shown by Čech [3]. Theorems 2.5.7 and 2.5.8 were established by Lavrentieff [7].

The concept of paracompactness was introduced by Dieudonné [4]. In [2], Bing introduced the concept of collectionwise normality and showed the collectionwise normality of paracompact spaces (Theorem 2.6.1). The equivalence of (b) and (c) in Theorem 2.6.3 was proved by Tukey [17], where he called spaces satisfying condition (c) *fully normal spaces*. The equivalence of (a) and (c) and the equivalence of (a), (d), and (e) were respectively proved by Stone [16] and Michael [10]. Theorem 2.6.5 on local properties was established by Michael [11]. Lemma 2.7.1 appeared in [8]. Theorem 2.7.2 and Proposition 2.7.4 were also established by Michael [11]. The simple proof of Proposition 2.7.4 presented here is due to Mather [9]. Theorem 2.7.6 was proved by Dieudonné [4]. These notes are based on historical and bibliographic notes in Engelking's book, listed above.

In some literature, it is mentioned that the direct limit of a closed tower of Hausdorff spaces need not be Hausdorff. The author could not find such an example in the literature. Example 2.10.3 is due to H. Ohta.

References

1. P. Alexandroff, P. Urysohn, Une condition nécessaire et suffisante pour qu'une classe (L) soit une classe (B). C.R. Acad. Sci. Paris Sér. A-B **177**, 1274–1276 (1923)
2. R.H. Bing, Metrization of topological spaces. Can. J. Math. **3**, 175–186 (1951)
3. E. Čech, On bicompact spaces. Ann. Math. **38**, 823–844 (1937)
4. J. Dieudonné, Une généralisation des espaces compacts. J. Math. Pures Appl. **23**, 65–76 (1944)
5. A.H. Frink, Distance functions and the metrization problems. Bull. Am. Math. Soc. **43**, 133–142 (1937)
6. F. Hausdorff, *Grundzüge der Mengenlehre* (Teubner, Leipzig, 1914)
7. M. Lavrentieff, Contribution á la théorie des ensembles homéomorphes. Fund. Math. **6**, 149–160 (1924)
8. S. Lefschetz, *Algebraic Topology*. American Mathematical Society Colloquim Publications **27** (American Mathematical Society, New York, 1942)
9. M.R. Mather, Paracompactness and partition of unity, Ph.D. thesis, Cambridge University, 1965
10. E. Michael, A note on paracompact spaces. Proc. Am. Math. Soc. **4**, 831–838 (1953)
11. E. Michael, Local properties of topological spaces. Duke Math. J. **21**, 163–171 (1954)
12. J. Nagata, On a necessary and sufficient condition of metrizability. J. Inst. Polyt. Osaka City Univ. **1**, 93–100 (1950)
13. M.E. Rudin, A new proof that metric spaces are paracompact. Proc. Am. Math. Soc. **20**, 603 (1969)
14. B.M. Scott, A "more topological" proof of the Tietze–Urysohn Theorem. Am. Math. Mon. **85**, 192–193 (1978)
15. Ju.M. Smirnov, On the metrization of topological spaces (Russian). Uspekhi Mat. Nauk **6**(6), 100–111 (1951)
16. A.H. Stone, Paracompactness and product spaces. Bull. Am. Math. Soc. **54**, 977–982 (1948)
17. J.W. Tukey, *Convergence and Uniformity in Topology*. Annals of Mathematics Studies, **2** (Princeton University Press, Princeton, 1940)
18. P. Urysohn, Über die Metrisation der kompakten topologischen Räume. Math. Ann. **92**, 275–293 (1924)
19. D.G. Wright, Tychonoff's theorem. Proc. Am. Math. Soc. **120**, 985–987 (1994)

Chapter 3
Topology of Linear Spaces and Convex Sets

In this chapter, several basic results on topological linear spaces and convex sets are presented. We will characterize finite-dimensionality, metrizability, and normability of topological linear spaces. Among the important results are the Hahn–Banach Extension Theorem, the Separation Theorem, the Closed Graph Theorem, and the Open Mapping Theorem. We will also prove the Michael Selection Theorem, which will be applied in the proof of the Bartle–Graves Theorem.

3.1 Flats and Affine Functions

In this section, we present the basic properties of flats and affine functions. Let E be a linear space (over $\mathbb{R}$). We call $F \subset E$ a **flat**[1] if the straight line through every distinct two points of F is contained in F, i.e.,

$$(1-t)x + ty \in F \quad \text{for each } x, y \in F \text{ and } t \in \mathbb{R}.$$

Evidently, the intersection and the product of flats are also flats. We have the following characterization of flats:

Proposition 3.1.1. *Let E be a linear space. For each non-empty subset $F \subset E$, the following conditions are equivalent:*

(a) *F is a flat;*
(b) *For each $n \in \mathbb{N}$, if $v_1, \ldots, v_n \in F$ and $\sum_{i=1}^n t_i = 1$, then $\sum_{i=1}^n t_i v_i \in F$;*
(c) *$F - x$ is a linear subspace of E for any $x \in F$;*
(d) *$F - x_0$ is a linear subspace of E for some $x_0 \in E$.*

---

[1] A flat is also called an **affine set**, a **linear manifold**, or a **linear variety**.

K. Sakai, *Geometric Aspects of General Topology*, Springer Monographs in Mathematics, DOI 10.1007/978-4-431-54397-8_3, © Springer Japan 2013

Proof. By induction on $n \in \mathbb{N}$, we can obtain (a) $\Rightarrow$ (b). Condition (c) follows from the case $n = 3$ of (b) because, for each $x, y, z \in F$ and $a, b \in \mathbb{R}$,

$$a(y - x) + b(z - x) + x = (1 - a - b)x + ay + bz.$$

To see (c) $\Rightarrow$ (a), let $x, y \in F$ and $t \in \mathbb{R}$. Since $F - x$ is a linear subspace of E by (c), we have $t(y - x) \in F - x$, which means $(1 - t)x + ty \in F$. The implication (c) $\Rightarrow$ (d) is obvious.

(d) $\Rightarrow$ (c): It suffices to show that if $F - x_0$ is a linear subspace of E, then $F - x = F - x_0$ for any $x \in F$. For every $z \in F$, we have

$$z - x = (z - x_0) - (x - x_0) \in F - x_0.$$

Here, take $z' \in F$ so that $(z - x_0) + (x - x_0) = z' - x_0$. Then, we have

$$z - x_0 = (z' - x_0) - (x - x_0) = z' - x \in F - x.$$

Consequently, we have $F - x = F - x_0$. □

In the proof of the implication (d) $\Rightarrow$ (c), we actually proved the following:

Corollary 3.1.2. *Let F be a flat in a linear space E. Then, $F - x = F - y$ for any $x, y \in F$.* □

A maximal proper flat $H \subsetneq E$ is called a **hyperplane** in E. The following proposition shows the relationship between hyperplanes and linear functionals.

Proposition 3.1.3. *Let E be a linear space.*

(1) *For each hyperplane $H \subset E$, there is a linear functional $f : E \to \mathbb{R}$ such that $H = f^{-1}(s)$ for some $s \in \mathbb{R}$;*
(2) *For each non-trivial linear functional $f : E \to \mathbb{R}$ and $s \in \mathbb{R}$, $f^{-1}(s)$ is a hyperplane in E;*
(3) *For linear functionals $f_1, f_2 : E \to \mathbb{R}$, if $f_1^{-1}(s_1) = f_2^{-1}(s_2)$ for some $s_1, s_2 \in \mathbb{R}$, then $f_2 = rf_1$ for some $r \in \mathbb{R}$.*

Proof. (1): For a given $x_0 \in H$, $H_0 = H - x_0$ is a maximal proper linear subspace of E (Proposition 3.1.1). Let $x_1 \in E \setminus H_0$. For each $x \in E$, there exists a unique $t \in \mathbb{R}$ such that $x - tx_1 \in H_0$. Indeed, $E = H_0 + \mathbb{R}x_1$ because of the maximality of H_0. Hence, we can write $x = z + tx_1$ for some $z \in H_0$ and $t \in \mathbb{R}$. Then, $x - tx_1 \in H_0$. Moreover, if $x - t'x_1 \in H_0$ and $t' \in \mathbb{R}$, then $(t - t')x_1 \in H_0$. Since $x_1 \notin H_0$, it follows that $t = t'$. Therefore, we have a function $f : E \to \mathbb{R}$ such that $x - f(x)x_1 \in H_0$. For each $x, y \in E$ and $a, b \in \mathbb{R}$,

$$(ax + by) - (af(x) + bf(y))x_1 = a(x - f(x)x_1) + b(y - f(y)x_1) \in H_0,$$

which means $f(ax+by) = af(x)+bf(y)$, i.e., f is linear. Observe that $f^{-1}(0) = H_0 = H - x_0$, hence it follows that $H = f^{-1}(f(x_0))$.

(2): From the non-triviality of f, it follows that $f(E) = \mathbb{R}$, and hence $\emptyset \subsetneqq f^{-1}(s) \subsetneqq E$. A simple calculation shows that $f^{-1}(s)$ is a flat. To prove the maximality, let $F \subset E$ be a flat with $f^{-1}(s) \subsetneqq F$. Take $x_0 \in f^{-1}(s)$ and $x_1 \in F \setminus f^{-1}(s)$. Since $f(x_1) \neq f(x_0)$ and F is a flat, it follows that $f(F) = \mathbb{R}$. For each $x \in E$, we can choose $y \in F \setminus f^{-1}(s)$ so that $f(y) \neq f(x)$. Note that $s = tf(x)+(1-t)f(y)$ for some $t \in \mathbb{R}\setminus\{0\}$. Let $z = tx+(1-t)y \in f^{-1}(s) \subset F$. Then, $x = (1-t^{-1})y + t^{-1}z \in F$. Accordingly, we have $F = E$.

(3): When $f_1^{-1}(s_1) = f_2^{-1}(s_2) = \emptyset$, both f_1 and f_2 are trivial (i.e., $f_1(E) = f_2(E) = \{0\}$), and hence $f_1 = f_2$. If $f_1^{-1}(s_1) = f_2^{-1}(s_2) \neq \emptyset$, take $x_0 \in f_1^{-1}(s_1) = f_2^{-1}(s_2)$. Then, it follows that

$$f_1^{-1}(0) = f_1^{-1}(s_1) - x_0 = f_2^{-1}(s_2) - x_0 = f_2^{-1}(0).$$

Let $H_0 = f_1^{-1}(0) = f_2^{-1}(0)$ and $x_1 \in E \setminus H_0$. Analogous to (1), each $x \in E$ can be uniquely written as $x = y + tx_1$, where $y \in H_0$ and $t \in \mathbb{R}$. Then, $f_1(x) = tf_1(x_1)$ and $f_2(x) = tf_2(x_1)$, hence $f_2(x) = f_1(x)f_1(x_1)^{-1}f_2(x_1)$. Let $r = f_1(x_1)^{-1}f_2(x_1)$. It follows that $f_2 = rf_1$. □

It is said that finitely many distinct points $v_1, \dots, v_n \in E$ are **affinely** (or **geometrically**) **independent** provided that, for $t_1, \dots, t_n \in \mathbb{R}$,

$$\sum_{i=1}^{n} t_i v_i = \mathbf{0}, \sum_{i=1}^{n} t_i = 0 \Rightarrow t_1 = \cdots = t_n = 0,$$

i.e., $v_1 - v_n, \dots, v_{n-1} - v_n$ are linearly independent. In this case, the subset $\{v_1, \dots, v_n\} \subset E$ is also said to be **affinely** (or **geometrically**) **independent**. An (infinite) subset $A \subset E$ is said to be **affinely** (or **geometrically**) **independent** if every finite subset of A is affinely independent. This condition is equivalent to the condition that $(A - v) \setminus \{\mathbf{0}\}$ is linearly independent for some/any $v \in A$.[2]

The smallest flat containing $A \subset E$ is called the **flat hull**[3] of A and is denoted by $\operatorname{fl} A$. Then, $\mathbb{R}^n = \operatorname{fl}\{\mathbf{0}, \mathbf{e}_1, \dots, \mathbf{e}_n\}$, where $\{\mathbf{e}_1, \dots, \mathbf{e}_n\}$ is the canonical orthonormal basis for $\mathbb{R}^n$ (i.e., $\mathbf{e}_i(i) = 1$ and $\mathbf{e}_i(j) = 0$ for $j \neq i$). Observe that

$$\operatorname{fl}\{v_1, \dots, v_n\} = \left\{\textstyle\sum_{i=1}^n t_i v_i \;\middle|\; \sum_{i=1}^n t_i = 1\right\} \text{ and}$$
$$\operatorname{fl} A = \bigcup\left\{\operatorname{fl}\{x_1, \dots, x_n\} \;\middle|\; n \in \mathbb{N},\ x_1, \dots, x_n \in A\right\}.$$

[2] The phrase "for some/any" means that we can choose one of "some" or "any" in the sentence. By this choice, we have two different conditions. The condition using "some" is weaker than the condition using "any" in general. However, these two conditions can be equivalent in a certain situation.

[3] The flat hull is also called the **affine hull**.

By Zorn's Lemma, every non-empty subset $A \subset E$ contains a maximal affinely independent subset $A_0 \subset A$. Then, $\mathrm{fl}\, A_0 = \mathrm{fl}\, A$ and each $x \in \mathrm{fl}\, A$ can be uniquely written as $x = \sum_{i=1}^n t_i v_i$, where $v_1, \dots, v_n \in A_0$ and $t_1, \dots, t_n \in \mathbb{R} \setminus \{0\}$ such that $\sum_{i=1}^n t_i = 1$. In fact, for some/any $v \in A_0$, $(A_0 - v) \setminus \{\mathbf{0}\}$ $(= (A_0 \setminus \{v\}) - v)$ is a Hamel basis for the linear subspace $\mathrm{fl}\, A - v$ $(= \mathrm{fl}\, A_0 - v)$ of E.

The **dimension** of a flat $F \subset E$ is denoted by $\dim F$, and is defined by the dimension of the linear space $F - x$ for some/any $x \in F$, i.e., $\dim F = \dim(F - x)$. When $\dim F = n$ (resp. $\dim F < \infty$ or $\dim F = \infty$), it is said that F is ***n*-dimensional** (resp. **finite-dimensional** (abbrev. **f.d.**) or **infinite-dimensional** (abbrev. **i.d.**)). Therefore, every n-dimensional flat $F \subset E$ contains $n + 1$ points $v_1, \dots, v_{n+1}$ such that $F = \mathrm{fl}\{v_1, \dots, v_{n+1}\}$. In this case, $v_1, \dots, v_{n+1}$ are affinely independent. Conversely, if $F = \mathrm{fl}\{v_1, \dots, v_{n+1}\}$ for some $n+1$ affinely independent points $v_1, \dots, v_{n+1} \in F$, then $\dim F = n$.

Let F and F' be flats in linear spaces E and E', respectively. A function $f : F \to F'$ is said to be **affine** if it satisfies the following condition:

$$f((1-t)x + ty) = (1-t)f(x) + tf(y) \quad \text{for each } x, y \in F \text{ and } t \in \mathbb{R},$$

which is equivalent to the following:

$$f\left(\textstyle\sum_{i=1}^n t_i v_i\right) = \sum_{i=1}^n t_i f(v_i)$$

$$\text{for each } n \in \mathbb{N},\ v_i \in F,\ t_i \in \mathbb{R} \text{ with } \sum_{i=1}^n t_i = 1.$$

Recall that $F \subset E$ is a flat if and only if $F - x_0$ is a linear subspace of E for some/any $x_0 \in F$ (Proposition 3.1.1).

Proposition 3.1.4. *Let $f : F \to F'$ be a function between flats F and F' in linear spaces E and E', respectively. In order that f is affine, it is necessary and sufficient that the following $f^{x_0} : F - x_0 \to F' - f(x_0)$ is linear for some/any $x_0 \in F$:*

$$f^{x_0}(x) = f(x + x_0) - f(x_0) \quad \textit{for each } x \in F - x_0.$$

Proof. (*Necessity*) For each $x, y \in F - x_0$ and $a, b \in \mathbb{R}$,

$$\begin{aligned} f^{x_0}(ax + by) &= f(ax + by + x_0) - f(x_0) \\ &= f(a(x + x_0) + b(y + x_0) + (1 - a - b)x_0) - f(x_0) \\ &= af(x + x_0) + bf(y + x_0) + (1 - a - b)f(x_0) - f(x_0) \\ &= a(f(x + x_0) - f(x_0)) + b(f(y + x_0) - f(x_0)) \\ &= af^{x_0}(x) + bf^{x_0}(y). \end{aligned}$$

(*Sufficiency*) For each $x, y \in F$ and $t \in \mathbb{R}$,

$$\begin{aligned} f((1-t)x+ty) &= f^{x_0}((1-t)x+ty-x_0)+f(x_0) \\ &= f^{x_0}((1-t)(x-x_0)+t(y-x_0))+f(x_0) \\ &= (1-t)f^{x_0}(x-x_0)+tf^{x_0}(y-x_0)+f(x_0) \\ &= (1-t)(f^{x_0}(x-x_0)+f(x_0))+t(f^{x_0}(y-x_0)+f(x_0)) \\ &= (1-t)f(x)+tf(y). \end{aligned}$$

□

Proposition 3.1.5. *Let A be a non-empty affinely independent subset of a linear space E. Then, every function $g : A \to E'$ to another linear space E' uniquely extends to an affine function $\tilde{g} : \operatorname{fl} A \to E'$ such that $\tilde{g}(\operatorname{fl} A) = \operatorname{fl} g(A)$. Accordingly, every affine function f defined on $F = \operatorname{fl} A$ is uniquely determined by $f|A$ and the image $f(F)$ is a flat.*

Proof. Let $F' = \operatorname{fl} g(A)$ and take $v_0 \in A$. Since $(A \setminus \{v_0\}) - v_0$ is a Hamel basis of the linear subspace $\operatorname{fl} A - v_0$ of E, we have the unique linear function $h : \operatorname{fl} A - v_0 \to F' - g(v_0)$ such that

$$h(v-v_0) = g(v) - g(v_0) \quad \text{for each } v \in A \setminus \{v_0\}.$$

Then, g uniquely extends to the affine function $\tilde{g} : \operatorname{fl} A \to F'$ defined by

$$\tilde{g}(x) = h(x-v_0) + g(v_0) \quad \text{for each } x \in \operatorname{fl} A.$$

It is easy to see that $\tilde{g}(\operatorname{fl} A) = \operatorname{fl} g(A)$. □

Additional Properties of Flats and Affine Functions 3.1.6.

In the following, let E and E' be linear spaces and $f : F \to E'$ be a function of a flat F in E.

(1) If f is affine and F' is a flat in E', then $f(F)$ and $f^{-1}(F')$ are flats in E' and E, respectively.
(2) A function f is affine if and only if the graph $\operatorname{Gr}(f) = \{(x, f(x)) \mid x \in F\}$ of f is a flat in $E \times E'$.

3.2 Convex Sets

In this section, we introduce the basic concepts of convex sets. A subset $C \subset E$ is said to be **convex** if the line segment with the end ponts in C is contained in C, i.e.,

$$(1-t)x + ty \in C \quad \text{for each } x, y \in C \text{ and } t \in \mathbf{I}.$$

By induction on n, it can be proved that every convex set $C \subset E$ satisfies the following condition:

$$\sum_{i=1}^{n} z(i)v_i \in C \quad \text{for each } n \in \mathbb{N},\ v_i \in C \text{ and } z \in \Delta^{n-1},$$

where $\Delta^{n-1} = \{z \in \mathbf{I}^n \mid \sum_{i=1}^{n} z(i) = 1\}$ is the standard $(n-1)$-simplex. The following is easy:

- If $A, B \subset E$ are convex, then $aA + bB$ is also convex for each $a, b \in \mathbb{R}$.

The **dimension** of a convex set $C \subset E$ is defined by the dimension of the flat hull $\mathrm{fl}\, C$, i.e., $\dim C = \dim \mathrm{fl}\, C$. Concerning the flat hull of a convex set, we have the following proposition:

Proposition 3.2.1. *For each convex set $C \subset E$,*

$$\mathrm{fl}\, C = \big\{(1-t)x + ty \bigm| x, y \in C,\ t \in \mathbb{R}\big\}.$$

Proof. Each $z \in \mathrm{fl}\, C$ can be written $z = \sum_{i=1}^{n} t_i x_i$, where $x_i \in C$ and $\sum_{i=1}^{n} t_i = 1$. We may assume that $t_1 \leq \cdots \leq t_n \in \mathbb{R} \setminus \{0\}$. If $t_1 \geq 0$ then $z \in C$. Otherwise, $t_k < 0$ and $t_{k+1} > 0$ for some $k = 1, \ldots, n-1$. Then, we have $t = \sum_{i=1}^{n-k} t_{k+i} > 0$, where $1 - t = \sum_{i=1}^{k} t_i < 0$. Let

$$x = \sum_{i=1}^{k} (1-t)^{-1} t_i x_i, \quad y = \sum_{i=1}^{n-k} t^{-1} t_{k+i} x_{k+i} \in C.$$

Then, $z = (1-t)x + ty$. Accordingly, we have

$$\mathrm{fl}\, C \subset \big\{(1-t)x + ty \bigm| x, y \in C,\ t \in \mathbb{R}\big\}.$$

The converse inclusion is obvious. □

The smallest convex set containing $A \subset E$ is called the **convex hull** of A and is denoted by $\langle A \rangle$. We simply write $\langle v_1, \ldots, v_n \rangle = \langle \{v_1, \ldots, v_n\} \rangle$. Then, $\Delta^{n-1} = \langle \mathbf{e}_1, \ldots, \mathbf{e}_n \rangle$. Observe that

$$\langle v_1, \ldots, v_n \rangle = \big\{ \textstyle\sum_{i=1}^{n} z(i) v_i \bigm| z \in \Delta^{n-1} \big\} \text{ and}$$
$$\langle A \rangle = \bigcup \big\{ \langle x_1, \ldots, x_n \rangle \bigm| n \in \mathbb{N},\ x_1, \ldots, x_n \in A \big\}.$$

For each two non-empty subsets $A, B \subset E$,

$$\langle A \cup B \rangle = \big\{(1-t)x + ty \bigm| x \in \langle A \rangle,\ y \in \langle B \rangle,\ t \in \mathbf{I}\big\} \text{ and}$$
$$\langle aA + bB \rangle = a\langle A \rangle + b\langle B \rangle \text{ for } a, b \in \mathbb{R}.$$

The second equality can be proved as follows: Because $a\langle A\rangle + b\langle B\rangle$ is convex and $aA + bB \subset a\langle A\rangle + b\langle B\rangle$, we have $\langle aA + bB\rangle \subset a\langle A\rangle + b\langle B\rangle$. To show that $a\langle A\rangle + b\langle B\rangle \subset \langle aA+bB\rangle$, let $x \in \langle A\rangle$ and $y \in \langle B\rangle$. Then, $x = \sum_{i=1}^n t_i x_i$ and $y = \sum_{j=1}^m s_j y_j$ for some $x_i \in A$, $y_j \in B$, and $t_i, s_j > 0$ with $\sum_{i=1}^n t_i = \sum_{j=1}^m s_j = 1$. Since $ax_i + by_j \in aA+bB$ and $\sum_{i=1}^n \sum_{j=1}^m t_i s_j = 1$, it follows that

$$\begin{aligned} ax + by &= \sum_{i=1}^n t_i(ax_i + by) = \sum_{i=1}^n t_i\Big(\sum_{j=1}^m s_j(ax_i + by_j)\Big) \\ &= \sum_{i=1}^n \sum_{j=1}^m t_i s_j(ax_i + by_j) \in \langle aA + bB\rangle. \end{aligned}$$

Let C and C' be non-empty convex sets in the linear spaces E and E', respectively. A function $f : C \to C'$ is said to be **affine** (or **linear in the affine sense**) provided

$$f((1-t)x + ty) = (1-t)f(x) + tf(y) \quad \text{for each } x, y \in C \text{ and } t \in \mathbf{I}.$$

As in the definition of a flat, **I** can be replaced by $\mathbb{R}$, i.e.,

$$\begin{aligned} &x, y \in C,\ t \in \mathbb{R},\ (1-t)x + ty \in C \\ &\quad\Rightarrow f((1-t)x + ty) = (1-t)f(x) + tf(y). \end{aligned}$$

Indeed, let $z = (1-t)x + ty \in C$ in the above expression. When $t < 0$, consider

$$x = \frac{1}{1-t}z + \frac{-t}{1-t}y,\ \frac{1}{1-t} \in \mathbf{I},\ \frac{-t}{1-t} = 1 - \frac{1}{1-t}.$$

When $t > 1$, consider

$$y = \frac{1}{t}z + \frac{t-1}{t}x,\ \frac{1}{t} \in \mathbf{I},\ \frac{t-1}{t} = 1 - \frac{1}{t}.$$

As is easily seen, $f : C \to C'$ is affine if and only if

$$f\big(\textstyle\sum_{i=1}^n z(i)v_i\big) = \displaystyle\sum_{i=1}^n z(i)f(v_i) \text{ for each } n \in \mathbb{N},\ v_i \in C \text{ and } z \in \Delta^{n-1},$$

which is equivalent to the following:

$$v_i \in C,\ t_i \in \mathbb{R},\ \sum_{i=1}^n t_i v_i \in C,\ \sum_{i=1}^n t_i = 1 \Rightarrow f\big(\textstyle\sum_{i=1}^n t_i v_i\big) = \displaystyle\sum_{i=1}^n t_i f(v_i).$$

For every affine function $f : C \to E'$ of a convex set $C \subset E$ into another linear space E', the image $f(C)$ is also convex.

Proposition 3.2.2. *Let C and D be non-empty convex sets in the linear spaces E and E', respectively. Every affine function $f : C \to D$ uniquely extends to an affine function $\tilde{f} : \mathrm{fl}\, C \to \mathrm{fl}\, D$. Moreover, if f is injective (or surjective) then so is $\tilde{f}$.*

Proof. Let C_0 be a maximal affinely independent subset of C. Then, $\mathrm{fl}\, C = \mathrm{fl}\, C_0$. Due to Proposition 3.1.5, $f|C_0$ uniquely extends to an affine function $\tilde{f} : \mathrm{fl}\, C \to \mathrm{fl}\, D$. From the above remark, we can see that $\tilde{f}|C = f$.

If f is injective, we show that $\tilde{f}$ is also injective. By the definition of $\tilde{f}$ in the proof of Proposition 3.1.5, it suffices to show that $f(C_0)$ is affinely independent. Assume that $f(C_0)$ is not affinely independent, i.e., there are distinct points $v_1, \ldots, v_n \in C_0$ and $t_1, \ldots, t_n \in \mathbb{R} \setminus \{0\}$ such that $\sum_{i=1}^n t_i f(v_i) = \mathbf{0}$ and $\sum_{i=1}^n t_i = 0$. Without loss of generality, it can be assumed that $t_1, \ldots, t_k > 0$ and $t_{k+1}, \ldots, t_n < 0$. Note that $1 < k < n$ and $\sum_{i=1}^k t_i = -\sum_{j=k+1}^n t_j > 0$. Let

$$x = \sum_{i=1}^{k} \frac{t_i}{s} v_i \quad \text{and} \quad y = \sum_{j=k+1}^{n} -\frac{t_j}{s} v_j, \quad \text{where } s = \sum_{i=1}^{k} t_i > 0.$$

Then, $x, y \in C$ and $f(x) = f(y)$ because

$$f(x) - f(y) = \frac{1}{s} \sum_{i=1}^{n} t_i f(v_i) = \mathbf{0}.$$

Since f is injective, we have $x = y$. Hence, it follows that $\sum_{i=1}^k t_i v_i = -\sum_{j=k+1}^n t_j v_j$, i.e., $\sum_{i=1}^n t_i v_i = \mathbf{0}$. Because C_0 is affinely independent, $t_1 = \cdots = t_n = 0$, which is a contradiction.

Finally, we show that if f is surjective then so is $\tilde{f}$. By Proposition 3.2.1, each $z \in \mathrm{fl}\, D$ can be written as follows:

$$z = (1-t)y + ty', \ y, y' \in D, \ t \in \mathbb{R}.$$

Since f is surjective, we have $x, x' \in C$ such that $f(x) = y$ and $f(x') = y'$. Then, $(1-t)x + tx' \in \mathrm{fl}\, C$ and

$$\tilde{f}((1-t)x + tx') = (1-t)y + ty' = z.$$

Therefore, $\tilde{f}$ is also surjective. □

Let C be a convex set in a linear space E. The following set is called the **radial interior** of C:

$$\mathrm{rint}\, C = \big\{ x \in C \bigm| \forall y \in C, \ \exists \delta > 0 \ \text{such that} \ (1+\delta)x - \delta y \in C \big\}.^4$$

---

[4]In Köthe's book, $\mathrm{rint}\, C$ is denoted by C^i and called the **algebraic kernel** of C.

In the case $C = \langle v_1, \ldots, v_n \rangle$, observe that

$$\operatorname{rint}\langle v_1, \ldots, v_n \rangle = \left\{ \sum_{i=1}^n z(i) v_i \;\middle|\; z \in \Delta^{n-1} \cap (0, \infty)^n \right\}.$$

Indeed, let $x_0 = \sum_{i=1}^n n^{-1} v_i \in \langle v_1, \ldots, v_n \rangle$. For each $x \in \operatorname{rint}\langle v_1, \ldots, v_n \rangle$, we have $y \in \langle v_1, \ldots, v_n \rangle$ such that $x \in \langle x_0, y \rangle$, i.e., $x = (1-t)x_0 + ty$ for some $t \in (0,1)$. Then, $y = \sum_{i=1}^n z(i) v_i$ for some $z \in \Delta^{n-1}$. It follows that $x = \sum_{i=1}^n ((1-t)n^{-1} + tz(i)) v_i$, where $\sum_{i=1}^n ((1-t)n^{-1} + tz(i)) = 1$ and $(1-t)n^{-1} + tz(i) > 0$ for all $i = 1, \ldots, n$. Thus, x is a point of the rightside set. Conversely, it is straightforward to prove that each point of the rightside set belongs to $\langle v_1, \ldots, v_n \rangle$.

In particular, $\operatorname{rint}\langle v_1, v_2 \rangle = \{(1-t)v_1 + tv_2 \mid 0 < t < 1\}$, and hence $\operatorname{rint}\langle v_1, v_2 \rangle = \langle v_1, v_2 \rangle \setminus \{v_1, v_2\}$ if $v_1 \neq v_2$. The radial interior of C can also be defined as

$$\operatorname{rint} C = \left\{ x \in C \;\middle|\; \forall y \in C,\ \exists z \in C \text{ such that } x \in \operatorname{rint}\langle y, z \rangle \right\}.$$

For each $x \in C$, the following subset $C_x \subset C$ is called the **face** of C at x:

$$\begin{aligned} C_x &= \left\{ y \in C \;\middle|\; \exists \delta > 0 \text{ such that } (1+\delta)x - \delta y \in C \right\} \\ &= \left\{ y \in C \;\middle|\; \exists z \in C \text{ such that } x \in \operatorname{rint}\langle y, z \rangle \right\}.^5 \end{aligned}$$

By an easy observation, we have

$$\operatorname{rint} C = \{ x \in C \mid C_x = C \}, \quad \text{i.e.,} \quad x \in \operatorname{rint} C \Leftrightarrow C_x = C.$$

When $C_x = \{x\}$, we call x an **extreme point** of C. It is said that $x \in E$ is **linearly accessible** from C if there is some $y \in C$ such that

$$\operatorname{rint}\langle x, y \rangle \subset C \quad (\text{i.e., } \langle x, y \rangle \setminus \{x\} \subset C).$$

The **radial closure** $\operatorname{rcl} C$ of C is the set of all linearly accessible points from C.[6] It should be noted that $\operatorname{rcl} C \subset \operatorname{fl} C$ by Proposition 3.2.1, hence $\operatorname{fl} \operatorname{rcl} C = \operatorname{fl} C$. Consequently, we have the following inclusions:

$$\operatorname{rint} C \subset C \subset \operatorname{rcl} C \subset \operatorname{fl} C.$$

The set $\partial C = \operatorname{rcl} C \setminus \operatorname{rint} C$ is called the **radial boundary** of C.

Remark 1. Note that $A \subset B$ implies $\operatorname{rcl} A \subset \operatorname{rcl} B$, but it does not imply $\operatorname{rint} A \subset \operatorname{rint} B$. For example, consider $A = \mathbf{I}^n \times \{\mathbf{0}\} \subset B = \mathbf{I}^{n+1}$. Then, $A \cap \operatorname{rint} B = \emptyset$.

---

[5] The face C_x is a little differently defined than the **supporting facet** of C through x in Köthe's book.

[6] In Köthe's book, $\operatorname{rcl} C$ is denoted by C^a and called the **algebraic hull** of C.

For the Hilbert cube $\boldsymbol{Q} = [-1,1]^{\mathbb{N}}$, we have

$$\operatorname{rint} \boldsymbol{Q} = \{x \in \boldsymbol{Q} \mid \sup_{i\in\mathbb{N}} |x(i)| < 1\} \subsetneqq (-1,1)^{\mathbb{N}}.$$

Observe that $\operatorname{rint}[-1,1]^{\mathbb{N}}_f = (-1,1)^{\mathbb{N}}_f$ but $\operatorname{rint} \mathbf{I}^{\mathbb{N}}_f = \emptyset$, where

$$[-1,1]^{\mathbb{N}}_f = \mathbb{R}^{\mathbb{N}}_f \cap [-1,1]^{\mathbb{N}},\ (-1,1)^{\mathbb{N}}_f = \mathbb{R}^{\mathbb{N}}_f \cap (-1,1)^{\mathbb{N}}, \text{ and } \mathbf{I}^{\mathbb{N}}_f = \mathbb{R}^{\mathbb{N}}_f \cap \mathbf{I}^{\mathbb{N}}.$$ [7]

As is easily observed, $\mathbf{I}^{\mathbb{N}}_f = \operatorname{rcl}(\mathbf{I}^{\mathbb{N}}_f \setminus \{\mathbf{0}\})$. It will be shown in Remark 3 that $\mathbf{I}^{\mathbb{N}}_f \setminus \{\mathbf{0}\} = \operatorname{rcl} C$ for some convex set $C \subset \mathbb{R}^{\mathbb{N}}_f$.

Remark 2. The unit closed ball $\mathbf{B}_{\boldsymbol{c}_0}$ of the Banach space $\boldsymbol{c}_0$ has no extreme points. In fact, every $x \in \mathbf{B}_{\boldsymbol{c}_0}$ is the midpoint of two distinct points $y, z \in \mathbf{B}_{\boldsymbol{c}_0}$, i.e., $x = \frac{1}{2}y + \frac{1}{2}z$. For example, choose $n \in \mathbb{N}$ so that $|x(n)| < \frac{1}{2}$ and let $y, z \in \mathbf{B}_{\boldsymbol{c}_0}$ such that $y(i) = z(i) = x(i)$ for $i \neq n$, $y(n) = x(n) + \frac{1}{2}$, and $z(n) = x(n) - \frac{1}{2}$.

Proposition 3.2.3. *Let $C \subset E$ be a convex set. If $x \in \operatorname{rint} C$, $y \in \operatorname{rcl} C$, and $0 \le t < 1$, then $(1-t)x + ty \in \operatorname{rint} C$, i.e., $\langle x, y\rangle \setminus \{y\} \subset \operatorname{rint} C$.*

Proof. For each $z \in C$, we have to find $v \in C$ and $0 < s < 1$ such that

$$(1-t)x + ty = (1-s)z + sv \in \operatorname{rint}\langle z, v\rangle.$$

Take $w \in C$ so that $\operatorname{rint}\langle w, y\rangle \subset C$, and choose $0 < r < 1$ so that

$$z' = (1+r)x - rz,\ w' = (1+r)x - rw \in C.$$

The desired v is to be written as

$$v = t_1 y + t_2 w + t_3 w' + t_4 z' = (t_1+t_2)u + (t_3+t_4)u' \in C,$$

where $t_1 + t_2 + t_3 + t_4 = 1$, $t_1, t_2, t_3, t_4 > 0$,

$$u = \frac{t_1}{t_1+t_2}y + \frac{t_1}{t_1+t_2}w,\ u' = \frac{t_3}{t_3+t_4}w' + \frac{t_4}{t_3+t_4}z' \in C.$$

Then, we have

$$\begin{aligned}(1-s)z + sv &= (1-s)z + s(t_1 y + t_2 w + t_3 w' + t_4 z')\\ &= st_1 y + s(t_2 - t_3 r)w + s(t_3+t_4)(1+r)x + (1 - s - st_4 r)z.\end{aligned}$$

To obtain $(1-s)z + sv = (1-t)x + ty$, it is enough to find $t_1, t_2, t_3, t_4 > 0$ and $0 < s < 1$ satisfying the simultaneous equations: $st_1 = t$, $t_2 = t_3 r$, $s(t_3+t_4)(1+r) =$

[7]It is known that $[-1,1]^{\mathbb{N}}_f \approx \mathbf{I}^{\mathbb{N}}_f$.

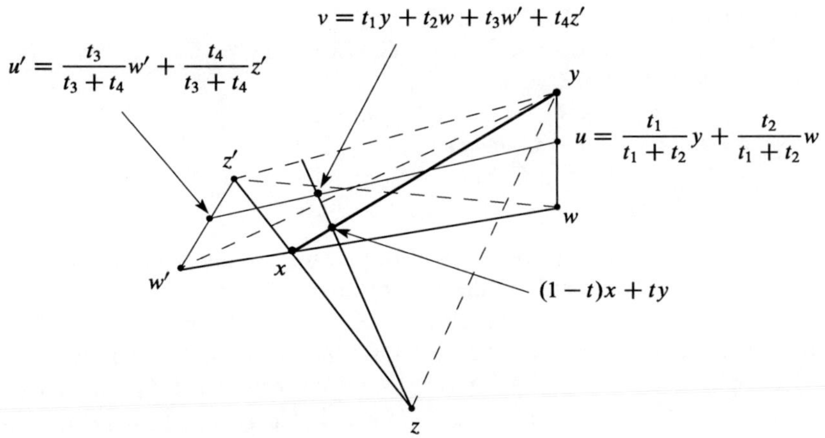

Fig. 3.1 $(1-t)x + ty \in \operatorname{rint} C$

$1-t$, and $1-s = st_4r$, i.e.,

$$(*) \quad t_1 = \frac{t}{s},\ t_4 = \frac{1-s}{rs},\ t_3 = \frac{1}{r} - \frac{1+rt}{(1+r)rs},\ t_2 = 1 - \frac{1+rt}{(1+r)s}.$$

Since $t_1, t_4 < 1$ and $0 < t_2$ $(< t_3)$, it is necessary to satisfy

$$\max\left\{t,\ \frac{1}{1+r},\ \frac{1+rt}{1+r}\right\} < s < 1.$$

We can take such an s because the left side of the above inequality is less than 1. Then, we can define $t_1, t_2, t_3, t_4 > 0$ as in $(*)$, which satisfies $t_1 + t_2 + t_3 + t_4 = 1$. Thus, we have the desired $v = t_1y + t_2w + t_3w' + t_4z' \in C$ — Fig. 3.1. □

Although we verified in Remark 1 that $A \subset B$ does not imply $\operatorname{rint} A \subset \operatorname{rint} B$ in general, we do have the following corollary:

Corollary 3.2.4. *Let A and B be non-empty convex sets in E. If $A \subset B$ and $A \cap \operatorname{rint} B \neq \emptyset$, then $\operatorname{rint} A \subset \operatorname{rint} B$.*

Proof. Let $x \in A \cap \operatorname{rint} B$. For each $y \in \operatorname{rint} A$, we have $z \in A$ such that $y \in \operatorname{rint}\langle x, z\rangle$. Since $\operatorname{rint}\langle x, z\rangle \subset \operatorname{rint} B$ by Proposition 3.2.3, it follows that $y \in \operatorname{rint} B$. □

Proposition 3.2.5. *For each convex set $C \subset E$, the following statements hold:*

(1) *Both* $\operatorname{rint} C$ *and* $\operatorname{rcl} C$ *are convex;*
(2) $\operatorname{rint}\operatorname{rint} C = \operatorname{rint} C \subset \operatorname{rint}\operatorname{rcl} C$;
(3) $\operatorname{rint} C \neq \emptyset \Rightarrow \operatorname{rint}\operatorname{rcl} C = \operatorname{rint} C,\ \operatorname{rcl}\operatorname{rint} C = \operatorname{rcl}\operatorname{rcl} C = \operatorname{rcl} C,$
in which case $\partial \operatorname{rint} C = \partial \operatorname{rcl} C = \partial C$;

(4) $\operatorname{rint} C \neq \emptyset \Rightarrow \operatorname{fl} C = \operatorname{fl} \operatorname{rint} C$;
(5) $\operatorname{rint} C \neq \emptyset$, $\operatorname{rcl} C = \operatorname{fl} C \Rightarrow \operatorname{rint} C = C = \operatorname{fl} C$;
(6) $\partial C \neq \emptyset \Leftrightarrow \emptyset \neq C \subsetneqq \operatorname{fl} C$;
(7) *C_x is convex and $C_x = C \cap \operatorname{fl} C_x$ for $x \in C$;*
(8) *$x \in \operatorname{rint} C_x$ for $x \in C$, hence $(C_x)_x = C_x$;*
(9) *$(C_x)_y = C_y$ for $x \in C$ and $y \in C_x$;*
(10) *$C_x = C_y$ for $x \in C$ and $y \in \operatorname{rint} C_x$.*

Proof. (1): To prove the convexity of $\operatorname{rint} C$, we can apply Proposition 3.2.3. It is now quite straightforward to show the convexity of $\operatorname{rcl} C$.

(2): To show $\operatorname{rint} C \subset \operatorname{rint}\operatorname{rint} C$, we can apply Proposition 3.2.3. Because $\operatorname{rint}(\operatorname{rint} C) \subset \operatorname{rint} C$ by Corollary 3.2.4, we have $\operatorname{rint}\operatorname{rint} C = \operatorname{rint} C$.

For each $x \in \operatorname{rint} C$ and $y \in \operatorname{rcl} C$, $\frac{1}{2}x + \frac{1}{2}y \in \operatorname{rint} C$ by Proposition 3.2.3. Then, we have $\delta > 0$ such that $(1+\delta)x - \delta(\frac{1}{2}x + \frac{1}{2}y) \in C$, i.e., $(1 + \frac{1}{2}\delta)x - \frac{1}{2}\delta y \in C$. Hence, $x \in \operatorname{rint}\operatorname{rcl} C$.

(3): Let $x_0 \in \operatorname{rint} C$. For each $x \in \operatorname{rint}\operatorname{rcl} C$, we have $y \in \operatorname{rcl} C$ such that $x \in \operatorname{rint}\langle x_0, y\rangle$, which implies that $x \in \operatorname{rint} C$ by Proposition 3.2.3. Combining this with (2) yields $\operatorname{rint}\operatorname{rcl} C = \operatorname{rint} C$.

We now have $x_0 \in \operatorname{rint} C = \operatorname{rint}\operatorname{rcl} C$. If $x \in \operatorname{rcl}\operatorname{rcl} C$, then $\operatorname{rint}\langle x_0, x\rangle \subset \operatorname{rint}\operatorname{rcl} C = \operatorname{rint} C$ by Proposition 3.2.3, which means that $x \in \operatorname{rcl}\operatorname{rint} C$. Since $\operatorname{rcl}\operatorname{rint} C \subset \operatorname{rcl} C \subset \operatorname{rcl}\operatorname{rcl} C$, we have $\operatorname{rcl}\operatorname{rint} C = \operatorname{rcl} C = \operatorname{rcl}\operatorname{rcl} C$.

(4): Let $x_0 \in \operatorname{rint} C$. For each $x \in C$, $\frac{1}{2}x + \frac{1}{2}x_0 \in \operatorname{fl}\operatorname{rint} C$ by Proposition 3.2.3. Then, it follows from Proposition 3.2.1 that $x = 2(\frac{1}{2}x + \frac{1}{2}x_0) - x_0 \in \operatorname{fl}\operatorname{rint} C$. Accordingly, we have $C \subset \operatorname{fl}\operatorname{rint} C$, which implies $\operatorname{fl} C \subset \operatorname{fl}\operatorname{rint} C$. Since $\operatorname{fl}\operatorname{rint} C \subset \operatorname{fl} C$, we have $\operatorname{fl} C = \operatorname{fl}\operatorname{rint} C$.

(5): Let $x_0 \in \operatorname{rint} C$. For each $x \in \operatorname{fl} C$, $2x - x_0 \in \operatorname{fl} C = \operatorname{rcl} C$. Then, $x = \frac{1}{2}x_0 + \frac{1}{2}(2x - x_0) \in \operatorname{rint} C \subset C$ by Proposition 3.2.3.

(6): Assume $\emptyset \neq C \subsetneqq \operatorname{fl} C$. Then, we have $x \in \operatorname{fl} C \setminus C$, which can be written as $x = (1+t)y - tz$ for some $y \neq z \in C$ and $t > 0$ by Proposition 3.2.1. Let

$$s = \inf\{t > 0 \mid (1+t)y - tz \notin C\} \geq 0.$$

Then, $(1+s)y - sz \in \operatorname{rcl} C \setminus \operatorname{rint} C = \partial C$.

When $C = \operatorname{fl} C$, i.e., C is a flat, we have $\operatorname{rcl} C = \operatorname{rint} C = C$ by definition, which means $\partial C = \emptyset$. Therefore, $\partial C \neq \emptyset$ implies $\emptyset \neq C \subsetneqq \operatorname{fl} C$.

(7): First, we show that C_x is convex. For each $y, z \in C_x$, we can choose $\delta > 0$ so that $(1+\delta)x - \delta y \in C$ and $(1+\delta)x - \delta z \in C$. Then, for each $t \in \mathbf{I}$,

$$\begin{aligned}(1+\delta)x &- \delta\big((1-t)y + tz\big) \\ &= (1-t)\big((1+\delta)x - \delta y\big) + t\big((1+\delta)x - \delta z\big) \in C,\end{aligned}$$

which means $(1-t)y + tz \in C_x$.

Because $C_x \subset C \cap \operatorname{fl} C_x$, it remains to show $C \cap \operatorname{fl} C_x \subset C_x$. By Proposition 3.2.1, each $y \in C \cap \operatorname{fl} C_x$ can be written as $y = (1-t)y' + ty''$ for some $y', y'' \in C_x$

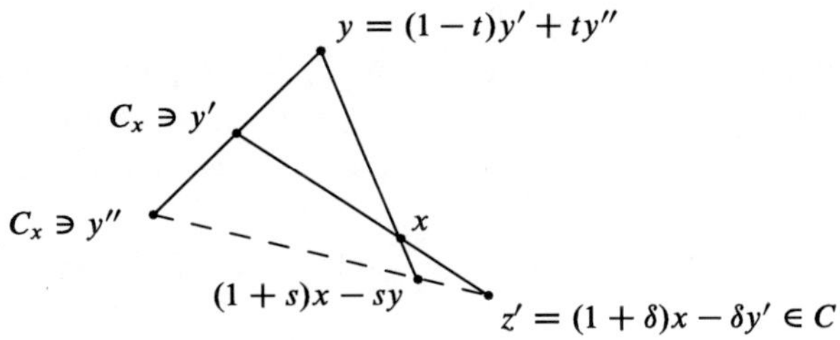

Fig. 3.2 $C \cap \operatorname{fl} C_x \subset C_x$

and $t \in \mathbb{R}$. Because of the convexity of C_x, we have $y \in C_x$ if $t \in \mathbf{I}$. Then, we may assume that $t < 0$ (if $t > 1$, exchange y' with y''). We have $\delta > 0$ such that $z' = (1+\delta)x - \delta y' \in C$. Observe that

$$\begin{aligned}(1+s)x - sy &= (1+s)\left(\frac{\delta}{1+\delta}y' + \frac{1}{1+\delta}z'\right) - s\big((1-t)y' + ty''\big)\\ &= \left(\frac{(1+s)\delta}{1+\delta} - s(1-t)\right)y' + \frac{1+s}{1+\delta}z' - sty''.\end{aligned}$$

Let $s = \delta/(1-t-t\delta) > 0$. Then, since $1+s = (1-t)(1+\delta)/(1-t-t\delta)$, it follows that

$$(1+s)x - sy = \frac{1-t}{1-t-t\delta}z' + \frac{-t\delta}{1-t-t\delta}y'' \in C,$$

which implies that $y \in C_x$ (Fig. 3.2).

(8): From the definition of $\operatorname{rint} C_x$, it easily follows that $x \in \operatorname{rint} C_x$.

(9): Because $C_x \subset C$, we have $(C_x)_y \subset C_y$ by definition. We will show that $C_y \subset C_x$, which implies $C_y = (C_y)_y \subset (C_x)_y$ by (8) and the definition. For each $z \in C_y$, choose $\delta_1 > 0$ so that $u = (1+\delta_1)y - \delta_1 z \in C$. On the other hand, since $y \in C_x$, we have $\delta_2 > 0$ such that $v = (1+\delta_2)x - \delta_2 y \in C$. Then,

$$\frac{(1+\delta_1)(1+\delta_2)}{1+\delta_1+\delta_2}x - \frac{\delta_1\delta_2}{1+\delta_1+\delta_2}z = \frac{1+\delta_1}{1+\delta_1+\delta_2}v + \frac{\delta_2}{1+\delta_1+\delta_2}u \in C,$$

which means that $z \in C_x$.

(10): Since $y \in \operatorname{rint} C_x$, we have $(C_x)_y = C_x$. On the other hand, $(C_x)_y = C_y$ by (9). □

Remark 3. It should be noted that, in general, $\operatorname{rcl}\operatorname{rcl} C \neq \operatorname{rcl} C$. For example, let C be the convex set in $\mathbb{R}^{\mathbb{N}}_f$ defined as follows:

$$C = \Big\{x \in \mathbf{I}^{\mathbb{N}}_f \;\Big|\; \exists k \in \mathbb{N} \text{ such that } \textstyle\sum_{i\in\mathbb{N}} x(i) \geq k^{-1},\\ x(i) \neq 0 \text{ at least } k \text{ many } i \in \mathbb{N}\Big\}.$$

It is easy to see that $\mathbf{0} \notin \operatorname{rcl} C$, i.e., $\operatorname{rcl} C \subset \mathbf{I}_f^{\mathbb{N}} \setminus \{\mathbf{0}\}$. For each $x \in \mathbf{I}_f^{\mathbb{N}} \setminus \{\mathbf{0}\}$, choose $k \in \mathbb{N}$ so that $k^{-1} \le \sum_{i\in\mathbb{N}} x(i)$, and let $y \in C$ such that $y(i) = k^{-2}$ for $i \le k$ and $y(i) = 0$ for $i > k$. If $0 < t \le 1$, then $(1-t)x + ty \in C$ because $(1-t)x(i) + ty(i) \neq 0$ for at least k many $i \in \mathbb{N}$ and

$$\sum_{i\in\mathbb{N}} \big((1-t)x(i) + ty(i)\big) = (1-t)\sum_{i\in\mathbb{N}} x(i) + t\sum_{i\in\mathbb{N}} y(i) \ge k^{-1}.$$

Therefore, $\operatorname{rcl} C = \mathbf{I}_f^{\mathbb{N}} \setminus \{\mathbf{0}\}$. As observed in Remark 1, $\operatorname{rcl}\big(\mathbf{I}_f^{\mathbb{N}} \setminus \{\mathbf{0}\}\big) = \mathbf{I}_f^{\mathbb{N}}$. Hence, we have $\operatorname{rcl}\operatorname{rcl} C \neq \operatorname{rcl} C$. It should also be noted that $\operatorname{rint} C = \emptyset$.

In the finite-dimensional case, we have the following proposition:

Proposition 3.2.6. *Every non-empty finite-dimensional convex set C has a non-empty radial interior, i.e., $\operatorname{rint} C \neq \emptyset$, and therefore*

$$\operatorname{rcl}\operatorname{rint} C = \operatorname{rcl}\operatorname{rcl} C = \operatorname{rcl} C \quad \textit{and} \quad \partial\operatorname{rint} C = \partial\operatorname{rcl} C = \partial C.$$

Proof. We have a maximal affinely independent finite subset $\{v_1, \dots, v_n\} \subset C$. Then, $v_0 = \sum_{i=1}^n n^{-1} v_i \in \operatorname{rint} C$. Indeed, since $C \subset \operatorname{fl}\{v_1, \dots, v_n\}$, each $x \in C$ can be written as $x = \sum_{i=1}^n t_i v_i$, where $\sum_{i=1}^n t_i = 1$. Observe that

$$\begin{aligned}(1+\delta)v_0 - \delta x &= (1+\delta)\sum_{i=1}^n n^{-1} v_i - \delta \sum_{i=1}^n t_i v_i \\ &= \sum_{i=1}^n (n^{-1} + \delta(n^{-1} - t_i))v_i.\end{aligned}$$

When $v_0 \neq x$, we have $s = \min\{n^{-1} - t_i \mid i=1,\dots,n\} < 0$. Let $\delta = 1/(-sn) > 0$. Then, $n^{-1} + \delta(n^{-1} - t_i) \ge 0$ for every $i = 1, \dots, n$, which implies that $(1+\delta)v_0 - \delta x \in C$. □

Additional Results for Convex Sets 3.2.7.

(1) For every two convex sets C and D,

$$(C \cap D)_x = C_x \cap D_x \quad \text{for each } x \in C \cap D.$$

(2) For every two convex sets C and D with $\operatorname{rint} C \cap \operatorname{rint} D \neq \emptyset$,

$$\operatorname{rint}(C \cap D) = \operatorname{rint} C \cap \operatorname{rint} D.$$

In general, $\operatorname{rint} C \cap \operatorname{rint} D \subset \operatorname{rint}(C \cap D)$.

Sketch of Proof. To show that $\operatorname{rint}(C \cap D) \subset \operatorname{rint} C \cap \operatorname{rint} D$, let $x_0 \in \operatorname{rint} C \cap \operatorname{rint} D$. For each $x \in \operatorname{rint}(C \cap D)$, take $y \in C \cap D$ so that $x \in \operatorname{rint}\langle x_0, y\rangle$. Since $\operatorname{rint}\langle x_0, y\rangle \subset \operatorname{rint} C$ by Proposition 3.2.3, it follows that $x \in \operatorname{rint} C$. Hence, $\operatorname{rint}(C \cap D) \subset \operatorname{rint} C$. Similarly, we have $\operatorname{rint}(C \cap D) \subset \operatorname{rint} D$.

(3) Let C and D be convex sets in the linear spaces E and E', respectively. Then, $C \times D$ is also convex,

$$\operatorname{rint}(C \times D) = \operatorname{rint} C \times \operatorname{rint} D \quad \text{and} \quad \operatorname{rcl}(C \times D) = \operatorname{rcl} C \times \operatorname{rcl} D.$$

Moreover, $(C \times D)_{(x,y)} = C_x \times D_y$ for each $(x, y) \in C \times D$.

(4) Let $f : C \to E'$ be an affine function of a convex set C in a linear space E into another linear space E', and D be a convex set in E'. Then, $f(C)$ and $f^{-1}(D)$ are convex and

$$f^{-1}(D)_x = C_x \cap f^{-1}(D_{f(x)}) \ \text{ for each } x \in f^{-1}(D)\ (\subset C).$$

In particular, $C_x \subset f^{-1}(f(C)_{f(x)})$ (i.e., $f(C_x) \subset f(C)_{f(x)}$) for each $x \in C$. When f is injective, $f(C_x) = f(C)_{f(x)}$ for each $x \in C$.

Sketch of Proof. It is easy to see that $f(f^{-1}(D)_x) \subset D_{f(x)}$, hence $f^{-1}(D)_x \subset f^{-1}(D_{f(x)})$. Also, $f^{-1}(D)_x \subset C_x$ because $f^{-1}(D) \subset C$. Accordingly, $f^{-1}(D)_x \subset C_x \cap f^{-1}(D_{f(x)})$. To prove the converse inclusion, for each $y \in f^{-1}(D_{f(x)}) \cap C_x$, choose $\delta > 0$ so that $(1+\delta)f(x) - \delta f(y) \in D$ and $(1+\delta)x - \delta y \in C$. Then, $(1+\delta)x - \delta y \in f^{-1}(D)$.

(5) For every (bounded) subset A of a normed linear space $E = (E, \|\cdot\|)$, the following hold:

(i) $\|x - y\| \le \sup_{z\in A} \|x - z\|$ for each $x \in E$ and $y \in \langle A\rangle$;
(ii) $\operatorname{diam}\langle A\rangle = \operatorname{diam} A$.

Sketch of Proof. (i): Write $y = \sum_{i=1}^n z(i)x_i$ for some $x_1, \ldots, x_n \in A$ and $z \in \Delta^{n-1}$.
(ii): For each $x, y \in \langle A\rangle$,

$$\|x - y\| \le \sup_{z\in A} \|x - z\| \le \sup_{z\in A} \sup_{z'\in A} \|z - z'\| = \operatorname{diam} A.$$

Remark 4. In (2) above, $\operatorname{rint}(C \cap D) \ne \operatorname{rint} C \cap \operatorname{rint} D$ in general. Consider the case that $C \cap D \ne \emptyset$ but $\operatorname{rint} C \cap \operatorname{rint} D = \emptyset$.

In (4) above, $f(C_x) \ne f(C)_{f(x)}$ in general. For instance, let $C = \{(s,t) \in \mathbb{R}^2 \mid |s| \le t \le 1\} \subset \mathbb{R}^2$. Then, $\operatorname{pr}_1(C) = [-1, 1]$, $\operatorname{pr}_1(C_0) = \{0\}$, and $\operatorname{pr}_1(C)_0 = \operatorname{pr}_1(C)$.

3.3 The Hahn–Banach Extension Theorem

We now prove the Hahn–Banach Extension Theorem and present a relationship between the sublinear functionals and the convex sets.

Let E be a linear space. A functional $p : E \to \mathbb{R}$ is **sublinear** if it satisfies the following conditions:

(SL_1) $p(x + y) \le p(x) + p(y)$ for each $x, y \in E$, and

(SL_2) $p(tx) = tp(x)$ for each $x \in E$ and $t > 0$.

Note that if $p : E \to \mathbb{R}$ is sublinear then $p(\mathbf{0}) = 0$ and $-p(-x) \leq p(x)$. For each $x, y \in E$ and $t \in \mathbf{I}$,

$$p((1-t)x + ty) \leq (1-t)p(x) + tp(y).$$

When $p : E \to \mathbb{R}$ is a non-negative sublinear functional, $p^{-1}([0, r)) = rp^{-1}([0, 1))$ and $p^{-1}([0, r]) = rp^{-1}(\mathbf{I})$ are convex for each $r > 0$.

In the following Hahn–Banach Extension Theorem, no topological concepts appear (even in the proof). Nevertheless, this theorem is very important in the study of topological linear spaces.

Theorem 3.3.1 (Hahn–Banach Extension Theorem). *Let $p : E \to \mathbb{R}$ be a sublinear functional of a linear space E and F be a linear subspace of E. If $f : F \to \mathbb{R}$ is a linear functional such that $f(x) \leq p(x)$ for every $x \in F$, then f extends to a linear functional $\tilde{f} : E \to \mathbb{R}$ such that $\tilde{f}(x) \leq p(x)$ for every $x \in E$.*

Proof. Let $\mathcal{F}$ be the collection of all linear functionals $f' : F' \to \mathbb{R}$ of a linear subspace $F' \subset E$ such that $F \subset F'$, $f'|F = f$, and $f'(x) \leq p(x)$ for every $x \in F'$. For $f', f'' \in \mathcal{F}$, we define $f' \leq f''$ if f'' is an extension of f'. Then, $\mathcal{F} = (\mathcal{F}, \leq)$ is an inductive ordered set, i.e., every totally ordered subset of $\mathcal{F}$ is upper bounded. By Zorn's Lemma, $\mathcal{F}$ has a maximal element $f_0 : F_0 \to \mathbb{R}$. It suffices to show that $F_0 = E$.

Assume that $F_0 \neq E$. Taking $x_1 \in E \setminus F_0$, we have a linear subspace $F_1 = F_0 + \mathbb{R}x_1 \supsetneq F_0$. We show that f_0 has a linear extension $f_1 : F_1 \to \mathbb{R}$ in $\mathcal{F}$, which contradicts the maximality of f_0. By assigning x_1 to $\alpha \in \mathbb{R}$, f_1 can be defined, i.e., $f_1(x + tx_1) = f_0(x) + t\alpha$ for $x \in F_0$ and $t \in \mathbb{R}$. In order that $f_1 \in \mathcal{F}$, we have to choose α so that for every $x \in F_0$ and $t > 0$,

$$f_0(x) + t\alpha \leq p(x + tx_1) \text{ and } f_0(x) - t\alpha \leq p(x - tx_1).$$

Dividing by t, we obtain the following equivalent condition:

$$f_0(y) - p(y - x_1) \leq \alpha \leq p(y + x_1) - f_0(y) \text{ for every } y \in F_0.$$

Hence, such an $\alpha \in \mathbb{R}$ exists if

$$\sup\{f_0(y) - p(y - x_1) \mid y \in F_0\} \leq \inf\{p(y + x_1) - f_0(y) \mid y \in F_0\}.$$

This inequality can be proved as follows: for each $y, y' \in F_0$,

$$f_0(y) + f_0(y') = f_0(y + y') \leq p(y + y') \leq p(y - x_1) + p(y' + x_1),$$

hence $f_0(y) - p(y - x_1) \leq p(y' + x_1) - f_0(y')$, which implies the desired inequality. □

Let F be a flat in a linear space E and $A \subset F$. The following set is called the **core** of A in F:

$$\operatorname{core}_F A = \{x \in A \mid \forall y \in F,\ \exists \delta > 0 \text{ such that } |t| \le \delta \Rightarrow (1-t)x + ty \in A\},$$

where $|t| \le \delta$ can be replaced by $-\delta \le t \le 0$ (or $0 \le t \le \delta$). Each point of $\operatorname{core}_F A$ is called a **core point** of A in F. In the case that A is convex,

$$\begin{aligned} x \in \operatorname{core}_F A &\Leftrightarrow \forall y \in F,\ \exists \delta > 0 \text{ such that } (1+\delta)x - \delta y \in A \\ &\Leftrightarrow \forall y \in F,\ \exists \delta > 0 \text{ such that } (1-\delta)x + \delta y \in A. \end{aligned}$$

When $F = E$, we can omit the phrase "in E" and simply write $\operatorname{core} A$ by removing the subscript E. By definition, $A \subset B \subset F$ implies $\operatorname{core}_F A \subset \operatorname{core}_F B$. We also have the following fact:

Fact. *For each $A \subset F$, $\operatorname{core}_F A \neq \emptyset$ if and only if $\operatorname{fl} A = F$.*

Indeed, the "if" part is trivial. To show the "only if" part, let $x \in \operatorname{core}_F A$. For each $y \in F$, we have $\delta > 0$ such that $z = (1+\delta)x - \delta y \in A$. Then, $y = \delta^{-1}(1+\delta)x - \delta^{-1}z \in \operatorname{fl} A$. Note that $\operatorname{fl} A \subset F$ because $A \subset F$. Consequently, $\operatorname{fl} A = F$.

Proposition 3.3.2. *For every convex set $A \subset E$, $\operatorname{core}_{\operatorname{fl} A} A = \operatorname{rint} A$, which is also convex. Hence, $\operatorname{core} A \neq \emptyset$ implies $\operatorname{core} A = \operatorname{rint} A$ and $\operatorname{core}\operatorname{core} A = \operatorname{core} A$.*

Proof. Because $\operatorname{core}_{\operatorname{fl} A} A \subset \operatorname{rint} A$ by definition, it suffices to show that $\operatorname{rint} A \subset \operatorname{core}_{\operatorname{fl} A} A$. For each $x \in \operatorname{rint} A$ and $y \in \operatorname{fl} A$, we need to find some $s > 0$ such that $(1+s)x - sy \in A$. This can be done using the same proof of the inclusion $C \cap \operatorname{fl} C_x \subset C_x$ in Proposition 3.2.5(7). □

Remark 5. When A is a finite-dimensional convex set, $\operatorname{core}_F A \neq \emptyset$ if and only if $F = \operatorname{fl} A$ according to Propositions 3.3.2 and 3.2.6. However, this does not hold for an infinite-dimensional convex set. For example, consider the convex set $\mathbf{I}^{\mathbb{N}}_f$ in $\mathbb{R}^{\mathbb{N}}$. Then, $\mathbb{R}^{\mathbb{N}}_f = \operatorname{fl} \mathbf{I}^{\mathbb{N}}_f$ and $\operatorname{core}_{\mathbb{R}^{\mathbb{N}}_f} \mathbf{I}^{\mathbb{N}}_f = \operatorname{rint} \mathbf{I}^{\mathbb{N}}_f = \emptyset$.

With regard to convex sets defined by a non-negative sublinear functional, we have the following proposition:

Proposition 3.3.3. *Let $p : E \to \mathbb{R}$ be a non-negative sublinear functional of a linear subspace E. Then,*

$$p^{-1}([0,1)) = \operatorname{core} p^{-1}([0,1)) = \operatorname{core} p^{-1}(\mathbf{I}).$$

Proof. The inclusion $\operatorname{core} p^{-1}([0,1)) \subset \operatorname{core} p^{-1}(\mathbf{I})$ is obvious.

Let $x \in p^{-1}([0,1))$. For each $y \in E$, we can choose $\delta > 0$ so that $\delta p(x-y) < 1 - p(x)$. Then,

$$0 \le p((1+\delta)x - \delta y) = p(x + \delta(x-y)) \le p(x) + \delta p(x-y) < 1,$$

i.e., $x \in \operatorname{core} p^{-1}([0,1))$. Hence, $p^{-1}([0,1)) \subset \operatorname{core} p^{-1}([0,1))$.

If $p(x) \ge 1$, then $x \notin \operatorname{core} p^{-1}(\mathbf{I})$ because

$$p((1+t)x - t\mathbf{0}) = (1+t)p(x) > 1 \quad \text{for any } t > 0.$$

This means that $\operatorname{core} p^{-1}(\mathbf{I}) \subset p^{-1}([0,1))$. □

For each $A \subset E$ with $\mathbf{0} \in \operatorname{core} A$, the **Minkowski functional** $p_A : E \to \mathbb{R}_+$ can be defined as follows:

$$p_A(x) = \inf\{s > 0 \mid x \in sA\} = \inf\{s > 0 \mid s^{-1}x \in A\}.$$

Then, for each $x \in E$ and $t > 0$,

$$\begin{aligned} p_A(tx) &= \inf\{s > 0 \mid s^{-1}tx \in A\} = \inf\{ts > 0 \mid (ts)^{-1}tx \in A\} \\ &= t \inf\{s > 0 \mid s^{-1}x \in A\} = tp_A(x), \end{aligned}$$

i.e., p_A satisfies (SL_2). In the above, $p_A(tx) = p_{t^{-1}A}(x)$. Then, it follows that $p_{t^{-1}A} = tp_A$ for each $t > 0$. Replacing t by t^{-1}, we have

$$p_{tA} = t^{-1}p_A \quad \text{for each } t > 0.$$

If $A \subset E$ is convex, the Minkowski functional p_A has the following desirable properties:

Proposition 3.3.4. *Let $A \subset E$ be a convex set with $\mathbf{0} \in \operatorname{core} A$. Then, the Minkowski functional p_A is sublinear and*

$$\operatorname{rint} A = \operatorname{core} A = p_A^{-1}([0,1)) \subset A \subset p_A^{-1}(\mathbf{I}) = \operatorname{rcl} A,$$

so $\partial A = p_A^{-1}(1)$. Moreover,

$$p_A(x) = 0 \Leftrightarrow \mathbb{R}_+ x \subset A.$$

In order that p_A is a norm on E, it is necessary and sufficient that $\mathbb{R}_+ x \not\subset A$ if $x \ne \mathbf{0}$ and $tA \subset A$ if $|t| < 1$.

Proof. First, we prove that p_A is sublinear. As already observed, p_A satisfies (SL_2). To show that p_A satisfies (SL_1), let $x, y \in E$. Since A is convex, we have

$$s^{-1}x,\ t^{-1}y \in A \Rightarrow (s+t)^{-1}(x+y) = \frac{s}{s+t}s^{-1}x + \frac{t}{s+t}t^{-1}y \in A,$$

which implies that $p_A(x+y) \le p_A(x) + p_A(y)$.

The first equality $\operatorname{rint} A = \operatorname{core} A$ has been stated in Proposition 3.3.2. It easily follows from the definitions that $\operatorname{core} A \subset p_A^{-1}([0,1)) \subset A \subset p_A^{-1}(\mathbf{I})$ and $p_A^{-1}(1) \subset \operatorname{rcl} A$, so $p_A^{-1}(\mathbf{I}) \subset \operatorname{rcl} A$. By Propositions 3.3.2 and 3.3.3, we have

$$\operatorname{core} A = \operatorname{core}\operatorname{core} A \subset \operatorname{core} p_A^{-1}([0,1)) = p_A^{-1}([0,1)) \subset \operatorname{core} A,$$

which means the second equality $\operatorname{core} A = p_A^{-1}([0,1))$. To obtain the third equality $p_A^{-1}(\mathbf{I}) = \operatorname{rcl} A$, it remains to show that $\operatorname{rcl} A \subset p_A^{-1}(\mathbf{I})$. Let $x \in \operatorname{rcl} A$. Since $\mathbf{0} \in \operatorname{rint} A$, it follows from Proposition 3.2.3 that $s^{-1}x \in \operatorname{rint} C \subset C$ for each $s > 1$, which implies that $p_A(x) \leq 1$, i.e., $x \in p_A^{-1}(\mathbf{I})$.

By definition, $p_A(x) = 0$ if and only if $tx \in A$ for an arbitrarily large $t > 0$, which means that $\mathbb{R}_+ x \subset A$ because A is convex.

Because p_A is sublinear, p_A is a norm if and only if $p_A(x) \neq 0$ and $p_A(x) = p_A(-x)$ for every $x \in E \setminus \{\mathbf{0}\}$. Because $p_A(x) \neq 0$ if and only if $\mathbb{R}_+ x \not\subset A$, it remains to show that $p_A(x) = p_A(-x)$ for every $x \in E \setminus \{\mathbf{0}\}$ if and only if $tA \subset A$ whenever $|t| < 1$.

Assume that $p_A(x) = p_A(-x)$ for each $x \in E$. If $x \in A$ and $|t| < 1$ then $p_A(tx) = p_A(|t|x) = |t|p_A(x) < 1$, which implies that $tx \in A$. Hence, $tA \subset A$ whenever $|t| < 1$.

Conversely, assume that $tA \subset A$ whenever $|t| < 1$. For each $s > p_A(x)$, $r^{-1}x \in A$ for some $0 < r < s$, and we have $s^{-1}(-x) = (-s^{-1}r)r^{-1}x \in A$, hence $p_A(-x) \leq p_A(x)$. Replacing x with $-x$, we have $p_A(x) \leq p_A(-x)$. Therefore, $p_A(x) = p_A(-x)$. □

When the Minkowski functional p_A is a norm on E, we call it the **Minkowski norm**. In this case, $\operatorname{rcl} A$, $\operatorname{rint} A$, and ∂A are the unit closed ball, the unit open ball, and the unit sphere, respectively, of the normed linear space $E = (E, p_A)$. Then, $\operatorname{rcl} A$ and $\operatorname{rint} A$ are symmetric about $\mathbf{0}$, i.e., $\operatorname{rcl} A = -\operatorname{rcl} A$ and $\operatorname{rint} A = -\operatorname{rint} A$. We should note that a convex set $A \subset E$ is symmetric about $\mathbf{0}$ if and only if $tA \subset A$ whenever $|t| \leq 1$ (in the next section, A is said to be circled).

A subset $W \subset E$ is called a **wedge** if $x + y \in W$ for each $x, y \in W$ and $tx \in W$ for each $x \in W, t \geq 0$, or equivalently, W is convex and $tW \subset W$ for every $t \geq 0$. Note that if $A \subset E$ is convex then $\mathbb{R}_+ A$ is a wedge. For a wedge $W \subset E$, the following statements are true:

(1) $\mathbf{0} \in \operatorname{core} W \Leftrightarrow W = E$;
(2) $W \neq E,\ x \in \operatorname{core} W \Rightarrow -x \notin W$.

A **cone** $C \subset E$ is a wedge with $C \cap (-C) = \{\mathbf{0}\}$. Each translation of a cone is also called a **cone**.

Using the Hahn–Banach Extension Theorem, we can prove the following separation theorem:

Theorem 3.3.5 (Separation Theorem). *Let A and B be convex sets in E such that $\operatorname{core} A \neq \emptyset$ and $(\operatorname{core} A) \cap B = \emptyset$. Then, there exists a linear functional $f : E \to \mathbb{R}$ such that $f(x) < f(y)$ for every $x \in \operatorname{core} A$ and $y \in B$, and $\sup f(A) \leq \inf f(B)$.*

Proof. Recall that $\operatorname{core} A = \operatorname{rint} A$ (Proposition 3.3.2). For a linear functional $f : E \to \mathbb{R}$, if $f(x) < f(y)$ for every $x \in \operatorname{core} A$ and $y \in B$, then $\sup f(A) \leq \inf f(B)$. Indeed, let $x \in A$, $y \in B$, $v \in \operatorname{core} A$, and $0 \leq t < 1$. Since $(1-t)v+tx \in \operatorname{core} A$ by Proposition 3.2.3, we have

$$(1-t)f(v) + tf(x) = f((1-t)v + tx) < f(y),$$

where the left side tends to $f(x)$ as $t \to 1$, and hence $f(x) \leq f(y)$.

Note that $W = \mathbb{R}_+(A - B)$ is a wedge. Moreover, $(\operatorname{core} A) - B \subset \operatorname{core} W$. Indeed, let $x \in \operatorname{core} A$ and $y \in B$. For each $z \in E$, choose $\delta > 0$ so that $(1+\delta)x - \delta(y+z) \in A$. Then,

$$(1+\delta)(x-y) - \delta z = (1+\delta)x - \delta(y+z) - y \in A - B \subset W.$$

Therefore, it suffices to construct a linear functional $f : E \to \mathbb{R}$ such that $f(\operatorname{core} W) \subset (-\infty, 0)$.

Now, we shall show that $W \cap (B - \operatorname{core} A) = \emptyset$. Assume that there exist $x_0 \in A$, $x_1 \in \operatorname{core} A$, $y_0, y_1 \in B$, and $t_0 \geq 0$ such that $t_0(x_0 - y_0) = y_1 - x_1$. Note that $\operatorname{rint}\langle x_0, x_1\rangle \subset \operatorname{rint} A = \operatorname{core} A$ by Proposition 3.2.3. Hence,

$$\frac{t_0}{t_0+1}x_0 + \frac{1}{t_0+1}x_1 = \frac{t_0}{t_0+1}y_0 + \frac{1}{t_0+1}y_1 \in (\operatorname{core} A) \cap B,$$

which contradicts the fact that $(\operatorname{core} A) \cap B = \emptyset$.

Take $v_0 \in (\operatorname{core} A) - B \subset \operatorname{core} W$. Then, note that $-v_0 \notin W$. For each $x \in E$, we have $\delta > 0$ such that $(1+\delta)v_0 - \delta(-x) \in W$, which implies $x + \delta^{-1}(1+\delta)v_0 \in W$. Then, we can define $p : E \to \mathbb{R}$ by

$$p(x) = \inf\{t \geq 0 \mid x + tv_0 \in W\}.$$

Because W is a wedge, we see that p is sublinear. Since $-v_0 \notin W$, it follows that $p(s(-v_0)) = s$ and $p(sv_0) = 0$ for every $s \geq 0$. Applying the Hahn–Banach Extension Theorem 3.3.1, we can obtain a linear functional $f : E \to \mathbb{R}$ such that $f(s(-v_0)) = s$ for each $s \in \mathbb{R}$ and $f(x) \leq p(x)$ for every $x \in E$ (see Fig. 3.3). For each $z \in \operatorname{core} W$, we have $\delta > 0$ such that $(1+\delta)z - \delta(z+v_0) \in W$, i.e., $z - \delta v_0 \in W$. Accordingly, $(z - \delta v_0) + tv_0 \in W$ for every $t \geq 0$, which means $p(z - \delta v_0) = 0$. Thus, we have

$$f(z) < f(z) + \delta = f(z - \delta v_0) \leq p(z - \delta v_0) = 0. \qquad \square$$

Remark 6. Using the Hahn–Banach Extension Theorem, we have proved the Separation Theorem. Conversely, the Hahn–Banach Extension Theorem can be derived from the Separation Theorem. Indeed, under the assumption of the Hahn–Banach Extension Theorem 3.3.1, we define

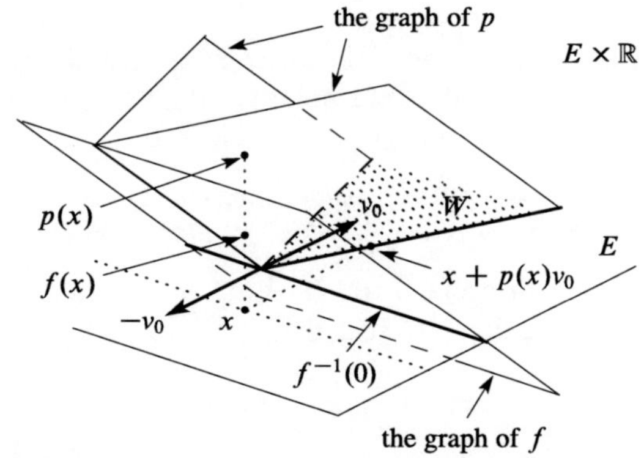

Fig. 3.3 The graphs of p and f

$$A = \{(x,t) \in E \times \mathbb{R} \mid t > p(x)\} \text{ and } B = \{(x, f(x)) \in E \times \mathbb{R} \mid x \in F\},$$

where $B = \mathrm{Gr}(f)$ is the graph of f. Then, A and B are disjoint convex sets in $E \times \mathbb{R}$. It is straightforward to show that $\operatorname{core} A = A \neq \emptyset$. By the Separation Theorem 3.3.5, we have a linear functional $\varphi : E \times \mathbb{R} \to \mathbb{R}$ such that $A \subset \varphi^{-1}((-\infty, r])$ and $B \subset \varphi^{-1}([r, \infty))$ for some $r \in \mathbb{R}$. Then, $r \leq 0$ because $0 = \varphi(0,0) \in \varphi(B)$. If $\varphi(z) < 0$ for some $z \in B$, then $\varphi(tz) = t\varphi(z) < r$ for sufficiently large $t > 0$. This is a contradiction because $tz \in B$. If $\varphi(z) > 0$ for some $z \in B$, then $-z \in B$ and $\varphi(-z) = -\varphi(z) < 0$, which is a contradiction. Therefore, $B \subset \varphi^{-1}(0)$. Note that $\varphi(0,1) < 0$ because $(0,1) \in A$. Since $\varphi(x,t) = \varphi(x,0) + t\varphi(0,1)$ for each $x \in E$, we have $\varphi(\{x\} \times \mathbb{R}) = \mathbb{R}$. Observe that $(\{x\} \times \mathbb{R}) \cap \varphi^{-1}(0)$ is a singleton. Then, f extends to the linear functional $\tilde{f} : E \to \mathbb{R}$ whose graph is $\varphi^{-1}(0)$, i.e., $(x, \tilde{f}(x)) \in \varphi^{-1}(0)$ for each $x \in E$. Since $\varphi^{-1}(0) \subset (E \times \mathbb{R}) \setminus A$, it follows that $\tilde{f}(x) \leq p(x)$ for every $x \in E$.

The Separation Theorem 3.3.5 can also be obtained as a corollary of the following two theorems, where we do not use the Hahn–Banach Extension Theorem 3.3.1.

Theorem 3.3.6. *For each pair of disjoint non-empty convex sets $A, B \subset E$, there exists a pair of disjoint convex sets $\widetilde{A}, \widetilde{B} \subset E$ such that $A \subset \widetilde{A}$, $B \subset \widetilde{B}$, and $\widetilde{A} \cup \widetilde{B} = E$.*

Proof. Let $\mathcal{P}$ be the collection of pairs (C, D) of disjoint convex sets such that $A \subset C$ and $B \subset D$. For $(C, D), (C', D') \in \mathcal{P}$, we define $(C, D) \leq (C', D')$ if $C \subset C'$ and $D \subset D'$. Then, it is easy to see that $\mathcal{P} = (\mathcal{P}, \leq)$ is an inductive ordered set. Due to Zorn's Lemma, $\mathcal{P}$ has a maximal element $(\widetilde{A}, \widetilde{B})$.

To show that $\widetilde{A} \cup \widetilde{B} = E$, assume the contrary, i.e., there exists a point $v_0 \in E \setminus (\widetilde{A} \cup \widetilde{B})$. By the maximality of $(\widetilde{A}, \widetilde{B})$, we can obtain two points

$$x \in \widetilde{A} \cap \langle \widetilde{B} \cup \{v_0\} \rangle \text{ and } y \in \widetilde{B} \cap \langle \widetilde{A} \cup \{v_0\} \rangle.$$

Then, $x \in \langle v_0, y_1\rangle$ for some $y_1 \in \widetilde{B}$ and $y \in \langle v_0, x_1\rangle$ for some $x_1 \in \widetilde{A}$. Note that $x \in \operatorname{rint}\langle v_0, y_1\rangle$ and $y \in \operatorname{rint}\langle v_0, x_1\rangle$. Consider the triangle $\langle v_0, x_1, y_1\rangle$. It is easy to see that $\langle x_1, x\rangle$ and $\langle y_1, y\rangle$ meet at a point v_1. Since $\langle x_1, x\rangle \subset \widetilde{A}$ and $\langle y_1, y\rangle \subset \widetilde{B}$, it follows that $v_1 \in \widetilde{A} \cap \widetilde{B}$, which is a contradiction. □

Theorem 3.3.7. *For each pair of disjoint non-empty convex sets $C, D \subset E$ with $C \cup D = E$, $\operatorname{rcl} C \cap \operatorname{rcl} D$ is a hyperplane if $\operatorname{rcl} C \cap \operatorname{rcl} D \neq E$.*

Proof. First, we show that $\operatorname{rcl} C \cap \operatorname{rcl} D = \partial C = \partial D$. To prove that $\partial C \subset \partial D$, let $x \in \partial C$. It suffices to find $y \in C$ such that

$$(1-t)x + ty \in C \quad \text{for } 0 < t \le 1 \text{ and}$$
$$(1+t)x - ty \in E \setminus C = D \quad \text{for } t > 0.$$

To this end, take $y', y'' \in C$ such that $(1-t)x + ty' \in C$ for $0 < t \le 1$ and $(1+t)x - ty'' \notin C$ for $t > 0$. Then, $y = \frac{1}{2}y' + \frac{1}{2}y'' \in C$ is the desired point. Indeed, for each $0 < t \le 1$,

$$\begin{aligned}(1-t)x + ty &= (1-t)x + \tfrac{1}{2}ty' + \tfrac{1}{2}ty'' \\ &= \left(1 - \tfrac{1}{2}t\right)\left(\frac{1-t}{1-\frac{1}{2}t}x + \frac{\frac{1}{2}t}{1-\frac{1}{2}t}y'\right) + \tfrac{1}{2}ty'' \in C.\end{aligned}$$

Moreover, note that

$$\begin{aligned}&(1-s)((1+t)x - ty) + sy' \\ &\qquad = (1-s)(1+t)x - \tfrac{1}{2}(1-s)ty' - \tfrac{1}{2}(1-s)ty'' + sy'.\end{aligned}$$

For each $t > 0$, let $s = t/(2+t) \in (0,1)$. Then, $(1-s)t = 2s$. Therefore, we have

$$(1-s)((1+t)x - ty) + sy' = (1+s)x - sy'' \notin C,$$

which means that $(1+t)x - ty \notin C$ (Fig. 3.4). Similarly, we have $\partial D \subset \partial C$. Hence, $\partial C = \partial D$. Since $\operatorname{rint} C \cap \operatorname{rint} D = \emptyset$, it follows that $\operatorname{rcl} C \cap \operatorname{rcl} D = \partial C = \partial D$.

Next, we show that ∂C is a flat. It suffices to show that if $x, y \in \partial C$ and $t > 0$, then $x' = (1+t)x - ty \in \partial C$. If $x' \notin \partial C$, then $x' \in \operatorname{rint} C$ or $x' \in \operatorname{rint} D$. In this case, $x \in \operatorname{rint}\langle x', y\rangle \subset \operatorname{rint} C$ or $x \in \operatorname{rint}\langle x', y\rangle \subset \operatorname{rint} D$ by Proposition 3.2.3. This is a contradiction. Therefore, $x' \in \partial C$.

It remains to show that if $\partial C \neq E$ then ∂C is a hyperplane. We have $v \in E \setminus \partial C$. It suffices to prove that $E = \operatorname{fl}(\partial C \cup \{v\})$. Without loss of generality, we may assume that $v \in \operatorname{rint} C$. On the other hand, $\partial C \neq \emptyset$ because $C \neq E$. Let $z \in \partial C$. Then, $w = z - (v - z) = 2z - v \in \operatorname{rint} D$. Otherwise, $w \in \operatorname{rcl} C$, from which, using Proposition 3.2.3, it would follow that $z = \frac{1}{2}v + \frac{1}{2}w \in \operatorname{rint}\langle v, w\rangle \subset \operatorname{rint} C$, which is a contradiction.

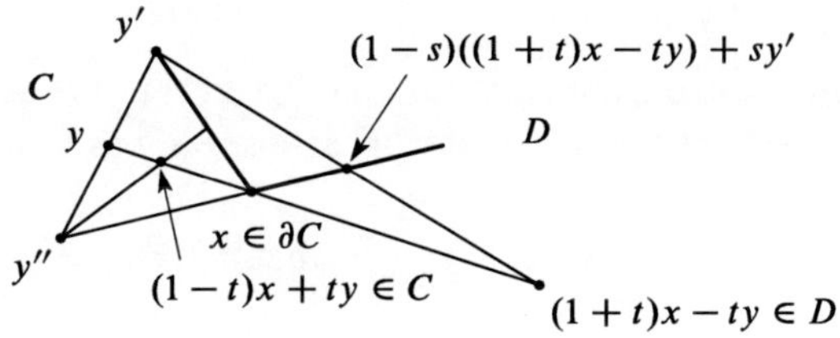

Fig. 3.4 $\partial C \subset \partial D$

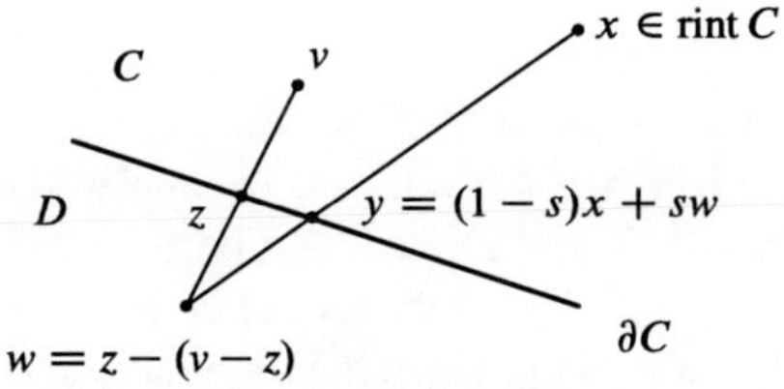

Fig. 3.5 The case $x \in \operatorname{rint} C$

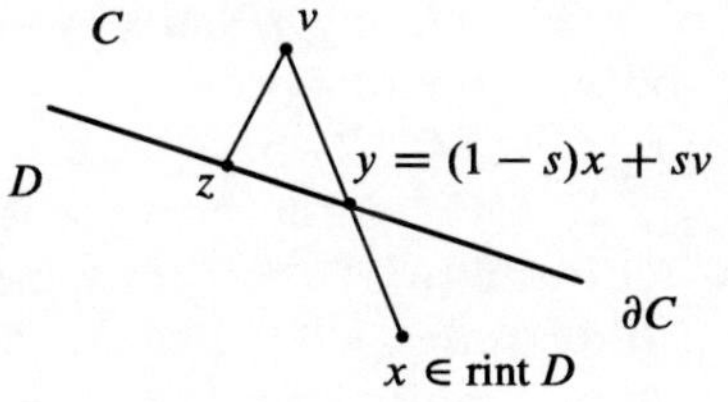

Fig. 3.6 The case $x \in \operatorname{rint} D$

For each $x \in E \setminus \partial C$, $x \in \operatorname{rint} C$ or $x \in \operatorname{rint} D$. When $x \in \operatorname{rint} C$, let

$$s = \sup \{ t \in \mathbf{I} \mid (1-t)x + tw \in C \}.$$

Refer to Fig. 3.5. Then, $y = (1-s)x + sw \in \partial C$, which implies that

$$x = \frac{1}{1-s} y - \frac{2s}{1-s} z + \frac{s}{1-s} v \in \mathrm{fl}(\partial C \cup \{v\}).$$

In the case that $x \in \operatorname{rint} D$, let

$$s = \sup \{ t \in \mathbf{I} \mid (1-t)x + tv \in D \}.$$

Now, refer to Fig. 3.6. Then, $y = (1-s)x + sv \in \partial D = \partial C$, which implies that

$$x = \frac{1}{1-s} y + \frac{-s}{1-s} v \in \mathrm{fl}(\partial C \cup \{v\}).$$

Consequently, it follows that $E = \text{fl}(\partial C \cup \{v\})$. □

Remark 7. In the above, the condition $\text{rcl}\, C \cap \text{rcl}\, D \neq E$ is necessary. For example, define the convex set C in the linear space $\mathbb{R}^{\mathbb{N}}_f$ as follows:

$$C = \{x \in \mathbb{R}^{\mathbb{N}}_f \mid n = \max\{i \mid x(i) \neq 0\} \Rightarrow x(n) > 0\}.$$

Let $D = \mathbb{R}^{\mathbb{N}}_f \setminus C = (-C) \setminus \{\mathbf{0}\}$. Then, D is also convex. As is easily observed, $\text{rcl}\, C = \text{rcl}\, D = \mathbb{R}^{\mathbb{N}}_f$, hence $\text{rcl}\, C \cap \text{rcl}\, D = \mathbb{R}^{\mathbb{N}}_f$.

The Separation Theorem 3.3.5 can also be obtained as a corollary of Theorems 3.3.6 and 3.3.7. In fact, let $A, B \subset E$ be convex sets with $\text{core}\, A \neq \emptyset$ and $(\text{core}\, A) \cap B = \emptyset$. Then, $\text{core}\, A = \text{rint}\, A$ is convex. We apply Theorem 3.3.6 to obtain disjoint non-empty convex sets C and D such that $\text{core}\, A \subset C$, $B \subset D$, and $C \cup D = E$. Observe that $\text{core}\, A \cap \text{rcl}\, D = \emptyset$, hence $\text{rcl}\, D \neq E$. It follows from Theorem 3.3.7 that $\text{rcl}\, C \cap \text{rcl}\, D$ is a hyperplane. Then, we have a linear functional $f : E \to \mathbb{R}$ such that $\text{rcl}\, C \cap \text{rcl}\, D = f^{-1}(s)$ for some $s \in \mathbb{R}$ (Proposition 3.1.3(1)). Since $\text{core}\, A \subset E \setminus f^{-1}(s)$, we have $\text{core}\, A \subset f^{-1}((s, \infty))$ or $\text{core}\, A \subset f^{-1}((-\infty, s))$. If $\text{core}\, A \subset f^{-1}((s, \infty))$, by replacing f and s by $-f$ and $-s$, it can be assumed that $\text{core}\, A \subset f^{-1}((-\infty, s))$.

We now show that $\text{rcl}\, C \subset f^{-1}((-\infty, s])$. Let $x \in \text{core}\, A$ ($\subset \text{rint}\, C$). Then, $x \in \text{rint}\, C$ and $f(x) < s$. If $f(y) > s$ for some $y \in \text{rcl}\, C$, we have $z \in \text{rint}\langle x, y\rangle \cap f^{-1}(s)$. Because $z \in \text{rcl}\, D$, $\text{rint}\langle w, z\rangle \subset D$ for some $w \in D$. On the other hand, $z \in \text{rint}\langle x, y\rangle \subset \text{rint}\, C$ (Proposition 3.2.3). Because $\text{rint}\, C = \text{core}\, C$, $\langle v, z\rangle \subset C = E \setminus D$ for some $v \in \text{rint}\langle w, z\rangle$, which is a contradiction.

Since $C \subset f^{-1}((-\infty, s])$, it follows that $D \supset f^{-1}((s, \infty))$. Observe that $\text{rint}\, D \supset f^{-1}((s, \infty))$. So, we have $x \in \text{rint}\, D$ and $f(x) > s$. Likewise for $\text{rcl}\, D$, we can show that $\text{rcl}\, D \subset f^{-1}([s, \infty))$. Accordingly, we have

$$\text{rcl}\, C = f^{-1}((-\infty, s]) \text{ and } \text{rcl}\, D = f^{-1}([s, \infty)).$$

Since $\text{core}\, A \subset f^{-1}((-\infty, s))$ and $B \subset f^{-1}([s, \infty))$, we have the desired result.

3.4 Topological Linear Spaces

A **topological linear space** E is a linear space with a topology such that the algebraic operations of addition $(x, y) \mapsto x + y$ and scalar multiplication $(t, x) \mapsto tx$ are continuous.[8] Every linear space E has such a topology. In fact,

---

[8]Here, we only consider linear spaces over $\mathbb{R}$. Recall that topological spaces are assumed to be **Hausdorff**. For topological linear spaces (more generally for topological groups), it suffices to assume axiom T_0, which implies **regularity** (Proposition 3.4.2 and its footnote). The continuity of scalar multiplication implies the continuity of the operation $x \mapsto -x$ because $(-1)x = -x$.

E has a Hamel basis B. As a linear subspace of the product space $\mathbb{R}^B$, $\mathbb{R}^B_f$ is a topological linear space that is linearly isomorphic to E by the linear isomorphism $\varphi : \mathbb{R}^B_f \to E$ defined by $\varphi(x) = \sum_{v \in B} x(v)v$. Then, φ induces a topology that makes E a topological linear space. In the next section, it will be seen that if E is finite-dimensional, then such a topology is unique. However, an infinite-dimensional linear space has various topologies for which the algebraic operations are continuous.

In the following proposition, we present the basic properties of a neighborhood basis at $\mathbf{0}$ in a topological linear space.

Proposition 3.4.1. *Let E be a topological linear space and $\mathcal{U}$ be a neighborhood basis at $\mathbf{0}$ in E. Then, $\mathcal{U}$ has the following properties:*

(1) *For each $U, V \in \mathcal{U}$, there is some $W \in \mathcal{U}$ such that $W \subset U \cap V$;*
(2) *For each $U \in \mathcal{U}$, there is some $V \in \mathcal{U}$ such that $V + V \subset U$;*
(3) *For each $U \in \mathcal{U}$, there is some $V \in \mathcal{U}$ such that $[-1, 1]V \subset U$;*
(4) *For each $x \in E$ and $U \in \mathcal{U}$, there is some $a > 0$ such that $x \in aU$;*
(5) $\bigcap \mathcal{U} = \{\mathbf{0}\}$.

Conversely, let E be a linear space with $\mathcal{U}$ a collection of subsets satisfying these conditions. Then, E has a topology such that addition and scalar multiplication are continuous and $\mathcal{U}$ is a neighborhood basis at $\mathbf{0}$.

Sketch of Proof. Property (1) is trivial; (2) comes from the continuity of addition at $(0, 0) \in E \times E$; (3) is obtained by the continuity of scalar multiplication at each $(t, 0) \in [-1, 1] \times E$ and the compactness of $[-1, 1]$; (4) follows from the continuity of scalar multiplication at $(0, x) \in \mathbb{R} \times E$; the Hausdorffness of E implies (5).

Given $\mathcal{U}$ with these properties, an open set in E is defined as a subset $W \subset E$ satisfying the condition that, for each $x \in W$, there is some $U \in \mathcal{U}$ such that $x + U \subset W$. (Verify the axioms of open sets, i.e., the intersection of finite open sets is open; every union of open sets is open.)

For each $x \in E$ and $U \in \mathcal{U}$, $x + U$ is a neighborhood of x in this topology.[9] Indeed, let

$$W = \{y \in E \mid \exists V \in \mathcal{U} \text{ such that } y + V \subset x + U\}.$$

Then, $x \in W \subset x + U$ because of (5). For each $y \in W$, we have $V \in \mathcal{U}$ such that $y + V \subset x + U$. Take $V' \in \mathcal{U}$ so that $V' + V' \subset V$ as in (2). Then, $y + V' \subset W$ because $(y + y') + V' \subset y + V \subset x + U$ for every $y' \in V'$. Therefore, W is open in E, so $x + U$ is a neighborhood of x in E. By the definition of the topology, $\{x + U \mid U \in \mathcal{U}\}$ is a neighborhood basis at x. In particular, $\mathcal{U}$ is a neighborhood basis at $\mathbf{0}$.

Since $\{x + U \mid U \in \mathcal{U}\}$ is a neighborhood basis at x, the continuity of addition follows from (2). Using (3), we can show that the operation $x \mapsto -x$ is continuous.

For scalar multiplication, let $x \in E$, $\alpha \in \mathbb{R}$, and $U \in \mathcal{U}$. Because of the continuity of $x \mapsto -x$, it can be assumed that $\alpha \geq 0$. Then, we can write $\alpha = n + t$, where $n \in \omega$ and $0 \leq t < 1$. Using (2) inductively, we can find $V_1 \supset \cdots \supset V_n \supset V_{n+1}$ in $\mathcal{U}$ such that

[9] If E is a topological linear space, $x + U$ is a neighborhood of $x \in E$ for any neighborhood U of $\mathbf{0}$.

$$V_1 + \cdots + V_n + (V_{n+1} + V_{n+1}) \subset U.$$

By (3), we have $W \in \mathcal{U}$ such that $[-1, 1]W \subset V_{n+1}$. Then, $x \in rW$ for some $r > 0$ by (4). Choose $\delta > 0$ so that $\delta < \min\{1/r, 1-t\}$. Let $y \in x + W$ and $|\alpha - \beta| < \delta$. Then, we can write $\beta = n + s$, where $t - \delta < s < t + \delta$. It follows that

$$\begin{aligned}\beta y - \alpha x &= (n+s)y - (n+t)x = n(y-x) + s(y-x) + (s-t)x \\ &\in nW + [-1,1]W + \delta[-1,1](rW) \\ &\subset nV_{n+1} + V_{n+1} + V_{n+1} \\ &\subset \underbrace{V_{n+1} + \cdots + V_{n+1}}_{n+2 \text{ many}} \subset U,\end{aligned}$$ [10]

hence $\beta y \in \alpha x + U$.

To see the Hausdorffness, let $x \neq y \in E$. By (5), we have $U \in \mathcal{U}$ such that $x - y \notin U$. By (2) and (3), we can find $V \in \mathcal{U}$ such that $V - V \subset U$. Then, $x + V$ and $y + V$ are neighborhoods of x and y, respectively. Observe that $(x + V) \cap (y + V) = \emptyset$.

It is said that $A \subset E$ is **circled** if $tA \subset A$ for every $t \in [-1, 1]$. It should be noted that the closure of a circled set A is also circled.

Indeed, let $x \in \operatorname{cl} A$ and $t \in [-1, 1]$. If $t = 0$, then $tx = 0 \in A \subset \operatorname{cl} A$. When $t \neq 0$, for each neighborhood U of tx in E, since $t^{-1}U$ is a neighborhood of $t^{-1}x$, $t^{-1}U \cap A \neq \emptyset$, which implies that $U \cap tA \neq \emptyset$. Because $tA \subset A$, $U \cap A \neq \emptyset$. Thus, it follows that $tx \in \operatorname{cl} A$.

In (3) above, $W = [-1, 1]V$ is a neighborhood of $\mathbf{0} \in E$ that is circled, i.e., $tW \subset W$ for every $t \in [-1, 1]$. Consequently, (3) is equivalent to the following condition:

(3)' $\mathbf{0} \in E$ has a neighborhood basis consisting of circled (open) sets.

A **topological group** G is a group with a topology such that the algebraic operations of multiplication $(x, y) \mapsto xy$ and taking inverses $x \mapsto x^{-1}$ are both continuous.[11] Then, G is **homogeneous**, that is, for each distinct $x_0, x_1 \in G$, there is a homeomorphism $h : G \to G$ such that $h(x_0) = x_1$. Such an h can be defined by $h(x) = x_0x^{-1}x_1$, where not only $h(x_0) = x_1$ but also $h(x_1) = x_0$. Every topological linear space is a topological group with respect to addition, so it is homogeneous.

Proposition 3.4.2. *Every topological group G has a closed neighborhood basis at each $g \in G$, i.e., it is regular.*[12] *For a topological linear space E, $\mathbf{0} \in E$ has a circled closed neighborhood basis.*

---

[10] It should be noted that, in general, $2V \subset V + V$ but $V + V \not\subset 2V$.

[11] These two operations are continuous if and only if the operation $(x, y) \to x^{-1}y$ is continuous.

[12] A topological group G is assumed to be **Hausdorff**, but it suffices to assume axiom T_0. In fact, axiom T_0 implies T_1 for a topological group G because of the homogeneity of G.

Sketch of Proof. Each neighborhood U of the unit $\mathbf{1} \in G$ contains a neighborhood V of $\mathbf{1}$ such that $V^{-1}V \subset U$. For each $x \in \operatorname{cl} V$, we have $y \in Vx \cap V$. Consequently, $x \in V^{-1}y \subset V^{-1}V \subset U$, so we have $\operatorname{cl} V \subset U$.

For the additional statement, recall that if V is circled then $\operatorname{cl} V$ is also circled.

Proposition 3.4.3. *Let G be a topological group and H be a subgroup of G.*

(1) *If H is open in G then H is closed in G.*
(2) *The closure $\operatorname{cl} H$ of H is a subgroup of G.*

Sketch of Proof. (1): For each $x \in G \setminus H$, Hx is an open neighborhood of x in G and $Hx \subset G \setminus H$.

(2): For each $x, y \in \operatorname{cl} H$, show that $x^{-1}y \in \operatorname{cl} H$, i.e., each neighborhood W of $x^{-1}y$ meets H. To this end, choose neighborhoods U and V of x and y, respectively, so that $U^{-1}V \subset W$.

Due to Proposition 3.4.3(1), a connected topological group G has no open subgroups except for G itself. Observe that every topological linear space E is path-connected. Consequently, E has no open linear subspaces except for E itself, i.e., every proper linear subspace of E is not open in E.

The continuity of linear functionals is characterized as follows:

Proposition 3.4.4. *Let E be a topological linear space. For a linear functional $f : E \to \mathbb{R}$ with $f(E) \neq \{0\}$, the following are equivalent:*

(a) *f is continuous;*
(b) *$f^{-1}(0)$ is closed in E;*
(c) *$f^{-1}(0)$ is not dense in E;*
(d) *$f(V)$ is bounded for some neighborhood V of $\mathbf{0} \in E$.*

Proof. The implication (a) $\Rightarrow$ (b) is obvious, and (b) $\Rightarrow$ (c) follows from $f(E) \neq \{0\}$ (i.e., $f^{-1}(0) \neq E$).

(c) $\Rightarrow$ (d): We have $x \in E$ and a circled neighborhood V of $\mathbf{0} \in E$ such that $(x+V) \cap f^{-1}(0) = \emptyset$. Then, $f(V)$ is bounded. Indeed, if $f(V)$ is unbounded, then there is some $z \in V$ such that $|f(z)| > |f(x)|$. In this case, $f(tz) = tf(z) = -f(x)$ for some $t \in [-1, 1]$, which implies that $-f(x) \in f(V)$. It follows that $0 \in f(x) + f(V) = f(x + V)$, which contradicts the fact that $(x + V) \cap f^{-1}(0) = \emptyset$.

(d) $\Rightarrow$ (a): For each $\varepsilon > 0$, we have $n \in \mathbb{N}$ such that $f(V) \subset (-n\varepsilon, n\varepsilon)$. Then, $n^{-1}V$ is a neighborhood of $\mathbf{0}$ in E and $f(n^{-1}V) \subset (-\varepsilon, \varepsilon)$. Therefore, f is continuous at $\mathbf{0} \in E$. Since f is linear, it follows that f is continuous at every point of E. □

Proposition 3.4.5. *Let E be a topological linear space and $A, B \subset E$.*

(1) *If B is open in E then $A + B$ is open in E.*
(2) *If A is compact and B is closed in E then $A + B$ is also closed in E.*

Sketch of Proof. (1): Note that $A + B = \bigcup_{x \in A}(x + B)$.

(2): To show that $E \setminus (A+B)$ is open in E, let $z \in E \setminus (A+B)$. For each $x \in A$, because $z-x \in E \setminus B$, we have open neighborhoods U_x, V_x of x, z in E such that $V_x - U_x \subset E \setminus B$. Since A is compact, $A \subset \bigcup_{i=1}^{n} U_{x_i}$ for some $x_1, \dots, x_n \in A$. Then, $V = \bigcap_{i=1}^{n} V_{x_i}$ is an open neighborhood of z in E. We can show that $V \cap (A+B) = \emptyset$, i.e., $V \subset E \setminus (A+B)$.

Remark 8. In (2) above, we cannot assert that $A + B$ is closed in E even if both A and B are closed and convex in E. For example, $A = \mathbb{R} \times \{0\}$ and $B = \{(x, y) \in \mathbb{R}^2 \mid x > 0,\ y \geq x^{-1}\}$ are closed convex sets in $\mathbb{R}^2$, but $A + B = \mathbb{R} \times (0, \infty)$ is not closed in $\mathbb{R}^2$.

Proposition 3.4.6. *Let F be a closed linear subspace of a topological linear space E. Then, the quotient linear space E/F with the quotient topology is also a topological linear space, and the quotient map $q : E \to E/F$ (i.e., $p(x) = x + F \in E/F$) is open, hence if $\mathcal{U}$ is a neighborhood basis at $\mathbf{0}$ in E, then $q(\mathcal{U}) = \{q(U) \mid U \in \mathcal{U}\}$ is a neighborhood basis $\mathbf{0}$ in E/F.*

Sketch of Proof. Apply Proposition 3.4.5(1) to show that the quotient map $q : E \to E/F$ is open. Then, in the diagrams below, $q \times q$ and $q \times \mathrm{id}_{\mathbb{R}}$ are open, so they are quotient maps:

$$\begin{array}{ccc} E \times E & \longrightarrow & E \\ {\scriptstyle q\times q}\downarrow & & \downarrow{\scriptstyle q} \\ E/F \times E/F & \longrightarrow & E/F, \end{array} \qquad \begin{array}{ccc} E \times \mathbb{R} & \longrightarrow & E \\ {\scriptstyle q\times \mathrm{id}_{\mathbb{R}}}\downarrow & & \downarrow{\scriptstyle q} \\ E/F \times \mathbb{R} & \longrightarrow & E/F. \end{array}$$

Accordingly, the continuity of addition and scalar multiplication are clear. Note that E/F is Hausdorff if and only if F is closed in E.

For convex sets in a topological linear space, we have the following:

Proposition 3.4.7. *For each convex set C in a topological linear space E, the following hold:*

(1) $\mathrm{cl}\, C$ *is convex and* $\mathrm{rcl}\, C \subset \mathrm{cl}\, C$, *hence* $\mathrm{rcl}\, C = C$ *if C is closed in E;*
(2) $\mathrm{int}_F\, C = \emptyset$ *for any flat F with* $\mathrm{fl}\, C \subsetneq F$;
(3) $\mathrm{int}_{\mathrm{fl}\, C}\, C \neq \emptyset$ *implies* $\mathrm{int}_{\mathrm{fl}\, C}\, C = \mathrm{core}_{\mathrm{fl}\, C}\, C = \mathrm{rint}\, C$.

Proof. By the definition and the continuity of algebraic operations, we can easily obtain (1). For (2), observe $\mathrm{int}_F\, C \subset \mathrm{core}_F\, C$. If $\mathrm{int}_F\, C \neq \emptyset$ then $\mathrm{fl}\, C = F$ by the Fact stated in the previous section.

(3): Due to Proposition 3.3.2, $\mathrm{core}_{\mathrm{fl}\, C}\, C = \mathrm{rint}\, C$. Note that $\mathrm{int}_{\mathrm{fl}\, C}\, C \subset \mathrm{core}_{\mathrm{fl}\, C}\, C$. Without loss of generality, we may assume that $\mathbf{0} \in \mathrm{int}_{\mathrm{fl}\, C}\, C$. Then, for each $x \in \mathrm{rint}\, C$, we can find $0 < s < 1$ such that $x \in sC$. Since $(1 - s)C$ is a neighborhood of $\mathbf{0} = x - x$ in $\mathrm{fl}\, C$, we have a neighborhood U of x in $\mathrm{fl}\, C$ such that $U - x \subset (1 - s)C$. Then, it follows that $U \subset (1 - s)C + sC = C$. Therefore, $x \in \mathrm{int}_{\mathrm{fl}\, C}\, C$. □

Remark 9. In the above, we cannot assert any one of $\mathrm{cl}\, C = \mathrm{rcl}\, C$, $\mathrm{int}_{\mathrm{fl}\, C}\, C = \mathrm{core}_{\mathrm{fl}\, C}\, C$, or $\mathrm{int}_{\mathrm{fl}\, C}\, C \neq \emptyset$. For example, $[-1, 1]^{\mathbb{N}}_f$ is a convex set in $\mathbb{R}^{\mathbb{N}}$ such that

$\operatorname{rcl}[-1,1]_f^{\mathbb{N}} = [-1,1]_f^{\mathbb{N}}$ but $\operatorname{cl}[-1,1]_f^{\mathbb{N}} = [-1,1]^{\mathbb{N}}$. Note that $\operatorname{fl}[-1,1]_f^{\mathbb{N}} = \mathbb{R}_f^{\mathbb{N}}$. Regard $[-1,1]_f^{\mathbb{N}}$ as a convex set in $\mathbb{R}_f^{\mathbb{N}}$. Then,

$$\operatorname{int}_{\mathbb{R}_f^{\mathbb{N}}}[-1,1]_f^{\mathbb{N}} = \emptyset \text{ but } \operatorname{core}_{\mathbb{R}_f^{\mathbb{N}}}[-1,1]_f^{\mathbb{N}} = \operatorname{rint}[-1,1]_f^{\mathbb{N}} = (-1,1)_f^{\mathbb{N}}.$$

By Proposition 3.4.7(1), if A is a subset of a topological linear space E, then $\operatorname{cl}\langle A\rangle$ is the smallest closed convex set containing A, which is called the **closed convex hull** of A.

Remark 10. In general, $\langle A\rangle$ is not closed in E even if A is compact. For example, let $A = \{a_n \mid n \in \omega\} \subset \ell_1$, where $a_0(i) = 2^{-i}$ for every $i \in \mathbb{N}$ and, for each $n \in \mathbb{N}$, $a_n(i) = 2^{-i}$ if $i \leq n$ and $a_n(i) = 0$ if $i > n$. Then, A is compact and $\langle A\rangle = \bigcup_{n\in\mathbb{N}}\langle a_0, a_1, \ldots, a_n\rangle$. For each $n \in \mathbb{N}$, let

$$x_n = 2^{-n}a_0 + 2^{-1}a_1 + \cdots + 2^{-n}a_n \in \langle a_0, a_1, \ldots, a_n\rangle.$$

Then, $x_n(i) = 2^{-2i+1}$ if $i \leq n$ and $x_n(i) = 2^{-n-i}$ if $i > n$. Hence, $(x_n)_{n\in\mathbb{N}}$ converges to $x_0 \in \ell_1$, where $x_0(i) = 2^{-2i+1}$ for each $i \in \mathbb{N}$. However, $x_0 \notin \langle A\rangle$. Otherwise, $x_0 \in \langle a_0, a_1, \ldots, a_n\rangle$ for some $n \in \mathbb{N}$, where we can write

$$x_0 = \sum_{i=0}^{n} z(i+1)a_i,\ z \in \Delta^n.$$

Then, we have the following:

$$z(1)a_0(n+1) = x_0(n+1) = 2^{-2n-1} = 2^{-n}a_0(n+1) \text{ and}$$
$$z(1)a_0(n+2) = x_0(n+2) = 2^{-2n-3} = 2^{-n-1}a_0(n+2),$$

hence $z(1) = 2^{-n}$ and $z(1) = 2^{-n-1}$. This is a contradiction. Therefore, $\langle A\rangle$ is not closed in ℓ_1.

The following is the topological version of the Separation Theorem 3.3.5:

Theorem 3.4.8 (SEPARATION THEOREM). *Let A and B be convex sets in a topological linear space E such that* $\operatorname{int} A \neq \emptyset$ *and* $(\operatorname{int} A) \cap B = \emptyset$. *Then, there is a continuous linear functional $f : E \to \mathbb{R}$ such that $f(x) < f(y)$ for each $x \in \operatorname{int} A$ and $y \in B$, and* $\sup f(A) \leq \inf f(B)$.

Proof. First, $\operatorname{int} A \neq \emptyset$ implies $\operatorname{core} A = \operatorname{int} A \neq \emptyset$ by Proposition 3.4.7(3). Then, by the Separation Theorem 3.3.5, we have a linear functional $f : E \to \mathbb{R}$ such that $f(x) < f(y)$ for every $x \in \operatorname{int} A$ and $y \in B$, and $\sup f(A) \leq \inf f(B)$. Note that $B - \operatorname{int} A$ is open in E and $f(z) > 0$ for every $z \in B - \operatorname{int} A$. Thus, $f^{-1}(0)$ is not dense in E. Therefore, f is continuous by Proposition 3.4.4. □

A topological linear space E is **locally convex** if $\mathbf{0} \in E$ has a neighborhood basis consisting of (open) convex sets; equivalently, open convex sets make up an open basis for E. It follows from Proposition 3.4.6 that for each locally convex topological space E and each closed linear subspace $F \subset E$, the quotient linear space E/F is also locally convex. For locally convex topological linear spaces, we have the following separation theorem:

Theorem 3.4.9 (STRONG SEPARATION THEOREM). *Let A and B be disjoint closed convex sets in a locally convex topological linear space E. If at least one of A and B is compact, then there is a continuous linear functional $f : E \to \mathbb{R}$ such that* $\sup f(A) < \inf f(B)$.

Proof. By Proposition 3.4.5(2), $B - A$ is closed in E. Since $A \cap B = \emptyset$, it follows that $\mathbf{0} \notin B - A$. Choose an open convex neighborhood U of $\mathbf{0}$ so that $U \cap (B - A) = \emptyset$. By the Separation Theorem 3.4.8, we have a nontrivial continuous linear functional $f : E \to \mathbb{R}$ such that $\sup f(U) \leq \inf f(B - A)$. Then, $\sup f(A) + \sup f(U) \leq \inf f(B)$, where $\sup f(U) > 0$ by the non-triviality of f. Thus, we have the result. □

As a particular case, we have the following:

Corollary 3.4.10. *Let E be a locally convex topological linear space. For each pair of distinct points $x, y \in E$, there exists a continuous linear functional $f : E \to \mathbb{R}$ such that $f(x) \neq f(y)$.* □

Concerning the continuity of sublinear functionals, we have the following:

Proposition 3.4.11. *Let $p : E \to \mathbb{R}$ be a non-negative sublinear functional of a topological linear space E. Then, p is continuous if and only if $p^{-1}([0,1))$ is a neighborhood of $\mathbf{0} \in E$.*

Proof. The "only if" part follows from $p^{-1}([0,1)) = p^{-1}((-1,1))$. To see the "if" part, let $\varepsilon > 0$. Since $p^{-1}([0,\varepsilon)) = \varepsilon p^{-1}([0,1))$ is a neighborhood of $\mathbf{0} \in E$, each $x \in E$ has the following neighborhood:

$$U = \left(x + p^{-1}([0,\varepsilon))\right) \cap \left(x - p^{-1}([0,\varepsilon))\right).$$

For each $y \in U$, since $p(y - x) < \varepsilon$ and $p(x - y) < \varepsilon$, it follows that

$$p(y) \leq p(y - x) + p(x) < p(x) + \varepsilon \text{ and}$$
$$p(y) \geq p(x) - p(x - y) > p(x) - \varepsilon,$$

which means that p is continuous at x. □

For each convex set $C \subset E$ with $\mathbf{0} \in \operatorname{int} C$, we have $\operatorname{int} C = \operatorname{core} C = p_C^{-1}([0,1))$ by Propositions 3.3.4 and 3.4.7(3). Then, the following is obtained from Proposition 3.4.11.

Corollary 3.4.12. *Let E be a topological linear space. For each convex set $C \subset E$ with $\mathbf{0} \in \operatorname{int} C$, the Minkowski functional $p_C : E \to \mathbb{R}$ is continuous. Moreover, $p_C^{-1}([0,1)) = \operatorname{int} C = \operatorname{rint} C$ and $p_C^{-1}(\mathbf{I}) = \operatorname{cl} C = \operatorname{rcl} C$, hence $p_C^{-1}(1) = \operatorname{bd} C = \partial C$.* □

The boundedness is a metric concept, but it can be extended to subsets of a topological linear space E. A subset $A \subset E$ is **topologically bounded**[13] provided that, for each neighborhood U of $\mathbf{0} \in E$, there exists some $r > 0$ such that $A \subset rU$. If $A \subset E$ is topologically bounded and $B \subset A$, then B is also topologically bounded. Recall that every neighborhood U of $\mathbf{0} \in E$ contains a circled neighborhood V of $\mathbf{0} \in E$ (cf. Proposition 3.4.1(3)). Since $sV \subset tV$ for $0 < s < t$, it is easy to see that every compact subset of E is topologically bounded. When E is a normed linear space, $A \subset E$ is topologically bounded if and only if A is bounded in the metric sense. Applying Minkowski functionals, we can show the following:

Theorem 3.4.13. *Let E be a topological linear space. Each pair of topologically bounded closed convex sets $C, D \subset E$ with $\operatorname{int} C \neq \emptyset$ and $\operatorname{int} D \neq \emptyset$ are homeomorphic to each other by a homeomorphism of E onto itself, hence $(C, \operatorname{bd} C) \approx (D, \operatorname{bd} D)$ and $\operatorname{int} C \approx \operatorname{int} D$.*

Proof. Without loss of generality, we may assume that $\mathbf{0} \in \operatorname{int} C \cap \operatorname{int} D$. Let p_C and p_D be the Minkowski functionals for C and D, respectively. By the topological boundedness of C and D, it is easy to see that $p_C(x),\ p_D(x) > 0$ for every $x \in E \setminus \{\mathbf{0}\}$. Then, we can define maps $\varphi, \psi : E \to E$ as follows: $\varphi(\mathbf{0}) = \psi(\mathbf{0}) = \mathbf{0}$,

$$\varphi(x) = \frac{p_C(x)}{p_D(x)} x \quad \text{and} \quad \psi(x) = \frac{p_D(x)}{p_C(x)} x \quad \text{for each } x \in E \setminus \{\mathbf{0}\}.$$

It follows from the continuity of p_C and p_D (Corollary 3.4.12) that φ and ψ are continuous at each $x \in E \setminus \{\mathbf{0}\}$.

To verify the continuity of φ at $\mathbf{0} \in E$, let U be a neighborhood of $\mathbf{0} \in E$. Since D is topologically bounded and C is a neighborhood of $\mathbf{0}$, there is an $r > 0$ such that $D \subset rC$. Then, $p_C(x) \le r p_D(x)$ for every $x \in E$. Choose a circled neighborhood V of $\mathbf{0} \in E$ so that $rV \subset U$. Then, $\varphi(V) \subset U$. Indeed, for each $x \in V \setminus \{\mathbf{0}\}$,

$$\varphi(x) = \frac{p_C(x)}{p_D(x)} x \in \frac{p_C(x)}{p_D(x)} V \subset rV \subset U.$$

Similarly, ψ is continuous at $\mathbf{0} \in E$.

For each $x \in E \setminus \{\mathbf{0}\}$, since $\varphi(x) \neq \mathbf{0}$,

[13]Usually, we say simply *bounded* but here add *topologically* in order to distinguish the metric sense. It should be noted that every metrizable space has an admissible bounded metric.

$$\psi\varphi(x) = \frac{p_D(\varphi(x))}{p_C(\varphi(x))}\varphi(x) = \frac{\dfrac{p_C(x)}{p_D(x)}p_D(x)}{\dfrac{p_C(x)}{p_D(x)}p_C(x)} \cdot \frac{p_C(x)}{p_D(x)}x = x.$$

Hence, $\psi\varphi = \mathrm{id}$. Similarly, $\varphi\psi = \mathrm{id}$. Therefore, φ is a homeomorphism with $\varphi^{-1} = \psi$. Moreover, observe that $\varphi(C) \subset D$ and $\psi(D) \subset C$, hence $\varphi(C) = D$. Thus, we have the result. □

The norm of a normed linear space E is the Minkowski functional for the unit closed ball $\mathbf{B}_E$ of E. Since $\mathrm{bd}\,\mathbf{B}_E$ is the unit sphere $\mathbf{S}_E$ of E, we have the following:

Corollary 3.4.14. *Let $E = (E, \|\cdot\|)$ be a normed linear space. For every bounded closed convex set $C \subset E$ with $\mathrm{int}\,C \neq \emptyset$, the pair $(C, \mathrm{bd}\,C)$ is homeomorphic to the pair $(\mathbf{B}_E, \mathbf{S}_E)$ of the unit closed ball and the unit sphere of E.* □

It is easy to see that every normed linear space $E = (E, \|\cdot\|)$ is homeomorphic to the unit open ball $\mathrm{B}(\mathbf{0}, 1) = \mathbf{B}_E \setminus \mathbf{S}_E$ of E.

In fact, the following are homeomorphisms (each of them is the inverse of the other):

$$E \ni x \mapsto \frac{1}{1 + \|x\|}x \in \mathrm{B}(\mathbf{0}, 1); \quad \mathrm{B}(\mathbf{0}, 1) \ni y \mapsto \frac{1}{1 - \|y\|}y \in E.$$

By applying the Minkowski functional, this can be extended as follows:

Theorem 3.4.15. *Every open convex set V in a topological linear space E is homeomorphic to E itself.*

Proof. Without loss of generality, it can be assumed that $\mathbf{0} \in \mathrm{int}\,V = V$. Then, we have $V = \mathrm{int}\,V = p_V^{-1}([0, 1))$ by Corollary 3.4.12. Using the Minkowski functional p_V, we can define maps $\varphi : V \to E$ and $\psi : E \to V$ as follows:

$$\varphi(x) = \frac{1}{1 - p_V(x)}x \quad \text{for } x \in V; \quad \psi(y) = \frac{1}{1 + p_V(y)}y \quad \text{for } y \in E.$$

Observe that $\psi\varphi = \mathrm{id}_V$ and $\varphi\psi = \mathrm{id}_E$. This means that φ is a homeomorphism with $\psi = \varphi^{-1}$. □

3.5 Finite-Dimensionality

Here, we prove that every finite-dimensional linear space has the unique topology that is compatible with the algebraic operations, and that a topological linear space is finite-dimensional if and only if it is locally compact.

First, we show the following proposition:

Proposition 3.5.1. *Every finite-dimensional flat F in an arbitrary linear space E has the unique (Hausdorff) topology such that the following operation is continuous:*

$$F \times F \times \mathbb{R} \ni (x, y, t) \mapsto (1-t)x + ty \in F.$$

With respect to this topology, every affine bijection $f : \mathbb{R}^n \to F$ is a homeomorphism, where $n = \dim F$. Then, F is affinely homeomorphic to $\mathbb{R}^n$. Moreover, if E is a topological linear space then F is closed in E.

Proof. As mentioned at the beginning of Sect. 3.4, E has a topology that makes E a topological linear space. With respect to the topology of F inherited from this topology, the above operation is continuous.

Note that there exists an affine bijection $f : \mathbb{R}^n \to F$, where $\dim F = n$. We shall show that any affine bijection $f : \mathbb{R}^n \to F$ is a homeomorphism with respect to any other topology of F such that the above operation is continuous, which implies that such a topology is unique and F is affinely homeomorphic to $\mathbb{R}^n$.

Since f is affine, we have

$$f(z) = \left(1 - \sum_{i=1}^{n} z(i)\right) f(\mathbf{0}) + \sum_{i=1}^{n} z(i) f(\mathbf{e}_i) \quad \text{for each } z \in \mathbb{R}^n.$$

Note that the following function is continuous:

$$\mathbb{R}^n \ni z \mapsto \left(1 - \sum_{i=1}^{n} z(i), z(1), \ldots, z(n)\right) \in \operatorname{fl} \Delta^n \subset \mathbb{R}^{n+1}.$$

Then, the continuity of f follows from the claim:

Claim. Given $v_1, \ldots, v_k \in F$, $k \leq n$, the following function is continuous:

$$\varphi_k : \operatorname{fl} \Delta^{k-1} \ni z \mapsto \sum_{i=1}^{k} z(i) v_i \in F.$$

Since $\operatorname{fl} \Delta^0 = \Delta^0$ is a singleton, the continuity of φ_1 is obvious. Assuming the continuity of φ_k, we shall show the continuity of φ_{k+1}. Let $\psi : \operatorname{fl} \Delta^{k-1} \times \mathbb{R} \to \operatorname{fl} \Delta^k$ be the map defined by $\psi(z, t) = ((1-t)z, t)$. Observe that

$$\varphi_{k+1}\psi(z, t) = (1-t) \sum_{i=1}^{k} z(i) v_i + t v_{k+1} = (1-t)\varphi_k(z) + t v_{k+1}.$$

From the property of the topology of F and the continuity of φ_k, it follows that $\varphi_{k+1}\psi$ is continuous. For each $i = 1, \ldots, k+1$, let $p_i = \operatorname{pr}_i | \operatorname{fl} \Delta^k : \operatorname{fl} \Delta^k \to \mathbb{R}$ be the restriction of the projection onto the i-th factor. Note that

$$\psi|\operatorname{fl}\Delta^{k-1}\times(\mathbb{R}\setminus\{1\}):\operatorname{fl}\Delta^{k-1}\times(\mathbb{R}\setminus\{1\})\to\operatorname{fl}\Delta^{k}\setminus p_{k+1}^{-1}(1)$$

is a homeomorphism. Hence, $\varphi_{k+1}|\operatorname{fl}\Delta^k\setminus p_{k+1}^{-1}(1)$ is continuous. Replacing the $(k+1)$-th coordinates with the i-th coordinates, we can see the continuity of $\varphi_{k+1}|\operatorname{fl}\Delta^k\setminus p_i^{-1}(1)$. Since $\operatorname{fl}\Delta^k=\bigcup_{i=1}^{k+1}(\operatorname{fl}\Delta^k\setminus p_i^{-1}(1))$, it follows that φ_{k+1} is continuous. Thus, the claim can be obtained by induction.

It remains to show the openness of f. On the contrary, assume that f is not open. Then, we have $x\in\mathbb{R}^n$ and $\varepsilon>0$ such that $f(\mathrm{B}(x,\varepsilon))$ is not a neighborhood of $f(x)$ in F. Since $\operatorname{bd}\mathrm{B}(x,\varepsilon)$ is a bounded closed set of $\mathbb{R}^n$, it is compact, hence $f(\operatorname{bd}\mathrm{B}(x,\varepsilon))$ is closed in F. Then, $F\setminus f(\operatorname{bd}\mathrm{B}(x,\varepsilon))$ is a neighborhood of $f(x)$ in F. Using the compactness of $\mathbf{I}$, we can find an open neighborhood U of $f(x)$ in F such that

$$(1-t)f(x)+tU\subset F\setminus f(\operatorname{bd}\mathrm{B}(x,\varepsilon))\quad\text{for every } t\in\mathbf{I}.$$

Then, $U\cap f(\operatorname{bd}\mathrm{B}(x,\varepsilon))=\emptyset$. Since $f(\mathrm{B}(x,\varepsilon))$ is not a neighborhood of $f(x)$, it follows that $U\not\subset f(\mathrm{B}(x,\varepsilon))$, and so we can take a point $y\in U\setminus f(\overline{\mathrm{B}}(x,\varepsilon))$. Now, we define a linear path $g:\mathbf{I}\to\mathbb{R}^n$ by $g(t)=(1-t)x+tf^{-1}(y)$. Since f is affine and $y\in U$, it follows that

$$fg(t)=(1-t)f(x)+ty\in F\setminus f(\operatorname{bd}\mathrm{B}(x,\varepsilon))\quad\text{for every } t\in\mathbf{I}.$$

Since f is a bijection, we have

$$g(\mathbf{I})\subset\mathbb{R}^n\setminus\operatorname{bd}\mathrm{B}(x,\varepsilon)=\mathrm{B}(x,\varepsilon)\cup(\mathbb{R}^n\setminus\overline{\mathrm{B}}(x,\varepsilon)).$$

Then, $g(\mathbf{0})=x\in\mathrm{B}(x,\varepsilon)$ and $g(1)=f^{-1}(y)\in\mathbb{R}^n\setminus\overline{\mathrm{B}}(x,\varepsilon)$, which contradicts the connectedness of $\mathbf{I}$. Thus, f is open.

In the case when E is a topological linear space, to prove that F is closed in E, take a point $x\in E\setminus F$ and consider the flat $F_x=\operatorname{fl}(F\cup\{x\})$. It is easy to construct an affine bijection $f:\mathbb{R}^{n+1}\to F_x$ such that $f(\mathbb{R}^n\times\{\mathbf{0}\})=F$. As we saw in the above, f is a homeomorphism, hence F is closed in F_x. Since $F_x\setminus F$ is open in F_x, we have an open set U in E such that $U\cap F_x=F_x\setminus F$. Then, U is a neighborhood of x in E and $U\subset E\setminus F$. Therefore, $E\setminus F$ is open in E, that is, F is closed in E. □

If a linear space E has a topology such that the operation

$$E\times E\times\mathbb{R}\ni(x,y,t)\mapsto(1-t)x+ty\in E$$

is continuous, then scalar multiplication and addition are also continuous with this topology because they can be written as follows:

$$E\times\mathbb{R}\ni(x,t)\mapsto tx=(1-t)\mathbf{0}+tx\in E;$$
$$E\times E\ni(x,y)\mapsto x+y=2\left(\tfrac{1}{2}x+\tfrac{1}{2}y\right)\in E.$$

Then, the following is obtained by Proposition 3.5.1:

Corollary 3.5.2. *Every finite-dimensional linear space E has the unique (Hausdorff) topology compatible with the algebraic operations (addition and scalar multiplication), and then it is linearly homeomorphic to $\mathbb{R}^n$, where $n = \dim E$.* □

Moreover, we have the following:

Corollary 3.5.3. *Let E be a topological linear space and F a finite-dimensional flat in another topological linear space. Then, every affine function $f : F \to E$ is continuous, and if f is injective then f is a closed embedding.*

Proof. By Proposition 3.5.1, F can be replaced with $\mathbb{R}^n$, where $n = \dim F$. Then, we can write

$$f(x) = \left(1 - \sum_{i=1}^{n} x(i)\right) f(\mathbf{0}) + \sum_{i=1}^{n} x(i) f(\mathbf{e}_i) \text{ for each } x \in \mathbb{R}^n,$$

where $\mathbf{e}_1, \ldots, \mathbf{e}_n$ is the canonical orthonormal basis for $\mathbb{R}^n$. Thus, the continuity of f is obvious. Since $f(\mathbb{R}^n)$ is a finite-dimensional flat in E, $f(\mathbb{R}^n)$ is closed in E by Proposition 3.5.1. If f is injective then $f : \mathbb{R}^n \to f(\mathbb{R}^n)$ is an affine bijection, which is a homeomorphism by Proposition 3.5.1. Hence, f is a closed embedding. □

Combining Proposition 3.2.2 and Corollary 3.5.3, we have

Corollary 3.5.4. *Let E be a topological linear space and C a finite-dimensional convex set in another topological linear space. Then, every affine function $f : C \to E$ is continuous. Moreover, if f is injective then f is an embedding.* □

For finite-dimensional convex sets in a linear space, we have the following:

Proposition 3.5.5. *Let C be a finite-dimensional convex set in an arbitrary linear space E. Then, $\operatorname{rint} C = \operatorname{int}_{\operatorname{fl} C} C$ with respect to the unique topology for $\operatorname{fl} C$ as in Proposition 3.5.1.*

Proof. We may assume that E is a topological linear space. By Proposition 3.4.7(3), it suffices to show that $\operatorname{int}_{\operatorname{fl} C} C \neq \emptyset$. We have affinely independent $v_0, v_1, \ldots, v_n \in C$ with $\operatorname{fl} C = \operatorname{fl}\{v_0, v_1, \ldots, v_n\}$, where $n = \dim C$. We have an affine bijection $f : \mathbb{R}^n \to \operatorname{fl} C$ such that $f(\mathbf{0}) = v_0$, $f(\mathbf{e}_1) = v_1, \ldots, f(\mathbf{e}_n) = v_n$. Then, f is a homeomorphism by Proposition 3.5.1, hence

$$\operatorname{int}_{\operatorname{fl} C} C \supset \operatorname{int}_{\operatorname{fl} C} \langle v_0, v_1, \ldots, v_n \rangle = f(\operatorname{int}_{\mathbb{R}^n} \langle \mathbf{0}, \mathbf{e}_1, \ldots, \mathbf{e}_n \rangle) \neq \emptyset. \qquad \square$$

Note that every compact set in a topological linear space is topologically bounded and closed. For an n-dimensional convex set C in a linear space, the flat hull $\operatorname{fl} C$ is affinely isomorphic to $\mathbb{R}^n$. Combining Propositions 3.5.1 and 3.5.5 with Corollary 3.4.14, we have the following:

Corollary 3.5.6. *For every n-dimensional compact convex set C in an arbitrary topological linear space E, the pair $(C, \partial C)$ is homeomorphic to the pair $(\mathbf{B}^n, \mathbf{S}^{n-1})$ of the unit closed n-ball and the unit $(n-1)$-sphere.* □

Remark 11. It should be noted that every bounded closed set in Euclidean space $\mathbb{R}^n$ is compact. More generally, we can prove the following:

Proposition 3.5.7. *Let E be an arbitrary topological linear space and $A \subset E$ with* $\dim \mathrm{fl}\, A < \infty$*. Then, A is compact if and only if A is topologically bounded and closed in E.*

Sketch of Proof. Using Proposition 3.5.1, this can be reduced to the case of $\mathbb{R}^n$.

The following convex version of Proposition 3.5.1 is not trivial.

Proposition 3.5.8. *Let C be an n-dimensional convex set in an arbitrary linear space E. If (1) C is the convex hull of a finite set*[14] *or (2)* $C = \mathrm{rint}\, C$*, then C has the unique (Hausdorff) topology such that the following operation is continuous:*

$$C \times C \times \mathbf{I} \ni (x, y, t) \mapsto (1-t)x + ty \in C.$$

In case (1), $\mathrm{rcl}\, C = C$ *and* $(C, \partial C) \approx (\mathbf{B}^n, \mathbf{S}^{n-1})$*; in case (2),* $C \approx \mathbb{R}^n$*.*

Proof. Like Proposition 3.5.1, it suffices to see the uniqueness and the additional statement. To this end, suppose that C has such a topology, but it is unknown whether this is induced from a topology of $\mathrm{fl}\, C$ or not.

Case (1): Let $C = \langle v_1, \ldots, v_k \rangle$ and define $f : \Delta^{k-1} \to C$ by $f(z) = \sum_{i=1}^k z(i)v_i$. In the same way as for the claim in the proof of Proposition 3.5.1, we can see that the continuity of the operation above induces the continuity of f. Since Δ^{k-1} is compact, f is a closed map, hence it is quotient. Thus, the topology of C is unique and C is compact with respect to this topology. Giving any topology on E so that E is a topological linear space, we have $\mathrm{rcl}\, C = C$ by Proposition 3.4.7(i) and $(C, \partial C) \approx (\mathbf{B}^n, \mathbf{S}^{n-1})$ by Corollary 3.5.6.

Case (2): Let $f : \mathbb{R}^n \to \mathrm{fl}\, C$ be an affine bijection, where $n = \dim \mathrm{fl}\, C = \dim C$. Since $D = f^{-1}(C)$ is an n-dimensional convex set in $\mathbb{R}^n$, $D = \mathrm{rint}\, D = \mathrm{int}\, D$ is open in $\mathbb{R}^n$ by Proposition 3.5.5, hence $D \approx \mathbb{R}^n$ by Proposition 3.4.15. Then, it suffices to show that $f|D : D \to C$ is a homeomorphism. For each $x \in D$, choose $\delta > 0$ so that $x + \delta\mathbf{B}^n = \overline{\mathrm{B}}(x, \delta) \subset D$. Let $v_0 = x - \delta\hat{\Delta}^{n-1}$, where $\hat{\Delta}^{n-1}$ is the barycenter of the standard $(n-1)$-simplex $\Delta^{n-1} = \langle \mathbf{e}_1, \mathbf{e}_2, \ldots, \mathbf{e}_n \rangle \subset \mathbb{R}^n$. For each $i = 1, \ldots, n$, let $v_i = x + \delta\mathbf{e}_i$. Then, $v_0, v_1, \ldots, v_n$ are affinely independent and

$$x \in \mathrm{int}_{\mathbb{R}^n} \langle v_0, v_1, \ldots, v_n \rangle \subset x + \delta\mathbf{B}^n \subset D,$$

hence $\langle v_0, v_1, \ldots, v_n \rangle$ is a neighborhood of x in D. On the other hand, we have the affine homeomorphism $\varphi : \Delta^n \to \langle v_0, v_1, \ldots, v_n \rangle$ defined by $\varphi(z) = \sum_{i=0}^n z(i +$

[14] In this case, C is called a **cell** or a **(convex) linear cell** (cf. Sect. 4.1).

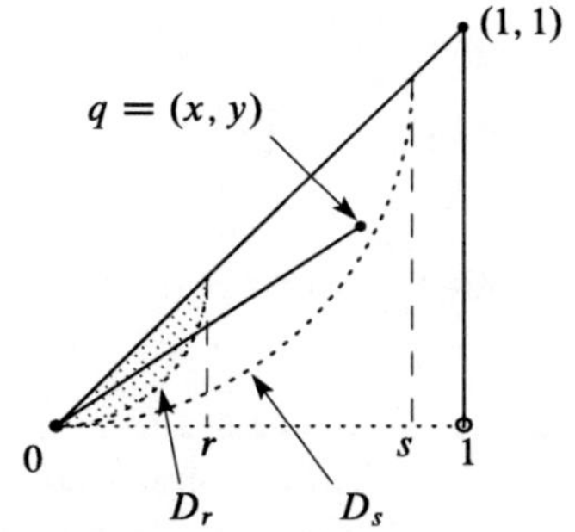

Fig. 3.7 The continuity of the operation at $(\mathbf{0}, q, 0) \in C \times C \times \mathbf{I}$

$1)v_i$. Since $f\varphi(z) = \sum_{i=0}^{n} z(i+1)f(v_i)$, the continuity of the operation above implies that of $f\varphi$, hence $f|\langle v_0, v_1, \ldots, v_n\rangle$ is continuous at x. Then, it follows that $f|D$ is continuous at x.

Since D is open in $\mathbb{R}^n$, we can apply the same argument as in the proof of Proposition 3.5.1 to prove that $f|D : D \to C$ is open. Consequently, $f|D : D \to C$ is a homeomorphism. □

Remark 12. For an arbitrary finite-dimensional convex C, Proposition 3.5.8 does not hold in general. For example, let

$$C = \{\mathbf{0}\} \cup \{(x, y) \in (0, 1]^2 \mid x \geq y\} \subset \mathbb{R}^2.$$

Then, C is a convex set that has a finer topology than usual such that the operation in Proposition 3.5.8 is continuous. Such a topology is generated by open sets in the usual topology and the following sets:

$$D_r = \{\mathbf{0}\} \cup (\mathrm{B}((0, r), r) \cap C), \ r > 0.$$

Note that this topology induces the same relative topology on $C \setminus \{\mathbf{0}\}$ as usual. Since $D_{\varepsilon/\sqrt{2}} \subset \mathrm{B}(\mathbf{0}, \varepsilon)$ for each $\varepsilon > 0$, $\{D_r \mid r > 0\}$ is a neighborhood basis at $\mathbf{0} \in C$ with respect to this topology.

We shall show that the operation

$$C \times C \times \mathbf{I} \ni (p, q, t) \mapsto (1-t)p + tq \in C$$

is continuous at $(p, q, t) \in C \times C \times \mathbf{I}$. If $(1-t)p + tq \neq \mathbf{0}$, it follows from the continuity with respect to the usual topology. The continuity at $(0, 0, t)$ follows from the convexity of D_r, $r > 0$.

To see the continuity at $(\mathbf{0}, q, 0)$ $(q \neq \mathbf{0})$, let $q = (x, y)$, where $0 < y \leq x \leq 1$. Choose $s > 0$ so that $q \in D_s$ (i.e., $s > (x^2 + y^2)/2y$). For each $0 < r < \min\{1, s\}$, let $0 \leq t \leq r/2s$, $p' \in D_{r/2}$, and $q' \in D_s$ (Fig. 3.7). Observe that

$$\frac{1-t}{r/s-t}(r/s)p' \in \frac{1-t}{r/s-t}(r/s)D_{r/2} \subset \frac{1}{r/2s}(r/s)D_{r/2} = D_r$$

and $(r/s)q' \in (r/s)D_s = D_r$. Since D_r is convex, it follows that

$$(1-t)p' + tq' = \left(1 - \frac{t}{r/s}\right)\frac{1-t}{r/s-t}(r/s)p' + \frac{t}{r/s}(r/s)q' \in D_r.$$

Thus, the operation is continuous at $(0, q, 0)$. The continuity at $(p, \mathbf{0}, 1)$ $(p \neq \mathbf{0})$ is the same.

A subset A of a topological linear space E is **totally bounded** provided, for each neighborhood U of $\mathbf{0} \in E$, there exists some finite set $M \subset E$ such that $A \subset M + U$. In this definition, M can be taken as a subset of A.

Indeed, for each neighborhood U of $\mathbf{0} \in E$, we have a circled neighborhood V such that $V + V \subset U$. Then, $A \subset M + V$ for some finite set $M \subset E$, where it can be assumed that $(x + V) \cap A \neq \emptyset$ for every $x \in M$. For each $x \in M$, choose $a_x \in A$ so that $a_x \in x + V$. Then, $x \in a_x - V = a_x + V$. It follows that $A \subset \bigcup_{x\in M}(x+V) \subset \bigcup_{x\in M}(a_x+V+V) \subset \bigcup_{x\in M}(a_x + U)$.

If $A \subset E$ is totally bounded and $B \subset A$, then B is also totally bounded. It is easy to see that every compact subset of E is totally bounded and every totally bounded subset of E is topologically bounded. In other words, we have:

$$\boxed{\text{compact}} \Rightarrow \boxed{\text{totally bounded}} \Rightarrow \boxed{\text{topologically bounded}}$$

For topological linear spaces, the finite-dimensionality can be simply characterized as follows:

Theorem 3.5.9. *Let E be a topological linear space. The following are equivalent:*

(a) *E is finite-dimensional;*
(b) *E is locally compact;*
(c) *$\mathbf{0} \in E$ has a totally bounded neighborhood in E.*

Proof. Since each n-dimensional topological linear space is linearly homeomorphic to $\mathbb{R}^n$ (Corollary 3.5.2), we have (a) $\Rightarrow$ (b). Since every compact subset of E is totally bounded, the implication (b) $\Rightarrow$ (c) follows.

(c) $\Rightarrow$ (a): Let U be a totally bounded neighborhood of $\mathbf{0} \in E$. By Proposition 3.4.1, we have a circled neighborhood V of $\mathbf{0}$ such that $V + V \subset U$. Then, V is also totally bounded. First, we show the following:

Claim. For each closed linear subspace $F \subsetneq E$, there is some $x \in U$ such that $(x + V) \cap F = \emptyset$.

Contrary to the claim, suppose that $(x + V) \cap F \neq \emptyset$ for every $x \in U$. Since $V = -V$, it follows that $U \subset F + V$, so we have $V + V \subset F + V$. If $(n-1)V \subset F + V$ then

$$nV \subset (n-1)V + V \subset F + V + V \subset F + F + V = F + V.$$

By induction, we have $nV \subset F + V$ for every $n \in \mathbb{N}$, which implies that $V \subset \bigcap_{n\in\mathbb{N}}(F + n^{-1}V)$.

Take $z \in E \setminus F$. Since F is closed in E, we have a circled neighborhood W of $\mathbf{0} \in E$ such that $W \subset V$ and $(z+W) \cap F = \emptyset$. The total boundedness of V implies the topological boundedness, hence $V \subset mW$ for some $m \in \mathbb{N}$. On the other hand, $k^{-1}z \in V$ for some $k \in \mathbb{N}$. Since $k^{-1}z \in V \subset F + (km)^{-1}V$, it follows that $z \in F + m^{-1}V \subset F + W$. This contradicts the fact that $(z + W) \cap F = \emptyset$.

Now, assume that E is infinite-dimensional. Let $v_1 \in U \setminus \{\mathbf{0}\}$ and $F_1 = \mathbb{R}v_1$. Then, F_1 is closed in E (Proposition 3.5.1) and $F_1 \neq E$. Applying the claim above, we have $v_2 \in U$ such that $(v_2 + V) \cap F_1 = \emptyset$. Note that $v_2 \notin v_1 + V$. Let $F_2 = \mathbb{R}v_1 + \mathbb{R}v_2$. Since F_2 is closed in E (Proposition 3.5.1) and $F_2 \neq E$, we can again apply the claim to find $v_3 \in U$ such that $(v_3 + V) \cap F_2 = \emptyset$. Then, note that $v_3 \notin v_i + V$ for $i = 1, 2$. By induction, we have $v_n \in U$, $n \in \mathbb{N}$, such that $v_n \notin v_i + V$ for $i < n$. Then, $\{v_n \mid n \in \mathbb{N}\}$ is not totally bounded. This is a contradiction. Consequently, E is finite-dimensional. □

By Theorem 3.5.9, every infinite-dimensional topological linear space is not locally compact.

3.6 Metrizability and Normability

In this section, we prove metrization and normability theorems for topological linear spaces. The metrizability of a topological linear space has the following very simple characterization:

Theorem 3.6.1. *A topological linear space E is metrizable if and only if $\mathbf{0} \in E$ has a countable neighborhood basis.*

In a more general setting, we shall prove a stronger result. A metric d on a group G is said to be **left** (resp. **right**) **invariant** if $d(x, y) = d(zx, zy)$ (resp. $d(x, y) = d(xz, yz)$) for each $x, y, z \in G$; equivalently, $d(x, y) = d(x^{-1}y, \mathbf{1})$ (resp. $d(x, y) = d(xy^{-1}, \mathbf{1})$) for each $x, y \in G$. When both of two metrics d and d' on a group G are left (or right) invariant, they are uniformly equivalent to each other if and only if they induce the same topology. It is said that d is **invariant** if it is left and right invariant. Every invariant metric d on a group G induces the topology on G that makes G a topological group. In fact,

$$d(x, y) = d(x^{-1}xy^{-1}, x^{-1}yy^{-1}) = d(y^{-1}, x^{-1}) = d(x^{-1}, y^{-1}) \text{ and}$$
$$d(xy, x'y') \le d(xy, x'y) + d(x'y, x'y') = d(x, x') + d(y, y').$$

It is easy to verify that a left (or right) invariant metric d on a group G is invariant if $d(x, y) = d(x^{-1}, y^{-1})$ for each $x, y \in G$. Theorem 3.6.1 comes from the following:

Theorem 3.6.2. *For a topological group G, the following are equivalent:*

(a) *G is metrizable;*
(b) *The unit $\mathbf{1} \in G$ has a countable neighborhood basis;*
(c) *G has an admissible bounded left invariant (right invariant) metric.*

Proof. Since the implications (a) $\Rightarrow$ (b) and (c) $\Rightarrow$ (a) are obvious, it suffices to show the implication (b) $\Rightarrow$ (c).

(b) $\Rightarrow$ (c):[15] We shall construct a left invariant metric $\rho \in \mathrm{Metr}(G)$. Then, a right invariant metric $\rho' \in \mathrm{Metr}(G)$ can be defined by $\rho'(x, y) = \rho(x^{-1}, y^{-1})$. By condition (b), we can find an open neighborhood basis $\{V_n \mid n \in \mathbb{N}\}$ at $\mathbf{1} \in G$ such that

$$V_n^{-1} = V_n \text{ and } V_{n+1}V_{n+1}V_{n+1} \subset V_n \text{ for each } n \in \mathbb{N}.^{16}$$

Let $V_0 = G$, and define

$$p(x) = \inf\{2^{-i} \mid x \in V_i\} \in \mathbf{I} \text{ for each } x \in G.$$

Since $V_n = V_n^{-1}$ for each $n \in \mathbb{N}$, it follows that $p(x) = p(x^{-1})$ for every $x \in G$. Note that $\bigcap_{n\in\omega} V_n = \{\mathbf{1}\}$.[17] Hence, for every $x \in G$,

$$p(x) = 0 \Leftrightarrow x = \mathbf{1}.$$

By induction on n, we shall prove the following:

$$(*)\ \ p(x_0^{-1}x_n) \leq 2\sum_{i=1}^{n} p(x_{i-1}^{-1}x_i) \text{ for each } x_0, x_1, \ldots, x_n \in G.^{18}$$

The case $n = 1$ is obvious. Assume $(*)$ for $m < n$. If $\sum_{i=1}^{n} p(x_{i-1}^{-1}x_i) = 0$ or $\sum_{i=1}^{n} p(x_{i-1}^{-1}x_i) \geq \frac{1}{2}$, it is trivial. When $2^{-k-1} \leq \sum_{i=1}^{n} p(x_{i-1}^{-1}x_i) < 2^{-k}$ for some $k \in \mathbb{N}$, choose $1 \leq m \leq n$ so that

$$\sum_{i=1}^{m-1} p(x_{i-1}^{-1}x_i) < 2^{-k-1} \text{ and } \sum_{i=m+1}^{n} p(x_{i-1}^{-1}x_i) < 2^{-k-1}.$$

---

[15]The idea of the proof is the same as that of Theorem 2.4.1 (b) $\Rightarrow$ (a).

[16]Note that $\{V_n x \mid n \in \mathbb{N}\}$ is an open neighborhood basis at $x \in G$. For each $x, y \in G$ and $n \in \mathbb{N}$, $V_{n+1}x \cap V_{n+1}y \neq \emptyset$ implies $V_{n+1}y \subset V_n x$. Indeed, $ux = vy$ for some $u, v \in V_{n+1}$, hence $V_{n+1}y = V_{n+1}v^{-1}ux \subset V_n x$. Thus, the metrizability of G can be obtained by the Frink Metrization Theorem 2.4.1. On the other hand, $\mathcal{V}_n = \{V_n x \mid x \in G\} \in \mathrm{cov}(G)$ and $\mathrm{st}\,\mathcal{V}_{n+1} \prec \mathcal{V}_n$. Indeed, $\mathrm{st}(V_{n+1}x, \mathcal{V}_{n+1}) \subset V_n x$. Thus, the metrizability of G can also be obtained by Corollary 2.4.4.

[17]It is assumed that G is Hausdorff.

[18]For each $x, y \in G$, let $\delta(x, y) = p(x^{-1}y)$. Then, this inequality is simply the one given in the sketch of the direct proof for Corollary 2.4.4.

Note that $p(x_{m-1}^{-1}x_m) < 2^{-k}$. By the inductive assumption, $p(x_0^{-1}x_{m-1}) < 2^{-k}$ and $p(x_m^{-1}x_n) < 2^{-k}$. Then, $x_0^{-1}x_{m-1},\ x_{m-1}^{-1}x_m,\ x_m^{-1}x_n \in V_{k+1}$. Since $V_{k+1}V_{k+1}V_{k+1} \subset V_k$, it follows that $x_0^{-1}x_n \in V_k$, hence

$$p(x_0^{-1}x_n) \leq 2^{-k} \leq 2\sum_{i=1}^{n} p(x_{i-1}^{-1}x_i).$$

Now, we can define a metric ρ on G as follows:

$$\rho(x,y) = \inf\left\{\textstyle\sum_{i=1}^{n} p(x_{i-1}^{-1}x_i) \;\middle|\; n \in \mathbb{N},\ x_i \in G,\ x_0 = x,\ x_n = y\right\}.$$

By the definition, ρ is left invariant. Note that $\rho(x,y) \leq p(x^{-1}y) \leq 1$. Then, $x^{-1}y \in V_n$ implies $\rho(x,y) \leq p(x^{-1}y) \leq 2^{-n} < 2^{-n+1}$, which means $xV_n \subset \mathrm{B}_\rho(x, 2^{-n+1})$. On the other hand, if $\rho(x,y) < 2^{-n}$ then $p(x^{-1}y) \leq 2\rho(x,y) < 2^{-n+1}$ by (∗), which implies $x^{-1}y \in V_n$. Thus, $\mathrm{B}_\rho(x, 2^{-n}) \subset xV_n$. Therefore, ρ is admissible. □

In the above proof, a right invariant metric $\rho \in \mathrm{Metr}(G)$ can be directly defined as follows:

$$\rho(x,y) = \inf\left\{\textstyle\sum_{i=1}^{n} p(x_{i-1}x_i^{-1}) \;\middle|\; n \in \mathbb{N},\ x_i \in G,\ x_0 = x,\ x_n = y\right\}.$$

Every metrizable topological linear space E has an admissible (bounded) metric ρ that is not only invariant but also satisfies the following:

(♯) $|t| \leq 1 \Rightarrow \rho(tx, \mathbf{0}) \leq \rho(x, \mathbf{0})$.

To verify this, let us recall how to define the metric ρ in the above proof. Taking a neighborhood basis $\{V_n \mid n \in \mathbb{N}\}$ at $\mathbf{0} \in E$ so that $V_n = -V_n$ and $V_{n+1} + V_{n+1} + V_{n+1} \subset V_n$ for each $n \in \mathbb{N}$, we define the admissible invariant metric ρ as follows:

$$\rho(x,y) = \inf\left\{\textstyle\sum_{i=1}^{n} p(x_i - x_{i-1}) \;\middle|\; n \in \mathbb{N},\ x_i \in E,\ x_0 = x,\ x_n = y\right\},$$

where $p(x) = \inf\{2^{-i} \mid x \in V_i\}$. Since E is a topological linear space, the condition that $V_n = -V_n$ can be replaced by a stronger condition that V_n is circled, i.e., $tV_n \subset V_n$ for $t \in [-1,1]$. Then, $p(tx) \leq p(x)$ for each $x \in E$ and $t \in [-1,1]$, which implies that $\rho(tx, \mathbf{0}) \leq \rho(x, \mathbf{0})$ for each $x \in E$ and $t \in [-1,1]$.

Let d be an invariant metric on a linear space E. Addition on a linear space E is clearly continuous with respect to d. On the other hand, scalar multiplication on E is continuous with respect to d if and only if d satisfies the following three conditions:

(i) $d(x_n, \mathbf{0}) \to 0 \;\Rightarrow\; \forall t \in \mathbb{R},\ d(tx_n, \mathbf{0}) \to 0$;
(ii) $t_n \to 0 \;\Rightarrow\; \forall x \in E,\ d(t_n x, \mathbf{0}) \to 0$;
(iii) $d(x_n, \mathbf{0}) \to 0,\ t_n \to 0 \;\Rightarrow\; d(t_n x_n, \mathbf{0}) \to 0$.

Indeed, the "only if" part is trivial. To show the "if" part, observe

$$t_n x_n - tx = (t_n - t)(x_n - x) + t(x_n - x) + (t_n - t)x.$$

Since d is invariant, it follows that

$$\begin{aligned} d(t_n x_n, tx) &= d((t_n - t)(x_n - x) + t(x_n - x) + (t_n - t)x, \mathbf{0}) \\ &\le d((t_n - t)(x_n - x), \mathbf{0}) + d(t(x_n - x), \mathbf{0}) + d((t_n - t)x, \mathbf{0}), \end{aligned}$$

where $d(t_n x_n, tx) \to 0$ if $t_n \to t$ and $d(x_n, x) \to 0$. Thus, the above three conditions imply the continuity of scalar multiplication on E with respect to d.

It should be remarked that condition (♯) implies condition (iii).

An invariant metric d on E satisfying these conditions is called a **linear metric**. A linear space with a linear metric is called a **metric linear space**. Then, every metric linear space is a metrizable topological linear space. Conversely, we have the following fact:

Fact. *Every admissible invariant metric for a metrizable topological linear space is a linear metric.*

For subsets of a metric linear space, the total boundedness coincides with that in the metric sense. On the other hand, the topological boundedness does not coincide with the metric boundedness. In fact, every metrizable topological linear space E has an admissible bounded invariant metric. For instance, given an admissible invariant metric d for E, the following are admissible bounded invariant metrics:

$$\min\{1, d(x,y)\}, \quad \frac{d(x,y)}{1 + d(x,y)}.$$

For a linear metric ρ on E with the condition (♯), the functional $E \ni x \mapsto \rho(x, \mathbf{0}) \in \mathbb{R}$ is called an ***F*-norm**. In other words, a functional $\|\cdot\| : E \to \mathbb{R}$ on a linear space E is called an ***F*-norm** if it satisfies the following conditions:

(F_1) $\|x\| \ge 0$ for every $x \in E$;
(F_2) $\|x\| = 0 \Rightarrow x = \mathbf{0}$;
(F_3) $|t| \le 1 \Rightarrow \|tx\| \le \|x\|$ for every $x \in E$;
(F_4) $\|x + y\| \le \|x\| + \|y\|$ for every $x, y \in E$;
(F_5) $\|x_n\| \to 0 \Rightarrow \|tx_n\| \to 0$ for every $t \in \mathbb{R}$;
(F_6) $t_n \to 0 \Rightarrow \|t_n x\| \to 0$ for every $x \in E$.

Conditions (F_3), (F_5), and (F_6) correspond to conditions (♯), (i), and (ii), respectively. The converse of (F_2) is true because $\|\mathbf{0}\| = 0$ by (F_6). Then, $\|x\| = 0$ if and only if $x = \mathbf{0}$. Condition (F_3) implies that $\|-x\| = \|x\|$ for every $x \in E$. Furthermore, conditions (F_3) and (F_4) imply condition (F_5). Indeed, using (F_4) inductively, we have $\|nx\| \le n\|x\|$ for every $n \in \mathbb{N}$. Each $t \in [0, \infty)$ can be written as $t = [t] + s$ for some $s \in [0, 1)$, where $[t]$ is the greatest integer $\le t$. Since $\|sx\| \le \|x\|$ by (F_3), it follows that $\|tx\| \le ([t] + 1)\|x\|$. Because $\|-x\| = \|x\|$, $\|tx\| \le ([|t|] + 1)\|x\|$ for every $t \in \mathbb{R}$. This implies condition (F_5). Thus, condition (F_5) is unnecessary.

A linear space E given an F-norm $\|\cdot\|$ is called an ***F*-normed linear space**. Every norm is an F-norm, hence every normed linear space is an F-normed space. An F-norm $\|\cdot\|$ induces the linear metric $d(x,y) = \|x - y\|$. Then, every F-normed linear space is a metric linear space. An F-norm on a topological linear space E is said to be **admissible** if it induces the topology for E. As we saw above, if E is metrizable, then E has an admissible invariant metric ρ satisfying ($\sharp$), which induces the F-norm. Therefore, we have the following:

Theorem 3.6.3. *A topological linear space has an admissible F-norm if and only if it is metrizable.* □

For each metrizable topological linear space, there exists an F-norm with the following stronger condition than (F_3):

(F_3^*) $x \neq \mathbf{0}$, $|t| < 1 \Rightarrow \|tx\| < \|x\|$,

which implies that $\|sx\| < \|tx\|$ for each $x \neq \mathbf{0}$ and $0 < s < t$. The following proposition guarantees the existence of an F-norm with the condition (F_3^*):

Proposition 3.6.4. *Every (completely) metrizable topological linear space E has an admissible invariant (complete) metric d such that $d(tx, \mathbf{0}) < d(x, \mathbf{0})$ if $x \neq \mathbf{0}$ and $|t| < 1$, which induces an admissible F-norm satisfying (F_3^*). If an admissible invariant metric ρ for E is given, d can be chosen so that $d \geq \rho$ (hence, if ρ is complete, then so is d). Moreover, if ρ is bounded, d can be chosen to be bounded.*

Proof. Given an admissible (bounded) invariant metric ρ for E, we define $d_1(x, y) = \sup_{0<s\leq 1} \rho(sx, sy)$. Then, d_1 is an invariant metric on E with $d_1 \geq \rho$ (if ρ is bounded then so is d_1). For each $\varepsilon > 0$, since the scalar multiplication $E \times \mathbb{R} \ni (x, s) \to sx \in E$ is continuous at $(\mathbf{0}, s)$ and $\mathbf{I}$ is compact, we can find $\delta > 0$ such that $\rho(x, \mathbf{0}) < \delta$ implies $\rho(sx, \mathbf{0}) < \varepsilon$ for every $s \in \mathbf{I}$, hence $\rho(x, y) < \delta$ implies $d_1(x, y) = \sup_{0<s\leq 1} \rho(sx, sy) \leq \varepsilon$. Thus, d_1 is uniformly equivalent to ρ. In particular, d_1 is admissible. For $r > 0$, we define an admissible invariant metric d_r for E by $d_r(x, y) = d_1(rx, ry)$ $(= \sup_{0<s\leq r} \rho(sx, sy))$. Observe that $d_r(tx, \mathbf{0}) \leq d_r(x, \mathbf{0})$ for each $x \in E$ and $t \in \mathbf{I}$.

Now, let $\mathbb{Q} \cap (0, 1] = \{r_n \mid n \in \mathbb{N}\}$, where $r_1 = 1$. We define $d(x, y) = \sum_{n\in\mathbb{N}} 2^{-n+1} d_{r_n}(x, y)$. Then, d is an invariant metric on E and

$$\rho(x, y) \leq d_1(x, y) \leq d(x, y) \leq 2d_1(x, y),$$

hence d is admissible (if ρ is bounded then so is d). It also follows that $d(tx, \mathbf{0}) \leq d(x, \mathbf{0})$ for each $x \in E$ and $t \in \mathbf{I}$. It remains to show that $d(tx, \mathbf{0}) \neq d(x, \mathbf{0})$ for each $x \in E \setminus \{\mathbf{0}\}$ and $0 < t < 1$. Since $\mathbb{Q} \cap (0, 1)$ is dense in $(0, 1)$, it suffices to show that $d(tx, \mathbf{0}) \neq d(x, \mathbf{0})$ for each $x \in E \setminus \{\mathbf{0}\}$ and $t \in \mathbb{Q} \cap (0, 1)$. Assume that there exists some $x \in E \setminus \{\mathbf{0}\}$ and $t \in \mathbb{Q} \cap (0, 1)$ such that $d(tx, \mathbf{0}) = d(x, \mathbf{0})$. Note that $d_r(tx, \mathbf{0}) = d_r(x, \mathbf{0})$ for each $r \in \mathbb{Q} \cap (0, 1)$. Then, it follows that

$$\begin{aligned} d_t(x, \mathbf{0}) &= d_t(tx, \mathbf{0}) = d_{t^2}(x, \mathbf{0}) = d_{t^2}(tx, \mathbf{0}) \\ &= d_{t^3}(x, \mathbf{0}) = d_{t^3}(tx, \mathbf{0}) = \cdots, \end{aligned}$$

so $d_t(x,\mathbf{0}) = d_{t^{n+1}}(x,\mathbf{0}) = d_t(t^n x,\mathbf{0})$ for every $n \in \mathbb{N}$. Since $\lim_{n\to\infty} t^n = 0$, it follows that $d_t(x,\mathbf{0}) = \lim_{n\to\infty} d_t(t^n x,\mathbf{0}) = 0$, hence $x = \mathbf{0}$, which is a contradiction. □

The topological linear space $\mathbb{R}^{\mathbb{N}} = \boldsymbol{s}$ (the space of sequences) has the following admissible F-norms:

$$\sup_{i\in\mathbb{N}} \min\{1/i, |x(i)|\}, \ \sum_{i\in\mathbb{N}} \min\{2^{-i}, |x(i)|\}, \ \sum_{i\in\mathbb{N}} \frac{2^{-i}|x(i)|}{1+|x(i)|}, \ \ldots.$$

The first two do not satisfy condition (F_3^*), but the third does.

We now consider the completion of metric linear spaces (cf. 2.3.10).

Proposition 3.6.5. *Let G be a topological group such that the topology is induced by an invariant metric d. The completion $\widetilde{G} = (\widetilde{G}, \tilde{d})$ of (G, d) is a group such that G is its subgroup and $\tilde{d}$ is invariant. Similarly, the completion of a metric (F-normed or normed) linear space E is a metric (F-normed or normed) linear space containing E as a linear subspace.*

Proof. We define the algebraic operations on $\widetilde{G}$ as follows: for each $x, y \in \widetilde{G}$, choose sequences $(x_i)_{i\in\mathbb{N}}$ and $(y_i)_{i\in\mathbb{N}}$ in G so as to converge to x and y, respectively. Since d is invariant, $(x_i y_i)_{i\in\mathbb{N}}$ and (x_i^{-1}) are Cauchy sequences in G. Then, define xy and x^{-1} as the limits of $(x_i y_i)_{i\in\mathbb{N}}$ and $(x_i^{-1})_{i\in\mathbb{N}}$, respectively. It is easily verified that these are well-defined. Since $\tilde{d}(x,y) = \lim_{i\to\infty} d(x_i, y_i)$, it is also easy to see that $\tilde{d}$ is invariant, which implies the continuity of the algebraic operations $(x,y) \mapsto xy$ and $x \mapsto x^{-1}$.

For the completion $\widetilde{E}$ of a metric linear space E, we can define not only addition but also scalar multiplication in the same way. To see the continuity of scalar multiplication, let $x \in \widetilde{E}$ and $t \in \mathbb{R}$. Choose a sequence $(x_i)_{i\in\mathbb{N}}$ in E so as to converge to x. For each $\varepsilon > 0$, we can choose $\delta_0 > 0$ (depending on t) so that

$$z \in E,\ d(z,\mathbf{0}) < \delta_0,\ |t - t'| < \delta_0 \Rightarrow d(t'z,\mathbf{0}) < \varepsilon/4.$$

Then, we have $n_0 \in \mathbb{N}$ such that $d(x_n, x_{n_0}) < \delta_0$ for every $n \geq n_0$. Choose $\delta_1 > 0$ so that $\delta_1 < \delta_0$ and

$$|s| < \delta_1 \Rightarrow d(sx_{n_0},\mathbf{0}) < \varepsilon/4.$$

Now, let $x' \in \widetilde{E}$ and $t' \in \mathbb{R}$ such that $\tilde{d}(x,x') < \delta_0$ and $|t - t'| < \delta_1$. Take a sequence $(x_i')_{i\in\mathbb{N}}$ in E so as to converge to x' and choose $n_1 \in \mathbb{N}$ so that $n_1 \geq n_0$ and $d(x_n, x_n') < \delta_0$ for every $n \geq n_1$. Then, for every $n \geq n_1$, it follows that

$$\begin{aligned} d(tx_n, t'x_n') &\leq d(tx_n, tx_{n_0}) + d(tx_{n_0}, t'x_{n_0}) + d(t'x_{n_0}, t'x_n) + d(t'x_n, t'x_n') \\ &= d(t(x_n - x_{n_0}),\mathbf{0}) + d((t-t')x_{n_0},\mathbf{0}) \\ &\quad + d(t'(x_{n_0} - x_n),\mathbf{0}) + d(t'(x_n - x_n'),\mathbf{0}) \\ &< \varepsilon/4 + \varepsilon/4 + \varepsilon/4 + \varepsilon/4 = \varepsilon. \end{aligned}$$

When E is an F-normed (or normed) linear space, it is easy to see that the F-norm (or norm) for E naturally extends to $\widetilde{E}$. □

Concerning the completeness of admissible invariant metrics, we have the following:

Theorem 3.6.6. *Let G be a completely metrizable topological group. Every admissible invariant metric for G is complete. In particular, a metric linear space is complete if it is absolutely G_δ (i.e., completely metrizable).*

Proof. Let d be an admissible invariant metric for G and $\widetilde{G}$ be the completion of (G, d). Note that $\widetilde{G}$ is a topological group by Proposition 3.6.5. It suffices to show that $\widetilde{G} = G$. Since G is completely metrizable, G is a dense G_δ-set in $\widetilde{G}$ (Theorem 2.5.2), hence we can write $\widetilde{G} \setminus G = \bigcup_{n\in\mathbb{N}} F_n$, where each F_n is a nowhere dense closed set in $\widetilde{G}$. Assume $\widetilde{G} \setminus G \neq \emptyset$ and take $x_0 \in \widetilde{G} \setminus G$. Since $x_0 x \in \widetilde{G} \setminus G$ for every $x \in G$, it follows that $G \subset \bigcup_{n\in\mathbb{N}} x_0^{-1} F_n$, where each $x_0^{-1} F_n$ is also a nowhere dense closed set in $\widetilde{G}$. Then, we have

$$\widetilde{G} = \bigcup_{n\in\mathbb{N}} F_n \cup \bigcup_{n\in\mathbb{N}} x_0^{-1} F_n,$$

which is the countable union of nowhere dense closed sets. This contradicts the complete metrizability of $\widetilde{G}$ (the Baire Category Theorem 2.5.1). □

Corollary 3.6.7. *Let G be a metrizable topological group. Every completely metrizable Abelian subgroup H of G is closed in G. Hence, in a metrizable topological linear space, every completely metrizable linear subspace is closed.*

Proof. By Theorem 3.6.2, G has an admissible left invariant metric d. Because H is an Abelian subgroup of G, the restriction of d on H is an admissible invariant metric for H, which is complete by Theorem 3.6.6. Hence, it follows that H is closed in G. □

It is said that an F-norm (or an F-normed space) is **complete** if the metric induced by the F-norm is complete. It should be noted that every metrizable topological linear space has an admissible F-norm (Proposition 3.6.4) and that every admissible F-norm for a completely metrizable topological linear space is complete (Theorem 3.6.6). A completely metrizable topological linear space (or a complete F-normed linear space) is called an ***F*-space**. A **Fréchet space** is a locally convex F-space, that is, a completely metrizable locally convex topological linear space. Every Banach space is a Fréchet space, but the converse does not hold. In fact, $\mathbf{s} = \mathbb{R}^{\mathbb{N}}$ is a Fréchet space but it is not normable (Proposition 1.2.1).

Concerning the quotient of an F-normed (or normed) linear space, we have the following:

Proposition 3.6.8. *Let $E = (E, \|\cdot\|)$ be an F-normed (or normed) linear space and F a closed linear subspace of E. Then, the quotient space E/F has the admissible F-norm (or norm) $|||\xi||| = \inf_{x\in\xi} \|x\|$, where if $\|\cdot\|$ is complete then so is $|||\cdot|||$. Hence, if E is (completely) metrizable or (completely) normable then so is E/F.*

Proof. It is easy to see that $|||\cdot|||$ is an F-norm (or a norm). It should be noted that the closedness of F is necessary for condition (F_2). Let $q : E \to E/F$ be the natural linear surjection, i.e., $q(x) = x + F$. Then, for each $\varepsilon > 0$,

$$\{q(x) \mid \|x\| < \varepsilon\} = \{\xi \in E/F \mid |||\xi||| < \varepsilon\},$$

which means that $q : E \to (E/F, |||\cdot|||)$ is open and continuous, so it is a quotient map. Then, $|||\cdot|||$ induces the quotient topology, i.e., $|||\cdot|||$ is admissible for the quotient topology. It also follows that if E is locally convex then so is E/F.

We should remark the following fact:

Fact. $|||\xi - \xi'||| = \inf\{\|x - x'\| \mid x' \in \xi'\}$ *for each* $x \in \xi$.

Indeed, the left side is not greater than the right side by definition. For each $x, y \in \xi$ and $y' \in \xi'$,

$$\|y - y'\| = \|x - (y' + x - y)\| \geq \inf\{\|x - x'\| \mid x' \in \xi'\}$$

because $y' + x - y \in \xi'$. Thus, the left side is not less than the right side.

We shall show that if $\|\cdot\|$ is complete then so is $|||\cdot|||$. To see the completeness of $|||\cdot|||$, it suffices to prove that each Cauchy sequence $(\xi_i)_{i\in\mathbb{N}}$ in E/F contains a convergent subsequence. Then, by replacing $(\xi_i)_{i\in\mathbb{N}}$ with its subsequence, we may assume that $|||\xi_i - \xi_{i+1}||| < 2^{-i}$ for each $i \in \mathbb{N}$. Using the fact above, we can inductively choose $x_i \in \xi_i$ so that $\|x_i - x_{i+1}\| < 2^{-i}$. Then, $(x_i)_{i\in\mathbb{N}}$ is a Cauchy sequence in E, which converges to some $x \in E$. It follows that $(\xi_i)_{i\in\mathbb{N}}$ converges to some $x + F$. □

In the above, E/F is called the **quotient F-normed** (or **normed**) **linear space** with the F-norm (or norm) $|||\cdot|||$, which is called the **quotient F-norm** (or **norm**). Note that E/F is locally convex if so is E. If E is a Banach space, a Fréchet space, or an F-space, then so is E/F for any closed linear subspace F of E.

Recall that $A \subset E$ is topologically bounded if, for each neighborhood U of $\mathbf{0} \in E$, there exists some $r \in \mathbb{R}$ such that $A \subset rU$.

Theorem 3.6.9. *A topological linear space E is normable if and only if there is a topologically bounded convex neighborhood of $\mathbf{0} \in E$.*

Proof. The "only if" part is trivial. To see the "if" part, let V be a topologically bounded convex neighborhood of $\mathbf{0} \in E$. Then, $W = V \cap (-V)$ is a topologically bounded circled convex neighborhood of $\mathbf{0} \in E$. Hence, the Minkowski functional p_W is a norm on E by Proposition 3.3.4. By Corollary 3.4.12,

$$\{x \in E \mid p_W(x) < \varepsilon\} = \varepsilon p_W^{-1}([0, 1)) = \varepsilon \operatorname{int} W \quad \text{for each } \varepsilon > 0.$$

For each neighborhood U of $\mathbf{0} \in E$, we can choose $r > 0$ such that $W \subset rU$. Then,

$$\{x \in E \mid p_W(x) < r^{-1}\} = r^{-1} \operatorname{int} W \subset r^{-1} W \subset U,$$

hence p_W induces the topology for E. □

For the local convexity, we have the following:

Theorem 3.6.10. *A (metrizable) topological linear space E is locally convex if and only if E is linearly homeomorphic to a linear subspace of the (countable) product $\prod_{\lambda\in\Lambda} E_\lambda$ of normed linear spaces E_λ.*

Proof. As is easily observed, the product of locally convex topological linear spaces is locally convex, and so is any linear subspace of a locally convex topological linear space. Moreover, the countable product of metrizable spaces is metrizable. Then, the "if" part follows.

We show the "only if" part. By the local convexity, E has a neighborhood basis $\{V_\lambda \mid \lambda \in \Lambda\}$ of $\mathbf{0} \in E$ consisting of circled closed convex sets (cf. Proposition 3.4.2), where $\operatorname{card} \Lambda = \aleph_0$ if E is metrizable (Theorem 3.6.1). For each $\lambda \in \Lambda$, let F_λ be a maximal linear subspace of E contained in V_λ. (The existence of F_λ is guaranteed by Zorn's Lemma.) Then, F_λ is closed in E. Let $q_\lambda : E \to E/F_\lambda$ be the natural linear surjection, where we do not give the quotient topology to E/F_λ but we want to define a norm on E/F_λ.

Observe that $q_\lambda(V_\lambda)$ is a circled convex set in E/F_λ and $\mathbf{0} \in \operatorname{core} q_\lambda(V_\lambda)$. Moreover, $\mathbb{R}_+\xi \not\subset q_\lambda(V_\lambda)$ for each $\xi \in (E/F_\lambda) \setminus \{\mathbf{0}\}$. Indeed, take $x \in E \setminus F_\lambda$ so that $q_\lambda(x) = \xi$. By the maximality of F_λ, $\mathbb{R}x + F_\lambda \not\subset V_\lambda$, i.e., $tx + y \notin V_\lambda$ for some $t \in \mathbb{R}$ and $y \in F_\lambda$, where we can take $t > 0$ because V_λ is circled. For each $z \in F_\lambda$,

$$tx + y = \tfrac{1}{2}(2tx + z) + \tfrac{1}{2}(2y - z).$$

Since $2y - z \in F_\lambda \subset V_\lambda$, it follows that $2tx + z \notin V_\lambda$. Then, $2t\xi = q_\lambda(2tx) \notin q_\lambda(V_\lambda)$.

By Proposition 3.3.4, the Minkowski functional $p_\lambda = p_{q_\lambda(V_\lambda)} : E/F_\lambda \to \mathbb{R}$ for $q_\lambda(V_\lambda)$ is a norm. Thus, we have a normed linear space $E_\lambda = (E/F_\lambda, p_\lambda)$. Observe that

$$\begin{aligned}\mathbf{0} \in \operatorname{int} V_\lambda = \operatorname{core} V_\lambda &\subset q_\lambda^{-1}(\operatorname{core} q_\lambda(V_\lambda)) \\ &= q_\lambda^{-1}\big(p_{q_\lambda(V_\lambda)}^{-1}([0,1))\big) = (p_\lambda q_\lambda)^{-1}([0,1)).\end{aligned}$$

By Proposition 3.4.11, the sublinear functional $p_\lambda q_\lambda : E \to \mathbb{R}$ is continuous, which implies that $q_\lambda : E \to E_\lambda$ is continuous.

Let $h : E \to \prod_{\lambda\in\Lambda} E_\lambda$ be the linear map[19] defined by $h(x) = (q_\lambda(x))_{\lambda\in\Lambda}$. If $x \neq \mathbf{0} \in E$ then $x \notin V_\lambda$ (so $x \notin F_\lambda$) for some $\lambda \in \Lambda$, which implies $q_\lambda(x) \neq 0$, hence $h(x) \neq \mathbf{0}$. Thus, h is a continuous linear injection. To see that h is an embedding, it suffices to show that

$$h(V_\lambda) \supset h(E) \cap \operatorname{pr}_\lambda^{-1}(p_\lambda^{-1}([0, \tfrac{1}{2}))) \quad \text{for each } \lambda \in \Lambda.$$

[19]That is, a continuous linear function.

If $p_\lambda(\mathrm{pr}_\lambda h(x)) < \frac{1}{2}$ then

$$q_\lambda(2x) = \mathrm{pr}_\lambda h(2x) \in p_\lambda^{-1}([0,1)) \subset q_\lambda(V_\lambda),$$

hence $2x - y \in F_\lambda$ for some $y \in V_\lambda$. Then, it follows that

$$x = \tfrac{1}{2}(2x - y) + \tfrac{1}{2}y \in V_\lambda,$$

so $h(x) \in h(V_\lambda)$. This completes the proof. □

Combining Theorem 3.6.10 with Proposition 3.6.5 and Corollary 3.6.7, we have the following:

Corollary 3.6.11. *A topological linear space E is a Fréchet space if and only if E is linearly homeomorphic to a closed linear subspace of the countable product $\prod_{i \in \mathbb{N}} E_i$ of Banach spaces E_i.* □

3.7 The Closed Graph and Open Mapping Theorems

This section is devoted to two very important theorems, the Closed Graph Theorem and the Open Mapping Theorem. They are proved using the Baire Category Theorem 2.5.1.

Theorem 3.7.1 (Closed Graph Theorem). *Let E and F be completely metrizable topological linear spaces and $f : E \to F$ be a linear function. If the graph of f is closed in $E \times F$, then f is continuous.*

Proof. It suffices to show the continuity of f at $\mathbf{0} \in E$. Let d and ρ be admissible complete invariant metrics for E and F, respectively (cf. Proposition 3.6.4).

First, we show that for each $\varepsilon > 0$, there is some $\delta(\varepsilon) > 0$ such that $\mathrm{B}_d(\mathbf{0}, \delta(\varepsilon)) \subset \mathrm{cl}\, f^{-1}(\mathrm{B}_\rho(\mathbf{0}, \varepsilon))$. Since $F = \bigcup_{n\in\mathbb{N}} n\mathrm{B}_\rho(\mathbf{0}, \varepsilon/2)$ and f is linear, it follows that $E = \bigcup_{n\in\mathbb{N}} nf^{-1}(\mathrm{B}_\rho(\mathbf{0}, \varepsilon/2))$. By the Baire Category Theorem 2.5.1, $\mathrm{int}\,\mathrm{cl}\, nf^{-1}(\mathrm{B}_\rho(\mathbf{0}, \varepsilon/2)) \neq \emptyset$ for some $n \in \mathbb{N}$, which implies that $\mathrm{int}\,\mathrm{cl}\, f^{-1}(\mathrm{B}_\rho(\mathbf{0}, \varepsilon/2)) \neq \emptyset$. Let $z \in \mathrm{int}\,\mathrm{cl}\, f^{-1}(\mathrm{B}_\rho(\mathbf{0}, \varepsilon/2))$ and choose $\delta(\varepsilon) > 0$ so that

$$z + \mathrm{B}_d(\mathbf{0}, \delta(\varepsilon)) = \mathrm{B}_d(z, \delta(\varepsilon)) \subset \mathrm{cl}\, f^{-1}(\mathrm{B}_\rho(\mathbf{0}, \varepsilon/2)).$$

Then, it follows that

$$\mathrm{B}_d(\mathbf{0}, \delta(\varepsilon)) \subset \mathrm{cl}\, f^{-1}(\mathrm{B}_\rho(\mathbf{0}, \varepsilon/2)) - z \subset \mathrm{cl}\, f^{-1}(\mathrm{B}_\rho(\mathbf{0}, \varepsilon)).$$

The second inclusion can be proved as follows: for each $y \in \mathrm{cl}\, f^{-1}(\mathrm{B}_\rho(\mathbf{0}, \varepsilon/2))$ and $\eta > 0$, we have $y', z' \in f^{-1}(\mathrm{B}_\rho(\mathbf{0}, \varepsilon/2))$ such that $d(y, y'), d(z, z') < \eta/2$, which implies $d(y - z, y' - z') < \eta$. Observe that

$$\rho(f(y' - z'), \mathbf{0}) = \rho(f(y'), f(z')) \leq \rho(f(y'), \mathbf{0}) + \rho(f(z'), \mathbf{0}) < \varepsilon,$$

which means $y' - z' \in f^{-1}(\mathrm{B}_\rho(\mathbf{0}, \varepsilon))$. Therefore, $y - z \in \operatorname{cl} f^{-1}(\mathrm{B}_\rho(\mathbf{0}, \varepsilon))$.

Now, for each $\varepsilon > 0$ and $x \in \mathrm{B}_d(\mathbf{0}, \delta(\varepsilon/2))$, we can inductively choose $x_n \in E$, $n \in \mathbb{N}$, so that $x_n \in f^{-1}(\mathrm{B}_\rho(\mathbf{0}, 2^{-n}\varepsilon))$ and

$$d\big(x, \textstyle\sum_{i=1}^n x_i\big) = d\big(x - \sum_{i=1}^n x_i,\, \mathbf{0}\big) < \min\big\{2^{-n}, \delta(2^{-n-1}\varepsilon)\big\}.$$

Indeed, if $x_1, \ldots, x_{n-1}$ have been chosen, then

$$x - \sum_{i=1}^{n-1} x_i \in \mathrm{B}_d(\mathbf{0}, \delta(2^{-n}\varepsilon)) \subset \operatorname{cl} f^{-1}(\mathrm{B}_\rho(\mathbf{0}, 2^{-n}\varepsilon)),$$

hence we can choose $x_n \in f^{-1}(\mathrm{B}_\rho(\mathbf{0}, 2^{-n}\varepsilon))$ so that

$$d\big(x, \textstyle\sum_{i=1}^n x_i\big) = d\big(x - \sum_{i=1}^{n-1} x_i,\, x_n\big) < \min\big\{2^{-n}, \delta(2^{-n-1}\varepsilon)\big\}.$$

Since $\rho(f(x_n), \mathbf{0}) < 2^{-n}\varepsilon$ for each $n \in \mathbb{N}$, it follows that $(f(\sum_{i=1}^n x_i))_{n\in\mathbb{N}}$ is a Cauchy sequence, which converges to some $y \in F$. For each $n \in \mathbb{N}$,

$$\rho\big(f(\textstyle\sum_{i=1}^n x_i),\, \mathbf{0}\big) \le \displaystyle\sum_{i=1}^n \rho(f(x_i),\, \mathbf{0}) < \sum_{i=1}^n 2^{-i}\varepsilon < \varepsilon,$$

hence $y \in \overline{\mathrm{B}}_\rho(\mathbf{0}, \varepsilon)$. On the other hand, $\sum_{i=1}^n x_i$ converges to x. Since the graph of f is closed in $E \times F$, the point (x, y) belongs to the graph of f, which means $f(x) = y \in \overline{\mathrm{B}}_\rho(\mathbf{0}, \varepsilon)$. Thus, we have $f(\mathrm{B}_d(\mathbf{0}, \delta(\varepsilon/2))) \subset \overline{\mathrm{B}}_\rho(\mathbf{0}, \varepsilon)$. Therefore, f is continuous. □

Corollary 3.7.2. *Let E and F be completely metrizable topological linear spaces. Then, every continuous linear isomorphism $f : E \to F$ is a homeomorphism.*

Proof. In general, the continuity of f implies the closedness of the graph of f in $E \times F$. By changing coordinates, the graph of f can be regarded as the graph of f^{-1}. Then, it follows that the graph of f^{-1} is closed in $F \times E$, which implies the continuity of f^{-1} by Theorem 3.7.1. □

Theorem 3.7.3 (Open Mapping Theorem). *Let E and F be completely metrizable topological linear spaces. Then, every continuous linear surjection $f : E \to F$ is open.*

Proof. Since $f^{-1}(\mathbf{0})$ is a closed linear subspace of E, the quotient linear space $E/f^{-1}(\mathbf{0})$ is completely metrizable by Proposition 3.6.8. Then, f induces the continuous linear isomorphism $\tilde{f} : E/f^{-1}(\mathbf{0}) \to F$. By Corollary 3.7.2, $\tilde{f}$ is a homeomorphism. Note that the quotient map $q : E \to E/f^{-1}(\mathbf{0})$ is open. Indeed, for every open set U in E, $q^{-1}(q(U)) = U + f^{-1}(\mathbf{0})$ is open in E, which means that $q(U)$ is open in $E/f^{-1}(\mathbf{0})$. Hence, f is also open. □

Note. In the above, the Closed Graph Theorem is first proved and then the Open Mapping Theorem is obtained as a corollary of the Closed Graph Theorem. Conversely, we can directly prove the Open Mapping Theorem and then obtain the Closed Graph Theorem as a corollary of the Open Mapping Theorem.

Direct Proof of the Open Mapping Theorem. Let d and ρ be admissible complete invariant metrics for E and F, respectively.

First, we show that for each $\varepsilon > 0$, there is some $\delta(\varepsilon) > 0$ such that $\mathrm{B}_\rho(\mathbf{0}, \delta(\varepsilon)) \subset \mathrm{cl}\, f(\mathrm{B}_d(\mathbf{0}, \varepsilon))$. Since $E = \bigcup_{n\in\mathbb{N}} n\mathrm{B}_d(\mathbf{0}, \varepsilon/2)$, it follows that $F = f(E) = \bigcup_{n\in\mathbb{N}} nf(\mathrm{B}_d(\mathbf{0}, \varepsilon/2))$. By the Baire Category Theorem 2.5.1, $\mathrm{int}\,\mathrm{cl}\, nf(\mathrm{B}_d(\mathbf{0}, \varepsilon/2)) \neq \emptyset$ for some $n \in \mathbb{N}$, which implies that $\mathrm{int}\,\mathrm{cl}\, f(\mathrm{B}_d(\mathbf{0}, \varepsilon/2)) \neq \emptyset$. Let $z \in \mathrm{int}\,\mathrm{cl}\, f(\mathrm{B}_d(\mathbf{0}, \varepsilon/2))$ and choose $\delta(\varepsilon) > 0$ so that

$$z + \mathrm{B}_\rho(\mathbf{0}, \delta(\varepsilon)) = \mathrm{B}_\rho(z, \delta(\varepsilon)) \subset \mathrm{cl}\, f(\mathrm{B}_d(\mathbf{0}, \varepsilon/2)).$$

Then, it follows that

$$\mathrm{B}_\rho(\mathbf{0}, \delta(\varepsilon)) \subset \mathrm{cl}\, f(\mathrm{B}_d(\mathbf{0}, \varepsilon/2)) - z \subset \mathrm{cl}\, f(\mathrm{B}_d(\mathbf{0}, \varepsilon)),$$

where the second inclusion can be seen as follows: for $y \in \mathrm{cl}\, f(\mathrm{B}_d(\mathbf{0}, \varepsilon/2))$ and $\eta > 0$, choose $y', z' \in \mathrm{B}_d(\mathbf{0}, \varepsilon/2)$ so that $\rho(y, f(y')), \rho(z, f(z')) < \eta/2$. Then, observe that $\rho(y - z, f(y' - z')) < \eta$ and $d(y' - z', \mathbf{0}) = d(y', z') < \varepsilon$, hence $y - z \in \mathrm{cl}\, f(\mathrm{B}_d(\mathbf{0}, \varepsilon))$.

Next, we prove that $\mathrm{cl}\, f(\mathrm{B}_d(\mathbf{0}, \varepsilon/2)) \subset f(\overline{\mathrm{B}}_d(\mathbf{0}, \varepsilon))$ for each $\varepsilon > 0$. For each $y \in \mathrm{cl}\, f(\mathrm{B}_d(\mathbf{0}, \varepsilon/2))$, choose $x_1 \in \mathrm{B}_d(\mathbf{0}, \varepsilon/2)$ so that

$$\rho(y, f(x_1)) < \min\{2^{-1}, \delta(2^{-2}\varepsilon)\}.$$

By induction, we can choose $x_n \in \mathrm{B}_d(\mathbf{0}, 2^{-n}\varepsilon)$, $n \in \mathbb{N}$, so that

$$\rho\big(y, f\big(\textstyle\sum_{i=1}^n x_i\big)\big) = \rho\big(y - \sum_{i=1}^n f(x_i), \mathbf{0}\big) < \min\{2^{-n}, \delta(2^{-n-1}\varepsilon)\}.$$

Indeed, if $x_1, \ldots, x_{n-1}$ have been chosen, then

$$y - \sum_{i=1}^{n-1} f(x_i) \in \mathrm{B}_\rho(\mathbf{0}, \delta(2^{-n}\varepsilon)) \subset \mathrm{cl}\, f(\mathrm{B}_d(\mathbf{0}, 2^{-n}\varepsilon)),$$

hence we can choose $x_n \in \mathrm{B}_d(\mathbf{0}, 2^{-n}\varepsilon)$ so that

$$\begin{aligned}\rho\big(y, f\big(\textstyle\sum_{i=1}^n x_i\big)\big) &= \rho\big(y - \textstyle\sum_{i=1}^{n-1} f(x_i), f(x_n)\big)\\ &< \min\{2^{-n}, \delta(2^{-n-1}\varepsilon)\}.\end{aligned}$$

Since $(\sum_{i=1}^n x_i)_{n\in\mathbb{N}}$ is a Cauchy sequence in E, it converges to some $x \in E$. On the other hand, $(f(\sum_{i=1}^n x_i))_{n\in\mathbb{N}}$ converges to y. By the continuity of f, we have $f(x) = y$. For each $n \in \mathbb{N}$,

$$d\big(\textstyle\sum_{i=1}^n x_i, \mathbf{0}\big) \leq \displaystyle\sum_{i=1}^n d(x_i, \mathbf{0}) < \sum_{i=1}^n 2^{-i}\varepsilon < \varepsilon,$$

hence $x \in \overline{\mathrm{B}}_d(\mathbf{0}, \varepsilon)$. Thus, it follows that $\mathrm{cl}\, f(\mathrm{B}_d(\mathbf{0}, \varepsilon/2)) \subset f(\overline{\mathrm{B}}_d(\mathbf{0}, \varepsilon))$.

To see that f is open, let U be an open set in E. For each $x \in U$, choose $\varepsilon > 0$ so that $\overline{\mathrm{B}}_d(\mathbf{0}, \varepsilon) \subset -x + U$. Since

$$\mathrm{B}_\rho(\mathbf{0}, \delta(\varepsilon/2)) \subset \mathrm{cl}\, f(\mathrm{B}_d(\mathbf{0}, \varepsilon/2)) \subset f(\overline{\mathrm{B}}_d(\mathbf{0}, \varepsilon)) \subset -f(x) + f(U),$$

it follows that $B_\rho(f(x), \delta(\varepsilon/2)) \subset f(U)$. Hence, $f(U)$ is open in F. □

Now, using the Open Mapping Theorem, we shall prove the Closed Graph Theorem.

Proof of the Closed Graph Theorem. The product space $E \times F$ is a completely metrizable topological linear space. The graph G of f is a linear subspace of $E \times F$ that is completely metrizable because it is closed in $E \times F$. Since $p = \mathrm{pr}_E|G : G \to E$ is a homeomorphism by the Open Mapping Theorem, $f = \mathrm{pr}_F \circ p^{-1}$ is continuous. □

Remark 13. In both the Closed Graph Theorem and the Open Mapping Theorem, the completeness is essential. Let $E = (\ell_1, \|\cdot\|_2)$, where $\ell_1 \subset \ell_2$ as sets and $\|\cdot\|_2$ is the norm inherited from ℓ_2. Then, E is not completely metrizable. Indeed, if so, it would be closed in ℓ_2 by Corollary 3.6.7, but E is dense in ℓ_2 and $E \neq \ell_2$. The linear bijection $f = \mathrm{id} : \ell_1 \to E$ is continuous, but is not a homeomorphism, so it is not an open map. It follows from the continuity of f that the graph of f is closed in $\ell_1 \times E$, hence the graph of f^{-1} is closed in $E \times \ell_1$. However, $f^{-1} : E \to \ell_1$ is not continuous.

3.8 Continuous Selections

Let X and Y be spaces and $\varphi : X \to \mathfrak{P}(Y)$ be a set-valued function, where $\mathfrak{P}(Y)$ is the power set of Y. We denote $\mathfrak{P}_0(Y) = \mathfrak{P}(Y) \setminus \{\emptyset\}$. A **(continuous) selection** for φ is a map $f : X \to Y$ such that $f(x) \in \varphi(x)$ for each $x \in X$. For a topological linear space Y, we denote by $\mathrm{Conv}(Y)$ the set of all non-empty convex sets in Y. In this section, we consider the problem of when a convex-valued function $\varphi : X \to \mathrm{Conv}(Y)$ has a selection.

It is said that $\varphi : X \to \mathfrak{P}(Y)$ is **lower semi-continuous (l.s.c.)** (resp. **upper semi-continuous (u.s.c.)**) if, for each open set V in Y,

$$\{x \in X \mid \varphi(x) \cap V \neq \emptyset\} \text{ (resp. } \{x \in X \mid \varphi(x) \subset V\}) \text{ is open in } X;$$

equivalently, for each open set V in Y and $x_0 \in X$ such that $\varphi(x_0) \cap V \neq \emptyset$ (resp. $\varphi(x_0) \subset V$), there exists a neighborhood U of x_0 in X such that $\varphi(x) \cap V \neq \emptyset$ (resp. $\varphi(x) \subset V$) for every $x \in U$. We say that φ is **continuous** if $\varphi : X \to \mathfrak{P}(Y)$ is l.s.c. and u.s.c. The continuity of φ coincides with that in the usual sense when $\mathfrak{P}(Y)$ is regarded as a space with the topology generated by the following sets:

$$U^- = \{A \in \mathfrak{P}(Y) \mid A \cap U \neq \emptyset\} \text{ and } U^+ = \{A \in \mathfrak{P}(Y) \mid A \subset U\},$$

where U is non-empty and open in Y. This topology is called the **Vietoris topology**, where $\emptyset$ is isolated because $\{\emptyset\} = \emptyset^+$ ($\emptyset \notin U^-$ for any open set U in Y). The Vietoris topology has an open basis consisting of the following sets: $\mathrm{V}(\emptyset) = \{\emptyset\}$ and

$$\mathrm{V}(U_1, \ldots, U_n) = \left\{A \subset Y \mid A \subset \textstyle\bigcup_{i=1}^n U_i,\ \forall i = 1, \ldots, n,\ A \cap U_i \neq \emptyset\right\}$$

$$= \left(\bigcup_{i=1}^{n} U_i\right)^+ \cap \bigcap_{i=1}^{n} U_i^-,$$

where $n \in \mathbb{N}$ and $U_1, \ldots, U_n$ are open in Y. In fact, $U^- = \mathrm{V}(U, X)$ and $U^+ = \mathrm{V}(U) \cup \mathrm{V}(\emptyset)$. The subspace $\mathrm{F}_1(Y) = \{\{y\} \mid y \in Y\}$ of $\mathfrak{P}_0(Y)$ consisting of all singletons is homeomorphic to Y because $U^+ \cap \mathrm{F}_1(Y) = U^- \cap \mathrm{F}_1(Y) = \mathrm{F}_1(U)$ for each open set U in Y. It should be noted that $\mathfrak{P}_0(Y)$ with the Vietoris topology is not T_1 in general.

For example, the space $\mathfrak{P}_0(\mathbf{I})$ is not T_1. Indeed, for any neighborhood of $\mathcal{U}$ of $\mathbf{I} \in \mathfrak{P}_0(\mathbf{I})$, there are open sets $U_1, \ldots, U_n$ in $\mathbf{I}$ such that $\mathbf{I} \in \mathrm{V}(U_1, \ldots, U_n) \subset \mathcal{U}$. Then, $D \in \mathrm{V}(U_1, \ldots, U_n) \subset \mathcal{U}$ for every dense subset $D \subset \mathbf{I}$. In particular, $\mathbf{I} \cap \mathbb{Q} \in \mathcal{U}$.

The subspace $\mathrm{Comp}(Y)$ of $\mathfrak{P}(Y)$ consisting of all non-empty compact sets is Hausdorff.[20] Indeed, for each $A \neq B \in \mathrm{Comp}(Y)$, we may assume that $A \setminus B \neq \emptyset$. Take $y_0 \in B \setminus A$. Because of the compactness of A, we have disjoint open sets U and V in Y such that $A \subset U$ and $y_0 \in V$. Then, $A \in U^+$, $B \subset V^-$, and $U^+ \cap V^- \neq \emptyset$. It will be prove that $\mathrm{Comp}(Y)$ is metrizable if Y is metrizable (Proposition 5.12.4). Moreover, $\mathrm{Cld}(Y)$ is metrizable if and only if Y is compact and metrizable (cf. Note after Proposition 5.12.4).

By the same argument as above, it follows that if Y is regular then the subspace $\mathrm{Cld}(Y)$ of $\mathfrak{P}(Y)$ consisting of all non-empty closed sets is Hausdorff. One should note that the converse is also true, that is, if $\mathrm{Cld}(Y)$ is Hausdorff then Y is regular. When Y is not regular, we have a closed set $A \subset Y$ and $y_0 \in Y \setminus A$ such that if U and V are open sets with $A \subset U$ and $y_0 \in V$ then $U \cap V \neq \emptyset$. Let $B = A \cup \{y_0\} \in \mathrm{Cld}(Y)$ and let $U_1, \ldots, U_n, U'_1, \ldots, U'_{n'}$ be open sets in Y such that

$$A \in \mathrm{V}(U_1, \ldots, U_n) \quad \text{and} \quad B \in \mathrm{V}(U'_1, \ldots, U'_{n'}).$$

Let $U_0 = \bigcap\{U'_i \mid U'_i \cap A = \emptyset\}$. Since $y_0 \in U_0$, we have $y_1 \in U_0 \cap \bigcup_{i=1}^{n} U_i$. It follows that

$$A \cup \{y_1\} \in \mathrm{V}(U_1, \ldots, U_n) \cap \mathrm{V}(U'_1, \ldots, U'_{n'}).$$

Thus, $\mathrm{Cld}(Y)$ is not Hausdorff.

Proposition 3.8.1. *For a function $g : Y \to X$, the set-valued function $g^{-1} : X \to \mathfrak{P}(Y)$ is l.s.c. (resp. u.s.c.) if and only if g is open (resp. closed).*

Proof. This follows from the fact that, for $V \subset Y$,

$$\{x \in X \mid g^{-1}(x) \cap V \neq \emptyset\} = g(V) \quad \text{and}$$

$$\{x \in X \mid g^{-1}(x) \subset V\} = X \setminus \{x \in X \mid g^{-1}(x) \cap (X \setminus V) \neq \emptyset\}$$
$$= X \setminus g(X \setminus V).$$

□

[20]Recall that Y is assumed to be Hausdorff.

Because of the following proposition, we consider the selection problem for l.s.c. set-valued functions.

Proposition 3.8.2. *Let $\varphi : X \to \mathfrak{P}_0(Y)$ be a set-valued function. Assume that, for each $x_0 \in X$ and $y_0 \in \varphi(x_0)$, there exists a neighborhood U of x_0 in X and a selection $f : U \to Y$ for $\varphi|U$ such that $f(x_0) = y_0$. Then, φ is l.s.c.*

Proof. Let V be an open set in Y and $x_0 \in X$ such that $\varphi(x_0) \cap V \neq \emptyset$. Take any $y_0 \in \varphi(x_0) \cap V$. From the assumption, there is a neighborhood U of x_0 in X with a selection $f : U \to Y$ for $\varphi|U$ such that $f(x_0) = y_0$. Then, $f^{-1}(V)$ is a neighborhood of x_0 in X and $f(x) \in \varphi(x) \cap V$ for each $x \in f^{-1}(V)$. □

Lemma 3.8.3. *Let $\varphi, \psi : X \to \mathfrak{P}(Y)$ be set-valued functions such that $\operatorname{cl}\varphi(x) = \operatorname{cl}\psi(x)$ for each $x \in X$. If φ is l.s.c. then so is ψ.*

Sketch of Proof. This follows from the fact that, for each open set V in Y and $B \subset Y$, $V \cap B \neq \emptyset$ if and only if $V \cap \operatorname{cl} B \neq \emptyset$.

Lemma 3.8.4. *Let $\varphi : X \to \mathfrak{P}(Y)$ be l.s.c., A be a closed set in X, and $f : A \to Y$ be a selection for $\varphi|A$. Define $\psi : X \to \mathfrak{P}(Y)$ by*

$$\psi(x) = \begin{cases} \{f(x)\} & \text{if } x \in A, \\ \varphi(x) & \text{otherwise.} \end{cases}$$

Then, ψ is also l.s.c.

Proof. For each open set V in Y, $f^{-1}(V)$ is open in A and

$$f^{-1}(V) \subset \{x \in X \mid \varphi(x) \cap V \neq \emptyset\},$$

where the latter set is open in X because φ is l.s.c. Then, we can choose an open set U in X so that $f^{-1}(V) = U \cap A$ and $U \subset \{x \in X \mid \varphi(x) \cap V \neq \emptyset\}$. Observe that

$$\{x \in X \mid \psi(x) \cap V \neq \emptyset\} = U \cup (\{x \in X \mid \varphi(x) \cap V \neq \emptyset\} \setminus A).$$

Thus, it follows that ψ is l.s.c. □

For each $W \subset Y^2$ and $y_0 \in Y$, we denote

$$W(y_0) = \{y \in Y \mid (y_0, y) \in W\}.$$

If W is a neighborhood of the diagonal $\Delta_Y = \{(y, y) \mid y \in Y\}$ in Y^2, then $W(y_0)$ is a neighborhood of y_0 in Y.

Lemma 3.8.5. *Let $\varphi : X \to \mathfrak{P}(Y)$ be l.s.c., $f : X \to Y$ be a map, and W be a neighborhood of Δ_Y in Y^2. Define a set-valued function $\psi : X \to \mathfrak{P}(Y)$ by $\psi(x) = \varphi(x) \cap W(f(x))$ for each $x \in X$. Then, ψ is l.s.c.*

Proof. Let V be an open set in Y and $x_0 \in X$ such that $\psi(x_0) \cap V \neq \emptyset$. Take any $y_0 \in \varphi(x_0) \cap W(f(x_0)) \cap V$. Since $(f(x_0), y_0) \in W$, there are open sets V_1 and V_2 in Y such that $(f(x_0), y_0) \in V_1 \times V_2 \subset W$. Then, x_0 has the following open neighborhood in X:

$$U = f^{-1}(V_1) \cap \{x \in X \mid \varphi(x) \cap V_2 \cap V \neq \emptyset\}.$$

For each $x \in U$, we have $y \in \varphi(x) \cap V_2 \cap V$. Since $(f(x), y) \in V_1 \times V_2 \subset W$, it follows that $y \in \varphi(x) \cap W(f(x)) \cap V$, hence $\psi(x) \cap V \neq \emptyset$. Therefore, ψ is l.s.c. □

Let E be a linear space. The set of all non-empty convex sets in E is denoted by $\operatorname{Conv}(E)$. Recall that $\langle A \rangle$ denotes the convex hull of $A \subset E$.

Lemma 3.8.6. *Let E be a topological linear space and $\varphi : X \to \mathfrak{P}_0(E)$ be an l.s.c. set-valued function. Define a convex-valued function $\psi : X \to \operatorname{Conv}(E)$ by $\psi(x) = \langle \varphi(x) \rangle$ for each $x \in X$. Then, ψ is also l.s.c.*

Proof. Let V be an open set in E and $x_0 \in X$ such that $\psi(x_0) \cap V \neq \emptyset$. Choose any $y_0 = \sum_{i=1}^n t_i y_i \in \psi(x_0) \cap V$, where $y_1, \ldots, y_n \in \varphi(x_0)$ and $t_1, \ldots, t_n \geq 0$ with $\sum_{i=1}^n t_i = 1$. Then, each y_i has an open neighborhood V_i such that $t_1 V_1 + \cdots + t_n V_n \subset V$. Since φ is l.s.c.,

$$U = \bigcap_{i=1}^n \{x \in X \mid \varphi(x) \cap V_i \neq \emptyset\}$$

is an open neighborhood of x_0 in X. For each $x \in U$, let $z_i \in \varphi(x) \cap V_i$, $i = 1, \ldots, n$. Then, $\sum_{i=1}^n t_i z_i \in \psi(x) \cap V$, hence $\psi(x) \cap V \neq \emptyset$. Therefore, ψ is l.s.c. □

Lemma 3.8.7. *Let X be paracompact, E be a topological linear space, and $\varphi : X \to \operatorname{Conv}(E)$ be an l.s.c. convex-valued function. Then, for each convex open neighborhood V of $\mathbf{0}$ in E, there exists a map $f : X \to E$ such that $f(x) \in \varphi(x) + V$ for each $x \in X$.*

Proof. For each $y \in E$, let

$$U_y = \{x \in X \mid \varphi(x) \cap (y - V) \neq \emptyset\}.$$

Since φ is l.s.c., we have $\mathcal{U} = \{U_y \mid y \in E\} \in \operatorname{cov}(X)$. From paracompactness, X has a locally finite partition of unity $(f_\lambda)_{\lambda \in \Lambda}$ subordinated to $\mathcal{U}$. For each $\lambda \in \Lambda$, choose $y_\lambda \in E$ so that $\operatorname{supp} f_\lambda \subset U_{y_\lambda}$. We define a map $f : X \to E$ by $f(x) = \sum_{\lambda \in \Lambda} f_\lambda(x) y_\lambda$. If $f_\lambda(x) \neq 0$ then $x \in U_{y_\lambda}$, which means that $\varphi(x) \cap (y_\lambda - V) \neq \emptyset$, i.e., $y_\lambda \in \varphi(x) + V$. Since each $\varphi(x) + V$ is convex, $f(x) \in \varphi(x) + V$. □

Now, we can prove the following:

Theorem 3.8.8 (MICHAEL SELECTION THEOREM). *Let X be a paracompact space and $E = (E, d)$ be a locally convex metric linear space.*[21] *Then, every l.s.c. convex-valued function $\varphi : X \to \mathrm{Conv}(E)$ admits a selection if each $\varphi(x)$ is d-complete. Moreover, if A is a closed set in X then each selection $f : A \to E$ for $\varphi|A$ can extend to a selection $\tilde{f} : X \to E$ for φ.*

Proof. Let $\{V_i \mid i \in \mathbb{N}\}$ be a neighborhood basis of $\mathbf{0}$ in E such that each V_i is symmetric, convex, and $\operatorname{diam} V_i < 2^{-(i+1)}$. By induction, we construct maps $f_i : X \to E$, $i \in \mathbb{N}$, so that, for each $x \in X$ and $i \in \mathbb{N}$,

(1) $f_i(x) \in \varphi(x) + V_i$ and
(2) $d(f_{i+1}(x), f_i(x)) < 2^{-i}$.

The existence of f_1 is guaranteed by Lemma 3.8.7. Assume we have maps $f_1, \ldots, f_n$ satisfying (1) and (2). Define $\psi : X \to \mathrm{Conv}(E)$ by

$$\psi(x) = \varphi(x) \cap (f_n(x) + V_n) \quad \text{for each } x \in X.$$

Since V_n is symmetric, we have $\psi(x) \neq \emptyset$ by (1). Consider the neighborhood $W = \{(x, y) \in E^2 \mid y - x \in V_n\}$ of Δ_E in E^2. Then, $W(f_n(x)) = f_n(x) + V_n$. By Lemma 3.8.5, ψ is l.s.c. We can apply Lemma 3.8.7 to obtain a map $f_{n+1} : X \to E$ such that

$$f_{n+1}(x) \in \psi(x) + V_{n+1} \quad \text{for each } x \in X.$$

Then, as is easily observed, f_{n+1} satisfies (1) and (2). Thus, we have the desired sequence of maps f_i, $i \in \mathbb{N}$.

Using maps $f_i : X \to E$, $i \in \mathbb{N}$, we shall define a selection $f : X \to E$ for φ. For each $x \in X$ and $i \in \mathbb{N}$, we have $x_i \in \varphi(x)$ such that $d(f_i(x), x_i) < 2^{-(i+1)}$ by (1). Then, $(x_i)_{i \in \mathbb{N}}$ is Cauchy in $\varphi(x)$. Since $\varphi(x)$ is complete, $(x_i)_{i \in \mathbb{N}}$ converges to $f(x) \in \varphi(x)$. Thus, we have $f : X \to E$. Note that $(f_i)_{i \in \mathbb{N}}$ uniformly converges to f, so f is continuous. Hence, f is a selection for φ.

For the additional statement, apply Lemma 3.8.4. □

Concerning factors of a metric linear space, we have the following:

Corollary 3.8.9 (BARTLE–GRAVES–MICHAEL). *Let E be a locally convex metric linear space and F be a linear subspace of E that is complete (so a Fréchet space). Then, $E \approx F \times E/F$. In particular, $E \approx \mathbb{R} \times G$ for some metric linear space G.*

Proof. Note that the quotient space E/F is metrizable (Proposition 3.6.8) and the natural map $g : E \to E/F$ is open, hence $g^{-1} : E/F \to \mathrm{Conv}(E)$ is l.s.c. by Proposition 3.8.1. Since $g^{-1}g(x) = x + F$ is complete for each $x \in E$, we apply the Michael Selection Theorem 3.8.8 to obtain a map $f : E/F \to E$ that is a

[21] Recall that a metric linear space is a linear space with a linear metric (cf. Sect. 3.5).

selection for g^{-1}, i.e., $gf = \mathrm{id}$. Then, $x - fg(x) \in F$ for each $x \in E$. Hence, a homeomorphism $h : E \to F \times (E/F)$ can be defined by

$$h(x) = (x - fg(x), g(x)) \quad \text{for each } x \in E.$$

In fact, $h^{-1}(y, z) = y + f(z)$ for each $(y, z) \in F \times E/F$. □

By combining the Michael Selection Theorem 3.8.8 and the Open Mapping Theorem 3.7.3, the following Bartle–Graves Theorem can be obtained as a corollary:

Theorem 3.8.10 (Bartle–Graves). *Let E and F be Fréchet spaces and $f : E \to F$ be a continuous linear surjection. Then, there is a map $g : F \to E$ such that $fg = \mathrm{id}$. Therefore, $E \approx F \times \ker f$ by the homeomorphism h defined as follows:*

$$h(x) = (f(x), x - gf(x)) \quad \textit{for each } x \in E. \quad \square$$

We show that each Banach space is a (topological) factor of $\ell_1(\Gamma)$. To this end, we need the following:

Theorem 3.8.11 (Banach–Mazur, Klee). *For every Banach space E, there is a continuous linear surjection $q : \ell_1(\Gamma) \to E$, where* $\operatorname{card} \Gamma = \operatorname{dens} E$.

Proof. The unit closed ball $\mathbf{B}_E$ of E has a dense set $\{\mathbf{e}_\gamma \mid \gamma \in \Gamma\}$. Since $\sum_{\gamma \in \Gamma} |x(\gamma)| = \|x\| < \infty$ for each $x \in \ell_1(\Gamma)$ and E is complete, we can define a linear map $q : \ell_1(\Gamma) \to E$ as follows:

$$q(x) = \sum_{\gamma \in \Gamma} x(\gamma)\mathbf{e}_\gamma \quad \text{for each } x \in E.$$[22]

Since $\|q(x)\| \leq \sum_{\gamma \in \Gamma} |x(\gamma)| = \|x\|$, it follows that q is continuous.

To see that q is surjective, it suffices to show $\mathbf{B}_E \subset q(\ell_1(\Gamma))$. For each $y \in \mathbf{B}_E$, we can inductively choose $\mathbf{e}_{\gamma_i}$, $i \in \mathbb{N}$, so that $\gamma_i \neq \gamma_j$ if $i \neq j$, and

$$\|y - \mathbf{e}_{\gamma_1}\| < 2^{-1}, \quad \|y - \mathbf{e}_{\gamma_1} - 2^{-1}\mathbf{e}_{\gamma_2}\| < 2^{-2},$$
$$\|y - \mathbf{e}_{\gamma_1} - 2^{-1}\mathbf{e}_{\gamma_2} - 2^{-2}\mathbf{e}_{\gamma_3}\| < 2^{-3}, \ \ldots.$$

We have $x \in \ell_1(\Gamma)$ defined by

$$x(\gamma) = \begin{cases} 2^{1-i} & \text{if } \gamma = \gamma_i, \\ 0 & \text{otherwise.} \end{cases}$$

Then, it follows that $y = \sum_{i=1}^{\infty} 2^{1-i}\mathbf{e}_{\gamma_i} = q(x)$. This completes the proof. □

---

[22]See Proposition 1.2.3.

As a combination of the Bartle–Graves Theorem 3.8.10 and Theorem 3.8.11 above, we have the following:

Corollary 3.8.12. *For any Banach space E, there exists a Banach space F such that $E \times F \approx \ell_1(\Gamma)$, where* $\operatorname{card} \Gamma = \operatorname{dens} E$. □

In the Michael Selection Theorem 3.8.8, the paracompactness of X is necessary. Actually, we have the following characterization:

Theorem 3.8.13. *A space X is paracompact if and only if the following holds for any Banach space E: if $\varphi : X \to \operatorname{Conv}(E)$ is l.s.c. and each $\varphi(x)$ is closed, then φ has a selection.*

Proof. Since the "only if" part is simply Theorem 3.8.8, it suffices to prove the "if" part. For each $\mathcal{U} \in \operatorname{cov}(X)$, we define $\varphi : X \to \mathfrak{P}_0(\ell_1(\mathcal{U}))$ as follows:

$$\varphi(x) = \{z \in \ell_1(\mathcal{U}) \mid \|z\| = 1, \ \forall U \in \mathcal{U}, \ z(U) \geq 0, z(U) = 0 \ \text{ if } x \notin U\}.$$

Clearly, each $\varphi(x)$ is a closed convex set.

To see that φ is l.s.c., let W be an open set in $\ell_1(\mathcal{U})$ and $z \in \varphi(x) \cap W$. Choose $\delta > 0$ so that $\mathrm{B}(z, 2\delta) \subset W$. Then, we have $V_1, \ldots, V_n \in \mathcal{U}[x]$ such that $\sum_{i=1}^n z(V_i) > 1 - \delta$, where $\bigcap_{i=1}^n V_i$ is a neighborhood of x in X. We define $z' \in \ell_1(\mathcal{U})$ as follows:

$$z'(V_i) = \frac{z(V_i)}{\sum_{j=1}^n z(V_j)} \quad \text{and} \quad z'(U) = 0 \ \text{ for } U \neq V_1, \ldots, V_n.$$

It is easy to see that $z' \in \varphi(x') \cap W$ for every $x' \in \bigcap_{i=1}^n V_i$. Thus, φ is l.s.c.

By the assumption, φ has a selection $f : X \to \ell_1(\mathcal{U})$. For each $U \in \mathcal{U}$, let $f_U : X \to \mathbf{I}$ be the map defined by $f_U(x) = f(x)(U)$ for $x \in X$. Then, $(f_U)_{U \in \mathcal{U}}$ is a partition of unity such that $f_U^{-1}((0, 1]) \subset U$ for every $U \in \mathcal{U}$. The result follows from Theorem 2.7.5. □

Remark 14. Let $g, h : X \to \mathbb{R}$ be real-valued functions on a space X such that g is u.s.c., h is l.s.c., and $g(x) \leq h(x)$ for each $x \in X$. We define the convex-valued function $\varphi : X \to \operatorname{Conv}(\mathbb{R})$ by $\varphi(x) = [g(x), h(x)]$ for each $x \in X$. Then, φ is l.s.c. Indeed, for each open set V in $\mathbb{R}$, let $\varphi(x) \cap V \neq \emptyset$. Take $y \in \varphi(x) \cap V$ and $a < y < b$ so that $[a, b] \subset V$. Since g is u.s.c. and h is l.s.c., x has a neighborhood U in X such that $x' \in U$ implies $g(x') < b$ and $h(x') > a$. Since $g(x') \leq h(x')$, it follows that

$$\varphi(x') \cap V \supset [g(x'), h(x')] \cap [a, b] = [\max\{a, g(x')\}, \min\{b, h(x')\}] \neq \emptyset.$$

Now, we can apply the Michael Selection Theorem 3.8.8 to obtain a map $f : X \to \mathbb{R}$ such that $g(x) \leq f(x) \leq h(x)$ for each $x \in X$. This is analogous to Theorem 2.7.6.

3.9 Free Topological Linear Spaces

The **free topological linear space** over a space X is a topological linear space $L(X)$ that contains X as a subspace and has the following extension property:

(LE) For an arbitrary topological linear space F, every map $f : X \to F$ of X uniquely extends to a linear map[23] $\tilde{f} : L(X) \to F$.

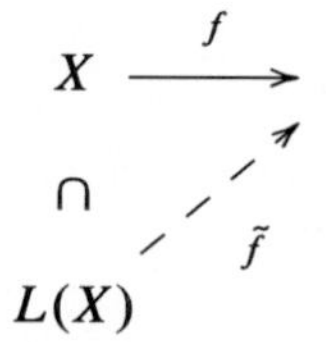

If such a space $L(X)$ exists, then it is uniquely determined up to linear homeomorphism, that is, if E is a topological linear space that contains X and has the property (LE), then E is linearly homeomorphic to $L(X)$.

Indeed, there exist linear maps $\varphi : L(X) \to E$ and $\psi : E \to L(X)$ such that $\varphi|X = \psi|X = \mathrm{id}_X$. Since $\mathrm{id}_{L(X)}$ is a linear map extending id_X, it follows from the uniqueness that $\psi\varphi = \mathrm{id}_{L(X)}$. Similarly, we have $\varphi\psi = \mathrm{id}_E$. Therefore, φ is a linear homeomorphism with $\psi = \varphi^{-1}$.

Lemma 3.9.1. *If X is a Tychonoff space,*

(1) *X is a Hamel basis for $L(X)$;*
(2) *$L(X)$ is regular.*

Proof. (1): First, let F be the linear span of X. Applying (LE), we have a linear map $r : L(X) \to F$ such that $r|X = \mathrm{id}_X$. Since $r : L(X) \to L(X)$ is a linear map extending id_X, we have $r = \mathrm{id}_{L(X)}$, which implies $F = L(X)$, that is, $L(X)$ is generated by X.

To see that X is linearly independent in $L(X)$, let $x_1, \ldots, x_n \in X$, where $x_i \neq x_j$ if $i \neq j$. For each $i = 1, \ldots, n$, there is a map $f_i : X \to \mathbf{I}$ such that $f_i(x_i) = 1$ and $f_i(x_j) = 0$ for $j \neq i$. Let $f : X \to \mathbb{R}^n$ be the map defined by $f(x) = (f_1(x), \ldots, f_n(x))$. Then, by (LE), f extends to a linear map $\tilde{f} : L(X) \to \mathbb{R}^n$, where $\tilde{f}(x_i) = f(x_i) = \mathbf{e}_i$ for each $i = 1, \ldots, n$. Since $\mathbf{e}_1, \ldots, \mathbf{e}_n$ is linearly independent in $\mathbb{R}^n$, it follows that $x_1, \ldots, x_n \in X$ is linearly independent in $L(X)$.

(2): Due to the Fact in Sect. 3.4 and Proposition 3.4.2, it suffices to show that $\{\mathbf{0}\}$ is closed in $L(X)$. Each $z \in L(X) \setminus \{\mathbf{0}\}$ can be uniquely represented as follows:

$$z = \sum_{i=1}^{n} t_i x_i, \; x_i \in X, \; t_i \in \mathbb{R} \setminus \{0\},$$

[23]That is, a continuous linear function.

where $x_i \neq x_j$ if $i \neq j$. There is a map $f : X \to \mathbf{I}$ such that $f(x_1) = 1$ and $f(x_i) = 0$ for each $i = 2, \ldots, n$. By (LE), f extends to a linear map $\tilde{f} : L(X) \to \mathbb{R}$. Then, $\tilde{f}(z) = t_1 f(x_1) = t_1 \neq 0 = \tilde{f}(\mathbf{0})$. Hence, $\tilde{f}^{-1}(\mathbb{R} \setminus \{0\})$ is an open neighborhood of z in $L(X)$ that misses $\mathbf{0}$. □

Remark 15. In the definition of a free topological linear space $L(X)$, specify a map $\eta : X \to L(X)$ instead of assuming $X \subset L(X)$ and replace the property (LE) with the following universality:

(*) For each map $f : X \to F$ of X to an arbitrary topological linear space F, there exists a unique linear map $\tilde{f} : L(X) \to F$ such that $\tilde{f}\eta = f$.

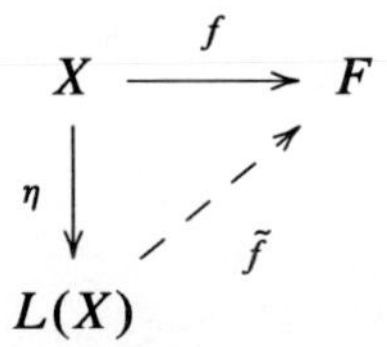

Then, we can show that η is an embedding if X is a Tychonoff space.

To see that η is injective, let $x \neq y \in X$. Then, there is a map $f : X \to \mathbf{I}$ with $f(x) = 0$ and $f(y) = 1$. By (*), we have a linear map $\tilde{f} : L(X) \to \mathbb{R}$ such that $\tilde{f}\eta = f$. Then, observe $\eta(x) \neq \eta(y)$.

To show that $\eta : X \to \eta(X)$ is open, let U be an open set in X. For each $x \in U$, there is a map $g : X \to \mathbf{I}$ such that $g(x) = 0$ and $g(X \setminus U) = 1$. By (*), we have a linear map $\tilde{g} : L(X) \to \mathbb{R}$ such that $\tilde{g}\eta = g$. Then, $V = \tilde{g}^{-1}((-\frac{1}{2}, \frac{1}{2}))$ is an open neighborhood $\eta(x)$ in $L(X)$. Since $\eta^{-1}(V) = g^{-1}([0, \frac{1}{2})) \subset U$, it follows that $V \cap \eta(X) \subset \eta(U)$, hence $\eta(U)$ is a neighborhood of $\eta(x)$ in $\eta(X)$. This means that $\eta(U)$ is open in $\eta(X)$. Thus, $\eta : X \to \eta(X)$ is open.

Since η is an embedding, X can be identified with $\eta(X)$, which is a subspace of $L(X)$. Then, (*) is equivalent to (LE). Here, it should be noted that the uniqueness of $\tilde{f}$ in (*) is not used to prove that η is an embedding. Moreover, the linear map $\tilde{f}$ in (*) is unique if and only if $L(X)$ is generated by $\eta(X)$. (For the "only if" part, refer to the proof of Lemma 3.9.1(1).)

Theorem 3.9.2. *For every Tychonoff space X, there exists the free topological linear space $L(X)$ over X.*

Proof. There exists a collection $\mathcal{F} = \{f_\lambda : X \to F_\lambda \mid \lambda \in \Lambda\}$ such that, for an arbitrary topological linear space F and each continuous map $f : X \to F$, there exist $\lambda \in \Lambda$ and a linear embedding $\varphi : F_\lambda \to F$ such that $\varphi f_\lambda = f$.

Indeed, for each cardinal $\tau \leq \operatorname{card} X$, let $\mathfrak{T}_\tau$ be the topologies $\mathcal{T}$ on $\mathbb{R}^\tau_f$ such that $(\mathbb{R}^\tau_f, \mathcal{T})$ is a topological linear space. Then, the desired collection is

$$\mathcal{F} = \bigcup_{\tau \leq \operatorname{card} X} \bigcup_{\mathcal{T} \in \mathfrak{T}_\tau} \mathrm{C}(X, (\mathbb{R}^\tau_f, \mathcal{T})).$$

Consequently, for an arbitrary topological linear space F and each continuous map $f : X \to F$, let $\tau = \operatorname{card} f(X) \leq \operatorname{card} X$. The linear span F' of $f(X)$ is linearly homeomorphic to $(\mathbb{R}^\tau_f, \mathcal{T})$ for some $\mathcal{T} \in \mathfrak{T}_\tau$. Let $\psi : F' \to (\mathbb{R}^\tau_f, \mathcal{T})$ be a linear homeomorphism. Accordingly, we have $g = \psi f \in \mathrm{C}(X, (\mathbb{R}^\tau_f, \mathcal{T}))$, and thus $f = \psi^{-1} g$.

The product space $\prod_{\lambda \in \Lambda} F_\lambda$ is a topological linear space. Let $\eta : X \to \prod_{\lambda \in \Lambda} F_\lambda$ be the map defined by $\eta(x) = (f_\lambda(x))_{\lambda \in \Lambda}$. We define $L(X)$ as the linear span of $\eta(X)$ in $\prod_{\lambda \in \Lambda} F_\lambda$. Then, $(L(X), \eta)$ satisfies the condition (*) in the above remark. In fact, for an arbitrary topological linear space F and each map $f : X \to F$, there exists $\lambda \in \Lambda$ and a linear embedding $\varphi : F_\lambda \to F$ such that $\varphi f_\lambda = f$. Consequently, we have a linear map $\tilde{f} = \varphi \mathrm{pr}_\lambda | L(X) : L(X) \to F$ and

$$\tilde{f} \eta(x) = \varphi \mathrm{pr}_\lambda \eta(x) = \varphi f_\lambda(x) = f(x) \quad \text{for every } x \in X.$$

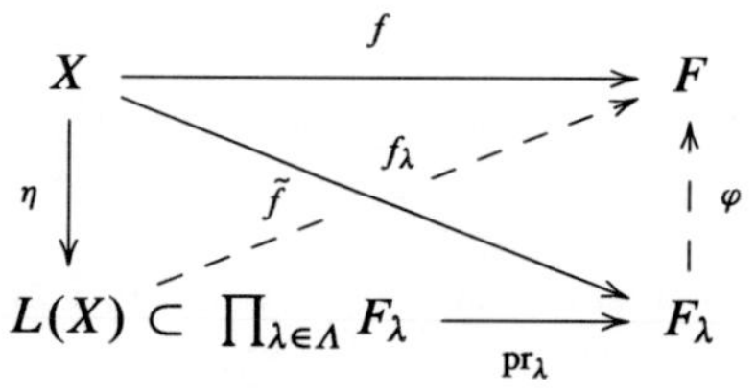

Because $L(X)$ is generated by $\eta(X)$, a linear map $\tilde{f} : L(X) \to F$ is uniquely determined by the condition that $\tilde{f}\eta = f$. As observed in the above remark, η is an embedding, hence X can be identified with $\eta(X)$. Then, $L(X)$ satisfies (LE), i.e., $L(X)$ is the free topological linear space over X. □

Let X and Y be Tychonoff spaces. For each map $f : X \to Y$, we have a unique linear map $f_\natural : L(X) \to L(Y)$ that is an extension of f by (LE).

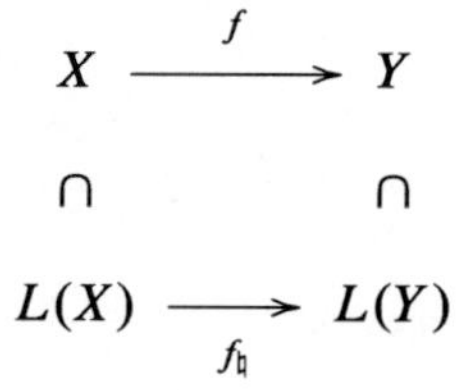

This is functorial, i.e., $(gf)_\natural = g_\natural f_\natural$ for every pair of maps $f : X \to Y$ and $g : Y \to Z$, and $\mathrm{id}_{L(X)} = (\mathrm{id}_X)_\natural$. Accordingly, we have a covariant functor from the category of Tychonoff spaces into the category of topological linear spaces. Consequently, every homeomorphism $f : X \to Y$ extends to a linear homeomorphism $f_\natural : L(X) \to L(Y)$.

In Sect. 7.12, we will construct a metrizable linear space that is not an absolute extensor for metrizable spaces. The free topological linear space $L(X)$ over a compactum X has an important role in the construction. The topological and geometrical structures of $L(X)$ will be studied in Sect. 7.11.

Notes for Chap. 3

There are lots of good textbooks for studying topological linear spaces. The following classical book of Köthe is still a very good source on this subject. The textbook by Kelly and Namioka is also recommended by many people. Besides these two books, the textbook by Day is a good reference for normed linear spaces as is Valentine's book for convex sets. Concerning non-locally convex F-spaces and Roberts' example (a compact convex set with no extreme points), one can refer to the book by Kalton, Peck and Roberts.

- G. Köthe, *Topological Vector Spaces, I,* English edition, GMW **159** (Springer-Verlag, New York, 1969)
- J.L. Kelly and I. Namioka, *Linear Topological Spaces,* Reprint edition, GTM **36** (Springer-Verlag, New York, 1976)
- M.M. Day, *Normed Linear Spaces*, 3rd edition, EMG **21** (Springer-Verlag, Berlin, 1973)
- F.A. Valentine, *Convex Sets* (McGraw-Hill Inc., 1964); Reprint of the 1964 original (R.E. Krieger Publ. Co., New York, 1976)
- N.J. Kalton, N.T. Peck and J.W. Roberts, *An F-space Sampler*, London Math. Soc. Lecture Note Ser. **89** (Cambridge Univ. Press, Cambridge, 1984)

For a systematic and comprehensive study on continuous selections, refer to the following book by Repovš and Semenov, which is written in instructive style.

- D. Repovš and P.V. Semenov, *Continuous Selections of Multivalued Mappings*, MIA **455** (Kluwer Acad. Publ., Dordrecht, 1998)

In Theorem 3.6.4, the construction of a metric d from d_0 is due to Eidelheit and Mazur [1].

The results of Sect. 3.8 are contained in the first part of Michael's paper [2], which consists of three parts. For the finite-dimensional case, refer to the second and third parts of [2] (cf. [3]) and the book of Repovš and Semenov. The finite-dimensional case is deeply related with the concept discussed in Sect. 6.11 but will not be treated in this book. The 0-dimensional case will be treated in Sect. 7.2.

References

1. M. Eidelheit, S. Mazur, Eine Bemerkung über die Räume vom Typus (F). Stud. Math. **7**, 159–161 (1938)
2. E. Michael, Continuous selections, I. Ann. Math. **63**, 361–382 (1956); Continuous selections, II. Ann. Math. **64**, 562–580 (1956); Continuous selections, III. Ann. Math. **65**, 375–390 (1957)
3. E. Michael, A generalization of a theorem on continuous selections. Proc. Am. Math. Soc. **105**, 236–243 (1989)

Chapter 4
Simplicial Complexes and Polyhedra

In this chapter, we introduce and demonstrate the basic concepts and properties of simplicial complexes. The importance and usefulness of simplicial complexes lies in the fact that they can be used to approximate and explore (topological) spaces. A polyhedron is the underlying space of a simplicial complex, which has two typical topologies, the so-called weak (Whitehead) topology and the metric topology. The paracompactness of the weak topology will be shown. We show that every completely metrizable space can be represented as the inverse limit of locally finite-dimensional polyhedra with the metric topology. In addition, we give a proof of the Whitehead–Milnor Theorem on the homotopy type of simplicial complexes. We also prove that a map between polyhedra is a homotopy equivalence if it induces isomorphisms between their homotopy groups.

This chapter is based on Chaps. 2 and 3. In particular, we employ the theory of convex sets and the related concepts discussed in Chap. 3.

4.1 Simplexes and Cells

Let E be a linear space. The convex hull $\sigma = \langle v_1, \dots, v_n \rangle$ of finitely many affinely independent points $v_1, \dots, v_n \in E$ is called a **simplex**. Each v_i is called a **vertex** of σ, and $n - 1$ is called the **dimension** of σ, written as $\dim \sigma = n - 1$. An n-dimensional simplex is called an ***n*-simplex**. A 0-simplex is a singleton and a 1-simplex is a line segment. Note that the affine image of a simplex σ (i.e., the image $f(\sigma)$ of an affine function $f : \sigma \to E'$ of σ into a linear space E') is, in general, not a simplex.

The convex hull $C = \langle A \rangle$ of a non-empty finite subset $A \subset E$ is called a **cell** (or a **linear cell**),[1] where the dimension of C is defined as the dimension of the flat

[1] More precisely, it is called a **convex linear cell**.

K. Sakai, *Geometric Aspects of General Topology*, Springer Monographs in Mathematics, DOI 10.1007/978-4-431-54397-8_4, © Springer Japan 2013

hull fl C, i.e., $\dim C = \dim \mathrm{fl}\, A < \infty$ (cf. Sect. 3.2). An n-dimensional cell is called an **n-cell** (or a **linear n-cell**). Obviously, every simplex is a cell. The affine image of a cell is always a cell and every cell is the affine image of some simplex. A 0-cell and a 1-cell are the same as a 0-simplex and a 1-simplex, respectively. If $\mathrm{card}\, A = n$ and $\dim\langle A\rangle = n - 1$, then A is affinely independent, hence $\langle A\rangle$ is a simplex. When $\langle v_1, \dots, v_{n+1}\rangle$ is an n-simplex, it follows that $v_1, \dots, v_{n+1}$ are affinely independent, hence they are vertices of the simplex.

The *radial* interior and the *radial* boundary of a cell C are simply called the **interior** and the **boundary** of C. Recall that they are defined without topology, that is,

$$\mathrm{rint}\, C = \big\{x \in C \bigm| \forall y \in C,\ \exists \delta > 0 \text{ such that } (1+\delta)x - \delta y \in C\big\}$$

and $\partial C = C \setminus \mathrm{rint}\, C$ (cf. Sect. 3.2).[2] For $x \neq y \in E$,

$$\partial\langle x, y\rangle = \{x, y\} \text{ and } \mathrm{rint}\langle x, y\rangle = \langle x, y\rangle \setminus \{x, y\}.$$

Then, we can also write as follows:

$$\mathrm{rint}\, C = \big\{x \in C \bigm| \forall y \in C,\ \exists z \in C \text{ such that } x \in \mathrm{rint}\langle y, z\rangle\big\}.$$

According to Proposition 3.5.1, the flat hull fl C has the unique topology such that the following operation is continuous:

$$\mathrm{fl}\, C \times \mathrm{fl}\, C \times \mathbb{R} \ni (x, y, t) \mapsto (1-t)x + ty \in \mathrm{fl}\, C,$$

and fl C is affinely homeomorphic to $\mathbb{R}^n$, where $n = \dim C = \dim \mathrm{fl}\, C$. With respect to this topology, as is shown in Proposition 3.5.5,

$$\mathrm{rint}\, C = \mathrm{int}_{\mathrm{fl}\, C}\, C \text{ and } \partial C = \mathrm{bd}_{\mathrm{fl}\, C}\, C.$$

Moreover, $(C, \partial C) \approx (\mathbf{B}^n, \mathbf{S}^{n-1})$ (Corollary 3.5.6). For every cell (or simplex) C, we always consider the topology to be inherited from this unique topology of fl C.

In fact, as seen in Proposition 3.5.8, C itself has the unique topology such that the following operation is continuous:

$$C \times C \times \mathbf{I} \ni (x, y, t) \mapsto (1-t)x + ty \in C.$$

[2]According to Proposition 3.5.8, C itself is equal to the radial closure

$$\mathrm{rcl}\, C = \{x \in E \mid \exists y \in C \text{ such that } \forall t \in \mathbf{I},\ (1+t)x - ty \in C\}.$$

For the standard n-simplex $\Delta^n \subset \mathbb{R}^{n+1}$,

$$\operatorname{rint} \Delta^n = \{z \in \Delta^n \mid z(i) > 0 \text{ for every } i = 1, \ldots, n+1\} \text{ and}$$
$$\partial \Delta^n = \{z \in \Delta^n \mid z(i) = 0 \text{ for some } i = 1, \ldots, n+1\}.$$

For an n-simplex $\sigma = \langle v_1, \ldots, v_{n+1} \rangle \subset E$, there exists the *natural* affine homeomorphism $\delta_\sigma : \operatorname{fl} \Delta^n \to \operatorname{fl} \sigma$ defined by

$$\delta_\sigma(z) = \sum_{i=1}^{n+1} z(i) v_i \quad \text{for each } z \in \operatorname{fl} \Delta^n.$$

Then, $\sigma = \delta_\sigma(\Delta^n)$, $\operatorname{rint} \sigma = \delta_\sigma(\operatorname{rint} \Delta^n)$, and $\partial\sigma = \delta_\sigma(\partial \Delta^n)$. The **barycenter** $\hat{\sigma}$ of σ is defined as follows:

$$\hat{\sigma} = \delta_\sigma\left(\frac{1}{n+1}, \ldots, \frac{1}{n+1}\right) = \sum_{i=1}^{n+1} \frac{1}{n+1} \cdot v_i.$$

The homeomorphism h_σ is not unique because it is depend on the the order of $\sigma^{(0)}$, but the barycenter $\hat{\sigma}$ is independent from the order of $\sigma^{(0)}$ and uniquely determined.

When $v_1, \ldots, v_{n+1}$ are not affinely independent, the map defined as δ_σ is not a homeomorphism, but we do have the following result:

Proposition 4.1.1. *For every finite subset* $A = \{v_1, \ldots, v_n\} \subset E$,

$$\operatorname{rint}\langle A \rangle = \left\{\textstyle\sum_{i=1}^n z(i) v_i \mid z \in \operatorname{rint} \Delta^{n-1}\right\}.$$

Proof. Take $z_0 \in \operatorname{rint} \Delta^{n-1}$ and let $x_0 = \sum_{i=1}^n z_0(i) v_i \in \langle A \rangle$. For each $x \in \operatorname{rint}\langle A \rangle$, we have $x_1 \in \langle A \rangle$ and $0 < \delta < 1$ such that $x = (1-\delta)x_0 + \delta x_1$. Write $x_1 = \sum_{i=1}^n z_1(i) v_i$, $z_1 \in \Delta^{n-1}$. Then, $z = (1-\delta)z_0 + \delta z_1 \in \operatorname{rint} \Delta^{n-1}$ (Proposition 3.2.3) and

$$x = (1-\delta)\sum_{i=1}^n z_0(i) v_i + \delta \sum_{i=1}^n z_1(i) v_i = \sum_{i=1}^n z(i) v_i.$$

Conversely, for each $z \in \operatorname{rint} \Delta^{n-1}$, we show $\sum_{i=1}^n z(i) v_i \in \operatorname{rint}\langle A \rangle$. Each $y \in \langle A \rangle$ can be written as $y = \sum_{i=1}^n z_0(i) v_i$ for some $z_0 \in \Delta^{n-1}$. On the other hand, we have $z_1 \in \Delta^{n-1}$ and $0 < \delta < 1$ such that $z = (1-\delta)z_0 + \delta z_1$. Let $y_1 = \sum_{i=1}^n z_1(i) v_i \in \langle A \rangle$. Then, it follows that

$$\sum_{i=1}^n z(i) v_i = (1-\delta)\sum_{i=1}^n z_0(i) v_i + \delta \sum_{i=1}^n z_1(i) v_i = (1-\delta)y + \delta y_1.$$

This means that $\sum_{i=1}^n z(i) v_i \in \operatorname{rint}\langle A \rangle$. □

Proposition 4.1.2. *Each cell* $C \subset E$ *has the smallest finite set* $C^{(0)}$ *such that* $\langle C^{(0)} \rangle = C$ *(i.e.,* $\langle A \rangle = C \Rightarrow C^{(0)} \subset A$*). In addition,* $C^{(0)} \subset \partial C$ *if* $\dim C > 0$.

Proof. By the definition of a cell, we can easily find a minimal finite set $C^{(0)}$ such that $\langle C^{(0)}\rangle = C$, i.e., $\langle B\rangle \neq C$ if $B \subsetneqq C^{(0)}$. We have to show that $\langle A\rangle = C$ implies $C^{(0)} \subset A$. Assume that $C^{(0)} \not\subset A$. Let $C^{(0)} = \{v_1, \dots, v_n\}$, where $v_1 \notin A$ and $v_i \neq v_j$ if $i \neq j$. We can write

$$v_1 = \sum_{i=1}^{m} z(i)x_i,\ x_1, \dots, x_m \in A,\ z \in \operatorname{rint} \Delta^{m-1},$$

where $x_i \neq x_j$ if $i \neq j$. Since $v_1 \notin A$, it follows that $m \geq 2$. Define

$$v' = (z(1) + \varepsilon)x_1 + (z(2) - \varepsilon)x_2 + \sum_{i=3}^{m} z(i)x_i,$$

$$v'' = (z(1) - \varepsilon)x_1 + (z(2) + \varepsilon)x_2 + \sum_{i=3}^{m} z(i)x_i,$$

where $\varepsilon > 0$ is chosen so that $z(1) \pm \varepsilon,\ z(2) \pm \varepsilon \in (0, 1)$. Then, $v' \neq v''$ because $v' - v'' = 2\varepsilon(x_1 - x_2) \neq 0$. Since $v', v'' \in C$, we can write

$$v' = \sum_{i=1}^{n} z'(i)v_i,\ v'' = \sum_{i=1}^{n} z''(i)v_i,\ z', z'' \in \Delta^{n-1},$$

and therefore

$$v_1 = \frac{1}{2}v' + \frac{1}{2}v'' = \sum_{i=1}^{n} \left(\frac{1}{2}z'(i) + \frac{1}{2}z''(i)\right) v_i.$$

Recall that $v_1 \notin \langle v_2, \dots, v_n\rangle$. Then, it follows that $\frac{1}{2}z'(1) + \frac{1}{2}z''(1) = 1$. Since $z'(1), z''(1) \in \mathbf{I}$, we have $z'(1) = z''(1) = 1$. Hence, $v' = v'' = v_1$, which is a contradiction. Thus, $C^{(0)}$ is the smallest finite set such that $\langle C^{(0)}\rangle = C$.

The additional assertion easily follows from Proposition 4.1.1 and the minimality of $C^{(0)}$. □

In Proposition 4.1.2, each point of $C^{(0)}$ is called a **vertex** of C; namely, $C^{(0)}$ is the set of vertices of C. Note that if $\sigma = \langle v_1, \dots, v_{n+1}\rangle$ is an n-simplex then $\sigma^{(0)} = \{v_1, \dots, v_{n+1}\}$. Thus, we have the following:

Corollary 4.1.3. *A cell $C \subset E$ is a simplex if and only if $C^{(0)}$ is affinely independent.* □

It is said that two simplexes σ and τ are **joinable** (or σ is **joinable** to τ) if $\sigma \cap \tau = \emptyset$ and $\sigma^{(0)} \cup \tau^{(0)}$ is affinely independent. In this case, $\langle \sigma^{(0)} \cup \tau^{(0)}\rangle$ is a simplex of dimension $\dim \sigma + \dim \tau + 1$, which is denoted by $\sigma\tau$ and called the **join** of σ and τ. When $\sigma = \{v\}$, the simplex $\{v\}\tau$ is simply denoted by $v\tau$.

The face of a cell C at $x \in C$ is defined as in Sect. 3.2, i.e.,

$$C_x = \{y \in C \mid \exists \delta > 0 \text{ such that } (1 + \delta)x - \delta y \in C\}.$$

Recall $\operatorname{rint} C = \{x \in C \mid C_x = C\}$, hence $\partial C = \{x \in C \mid C_x \neq C\}$. Moreover, $x \in \operatorname{rint} C_x$ (3.2.5(8)) and $C_x = C_y$ for every $y \in \operatorname{rint} C_x$ (3.2.5(10)). Recall that an extreme point of C is a point $x \in C$ such that $C_x = \{x\}$.

Proposition 4.1.4. *For each cell $C \subset E$ and $x \in C$, the following hold:*

(1) *C_x is a cell with $C_x^{(0)} = C^{(0)} \cap C_x$;*
(2) *x is a vertex of C if and only if it is an extreme point of C,*
i.e., $x \in C^{(0)} \Leftrightarrow C_x = \{x\}$.

Proof. (1): To see that C_x is a cell, it suffices to show that $C_x = \langle C^{(0)} \cap C_x \rangle$. Since C_x is convex (Proposition 3.2.5(7)), we have $\langle C^{(0)} \cap C_x \rangle \subset C_x$. Each $y \in C_x$ can be written as

$$y = \sum_{i=1}^{n} z(i) v_i, \; v_1, \ldots, v_n \in C^{(0)}, \; z \in \operatorname{rint} \Delta^{n-1}.$$

Choose $\delta \in (0,1)$ so that $(1+\delta)x - \delta y \in C$. For each $i = 1, \ldots, n$, let

$$x_i = (1 - \delta + \delta z(i))x + \sum_{j \neq i} \delta z(j) v_j \in C.$$

Then, it follows that

$$\left(1 + \frac{1}{2}\delta z(i)\right) x - \frac{1}{2}\delta z(i) v_i = \frac{1}{2}\big((1+\delta)x - \delta y\big) + \frac{1}{2}x_i \in C,$$

which means that $v_i \in C_x$. Hence, $y \in \langle C^{(0)} \cap C_x \rangle$.

Since $C_x = \langle C^{(0)} \cap C_x \rangle$, it follows that $C_x^{(0)} \subset C^{(0)} \cap C_x$ and

$$\begin{aligned}
\langle C_x^{(0)} \cup (C^{(0)} \setminus C_x) \rangle &= \langle C_x \cup (C^{(0)} \setminus C_x) \rangle \\
&= \big\langle (C^{(0)} \cap C_x) \cup (C^{(0)} \setminus C_x) \big\rangle \\
&= \langle C^{(0)} \rangle = C.
\end{aligned}$$

The latter implies $C^{(0)} \cap C_x \subset C_x^{(0)}$. Hence, $C^{(0)} \cap C_x = C_x^{(0)}$.

(2): If $C_x = \{x\}$ then $x \in C_x^{(0)} \subset C^{(0)}$ by (1). Conversely, if $x \in C^{(0)}$ then $x \in C^{(0)} \cap C_x = C_x^{(0)}$ by (1). Since $x \in \operatorname{rint} C_x$ (Proposition 3.2.5(8)), we have $\dim C_x = 0$ by Proposition 4.1.2, which implies $C_x = \{x\}$. $\square$

A cell D is call a **face** of a cell C (denoted by $D \leq C$ or $C \geq D$) if $D = C_x$ for some $x \in C$. If $D \leq C$ and $D \neq C$, D is called a **proper face** of C (denoted by $D < C$ or $C > D$). An n-dimensional face is called an ***n*-face**. A face of a simplex σ is also a simplex (cf. Proposition 4.1.4(1)) and $\langle v_1, \ldots, v_k \rangle \leq \sigma$ for $v_1, \ldots, v_k \in \sigma^{(0)}$. Note that σ is the join $\sigma_0\sigma_1$ of any two disjoint faces σ_0 and σ_1 with $\sigma^{(0)} = \sigma_0^{(0)} \cup \sigma_1^{(0)}$, where σ_i is called the **opposite face** of σ to σ_{1-i} $(i = 0, 1)$. Moreover, it follows that $\tau_0\tau_1 \leq \sigma = \sigma_0\sigma_1$ for each $\tau_i \leq \sigma_i$ $(i = 0, 1)$.

Proposition 4.1.5. *For simplexes σ and τ,*

$$\tau \leq \sigma \Leftrightarrow \tau^{(0)} \subset \sigma^{(0)}.$$

Proof. The implication $\Rightarrow$ is a direct consequence of Proposition 4.1.4(1). If $\tau^{(0)} \subset \sigma^{(0)}$ then $\tau \subset \sigma$. Take any $x \in \operatorname{rint} \tau$. Then, $\tau \subset \sigma_x$ by the definition of σ_x. For each $y \in \sigma \setminus \tau$, we have distinct $v_1, \ldots, v_n \in \sigma^{(0)}$ and $z \in \operatorname{rint} \Delta^{n-1}$ such that $y = \sum_{i=1}^n z(i) v_i$ and $v_1 \notin \tau^{(0)}$. Then, $\tau \cap \operatorname{rint}\langle y, y' \rangle = \emptyset$ for any $y' \in \sigma$, hence $y \notin \sigma_x$. Thus, we have $\tau = \sigma_x$. □

Proposition 4.1.6. *For each cell $C \subset E$, the following hold:*

(1) $\emptyset \neq A \subset C^{(0)} \Rightarrow \langle A \rangle^{(0)} = A$;
(2) *If $A \subset C^{(0)}$ is not a singleton then $C^{(0)} \cap \operatorname{rint}\langle A \rangle = \emptyset$;*
(3) $D \leq C \Rightarrow D = C \cap \operatorname{fl} D$ *and* $D^{(0)} = C^{(0)} \cap D = C^{(0)} \cap \operatorname{fl} D$;
(4) $D \leq C \Rightarrow D_x = C_x$ *for each* $x \in D$, *hence* $D' \leq C$ *for each* $D' \leq D$;
(5) $D \leq C \Rightarrow D = C_x$ *for each* $x \in \operatorname{rint} D$;
(6) $D,\ D' \leq C,\ D \cap \operatorname{rint} D' \neq \emptyset \Rightarrow D' \leq D$;
(7) $D < C \Rightarrow \dim D < \dim C$;
(8) $\partial C = \bigcup\{D \mid D < C\} = \bigcup\{D \mid D < C,\ \dim D = \dim C - 1\}$.

Proof. By virtue of Propositions 3.2.5(7) and 4.1.4(1), we have (3). For (4) and (5), we refer to Propositions 3.2.5(9) and 3.2.5(10), respectively. It is easy to obtain (6) from (4) and (5). For (7), it follows from (3) that $D < C$ implies $\operatorname{fl} D \subsetneqq \operatorname{fl} C$, so $\dim D < \dim C$. Statements (1), (2), and (8) remain to be proved.

(1): First, note that $\langle A \rangle^{(0)} \subset A \subset C^{(0)}$. Let $B = (C^{(0)} \setminus A) \cup \langle A \rangle^{(0)} \subset C^{(0)}$. Since $A \subset \langle A \rangle \subset \langle B \rangle$, we have $C^{(0)} \subset \langle B \rangle \subset C$, hence $\langle B \rangle = C$. Therefore, $B = C^{(0)}$ by Proposition 4.1.2. This means $\langle A \rangle^{(0)} = A$.

(2): Assume that $A \subset C^{(0)}$ contains at least two vertices and $\operatorname{rint}\langle A \rangle$ contains some $v \in C^{(0)}$. Since $A \setminus \{v\} \neq \emptyset$, it follows from Proposition 4.1.1 that $v \in \langle A \setminus \{v\} \rangle$. This implies $\langle C^{(0)} \rangle = \langle C^{(0)} \setminus \{v\} \rangle$, which contradicts the definition of $C^{(0)}$. Thus, $C^{(0)} \cap \operatorname{rint}\langle A \rangle = \emptyset$.

(8): Since $\partial C = \{x \in C \mid C_x \neq C\}$, we have the first equality. To prove the second equality, it suffices to show that each $x \in \partial C$ is contained in an $(n-1)$-face of C, where $n = \dim C$. Let D be a maximal proper face of C containing x. Then, $\dim D \leq n-1$ by (7). Assume $\dim D < n-1$. Since $\operatorname{rint} D$ misses any other proper face of C by (6), we have $\partial C \setminus \operatorname{rint} D = \bigcup\{D' \mid D \neq D' < C\}$, which is a compact set in the flat $\operatorname{fl} C$ given the unique topology (Proposition 3.5.1). Take $x_0 \in \operatorname{rint} D$. Since $\operatorname{fl} C$ is affinely homeomorphic to $\mathbb{R}^n$, x_0 has a convex neighborhood V in $\operatorname{fl} C$ such that $V \cap (\partial C \setminus \operatorname{rint} D) = \emptyset$. Since $x_0 \in \operatorname{rint} D \subset \partial C$, we can find $x_1 \in C \setminus D$ such that $(1+t)x_0 - t x_1 \notin C \cup \operatorname{fl} D$ for every $t > 0$. Choosing $s, t > 0$ sufficiently small, we have

$$y = (1-s)x_0 + s x_1 \in V \cap \operatorname{rint} C,\ z = (1+t)x_0 - t x_1 \in V \setminus (C \cup \operatorname{fl} D).$$

Consequently, $x_0 \in \operatorname{rint}\langle y, z\rangle$. Note that

$$\dim \operatorname{fl}(D \cup \{y\}) \leq n - 1 < n = \dim \operatorname{fl} C.$$

Hence, there exists a $v \in (V \cap C) \setminus \operatorname{fl}(D \cup \{y\})$. Since $z \in \operatorname{fl}(D \cup \{y\})$ and $v \notin \operatorname{fl}(D \cup \{y\})$, it follows that $\langle v, z\rangle \cap \operatorname{fl} D = \emptyset$, so $\langle v, z\rangle \cap D = \emptyset$. On the other hand, $\langle v, z\rangle \cap \partial C \neq \emptyset$ because $v \in C$ and $z \notin C$. Since $\langle v, z\rangle \subset V$ and $V \cap \partial C \subset \operatorname{rint} D$, we have $\langle v, z\rangle \cap D \neq \emptyset$, which is a contradiction. Therefore, $\dim D = n - 1$. □

Proposition 4.1.7. *For each n-cell $C \subset E$ and each k-face $D < C$ ($k < n$), there exist faces $D = D_k < D_{k+1} < \cdots < D_n = C$ such that* $\dim D_i = i$ *for $k \leq i \leq n$.*

Proof. The case $n - k = 1$ is obvious. When $n - k > 1$, let $x \in \operatorname{rint} D$. By Proposition 4.1.6(8), we have $(n-1)$-face $C' < C$ such that $x \in C'$. Then, $C'_x = C_x = D$ by Propositions 4.1.6(4) and (5). Hence, $D < C'$. The result can be obtained by induction. □

Using affine functionals, we characterize cells as follows:

Proposition 4.1.8. *Let $\emptyset \neq C \subset E$ be non-degenerate. In order for C to be a cell, it is necessary and sufficient that* $\dim \operatorname{fl} C < \infty$, *$x + \mathbb{R}_+(y - x) \not\subset C$ for each pair of distinct points $x, y \in C$, and there are finitely many non-constant affine functionals $f_1, \ldots, f_k : \operatorname{fl} C \to \mathbb{R}$ such that $C = \bigcap_{i=1}^k f_i^{-1}(\mathbb{R}_+)$.*

Proof. (*Necessity*) Let $n = \dim C$. By virtue of Proposition 4.1.6(8), we can write $\partial C = \bigcup_{i=1}^k D_i$, where each D_i is an $(n-1)$-face of C. Because $\operatorname{fl} D_i$ is a hyperplane in $\operatorname{fl} C$, there is an affine functional $f_i : \operatorname{fl} C \to \mathbb{R}$ such that $\operatorname{fl} D_i = f_i^{-1}(0)$ (Proposition 3.1.3(1)). Then, $C \subset f_i^{-1}(\mathbb{R}_+)$ or $C \subset f_i^{-1}(-\mathbb{R}_+)$. Replacing f_i with $-f_i$ if $C \subset f_i^{-1}(-\mathbb{R}_+)$, we may assume that $C \subset f_i(\mathbb{R}_+)$ for every $i = 1, \ldots, k$. Thus, we have $C \subset \bigcap_{i=1}^k f_i^{-1}(\mathbb{R}_+)$. Suppose that there is a $z \in \bigcap_{i=1}^k f_i^{-1}(\mathbb{R}_+) \setminus C$. By taking $y \in \operatorname{rint} C$, we have $x \in \operatorname{rint}\langle y, z\rangle \cap \partial C$. Then, x is contained in some D_i. Since $f_i(x) = 0$ and $f_i(y) > 0$, it follows that $f_i(z) < 0$, which is a contradiction. Therefore, $C = \bigcap_{i=1}^k f_i^{-1}(\mathbb{R}_+)$.

(*Sufficiency*) First, note that $C = \bigcap_{i=1}^k f_i^{-1}(\mathbb{R}_+)$ is convex and $\operatorname{rint} C = \operatorname{core}_{\operatorname{fl} C} C = \bigcap_{i=1}^k f_i^{-1}((0, \infty))$, hence $\partial C = C \setminus \operatorname{rint} C = \bigcup_{i=1}^k (C \cap f_i^{-1}(0))$. Moreover, $C \cap f_{i_0}^{-1}(0) = \emptyset$ implies $C = \bigcap_{i \neq i_0} f_i^{-1}(\mathbb{R}_+)$, that is, $f_{i_0}(x) \geq 0$ for every $x \in \bigcap_{i \neq i_0} f_i^{-1}(\mathbb{R}_+)$. Indeed, assume that $f_{i_0}(x) < 0$ for some $x \in \bigcap_{i \neq i_0} f_i^{-1}(\mathbb{R}_+)$. Take any point $y \in C$. Because $f_{i_0}(y) \geq 0$, we have $z \in \langle x, y\rangle$ such that $f_{i_0}(z) = 0$. Then, $z \in \bigcap_{i=1}^k f_i^{-1}(\mathbb{R}_+) = C$, so $C \cap f_{i_0}^{-1}(0) \neq \emptyset$, which is a contradiction. Thus, we may assume that $C \cap f_i^{-1}(0) \neq \emptyset$ for every $i = 1, \ldots, k$.

Now, by induction on $n = \dim \operatorname{fl} C$, we shall show that C is a cell. For each $i = 1, \ldots, k$, let $D_i = C \cap f_i^{-1}(0) \neq \emptyset$. Then, as observed above, $\partial C = \bigcup_{i=1}^k D_i$. Since $\operatorname{fl} D_i \subset f_i^{-1}(0)$ and $\dim f_i^{-1}(0) = n - 1$ (Proposition 3.1.3(2)), each $D_i = \bigcap_{j \neq i} (f_j | \operatorname{fl} D_i)^{-1}(\mathbb{R}_+)$ is a cell by the inductive assumption. Thus, we have a finite set $A = \bigcup_{i=1}^k D_i^{(0)} \subset \partial C$. Consequently, $\partial C = \bigcup_{i=1}^k D_i \subset \langle A \rangle \subset C$. Take any point $v \in A \subset \partial C$. For each $x \in \operatorname{rint} C$, $v + \mathbb{R}_+(x - v) \not\subset C$, hence there is a $y \in \partial C$ such that $x \in \langle v, y \rangle$. Then, $x \in \langle A \rangle$ because $v, y \in \langle A \rangle$. Therefore, $C = \langle A \rangle$ is a cell. □

Later, we will use the following results, which are easily proved.

Additional Results for Cells 4.1.9.

(1) For each cell $C \subset E$ and each flat $F \subset E$ with $C \cap F \neq \emptyset$, the intersection $C \cap F$ is also a cell.
(2) For every two cells $C, D \subset E$ with $C \cap D \neq \emptyset$, the intersection $C \cap D$ is also a cell with $(C \cap D)_x = C_x \cap D_x$ for each $x \in C \cap D$. If $\operatorname{rint} C \cap \operatorname{rint} D \neq \emptyset$, then $\operatorname{rint}(C \cap D) = \operatorname{rint} C \cap \operatorname{rint} D$.
(3) Let $f : C \to E'$ be an affine map from a cell $C \subset E$ into another linear space E'. Then, $f^{-1}(D)$ is a cell for every cell $D \subset E'$ with $D \cap f(C) \neq \emptyset$, where $f^{-1}(D)_x = C_x \cap f^{-1}(D_{f(x)})$ for each $x \in f^{-1}(D)$. When f is injective, $f(C_x) = f(C)_{f(x)}$ for each $x \in C$.

Sketch of Proof. For the above three items, apply the characterization 4.1.8 (cf. Proposition 3.2.2 for (3)). The statements about faces in (2) and (3) are the same as 3.2.7(1) and (4), respectively.The statement about the radial interior in (2) is 3.2.7(2).

(4) For every two cells $C, D \subset E$, $C \times D$ is also a cell with $\operatorname{rint} C \times D = \operatorname{rint} C \times \operatorname{rint} D$ and $(C \times D)_{(x,y)} = C_x \times D_y$ for each $(x, y) \in C \times D$.

Sketch of Proof. Note that $C \times D = \langle C^{(0)} \times D^{(0)} \rangle$ and see 3.2.7(3).

4.2 Complexes and Subdivisions

Throughout this section, let E be a linear space. A collection K of cells in E is called a **cell complex**[3] if K satisfies the following two conditions:

(C1) If $C \in K$ and $D \leq C$ then $D \in K$;
(C2) For each $C, D \in K$ with $C \cap D \neq \emptyset$, $C \cap D \leq C$ (and $C \cap D \leq D$).

Under condition (C1), condition (C2) is equivalent to each of the following:

(C2′) For each $C, D \in K$, $C \cap \operatorname{rint} D \neq \emptyset$ implies $D \leq C$;
(C2″) For each $C, D \in K$, $C \neq D$ implies $\operatorname{rint} C \cap \operatorname{rint} D = \emptyset$ (equivalently, $\operatorname{rint} C \cap \operatorname{rint} D \neq \emptyset$ implies $C = D$).

Sketch of Proof. Since $C \leq D$ and $D \leq C$ imply $C = D$, we have (C2′) ⇒ (C2″). To see (C2) ⇒ (C2′), show that $C \cap \operatorname{rint} D \neq \emptyset$ implies $C \cap D = D$.

[3]More precisely, it is called a **(convex) linear cell complex**.

(C2″) ⇒ (C2): Assume $C \cap D \neq \emptyset$ and take a point $x \in \operatorname{rint}(C \cap D)$. Since $C_x, D_x \in K$ by (C1) and $x \in \operatorname{rint} C_x \cap \operatorname{rint} D_x$, we have $C_x = D_x$ by (C2″). It follows from 4.1.9(2) that

$$C \cap D = (C \cap D)_x = C_x \cap D_x = C_x \leq C.$$

For each n-cell C, we can define the following cell complexes (cf. Proposition 4.1.6(6)):

$$F(C) = \{D \mid D \leq C\} \quad \text{and} \quad F(\partial C) = \{D \mid D < C\}.$$

Condition (C1) means that $F(C) \subset K$ for each $C \in K$. A cell complex consisting of simplexes is called a **simplicial complex**. For a simplex σ, $F(\sigma)$ and $F(\partial\sigma)$ are simplicial complexes.

The next fact follows from (C2) and Proposition 4.1.6(3):

Fact. *For each* $C, D \in K$, $D \leq C \Leftrightarrow D^{(0)} \subset C^{(0)}$.

Let K be a cell (simplicial) complex. We call $K^{(0)} = \bigcup_{C \in K} C^{(0)}$ the **set of vertices**. It is said that K is **finite**, **infinite**, or **countable** according to $\operatorname{card} K$ (equivalently $\operatorname{card} K^{(0)}$). If $\operatorname{card} K$ is infinite, we have $\operatorname{card} K = \operatorname{card} K^{(0)}$.

Indeed, $K \ni C \mapsto C^{(0)} \in \operatorname{Fin}(K^{(0)})$ is an injection by Proposition 4.1.2 (or the above Fact). Then, $\operatorname{card} K^{(0)}$ is also infinite, hence it follows that

$$\operatorname{card} K \leq \operatorname{card} \operatorname{Fin}(K^{(0)}) = \operatorname{card} K^{(0)} \leq \operatorname{card} K.$$

The **dimension** of K is $\dim K = \sup_{C \in K} \dim C$. If $\dim K = \infty$, K is said to be **infinite-dimensional** (abbrev. **i.d.**). When $\dim K < \infty$, K is **finite-dimensional** (abbrev. **f.d.**). It is said that K is ***n*-dimensional** if $\dim K = n$. Note that every cell complex K with $\dim K \leq 1$ is simplicial.

The **polyhedron** $|K|$ of K is defined as follows:

$$|K| = \bigcup K = \bigcup_{C \in K} C = \bigcup_{C \in K} \operatorname{rint} C \ (\subset E).$$

Recall that each cell $C \in K$ is given the unique topology, as mentioned in the previous section, and if $\dim C = n$ then C with this topology is homeomorphic to the unit closed n-ball $\mathbf{B}^n$ (Proposition 3.5.8). The topology for $|K|$ is defined as follows:

$$U \subset |K| \text{ is open in } |K| \Leftrightarrow \forall C \in K,\ U \cap C \text{ is open in } C$$

$$\Big(\text{equiv. } A \subset |K| \text{ is closed in } |K| \Leftrightarrow \forall C \in K,\ A \cap C \text{ is closed in } C\Big).$$

This topology is called the **Whitehead** (or **weak**) **topology**. Then, $K^{(0)}$ is discrete in $|K|$. Each $C \in K$ is a closed subspace of $|K|$ because $C \cap D \leq D$ for any $D \in K$ with $C \cap D \neq \emptyset$. The following fact is used very often:

Fact. *For an arbitrary space X, each $f : |K| \to X$ is continuous if and only if $f|C$ is continuous for every $C \in K$.*

Remark 1. If V is a neighborhood of $x \in |K|$, then $V \cap C$ is a neighborhood x in C for every $C \in K[x]$. However, the converse does not hold. For example, let K be the 2-dimensional simplicial complex in $\mathbb{R}^{\mathbb{N}}_f$ defined as follows:

$$K = \{\mathbf{0}, \mathbf{e}_i, \langle \mathbf{0}, \mathbf{e}_i \rangle, \langle \mathbf{e}_1, \mathbf{e}_{i+1} \rangle, \langle \mathbf{0}, \mathbf{e}_1, \mathbf{e}_{i+1} \rangle \mid i \in \mathbb{N}\}.$$

We define $V = \bigcup_{i \in \mathbb{N}} \langle \mathbf{0}, 2^{-i}\mathbf{e}_1, 2^{-i}\mathbf{e}_{i+1} \rangle \subset |K|$. For each simplex $\sigma \in K[0]$, $V \cap \sigma$ is a neighborhood of $\mathbf{0}$ in σ. Nevertheless, V is not a neighborhood of $\mathbf{0}$ in $|K|$. Indeed, for each $i \in \mathbb{N}$,

$$(\operatorname{int}_{|K|} V) \cap \langle \mathbf{0}, \mathbf{e}_1, \mathbf{e}_{i+1} \rangle \subset \langle \mathbf{0}, 2^{-i}\mathbf{e}_1, 2^{-i}\mathbf{e}_{i+1} \rangle.$$

Hence, $(\operatorname{int}_{|K|} V) \cap \langle \mathbf{0}, \mathbf{e}_1 \rangle \subset \bigcap_{n \in \mathbb{N}} \langle \mathbf{0}, 2^{-i}\mathbf{e}_1 \rangle = \{\mathbf{0}\}$, which implies $(\operatorname{int}_{|K|} V) \cap \langle \mathbf{0}, \mathbf{e}_1 \rangle = \emptyset$. Thus, V is not a neighborhood of $\mathbf{0}$ in $|K|$.

Each $x \in |K|$ is contained in the interior of the unique cell $c_K(x) \in K$, which is called the **carrier** of x in K. In other words, $c_K(x)$ is the smallest cell of K containing x. If $x \in C \in K$ then $C_x = c_K(x)$. A cell $C \in K$ is said to be **principal** if C is not a proper face of any cell of K, that is, it is a maximal cell of K. A cell $C \in K$ is principal if and only if $\operatorname{int}_{|K|} C \neq \emptyset$. In general, $\operatorname{int}_{|K|} C \neq \operatorname{rint} C$ even if $C \in K$ is principal. If $\dim K = n$, then every n-cell of K is principal.

A cell complex L is called a **subcomplex** of a cell complex K if $L \subset K$. A subcollection $L \subset K$ is a subcomplex of K if and only if L satisfies condition (C1). Evidently, unions and intersections of subcomplexes of K are also subcomplexes of K. Every subcomplex of a simplicial complex is a simplicial complex. The ***n*-skeleton** of K is the subcomplex:

$$K^{(n)} = \{C \in K \mid \dim C \leq n\} \subset K.$$

The 0-skeleton is the set of vertices. For each cell $C \in K$, $F(C)$ and $F(\partial C)$ are subcomplexes of K and $F(\partial C) = F(C)^{(n-1)}$ if $n = \dim C$.

Proposition 4.2.1. *For every subcomplex L of a cell complex of K, $|L|$ is a closed subspace of $|K|$.*

Proof. As is easily observed, $A \cap |L|$ is closed in $|L|$ for each closed set A in $|K|$. Then, it suffices to show that every closed set A in $|L|$ is closed in $|K|$. For each $C \in K$,

$$A \cap C = A \cap C \cap |L| = \bigcup_{D \in L} A \cap C \cap D = \bigcup_{D \in L \cap F(C)} A \cap D.$$

For each $D \in L \cap F(C)$, D is a closed subspace of C and $A \cap D$ is closed in D, hence it is closed in C. Because $L \cap F(C)$ is finite, $A \cap C$ is closed in C. Therefore, A is closed in $|K|$. □

Proposition 4.2.2. *The polyhedron $|K|$ is perfectly normal.*

Proof. By definition, it is obvious that $|K|$ is T_1. For any disjoint closed sets $A, B \subset |K|$, it suffices to find a map $f : |K| \to \mathbf{I}$ such that $f^{-1}(0) = A$ and $f^{-1}(1) = B$. We will inductively construct maps $f_n : |K^{(n)}| \to \mathbf{I}$, $n \in \mathbb{N}$, so that $f_n^{-1}(0) = A \cap |K^{(n)}|$ and $f_n^{-1}(1) = B \cap |K^{(n)}|$. Then, the map f can be defined by $f||K^{(n)}| = f_n$ for each $n \in \mathbb{N}$.

Since $K^{(0)}$ is discrete, f_0 can be easily constructed. Assume that $f^{(n-1)}$ has been constructed. For each n-cell $C \in K$, we apply the Tietze Extension Theorem 2.2.2 to obtain a map $g_C : C \to \mathbf{I}$ such that $g_C|\partial C = f_{n-1}|\partial C$, $g_C(A \cap C) = \{0\}$, and $g_C(B \cap C) = \{1\}$. On the other hand, because $C \in K$ is metrizable (so perfectly normal), there is a map $h_C : C \to \mathbf{I}$ with

$$h_C^{-1}(0) = (A \cap C) \cup (B \cap C) \cup \partial C.$$

We define a map $f_C : C \to \mathbf{I}$ by

$$f_C(x) = (1 - h_C(x))g_C(x) + \frac{1}{2}h_C(x).$$

Then, $f_C|\partial C = f_{n-1}|\partial C$, $f_C^{-1}(0) = A \cap C$ and $f_C^{-1}(1) = B \cap C$. Hence, f_n can be defined by $f_n||K^{(n-1)}| = f_{n-1}$ and $f_n|C = f_C$ for every n-cell of K. □

A **full complex** (or **full simplicial complex**) is a simplicial complex K such that $K^{(0)}$ is affinely independent and $\langle v_1, \dots, v_n \rangle \in K$ for all finitely many distinct vertices $v_1, \dots, v_n \in K^{(0)}$. Every simplex is the polyhedron of a finite full complex. For each affinely independent set A in E, let $\Delta(A)$ denote the full complex with A the set of vertices (i.e., $\Delta(A)^{(0)} = A$). In the case when A is infinite, $|\Delta(A)|$ might be considered as an infinite-dimensional simplex. In fact, $|\Delta(\sigma^{(0)})| = \sigma$ for a simplex σ, where note that $\Delta(\sigma^{(0)}) = F(\sigma)$. An infinite full complex has no principal simplexes. For a simplicial complex K, if $K^{(0)}$ is affinely independent, then K is a subcomplex of the full complex $\Delta(K^{(0)})$.

On the other hand, it is said that a subcomplex L of a simplicial complex K is **full in** K or a **full subcomplex** of K if $\sigma \in L$ for any $\sigma \in K$ with $\sigma^{(0)} \subset L^{(0)}$, that is, L is a maximal subcomplex of K such that the set of vertices is $L^{(0)}$.[4] The n-skeleton $K^{(n)}$ is not full in K unless $K^{(n)} = K$. In general, a full subcomplex of a simplicial complex is not a full complex, but a full subcomplex of a full complex is always a full complex.

[4]It should be noted that although the same word **full** is used, *full subcomplex* and *full complex* are different concepts. The former is used in the relative sense, but the latter is in the absolute sense.

The following subcomplex of K is called the **star** at $C \in K$:

$$\mathrm{St}(C, K) = \{D \mid \exists D' \in K \text{ such that } C \leq D',\ D \leq D'\}.$$

Evidently, $\mathrm{St}(C, K) = F(C)$ (i.e., $|\mathrm{St}(C, K)| = C$) if and only if C is principal in K. We must not confuse $\mathrm{St}(C, K)$ with $\mathrm{st}(C, K)$.[5]

Note that $|\mathrm{St}(C, K)| = \mathrm{st}(\mathrm{rint}\, C, K)$ for each $C \in K$ but, in general, $|\mathrm{St}(C, K)| \subsetneqq \mathrm{st}(C, K)$. Observe that

$$\mathrm{st}(x, K) = |\mathrm{St}(c_K(x), K)| \text{ for each } x \in |K|.$$

If K is a simplicial complex, the **link** of $\sigma \in K$ can be defined as follows:

$$\begin{aligned}\mathrm{Lk}(\sigma, K) &= \mathrm{St}(\sigma, K) \setminus K[\sigma] = \{\tau \in \mathrm{St}(\sigma, K) \mid \tau \cap \sigma = \emptyset\} \\ &= \{\tau \in K \mid \tau\sigma \in K\}.\end{aligned}$$

Note that $\mathrm{Lk}(\sigma, K) = \emptyset$ if and only if σ is principal in K. For each non-principal simplex $\sigma \in K$, we have $|\mathrm{St}(\sigma, K)| = \bigcup_{\tau \in \mathrm{Lk}(\sigma,K)} \sigma\tau$.

We define the **open star** at $x \in |K|$ (with respect to K) as follows:

$$O_K(x) = |K| \setminus |K \setminus K[x]| = \bigcup_{C \in K[x]} \mathrm{rint}\, C,$$

where $K \setminus K[x]$ is a subcomplex of K, hence $O_K(x)$ is an open neighborhood of x in $|K|$. Since $O_K(x) \subset \mathrm{st}(x, K)$, it follows that $\mathrm{st}(x, K)$ $(= |\mathrm{St}(c_K(x), K)|)$ is a closed neighborhood of x in $|K|$. Note the following equivalences:

$$\begin{aligned} y \in O_K(x) &\Leftrightarrow c_K(y) \in K[x] \Leftrightarrow c_K(x) \leq c_K(y) \Leftrightarrow c_K(x)^{(0)} \subset c_K(y)^{(0)} \\ &\Leftrightarrow \forall v \in c_K(x)^{(0)},\ c_K(y) \in K[v] \Leftrightarrow \forall v \in c_K(x)^{(0)},\ y \in O_K(v). \end{aligned}$$

Therefore, we have

$$O_K(x) = \bigcap_{v \in c_K(x)^{(0)}} O_K(v).$$

Then, $|K|$ has the following open and closed covers:

$$\mathcal{O}_K = \{O_K(v) \mid v \in K^{(0)}\};\ \mathcal{S}_K = \{|\mathrm{St}(v, K)| \mid v \in K^{(0)}\},$$

where $\mathcal{O}_K^{\mathrm{cl}} = \mathcal{S}_K$. If K is a simplicial complex,

$$O_K(v) = |\mathrm{St}(v, K)| \setminus |\mathrm{Lk}(v, K)| \text{ for each } v \in K^{(0)}.$$

[5]For any $A \subset |K|$, we denote $K[A] = \{C \in K \mid C \cap A \neq \emptyset\}$ and $\mathrm{st}(A, K) = \bigcup K[A]$. When $A = \{x\}$, $K[x] = \{C \in K \mid x \in C\}$ and $\mathrm{st}(x, K) = \bigcup K[x]$. See Sects. 2.3 and 2.4 (cf. Sect. 1.1).

Proposition 4.2.3. *Let K be a simplicial complex and $v_1, \dots, v_n \in K^{(0)}$. Then, $\langle v_1, \dots, v_n \rangle \in K$ if and only if $\bigcap_{i=1}^{n} O_K(v_i) \neq \emptyset$.*

Proof. If $\sigma = \langle v_1, \dots, v_n \rangle \in K$, then $\bigcap_{i=1}^{n} O_K(v_i) \supset \operatorname{rint} \sigma \neq \emptyset$. Thus, we have the "only if" part. To prove the "if" part, assume that $\bigcap_{i=1}^{n} O_K(v_i)$ contains a point x. Then, $v_1, \dots, v_n \in c_K(x)^{(0)}$. Hence, $\langle v_1, \dots, v_n \rangle \in K$. □

Proposition 4.2.4. *Let K be a simplicial complex and L a subcomplex of K. Then, $\mathcal{O}_K||L| = \mathcal{O}_L$.*

Proof. Since $O_L(v) = O_K(v) \cap |L|$ for each $v \in L^{(0)}$, we have $\mathcal{O}_L \subset \mathcal{O}_K||L|$. To prove $\mathcal{O}_K||L| \subset \mathcal{O}_L$, let $v \in K^{(0)}$ with $O_K(v) \cap |L| \neq \emptyset$. We have a simplex $\sigma \in L$ such that $O_K(v) \cap \sigma \neq \emptyset$, which means that $v \in \sigma^{(0)} \subset L^{(0)}$. Then, $O_K(v) \cap |L| = O_L(v)$. Therefore, $\mathcal{O}_K||L| = \mathcal{O}_L$. □

A cell (or simplicial) complex K is said to be **locally finite**, **locally countable**, or **locally finite-dimensional** (abbrev. **l.f.d.**) according to whether the star $\operatorname{St}(v, K)$ at every $v \in K^{(0)}$ is finite, countable, or finite-dimensional, respectively. Every locally finite cell complex is l.f.d. Note that $K[O_K(v)] = K[v]$ and $\operatorname{St}(v, K) = \bigcup_{C \in K[v]} F(C)$ for every $v \in K^{(0)}$. Then, we have the following:

Proposition 4.2.5. *A cell complex K is locally finite (or locally countable) if and only if K is locally finite (or locally countable) as a collection of subsets in the space $|K|$.* □

For compact sets in $|K|$, we have the following:

Proposition 4.2.6. *Let K be a cell complex. Every compact set $A \subset |K|$ is contained in $|L|$ for some finite subcomplex $L \subset K$. Consequently, $|K|$ is compact if and only if K is finite.*

Proof. It suffices to show that $K_A = \{D \in K \mid A \cap \operatorname{rint} D \neq \emptyset\}$ is finite. For each $D \in K_A$, take $x_D \in A \cap \operatorname{rint} D$. Since $C \cap \{x_D \mid D \in K_A\}$ is finite for each $C \in K$ by (C2′), any subset of $\{x_D \mid D \in K_A\}$ is closed in $|K|$, hence $\{x_D \mid D \in K_A\}$ is discrete in $|K|$. Since A is compact, it follows that K_A is finite. □

When two cell complexes have the same polyhedron, the following proposition holds:

Proposition 4.2.7. *For each pair of cell complexes K_1 and K_2 with $|K_1| = |K_2|$ as sets, $|K_1| = |K_2|$ as spaces if and only if each cell of K_i is covered by finitely many cells of K_{3-i} for $i = 1, 2$.*

Proof. The intersection of a cell of K_1 and a cell of K_2 is also a cell by 4.1.9(2). This intersection has the unique topology mentioned in the previous section, hence it is a subspace of both spaces $|K_1|$ and $|K_2|$. If each cell of K_i is covered by finitely many cells of K_{3-i}, then every closed set in $|K_i|$ is also closed in $|K_{3-i}|$ for $i = 1, 2$. Thus, we have proved the "if" part. The "only if" part follows from Proposition 4.2.6. □

It is said that a cell complex K' is a **subdivision** of a cell complex K, or K' **subdivides** K, if the following conditions are satisfied:

(S1) Each cell of K is covered by finitely many cells of K';
(S2) $K' \prec K$ (i.e., each $C' \in K'$ is contained in some $C \in K$).

Due to Proposition 4.2.7,

- For every subdivision K' of a cell complex K, $|K'| = |K|$ as spaces.

Evidently, if K' is a subdivision of K and K'' is a subdivision of K', then K'' is a subdivision of K. A simplicial complex K' subdividing K is called a **simplicial subdivision** and denoted as follows:

$$K' \lhd K \text{ or } K \rhd K'.$$

Lemma 4.2.8. *Let K' be a subdivision of a cell complex K. For each $C \in K'$, let $D \in K$ be the smallest cell containing C. Then,* $\operatorname{rint} C \subset \operatorname{rint} D$.

Proof. Take any $x \in \operatorname{rint} C$. Then, $C \subset D_x$ by the definition of D_x. By the minimality, $D_x = D$, which means $x \in \operatorname{rint} D$. □

It should be noted that condition (S1) can be strengthened as follows:

Proposition 4.2.9. *Let K' be a subdivision of a cell complex K. For each $C \in K$, there are finitely many $D_1, \ldots, D_k \in K'$ such that $C = \bigcup_{i=1}^k \operatorname{rint} D_i = \bigcup_{i=1}^k D_i$.*

Proof. Because of condition (S1), we can find finitely many $D_1, \ldots, D_k \in K'$ such that $C \subset \bigcup_{i=1}^k \operatorname{rint} D_i \subset \bigcup_{i=1}^k D_i$, where it can be assumed that each $\operatorname{rint} D_i$ meets C. By Lemma 4.2.8, $\operatorname{rint} D_i \subset \operatorname{rint} C_i$ for some $C_i \in K$. Then, $C \cap \operatorname{rint} C_i \neq \emptyset$, which means $C_i \leq C$ by (C2′). Thus, we have $\bigcup_{i=1}^k D_i \subset \bigcup_{i=1}^k C_i \subset C$. □

With regard to subdivisions, we have the following:

Theorem 4.2.10. *Every cell complex K has a simplicial subdivision L with the same vertices, i.e., $K^{(0)} = L^{(0)}$.*

Proof. Give an order on $K^{(0)}$ so that $C^{(0)}$ has the maximum $v_C = \max C^{(0)}$ for each $C \in K$ (e.g., a total order). Let $L_0 = K^{(0)}$ and $L_1 = K^{(1)}$. Suppose that a simplicial subdivision $L_n \lhd K^{(n)}$ has been defined so that

(1) $L_n^{(0)} = K^{(0)}$ and $L_{n-1} \subset L_n$;
(2) $v_{c_K(\hat{\sigma})} \in \sigma^{(0)}$ for each $\sigma \in L_n$,

where $c_K(\hat{\sigma})$ is the carrier of the barycenter $\hat{\sigma}$ of σ in K (note that $c_K(\hat{\sigma}) \in K^{(n)}$ because $\hat{\sigma} \in |K^{(n)}|$). Let $C \in K$ be an $(n+1)$-cell. For each $\sigma \in L_n$ with $\sigma \subset \partial C$, we have $c_K(\hat{\sigma}) \leq C$, $\sigma \subset c_K(\hat{\sigma})$ by Lemma 4.2.8 and $v_{c_K(\hat{\sigma})} \in \sigma^{(0)}$ by the assumption. If $v_C \in c_K(\hat{\sigma})^{(0)}$ then $v_C = v_{c_K(\hat{\sigma})} \in \sigma^{(0)}$. When $v_C \not\in c_K(\hat{\sigma})^{(0)}$, since $c_K(\hat{\sigma})^{(0)} = C^{(0)} \cap \operatorname{fl} c_K(\hat{\sigma})$ (cf. 4.1.6(3)), it follows that $v_C \not\in \operatorname{fl} c_K(\hat{\sigma})$, so $v_C \not\in \operatorname{fl} \sigma$. Therefore, v_C is joinable to σ, that is, we have the simplex $v_C\sigma$ in C with $\sigma < v_C\sigma$. Now, we define

$$L_{n+1} = L_n \cup \{v_C\sigma \mid C \in K^{(n+1)} \setminus K^{(n)},\ \sigma \in L_n$$
$$\text{with } \sigma \subset \partial C \text{ and } v_C \notin c_K(\hat{\sigma})^{(0)}\}.$$

It is easy to verify that L_{n+1} is a simplicial complex and $L_{n+1} \lhd K^{(n+1)}$. By definition, L_{n+1} satisfies conditions (1) and (2).

By induction, we have simplicial subdivisions $L_n \lhd K^{(n)}$, $n \in \mathbb{N}$, such that $L_n^{(0)} = K^{(0)}$ and $L_{n-1} \subset L_n$. Then, $L = \bigcup_{n\in\mathbb{N}} L_n$ is a simplicial subdivision of K with $L^{(0)} = K^{(0)}$. □

In the above proof of Theorem 4.2.10, we give an order on the set $K^{(0)}$ of vertices such that $C^{(0)}$ has the maximum for each cell $C \in K$. A cell complex K with such an order on $K^{(0)}$ is called an **ordered cell complex**. If K is a simplicial complex, $\sigma^{(0)}$ has the maximum for each simplex $\sigma \in K$ if and only if $\sigma^{(0)}$ is totally ordered for each $\sigma \in K$. Thus, an **ordered simplicial complex** is a simplicial complex K with an order on $K^{(0)}$ such that $\sigma^{(0)}$ is totally ordered for each $\sigma \in K$.

Theorem 4.2.11. *Let K_1 and K_2 be cell complexes such that $|K_1| = |K_2|$ as spaces. Then, K_1 and K_2 have a common subdivision K. In addition, if K_0 is a subcomplex of both K_1 and K_2, then K_0 is also a subcomplex of K.*

This follows from Proposition 4.2.7 and the next proposition:

Proposition 4.2.12. *For each pair of cell complexes K_1 and K_2, the following K is a cell complex with $|K| = |K_1| \cap |K_2|$ (as sets):*

$$K = \{C \cap D \mid C \in K_1,\ D \in K_2 \text{ such that } C \cap D \neq \emptyset\}.$$

If K_0 is a subcomplex of both K_1 and K_2, then K_0 is also a subcomplex of K. Moreover, if each cell of K_i is covered by only finitely many cells of K_{3-i} for $i = 1, 2$, then $|K|$ is a closed subspace of both $|K_1|$ and $|K_2|$.

Proof. For each pair $C \in K_1$ and $D \in K_2$ with $C \cap D \neq \emptyset$, $C \cap D$ is a cell and $(C \cap D)_x = C_x \cap D_x$ for each $x \in C \cap D$ by 4.1.9(2). Thus, K satisfies (C1).

We will show that K satisfies (C2″), that is, for $C, C' \in K_1$ and $D, D' \in K_2$, the following implication holds:

$$\operatorname{rint}(C \cap D) \cap \operatorname{rint}(C' \cap D') \neq \emptyset \Rightarrow C \cap D = C' \cap D'.$$

Let $x \in \operatorname{rint}(C \cap D) \cap \operatorname{rint}(C' \cap D')$. Then, $(C \cap D)_x = C \cap D$ and $(C' \cap D')_x = C' \cap D'$. On the other hand, by Proposition 3.2.5(8),

$$x \in \operatorname{rint} C_x \cap \operatorname{rint} C'_x \cap \operatorname{rint} D_x \cap \operatorname{rint} D'_x.$$

Then, it follows from (C2″) that $C_x = C'_x$ and $D_x = D'_x$. Therefore, we can apply 4.1.9(2) to obtain the following equality:

$$C \cap D = (C \cap D)_x = C_x \cap D_x = C'_x \cap D'_x = (C' \cap D')_x = C' \cap D'.$$

By the definition of K, if K_0 is a subcomplex of both K_1 and K_2 then K_0 is also a subcomplex of K. Moreover, it is evident that the inclusions $|K| \subset |K_i|$, $i = 1, 2$, are continuous. When each cell of K_i is covered by only finitely many cells of K_{3-i} for $i = 1, 2$, it is easy to see that each closed set in $|K|$ is closed in $|K_1|$ and $|K_2|$. □

Combining Theorems 4.2.11 and 4.2.10, we have the following:

Corollary 4.2.13. *Let K_1 and K_2 be simplicial complexes such that $|K_1| = |K_2|$ as spaces. Then, K_1 and K_2 have a common simplicial subdivision K. Additionally, if K_0 is a subcomplex of both K_1 and K_2, then K_0 is also a subcomplex of K.* □

Proposition 4.2.14. *Let K be a cell complex and K' be a subdivision of K. Then, every subcomplex L of K is subdivided by the subcomplex $L' = \{C' \in K' \mid C' \subset |L|\}$ of K'.*

Proof. Obviously, L' is a subcomplex of K'. For each $C' \in L'$, let $C \in K$ be the smallest cell containing C'. On the other hand, we have $D \in L$ containing some $x \in \operatorname{rint} C'$. Since $\operatorname{rint} C' \subset \operatorname{rint} C$ by Lemma 4.2.8, it follows that $D \cap \operatorname{rint} C \neq \emptyset$, which implies $C \leq D$ by (C2′), hence $C \in L$. Thus, we have $L' \prec L$.

For each $C \in L$, we have finitely many $D'_1, \ldots, D'_k \in K'$ such that $C = \bigcup_{i=1}^k D'_i$ by Proposition 4.2.9. Since $D'_i \subset |L|$, it follows that $D'_i \in L'$. Thus, C is covered by finitely many cells of L'. □

Proposition 4.2.15. *Let K' be a subdivision of a cell complex K. Then, $O_{K'}(v') \subset O_K(v)$ for each $v' \in K'^{(0)}$ and $v \in c_K(v')^{(0)}$. Consequently, $\mathcal{O}_{K'} \prec \mathcal{O}_K$.*

Proof. For each $C \in K'$ with $v' \in C^{(0)}$, there is some $D \in K$ such that $\operatorname{rint} C \subset \operatorname{rint} D$ by Lemma 4.2.8. Then, $D \cap \operatorname{rint} c_K(v') \neq \emptyset$, which implies $c_K(v') \leq D$ by (C2′), hence $v \in c_K(v')^{(0)} \subset D^{(0)}$. Thus, $\operatorname{rint} C \subset \operatorname{rint} D \subset O_K(v)$. Therefore, $O_{K'}(v') \subset O_K(v)$. □

Some Topological Properties of Polyhedra 4.2.16.
Let K be a cell complex.

(1) $|K|$ is separable if and only if K is countable.

Sketch of Proof. Each $C \in K$ has a countable dense set D_C. If K is countable, then $D = \bigcup_{C \in K} D_C$ is a countable dense set in $|K|$.
For a countable subset $A \subset |K|$, the following is countable:

$$\{C \in K \mid \exists D \in K \text{ such that } C \leq D,\ A \cap \operatorname{rint} D \neq \emptyset\}.$$

If K is uncountable, we can find a cell $C_0 \in K$ such that $A \cap \operatorname{rint} C = \emptyset$ if $C \in K$ and $C_0 \leq C$. Let $x \in \operatorname{rint} C_0$. Then, $O_K(x) \cap A = \emptyset$, which means that A is not dense in $|K|$.

(2) The following are equivalent:

(a) K is locally finite;
(b) $|K|$ is locally compact;
(c) $|K|$ is metrizable;

(d) $|K|$ is first-countable.

Sketch of Proof. Because of Proposition 4.2.6, we have (a) $\Rightarrow$ (b). The implication (c) $\Rightarrow$ (d) is obvious. Since a space is metrizable if it is a locally finite union of metrizable closed subspaces (2.4.5(2)), the implication (a) $\Rightarrow$ (c) follows from Proposition 4.2.5.[6]
(b) $\Rightarrow$ (a): Let V be a compact neighborhood of $v \in K^{(0)}$. Due to Proposition 4.2.6, $V \subset |L|$ for some finite subcomplex $L \subset K$. If $C \in K$ and $v \in C^{(0)}$ then $V \cap C$ is a neighborhood of v in C. Since $V \cap \operatorname{rint} C \neq \emptyset$, it follows that $\operatorname{rint} C$ meets some $D \in L$, which implies $C \leq D$. Hence, $\mathrm{St}(v, K) \subset L$.
(d) $\Rightarrow$ (a): Assume that $\mathrm{St}(v, K)$ is infinite for some $v \in K^{(0)}$. Then, v is a vertex of distinct 1-cells (1-simplexes) A_n, $n \in \mathbb{N}$. Let $\{U_n \mid n \in \mathbb{N}\}$ be a neighborhood basis at v in $|K|$. For each $n \in \mathbb{N}$, choose $a_n \in \operatorname{rint} A_n \cap U_n$. Then, $|K| \setminus \{a_n \mid n \in \mathbb{N}\}$ is an open neighborhood of v in $|K|$ but it does not contain any U_n, which is a contradiction.

(3) The following are equivalent:

(a) K is countable and locally finite;
(b) $|K|$ is second-countable;
(c) $|K|$ is separable and metrizable.

Sketch of Proof. Combine (1) and (2) above.

(4) In general, $w(|K|) \neq \operatorname{dens}|K|$ (i.e., $w(|K|) > \operatorname{dens}|K|$).

Sketch of Proof. Let $K^{(0)} = \{v_i \mid i \in \omega\}$ and define $K = K^{(0)} \cup \{\langle v_0, v_i\rangle \mid i \in \mathbb{N}\}$. Then, $\operatorname{dens}|K| = \aleph_0$ by (1) above. However, $|K|$ is not first countable by (2), hence it is not second countable, that is, $w(|K|) > \aleph_0$.

(5) Let $f : X \to |K|$ be a map of a metrizable (more generally, first countable) space X. Then, each $x \in X$ has a neighborhood U_x in X such that $f(U_x) \subset |K_x|$ for some finite subcomplex K_x of K.

Sketch of Proof. If $x \in X$ does not have such a neighborhood, then we can find a sequence $(x_i)_{i\in\mathbb{N}}$ in X such that $\lim_{i\to\infty} x_i = x$ and $c_K(x_i) \neq c_K(x_j)$ if $i \neq j$. Because $\{f(x), f(x_i) \mid i \in \mathbb{N}\} \subset |K|$ is compact, this contradicts Proposition 4.2.6.

A **polyhedron** (or a **topological polyhedron**) is defined as a space P such that $P = |K|$ (or $P \approx |K|$) for some cell complex K. A subspace Q of a polyhedron (or a topological polyhedron) P is called a **subpolyhedron** of P if there exists a pair (K, L) of a cell complex and a subcomplex such that $P = |K|$ and $Q = |L|$ (or $(P, Q) \approx (|K|, |L|)$). Every subpolyhedron of P is closed in P according to Proposition 4.2.1. It follows from 4.2.16(2) that a (topological) polyhedron is metrizable if and only if it is locally compact. In general, for a (topological) polyhedron P, $w(P) \neq \operatorname{dens} P$ (cf. 4.2.16(4)). A **triangulation** of a polyhedron (or a topological polyhedron) P is a simplicial complex K such that $|K| = P$ (or $|K| \approx P$). Then, it is also said that P is **triangulated** by K or K **triangulates** P. According to Theorem 4.2.10, every (topological) polyhedron has a triangulation.

---

[6]This can be shown as follows: If K is a simplicial complex, (a) $\Rightarrow$ (c) will be proved in Theorem 4.5.6. Due to Theorem 4.2.10, every cell complex has a simplicial subdivision. Evidently, every subdivision of a locally finite cell complex is also locally finite. Thus, (a) $\Rightarrow$ (c) is also valid for every cell complex.

4.3 Product Complexes and Homotopy Extension

For two cell complexes K (in E) and L (in F), by virtue of 4.1.9(4), the **product cell complex** can be defined as the following cell complex (in $E \times F$):

$$K \times_c L = \{C \times D \mid C \in K,\ D \in L\}.$$

Note the following facts:

- $|K \times_c L| = |K| \times |L|$ as sets;
- $\mathrm{St}(C \times D, K \times_c L) = \mathrm{St}(C, K) \times_c \mathrm{St}(D, L)$ for each $C \in K$ and $D \in L$ (cf. 4.1.9(4));
- $c_{K \times_c L}(x, y) = c_K(x) \times c_L(y)$ for each $(x, y) \in |K| \times |L|$ (cf. 4.1.9(4));
- $O_{K \times_c L}(x, y) = O_K(x) \times O_L(y)$ as sets for each $(x, y) \in |K| \times |L|$.

The projections $\mathrm{pr}_1 : |K \times_c L| \to |K|$ and $\mathrm{pr}_2 : |K \times_c L| \to |L|$ are continuous, which means that the identity $\mathrm{id} : |K \times_c L| \to |K| \times |L|$ is continuous. When K and L are finite, $|K \times_c L| = |K| \times |L|$ as spaces because $|K \times_c L|$ is compact by Proposition 4.2.6. More generally, we can prove the following:

Theorem 4.3.1. *For each pair of cell complexes K and L, $|K \times_c L| = |K| \times |L|$ as spaces if (1) both K and L are locally countable or (2) one of K or L is locally finite.*

Proof. Since $\mathrm{id} : |K \times_c L| \to |K| \times |L|$ is continuous, it suffices to show the continuity of $\mathrm{id} : |K| \times |L| \to |K \times_c L|$ at each $(x, y) \in |K| \times |L|$. Choose $(v, u) \in K^{(0)} \times L^{(0)}$ so that $(x, y) \in O_K(v) \times O_L(u)$. Then,

$$O_{K \times_c L}(v, u) = O_K(v) \times O_L(u)$$

is an open neighborhood of (x, y) in both $|K \times_c L|$ and $|K| \times |L|$. Replacing K and L by $\mathrm{St}(v, K)$ and $\mathrm{St}(u, L)$, case (1) reduces to the case where K and L are countable and case (2) reduces to the case where L is finite.

Case (1): As noted above, we may assume that both K and L are countable. Then, K and L have towers $K_1 \subset K_2 \subset \cdots$ and $L_1 \subset L_2 \subset \cdots$ of finite subcomplexes such that $x \in |K_1|$, $y \in |L_1|$, $K = \bigcup_{n \in \mathbb{N}} K_n$, and $L = \bigcup_{n \in \mathbb{N}} L_n$. Observe that $|K|$, $|L|$, and $|K \times_c L|$ have the weak topologies determined by $\{|K_n| \mid n \in \mathbb{N}\}$, $\{|L_n| \mid n \in \mathbb{N}\}$, and $\{|K_n \times_c L_n| \mid n \in \mathbb{N}\}$, respectively. For each $n \in \mathbb{N}$, $|K_n \times_c L_n| = |K_n| \times |L_n|$ as spaces because K_n and L_n are finite. Then, for each neighborhood W of (x, y) in $|K \times_c L|$, we can inductively choose open neighborhoods U_n of x in $|K_n|$ and V_n of y in $|L_n|$ so that

$$\mathrm{cl}\, U_{n-1} \times \mathrm{cl}\, V_{n-1} \subset U_n \times V_n \subset \mathrm{cl}\, U_n \times \mathrm{cl}\, V_n \subset W \cap |K_n \times_c L_n|,$$

where $\mathrm{cl}\, U_n$ and $\mathrm{cl}\, V_n$ are compact. Observe that $U = \bigcup_{n\in\mathbb{N}} U_n$ and $V = \bigcup_{n\in\mathbb{N}} V_n$ are open neighborhoods of x and y in $|K|$ and $|L|$, respectively, and $U \times V \subset W$. Therefore, W is a neighborhood of (x, y) in $|K| \times |L|$.

Case (2): As noted above, it can be assumed that L is finite. Let W be a neighborhood of (x, y) in $|K \times_c L|$. For each $D \in L$, $W \cap (\{x\} \times D)$ is a neighborhood of (x, y) in $\{x\} \times D$ because $W \cap (c_K(x) \times D)$ is also in $c_K(x) \times D$. Thus, $W \cap (\{x\} \times |L|)$ is a neighborhood of (x, y) in $\{x\} \times |L|$. We can choose a neighborhood V of y in $|L|$ so that $\{x\} \times \mathrm{cl}\, V \subset W$, where $\mathrm{cl}\, V$ is compact because so is $|L|$. By induction on $n \in \omega$, for each n-cell $C \in K[x]$, we can choose an open neighborhood U_C of x in C so that $\mathrm{cl}\, U_C \times \mathrm{cl}\, V \subset W$ and $U_C \cap D = U_D$ for each $D \in F(\partial C)[x]$. In fact, $\bigcup_{D\in F(\partial C)[x]} \mathrm{cl}\, U_D \subset C$ is compact because $F(\partial C)[x]$ is finite. Then, we can find an open set U'_C in C such that

$$\bigcup_{D\in F(\partial C)[x]} \mathrm{cl}\, U_D \times \mathrm{cl}\, V \subset U'_C \times \mathrm{cl}\, V \subset \mathrm{cl}\, U'_C \times \mathrm{cl}\, V \subset W.$$

Therefore, $U_C = U'_C \setminus \bigcup_{D\in F(\partial C)[x]}(D \setminus U_D)$ is the desired neighborhood. Now, let $U = \bigcup_{C\in K[x]} U_C$. Then, U is a neighborhood of x in $\mathrm{st}(x, K)$ and $U \times V \subset W$. Since $\mathrm{st}(x, K)$ is a neighborhood of x in $|K|$, U is also a neighborhood of x in $|K|$. Hence, W is a neighborhood of (x, y) in $|K| \times |L|$. □

We denote $I = F(\mathbf{I})$ $(= \{0, 1, \mathbf{I}\})$, which is the cell complex with $|I| = \mathbf{I}$. It follows from Theorem 4.3.1 that $|K| \times \mathbf{I} = |K \times_c I|$ as spaces for every cell complex K. Due to the following proposition, the conditions given by Theorem 4.3.1 are essential.

Proposition 4.3.2. *There exist* 1*-dimensional cell complexes* K *and* L *with* $\mathrm{card}\, K^{(0)} = 2^{\aleph_0}$ *and* $\mathrm{card}\, L^{(0)} = \aleph_0$ *such that* K *is not locally countable,* L *is not locally finite, and* $|K \times_c L| \neq |K| \times |L|$ *as spaces.*

Proof. We define K and L in the linear spaces $\mathbb{R}^{\mathbb{N}^\mathbb{N}}$ and $\mathbb{R}^\mathbb{N}$ as follows:

$$K = \{\mathbf{0}, \mathbf{e}_a, \langle \mathbf{0}, \mathbf{e}_a\rangle \mid a \in \mathbb{N}^\mathbb{N}\} \text{ and } L = \{\mathbf{0}, \mathbf{e}_i, \langle \mathbf{0}, \mathbf{e}_i\rangle \mid i \in \mathbb{N}\},$$

where $\mathbf{e}_a \in \mathbb{R}^{\mathbb{N}^\mathbb{N}}$ and $\mathbf{e}_i \in \mathbb{R}^\mathbb{N}$ are the unit vectors (i.e., $\mathbf{e}_\gamma(\gamma) = 1$ and $\mathbf{e}_\gamma(\gamma') = 0$ for $\gamma' \neq \gamma$). For each $a \in \mathbb{N}^\mathbb{N}$ and $i \in \mathbb{N}$, let

$$y_{a,i} = a(i)^{-1}(\mathbf{e}_a, \mathbf{e}_i) \in \langle \mathbf{0}, \mathbf{e}_a\rangle \times \langle \mathbf{0}, \mathbf{e}_i\rangle \subset \mathbb{R}^{\mathbb{N}^\mathbb{N}} \times \mathbb{R}^\mathbb{N}.$$

Then, $Y = \{y_{a,i} \mid a \in \mathbb{N}^\mathbb{N}, i \in \mathbb{N}\}$ is closed in $|K \times_c L|$, where it should be noted that $(0, 0) \notin Y$.

To see that Y is not closed in $|K| \times |L|$, we show that $(\mathbf{0}, \mathbf{0}) \in \mathrm{cl}_{|K|\times|L|}\, Y$. Let U be a neighborhood of $\mathbf{0}$ in $|K|$ and V a neighborhood of $\mathbf{0}$ in $|L|$. For each $a \in \mathbb{N}^\mathbb{N}$, choose $\varepsilon_a > 0$ so that $[0, \varepsilon_a]\mathbf{e}_a \subset U$. For each $i \in \mathbb{N}$, choose $\delta_i > 0$ so that $[0, \delta_i]\mathbf{e}_i \subset V$. Then, we have $a_0 \in \mathbb{N}^\mathbb{N}$ such that $a_0(i)^{-1} < \min\{\delta_i, i^{-1}\}$ for

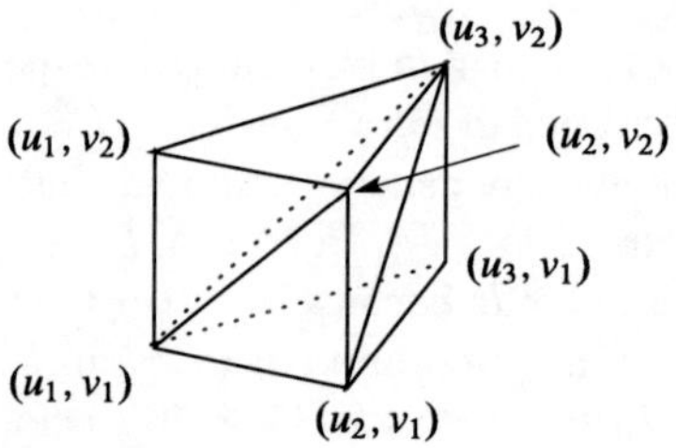

Fig. 4.1 A product simplicial complex

each $i \in \mathbb{N}$. Choose $i_0 \in \mathbb{N}$ so that $i_0^{-1} < \varepsilon_{a_0}$. Since $a_0(i_0)^{-1} < i_0^{-1} < \varepsilon_{a_0}$ and $a_0(i_0)^{-1} < \delta_{i_0}$, it follows that $y_{a_0,i_0} \in U \times V$, hence $(U \times V) \cap Y \neq \emptyset$. Thus, we have $(\mathbf{0}, \mathbf{0}) \in \mathrm{cl}_{|K|\times|L|} Y$. □

For simplicial complexes K and L, the product complex $K \times_c L$ is not simplicial but does have a simplicial subdivision with the same vertices by Theorem 4.2.10. From the proof of 4.2.10, such a simplicial subdivision of $K \times_c L$ can be obtained by giving an order on $(K \times_c L)^{(0)} = K^{(0)} \times L^{(0)}$ so that $K \times_c L$ is an ordered complex, that is, the set of vertices of each cell has the maximum. If K and L are ordered simplicial complexes, we can define an order on $K^{(0)} \times L^{(0)}$ as follows:

$$(u, v) \leq (u', v') \text{ if } u \leq u' \text{ and } v \leq v'.$$

By this order, $K \times_c L$ is an ordered cell complex. By $K \times_s L$, we denote the simplicial subdivision of $K \times_c L$ defined by using this order and call it the **product simplicial complex** of K and L (Fig. 4.1). In fact, $K \times_s L$ can be written as follows:

$$K \times_s L = \{\langle (u_1, v_1), \ldots, (u_k, v_k)\rangle \mid \exists \sigma \in K, \exists \tau \in L \text{ such that } u_1 \leq \cdots \leq u_k \in \sigma^{(0)},\ v_1 \leq \cdots \leq v_k \in \tau^{(0)}\}.$$

The above is the simplicial subdivision of $K \times_c L$ obtained by the procedure in the proof of Theorem 4.2.10. To show this, it suffices to verify that the simplicial subdivision of the n-skeleton $(K \times_c L)^{(n)}$ defined by this procedure can be written as follows:

$$\{\langle (u_1, v_1), \ldots, (u_k, v_k)\rangle \mid \exists \sigma \times \tau \in (K \times_c L)^{(n)} \text{ such that } u_1 \leq \cdots \leq u_k \in \sigma^{(0)},\ v_1 \leq \cdots \leq v_k \in \tau^{(0)}\}.$$

This can be proved by induction. According to the proof of Theorem 4.2.10, the simplicial subdivision of $(K \times_c L)^{(n+1)}$ is defined as the simplicial complex consisting of the simplexes $\langle (u_1, v_1), \ldots, (u_k, v_k), (u_{k+1}, v_{k+1})\rangle$, where $u_1 \leq \cdots \leq u_k \in \sigma_0^{(0)}$, $v_1 \leq \cdots \leq v_k \in \tau_0^{(0)}$ for some $\sigma_0 \times \tau_0 \in (K \times_c L)^{(n)}$ and (u_{k+1}, v_{k+1}) is the maximum vertex of the carrier $\sigma \times \tau \in K \times_c L$ of the barycenter of $\langle (u_1, v_1), \ldots, (u_k, v_k)\rangle$, where $\sigma \times \tau \in (K \times_c L)^{(n+1)}$. Since $\sigma \cap \mathrm{rint}\, \sigma_0 \neq \emptyset$ and $\tau \cap \mathrm{rint}\, \tau_0 \neq \emptyset$, we have $\sigma_0 \leq \sigma$ and $\tau_0 \leq \tau$, hence $u_1 \leq \cdots \leq u_k \leq u_{k+1} \in \sigma^{(0)}$ and $v_1 \leq \cdots \leq v_k \leq v_{k+1} \in \tau^{(0)}$.

Conversely, consider the simplex $\langle(u_1, v_1), \ldots, (u_k, v_k)\rangle$, where $u_1 \leq \cdots \leq u_k \in \sigma^{(0)}$, $v_1 \leq \cdots \leq v_k \in \tau^{(0)}$ for some $\sigma \times \tau \in (K \times_c L)^{(n+1)}$. We may assume that $k > 1$ and $(u_k, v_k) \neq (u_{k-1}, v_{k-1})$. Let σ' be the face of σ with the vertices $u_1, \ldots, u_{k-1}$ and τ' the face of τ with the vertices $v_1, \ldots, v_{k-1}$. Then, $\sigma' \times \tau' \in (K \times_c L)^{(n)}$, hence $\langle(u_1, v_1), \ldots, (u_{k-1}, v_{k-1})\rangle$ is a simplex of the simplicial subdivision of $(K \times_c L)^{(n)}$ by the inductive assumption. Since the barycenter of $\langle(u_1, v_1), \ldots, (u_{k-1}, v_{k-1})\rangle$ is contained in the cell $\sigma \times \tau$, $\langle(u_1, v_1), \ldots, (u_k, v_k)\rangle$ is a simplex of the simplicial subdivision of $(K \times_c L)^{(n+1)}$.

We now consider the following useful theorem:

Theorem 4.3.3 (HOMOTOPY EXTENSION THEOREM). *Let L be a subcomplex of a cell complex K and $h : |L| \times \mathbf{I} \to X$ a homotopy into an arbitrary space X. If h_0 extends to a map $f : |K| \to X$, then h extends to a homotopy $\bar{h} : |K| \times \mathbf{I} \to X$ with $\bar{h}_0 = f$. Moreover, if h is a $\mathcal{U}$-homotopy for an open cover $\mathcal{U}$ of X, then $\bar{h}$ can be taken as a $\mathcal{U}$-homotopy.*

Proof. Let $h : |L| \times \mathbf{I} \to X$ be a $\mathcal{U}$-homotopy such that h_0 extends to a map $f : |K| \to X$. For each $n \in \omega$, we define

$$K_n = L \cup K^{(n)} \text{ and } P_n = (K \times_c \{0\}) \cup (K_n \times_c I).$$

Then, K_n and P_n are subcomplexes of K and $K \times_c I$, respectively. Moreover, $|P_n|$ is a closed subspace of $|K| \times \mathbf{I}$ that contains $|L| \times \mathbf{I}$ as a subspace (cf. Theorem 4.3.1). Moreover, $|K| \times \mathbf{I} = \bigcup_{n \in \omega} |P_n|$ has the weak topology with respect to the tower $|P_0| \subset |P_1| \subset |P_2| \subset \cdots$.

We can define the map $g_0 : |P_0| \to X$ as follows:

$$g_0(x, t) = \begin{cases} h(x, t) & \text{for } (x, t) \in |L| \times \mathbf{I}, \\ f(x) & \text{for } (x, t) \in (|K| \times \{0\}) \cup (|K^{(0)}| \setminus |L|) \times \mathbf{I}. \end{cases}$$

It is obvious that $g_0||K_0| \times \mathbf{I}$ is a $\mathcal{U}$-homotopy. Assume that we have maps $g_i : |P_i| \to X, i < n$, such that $g_i||P_{i-1}| = g_{i-1}$ and $g_i||K_i| \times \mathbf{I}$ is a $\mathcal{U}$-homotopy. Let $C \in K \setminus L$ be an n-cell. By taking $v_C \in \operatorname{rint} C$, each $x \in C$ can be written as

$$x = (1 - s)y + sv_C, \ y \in \partial C, \ s \in \mathbf{I}.$$

We can choose $0 < \delta_C < 1$ so that $\{g_{n-1}(C(y)) \mid y \in \partial C\} \prec \mathcal{U}$, where

$$C(y) = \{y\} \times \mathbf{I} \cup \{((1 - s)y + sv_C, \ 0) \mid 0 \leq s \leq 2\delta_C\}.$$

See Fig. 4.2. We can define a map

$$r_C : C \times \mathbf{I} \to C \times \{0\} \cup \partial C \times \mathbf{I} \subset |P_{n-1}|$$

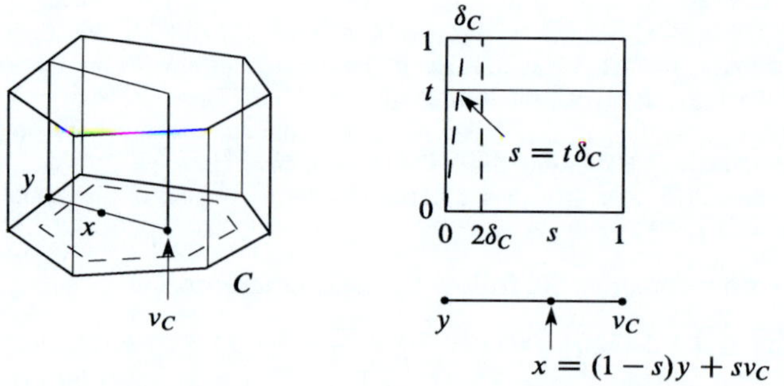

Fig. 4.2 $C \times \mathbf{I}$

as follows:

$$r_C((1-s)y + sv_C,\ t)$$
$$= \begin{cases} ((1-s)y + sv_C,\ 0) & \text{if } 2\delta_C \leq s \leq 1, \\ \left(\left(1 - \dfrac{2(s - t\delta_C)}{2-t}\right)y + \dfrac{2(s-t\delta_C)}{2-t}v_C,\ 0\right) & \text{if } t\delta_C \leq s \leq 2\delta_C, \\ (y,\ t - \delta_C^{-1}s) & \text{if } 0 \leq s \leq t\delta_C. \end{cases}$$

For each $x \in C$, $r_C(\{x\} \times \mathbf{I}) = \{(x, 0)\}$ or $r_C(\{x\} \times \mathbf{I}) \subset C(y)$, where $y \in \partial C$ with $x \in \langle y, v_C \rangle$. It follows that

$$\{g_{n-1}(r_C(\{x\} \times \mathbf{I})) \mid x \in C\} \prec \mathcal{U}.$$

Observe that $r_C|C \times \{0\} \cup \partial C \times \mathbf{I} = \mathrm{id}$. We can extend g_{n-1} to a map $g_n : |P_n| \to X$ defined by $g_n|C \times \mathbf{I} = g_{n-1}r_C$ for each n-cell $C \in K \setminus L$. Then, $g_n||K_n| \times \mathbf{I}$ is a $\mathcal{U}$-homotopy.

By induction, we can obtain maps $g_n : |P_n| \to X$, $n \in \omega$, such that $g_n||P_{n-1}| = g_{n-1}$ and $g_n||K_n| \times \mathbf{I}$ is a $\mathcal{U}$-homotopy. The desired $\mathcal{U}$-homotopy $\bar{h} : |K| \times \mathbf{I} \to X$ can be defined by $\bar{h}||P_n| = g_n$. □

For a cell complex L, two maps $f, g : X \to |L|$ are said to be **contiguous** (with respect to L) if f, g are L-close, that is, for each $x \in X$, there is some $C \in L$ such that $f(x), g(x) \in C$.

Proposition 4.3.4. *Let K and L be cell complexes. If two maps $f, g : |K| \to |L|$ are contiguous (with respect to L) then $f \simeq_L g$ by the homotopy $h : |K| \times \mathbf{I} \to |L|$ defined as follows:*

$$h(x,t) = (1-t)f(x) + tg(x) \ \textit{for each}\ (x,t) \in |K| \times \mathbf{I}.$$

Proof. Because f and g are contiguous, h is well-defined. We need to prove the continuity of h. Since $|K| \times \mathbf{I} = |K \times_c I|$ (Theorem 4.3.1), it suffices to show that $h|C \times \mathbf{I}$ is continuous for each $C \in K$. According to Proposition 4.2.6, $f(C) \cup g(C) \subset |L_0|$ for some finite subcomplex L_0 of L. For each pair $(D_1, D_2) \in L^2$, let $L(D_1, D_2)$ be the subcomplex of L consisting of faces of the minimal cell of L containing $D_1 \cup D_2$, where $L(D_1, D_2) = \emptyset$ if L has no cells containing $D_1 \cup D_2$. Then, $L_1 = \bigcup_{(D_1,D_2)\in L_0^2} L(D_1, D_2)$ is a finite subcomplex of L with $L_0 \subset L_1$ and $h(C \times \mathbf{I}) \subset |L_1|$. Note that the flat hull $\mathrm{fl}\,|L_1|$ is finite-dimensional. Due to Proposition 3.5.1, $\mathrm{fl}\,|L_1|$ has the unique topology for which the following operation is continuous:

$$\mathrm{fl}\,|L_1| \times \mathrm{fl}\,|L_1| \times \mathbf{I} \ni (x, y, t) \mapsto (1-t)x + ty \in \mathrm{fl}\,|L_1|.$$

The topology of $|L_1|$ is equal to the relative topology with respect to this topology. Indeed, due to Proposition 3.5.8, each $D \in L_1$ has the unique topology for which the operation $(x, y, t) \mapsto (1-t)x+ty$ is continuous. Hence, the inclusion $|L_1| \subset \mathrm{fl}\,|L_1|$ is continuous. By the compactness of $|L_1|$, this inclusion is a closed embedding. Then, it follows that $h|C \times \mathbf{I} : C \times \mathbf{I} \to |L_1|$ is continuous. □

In Proposition 4.3.4, h is called the **straight-line homotopy**.

Remark 2. Using the same arguments, we can prove that Proposition 4.3.4 is valid even if $|K|$ is replaced by a locally compact space X.[7] It will be shown in Sect. 4.9 that every two contiguous maps defined on an arbitrary space are homotopic, where the homotopy is not always given by the straight-line homotopy. In fact, there are some cases where the straight-line homotopy is not continuous. Such an example can be obtained by reforming the example given in Proposition 4.3.2.

Let $\Gamma = \mathbb{N}^{\mathbb{N}} \cup \mathbb{N}$. We define $K = \{0, \mathbf{e}_\gamma, \langle 0, \mathbf{e}_\gamma \rangle \mid \gamma \in \Gamma\}$ and

$$L = K \cup \left\{\langle \mathbf{e}_\gamma, \mathbf{e}_{\gamma'} \rangle,\ \langle 0, \mathbf{e}_\gamma, \mathbf{e}_{\gamma'} \rangle \;\middle|\; \gamma \neq \gamma' \in \Gamma\right\},$$

where $\mathbf{e}_\gamma \in \mathbb{R}^\Gamma$ is the unit vector in $\mathbb{R}^\Gamma$ (i.e., $\mathbf{e}_\gamma(\gamma) = 1$ and $\mathbf{e}_\gamma(\gamma') = 0$ for $\gamma' \neq \gamma$). Then, $K \subset L$ are cell complexes in $\mathbb{R}^\Gamma$ with $\dim K = 1$ and $\dim L = 2$. Let $f, g : |K|^2 \to |L|$ be maps defined by $f(x, y) = x$ and $g(x, y) = y$ for each $(x, y) \in |K|^2$, where $|K|^2 \neq |K \times_c K|$ as spaces (see the proof of Proposition 4.3.2). Evidently, these maps f, g are contiguous. We can define $h : |K|^2 \times \mathbf{I} \to |L|$ as follows:

$$h(x, y, t) = (1-t)f(x, y) + tg(x, y) = (1-t)x + ty.$$

We will prove that h is not continuous at $(0, 0, \frac{1}{2}) \in |K|^2 \times \mathbf{I}$. For each $a \in \mathbb{N}^{\mathbb{N}}$ and $i \in \mathbb{N}$, let

$$v_{a,i} = \frac{1}{2}a(i)^{-1}\mathbf{e}_a + \frac{1}{2}a(i)^{-1}\mathbf{e}_i \in \langle 0, \mathbf{e}_a, \mathbf{e}_i \rangle \subset |L|.$$

[7] More generally, it can be replaced by a k-space X. Indeed, to show the continuity of the straight-line homotopy h, it suffices to prove the continuity of $h|C \times \mathbf{I}$ for every compact set C in X.

As is easily observed, $U = |L| \setminus \{v_{a,i} \mid a \in \mathbb{N}^{\mathbb{N}}, i \in \mathbb{N}\}$ is an open neighborhood of $0 = h(0,0,\frac{1}{2})$ in $|L|$. Then, $h(W^2 \times \{\frac{1}{2}\}) \not\subset U$ for any neighborhood W of 0 in $|K|$. Indeed, for each $\gamma \in \Gamma$, choose $\delta_\gamma > 0$ so that $[0, \delta_\gamma]\mathbf{e}_\gamma \subset W$. We have $a_0 \in \mathbb{N}^{\mathbb{N}}$ such that $a_0(i) > \max\{i, \delta_i^{-1}\}$ for every $i \in \mathbb{N}$. Take $i_0 \in \mathbb{N}$ so that $i_0 > \delta_{a_0}^{-1}$. Since $a_0(i_0)^{-1} < i_0^{-1} < \delta_{a_0}$, we have $x_0 = a_0(i_0)^{-1}\mathbf{e}_{a_0} \in W$. On the other hand, since $a_0(i_0)^{-1} < \delta_{i_0}$, we have $y_0 = a_0(i_0)^{-1}\mathbf{e}_{i_0} \in W$. Then, it follows that

$$h\left(x_0, y_0, \frac{1}{2}\right) = \frac{1}{2}a_0(i_0)^{-1}\mathbf{e}_{a_0} + \frac{1}{2}a_0(i_0)^{-1}\mathbf{e}_{i_0} = v_{a_{i_0}, i_0} \notin U.$$

Therefore, h is not continuous at $(0,0,\frac{1}{2})$.

4.4 PL Maps and Simplicial Maps

Let K and L be cell complexes (in E and F, respectively). A map $f : |K| \to |L|$ is said to be **piecewise linear (PL)** if there is a subdivision K' of K such that f s affine on each cell $C \in K'$, i.e., $f|C : C \to F$ is affine. The continuity of f follows from Corollary 3.5.4. Then, the graph $G(f) = \{(x, f(x)) \mid x \in |K|\}$ of f is a closed subspace of the product space $|K| \times |L|$, and $\mathrm{pr}_{|K|}|G(f) : G(f) \to |K|$ is a homeomorphism whose inverse is the natural injection $i_f : |K| \to G(f)$ defined by $i_f(x) = (x, f(x))$ for each $x \in |K|$. Here, in general, $|K| \times |L| \neq |K \times_c L|$ as spaces (Proposition 4.3.2), but we have the following:

Lemma 4.4.1. *The topology of the graph $G(f)$ of each PL map $f : |K| \to |L|$ is equal to the one inherited from $|K \times_c L|$.*

Proof. Because $\mathrm{id} : |K \times_c L| \to |K| \times |L|$ is continuous, it suffices to show that $A \subset G(f)$ is closed in $|K| \times |L|$ if $A \cap (C \times D)$ is closed in $C \times D$ for each $C \in K$ and $D \in L$. Recall that $\mathrm{pr}_{|K|}|G(f) : G(f) \to |K|$ is a homeomorphism. Thus, we may show that $\mathrm{pr}_{|K|}(A)$ is closed in $|K|$. For each $C \in K$, since $f(C)$ is compact, we have $D_1, \dots, D_k \in L$ such that $f(C) \subset \bigcup_{i=1}^k D_i$. Since $\mathrm{pr}_{|K|}^{-1}(C) \cap G(f) = (C \times f(C)) \cap G(f)$, we have

$$A \cap \mathrm{pr}_{|K|}^{-1}(C) = A \cap (C \times f(C)) = \bigcup_{i=1}^{k} A \cap (C \times D_i).$$

Since each $A \cap (C \times D_i)$ is compact as a closed subset of $C \times D_i$, it follows that $A \cap \mathrm{pr}_{|K|}^{-1}(C)$ is compact. Hence, $\mathrm{pr}_{|K|}(A) \cap C$ is also compact, which implies that $\mathrm{pr}_{|K|}(A) \cap C$ is closed in C. Therefore, $\mathrm{pr}_{|K|}(A)$ is closed in $|K|$. □

We have the following characterization of PL maps:

Theorem 4.4.2. *Let K and L be cell complexes. A map $f : |K| \to |L|$ is PL if and only if the graph $G(f)$ of f is a polyhedron.*

Proof. To prove the "only if" part, let $i_f : |K| \to G(f)$ be the natural injection and K' be a subdivision of K such that f is affine on each cell in K'. Then, $i_f(K') = \{i_f(C) \mid C \in K'\}$ is a cell complex (cf. 4.1.9(3)). Note that $|i_f(K')| = G(f)$ as sets. For each $A \subset G(f)$,

$$\begin{aligned} A \text{ is closed in } G(f) &\Leftrightarrow \mathrm{pr}_{|K|}(A) \text{ is closed in } |K| = |K'| \\ &\Leftrightarrow \forall C \in K',\ \mathrm{pr}_{|K|}(A) \cap C \text{ is closed in } C \\ &\Leftrightarrow \forall C \in K',\ A \cap i_f(C) \text{ is closed in } i_f(C) \\ &\Leftrightarrow A \text{ is closed in } |i_f(K')|. \end{aligned}$$

Therefore, $|i_f(K')| = G(f)$ as spaces.

To prove the "if" part, let M be a cell complex with $|M| = G(f)$. Then, $\mathrm{pr}_{|K|}(M) = \{\mathrm{pr}_{|K|}(D) \mid D \in M\}$ is a cell complex (cf. 4.1.9(3)). Since $\mathrm{pr}_{|K|}|G(f) : G(f) = |M| \to |\mathrm{pr}_{|K|}(M)|$ is a homeomorphism, it follows that $|\mathrm{pr}_{|K|}(M)| = |K|$ as spaces. By Theorem 4.2.11, K and $\mathrm{pr}_{|K|}(M)$ have a common subdivision K'. For each $C \in K'$, $i_f(C)$ is a cell contained in some cell in M. Note that $\mathrm{pr}_{|K|}|i_f(C)$ is an affine homeomorphism, hence so is $i_f|C$. Since $\mathrm{pr}_{|L|}$ is affine on each cell in M, $f = \mathrm{pr}_{|L|} \circ i_f$ is affine on each cell in K', that is, f is PL. □

Lemma 4.4.3. *For every PL map $f : |K| \to |L|$, K has a subdivision K' such that f is affine on each cell $C \in K'$ and $f(K') \prec L$ (i.e., for each cell $C \in K'$, $f(C)$ is contained in some cell in L).*

Proof. By replacing K with a subdivision, we may assume that $f|C$ is affine for each $C \in K$. According to Theorem 4.4.2, there is a cell complex M such that $|M| = G(f)$, the graph of f. By Proposition 4.2.12, the following is a cell complex:

$$M' = \{C \cap D \mid C \in M,\ D \in K \times_c L \ \text{ such that } \ C \cap D \neq \emptyset\} \prec M.$$

Each $C \in M$ is covered by finitely many cells of $K \times_c L$ because it is compact in $|K \times_c L|$ by Lemma 4.4.1. Then, each cell of M is covered by finitely many cells of M'. Therefore, M' is a subdivision of M. We apply Theorem 4.2.11 to obtain a common subdivision K' of K and $\mathrm{pr}_{|K|}(M')$. Observe that $\mathrm{pr}_{|L|}(M') \prec L$ and $f = \mathrm{pr}_{|L|} \circ i_f$, where $i_f : |K| \to G(f)$ is the natural injection. Then, we have $f(K') \prec \mathrm{pr}_{|L|}(M') \prec L$. □

Using Lemma 4.4.3, we can easily prove the following:

Proposition 4.4.4. *The composition of PL maps is also PL.* □

Remark 3. In Proposition 4.3.4, if f and g are PL and $h : |K| \times \mathbf{I} \to |L|$ is the straight-line homotopy from f to g, then each $h_t : |K| \to |L|$ is PL. But, in general, $h : |K \times_c I| \to |L|$ is not PL. In fact, by Theorem 4.2.11, K has a subdivision K' such that both $f|C$ and $g|C$ are affine for each $C \in K'$. Then, $h_t|C$ is affine by

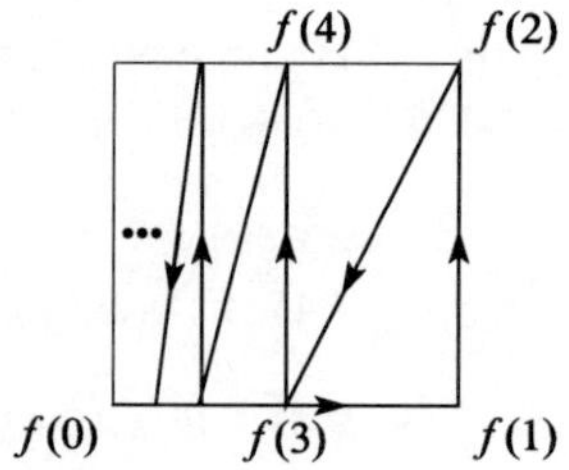

Fig. 4.3 The image of the PL map f

definition. As an example of the straight-line homotopy h being non-PL, consider the affine maps $f, g : \mathbf{I} \to \mathbf{I}^2$ defined by $f(s) = (s, 0)$ and $g(s) = (0, s)$. In this case, the straight-line homotopy h is defined by $h(s,t) = ((1-t)s, ts)$. Note that

$$h((1-t)(0,0) + t(1,1)) = h(t,t) = (t - t^2, t^2) \text{ for each } t \in \mathbf{I}.$$

Any cell complex K with $|K| = \mathbf{I}^2$ has a cell C such that $A = C \cap \langle (0,0), (1,1) \rangle$ is a non-degenerate line segment. Then, $h|C$ is not affine.

Remark 4. It should be remarked that the image of a PL map is, in general, not a polyhedron. In fact, let $f : \mathbb{R}_+ \to \mathbf{I}^2$ be the PL map defined as follows:

$$f(t) = \begin{cases} (t, 0) & \text{if } t \in \mathbf{I} = [0,1], \\ (2^{-i+1}, t - 2i + 1) & \text{if } t \in [2i-1, 2i], \\ (2^{-i}(2i + 2 - t), 2i + 1 - t) & \text{if } t \in [2i, 2i+1]. \end{cases}$$

Then, $f(\mathbb{R}_+)$ is not a polyhedron. Indeed, if $f(\mathbb{R}_+) = |L|$ for a cell complex L, then $f(2n-1)$, $n \in \omega$, should be vertices of L, which are contained in the compact set $f(\mathbf{I})$ (Fig. 4.3).

Let K and L be simplicial complexes. A function $f : |K| \to |L|$ is called a **simplicial map** from K to L (or with respect to K and L) if $f|\sigma$ is affine and $f(\sigma) \in L$ for each $\sigma \in K$, where $\dim f(\sigma) \leq \dim \sigma$. Evidently, $f(K^{(0)}) \subset L^{(0)}$ and $f(K) = \{f(\sigma) \mid \sigma \in K\}$ is a subcomplex of L. When $\sigma = \langle v_1, \ldots, v_n \rangle \in K$, we have $f(\sigma) = \langle f(v_1), \ldots, f(v_n) \rangle \in L$ and

$$f\left(\textstyle\sum_{i=1}^n t_i v_i\right) = \sum_{i=1}^n t_i f(v_i) \text{ for each } t_i \geq 0 \text{ with } \sum_{i=1}^n t_i = 1,$$

where it is possible that $f(v_i) = f(v_j)$ for some $i \neq j$. Every simplicial map $f : |K| \to |L|$ is PL, so it is continuous (Corollary 3.5.4). For a simplicial map from K to L, we may write $f : K \to L$. In fact, although it is actually a function from $|K|$ to $|L|$, f induces a function from K to L because $f(\sigma) \in L$ for each $\sigma \in K$. Note that the composition of simplicial maps and the restriction of a simplicial map to a subcomplex are also simplicial.

Proposition 4.4.5. *Let K and L be simplicial complexes. For a function $f_0 : K^{(0)} \to L^{(0)}$, the following are equivalent:*

(a) *f_0 extends to a simplicial map $f : K \to L$;*
(b) *$\langle f_0(\sigma^{(0)})\rangle \in L$ for each $\sigma \in K$;*
(c) *$\bigcap_{v\in\sigma^{(0)}} O_L(f_0(v)) \neq \emptyset$ for each $\sigma \in K$.*

In this case, the simplicial extension f of f_0 is unique.

Proof. The implication (a) $\Rightarrow$ (b) follows from the definition. By Proposition 4.2.3, we have (b) $\Leftrightarrow$ (c). It remains to show the implication (b) $\Rightarrow$ (a). For each $\sigma \in K$, the function $f_0|\sigma^{(0)}$ uniquely extends to an affine map $f_\sigma : \sigma \to \langle f_0(\sigma^{(0)})\rangle \subset |L|$. Because $f_\sigma|\sigma \cap \tau = f_\tau|\sigma \cap \tau$ for each $\sigma, \tau \in K$, we can define $f : |K| \to |L|$ by $f|\sigma = f_\sigma$. Then, f is simplicial with respect to K and L. The uniqueness of f follows from the uniqueness of f_σ, $\sigma \in K$. □

With regard to subdivisions of simplicial maps, we have the following:

Proposition 4.4.6. *Let K and L be simplicial complexes and $f : K \to L$ a simplicial map. For each simplicial subdivision $L' \lhd L$, there exists a simplicial subdivision $K' \lhd K$ such that $f : K' \to L'$ is simplicial.*

Proof. We define

$$K_0 = \{\sigma \cap f^{-1}(\tau) \mid \sigma \in K,\ \tau \in L',\ \tau \subset f(\sigma)\}.$$

Then, K_0 is a cell complex subdividing K. Indeed, let $\sigma \in K$ and $\tau \in L'$ with $\tau \subset f(\sigma)$ and $x \in \sigma \cap f^{-1}(\tau)$. By 4.1.9(3), $(f|\sigma)^{-1}(\tau) = \sigma \cap f^{-1}(\tau)$ is a cell with $(\sigma \cap f^{-1}(\tau))_x = \sigma_x \cap f^{-1}(\tau_{f(x)})$. Since $\operatorname{rint} f(\sigma_x) \cap \operatorname{rint} \tau_{f(x)} \neq \emptyset$, we have $\tau_{f(x)} \subset f(\sigma_x)$ because $L' \lhd L$. Thus, K_0 satisfies (C1). To show (C2″), let $\sigma, \sigma' \in K$ and $\tau, \tau' \in L'$ with $\tau \subset f(\sigma)$, $\tau' \subset f(\sigma')$, and

$$x \in \operatorname{rint}(\sigma \cap f^{-1}(\tau)) \cap \operatorname{rint}(\sigma' \cap f^{-1}(\tau'))\ \big(\subset \sigma \cap \sigma' \cap f^{-1}(\tau \cap \tau')\big).$$

Since $\operatorname{rint}\sigma_x \cap \operatorname{rint}\sigma'_x \neq \emptyset$ and $\operatorname{rint}\tau_{f(x)} \cap \operatorname{rint}\tau'_{f(x)} \neq \emptyset$, we have $\sigma_x = \sigma'_x$ and $\tau_{f(x)} = \tau'_{f(x)}$. Then, from 4.1.9(3), it follows that

$$\begin{aligned}\sigma \cap f^{-1}(\tau) &= (\sigma \cap f^{-1}(\tau))_x = \sigma_x \cap f^{-1}(\tau_{f(x)})\\ &= \sigma'_x \cap f^{-1}(\tau'_{f(x)}) = (\sigma' \cap f^{-1}(\tau'))_x = \sigma' \cap f^{-1}(\tau').\end{aligned}$$

Therefore, K_0 satisfies (C2″).

Let $\sigma \in K$ and $\tau \in L'$ with $\tau \subset f(\sigma)$. Then, $f(\sigma \cap f^{-1}(\tau)) = f(\sigma) \cap \tau = \tau$. For each $x \in \sigma \cap f^{-1}(\tau)$, since $\tau_{f(x)} \subset f(\sigma_x)$ as seen in the verification of (C1) above, it follows that

$$f((\sigma \cap f^{-1}(\tau))_x) = f(\sigma_x \cap f^{-1}(\tau_{f(x)})) = f(\sigma_x) \cap \tau_{f(x)} = \tau_{f(x)}.$$

In particular, if $v \in (\sigma \cap f^{-1}(\tau))^{(0)}$ then $f(v) \in \tau^{(0)}$. Consequently, we have $f(K_0^{(0)}) \subset L'^{(0)}$.

By Theorem 4.2.10, we have a simplicial subdivision $K' \lhd K_0$ such that $K'^{(0)} = K_0^{(0)}$. Then, $f(K'^{(0)}) \subset L'^{(0)}$. For each simplex $\sigma' \in K'$, we have $\sigma \in K$ and $\tau \in L'$ such that $\tau \subset f(\sigma)$, $\sigma' \subset \sigma \cap f^{-1}(\tau)$, and $\sigma'^{(0)} \subset (\sigma \cap f^{-1}(\tau))^{(0)}$. Since $f|\sigma'$ is affine and $f(\sigma'^{(0)}) \subset \tau^{(0)}$, it follows that $f(\sigma') \leq \tau$, hence $f(\sigma') \in L'$. Thus, $f : K' \to L'$ is simplicial. □

For homeomorphisms that are PL, we have the following:

Theorem 4.4.7. *Let K and L be cell complexes. If a homeomorphism $f : |K| \to |L|$ is PL, then the inverse $f^{-1} : |L| \to |K|$ is also PL and K has a subdivision K^* such that $f|C : C \to f(C)$ is an affine homeomorphism for each $C \in K^*$ and $L^* = \{f(C) \mid C \in K^*\}$ is a cell complex subdividing L.*

Proof. Because the graph $G(f)$ of f can be regarded as the graph of f^{-1} by changing the first and the second factors, the first assertion follows from Theorem 4.4.2.

Let $i_f : |K| \to G(f)$ and $i_{f^{-1}} : |L| \to G(f)$ be the natural injections, where $i_{f^{-1}}(y) = (f^{-1}(y), y)$ for each $y \in |L|$. Let K' be a subdivision of K such that f is affine on each cell $C \in K'$, and L' be a subdivision of L such that f^{-1} is affine on each cell $D \in L'$. As observed in the proof of Theorem 4.4.2, $i_f(K')$ and $i_{f^{-1}}(L')$ are cell complexes with $|i_f(K')| = |i_{f^{-1}}(L')| = G(f)$ as spaces. By virtue of Theorem 4.2.11, $i_f(K')$ and $i_{f^{-1}}(L')$ have a common subdivision M. Then, $K^* = \mathrm{pr}_{|K|}(M)$ and $L^* = \mathrm{pr}_{|L|}(M)$ are subdivisions of K' and L', respectively. In addition, $f(K^*) = \mathrm{pr}_{|L|} i_f(K^*) = \mathrm{pr}_{|L|}(M) = L^*$, that is, $L^* = \{f(C) \mid C \in K^*\}$. For each $C \in K^*$, $f|C = \mathrm{pr}_{|L|} \circ i_f|C : C \to f(C)$ is an affine homeomorphism. □

A **piecewise linear (PL) homeomorphism** is literally defined as a homeomorphism being PL. Due to Theorem 4.4.7, the inverse of a PL homeomorphism is also a PL homeomorphism. For cell complexes K and L, the polyhedra $|K|$ is **PL homeomorphic** to $|L|$ if there exists a PL homeomorphism $f : |K| \to |L|$.

Remark 5. Every PL bijection between compact polyhedra is a PL homeomorphism. However, a bijective PL map $f : |K| \to |L|$ is, in general, not a PL homeomorphism. For example, define $f : \mathbb{R}_+ \to \partial \mathbf{I}^2$ as follows:

$$f(t) = \begin{cases} (t, 0) & \text{if } t \in \mathbf{I} = [0, 1], \\ (1, t-1) & \text{if } t \in [1, 2], \\ (3-t, 1) & \text{if } t \in [2, 3], \\ (0, 2^{-n+2}(n+2-t)) & \text{if } t \in [n, n+1],\ n \geq 3. \end{cases}$$

Then, f is a PL bijection that is not a PL homeomorphism.

For simplicial complexes K and L, if a bijection $f : |K| \to |L|$ is simplicial with respect to K and L, then the inverse f^{-1} is also a simplicial map from L to K. A simplicial bijection is called a **simplicial isomorphism**. This is a homeomorphism, so it is also called a **simplicial homeomorphism**. Obviously, a simplicial isomorphism is a PL homeomorphism. The inverse of a simplicial isomorphism and the composition of simplicial isomorphisms are also simplicial isomorphisms. It is said that K is **simplicially isomorphic** to L (denoted by $K \equiv L$) if there exists a simplicial isomorphism $f : K \to L$. Obviously, $\equiv$ is an equivalence relation among simplicial complexes.

There exists a weaker equivalence relation among simplicial complexes. It is said that K is **combinatorially equivalent** to L (denoted by $K \cong L$) if they have simplicial subdivisions that are simplicially isomorphic. Then, $\cong$ is an equivalence relation among simplicial complexes.

It is obvious that $\cong$ is reflective and symmetric. To see that $\cong$ is transitive, let $K_1 \cong K_2$ and $K_2 \cong K_3$. Then, $K_1' \equiv K_2'$ for some $K_1' \lhd K_1$ and $K_2' \lhd K_2$, and $K_2'' \equiv K_3''$ for some $K_2'' \lhd K_2$ and $K_3'' \lhd K_3$. By virtue of Corollary 4.2.13, K_2' and K_2'' have a common simplicial subdivision K_2''', which induces $K_1''' \lhd K_1'$ and $K_3''' \lhd K_3''$ such that $K_1''' \equiv K_2'''$ and $K_3''' \equiv K_2'''$. Hence, $K_1''' \equiv K_3'''$, which means $K_1 \cong K_3$. Therefore, $\cong$ is an equivalence relation among simplicial complexes.

This fact also follows from Theorem 4.4.8 below, Theorem 4.4.7, and Proposition 4.4.4.

Theorem 4.4.8. *Two simplicial complexes K and L are combinatorially equivalent to each other if and only if $|K|$ and $|L|$ are PL homeomorphic to each other, that is, there exists a PL homeomorphism $f : |K| \to |L|$.*

Proof. If $K \cong L$, then K and L have simplicial subdivisions K' and L', respectively, such that $K' \equiv L'$, hence there is a simplicial isomorphism $f : K' \to L'$. Then, $f : |K| \to |L|$ is a PL homeomorphism.

Conversely, let $f : |K| \to |L|$ be a PL homeomorphism. By Theorem 4.4.7, there is a cell complex K' subdividing K such that $f|C : C \to f(C)$ is an affine homeomorphism for each $C \in K'$ and $L' = \{f(C) \mid C \in K'\}$ is a subdivision of L. By Theorem 4.2.10, we have a simplicial subdivision K'' of K' with the same vertices. Then, $L'' = \{f(\sigma) \mid \sigma \in K''\}$ is a simplicial subdivision of L. Observe that $f : K'' \to L''$ is a simplicial isomorphism. Thus, we have $K'' \equiv L''$, that is, $K \cong L$. □

For simplicial complexes K and L, the following implications are trivial:

$$K \equiv L \Rightarrow K \cong L \Rightarrow |K| \approx |L|.$$

Although it goes without saying that the converse of the first implication does not hold, the converse of the second does not either. It should be noted that $|K| = |L|$ implies $K \cong L$ by Theorems 4.2.11 and 4.2.10. The converse of the second implication is called **Hauptvermutung** (*the fundamental conjecture*). It took a long time to find finite simplicial complexes K and L such that $K \not\cong L$ but $|K| \approx |L|$.

It is known that this conjecture does not hold even if $|K|$ and $|L|$ are n-manifolds (i.e., there exists an n-manifold that has topological triangulations $K \not\cong L$).[8]

Remark 6. By Theorem 4.4.8, it might be expected that every PL map $f : |K| \to |L|$ is simplicial with respect to some simplicial subdivisions of K and L. However, this is not the case. For example, let K be the natural triangulation of $\mathbb{R}_+$, that is, $K = \omega \cup \{[n-1, n] \mid n \in \mathbb{N}\}$, and let $L = I = \{0, 1, \mathbf{I}\}$. We define a PL map $f : |K| \to |L|$ as follows:

$$f(2n) = 2^{-n-1} \text{ and } f(2n+1) = 1 - 2^{-n-1} \text{ for each } n \in \omega,$$

and f is affine on each $[n, n+1]$. Since every subdivision of K contains ω as vertices but every subdivision of L has only finitely many vertices, then f is not simplicial with respect to any simplicial subdivisions of K and L.

In Sect. 4.6, it will be proved that every *proper* PL map $f : |K| \to |L|$ is simplicial with respect to some simplicial subdivisions of K and L. According to Proposition 4.2.6, a (PL) map $f : |K| \to |L|$ is proper if and only if, for each $\sigma \in L$, there is a finite subcomplex $K_\sigma \subset K$ such that $f^{-1}(\sigma) \subset |K_\sigma|$.

4.5 The Metric Topology of Polyhedra

Let K be a simplicial complex. As shown in 4.2.16(2), $|K|$ is non-metrizable unless K is locally finite. In this section, we introduce the natural metric on the polyhedron $|K|$ that induces the same topology as the Whitehead topology if K is locally finite.

Each point $x \in |K|$ has the unique representation

$$x = \sum_{i=1}^{n+1} z(i)v_i, \; z \in \operatorname{rint} \Delta^n, \; c_K(x) = \langle v_1, \ldots, v_{n+1} \rangle.$$

For each $v \in K^{(0)}$, let

$$\beta_v^K(x) = \begin{cases} z(i) & \text{if } v = v_i, \; i = 1, \ldots, n+1, \\ 0 & \text{otherwise.} \end{cases}$$

Thus, we have maps $\beta_v^K : |K| \to \mathbf{I}$, $v \in K^{(0)}$, which are affine on each simplex of K. It follows from the definition that

$$\sum_{v \in K^{(0)}} \beta_v^K(x) = 1 \text{ for each } x \in |K|,$$

[8] An ***n*-manifold (without boundary)** is a paracompact space such that each point has an open neighborhood that is homeomorphic to an open set in $\mathbb{R}^n$.

where $\{v \in K^{(0)} \mid \beta_v^K(x) \neq 0\} = c_K(x)^{(0)}$, i.e., $\beta_v^K(x) > 0 \Leftrightarrow v \in c_K(x)^{(0)}$. Namely, $(\beta_v^K)_{v \in K^{(0)}}$ is a partition of unity on $|K|$ with $\operatorname{supp} \beta_v^K = |\operatorname{St}(v, K)|$. In fact, we have

$$(\beta_v^K)^{-1}((0, 1]) = O_K(v) \quad \text{for every } v \in K^{(0)}.$$

Then, each $x \in |K|$ is uniquely represented in the form

$$x = \sum_{v \in K^{(0)}} \beta_v^K(x) v,$$

where $\beta_v^K(x)$ is called the **barycentric coordinate** of x at v with respect to K. The injection $\beta^K : |K| \to \ell_1(K^{(0)})$ defined by $\beta^K(x)(v) = \beta_v^K(x)$ is called the **canonical representation** of K. Observe that $\beta^K(v) = \mathbf{e}_v$ is the unit vector of $\ell_1(K^{(0)})$. For each $\sigma \in K$, the restriction $\beta^K|\sigma$ is affine. It should be noted that $\beta^K(K) = \{\beta^K(\sigma) \mid \sigma \in K\}$ is a simplicial complex in the Banach space $\ell_1(K^{(0)})$ and $\beta^K : K \to \beta^K(K)$ is a simplicial isomorphism.

Now, we define the metric ρ_K on $|K|$ as follows:

$$\rho_K(x, y) = \|\beta^K(x) - \beta^K(y)\|_1 = \sum_{v \in K^{(0)}} |\beta_v^K(x) - \beta_v^K(y)|,$$

where $\| \cdot \|_1$ is the norm for $\ell_1(K^{(0)})$. Note that $\rho_K(v, v') = 2$ for each pair of distinct vertices $v, v' \in K^{(0)}$. The topology on $|K|$ induced by this metric ρ_K is called the **metric topology**. The space $|K|$ with this topology is denoted by $|K|_{\mathrm{m}}$. This space is homeomorphic to the subspace $\beta^K(|K|) = |\beta^K(K)|$ of the Banach space $\ell_1(K^{(0)})$ because β^K is an isometry. The space $|K|_{\mathrm{m}}$ (or the metric space $(|K|, \rho_K)$) is called a **metric polyhedron**. Note that $\ell_1(K^{(0)}) \subset \mathbb{R}^{K^{(0)}}$, and that the topology of $\beta^K(|K|)$ inherited from $\ell_1(K^{(0)})$ coincides with the one inherited from the product space $\mathbb{R}^{K^{(0)}}$ because $\beta^K(|K|)$ is contained in the unit sphere of $\ell_1(K^{(0)})$ (cf. Proposition 1.2.4). Hence, the metric topology on $|K|$ is the coarsest topology such that all $\beta_v^K : |K| \to \mathbf{I}$ $(v \in K^{(0)})$ are continuous. Then, we have the following:

Fact. *For an arbitrary space X, each $f : X \to |K|_{\mathrm{m}}$ is continuous if and only if $\beta_v^K f$ is continuous for every $v \in K^{(0)}$.*

Since every $\beta_v^K : |K| \to \mathbf{I}$ is continuous (with respect to the Whitehead topology), the identity $\mathrm{id}_{|K|} : |K| \to |K|_{\mathrm{m}}$ is continuous, hence the Whitehead topology and the metric topology are identical on each simplex of K.

The open star $O_K(v)$ at each $v \in K^{(0)}$ is open in $|K|_{\mathrm{m}}$ because it is simply $(\beta_v^K)^{-1}((0, 1])$ $(= |K| \setminus (\beta_v^K)^{-1}(0))$. Hence, $\mathcal{O}_K \in \operatorname{cov}(|K|_{\mathrm{m}})$. For each $x \in |K|$, we have

$$O_K(x) = \bigcup_{\sigma \in K[x]} \operatorname{rint} \sigma = \bigcap_{v \in c_K(x)^{(0)}} O_K(v) \subset |\operatorname{St}(c_K(x), K)|.$$

Hence, the open star $O_K(x)$ is an open neighborhood of x in $|K|_{\mathrm{m}}$. For each $0 < r \leq 1$, we have the following open neighborhood of x in $|K|_{\mathrm{m}}$:

$$O_K(x,r) = (1-r)x + rO_K(x) = \{(1-r)x + ry \mid y \in O_K(x)\}.$$

Then, $O_K(x,r) \subset \mathrm{B}_{\rho_K}(x, 2r)$. Indeed, for each $y \in O_K(x)$, since $c_K(x) \leq c_K(y)$ and β_v^K is affine on $c_K(y)$, it follows that

$$\begin{aligned}\rho_K((1-r)x + ry, x) &= \sum_{v \in K^{(0)}} \left|\beta_v^K((1-r)x + ry) - \beta_v^K(x)\right| \\ &= \sum_{v \in K^{(0)}} r\left|\beta_v^K(y) - \beta_v^K(x)\right| = r\rho_K(x,y) < 2r.\end{aligned}$$

As a consequence, we have the following:

Proposition 4.5.1. *Let K be a simplicial complex and $x \in |K|$. Then, $\{O_K(x,r) \mid 0 < r \leq 1\}$ is an open neighborhood basis of x in $|K|_{\mathrm{m}}$.* □

The following proposition can be easily proved:

Proposition 4.5.2. *For every subcomplex L of a simplicial complex K, the metric ρ_L is the restriction of ρ_K and $|L|_{\mathrm{m}}$ is a closed subspace of $|K|_{\mathrm{m}}$.* □

Moreover, we have:

Proposition 4.5.3. *For a simplicial complex K and each $x \in |K|$, the closure of $O_K(x)$ in $|K|_{\mathrm{m}}$ coincides with $|\operatorname{St}(c_K(x), K)|$. In particular, for a vertex $v \in K^{(0)}$, $|\operatorname{St}(v, K)|$ is the closure of $O_K(v)$ in $|K|_{\mathrm{m}}$.*

Proof. According to Proposition 4.5.2, $|\operatorname{St}(c_K(x), K)|$ is closed in $|K|_{\mathrm{m}}$. Then, it suffices to show that $(1-t)y + tx \in O_K(x)$ for each $y \in |\operatorname{St}(c_K(x), K)|$ and $0 < t \leq 1$. For each $v \in c_K(x)^{(0)}$, since $\beta_v^K(x) > 0$, it follows that

$$\beta_v^K((1-t)y + tx) = (1-t)\beta_v^K(y) + t\beta_v^K(x) > 0,$$
$$\text{i.e., } (1-t)y + tx \in (\beta_v^K)^{-1}((0,1]) = O_K(v).$$

Hence, $(1-t)y + tx \in \bigcap_{v \in c_K(x)^{(0)}} O_K(v) = O_K(x)$. □

Thus, with respect to the metric topology as well as the Whitehead topology, $\mathcal{S}_K = \mathcal{O}_K^{\mathrm{cl}}$ and $(\beta_v^K)_{v \in K^{(0)}}$ is a partition of unity on $|K|_{\mathrm{m}}$ with $\operatorname{supp} \beta_v^K = |\operatorname{St}(v, K)|$.

Using the metric topology instead of the Whitehead (weak) topology, Proposition 4.3.4 can be generalized as follows:

Proposition 4.5.4. *Let K be a simplicial complex and X an arbitrary space. If two maps $f, g : X \to |K|_{\mathrm{m}}$ are contiguous (with respect to K) then $f \simeq_K g$ by the straight-line homotopy $h : X \times \mathbf{I} \to |K|_{\mathrm{m}}$, that is,*

$$h(x,t) = (1-t)f(x) + tg(x) \ \textit{for each } (x,t) \in X \times \mathbf{I}.$$

Proof. It suffices to verify the continuity of h. This follows from the continuity of $\beta_v^K h : X \times \mathbf{I} \to \mathbf{I}$, $v \in K^{(0)}$, where

$$\beta_v^K h(x,t) = (1-t)\beta_v^K f(x) + t\beta_v^K g(x). \qquad \square$$

Let K and L be simplicial complexes. Each simplicial map $f : K \to L$ can be represented as follows:

$$\begin{aligned} f(x) &= f\big(\textstyle\sum_{v\in K^{(0)}} \beta_v^K(x)v\big) \\ &= \sum_{v\in K^{(0)}} \beta_v^K(x) f(v) = \sum_{u\in L^{(0)}} \sum_{v\in f^{-1}(u)} \beta_v^K(x)u, \end{aligned}$$

that is, $\beta_u^L(f(x)) = \sum_{v\in f^{-1}(u)} \beta_v^K(x)$ for each $x \in |K|$ and $u \in L^{(0)}$. Then, it is easy to show the following:

Proposition 4.5.5. *Let $f : K \to L$ be a simplicial map between simplicial complexes. Then, $\rho_L(f(x), f(y)) \leq \rho_K(x,y)$ for each $x, y \in |K|$, hence $f : |K|_{\mathrm{m}} \to |L|_{\mathrm{m}}$ is continuous. When f is injective, $f : (|K|, \rho_K) \to (|L|, \rho_L)$ is a closed isometry, so it is a closed embedding. Particularly, if f is bijective (i.e., f is a simplicial homeomorphism), it is a homeomorphism. As a consequence,*

$$K \equiv L \Rightarrow |K|_{\mathrm{m}} \approx |L|_{\mathrm{m}}. \qquad \square$$

For a finite simplicial complex K, since $|K|$ is compact (Proposition 4.2.6), it follows that $\mathrm{id}_{|K|} : |K| \to |K|_{\mathrm{m}}$ is a homeomorphism, that is, the metric topology of $|K|$ coincides with the Whitehead topology. More generally, we can prove the following theorem (cf. 4.2.16(2)):

Theorem 4.5.6. *For a simplicial complex K, the metric topology of $|K|$ coincides with the Whitehead topology (i.e., $|K|_{\mathrm{m}} = |K|$ as spaces) if and only if K is locally finite.*

Proof. If $|K|_{\mathrm{m}} = |K|$ as spaces then $|K|$ is metrizable, so K is locally finite by 4.2.16(2). To show the converse, let $\varphi = \mathrm{id} : |K| \to |K|_{\mathrm{m}}$. Assume that K is locally finite, that is, $\mathrm{St}(v, K)$ is finite for each $v \in K^{(0)}$. Then, each $\varphi\big| |\mathrm{St}(v,K)|$ is a homeomorphism, so $\varphi|O_K(v)$ is also a homeomorphism. Since $\mathcal{O}_K$ is an open cover of both $|K|_{\mathrm{m}}$ and $|K|$, it follows that φ is a homeomorphism. $\square$

Concerning subdivisions, we have the following result:

Proposition 4.5.7. *Let $K \rhd K'$ be simplicial complexes. Then, $\rho_K(x,y) \leq \rho_{K'}(x,y)$ for each $x, y \in |K| = |K'|$, hence $\mathrm{id} : |K'|_{\mathrm{m}} \to |K|_{\mathrm{m}}$ is continuous.*

Proof. For each $x \in |K| = |K'|$ and $v \in K^{(0)}$, we have

$$\beta_v^K(x) = \beta_v^K\big(\textstyle\sum_{w\in K'^{(0)}} \beta_w^{K'}(x)w\big) = \displaystyle\sum_{w\in K'^{(0)}} \beta_w^{K'}(x)\beta_v^K(w).$$

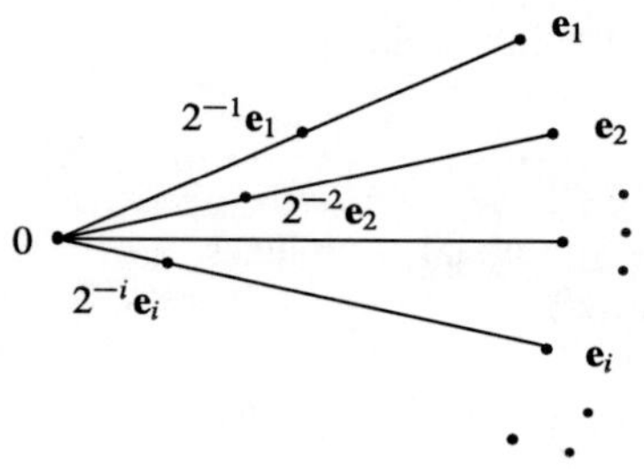

Fig. 4.4 $|K'|_{\mathrm{m}} \neq |K|_{\mathrm{m}}$

Then, for each $x, y \in |K| = |K'|$,

$$\begin{aligned}\rho_K(x,y) &= \sum_{v\in K^{(0)}} \left|\beta_v^K(x) - \beta_v^K(y)\right| \\ &\leq \sum_{v\in K^{(0)}} \sum_{w\in K'^{(0)}} |\beta_w^{K'}(x) - \beta_w^{K'}(y)|\beta_v^K(w) \\ &= \sum_{w\in K'^{(0)}} \left|\beta_w^{K'}(x) - \beta_w^{K'}(y)\right| = \rho_{K'}(x,y).\end{aligned}$$

□

In contrast to the Whitehead topology, $|K'|_{\mathrm{m}} \neq |K|_{\mathrm{m}}$ for some subdivision $K' \lhd K$. Such an example can be defined in ℓ_1 as follows:

$$K = \left\{\mathbf{0},\ \mathbf{e}_i,\ \langle 0, \mathbf{e}_i\rangle \;\middle|\; i \in \mathbb{N}\right\} \text{ and}$$
$$K' = \left\{\mathbf{0},\ 2^{-i}\mathbf{e}_i,\ \mathbf{e}_i,\ \langle \mathbf{0}, 2^{-i}\mathbf{e}_i\rangle,\ \langle 2^{-i}\mathbf{e}_i, \mathbf{e}_i\rangle \;\middle|\; i \in \mathbb{N}\right\} \lhd K,$$

where each $\mathbf{e}_i \in \ell_1$ is defined by $\mathbf{e}_i(i) = 1$ and $\mathbf{e}_i(j) = 0$ for $j \neq i$. Then, $|K|_{\mathrm{m}}$ is simply the hedgehog $J(\mathbb{N})$. The set $\{2^{-i}\mathbf{e}_i \mid i \in \mathbb{N}\}$ is closed in $|K'|_{\mathrm{m}}$ but $\lim_{i\to\infty} 2^{-i}\mathbf{e}_i = \mathbf{0}$ in $|K|_{\mathrm{m}}$ — Fig. 4.4.

For each simplicial map $f : K \to L$, both maps $f : |K| \to |L|$ and $f : |K|_{\mathrm{m}} \to |L|_{\mathrm{m}}$ are continuous (Proposition 4.5.5). Moreover, recall that every PL map $f : |K| \to |L|$ is continuous. However, a PL map $f : |K|_{\mathrm{m}} \to |L|_{\mathrm{m}}$ is not continuous even if f is bijective. In fact, consider $K' \lhd K$ in the above example and let $L = K$. We define a PL map $f : |K| \to |L|$ by $f(\mathbf{0}) = \mathbf{0}$, $f(\mathbf{e}_i) = \mathbf{e}_i$ and $f(2^{-i}\mathbf{e}_i) = \frac{1}{2}\mathbf{e}_i$, where $\frac{1}{2}\mathbf{e}_i$ is the barycenter of $\langle 0, \mathbf{e}_i\rangle$. Then, $f : |K|_{\mathrm{m}} \to |L|_{\mathrm{m}}$ is not continuous, because $2^{-i}\mathbf{e}_i \to \mathbf{0}$ in $|K|_{\mathrm{m}}$ but $f(2^{-i}\mathbf{e}_i) = \frac{1}{2}\mathbf{e}_i \not\to f(\mathbf{0}) = \mathbf{0}$ in $|L|_{\mathrm{m}}$.

It is inconvenient that the metric topology is changed by subdivisions and that PL maps are not continuous with respect to the metric topology. However, concerning product simplicial complexes, the metric topology has the advantage of the Whitehead topology.

Theorem 4.5.8. *For each pair of ordered simplicial complexes K and L,*

$$|K \times_s L|_{\mathrm{m}} = |K|_{\mathrm{m}} \times |L|_{\mathrm{m}} \text{ as spaces.}$$

Proof. The projections $\mathrm{pr}_1 : |K \times_s L|_{\mathrm{m}} \to |K|_{\mathrm{m}}$ and $\mathrm{pr}_2 : |K \times_s L|_{\mathrm{m}} \to |L|_{\mathrm{m}}$ are simplicial, so they are continuous. Therefore, $\mathrm{id} : |K \times_s L|_{\mathrm{m}} \to |K|_{\mathrm{m}} \times |L|_{\mathrm{m}}$ is continuous.

We will prove the continuity of $\mathrm{id} : |K|_{\mathrm{m}} \times |L|_{\mathrm{m}} \to |K \times_s L|_{\mathrm{m}}$ at each $(x, y) \in |K|_{\mathrm{m}} \times |L|_{\mathrm{m}}$. To this end, we need the data of $\beta^{K\times_s L}_{(u,v)}(x, y)$. Note that the carrier $c_{K\times_s L}(x, y)$ is contained in the cell $c_K(x) \times c_L(y)$. Let

$$c_K(x) = \langle u_1, \ldots, u_n \rangle,\ u_1 < \cdots < u_n \text{ and}$$
$$c_L(y) = \langle v_1, \ldots, v_m \rangle,\ v_1 < \cdots < v_m,$$

and define $0 = a_0 < a_1 < \cdots < a_n = 1$ and $0 = b_0 < b_1 < \cdots < b_m = 1$ as

$$a_k = \sum_{i=1}^{k} \beta^K_{u_i}(x) \text{ and } b_k = \sum_{i=1}^{k} \beta^L_{v_i}(y).$$

In this case, we can write

$$\{a_0, \ldots, a_n, b_0, \ldots, b_m\} = \{c_0, \ldots, c_\ell\}$$

such that $0 = c_0 < c_1 < \cdots < c_\ell = 1$, where $\max\{n, m\} \leq \ell < m + n$ and $\sum_{k=1}^{\ell}(c_k - c_{k-1}) = c_\ell - c_0 = 1$. For each $k = 1, \ldots, \ell$, let

$$a_{i(k)-1} < c_k \leq a_{i(k)} \text{ and } b_{j(k)-1} < c_k \leq b_{j(k)},$$

and define $(\bar{u}_k, \bar{v}_k) = (u_{i(k)}, v_{j(k)})$. Then, we have

$$\langle (\bar{u}_1, \bar{v}_1), \ldots, (\bar{u}_\ell, \bar{v}_\ell) \rangle \in K \times_s L,$$

which is the carrier of (x, y), and $\beta^{K\times_s L}_{(\bar{u}_k,\bar{v}_k)}(x, y) = c_k - c_{k-1}$ because

$$\begin{aligned}
&\sum_{k=1}^{\ell}(c_k - c_{k-1})(\bar{u}_k, \bar{v}_k) \\
&= \left(\sum_{i=1}^{n} \sum_{i(k)=i} (c_k - c_{k-1}) u_{i(k)}, \sum_{j=1}^{m} \sum_{j(k)=j} (c_k - c_{k-1}) v_{j(k)} \right) \\
&= \left(\sum_{i=1}^{n} (a_i - a_{i-1}) u_i, \sum_{j=1}^{m} (b_j - b_{j-1}) v_j \right) \\
&= \left(\sum_{i=1}^{n} \beta^K_{u_i}(x) u_i, \sum_{j=1}^{m} \beta^L_{v_j}(y) v_j \right) = (x, y).
\end{aligned}$$

For each $\varepsilon > 0$, choose $\delta > 0$ so that $2\delta < c_k - c_{k-1}$ for every $k = 1, \dots, \ell$ and $2(\ell + 1)\delta < \varepsilon$. Then, $\beta^K_{u_i}(x) > 2\delta$ for $i = 1, \dots, n$ and $\beta^L_{v_j}(y) > 2\delta$ for $j = 1, \dots, m$. Now, let $(x', y') \in |K| \times |L|$ with $\rho_K(x, x') < \delta$ and $\rho_L(y, y') < \delta$. To show that $\rho_{K\times_s L}((x, y), (x', y')) < \varepsilon$, let

$$c_K(x') = \langle u'_1, \cdots, u'_{n'}\rangle,\ u'_1 < \cdots < u'_{n'} \text{ and}$$
$$c_L(y') = \langle v'_1, \cdots, v'_{m'}\rangle,\ v'_1 < \cdots < v'_{m'}.$$

In the same way as x and y, define $a'_k = \sum_{i=1}^k \beta^K_{u'_i}(x')$, $b'_k = \sum_{i=1}^k \beta^L_{v'_i}(y')$, and write

$$\{a'_0, \dots, a'_{n'}, b'_0, \dots, b'_{m'}\} = \{c'_0, \dots, c'_{\ell'}\},$$

where $0 = c'_0 < c'_1 < \cdots < c'_{\ell'} = 1$. For each $k = 1, \dots, \ell'$, let $a'_{i'(k)-1} < c'_k \le a'_{i'(k)}$ and $b'_{j'(k)-1} < c'_k \le b'_{j'(k)}$, and define $(\bar{u}'_k, \bar{v}'_k) = (u'_{i'(k)}, v'_{j'(k)})$. Then,

$$\langle(\bar{u}'_1, \bar{v}'_1), \dots, (\bar{u}'_{\ell'}, \bar{v}'_{\ell'})\rangle = c_{K\times_s L}(x', y')$$

and $\beta^{K\times_s L}_{(\bar{u}'_k, \bar{v}'_k)}(x', y') = c'_k - c'_{k-1}$. For each $i = 1, \dots, n$,

$$\beta^K_{u_i}(x') \ge \beta^K_{u_i}(x) - |\beta^K_{u_i}(x) - \beta^K_{u_i}(x')| > 2\delta - \rho_K(x, x') > \delta > 0,$$

which implies that $u_1 < \cdots < u_n$ is a subsequence of $u'_1 < \cdots < u'_{n'}$, that is, we can take $1 \le p(1) < \cdots < p(n) \le n'$ to write $u_i = u'_{p(i)}$. Observe that

$$\begin{aligned} |a'_{p(k)} - a_k| &\le \sum_{i=1}^k |\beta^K_{u_i}(x) - \beta^K_{u_i}(x')| + \sum_{u\notin\{u_1,\dots,u_n\}} \beta^K_u(x') \\ &\le \rho_K(x, x') < \delta \quad \text{and} \\ |a'_{p(k)-1} - a_{k-1}| &\le \sum_{i=1}^{k-1} |\beta^K_{u_i}(x) - \beta^K_{u_i}(x')| + \sum_{u\notin\{u_1,\dots,u_n\}} \beta^K_u(x') \\ &\le \rho_K(x, x') < \delta. \end{aligned}$$

Similarly, we can take $1 \le q(1) < \cdots < q(m) \le m'$ to write $v_j = v'_{q(j)}$. Then,

$$|b'_{q(j)} - b_j| < \delta \text{ and } |b'_{q(j)-1} - b_{j-1}| < \delta.$$

On the other hand, for each $k = 1, \dots, \ell$, because $a_{i(k)-1} < c_k \le a_{i(k)}$ and $b_{j(k)-1} < c_k \le b_{j(k)}$, we have

$$c_k = \min\{a_{i(k)}, b_{j(k)}\} \text{ and } c_{k-1} = \max\{a_{i(k)-1}, b_{j(k)-1}\}.$$

Then, it follows that

$$b'_{q(j(k))-1} < b_{j(k)-1} + \delta \le c_{k-1} + \delta < c_k - \delta \le a_{i(k)} - \delta < a'_{p(i(k))}.$$

Similarly, $a'_{p(i(k))-1} < b'_{q(j(k))}$. Hence, there is some $r(k) = 1, \dots, \ell'$ such that

$$\begin{aligned} c'_{r(k)-1} &= \max\{a'_{p(i(k))-1}, b'_{q(j(k))-1}\} \\ &< c'_{r(k)} = \min\{a'_{p(i(k))}, b'_{q(j(k))}\}. \end{aligned}$$

This means that $p(i(k)) = i'(r(k))$ and $q(j(k)) = j'(r(k))$, which implies that

$$\begin{aligned} (\bar{u}'_{r(k)}, \bar{v}'_{r(k)}) &= (u'_{i'(r(k))}, v'_{j'(r(k))}) = (u'_{p(i(k))}, v'_{q(j(k))}) \\ &= (u_{i(k)}, v_{j(k)}) = (\bar{u}_k, \bar{v}_k) \end{aligned}$$

and $\beta^{K \times_s L}_{(\bar{u}_k, \bar{v}_k)}(x', y') = c'_{r(k)} - c'_{r(k)-1}$. Observe that

$$\begin{aligned} c_k - \delta &= \min\{a_{i(k)} - \delta, b_{j(k)} - \delta\} \\ &< c'_{r(k)} = \min\{a'_{p(i(k))}, b'_{q(j(k))}\} \\ &< c_k + \delta = \min\{a_{i(k)} + \delta, b_{j(k)} + \delta\} \quad \text{and} \\ c_{k-1} - \delta &= \max\{a_{i(k)-1} - \delta, b_{j(k)-1} - \delta\} \\ &< c'_{r(k)-1} = \max\{a'_{p(i(k))-1}, b'_{q(j(k))-1}\} \\ &< c_{k-1} + \delta = \max\{a_{i(k)-1} + \delta, b_{j(k)-1} + \delta\}. \end{aligned}$$

Moreover, it should be noted that

$$\sum_{i'(r)=i} (c'_r - c'_{r-1}) = a'_i - a'_{i-1} = \beta^K_{u'_i}(x')$$

$$\sum_{j'(r)=j} (c'_r - c'_{r-1}) = b'_j - b'_{j-1} = \beta^L_{v'_j}(y') \quad \text{and}$$

$$r \in \{r(1), \dots, r(\ell)\} \Leftrightarrow \begin{cases} i'(r) \in \{p(1), \dots, p(n)\}, \\ j'(r) \in \{q(1), \dots, q(m)\}. \end{cases}$$

Then, it follows that

$$\begin{aligned} \sum_{r \notin \{r(1),\dots,r(\ell)\}} (c'_r - c'_{r-1}) \le & \sum_{i \notin \{p(1),\dots,p(n)\}} \sum_{i'(r)=i} (c'_r - c'_{r-1}) \\ & + \sum_{j \notin \{q(1),\dots,q(m)\}} \sum_{j'(r)=j} (c'_r - c'_{r-1}) \end{aligned}$$

$$\begin{aligned}
&= \sum_{i \notin \{p(1),\dots,p(n)\}} \beta^K_{u'_i}(x') + \sum_{j \notin \{q(1),\dots,q(m)\}} \beta^L_{v'_j}(y') \\
&= \sum_{u \notin \{u_1,\dots,u_n\}} \beta^K_u(x') + \sum_{v \notin \{v_1,\dots,v_m\}} \beta^L_v(y') \\
&< \rho_K(x,x') + \rho_L(y,y') < 2\delta.
\end{aligned}$$

Consequently, we have the following estimation:

$$\begin{aligned}
\rho_{K\times_s L}((x,y),(x',y')) &= \sum_{k=1}^{\ell} \left|(c_k - c_{k-1}) - (c'_{r(k)} - c'_{r(k)-1})\right| \\
&\quad + \sum_{r \notin \{r(1),\dots,r(\ell)\}} (c'_r - c'_{r-1}) \\
&< \sum_{k=1}^{\ell} |c_k - c'_{r(k)}| + \sum_{k=1}^{\ell} |c_{k-1} - c'_{r(k)-1}| + 2\delta \\
&< 2(\ell+1)\delta < \varepsilon.
\end{aligned}$$

This completes the proof. □

For a simplicial complex K, we can characterize the complete metrizability of $|K|_{\mathrm{m}}$ as follows:

Theorem 4.5.9. *For a simplicial complex K, the following are equivalent:*

(a) *$|K|_{\mathrm{m}}$ is completely metrizable;*
(b) *K contains no infinite full complexes as subcomplexes;*
(c) *ρ_K is complete.*

Proof. Since (c) ⇒ (a) is obvious, we show (a) ⇒ (b) ⇒ (c).

(a) ⇒ (b): Assume that K contains an infinite full complex as a subcomplex. Then, we have a countably infinite full complex $L \subset K$. Because $|K|_{\mathrm{m}}$ is completely metrizable, its closed subspace $|L|_{\mathrm{m}}$ is also completely metrizable. However, $|L|$ is the union of countably many simplexes that have no interior points. This contradicts the Baire Category Theorem 2.5.1. Therefore, K contains no infinite full complexes.

(b) ⇒ (c): Let $(x_i)_{i\in\mathbb{N}}$ be a ρ_K-Cauchy sequence in $|K|_{\mathrm{m}}$. Since $\beta^K : (|K|, \rho_K) \to \ell_1(K^{(0)})$ is an isometry, $(\beta^K(x_i))_{i\in\mathbb{N}}$ is Cauchy in $\ell_1(K^{(0)})$, hence we have $\lambda = \lim_{i\to\infty} \beta^K(x_i) \in \ell_1(K^{(0)})$. Observe

$$\sum_{v\in K^{(0)}} \lambda(v) = \|\lambda\|_1 = \lim_{i\to\infty} \|\beta^K(x_i)\|_1 = 1.$$

Let $A = \{v \in K^{(0)} \mid \lambda(v) > 0\}$. Each finite subset $F \subset A$ is contained in $c_K(x_i)^{(0)}$ for sufficiently large $i \in \mathbb{N}$, hence $\langle F \rangle \in K$. Thus, K contains the full complex $\Delta(A)$ as a subcomplex. It follows from (b) that A is finite. Using the above argument, we have $\langle A \rangle \in K$, which means

$$x = \sum_{v \in A} \lambda(v)v \in \langle A \rangle \subset |K|.$$

Therefore, ρ_K is complete. □

There is another admissible metric on $|K|_{\mathrm{m}}$ that is widely adopted because it allows for easy estimates of distances.

Another Metric on a Polyhedron 4.5.10.

(1) For a simplicial complex K, the following metric is admissible for $|K|_{\mathrm{m}}$:

$$\begin{aligned}\rho_K^\infty(x, y) &= \|\beta^K(x) - \beta^K(y)\|_\infty \\ &= \sup_{v \in K^{(0)}} |\beta_v^K(x) - \beta_v^K(y)| \ \big(\leq \rho_K(x, y) \big),\end{aligned}$$

where $\|\cdot\|_\infty$ is the norm for $\ell_\infty(K^{(0)})$.

Sketch of Proof. For the continuity of $\mathrm{id} : (|K|, \rho_K^\infty) \to (|K|, \rho_K)$, see Proposition 1.2.4.

(2) Proposition 4.5.7 is not valid for the metric ρ_K^∞, that is, the inequality $\rho_K^\infty(x, y) \leq \rho_{K'}^\infty(x, y)$ does not hold for some $K' \lhd K$.

Example. Consider the following simplicial complexes in $\mathbb{R}$:

$$K = \{0, \pm 1, \langle 0, \pm 1 \rangle\}, \ K' = \{0, \pm\frac{1}{2}, \pm 1, \langle 0, \pm\frac{1}{2} \rangle, \langle \pm\frac{1}{2}, \pm 1 \rangle\}.$$

Then, $K' \lhd K$ but $\rho_K^\infty(-\frac{3}{4}, \frac{3}{4}) = \frac{3}{4} > \rho_{K'}^\infty(-\frac{3}{4}, \frac{3}{4}) = \frac{1}{2}$.

(3) For a simplicial complex K, the following are equivalent:

(a) K is finite-dimensional;
(b) ρ_K is uniformly equivalent to ρ_K^∞;
(c) ρ_K^∞ is complete.

Sketch of Proof. (a) ⇒ (b) and (c): For each $x, y \in |K|$,

$$\rho_K^\infty(x, y) \leq \rho_K(x, y) \leq 2(\dim K + 1)\rho_K^\infty(x, y).$$

Then, applying Theorem 4.5.9, we have (c).
(b) or (c) ⇒ (a): If $\dim K = \infty$, then we can inductively choose n-simplexes $\sigma_n \in K$, $n \in \mathbb{N}$, so that $\sigma_n \cap \sigma_m = \emptyset$ if $n \neq m$. Then, for any $n < m$, $\rho_K(\hat{\sigma}_n, \hat{\sigma}_m) = 2$ and $\rho_K^\infty(\hat{\sigma}_n, \hat{\sigma}_m) = 1/(n + 1)$. This is contrary to both (b) and (c).

4.6 Derived and Barycentric Subdivisions

In this section, we introduce derived subdivisions and barycentric subdivisions, and prove that proper PL maps are simplicial with respect to some simplicial subdivisions.

The following lemma is useful to construct simplicial subdivisions:

Lemma 4.6.1. *Let C be an n-cell with $u_0 \in \operatorname{rint} C$ and K' a simplicial subdivision of the cell complex $F(\partial C)$. Then,*

$$K'' = K' \cup \{u_0\} \cup \{u_0\sigma \mid \sigma \in K'\} \lhd F(C)$$

and K' is a full subcomplex of K''.

Proof. For each $\sigma \in K'$, choose $D < C$ so that $\sigma \subset D$. Since $u_0 \notin D = C \cap \operatorname{fl} D$, we have the join $u_0\sigma$, which is a simplex in C. Evidently, K'' satisfies (C1) and $K'' \prec F(C)$. To verify that K'' is a simplicial subdivision of $F(C)$, we have to show that K'' satisfies (C2″) and C is covered by simplexes of K''. On the other hand, it easily follows from the definition of K'' that K' is full in K''. Then, it suffices to show that each $x \in C$ is contained in the interior of a unique simplex of K''.

The case $x = u_0$ is obvious. The case $x \in \partial C$ follows from the fact that $K' \lhd F(\partial C)$ and $\operatorname{rint} u_0\sigma \cap \partial C = \emptyset$ for each $\sigma \in K'$. When $x \in \operatorname{rint} C$ and $x \neq u_0$, let

$$t_0 = \sup\{t > 0 \mid (1-t)u_0 + tx \in C\} > 1 \text{ and } y = (1-t_0)u_0 + t_0 x.$$

Then, $y \in \partial C$ and $x \in \langle u_0, y\rangle$. By Proposition 3.2.3, such a point $y \in \partial C$ is unique. Observe that the join $u_0 c_{K'}(y)$ is a unique simplex of K'' such that $x \in \operatorname{rint} u_0 c_{K'}(y)$, where $c_{K'}(y)$ is the carrier of y in K'. □

Proposition 4.6.2. *Let K be a cell complex and L' a simplicial subdivision of a subcomplex L of K. Given $v_C \in \operatorname{rint} C$ for each $C \in K \setminus L$, there exists a simplicial subdivision K' of K such that L' is a full subcomplex of K' and*

$$K'^{(0)} = L'^{(0)} \cup K^{(0)} \cup \{v_C \mid C \in K \setminus L\}.$$

Proof. For each $n \in \omega$, let $K_n = L \cup K^{(n)}$. Then, each K_n is a subcomplex of K and $K = \bigcup_{n\in\mathbb{N}} K_n$. Note that $K'_0 = L' \cup K^{(0)} \lhd K_0$ with $K_0'^{(0)} = L'^{(0)} \cup K^{(0)}$ and L' is full in K'_0. Assume that $K'_{n-1} \lhd K_{n-1}$ such that L' is a full subcomplex of K'_{n-1} and

$${K'_{n-1}}^{(0)} = K_0'^{(0)} \cup \{v_C \mid C \in K^{(n-1)} \setminus L\}.$$

For each $C \in K^{(n)} \setminus K_{n-1}$, let

$$K_{\partial C} = \{\sigma \in K'_{n-1} \mid \sigma \subset \partial C\} \lhd F(\partial C).$$

By Lemma 4.6.1, we have

$$K_C = K_{\partial C} \cup \{v_C\} \cup \{v_C\sigma \mid \sigma \in K_{\partial C}\} \lhd F(C)$$

and $K_{\partial C}$ is full in K_C. Thus, we have a simplicial complex

$$K'_n = K'_{n-1} \cup \bigcup_{C \in K^{(n)} \setminus K_{n-1}} K_C.$$

It is easy to see that $K'_n \lhd K_n$, L' is a full subcomplex in K'_n, and

$$K_n'^{(0)} = K_0'^{(0)} \cup \{v_C \mid C \in K^{(n)} \setminus L\}.$$

By induction, we have $K'_n \lhd K_n$, $n \in \mathbb{N}$, with the above conditions. Observe that $K' = \bigcup_{n\in\mathbb{N}} K'_n$ is the desired simplicial subdivision. □

When $L = \emptyset$ in Proposition 4.6.2 above, the obtained subdivision K' is called a **derived subdivision** of K. This is written as follows:

$$K' = \{\langle v_{C_1}, \dots, v_{C_n}\rangle \mid C_1 < \cdots < C_n \in K\},$$

and K' is an ordered simplicial complex with the natural order on $K'^{(0)}$ defined as follows: $v_C \leq v_D$ if $C \leq D$. If K is simplicial and $v_\sigma = \hat{\sigma}$ for each $\sigma \in K$, the derived subdivision K' of K is called the **barycentric subdivision** of K, which is denoted by $\operatorname{Sd} K$.

When L is simplicial and $L' = L$ in Proposition 4.6.2, we call K' a **derived subdivision** of K **relative to** L, where

$$\begin{aligned} K' = L \cup \{\langle v_{C_1}, \dots, v_{C_n}\rangle \mid C_1 < \cdots < C_n \in K \setminus L\} \\ \cup \{\langle v_1, \dots, v_m, v_{C_1}, \dots, v_{C_n}\rangle \mid \langle v_1, \dots, v_m\rangle \in L, \\ C_1 < \cdots < C_n \in K \setminus L \text{ with } \langle v_1, \dots, v_m\rangle < C_1\}. \end{aligned}$$

If K and L are simplicial and $v_\sigma = \hat{\sigma}$ for each $\sigma \in K \setminus L$, the derived subdivision of K relative to L is called the **barycentric subdivision** of K **relative to** L, which is denoted by $\operatorname{Sd}_L K$ (cf. Fig. 4.5). Note that L is a full subcomplex of $\operatorname{Sd}_L K$ by Proposition 4.6.2.

For simplicial complexes K and L, since the barycentric subdivisions $\operatorname{Sd} K$ and $\operatorname{Sd} L$ are ordered simplicial complexes, we can define the product simplicial complex $\operatorname{Sd} K \times_s \operatorname{Sd} L$, that consists of simplexes

$$\langle(\hat{\sigma}_1, \hat{\tau}_1), \dots, (\hat{\sigma}_n, \hat{\tau}_n)\rangle,\ \sigma_1 \leq \cdots \leq \sigma_n \in K,\ \tau_1 \leq \cdots \leq \tau_n \in L.$$

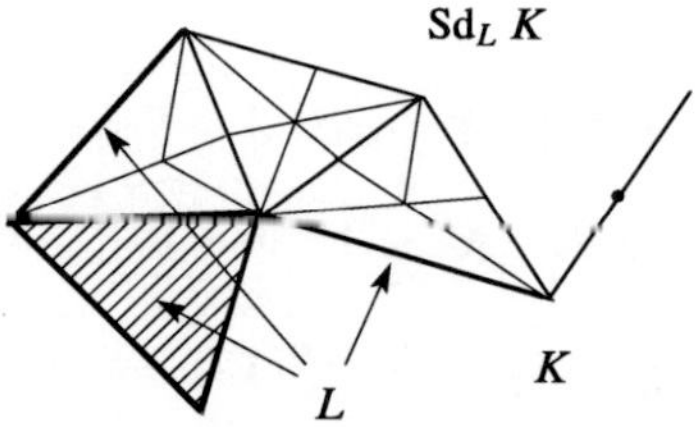

Fig. 4.5 Definition of $\mathrm{Sd}_L\,K$

On the other hand, by giving $v_{\sigma\times\tau} \in \mathrm{rint}(\sigma \times \tau)$ for each $\sigma \times \tau \in K \times_c L$, we can define a derived subdivision of the product cell complex $K \times_c L$, which consists of simplexes

$$\langle v_{\sigma_1\times\tau_1}, \ldots, v_{\sigma_n\times\tau_n}\rangle,\ \sigma_1 \times \tau_1 < \cdots < \sigma_n \times \tau_n \in K \times_c L.$$

If $v_{\sigma\times\tau} = (\hat{\sigma}, \hat{\tau})$ for each $\sigma \times \tau \in K \times_c L$, the derived subdivision of $K \times_c L$ is simply $\mathrm{Sd}\,K \times_s \mathrm{Sd}\,L$.

Applying these derived subdivisions, we prove the following:

Theorem 4.6.3. *Let K, K_0, and L be simplicial complexes such that $K_0 \subset K \cap L$, $|L| \subset |K|$, and $L \setminus K_0$ is finite. Then, K has a simplicial subdivision K' such that K_0 is a full subcomplex of K' and L is subdivided by some subcomplex of K'.*

Proof. We show the theorem by induction on $n = \mathrm{card}(L \setminus K_0)$. In the case $n = 0$, a derived subdivision K' of K relative to $L = K_0$ is the desired one. In the case $n > 0$, let σ be a maximal dimensional simplex of $L \setminus K_0$. Since σ is a principal simplex of L, $L_1 = L \setminus \{\sigma\}$ is a subcomplex of L. By the inductive assumption, we have $K' \lhd K$ such that K_0 is a full subcomplex of K' and $L_1' \lhd L_1$ for some $L_1' \subset K'$. Then, $\partial\sigma$ is triangulated by the subcomplex $L'_{\partial\sigma} = \{\tau \in L_1' \mid \tau \subset \partial\sigma\}$ of L_1' ($\subset K'$). Let

$$L'_\sigma = \big\{\tau \cap \sigma \ \big|\ \tau \in K' \ \text{ such that } \tau \cap \sigma \neq \emptyset\big\}.$$

Since $\tau \cap \sigma' \leq \tau$ for each $\sigma' < \sigma$ and $\tau \in K'$ with $\tau \cap \sigma' \neq \emptyset$, L'_σ is a cell complex by Proposition 4.2.12. Then, $|L'_\sigma| = \sigma$ and $L'_{\partial\sigma} \subset L'_\sigma$. It should be noted that if $\tau \cap \sigma \neq \emptyset$ but $\mathrm{rint}\,\tau \cap \sigma = \emptyset$, then $\tau \cap \sigma = \tau' \cap \sigma$ for some $\tau' < \tau$ with $\mathrm{rint}\,\tau' \cap \sigma \neq \emptyset$.

Now, consider the subcomplex $K_1' = \{\tau \in K' \mid \tau \cap \mathrm{rint}\,\sigma = \emptyset\}$ of K'. Then, $K_0 \cup L_1' \subset K_1'$ and $\sigma \cap |K_1'| = |L'_{\partial\sigma}| = \partial\sigma$. For each $\tau \in K' \setminus K_1'$, choose $v_\tau \in \mathrm{rint}\,\tau$ so that

$$\mathrm{rint}\,\tau \cap \mathrm{rint}\,\sigma \neq \emptyset \ \Rightarrow\ v_\tau \in \mathrm{rint}\,\tau \cap \mathrm{rint}\,\sigma,$$

where it should be noted that $\mathrm{rint}\,\tau \cap \mathrm{rint}\,\sigma = \emptyset$ implies $\mathrm{rint}\,\tau' \cap \mathrm{rint}\,\sigma \neq \emptyset$ for some $\tau' < \tau$ because $\tau \cap \mathrm{rint}\,\sigma \neq \emptyset$. Using these points v_τ, we define K'' as a derived

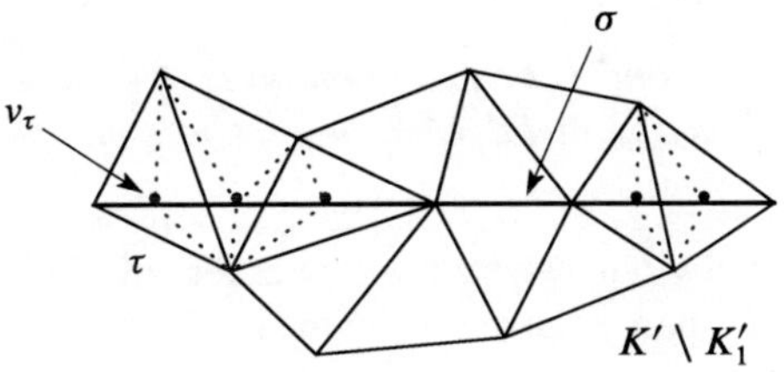

Fig. 4.6 Definition of K''

subdivision of K' relative to K'_1 (cf. Fig. 4.6). Then, $K_0 \cup L'_1 \subset K''$. Since K_0 is full in K'_1 and K'_1 is full in K'', it follows that K_0 is full in K''.

On the other hand, for each cell $C \in L'_\sigma \setminus L'_{\partial\sigma}$, let $v_C = v_{\tau_C}$, where $\tau_C \in K'$ is the unique cell such that $C = \tau_C \cap \sigma$ and $\operatorname{rint} \tau_C \cap \sigma \neq \emptyset$. Then, $v_C \in \operatorname{rint} C$. Indeed, since $\operatorname{rint} \tau_C \cap \operatorname{rint} \sigma \neq \emptyset$, we have $\operatorname{rint} C = \operatorname{rint} \tau_C \cap \operatorname{rint} \sigma$ by 4.1.9(2). Using these points v_C, we have the derived subdivision L''_σ of L'_σ relative to $L'_{\partial\sigma}$. Then, L''_σ is a triangulation of σ, which is simply the subcomplex of K'' consisting of simplexes with vertices in σ, that is, $L''_\sigma = \{\tau \in K'' \mid \tau \subset \sigma\}$. Thus, we have a subcomplex $L'' = L'_1 \cup L''_\sigma$ of K'' such that $L'' \lhd L$. □

To prove that proper PL maps are simplicial with respect to some simplicial subdivisions, we need the following:

Lemma 4.6.4. *Let $C_1, \ldots, C_n$ be cells contained in a cell C and K_0 be a triangulation of ∂C such that if $C_i \cap \partial C \neq \emptyset$ then $C_i \cap \partial C$ is triangulated by a subcomplex of K_0. Then, C has a triangulation K such that $K_0 \subset K$ and $C_1, \ldots, C_n$ are triangulated by subcomplexes of K.*

Proof. It suffices to prove the case $n = 1$. Indeed, assume that C has triangulations $K_1, \ldots, K_n$ such that $K_0 \subset K_i$ and C_i is triangulated by a subcomplex of K_i. By Corollary 4.2.13, we have a common simplicial subdivision K of $K_1, \ldots, K_n$ with $K_0 \subset K$, which is the desired triangulation of C.

The case $n = 1$ can be shown as follows: Consider the cell complex $F(C_1)$ and its subcomplex $L_1 = \{D \in F(C_1) \mid D \subset \partial C\}$, where $|L_1| = C_1 \cap \partial C$. Indeed, for each $x \in C_1 \cap \partial C$, take $D \leq C_1$ with $x \in \operatorname{rint} D$. Then, $D = D_x \subset C_x$. Since $x \notin \operatorname{rint} C$ implies $C_x < C$, we have $D \subset \partial C$, that is, $D \in L_1$. Note that K_0 gives the simplicial subdivision L'_1 of L_1. We can apply Proposition 4.6.2 to obtain a simplicial subdivision K_1 of $F(C_1)$ with $L'_1 \subset K_1$. Then, $K_1 \cup K_0$ is a triangulation of $C_1 \cup \partial C$. On the other hand, $F(C)$ has a simplicial subdivision K with $K_0 \subset K$ by Proposition 4.6.2. Applying Theorem 4.6.3, we can obtain a simplicial subdivision K'_1 of K such that $K_0 \subset K'_1$ and $K_1 \cup K_0$ is subdivided by a subcomplex of K'_1. □

Now, we shall show the following:

Proposition 4.6.5. *Let K and L be cell complexes. A proper map $f : |K| \to |L|$ is PL if and only if f is simplicial with respect to some simplicial subdivisions $K' \lhd K$ and $L' \lhd L$.*

Proof. The "if" part is obvious, where the properness of f is not necessary.

To see the "only if" part, replace K with a subdivision so that $f|C$ can be assumed to be affine for each $C \in K$. For each cell $D \in L$, let K_D be the smallest subcomplex of K such that $f^{-1}(D) \subset |K_D|$, which also can be defined as

$$K_D = \{C \in K \mid \exists C' \in K \text{ such that } C \leq C',\ \operatorname{rint} C' \cap f^{-1}(D) \neq \emptyset\}.$$

Since $f^{-1}(D)$ is compact by the properness of f, it follows that K_D is a finite subcomplex of K. According to the definition, $K_{D'} \subset K_D$ for $D' < D$.

By induction on $\dim D$, we can apply Lemma 4.6.4 to obtain a triangulation L_D of each $D \in L$ such that $f(C) \cap D$ is triangulated by a subcomplex of L_D for each $C \in K_D$ and $L_{D'} \subset L_D$ for every $D' < D$. Indeed, assume that $L_{D'}$ has been obtained for every $D' < D$. Then, $L_{\partial D} = \bigcup_{D'<D} L_{D'}$ is a triangulation of ∂D. Applying Lemma 4.6.4, we have a triangulation L_D of D such that $L_{\partial D} \subset L_D$ and $f(C) \cap D$ is triangulated by a subcomplex of L_D for every $C \in K_D$.

Now, we have a simplicial subdivision $L' = \bigcup_{D \in L} L_D$ of L, where $f(C)$ is triangulated by a subcomplex of L' for every $C \in K$. For each $C \in K$ and $\tau \in L'$ with $\tau \cap f(C) \neq \emptyset$, $C \cap f^{-1}(\tau) = (f|C)^{-1}(\tau)$ is a cell and $(f|C)^{-1}(\tau)_x = C_x \cap f^{-1}(\tau_{f(x)})$ for $x \in C \cap f^{-1}(\tau)$ by 4.1.9(3). By the analogy of Proposition 4.2.12, we can show that, for each $C, C' \in K$ and $\tau, \tau' \in L$,

$$\operatorname{rint}(C \cap f^{-1}(\tau)) \cap \operatorname{rint}(C' \cap f^{-1}(\tau')) \neq \emptyset \Rightarrow C \cap f^{-1}(\tau) = C' \cap f^{-1}(\tau').$$

Thus, the following is a cell complex:

$$\{C \cap f^{-1}(\tau) \mid C \in K,\ \tau \in L',\ C \cap f^{-1}(\tau) \neq \emptyset\},$$

which is a subdivision of K. By Theorem 4.2.10, we have a simplicial subdivision K' of this complex with the same vertices. Then, f is simplicial with respect to K' and L'. □

As we saw at the end of Sect. 4.4, the properness of f is essential in the "only if" part of Proposition 4.6.5. By Propositions 4.4.6 and 4.6.5, we have the following:

Corollary 4.6.6. *Let K_1, K_2, and K_3 be simplicial complexes. For each simplicial map $f : K_1 \to K_2$ and each proper PL map $g : |K_2| \to |K_3|$, there are simplicial subdivisions $K_1' \lhd K_1$ and $K_3' \lhd K_3$ such that $gf : K_1' \to K_3'$ is simplicial.*

Proof. Using Proposition 4.6.5, we can find simplicial subdivisions $K_2' \lhd K_2$ and $K_3' \lhd K_3$ such that $g : K_2' \to K_3'$ is simplicial. Then, by Proposition 4.4.6, K_1 has simplicial subdivisions $K_1' \lhd K_1$ such that $f : K_1' \to K_2'$ is simplicial, whence $gf : K_1' \to K_3'$ is simplicial. □

Remark 7. As observed, a PL map cannot be defined as a map $f : |K| \to |L|$ that is simplicial with respect to some simplicial subdivisions $K' \lhd K$ and $L' \lhd L$. If we adopted such a definition, then we could not assert that the composition of PL maps is PL. In fact, even if $f : K_1 \to K_2$ is a simplicial isomorphism and $g : |K_2| \to |K_3|$ is simplicial with respect to some simplicial subdivisions, the composition gf is not simplicial with respect to any simplicial subdivisions. For example, let $K = \{n,\ [n, n+1] \mid n \in \omega\}$ and $I = \{0,\ 1,\ \mathbf{I}\}$ be the natural triangulations of $[0, \infty)$ and $\mathbf{I}$, respectively. We define $K' \lhd K$ as follows:

$$K' = \{n,\ n + 2^{-(n+1)},\ [n, n + 2^{-(n+1)}],\ [n + 2^{-(n+1)}, n+1] \mid n \in \omega\}.$$

Let $f : K \to K'$ and $g : K \to I$ be the simplicial maps defined by

$$f(2n) = n,\ f(2n+1) = n + 2^{-(n+1)},\ g(2n) = 0 \text{ and } g(2n+1) = 1.$$

Then, $f : K \to K'$ is a simplicial isomorphism and $g : |K'| = |K| \to |I| = \mathbf{I}$ is a PL map. Observe that $gf(4n+1) = 2^{-(2n+1)}$ and $gf(4n+3) = 1 - 2^{-(2n+2)}$ for each $n \in \omega$, whence $gf(\omega)$ is infinite. Since every subdivision of K contains ω as vertices but every subdivision of I contains only finitely many vertices, gf is not simplicial with respect to any simplicial subdivisions of K and I.

As we saw in Sect. 4.5, a subdivision generally changes the metric topology but the barycentric subdivision does not.

Theorem 4.6.7. *For each simplicial complex K, $|\operatorname{Sd} K|_{\mathrm{m}} = |K|_{\mathrm{m}}$ as spaces.*

Proof. When K is finite, the result follows from Proposition 4.5.7 and the compactness of $|\operatorname{Sd} K|_{\mathrm{m}}$. We may assume that K is infinite. By Proposition 4.5.7, it suffices to show that $\mathrm{id} : |K|_{\mathrm{m}} \to |\operatorname{Sd} K|_{\mathrm{m}}$ is continuous. Let $x \in |K|$ and $k = \dim c_K(x)$. For each $\varepsilon > 0$, choose $\delta > 0$ so that $\delta < (2k+3)^{-1}\varepsilon$ and

$$\beta_v^K(x) \neq \beta_{v'}^K(x) \Rightarrow \delta < \frac{1}{2}|\beta_v^K(x) - \beta_{v'}^K(x)|.$$

The last condition implies that $\delta < \frac{1}{2}\beta_v^K(x)$ for every $v \in c_K(x)^{(0)}$ because $\beta_{v'}^K(x) = 0$ for some $v' \in K^{(0)}$. For each $y \in |K|$ with $\rho_K(x, y) < \delta$, the following hold:

$$\beta_v^K(x) > \beta_{v'}^K(x) \Rightarrow \beta_v^K(y) > \beta_{v'}^K(y)\,;$$
$$\beta_v^K(x) = 0 \Leftrightarrow \beta_v^K(y) < \delta.$$

Note that the first implication holds even if $\beta_{v'}^K(x) = 0$, hence $c_K(x) \leq c_K(y)$. Since $\beta_v^K(y) \leq \beta_{v'}^K(y)$ implies $\beta_v^K(x) \leq \beta_{v'}^K(x)$, we can write

$$c_K(x) = \langle v_0, \ldots, v_k \rangle,\ c_K(y) = \langle v_0, \ldots, v_n \rangle,\ k \leq n,$$
$$\beta_{v_0}^K(x) \geq \cdots \geq \beta_{v_k}^K(x) > 0 \text{ and } \beta_{v_0}^K(y) \geq \cdots \geq \beta_{v_n}^K(y) > 0.$$

For each $i = 0, \ldots, n$, let $\sigma_i = \langle v_0, \ldots, v_i \rangle$. Then, $\sigma_0 < \sigma_1 < \cdots < \sigma_n$, $\sigma_k = c_K(x)$, and $\sigma_n = c_K(y)$. Observe that

$$\begin{aligned} x &= \beta^K_{v_k}(x)v_k + \beta^K_{v_{k-1}}(x)v_{k-1} + \cdots + \beta^K_{v_1}(x)v_1 + \beta^K_{v_0}(x)v_0 \\ &= (k+1)\beta^K_{v_k}(x)\hat{\sigma}_k + k\big(\beta^K_{v_{k-1}}(x) - \beta^K_{v_k}(x)\big)\hat{\sigma}_{k-1} + \\ &\qquad \cdots + 2\big(\beta^K_{v_1}(x) - \beta^K_{v_2}(x)\big)\hat{\sigma}_1 + \big(\beta^K_{v_0}(x) - \beta^K_{v_1}(x)\big)\hat{\sigma}_0. \end{aligned}$$

Hence, $x \in \langle \hat{\sigma}_0, \ldots, \hat{\sigma}_k \rangle \in \operatorname{Sd} K$, $\beta^{\operatorname{Sd} K}_{\hat{\sigma}_k}(x) = (k+1)\beta^K_{v_k}(x)$ and

$$\beta^{\operatorname{Sd} K}_{\hat{\sigma}_i}(x) = (i+1)\big(\beta^K_{v_i}(x) - \beta^K_{v_{i+1}}(x)\big) \quad \text{for } i = 0, \ldots, k-1.$$

Similarly, we have $y \in \langle \hat{\sigma}_0, \ldots, \hat{\sigma}_n \rangle \in \operatorname{Sd} K$, $\beta^{\operatorname{Sd} K}_{\hat{\sigma}_n}(y) = (n+1)\beta^K_{v_n}(y)$ and

$$\beta^{\operatorname{Sd} K}_{\hat{\sigma}_i}(y) = (i+1)\big(\beta^K_{v_i}(y) - \beta^K_{v_{i+1}}(y)\big) \quad \text{for } i = 0, \ldots, n-1.$$

Then, it follows that

$$\begin{aligned} \rho_{\operatorname{Sd} K}(x, y) &= \sum_{i=0}^{k} \big|\beta^{\operatorname{Sd} K}_{\hat{\sigma}_i}(x) - \beta^{\operatorname{Sd} K}_{\hat{\sigma}_i}(y)\big| + \sum_{i=k+1}^{n} \beta^{\operatorname{Sd} K}_{\hat{\sigma}_i}(y) \\ &\le \sum_{i=0}^{k} (2i+1)\big|\beta^K_{v_i}(x) - \beta^K_{v_i}(y)\big| + (k+1)\beta^K_{v_{k+1}}(y) \\ &\qquad\qquad + (k+2)\beta^K_{v_{k+1}}(y) + \sum_{i=k+2}^{n} \beta^K_{v_i}(y) \\ &\le (2k+3)\sum_{i=0}^{n} \big|\beta^K_{v_i}(x) - \beta^K_{v_i}(y)\big| \\ &= (2k+3)\rho_K(x, y) < (2k+3)\delta < \varepsilon. \end{aligned}$$

Thus, $\mathrm{id} : |K|_{\mathrm{m}} \to |\operatorname{Sd} K|_{\mathrm{m}}$ is continuous. □

We have the following generalization of 4.2.16(1):

Proposition 4.6.8. *For each infinite cell complex K,* $\operatorname{dens}|K| = \operatorname{card} K^{(0)}$. *If K is simplicial,* $\operatorname{dens}|K| = \operatorname{dens}|K|_{\mathrm{m}} = \operatorname{card} K^{(0)}$.

Sketch of Proof. As in 4.2.16(1), we can construct a dense set D in $|K|$ with $\operatorname{card} D = \operatorname{card} K = \operatorname{card} K^{(0)}$, hence $\operatorname{dens}|K| \le \operatorname{card} K^{(0)}$. On the other hand, let K' be a derived subdivision of K. Then, $\{O_{K'}(v) \mid v \in K^{(0)}\}$ is a pair-wise disjoint collection of open sets in $|K|$, hence $\operatorname{card} K^{(0)} \le c(|K|) \le \operatorname{dens}|K|$.

If K is simplicial, the above D is also dense in $|K|_{\mathrm{m}}$. Moreover, $\{O_{\operatorname{Sd} K}(v) \mid v \in K^{(0)}\}$ is a pair-wise disjoint collection of open sets in $|K|_{\mathrm{m}}$.

It should be remarked that $w(|K|_{\mathrm{m}}) = \operatorname{dens}|K|_{\mathrm{m}} = \operatorname{card} K^{(0)}$ for every infinite simplicial complex K (see Sect. 1.1) but, as we saw in 4.2.16(4), $w(|K|) \neq \operatorname{card} K^{(0)}$ in general ($w(|K|) \geq \operatorname{dens}|K| = \operatorname{card} K^{(0)}$).

4.7 Small Subdivisions

In this section, it is proved that every simplicial complex K has an arbitrarily small simplicial subdivision, that is, for any $\mathcal{U} \in \operatorname{cov}(|K|)$, there is $K' \lhd K$ such that $\mathcal{S}_{K'} \prec \mathcal{U}$. As applications, we can prove the paracompactness of $|K|$ and the Simplicial (PL) Approximation Theorem.

We inductively define the ***n*-th barycentric subdivision** $\operatorname{Sd}^n K$ of K and the ***n*-th barycentric subdivision** $\operatorname{Sd}^n_L K$ **of** K **relative to** L as follows:

$$\operatorname{Sd}^n K = \operatorname{Sd}(\operatorname{Sd}^{n-1} K) \text{ and } \operatorname{Sd}^n_L K = \operatorname{Sd}_L(\operatorname{Sd}^{n-1}_L K),$$

where $\operatorname{Sd}^0 K = \operatorname{Sd}^0_L K = K$. In the following, we show that if K is finite then the size of $\operatorname{Sd}^m K$ becomes smaller as m gets larger. The following is a special case of 3.2.7(5)(the proof is easy):

Lemma 4.7.1. *For every cell C in a normed linear space $E = (E, \|\cdot\|)$, the following hold:*

(1) $\|x - y\| \leq \sup_{v \in C^{(0)}} \|x - v\|$ *for each $x \in E$ and $y \in C$;*
(2) $\operatorname{diam} C = \operatorname{diam} C^{(0)}$. □

Proposition 4.7.2. *For every simplicial complex K in a normed linear space E,* $\operatorname{mesh} K = \operatorname{mesh} K^{(1)}$.

Proof. By Lemma 4.7.1(2), we have $\operatorname{diam} \sigma = \max_{v,u \in \sigma^{(0)}} \operatorname{diam}\langle u, v\rangle$ for each $\sigma \in K$. Then, it follows that $\operatorname{mesh} K = \operatorname{mesh} K^{(1)}$. □

Lemma 4.7.3. *Let $f : \sigma \to E$ be an affine map of an n-simplex σ into a normed linear space E. Then,*

$$\operatorname{mesh} f(\operatorname{Sd} F(\sigma)) \leq \frac{n}{n+1} \cdot \operatorname{diam} f(\sigma).$$

Proof. Note that $f(\sigma)$ is a cell in E and $f(\sigma)^{(0)} \subset f(\sigma^{(0)})$. Let $\sigma'' < \sigma' \leq \sigma$. It follows from Lemma 4.7.1(1), (2) that

$$\begin{aligned} \|f(\hat{\sigma}') - f(\hat{\sigma}'')\| &\leq \max_{v \in \sigma''^{(0)}} \|f(\hat{\sigma}') - f(v)\| \\ &\leq \max_{v \in \sigma''^{(0)}} \sum_{u \in \sigma'^{(0)}} \frac{1}{\dim \sigma' + 1} \|f(u) - f(v)\| \end{aligned}$$

$$= \max_{v \in \sigma''^{(0)}} \sum_{u \in \sigma'^{(0)} \setminus \{v\}} \frac{1}{\dim \sigma' + 1} \|f(u) - f(v)\|$$

$$\leq \frac{\dim \sigma'}{\dim \sigma' + 1} \cdot \operatorname{diam} f(\sigma') \leq \frac{n}{n+1} \cdot \operatorname{diam} f(\sigma).$$

Then, we have the result by Proposition 4.7.2. □

The following is the special case of Lemma 4.7.3 when $f = \mathrm{id}_\sigma$:

Lemma 4.7.4. *For every n-simplex σ in a normed linear space E,*

$$\operatorname{mesh} \operatorname{Sd} F(\sigma) \leq \frac{n}{n+1} \cdot \operatorname{diam} \sigma.$$ □

For each simplicial complex K, we have the isometry $\beta^K : (|K|, \rho_K) \to \ell_1(K^{(0)})$. Then, we can regard $|K| \subset \ell_1(K^{(0)})$, and ρ_K is induced from the norm of $\ell_1(K^{(0)})$. Moreover, for an n-simplex $\sigma \in K$, $\rho_K(v, \hat{\sigma}) = 2n/(n+1)$ for every $v \in \sigma^{(0)}$. Hence, the following can be obtained from Lemma 4.7.4:

Proposition 4.7.5. *Let K be a simplicial complex. For any simplicial subdivision $K' \lhd K$, $\operatorname{mesh}_{\rho_K} K' = \operatorname{mesh}_{\rho_K} K'^{(1)}$. If $\dim K = m$ then*

$$\operatorname{mesh}_{\rho_K} \operatorname{Sd} K' \leq \frac{m}{m+1} \cdot \operatorname{mesh}_{\rho_K} K'.$$

In particular, if $K' = K$ and $\dim K = m$, we have $\operatorname{mesh}_{\rho_K} \operatorname{Sd} K = 2m/(m+1)$. □

Using Proposition 4.7.5 inductively, we have the following:

Theorem 4.7.6. *For every finite-dimensional simplicial complex K,*

$$\operatorname{mesh}_{\rho_K} \operatorname{Sd}^n K \leq 2\left(\frac{\dim K}{\dim K + 1}\right)^n \quad \textit{for every } n \in \mathbb{N}.$$

Hence, $\lim_{n\to\infty} \operatorname{mesh}_{\rho_K} \operatorname{Sd}^n K = 0$. □

Corollary 4.7.7. *For a finite simplicial complex K, each open cover $\mathcal{U}$ of $|K|_{\mathrm{m}}$ $(= |K|)$ is refined by $\mathcal{S}_{\operatorname{Sd}^n K}$ for some $n \in \mathbb{N}$.*

Proof. Since $|K|_{\mathrm{m}} = |K|$ is compact (Proposition 4.2.6 and Theorem 4.5.6) and $\dim K < \infty$, we can choose $n \in \mathbb{N}$ so large that $\operatorname{mesh}_{\rho_K} \operatorname{Sd}^n K$ is less than the Lebesgue number of $\mathcal{U}$. Then, we have the desired n. □

For infinite-dimensional simplicial complexes, we have the following:

Proposition 4.7.8. *If K is an infinite-dimensional simplicial complex, then*

$$\operatorname{mesh}_{\rho_K} \operatorname{Sd}^n K = 2 \ (= \operatorname{mesh}_{\rho_K} K) \ \textit{for every } n \in \mathbb{N}.$$

Proof. By induction on $n \in \mathbb{N}$, we shall show the following:

$(\star)_n$ for each $m \in \mathbb{N}$ and $\varepsilon > 0$, there exist a vertex $v \in K^{(0)}$ and an m-simplex $\sigma \in \mathrm{Lk}(v, \mathrm{Sd}^n K)$ such that $\beta_v^K(u) < \varepsilon$ for every $u \in \sigma^{(0)}$.

If $(\star)_n$ has been shown, then $\tau = \langle \{v\} \cup \sigma \rangle \in \mathrm{Sd}^n K$ and

$$\operatorname{diam} \tau \geq \max_{u \in \sigma^{(0)}} \rho_K(v, u) = \max_{u \in \sigma^{(0)}} 2(1 - \beta_v^K(u)) > 2(1 - \varepsilon).$$

Thus, it would follow that $\operatorname{mesh}_{\rho_K} \mathrm{Sd}^n K = 2$.

To see $(\star)_1$, for each $m \in \mathbb{N}$ and $\varepsilon > 0$, choose $k \in \mathbb{N}$ so that $(k+2)^{-1} < \varepsilon$. Since $\dim K = \infty$, K has simplexes $\sigma_1 < \cdots < \sigma_{m+1}$ with $\dim \sigma_i = k + i$. Let $v \in \sigma_1^{(0)}$. Then, we have an m-simplex $\sigma = \langle \hat{\sigma}_1, \ldots, \hat{\sigma}_{m+1} \rangle \in \mathrm{Lk}(v, \mathrm{Sd}\, K)$. Observe that $\beta_v^K(\hat{\sigma}_i) = (k + i + 1)^{-1} < \varepsilon$ for every $i = 1, \ldots, m+1$.

Now, we prove the implication $(\star)_n \Rightarrow (\star)_{n+1}$. For each $m \in \mathbb{N}$ and $\varepsilon > 0$, choose $k \in \mathbb{N}$ so that $(k+2)^{-1} < \varepsilon/2$. By $(\star)_n$, we have $v \in K^{(0)}$ and a $(k+m)$-simplex $\sigma = \langle v_1, \ldots, v_{k+m+1} \rangle \in \mathrm{Lk}(v, \mathrm{Sd}^n K)$ such that

$$\beta_v^K(v_i) < \varepsilon/2 \text{ for every } i = 1, \ldots, k+m+1.$$

For each $i = 1, \ldots, m+1$, let $\sigma_i = \langle v, v_1, \ldots, v_{k+i} \rangle \in \mathrm{Sd}^n K$. Since $v < \sigma_1 < \cdots < \sigma_{m+1}$, we have an m-simplex

$$\sigma = \langle \hat{\sigma}_1, \ldots, \hat{\sigma}_{m+1} \rangle \in \mathrm{Lk}(v, \mathrm{Sd}^{n+1} K).$$

Then, it follows that

$$\begin{aligned}\beta_v^K(\hat{\sigma}_i) &= \frac{1}{k+i+1}\left(1 + \sum_{j=1}^{k+i} \beta_v^K(v_j)\right) \\ &< \frac{1}{k+i+1}\left(1 + \frac{(k+i)\varepsilon}{2}\right) < \frac{1}{k+2} + \frac{\varepsilon}{2} < \varepsilon.\end{aligned}$$

This completes the proof. □

Remark 8. For the metric ρ_K^∞, we have the following:

- If $\dim K < \infty$ then $\operatorname{mesh}_{\rho_K^\infty} \mathrm{Sd}^n K \leq \left(\dfrac{\dim K}{\dim K + 1}\right)^n$ for every $n \in \mathbb{N}$;
- If $\dim K = \infty$ then $\operatorname{mesh}_{\rho_K^\infty} \mathrm{Sd}^n K = 1$ $(= \operatorname{mesh}_{\rho_K^\infty} K)$ for every $n \in \mathbb{N}$.

To construct small subdivisions of infinite simplicial complexes, the following is available:

Lemma 4.7.9. *Let K be a finite simplicial complex and L a subcomplex of K. Given an open neighborhood $U(v)$ of $|\mathrm{St}(v, L)|$ in $|K|_{\mathrm{m}}$ $(= |K|)$ for every $v \in L^{(0)}$, there exists a subdivision K' of K such that $L \subset K'$ and $|\mathrm{St}(v, K')| \subset U(v)$ for every $v \in L^{(0)}$.*

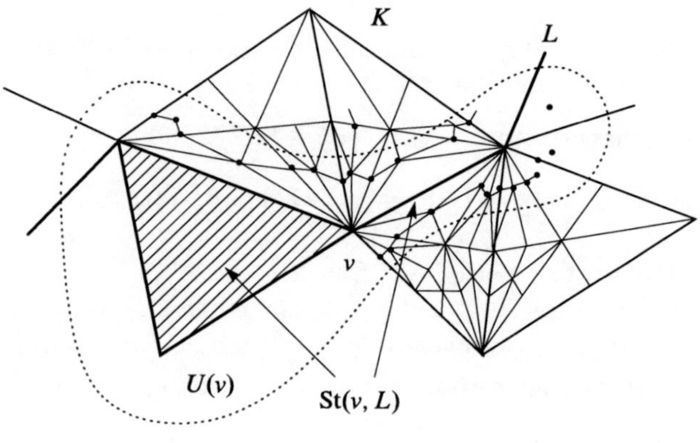

Fig. 4.7 Small subdivision

Proof. Replacing K by $\mathrm{Sd}_L K$, we may assume that L is full in K. Since $|\mathrm{St}(v, L)|$ and $|K| \setminus U(v)$ are disjoint closed sets in the compact metric space $(|K|, \rho_K)$, we have

$$\delta = \min_{v \in L^{(0)}} \mathrm{dist}_{\rho_K}(|\mathrm{St}(v, L)|,\ |K| \setminus U(v)) > 0.$$

For each $\sigma \in K \setminus L$, choose $v(\sigma) \in \mathrm{rint}\,\sigma$ so that

$$\sigma \cap |L| \neq \emptyset \Rightarrow \mathrm{dist}_{\rho_K}(v(\sigma), \sigma \cap |L|) < \delta.$$

Using the points $v(\sigma)$, we define K' as a derived subdivision of K relative to L, which is as desired (Fig. 4.7).

To see that $|\mathrm{St}(v, K')| \subset U(v)$ for each $v \in L^{(0)}$, let $x \in |\mathrm{St}(v, K')|$. Then, we have $\sigma_0 = \langle v_1, \ldots, v_m \rangle \in \mathrm{St}(v, L)$ and $\sigma_1 < \cdots < \sigma_n \in K \setminus L$ such that

$$v_1 = v,\ \sigma_0 < \sigma_1 \text{ and } x \in \langle v_1, \ldots, v_m, v(\sigma_1), \ldots, v(\sigma_n) \rangle \subset \sigma_n,$$

whence we can write

$$x = \sum_{i=1}^{m} z(i) v_i + \sum_{j=1}^{n} z(j+m) v(\sigma_j) \text{ for some } z \in \Delta^{m+n-1}.$$

For each $j = 1, \ldots, n$, since $\sigma_j \cap |L| \neq \emptyset$, we can choose $u_j \in \sigma_j \cap |L|$ so that $\rho_K(v(\sigma_j), u_j) < \delta$. Since L is full in K, $\sigma_i \cap |L| = \langle \sigma_i^{(0)} \cap |L| \rangle \in L$ for each $i = 1, \ldots, n$ and $\sigma_1 \cap |L| < \cdots < \sigma_n \cap |L|$, hence we have

$$y = \sum_{i=1}^{m} z(i) v_i + \sum_{j=1}^{n} z(j+m) u_j \in \sigma_n \cap |L| \subset |\mathrm{St}(v, L)|.$$

Since $\beta^K|\sigma_n$ is affine, it follows that

$$\begin{aligned}\rho_K(x,y) &= \left\|\beta^K(x) - \beta^K(y)\right\|_1 \\ &= \left\|\sum_{j=1}^n z(j+m)\beta^K(v(\sigma_j)) - \sum_{j=1}^n z(j+m)\beta^K(u_j)\right\|_1 \\ &\le \sum_{j=1}^n z(j+m)\left\|\beta^K(v(\sigma_j)) - \beta^K(u_j)\right\|_1 \\ &= \sum_{j=1}^n z(j+m)\rho_K(v(\sigma_j), u_j) < \delta.\end{aligned}$$

Thus, $\rho_K(x, |\operatorname{St}(v, L)|) < \delta$, which means that $x \in U(v)$. □

We can apply Lemma 4.7.9 above to strengthen Corollary 4.7.7 as follows:

Proposition 4.7.10. *Let K be a finite simplicial complex, L be a subcomplex of K, and $\mathcal{U}$ be an open cover of $|K|_{\mathrm{m}}$ $(= |K|)$ such that $\mathcal{S}_L \prec \mathcal{U}$. Then, K has a subdivision K' such that $L \subset K'$ and $\mathcal{S}_{K'} \prec \mathcal{U}$. Moreover, if $U(v) \in \mathcal{U}$ are given for all $v \in L^{(0)}$ so that $|\operatorname{St}(v, L)| \subset U(v)$, then $|\operatorname{St}(v, K')| \subset U(v)$ for all $v \in L^{(0)}$.*

Proof. As the additional statement, we assume that $|\operatorname{St}(v, L)| \subset U(v) \in \mathcal{U}$ for each $v \in L^{(0)}$. By Lemma 4.7.9, we have a subdivision K' of K such that $L \subset K'$ and $|\operatorname{St}(v, K')| \subset U(v)$ for each $v \in L^{(0)}$. By Corollary 4.7.7, $\mathcal{S}_{\operatorname{Sd}^m K'} \prec \mathcal{U}$ for some $m \in \mathbb{N}$. Then, $\operatorname{Sd}_L^m K'$ is the desired subdivision of K. Indeed, $|\operatorname{St}(v, \operatorname{Sd}_L^m K')| \subset |\operatorname{St}(v, K')| \subset U(v)$ for each $v \in L^{(0)}$. Let $v \in (\operatorname{Sd}_L^m K')^{(0)} \setminus L^{(0)}$. If $\operatorname{St}(v, \operatorname{Sd}_L^m K') \cap L = \emptyset$ then $|\operatorname{St}(v, \operatorname{Sd}_L^m K')| = |\operatorname{St}(v, \operatorname{Sd}^m K')|$, which is contained in some $U \in \mathcal{U}$. When $\operatorname{St}(v, \operatorname{Sd}_L^m K') \cap L \neq \emptyset$, we have $v' \in L^{(0)}$ such that $\langle v', v\rangle \in \operatorname{Sd}_L^m K'$, that is, $v' \in \sigma^{(0)}$ and $v = \hat{\sigma}$ for some $\sigma \in \operatorname{Sd}_L^{m-1} K'$. Then, it follows that

$$|\operatorname{St}(v, \operatorname{Sd}_L^m K')| \subset |\operatorname{St}(\sigma, \operatorname{Sd}_L^{m-1} K')| \subset |\operatorname{St}(v', K')| \subset U(v').$$

Therefore, $\mathcal{S}_{\operatorname{Sd}_L^m K'} \prec \mathcal{U}$. □

For a simplicial complex K, it is convenient to denote by $K(n)$ the set of all n-simplexes of K, that is, $K(n) = K^{(n)} \setminus K^{(n-1)}$. Now, we can prove the following:

Theorem 4.7.11 (J.H.C. Whitehead). *Let K be an arbitrary simplicial complex. For any open cover $\mathcal{U}$ of $|K|$, K has a simplicial subdivision K' such that $\mathcal{S}_{K'} \prec \mathcal{U}$.*

Proof. By induction, we shall construct subdivisions $K_n \lhd K^{(n)}$ and choose $U(v) \in \mathcal{U}$ for each $v \in K_n^{(0)} \setminus K_{n-1}^{(0)}$ so that $K_{n-1} \subset K_n$ and $|\operatorname{St}(v, K_n)| \subset U(v)$ for every $v \in K_n^{(0)} = \bigcup_{i \le n} K_i^{(0)}$. Then, $K' = \bigcup_{n\in\mathbb{N}} K_n$ would be the desired subdivision of K. Indeed, each $v \in K'^{(0)}$ belongs to some K_n, hence

$$|\operatorname{St}(v, K')| = \bigcup_{i \ge n} |\operatorname{St}(v, K_i)| \subset U(v) \in \mathcal{U}.$$

Now, assume that $K_{n-1} \lhd K^{(n-1)}$ has been constructed and $U(v) \in \mathcal{U}$ has been chosen for every $v \in K_{n-1}^{(0)}$ such that $|\operatorname{St}(v, K_{n-1})| \subset U(v)$. For each $\sigma \in K(n)$, let

$$K_{\partial\sigma} = \{\tau \in K_{n-1} \mid \tau \subset \partial\sigma\} \lhd F(\partial\sigma).$$

Applying Lemma 4.6.1, we have

$$K_\sigma = K_{\partial\sigma} \cup \{\hat{\sigma}\} \cup \{\hat{\sigma}\tau \mid \tau \in K_{\partial\sigma}\} \lhd F(\sigma).$$

By Proposition 4.7.10, we have $K'_\sigma \lhd K_\sigma$ such that $K_{\partial\sigma} \subset K'_\sigma$, $\mathcal{S}_{K'_\sigma} \prec \mathcal{U}$ and $|\operatorname{St}(v, K'_\sigma)| \subset U(v)$, for every $v \in K_{\partial\sigma}^{(0)}$. Then,

$$K_n = \bigcup_{\sigma \in K(n)} K'_\sigma \cup K_{n-1} \lhd K^{(n)}.$$

For each $v \in K_n^{(0)} \setminus K_{n-1}^{(0)}$, we have an n-simplex $\sigma \in K$ such that $v \in \operatorname{rint}\sigma$, whence $\operatorname{St}(v, K_n) = \operatorname{St}(v, K'_\sigma)$. Therefore, we can choose $U(v) \in \mathcal{U}$ so that $|\operatorname{St}(v, K_n)| \subset U(v)$. On the other hand, for every $v \in K_{n-1}^{(0)}$,

$$|\operatorname{St}(v, K_n)| = \bigcup\{|\operatorname{St}(v, K'_\sigma)| \mid v \in \sigma \in K(n)\} \cup |\operatorname{St}(v, K_{n-1})| \subset U(v).$$

This completes the proof. □

By Theorem 4.7.11 above, we have the following:

Corollary 4.7.12. *Every polyhedron is paracompact.*

Proof. It suffices to show that $|K|$ is paracompact for any simplicial complex K. For any open cover $\mathcal{U}$ of $|K|$, K has a subdivision K' such that $\mathcal{S}_{K'} \prec \mathcal{U}$. Then, $(\beta_v^{K'})_{v \in K'^{(0)}}$ is a partition of unity on $|K'| = |K|$ subordinated by $\mathcal{U}$. Hence, $|K|$ is paracompact by Theorem 2.7.5. □

Let K and L be simplicial complexes. A simplicial map $g : K \to L$ is called a **simplicial approximation** of a map $f : |K| \to |L|$ if each $g(x)$ is contained in the carrier $c_L(f(x))$ of $f(x)$ in L. Then, $f \simeq_L g$ by Proposition 4.3.4, which is realized by the straight-line homotopy. The definition of a **simplicial approximation** $g : K \to L$ of a map $f : |K|_{\mathrm{m}} \to |L|_{\mathrm{m}}$ is the same. Then, $f \simeq_L g$ by Proposition 4.5.4, which is also realized by the straight-line homotopy.

Lemma 4.7.13. *A simplicial map $g : K \to L$ is a simplicial approximation of a map $f : |K| \to |L|$ (or $f : |K|_{\mathrm{m}} \to |L|_{\mathrm{m}}$) if and only if $f(O_K(v)) \subset O_L(g(v))$ for every $v \in K^{(0)}$.*

Proof. First, assume that $g : K \to L$ is a simplicial approximation of f and let $v \in K^{(0)}$. For each $x \in O_K(v)$, we have $v \in c_K(x)^{(0)}$. Observe that $g(x) \in c_L(f(x)) \cap \operatorname{rint} g(c_K(x))$. Then, $g(c_K(x)) \leq c_L(f(x))$, hence $g(v) \in g(c_K(x))^{(0)} \subset c_L(f(x))^{(0)}$. Therefore, $f(x) \in \operatorname{rint} c_L(f(x)) \subset O_L(g(v))$.

Conversely, assume that $f(O_K(v)) \subset O_L(g(v))$ for every $v \in K^{(0)}$ and let $x \in |K|$. For every $v \in c_K(x)^{(0)}$, since $x \in \operatorname{rint} c_K(x) \subset O_K(v)$, we have $f(x) \in f(O_K(v)) \subset O_L(g(v))$, which means that $g(v) \in c_L(f(x))^{(0)}$. Thus, $g(c_K(x)) \leq c_L(f(x))$. Therefore, $g(x) \in c_L(f(x))$. □

As an application of Theorem 4.7.11, we have the following:

Theorem 4.7.14 (SIMPLICIAL APPROXIMATION). *Let K and L be simplicial complexes. Then, each map $f : |K| \to |L|$ has a simplicial approximation $g : K' \to L$ for some $K' \lhd K$.*

Proof. By Theorem 4.7.11, we have $K' \lhd K$ such that $\mathcal{S}_{K'} \prec f^{-1}(\mathcal{O}_L)$. Let $g' : K'^{(0)} \to L^{(0)}$ be a function such that $|\operatorname{St}(v, K')| \subset f^{-1}(O_L(g'(v)))$. For each $\sigma \in K'$,

$$\sigma \subset \bigcap_{v \in \sigma^{(0)}} |\operatorname{St}(v, K')| \subset \bigcap_{v \in \sigma^{(0)}} f^{-1}(O_L(g'(v))),$$

hence $\bigcap_{v \in \sigma^{(0)}} O_L(g'(v)) \neq \emptyset$. Due to Proposition 4.4.5, g' induces a simplicial map $g : K' \to L$. By Lemma 4.7.13, g is a simplicial approximation of f. □

Remark 9. We can easily generalize Theorem 4.7.14 as follows: Let K and $L_1, \ldots, L_n$ be simplicial complexes. For any maps $f_i : |K| \to |L_i|$, $i = 1, \ldots, n$, there exist a subdivision K' of K and simplicial maps $g_i : K' \to L_i$, $i = 1, \ldots, n$, such that each g_i is a simplicial approximation of f_i.

The following is a combination of Theorems 4.7.11 and 4.7.14:

Corollary 4.7.15 (PL APPROXIMATION THEOREM). *Let K and L be simplicial complexes and $f : |K| \to |L|$ be a map. For each open cover $\mathcal{U}$ of $|L|$, there is a PL map $g : |K| \to |L|$ that is $\mathcal{U}$-close to f.* □

4.8 Admissible Subdivisions

Let K be a simplicial complex and K' a simplicial subdivision of K. In general, the metric $\rho_{K'}$ is not admissible for $|K|_{\mathrm{m}}$, so the topology induced by $\rho_{K'}$ is different from the one induced by ρ_K (cf. Sect. 4.5). We call K' an **admissible subdivision** of K if the metric $\rho_{K'}$ is admissible for $|K|_{\mathrm{m}}$; equivalently, $|K'|_{\mathrm{m}} = |K|_{\mathrm{m}}$ as spaces. [9] For instance, the barycentric subdivision is admissible (Theorem 4.6.7) and, if K is locally finite, every subdivision of K is admissible (Theorem 4.5.6). In this section, we prove the metric topology version of Whitehead's Theorem 4.7.11 on small subdivisions. It should be remarked that $\operatorname{mesh}_{\rho_K} \operatorname{Sd}^n K = 2$ $(= \operatorname{mesh}_{\rho_K} K)$ for every $n \in \mathbb{N}$ if $\dim K = \infty$ (Proposition 4.7.8).

[9] In [7], such a subdivision is called a **proper subdivision**.

First, we give the following characterization of admissible subdivisions using open stars:

Lemma 4.8.1. *A simplicial subdivision K' of a simplicial complex K is admissible if and only if the open star $O_{K'}(v)$ of each vertex $v \in K'(0)$ is open in $|K|_{\mathrm{m}}$.*

Proof. The "only if" part follows from the fact that the open star $O_{K'}(v)$ is open in $|K'|_{\mathrm{m}}$. To see the "if" part, it suffices to show that $\mathrm{id} : |K|_{\mathrm{m}} \to |K'|_{\mathrm{m}}$ is continuous at each $x \in |K|$ (Proposition 4.5.7). Let U be a neighborhood of x in $|K'|_{\mathrm{m}}$. We can find $0 < r < 1$ such that $O_{K'}(x, r) \in U$ by Proposition 4.5.1. On the other hand, $O_{K'}(x) = \bigcap_{v \in c_{K'}(x)^{(0)}} O_{K'}(v)$ is open in $|K|_{\mathrm{m}}$ because so is each $O_{K'}(v)$. Again by Proposition 4.5.1, we can find $0 < s < 1$ such that $O_K(x, s) \subset O_{K'}(x)$. Then, it follows that

$$\begin{aligned} O_K(x, rs) &= (1 - rs)x + rsO_K(x) \\ &= (1 - r)x + r\big((1 - s)x + sO_K(x)\big) \\ &= (1 - r)x + rO_K(x, s) \\ &\subset (1 - r)x + rO_{K'}(x) = O_{K'}(x, r) \subset U, \end{aligned}$$

which means that U is a neighborhood of x in $|K|_{\mathrm{m}}$. □

For $A \subset |K|$, we introduce the following subcomplexes of K:

$$\begin{aligned} N(A, K) &= \{\sigma \in K \mid \exists \tau \in K[A] \text{ such that } \sigma \leq \tau\}, \\ C(A, K) &= K \setminus K[A] = \{\sigma \in K \mid \sigma \cap A = \emptyset\} \text{ and} \\ B(A, K) &= N(A, K) \cap C(A, K). \end{aligned}$$

If $A = |L|$ for a subcomplex $L \subset K$, we simply denote $N(L, K)$, $C(L, K)$, and $B(L, K)$ instead of $N(|L|, K)$, $C(|L|, K)$, and $B(|L|, K)$, respectively. Note that $N(\{v\}, K) = \mathrm{St}(v, K)$ for each $v \in K^{(0)}$ but $N(\sigma, K) \supsetneqq \mathrm{St}(\sigma, K)$ for each $\sigma \in K \setminus K^{(0)}$ in general. For each simplex $\sigma \in K$, $|N(\sigma, K)| = \mathrm{st}(\sigma, K)$ and $|\mathrm{St}(\sigma, K)| = \mathrm{st}(\mathrm{rint}\,\sigma, K) = \mathrm{st}(\hat{\sigma}, K)$. Moreover, note that each $x \in |K|$ is joinable to each simplex $\sigma \in \mathrm{St}(c_K(x), K) \cap C(x, K)$,[10] so we have the join $x\sigma$ contained in $|\mathrm{St}(c_K(x), K)|$.

Now, take $A \subset |K|$ so that $O_K(x) \cap O_K(x') = \emptyset$ if $x \neq x' \in A$ (i.e., K has no simplex containing more than one point of A). Then, the simplicial subdivision K_A of K can be defined as follows:

$$K_A = C(A, K) \cup \{x\sigma \mid x \in A,\ \sigma \in \mathrm{St}(c_K(x), K) \cap C(x, K)\}.$$

[10] In general, $\mathrm{St}(c_K(x), K) \cap C(x, K) \supsetneqq \mathrm{Lk}(c_K(x), K)$.

Observe that $K_A^{(0)} = A \cup K^{(0)}$, $C(A, K_A) = C(A, K)$, and $O_{K_A}(x) = O_K(x)$ for each $x \in A$. When $A = \{x\}$, we write $K_{\{x\}} = K_x$. The operation $K \to K_x$ (or K_x itself) is called a **starring** of K at x. A subdivision obtained by finite starrings is known as a **stellar subdivision**. In general, $(K_x)_y \neq (K_y)_x$ for distinct two points $x, y \in |K|$.

Lemma 4.8.2. *For each $w \in |K| \setminus K^{(0)}$, the starring K_w is an admissible subdivision of K.*

Proof. Due to Proposition 4.5.7, we need to show that $\mathrm{id} : |K|_{\mathrm{m}} \to |K_w|_{\mathrm{m}}$ is continuous. It suffices to prove that $\beta_v^{K_w} : |K|_{\mathrm{m}} \to \mathbf{I}$ is continuous for each $v \in K_w^{(0)}$. Using the barycentric coordinates with respect to K_w, each point $x \in |K|$ can be written as follows:

$$x = \beta_w^{K_w}(x)w + \sum_{u \in K^{(0)}} \beta_u^{K_w}(x)u.$$

Since $\beta_v^K(v) = 1$ and $\beta_v^K(u) = 0$ for each $u \in K^{(0)} \setminus \{v\}$, it follows that

$$\beta_v^K(x) = \beta_w^{K_w}(x)\beta_v^K(w) + \beta_v^{K_w}(x)$$
$$\big(\text{i.e., } \beta_v^{K_w}(x) = \beta_v^K(x) - \beta_w^{K_w}(x)\beta_v^K(w)\big).$$

Then, it is enough to verify the continuity of $\beta_w^{K_w} : |K|_{\mathrm{m}} \to \mathbf{I}$.

We shall show that

$$\beta_w^{K_w} = \min_{v \in c_K(w)^{(0)}} \beta_v^K(w)^{-1}\beta_v^K : |K|_{\mathrm{m}} \to \mathbf{I},$$

which implies that $\beta_w^{K_w}$ is continuous. For each $x \in |K|$, if $c_K(w) \not\leq c_K(x)$ then $\beta_w^{K_w}(x) = 0$ and $\min_{v \in c_K(w)^{(0)}} \beta_v^K(x)/\beta_v^K(w) = 0$. If $c_K(w) \leq c_K(x)$, let $v_0 \in c_K(w)^{(0)}$ such that

$$\beta_{v_0}^K(x)/\beta_{v_0}^K(w) = \min_{v \in c_K(w)^{(0)}} \beta_v^K(x)/\beta_v^K(w) \in (0, 1].$$

Let σ be the opposite face of $c_K(x)$ to v_0. Observe that

$$\begin{aligned}
&\frac{\beta_{v_0}^K(x)}{\beta_{v_0}^K(w)} + \sum_{v \in \sigma^{(0)}} \left(\beta_v^K(x) - \frac{\beta_{v_0}^K(x)}{\beta_{v_0}^K(w)}\beta_v^K(w)\right) \\
&\quad = \frac{\beta_{v_0}^K(x)}{\beta_{v_0}^K(w)} + \big(1 - \beta_{v_0}^K(x)\big) - \frac{\beta_{v_0}^K(x)}{\beta_{v_0}^K(w)}\big(1 - \beta_{v_0}^K(w)\big) = 1.
\end{aligned}$$

Then, we have

$$\frac{\beta_{v_0}^K(x)}{\beta_{v_0}^K(w)}w + \sum_{v \in \sigma^{(0)}} \left(\beta_v^K(x) - \frac{\beta_{v_0}^K(x)}{\beta_{v_0}^K(w)}\beta_v^K(w)\right)v \in w\sigma \in K_w,$$

which is simply x. Indeed,

$$\begin{aligned}
&\frac{\beta^K_{v_0}(x)}{\beta^K_{v_0}(w)}w + \sum_{v\in K^{(0)}\setminus\{v_0\}} \left(\beta^K_v(x) - \frac{\beta^K_{v_0}(x)}{\beta^K_{v_0}(w)}\beta^K_v(w)\right)v \\
&= \sum_{v\in K^{(0)}} \frac{\beta^K_{v_0}(x)}{\beta^K_{v_0}(w)}\beta^K_v(w)v + \sum_{v\in K^{(0)}\setminus\{v_0\}} \left(\beta^K_v(x) - \frac{\beta^K_{v_0}(x)}{\beta^K_{v_0}(w)}\beta^K_v(w)\right)v \\
&= \frac{\beta^K_{v_0}(x)}{\beta^K_{v_0}(w)}\beta^K_{v_0}(w)v_0 + \sum_{v\in\sigma^{(0)}} \beta^K_v(x)v = \sum_{v\in K^{(0)}} \beta^K_v(x)v = x.
\end{aligned}$$

Therefore, $\beta^{K_w}_w(x) = \beta^K_{v_0}(x)/\beta^K_{v_0}(w)$. □

Lemma 4.8.3. *Let K' and K'' be simplicial subdivisions of K such that $K'^{(0)}$ and $K''^{(0)}$ are discrete in $|K|_{\mathrm{m}}$. Then, K' and K'' have a common simplicial subdivision K''' such that $K'''^{(0)}$ is discrete in $|K|_{\mathrm{m}}$.*

Proof. Here, we use the following admissible metric on $|K|_{\mathrm{m}}$:

$$d(x,y) = \sqrt{\sum_{v\in K^{(0)}} \left(\beta^K_v(x) - \beta^K_v(y)\right)^2}.$$

Then, each n-simplex $\sigma \in K$ with this metric is isometric to the standard n-simplex of Euclidean space $\mathbb{R}^{n+1}$, so $\operatorname{diam}_d \sigma = \sqrt{2}$ if $n \neq 0$.

By Proposition 4.2.12, we have the following cell complex L, which is a common subdivision of K' and K'':

$$L = \left\{\sigma' \cap \sigma'' \;\middle|\; \sigma' \in K',\ \sigma'' \in K'' \text{ such that } \sigma' \cap \sigma'' \neq \emptyset\right\}.$$

By Theorem 4.2.10, L has a simplicial subdivision K''' such that $K'''^{(0)} = L^{(0)}$. Then, it suffices to show that $L^{(0)}$ is discrete in $|K|_{\mathrm{m}}$.

Let $x_0 \in |K|_{\mathrm{m}}$. Since $L^{(0)} \cap c_K(x_0)$ is finite, $K'^{(0)} \cup K''^{(0)}$ is discrete in X, and $c_K(x_0)$ is compact, we can find $0 < \delta < 1$ such that $B_d(x_0,\delta) \subset O(x_0,K)$,

$$\delta < \operatorname{dist}_d\left(c_K(x_0), \left(K'^{(0)} \cup K''^{(0)}\right) \setminus c_K(x_0)\right) \text{ and}$$
$$\delta < \min\left\{d(v,w) \;\middle|\; v \neq w \in \left(L^{(0)} \cap c_K(x_0)\right) \cup \{x_0\}\right\}.$$

We show that $d(x_0,v) \geq \delta^2/\sqrt{2}$ for every $v \in L^{(0)} \cap B_d(x_0,\delta) \setminus c_K(x_0)$, which implies that $B_d(x_0,\delta^2/\sqrt{2}) \cap (L^{(0)} \setminus \{x_0\}) = \emptyset$.

Since $v \in O(x_0,K) \setminus c_K(x_0)$, we have $c_K(x_0) < c_K(v)$. Since $c_K(v)$ is isometric to the standard simplex of Euclidean space, there exists the nearest point $u \in c_K(x_0)$ to v, i.e., $d(v,u) = \operatorname{dist}_d(v, c_K(x_0))$. Then, the line segment $\langle u,v\rangle$ is upright on $c_K(x_0)$. Since $v \in L^{(0)} \setminus (K'^{(0)} \cup K''^{(0)})$, it follows that $\{v\} = \sigma' \cap \sigma''$ for some

$\sigma' \in K' \setminus K'^{(0)}$ and $\sigma'' \in K'' \setminus K''^{(0)}$. Then, $\sigma'_0 = \sigma' \cap c_K(x_0) \neq \emptyset$ and $\sigma''_0 = \sigma'' \cap c_K(x_0) \neq \emptyset$. Otherwise,

$$d(x_0, v) \geq \mathrm{dist}_d \left(c_K(x_0), \left(K'^{(0)} \cup K''^{(0)}\right) \setminus c_K(x)\right) \geq \delta,$$

which is a contradiction. Let σ'_1 and σ''_1 be the faces of σ' and σ'' that are opposite σ'_0 and σ''_0, respectively. In other words, σ'_1 and σ''_1 are the simplexes spanned by the vertices σ' and σ'' that do not belong to σ'_0 and σ''_0, respectively. Then, we can write

$$v = (1-t')y' + t'z' = (1-t'')y'' + t''z'',$$

where $y' \in \sigma'_0$, $z' \in \sigma'_1$, $y'' \in \sigma''_0$, $z'' \in \sigma''_1$, and $t', t'' \in (0,1)$. Since $\sigma' \cap \sigma'' = \{v\}$ and $v \notin c_K(x_0)$, we have $\sigma'_0 \cap \sigma''_0 = \sigma' \cap \sigma'' \cap c_K(x_0) = \emptyset$, hence

$$d(y', u) + d(y'', u) \geq d(y', y'') \geq \mathrm{dist}_d(\sigma'_0, \sigma''_0) = \mathrm{dist}_d((\sigma'_0)^{(0)}, (\sigma''_0)^{(0)}) \geq \delta.$$

Then, $d(y', u) \geq \delta/2$ or $d(y'', u) \geq \delta/2$. We may assume that $d(y', u) \geq \delta/2$.

In the same way as above, let $x' \in c_K(x_0)$ be the nearest point to z', i.e., $d(z', x') = \mathrm{dist}_d(z', c_K(x_0)) > \delta$, where the line segment $\langle x', z' \rangle$ is upright on $c_K(x_0)$. Since the right triangle $x'y'z'$ is similar to the right triangle $uy'v$ and $d(x', y') \leq \mathrm{diam}_d\, c_K(x_0) = \sqrt{2}$, it follows that

$$d(x_0, v) \geq d(u, v) = \frac{d(x', z')}{d(x', y')} \cdot d(u, y') \geq \delta^2/\sqrt{2}.$$

This completes the proof. □

Theorem 4.8.4. *A simplicial subdivision K' of a simplicial complex K is admissible if and only if $K'^{(0)}$ is discrete in $|K|_{\mathrm{m}}$.*

Proof. Since $K'^{(0)}$ is discrete in $|K'|$, it suffices to show the "if" part. By virtue of Proposition 4.5.7, we need only show the continuity of $\mathrm{id} : |K|_{\mathrm{m}} \to |K'|_{\mathrm{m}}$ at each $w \in |K|$. By Lemma 4.8.3, there is a common subdivision K'' of K_w and K' such that $K''^{(0)}$ is discrete in $|K|$. Then, $\mathrm{id} : |K''|_{\mathrm{m}} \to |K'|_{\mathrm{m}}$ is continuous. It suffices to show the continuity of $\mathrm{id} : |K|_{\mathrm{m}} = |K_w|_{\mathrm{m}} \to |K''|_{\mathrm{m}}$ at w, where $w \in K_w^{(0)}$. Thus, we may assume that $w \in K^{(0)}$.

For each $x \in |K|$, observe that

$$\begin{aligned} \rho_K(x, w) &= \sum_{v \in K^{(0)}} \left|\beta_v^K(x) - \beta_v^K(w)\right| \\ &= 1 - \beta_w^K(x) + \sum_{v \in K^{(0)} \setminus \{w\}} \beta_v^K(x) = 2(1 - \beta_w^K(x)). \end{aligned}$$

For the same reason, we have $\rho_{K'}(x, w) = 2(1 - \beta_w^{K'}(x))$.

Let $\delta = \operatorname{dist}_{\rho_K}(w, K'^{(0)} \setminus \{w\}) > 0$. For each $\varepsilon > 0$, we shall show that if $\rho_K(x,w) < \delta\varepsilon/2$ then $\rho_{K'}(x,w) < \varepsilon$. For every $v \in K'^{(0)} \setminus \{w\}$, $\beta_w^K(v) \leq 1 - \delta/2$ because $2(1 - \beta_w^K(v)) = \rho_K(v,w) \geq \delta$. For each $x \in |K|$,

$$\beta_w^K(x) = \sum_{v \in K'^{(0)}} \beta_v^{K'}(x)\beta_w^K(v) \leq \beta_w^{K'}(x) + \sum_{v \in K'^{(0)} \setminus \{w\}} \beta_v^{K'}(x)(1 - \delta/2)$$

$$\leq \beta_w^{K'}(x) + (1 - \beta_w^{K'}(x))(1 - \delta/2) = \delta\beta_w^{K'}(x)/2 + 1 - \delta/2.$$

It follows that

$$\rho_{K'}(x,w)/2 = 1 - \beta_w^{K'}(x) \leq \frac{2(1 - \beta_w^K(x))}{\delta} = \rho_K(x,w)/\delta.$$

Therefore, $\rho_{K'}(x,w) < \varepsilon$. □

Combining Theorem 4.8.4 with Lemma 4.8.3, we have the following:

Corollary 4.8.5. *Every two admissible subdivisions of K have an admissible common subdivision.* □

Lemma 4.8.6. *Let K be a simplicial complex and L a finite-dimensional full subcomplex of K. Every simplicial subdivision B' of $B(L,K)$ extends to a simplicial subdivision N' of $N(L,K)$ such that $L \cup B' \subset N'$ and $N'^{(0)} = L^{(0)} \cup B'^{(0)}$.*

Proof. For each $\tau \in B'$, let $c_K(\hat{\tau})$ be the carrier of the barycenter of τ in K. Then, $c_K(\hat{\tau}) \in B(L,K)$ and $\operatorname{Lk}(c_K(\hat{\tau}), K) \cap L \neq \emptyset$. For each $\sigma \in \operatorname{Lk}(c_K(\hat{\tau}), K) \cap L$, we have $\sigma\tau \subset \sigma c_K(\hat{\tau}) \in K$. Then, we can define

$$N' = L \cup B' \cup \{\sigma\tau \mid \sigma \in \operatorname{Lk}(c_K(\hat{\tau}), K) \cap L,\ \tau \in B'\}.$$

Obviously, $N'^{(0)} = L^{(0)} \cup B'^{(0)}$. For each $x \in |N(L,K)| \setminus |L \cup B'|$, since L is full in K, we have $\sigma = c_K(x) \cap |L| \in L$. Let σ' be the opposite face of $c_K(x)$ from σ. Then, $\sigma' \in B(L,K)$. Since B' is a subdivision of $B(L,K)$, we have $\tau \in B'$ such that $c_K(\hat{\tau}) = \sigma'$ and $x \in \sigma\tau$. Thus, N' is a subdivision of $N(L,K)$. □

For $A \subset |K|$, let $\beta_A^K = \sum_{v \in K^{(0)} \cap A} \beta_v^K : |K| \to \mathbf{I}$. When A is a simplex $\sigma \in K$, we have $\sigma = (\beta_\sigma^K)^{-1}(1)$ and $(\beta_\sigma^K)^{-1}((0,1]) = \bigcup_{v \in \sigma^{(0)}} O_K(v)$.

Lemma 4.8.7. $(\beta_\sigma^K)^{-1}((1-r,1]) \subset \{y \in |K| \mid \operatorname{dist}_{\rho_K}(y,\sigma) < 2r\}$.

Proof. For each $y \in (\beta_\sigma^K)^{-1}((1-r,1])$, we have

$$x = \sum_{v \in \sigma^{(0)}} \frac{\beta_v^K(y)}{\beta_\sigma^K(y)} v \in \sigma.$$

Then, it follows that

$$\begin{aligned}\rho_K(x,y) &= \sum_{v\in K^{(0)}} |\beta_v^K(x) - \beta_v^K(y)| \\ &= \sum_{v\in\sigma^{(0)}} (\beta_v^K(x) - \beta_v^K(y)) + \sum_{v\in K^{(0)}\setminus\sigma^{(0)}} \beta_v^K(y) \\ &= 2\big(1 - \beta_\sigma^K(y)\big) < 2r.\end{aligned}$$

Therefore, $\mathrm{dist}_{\rho_K}(y,\sigma) < 2r$. □

The following is the metric topology version of Whitehead's Theorem 4.7.11 on small subdivisions:

Theorem 4.8.8 (Henderson–Sakai). *Let K be an arbitrary simplicial complex. For any open cover $\mathcal{U}$ of $|K|_{\mathrm{m}}$, K has an admissible subdivision K' such that $\mathcal{S}_{K'} \prec \mathcal{U}$.*

Proof. First, note that if a subdivision K' of K refines $\mathcal{U}$ then $\mathcal{S}_{K'} \prec \mathrm{st}\,\mathcal{U}$. Because every open cover of $|K|_{\mathrm{m}}$ has the open star-refinement, it suffices to construct an admissible subdivision K' of K that refines $\mathcal{U}$. We will inductively construct admissible subdivisions K_n of K, $n \geq 0$, so as to satisfy the following conditions:

(1) K_n is a subdivision of K_{n-1};
(2) $K_n||K^{(n-1)}| = K_{n-1}||K^{(n-1)}|$;
(3) $K_n[K^{(n)}] \prec \mathcal{U}$ (for simplicity, we write $K_n[K^{(n)}]$ instead of $K_n[|K^{(n)}|]$);
(4) $|C(K^{(n-1)}, K_n)| = |C(K^{(n-1)}, K_{n-1})|$,
or equivalently $|N(K^{(n-1)}, K_n)| = |N(K^{(n-1)}, K_{n-1})|$,

where $K_{-1} = \mathrm{Sd}\,K$ and $K^{(-1)} = \emptyset$. Condition (2) guarantees that $K' = \bigcup_{n\in\mathbb{N}} K_n||K^{(n)}|$ is a simplicial subdivision of K, where it should be noted that $K_0||K^{(0)}| = K^{(0)} \subset K_1||K^{(1)}|$. Then, $K' \prec \mathcal{U}$ by (3). Because each K_n is admissible, $K'^{(0)}||K^{(n)}| = K_n^{(0)}||K^{(n)}|$ is discrete in $|K|_{\mathrm{m}}$ by (2). Since $|C(K^{(n)}, K')| \subset |C(K^{(n)}, K_n)|$ by (2) and (4), $C(K^{(n)}, K')^{(0)}$ has no accumulation points in $|K^{(n)}|$. Then, it follows that $K'^{(0)}$ is discrete in $|K|_{\mathrm{m}}$, which means that K' is an admissible subdivision of K according to Theorem 4.8.4.

For each vertex $v \in K^{(0)}$, choose $1/2 < t_v < 1$ so that $(\beta_v^{\mathrm{Sd}^2 K})^{-1}([t_v, 1])$ is contained in some $U_v \in \mathcal{U}$ (Lemma 4.8.7). Dividing each $\sigma \in (\mathrm{Sd}^2 K)[v] \setminus \{v\}$ into two cells by $(\beta_v^{\mathrm{Sd}^2 K})^{-1}(t_v)$, we have a cell complex L subdividing $\mathrm{Sd}^2 K$, (cf. Fig. 4.8) that is,

$$\begin{aligned}L =& K^{(0)} \cup C(K^{(0)}, \mathrm{Sd}^2 K) \cup \big\{\sigma \cap (\beta_v^{\mathrm{Sd}^2 K})^{-1}(t_v), \sigma \cap (\beta_v^{\mathrm{Sd}^2 K})^{-1}([0,t_v]), \\ &\sigma \cap (\beta_v^{\mathrm{Sd}^2 K})^{-1}([t_v,1]) \;\big|\; \sigma \in (\mathrm{Sd}^2 K)[v] \setminus \{v\},\ v \in K^{(0)}\big\}.\end{aligned}$$

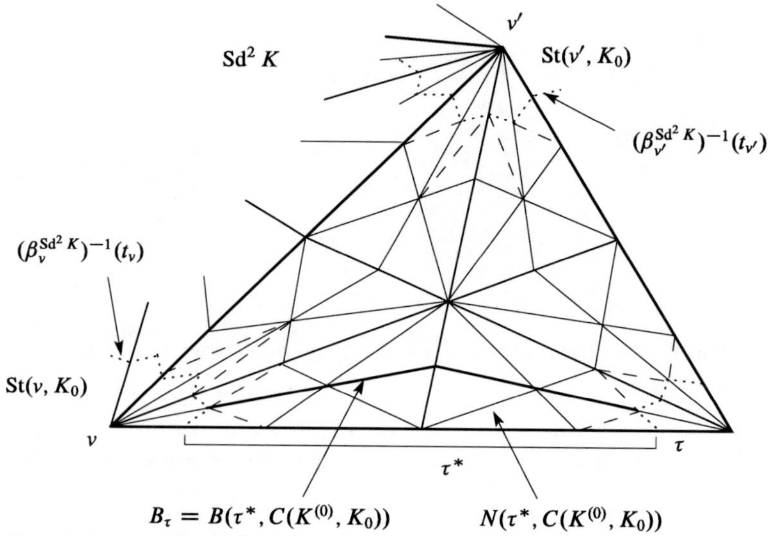

Fig. 4.8 The subdivision K_0 of $\mathrm{Sd}^2 K$

Then, $L^{(0)}$ is discrete in $|K|_{\mathrm{m}}$. Indeed, $L^{(0)}$ consists of the vertices $(\mathrm{Sd}^2 K)^{(0)}$ and the points

$$v_w = (1 - t_v)w + t_v v, \ v \in K^{(0)}, \ w \in \mathrm{Lk}(v, \mathrm{Sd}^2 K)^{(0)}.$$

Since $\mathrm{Sd}^2 K$ is an admissible subdivision of K, $(\mathrm{Sd}^2 K)^{(0)}$ is discrete in $|K|_{\mathrm{m}}$. On the other hand, $\{(\beta_v^{\mathrm{Sd}^2 K})^{-1}(t_v) \mid v \in K^{(0)}\}$ is discrete in $|K|_{\mathrm{m}}$. Then, it suffices to show that $\{v_w \mid w \in \mathrm{Lk}(v, \mathrm{Sd}^2 K)^{(0)}\}$ is discrete in $(\beta_v^{\mathrm{Sd}^2 K})^{-1}(t_v)$ for each $v \in K^{(0)}$. Note that the metric $\rho_{\mathrm{Sd}^2 K}$ is admissible for $|K|_{\mathrm{m}}$. For each $w, w' \in \mathrm{Lk}(v, \mathrm{Sd}^2 K)^{(0)}$,

$$\rho_{\mathrm{Sd}^2 K}(v_w, v_{w'}) = \beta_w^{\mathrm{Sd}^2 K}(v_w) + \beta_{w'}^{\mathrm{Sd}^2 K}(v_{w'}) = 2(1 - t_v).$$

Now, let K_0 be a simplicial subdivision of L with $K_0^{(0)} = L^{(0)}$ (cf. Fig. 4.8). Since $K_0^{(0)} = L^{(0)}$ is discrete in $|K|_{\mathrm{m}}$, K_0 is an admissible subdivision of K by Theorem 4.8.4. Observe

$$|\,\mathrm{St}(v, K_0)| = (\beta_v^{\mathrm{Sd}\,K})^{-1}([t_v, 1]) \subset U_v \ \text{ for } v \in K_0^{(0)}.$$

Then, K_0 satisfies condition (3).

Assume that K_{n-1} has been obtained. For each n-simplex $\tau \in K$, we define

$$\tau^* = \tau \cap |C(K^{(n-1)}, K_{n-1})|.$$

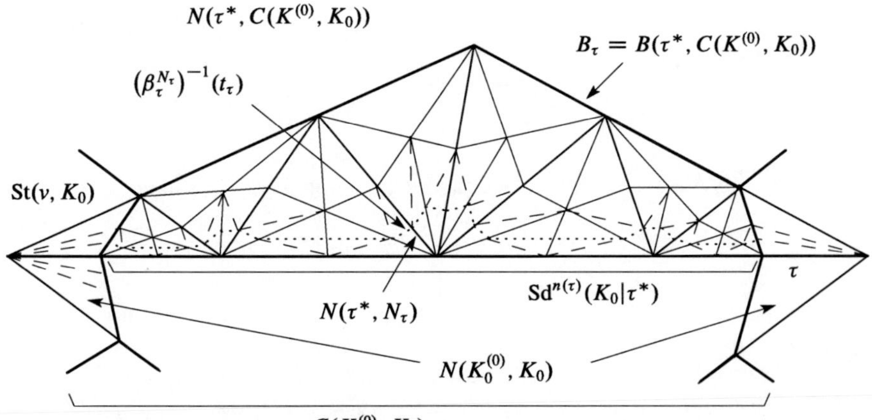

Fig. 4.9 The subdivision N_τ of $N(\tau, K_0)$

Note that $K_{n-1}|\tau^*$ is a triangulation of τ^*. We can choose $n(\tau) \in \mathbb{N}$ so that $\mathrm{Sd}^{n(\tau)}(K_{n-1}|\tau^*) \prec \mathcal{U}$. Let

$$B_\tau = B(\tau^*, C(K^{(n-1)}, K_{n-1})) \text{ and}$$
$$N_\tau = \mathrm{Sd}_{B_\tau}^{n(\tau)} N(\tau^*, C(K^{(n-1)}, K_{n-1})).$$

Then, N_τ is an admissible subdivision of $N(\tau^*, C(K^{(n-1)}, K_{n-1}))$, hence $|N_\tau|_{\mathrm{m}}$ is a subspace of $|K_{n-1}|_{\mathrm{m}} = |K|_{\mathrm{m}}$ (cf. Figs. 4.8 and 4.9). Moreover,

$$N_\tau|\tau^* = \mathrm{Sd}^{n(\tau)}(K_{n-1}|\tau^*) \prec \mathcal{U},$$

hence each $\sigma \in N_\tau|\tau^*$ is contained in some $U_\sigma \in \mathcal{U}$. By Lemma 4.8.7, $(\beta_\sigma^{N_\tau})^{-1}([t,1]) \subset U_\sigma$ for some $1/2 < t < 1$. Since $N_\tau|\tau^*$ is finite, we can find $1/2 < t_\tau < 1$ such that

$$\{(\beta_\sigma^{N_\tau})^{-1}([t_\tau,1]) \mid \sigma \in N_\tau|\tau^*\} \prec \mathcal{U}.$$

For each $\sigma \in N_\tau[\tau^*] \setminus N_\tau|\tau^*$, we have $\sigma \cap \tau^* \in N_\tau|\tau^*$ and $\beta_{\sigma\cap\tau^*}^{N_\tau}|\sigma = \beta_{\tau^*}^{N_\tau}|\sigma$. Dividing each $\sigma \in N_\tau[\tau^*] \setminus N_\tau|\tau^*$ into two cells by $(\beta_{\tau^*}^{N_\tau})^{-1}(t_\tau)$, we have a cell complex L_τ subdividing N_τ (cf. Fig. 4.9), that is,

$$\begin{aligned} L_\tau = N_\tau|\tau^* \cup C(\tau^*, N_\tau) \\ \cup \{\sigma \cap (\beta_{\tau^*}^{N_\tau})^{-1}(t_\tau),\ \sigma \cap (\beta_{\tau^*}^{N_\tau})^{-1}([0,t_\tau]), \\ \sigma \cap (\beta_{\tau^*}^{N_\tau})^{-1}([t_\tau,1]) \mid \sigma \in N_\tau[\tau^*] \setminus N_\tau|\tau^*\}. \end{aligned}$$

Then, $L_\tau^{(0)}$ is discrete in $|N_\tau|_{\mathrm{m}}$, so is discrete in $|K|_{\mathrm{m}}$. Indeed, $L_\tau^{(0)}$ consists of $N_\tau^{(0)}$ and the points

$$(1 - t_\tau)w + t_\tau v, \ v \in N_\tau^{(0)}|\tau^*, \ w \in \mathrm{Lk}(v, N_\tau)^{(0)} \setminus \tau^*,$$

where $N_\tau^{(0)}$ is discrete in $|N_\tau|_{\mathrm{m}}$. As can be easily observed, we have

$$\mathrm{dist}_{\rho_{N_\tau}}(N_\tau^{(0)}, (\beta_{\tau^*}^{N_\tau})^{-1}(t_\tau)) \geq \min\{2t_\tau, \ 2(1 - t_\tau)\}.$$

For each $v, v' \in N_\tau^{(0)}|\tau^*$, $w \in \mathrm{Lk}(v, N_\tau)^{(0)} \setminus \tau^*$, and $w' \in \mathrm{Lk}(v', N_\tau)^{(0)} \setminus \tau^*$, if $v \neq v'$ or $w \neq w'$ then

$$\rho_{N_\tau}((1 - t_\tau)w + t_\tau v, (1 - t_\tau)w' + t_\tau v') \geq \min\{2t_\tau, \ 2(1 - t_\tau)\}.$$

Now, for each $\tau \in K(n)$, let K_τ be a simplicial subdivision of L_τ with $K_\tau^{(0)} = L_\tau^{(0)}$. Observe

$$B_\tau = K_\tau \cap C(K^{(n)}, K_{n-1}) \ \text{ and } \ |B_\tau| = |K_\tau| \cap |C(K^{(n)}, K_{n-1})|.$$

Then, the following is a simplicial subdivision of $C(K^{(n-1)}, K_{n-1})$ (cf. Fig. 4.9):

$$C' = C(K^{(n)}, K_{n-1}) \cup \bigcup_{\tau \in K(n)} K_\tau.$$

According to Lemma 4.8.6, $N(K^{(n-1)}, K_{n-1})$ has a simplicial subdivision N' such that

$$N' \| B(K^{(n-1)}, K_{n-1})| = C' \| B(K^{(n-1)}, K_{n-1})| \ \text{ and }$$
$$N'^{(0)} = N(K^{(n-1)}, K_{n-1})^{(0)} \cup B'^{(0)}.$$

Then, $K_n = C' \cup B'$ is a simplicial subdivision of K_{n-1} such that

$$|N(K^{(n-1)}, K_{n-1})| = |N(K^{(n-1)}, K_n)|,$$

that is, K_n satisfies conditions (1) and (4). Note that

$$\begin{aligned} K_n^{(0)} &= N(K^{(n-1)}, K_{n-1})^{(0)} \cup C(K^{(n)}, K_{n-1})^{(0)} \cup \bigcup_{\tau \in K(n)} K_\tau^{(0)} \\ &= K_{n-1}^{(0)} \cup \bigcup_{\tau \in K(n)} N_\tau^{(0)}, \end{aligned}$$

which is discrete in $|K|_\mathrm{m}$. This means that K_n is an admissible subdivision of K by Theorem 4.8.4. By our construction, we have $K_n||K^{(n-1)}| = K_{n-1}||K^{(n-1)}|$, that is, K_n satisfies condition (2). Moreover, $K_n[K^{(n)}] \prec \mathcal{U}$ because

$$K_n[K^{(n-1)}] \prec K_{n-1}[K^{(n-1)}] \prec \mathcal{U} \text{ and}$$

$$K_n[K^{(n)}] \setminus K_n[K^{(n-1)}] \subset \bigcup_{\tau \in K(n)} N_\tau \prec \mathcal{U}.$$

Thus, K_n satisfies condition (3). This completes the proof. □

The next theorem can be proved by the same argument as in the proof of Theorem 4.7.14:

Theorem 4.8.9 (Simplicial Approximation). *Let K and L be simplicial complexes. Each map $f : |K|_\mathrm{m} \to |L|_\mathrm{m}$ has a simplicial approximation $g : K' \to L$ such that K' is an admissible subdivision of K, hence $g : |K|_\mathrm{m} \to |L|_\mathrm{m}$ is continuous.* □

Then, we have the following version of Corollary 4.7.15:

Corollary 4.8.10 (PL Approximation Theorem). *Let K and L be simplicial complexes and $f : |K|_\mathrm{m} \to |L|_\mathrm{m}$ a map. For each open cover $\mathcal{U}$ of $|L|_\mathrm{m}$, there is a simplicial map $g : K' \to L'$ with respect to admissible subdivisions K' and L' of K and L, respectively. In this case, $g : |K|_\mathrm{m} \to |L|_\mathrm{m}$ is continuous.* □

4.9 The Nerves of Open Covers

Let V be an arbitrary set. Recall $\mathrm{Fin}(V)$ is the collection of all non-empty finite subsets of V. An **abstract complex** $\mathcal{K}$ over V is a subcollection $\mathcal{K} \subset \mathrm{Fin}(V)$ satisfying the following condition:

(AC) if $A \in \mathcal{K}$ and $\emptyset \neq B \subset A$ then $B \in \mathcal{K}$.

A subcollection $\mathcal{L} \subset \mathcal{K}$ satisfying (AC) is called a **subcomplex** of $\mathcal{K}$. In particular, $\mathrm{Fin}(V)$ is an abstract complex and every abstract complex $\mathcal{K}$ over V is a subcomplex of $\mathrm{Fin}(V)$. For each $n \in \omega$, the ***n*-skeleton** $\mathcal{K}^{(n)}$ of $\mathcal{K}$ is defined by

$$\mathcal{K}^{(n)} = \{A \in \mathcal{K} \mid \operatorname{card} A \leq n + 1\},$$

where we regard $\mathcal{K}^{(0)} \subset V$, and so $\mathrm{Fin}(V)^{(0)} = V$. Each $\mathcal{K}^{(n)}$ is a subcomplex of $\mathcal{K}$. If $\mathcal{K} = \mathcal{K}^{(n)}$, we say that $\mathcal{K}$ is **at most *n*-dimensional** and write $\dim \mathcal{K} \leq n$. It is said that $\mathcal{K}$ is ***n*-dimensional** (written as $\dim \mathcal{K} = n$) if $\dim \mathcal{K} \leq n$ and $\dim \mathcal{K} \not\leq n - 1$. Note that every abstract complex $\mathcal{K}$ over V with $\dim \mathcal{K} \leq n$ is a subcomplex of $\mathrm{Fin}(V)^{(n)}$.

For any simplicial complex K, $\mathcal{K} = \{\sigma^{(0)} \mid \sigma \in K\} \subset \mathrm{Fin}(K^{(0)})$ is an abstract complex, which is called the **abstract complex defined by** K. Each $\mathcal{K}^{(n)}$ is defined by $K^{(n)}$. In particular, $\mathcal{K}^{(0)} = K^{(0)}$.

Conversely, each abstract complex $\mathcal{K}$ over V is defined by some simplicial complex. In fact, consider the linear space $\mathbb{R}^V_f$. By identifying each $v \in V$ with $\mathbf{e}_v \in \mathbb{R}^V_f$ defined by $\mathbf{e}_v(v) = 1$ and $\mathbf{e}_v(v') = 0$ if $v \neq v'$, we can regard V as a Hamel basis for $\mathbb{R}^V_f$. Then, $K = \{\langle A \rangle \mid A \in \mathcal{K}\}$ is a simplicial complex that defines $\mathcal{K}$. This K is called the **simplicial complex defined by** $\mathcal{K}$. Then each $K^{(n)}$ is defined by $\mathcal{K}^{(n)}$ and $K^{(0)} = \mathcal{K}^{(0)}$. Note that the full simplicial complex $\Delta(V)$ is the simplicial complex defined by $\mathrm{Fin}(V)$.

Remark 10. When V is a subset of a linear space E, for an abstract complex $\mathcal{K}$ over V, $K = \{\langle A \rangle \mid A \in \mathcal{K}\}$ is a simplicial complex that defines $\mathcal{K}$ if and only if each $A \in \mathcal{K}$ is affinely independent and

$$\langle A \rangle \cap \langle A' \rangle = \langle A \cap A' \rangle \quad \text{for each } A, A' \in \mathcal{K}.$$

In particular, K is a simplicial complex if $\mathcal{K}$ satisfies the condition:

($\sharp$) $A \cup A'$ is affinely independent for each $A, A' \in \mathcal{K}$.

The General Position Lemma states that there exists a countable (discrete) set V in $\mathbb{R}^{2n+1}$ such that each $2n + 2$ many points of V are affinely independent. This can be easily proved by using the Baire Category Theorem 2.5.1 and the fact that every hyperplane (= $2n$-dimensional flat) in $\mathbb{R}^{2n+1}$ is nowhere dense. The proof will be detailed in Sect. 5.8 (cf. Lemma 5.8.4). For such a set $V \subset \mathbb{R}^{2n+1}$, $\mathrm{Fin}(V)^{(n)}$ satisfies condition ($\sharp$) above. Therefore, for every abstract complex $\mathcal{K}$ over V with $\dim \mathcal{K} \leq n$, $K = \{\langle A \rangle \mid A \in \mathcal{K}\}$ is a simplicial complex that defines $\mathcal{K}$.

Remark 11. Note that every abstract complex $\mathcal{K}$ with $\dim \mathcal{K} \leq n$ is simplicially isomorphic to a subcomplex of $\mathrm{Fin}(V)^{(n)}$ for any set V with $\mathrm{card}\, V \geq \mathrm{card}\, \mathcal{K}^{(0)}$. Then, it follows that every countable complex K with $\dim K \leq n$ is simplicially isomorphic to a simplicial complex in $\mathbb{R}^{2n+1}$.

Remark 12. The barycentric subdivision $\mathrm{Sd}\, K$ of a simplicial complex K is simplicially isomorphic to the simplicial complex defined by the abstract complex

$$\big\{\{\sigma_1, \cdots, \sigma_n\} \bigm| \sigma_1 < \cdots < \sigma_n \in K\big\}.$$

Now, consider two abstract complexes $\mathcal{K}$ and $\mathcal{L}$ over V and W, respectively. Let K and L be the simplicial complexes defined by $\mathcal{K}$ and $\mathcal{L}$, respectively. Recall $\mathcal{K}^{(0)} = K^{(0)}$ and $\mathcal{L}^{(0)} = L^{(0)}$. Suppose that a function $\varphi : K^{(0)} \to L^{(0)}$ satisfies the following condition:

($*$) $A \in \mathcal{K}$ implies $\varphi(A) \in \mathcal{L}$.

Then, $\varphi : K^{(0)} \to L^{(0)}$ induces the simplicial map $f : K \to L$ with $f|K^{(0)} = \varphi$. Conversely, for any simplicial map $f : K \to L$, the restriction $\varphi = f|K^{(0)} : K^{(0)} \to L^{(0)}$ satisfies condition $(*)$ and f itself is the simplicial map induced by this φ. Such a function $\varphi : K^{(0)} \to L^{(0)}$ is also called a **simplicial map** from $\mathcal{K}$ to $\mathcal{L}$, and is written as $\varphi : \mathcal{K} \to \mathcal{L}$. If a bijection $\varphi : K^{(0)} \to L^{(0)}$ satisfies the condition that $A \in \mathcal{K}$ if and only if $\varphi(A) \in \mathcal{L}$, then φ induces the simplicial isomorphism $f : K \to L$ with $f|K^{(0)} = \varphi$, and any simplicial isomorphism $f : K \to L$ is induced by such a bijection $\varphi : K^{(0)} \to L^{(0)}$. Such a bijection $\varphi : K^{(0)} \to L^{(0)}$ is also called a **simplicial isomorphism** from $\mathcal{K}$ to $\mathcal{L}$. It is said that $\mathcal{K}$ is **simplicially isomorphic** to $\mathcal{L}$ (denoted by $\mathcal{K} \equiv \mathcal{L}$) if there is a simplicial isomorphism from $\mathcal{K}$ to $\mathcal{L}$.

For any open cover $\mathcal{U}$ of a space X, we define the abstract complex $\mathfrak{N}(\mathcal{U})$ over $\mathcal{U} \setminus \{\emptyset\}$ as follows:

$$\mathfrak{N}(\mathcal{U}) = \big\{\{U_1, \cdots, U_n\} \in \mathrm{Fin}(\mathcal{U}) \bigm| U_1 \cap \cdots \cap U_n \neq \emptyset\big\}.$$

The simplicial complex $N(\mathcal{U})$ defined by $\mathfrak{N}(\mathcal{U})$ is called the **nerve** of $\mathcal{U}$. A map $f : X \to |N(\mathcal{U})|$ (or $f : X \to |N(\mathcal{U})|_{\mathrm{m}}$) is called a **canonical map** for $\mathcal{U}$ if

$$f^{-1}(O_{N(\mathcal{U})}(U)) \subset U \quad \text{for each } U \in N(\mathcal{U})^{(0)} = \mathcal{U}.$$

Then, $f^{-1}(\mathcal{O}_{N(\mathcal{U})})$ is an open refinement of the open cover $\mathcal{U}$. Observe that every compact set in $|N(\mathcal{U})|$ meets $O_{N(\mathcal{U})}(U)$ for only finitely many $U \in \mathcal{U}$. Hence, if every $U \in \mathcal{U}$ has the compact closure in X, then each canonical map $f : X \to |N(\mathcal{U})|$ is proper.

Remark 13. For a subspace A of X and $\mathcal{U} \in \mathrm{cov}(X)$, we have $\mathcal{U}|A \in \mathrm{cov}(A)$. Assume that $U_1 \cap A \neq U_2 \cap A$ if $U_1 \neq U_2 \in \mathcal{U}[A]$. Then, by identifying each $U \cap A \in (\mathcal{U}|A) \setminus \{\emptyset\}$ with $U \in \mathcal{U}[A]$ $(\subset \mathcal{U} \setminus \{\emptyset\})$, the nerve $N(\mathcal{U}|A)$ can be regarded as the following subcomplex of the nerve $N(\mathcal{U})$:

$$\big\{\langle U_1, \ldots, U_n\rangle \in N(\mathcal{U}) \bigm| U_1 \cap \cdots \cap U_n \cap A \neq \emptyset\big\}.$$[11]

In this case, for each canonical map $f : X \to |N(\mathcal{U})|$ (or $f : X \to |N(\mathcal{U})|_{\mathrm{m}}$), the restriction $f|A : A \to |N(\mathcal{U}|A)|$ (or $f|A : A \to |N(\mathcal{U}|A)|_{\mathrm{m}}$) is a canonical map for $\mathcal{U}|A$. Indeed, for each $x \in A$, $c_{N(\mathcal{U})}(f(x))^{(0)} \subset \mathcal{U}[x] = (\mathcal{U}|A)[x]$, so $c_{N(\mathcal{U})}(f(x)) \in N(\mathcal{U}|A)$. Therefore, $f(A) \subset |N(\mathcal{U}|A)|$.

Due to Proposition 4.2.3, every simplicial complex K can be regarded as the nerve $N(\mathcal{O}_K)$ of the open cover $\mathcal{O}_K$ of $|K|$ (or $|K|_{\mathrm{m}}$) by identifying each $v \in K^{(0)}$ with $O_K(v) \in \mathcal{O}_K$. The identity $\mathrm{id} : |K| \to |K|$ is a canonical map for $\mathcal{O}_K$, where it should be noted that $\mathrm{id} : |K| \to |K|_{\mathrm{m}}$ and $\mathrm{id} : |K|_{\mathrm{m}} \to |K|_{\mathrm{m}}$ are also canonical maps for $\mathcal{O}_K$.

[11] This is not equal to $N(\mathcal{U}[A])$.

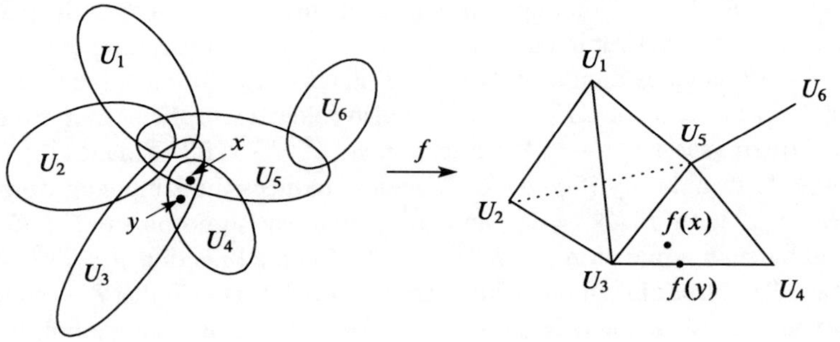

Fig. 4.10 A canonical map

We now have the following characterization of canonical maps:

Proposition 4.9.1. *For an open cover $\mathcal{U}$ of X, a map $f : X \to |N(\mathcal{U})|$ (or $f : X \to |N(\mathcal{U})|_m$) is a canonical map if and only if*

$$c_{N(\mathcal{U})}(f(x))^{(0)} \subset \mathcal{U}[x] \text{ for each } x \in X,$$

where $c_{N(\mathcal{U})}(f(x)) \in N(\mathcal{U})$ is the carrier of $f(x)$. If $\mathcal{U}[x]$ is finite, this condition is equivalent to $f(x) \in \langle \mathcal{U}[x] \rangle \in N(\mathcal{U})$ — Fig. 4.10.

Proof. To prove the "if" part, let $U \in \mathcal{U}$ and $x \in f^{-1}(O_{N(\mathcal{U})}(U))$. From the condition, it follows that $U \in c_{N(\mathcal{U})}(f(x))^{(0)} \subset \mathcal{U}[x]$, which means that $x \in U$. Therefore, $f^{-1}(O_{N(\mathcal{U})}(U)) \subset U$.

To show the "only if" part, let $x \in X$ and $U \in c_{N(\mathcal{U})}(f(x))^{(0)}$. Observe that $f(x) \in \operatorname{rint} c_{N(\mathcal{U})}(f(x)) \subset O_{N(\mathcal{U})}(U)$, hence $x \in f^{-1}(O_{N(\mathcal{U})}(U)) \subset U$. Thus, we have $c_{N(\mathcal{U})}(f(x))^{(0)} \subset \mathcal{U}[x]$. □

Proposition 4.9.1 yields the following:

Corollary 4.9.2. *Let $\mathcal{U}$ be an open cover of X. Then, any two canonical maps $f, g : X \to |N(\mathcal{U})|$ (or $f, g : X \to |N(\mathcal{U})|_m$) are contiguous.* □

For each open refinement $\mathcal{V}$ of $\mathcal{U} \in \operatorname{cov}(X)$, we have a simplicial map $\varphi : N(\mathcal{V}) \to N(\mathcal{U})$ such that $V \subset \varphi(V)$ for each $V \in \mathcal{V} = N(\mathcal{V})^{(0)}$. Such a simplicial map is called a **refining simplicial map**.

Corollary 4.9.3. *Let $\mathcal{U}$ and $\mathcal{V}$ be open covers of X with $\mathcal{V} \prec \mathcal{U}$ and $\varphi : N(\mathcal{V}) \to N(\mathcal{U})$ be a refining simplicial map. If $f : X \to |N(\mathcal{V})|$ (or $f : X \to |N(\mathcal{V})|_m$) is a canonical map for $\mathcal{V}$, then $\varphi f : X \to |N(\mathcal{U})|$ (or $\varphi f : X \to |N(\mathcal{U})|_m$) is also a canonical map for $\mathcal{U}$.*

Proof. For each $x \in X$,

$$c_{N(\mathcal{U})}(\varphi f(x))^{(0)} = \varphi\big(c_{N(\mathcal{V})}(f(x))^{(0)}\big) \subset \varphi(\mathcal{V}[x]) \subset \mathcal{U}[x]. \qquad \square$$

Concerning the existence of canonical maps, we have the following:

Theorem 4.9.4. *Every locally finite open cover $\mathcal{U}$ of a normal space X has a canonical map $f : X \to |N(\mathcal{U})|$ such that each point $x \in X$ has a neighborhood V_x with $f(V_x) \subset |K_x|$ for some finite subcomplex K_x of $N(\mathcal{U})$, so $f : X \to |N(\mathcal{U})|_{\mathrm{m}}$ is also a canonical map.*

Proof. According to Theorem 2.7.2, X has a partition of unity $(f_U)_{U\in\mathcal{U}}$ such that $\operatorname{supp} f_U \subset U$ for each $U \in \mathcal{U}$. We can define a map $f : X \to |N(\mathcal{U})|_{\mathrm{m}}$ as follows:

$$f(x) = \sum_{U\in\mathcal{U}} f_U(x) \cdot U \quad \left(\text{i.e., } \beta_U^{N(\mathcal{U})}(f(x)) = f_U(x)\right).$$

Observe that $f^{-1}(O_{N(\mathcal{U})}(U)) \subset \operatorname{supp} f_U \subset U$ for each $U \in \mathcal{U}$. Then, $f : X \to |N(\mathcal{U})|_{\mathrm{m}}$ is a canonical map for $N(\mathcal{U})$.

We need to verify the continuity of $f : X \to |N(\mathcal{U})|$ (with respect to the Whitehead topology). Each $x \in X$ has a neighborhood V_x such that $\mathcal{U}[V_x]$ is finite. We have the finite subcomplex K_x of $N(\mathcal{U})$ with $K_x^{(0)} = \mathcal{U}[V_x]$. Note that $f(V_x) \subset |K_x|$. Then, it follows that $f|V_x : V_x \to |K_x|_{\mathrm{m}} = |K_x|$ is continuous. Consequently, $f : X \to |N(\mathcal{U})|$ is continuous. □

Because every open cover of a paracompact space has a locally finite open refinement, the following corollary results from the combination of Theorem 4.9.4 and Corollary 4.9.3:

Corollary 4.9.5. *For every open cover $\mathcal{U}$ of a paracompact space X, there exists a canonical map $f : X \to |N(\mathcal{U})|$ such that $f : X \to |N(\mathcal{U})|_{\mathrm{m}}$ is also a canonical map.* □

Applying Corollary 4.9.5, we will prove the following:

Theorem 4.9.6. *For every simplicial complex K, the identity $\varphi = \mathrm{id} : |K| \to |K|_{\mathrm{m}}$ is a homotopy equivalence with a homotopy inverse $\psi : |K|_{\mathrm{m}} \to |K|$ such that $\psi\varphi \simeq_K \mathrm{id}$ and $\varphi\psi \simeq_K \mathrm{id}$, where $\psi\varphi \simeq_{\mathcal{O}_K} \mathrm{id}$ and $\varphi\psi \simeq_{\mathcal{O}_K} \mathrm{id}$ are also valid. These homotopies are realized by the straight-line homotopy.*

Proof. Consider K as the nerve $N(\mathcal{O}_K)$ of the open cover $\mathcal{O}_K \in \operatorname{cov}(|K|_{\mathrm{m}})$, where each vertex $v \in K^{(0)}$ is identified with the open star $O_K(v) \in \mathcal{O}_K$. By virtue of Corollary 4.9.5, we have a canonical map $\psi : |K|_{\mathrm{m}} \to |N(\mathcal{O}_K)| = |K|$. Then, $\psi\varphi, \mathrm{id} : |K| \to |K|$ are contiguous and $\varphi\psi, \mathrm{id} : |K|_{\mathrm{m}} \to |K|_{\mathrm{m}}$ are also contiguous by Corollary 4.9.2. Due to Propositions 4.3.4 and 4.5.4, $\psi\varphi \simeq_K \mathrm{id}$ and $\varphi\psi \simeq_K \mathrm{id}$ by the straight-line homotopy defined as

$$h(x,t) = (1-t)\psi\varphi(x) + tx = (1-t)\varphi\psi(x) + tx.$$

Since $c_K(\psi(x)) \leq c_K(x)$ for each $x \in |K|$, each $h(\{x\} \times \mathbf{I})$ is contained in not only $c_K(x)$ but also $O_K(v)$ for any $v \in c_K(\psi(x))^{(0)}$. □

Remark 14. In the above proof, let L be a subcomplex of a simplicial complex K. Then, $\mathcal{O}_L = \mathcal{O}_K||L|$ according to Proposition 4.2.4. As noted in Remark 13, identifying $O_L(v)$ with $O_K(v)$ for each $v \in L^{(0)}$, the nerve $N(\mathcal{O}_L)$ can be regarded as a subcomplex of the nerve of $N(\mathcal{O}_K)$, where the pair (K, L) can be identified with the pair $(N(\mathcal{O}_K), N(\mathcal{O}_L))$. Moreover, the restriction $\psi||L| : |L|_m \to |N(\mathcal{O}_L)| = |L|$ is also a canonical map for $\mathcal{O}_L$, which is a homotopy inverse of $\varphi||L| = \mathrm{id} : |L| \to |L|_m$. In this case, $(\psi||L|)(\varphi||L|) \simeq_L \mathrm{id}$ and $(\varphi||L|)(\psi||L|) \simeq_L \mathrm{id}$ are given by the straight-line homotopies, that is, the restrictions of the homotopies $\psi\varphi \simeq_K \mathrm{id}$ and $\varphi\psi \simeq_K \mathrm{id}$, respectively.

We can now generalize Proposition 4.3.4 as follows:

Proposition 4.9.7. *Let K be a simplicial complex and X an arbitrary space. If two maps $f, g : X \to |K|$ are contiguous then $f \simeq_K g$.*

Proof. Let $\varphi = \mathrm{id} : |K| \to |K|_m$. By virtue of Theorem 4.9.6, we have a map $\psi : |K|_m \to |K|$ such that $\psi\varphi \simeq_K \mathrm{id}$ by the straight-line homotopy $h : |K| \times \mathbf{I} \to |K|$, where $c_K(\psi(x)) \leq c_K(x)$ and $h(\{x\} \times \mathbf{I}) \subset c_K(x)$ for each $x \in |K|$ (see the proof of Theorem 4.9.6). On the other hand, by Proposition 4.5.4, we have $\varphi f \simeq_K \varphi g$, which is realized by the straight-line homotopy $h' : X \times \mathbf{I} \to |K|_m$. For each $x \in X$, choose $\sigma_x \in K$ so that $h'(\{x\} \times \mathbf{I}) \subset \sigma_x$. Observe that $\psi h'(\{x\} \times \mathbf{I}) \subset \psi(\sigma_x) \subset \sigma_x$, $h(\{f(x)\} \times \mathbf{I}) \subset c_K(f(x)) \subset \sigma_x$, and $h(\{g(x)\} \times \mathbf{I}) \subset c_K(g(x)) \subset \sigma_x$. Then, by connecting three homotopies $h(f \times \mathrm{id}_\mathbf{I})$, $\psi h'$, and $h(g \times \mathrm{id}_\mathbf{I})$, we can get a K-homotopy from f to g, hence $f \simeq_K g$. □

Combining Corollary 4.9.2 with Proposition 4.9.7 (or 4.5.4), we have the following corollary:

Corollary 4.9.8. *Let $\mathcal{U}$ be an open cover of a space X. Then, $f \simeq_{N(\mathcal{U})} g$ for any two canonical maps $f, g : X \to |N(\mathcal{U})|$ (or $f, g : X \to |N(\mathcal{U})|_m$).* □

An open cover $\mathcal{U}$ of a space X is said to be **star-finite** if $\mathcal{U}[U]$ is finite for each $U \in \mathcal{U}$, which is equivalent to the condition that the nerve $N(\mathcal{U})$ is locally finite. In fact, $\mathrm{St}(U, N(\mathcal{U}))^{(0)} = \mathcal{U}[U]$ for each $U \in \mathcal{U} = N(\mathcal{U})^{(0)}$. Thus, the star-finiteness of an open cover chraterizes the local finiteness of its nerve. On the other hand, the nerve $N(\mathcal{U})$ is locally finite-dimensional (l.f.d.) if and only if $\sup_{x \in U} \mathrm{card}\, \mathcal{U}[x] < \infty$ for each $U \in \mathcal{U}$. In this case, we have

$$\sup_{x \in U} \mathrm{card}\, \mathcal{U}[x] = \dim \mathrm{St}(U, N(\mathcal{U})) + 1.$$

Note that every star-finite open cover is locally finite and its nerve is locally finite-dimensional, and that if an open cover is locally finite or its nerve is locally finite-dimensional then it is point-finite, that is, we have the following implications:

$$\begin{array}{ccc} \boxed{\text{locally finite nerve} \equiv \text{star-finite}} & \Longrightarrow & \boxed{\text{locally finite}} \\ \Downarrow & & \Downarrow \\ \boxed{\text{locally finite-dimensional nerve}} & \Longrightarrow & \boxed{\text{point-finite}} \end{array}$$

In the above, the converse implications do not hold and there are no connections between the local finiteness of an open cover and the local finite-dimensionality of its nerve. In fact, $\mathcal{U} = \{\mathbb{R}, (n, \infty) \mid n \in \mathbb{N}\}$ is a locally finite open cover of $\mathbb{R}$ but the nerve $N(\mathcal{U})$ is not locally finite-dimensional. This example also shows that the converse implication in the top row of the figure does not hold. Since the cover is point-finite, the converse in the bottom row does not hold either. On the other hand, let $X = (\mathbb{N} \times \mathbf{I})/(\mathbb{N} \times \{0\})$ be the quotient space (or $X = J(\mathbb{N})$ the hedgehog (cf. Sect. 2.3)). Let $U_0 = X$ and, for each $n \in \mathbb{N}$, let $U_n = \{n\} \times (0, 1]$. Then, $\mathcal{U} = \{U_n \mid n \in \omega\}$ is an open cover with $\dim N(\mathcal{U}) = 1$, which is not locally finite in X. This shows that the converse implication on the left side of the figure does not hold. Since the cover is point-finite, the converse on the right side does not hold either.

Theorem 4.9.9. *Every open cover of a paracompact space has a locally finite σ-discrete open refinement with the locally finite-dimensional nerve.*

Proof. It suffices to show that, for a simplicial complex K, the open cover $\mathcal{O}_K \in \operatorname{cov}(|K|)$ has a locally finite σ-discrete open refinement $\mathcal{V}$ with the locally finite-dimensional nerve. In fact, every open cover $\mathcal{U}$ of a paracompact space X has a canonical map $f : X \to |N(\mathcal{U})|$ by Corollary 4.9.5. When $K = N(\mathcal{U})$,

$$f^{-1}(\mathcal{V}) \prec f^{-1}(\mathcal{O}_K) = f^{-1}(\mathcal{O}_{N(\mathcal{U})}) \prec \mathcal{U},$$

$f^{-1}(\mathcal{V})$ is locally finite σ-discrete and $N(f^{-1}(\mathcal{V}))$ is l.f.d. because $N(f^{-1}(\mathcal{V}))$ is simplicially isomorphic to the subcomplex $L \subset N(\mathcal{V})$ defined as follows:

$$L = \{\langle V_1, \ldots, V_n \rangle \mid \textstyle\bigcap_{i=1}^n f^{-1}(V_i) \neq \emptyset\}.$$

We will construct an open collection $\mathcal{V} = \{V_n(\sigma) \mid \sigma \in K^{(n)},\ n \in \omega\}$ satisfying the following conditions:

(1) $\{V_n(\sigma) \mid \sigma \in K^{(n)},\ \dim \sigma = i\}$ is discrete in $|K|$ for each $i \leq n$;
(2) $\operatorname{cl} V_n(\sigma) \subset O_{\operatorname{Sd} K}(\hat{\sigma}) \setminus \operatorname{cl} V_{n-2}$ for each $\sigma \in K$ and $n \geq \dim \sigma$;
(3) $|K^{(n)}| \cup \operatorname{cl} V_{n-1} \subset V_n$,

where

$$V_n = \bigcup \{V_i(\sigma) \mid \sigma \in K^{(i)},\ i \leq n\} \quad (V_{-1} = V_{-2} = \emptyset).$$

Then, $\mathcal{V} \in \operatorname{cov}(|K|)$ by (3), $\mathcal{V} \prec \mathcal{O}_{\operatorname{Sd} K} \prec \mathcal{O}_K$ by (2), and $\mathcal{V}$ is σ-discrete in $|K|$ by (1). Moreover, we can see that $\mathcal{V}$ is locally finite and $N(\mathcal{V})$ is l.f.d.

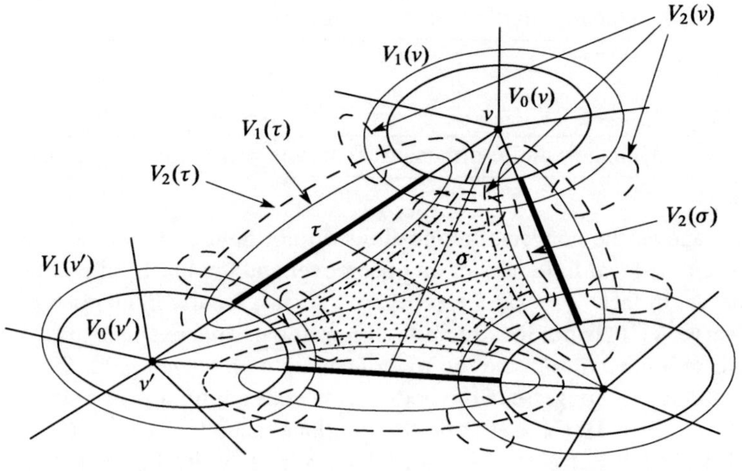

Fig. 4.11 $\mathcal{V} = \{V_n(\sigma) \mid n \in \omega,\ \sigma \in K^{(n)}\}$

The local finiteness of $\mathcal{V}$ can be shown as follows: Each $x \in |K|$ is contained in some $V_n \setminus \operatorname{cl} V_{n-2}$ by (3). From (2), it follows that

$$(V_n \setminus \operatorname{cl} V_{n-2}) \cap V_i(\sigma) = \emptyset \quad \text{for } i > n+1 \text{ and } i < n-1.$$

By (1), x has a neighborhood W in $V_n \setminus \operatorname{cl} V_{n-2}$ that meets at most one member of $\{V_i(\sigma) \mid \dim \sigma = k\}$ for each $i = n-1, n, n+1$ and each $k \leq i$. Consequently, W meets at most $3n+3$ of the points $V_i(\sigma)$.

To prove that $N(\mathcal{V})$ is l.f.d., let $n \in \omega$ and $\sigma \in K^{(n)}$. Due to (2), for each $x \in V_n(\sigma)$,

$$\{i \mid \exists \tau \in K \text{ such that } x \in V_i(\tau)\} = \{n-1, n\} \text{ or } \{n\} \text{ or } \{n, n+1\}.$$

For each $i \in \omega$, $V_i(\tau)$ is defined only if $\dim \tau \leq i$, so x is contained in at most $i+1$ of the open sets $V_i(\tau)$ by (1). Therefore, each $x \in V_n(\sigma)$ is contained in at most $2n+3$ of the open sets $V_i(\tau)$, that is, $\dim \operatorname{St}(V_n(\sigma), N(\mathcal{V})) \leq 2n+2$ (Fig. 4.11).

Now, let us construct $\mathcal{V}$. First, for each $v \in K^{(0)}$, choose an open set $V_0(v)$ in $|K|$ so that $v \in V_0(v) \subset \operatorname{cl} V_0(v) \subset O_{\operatorname{Sd} K}(v)$. Each $x \in |K|$ is contained in the open star $O_{\operatorname{Sd} K}(\hat{\sigma})$ for some $\sigma \in K$, which meets only finitely many $\operatorname{cl} V_0(v)$ because

$$O_{\operatorname{Sd} K}(\hat{\sigma}) \cap O_{\operatorname{Sd} K}(v) \neq \emptyset \ \Leftrightarrow\ \langle v, \hat{\sigma} \rangle \in \operatorname{Sd} K \ \Leftrightarrow\ v \in \sigma^{(0)}.$$

Then, $\{\operatorname{cl} V_0(v) \mid v \in K^{(0)}\}$ is a pair-wise disjoint locally finite collection of closed sets in $|K|$, which means that it is discrete in $|K|$. Therefore, $\{V_0(v) \mid v \in K^{(0)}\}$ is discrete in $|K|$.

Suppose that $V_i(\sigma)$ has been defined for $i \leq n-1$ and $\sigma \in K^{(i)}$ so as to satisfy (1), (2), and (3). Let V'_{n-1} be an open set in $|K|$ such that

$$|K^{(n-1)}| \cup \operatorname{cl} V_{n-2} \subset V'_{n-1} \subset \operatorname{cl} V'_{n-1} \subset V_{n-1}.$$

For each n-simplex $\sigma \in K$, since $|K^{(n-1)}| \subset V_{n-1}$, it follows that

$$\sigma \setminus V_{n-1} = \operatorname{rint} \sigma \setminus V_{n-1} \subset O_{\operatorname{Sd} K}(\hat{\sigma}) \setminus \operatorname{cl} V'_{n-1} \subset O_{\operatorname{Sd} K}(\hat{\sigma}) \setminus \operatorname{cl} V_{n-2}.$$

Then, we have an open set $V_n(\sigma)$ in $|K|$ such that

$$\sigma \setminus V_{n-1} \subset V_n(\sigma) \subset \operatorname{cl} V_n(\sigma) \subset O_{\operatorname{Sd} K}(\hat{\sigma}) \setminus \operatorname{cl} V'_{n-1}.$$

Each $x \in |K^{(n-1)}|$ is contained in the open set V'_{n-1}, which misses $\operatorname{cl} V_n(\sigma)$ for every n-simplex $\sigma \in K$. On the other hand, each $x \in |K| \setminus |K^{(n-1)}|$ is contained in the open star $O_{\operatorname{Sd} K}(\hat{\tau})$ for some $\tau \in K \setminus K^{(n-1)}$, which meets only finitely many $\operatorname{cl} V_n(\sigma)$ (where $\dim \sigma = n$) because

$$O_{\operatorname{Sd} K}(\hat{\tau}) \cap O_{\operatorname{Sd} K}(\hat{\sigma}) \neq \emptyset \;\Leftrightarrow\; \langle \hat{\sigma}, \hat{\tau} \rangle \in \operatorname{Sd} K \;\Leftrightarrow\; \sigma < \tau.$$

Then, $\{\operatorname{cl} V_n(\sigma) \mid \sigma \in K^{(n)},\ \dim \sigma = n\}$ is a pair-wise disjoint locally finite collection of closed sets in $|K|$, so it is discrete in $|K|$. Thus, $\{V_n(\sigma) \mid \sigma \in K^{(n)},\ \dim \sigma = n\}$ is discrete in $|K|$.[12]

For each $\sigma \in K^{(n-1)}$, observe that

$$\operatorname{cl} V_{n-1}(\sigma) \setminus V_{n-1} \subset O_{\operatorname{Sd} K}(\hat{\sigma}) \setminus \operatorname{cl} V'_{n-1} \subset O_{\operatorname{Sd} K}(\hat{\sigma}) \setminus \operatorname{cl} V_{n-2}.$$

Then, we have an open set $V_n(\sigma)$ in $|K|$ such that

$$\operatorname{cl} V_{n-1}(\sigma) \setminus V_{n-1} \subset V_n(\sigma) \subset \operatorname{cl} V_n(\sigma) \subset O_{\operatorname{Sd} K}(\hat{\sigma}) \setminus \operatorname{cl} V'_{n-1}.$$

By the same approach as above, we can see that $\{V_n(\sigma) \mid \sigma \in K^{(n)},\ \dim \sigma = i\}$ is discrete in $|K|$ for each $i < n$.

Since $\{V_i(\sigma) \mid \sigma \in K^{(i)},\ i \leq n-1\}$ is locally finite in $|K|$, we have

$$\operatorname{cl} V_{n-1} = \bigcup \{\operatorname{cl} V_i(\sigma) \mid \sigma \in K^{(i)},\ i \leq n-1\}.$$

Note that $\operatorname{cl} V_i(\sigma) \subset \operatorname{cl} V_{n-2} \subset V_{n-1}$ for $i \leq n-2$ and $\sigma \in K^{(i)}$. Hence,

$$\operatorname{bd} V_{n-1} \subset \bigcup_{\sigma \in K^{(n-1)}} \operatorname{cl} V_{n-1}(\sigma) \setminus V_{n-1} \subset \bigcup_{\sigma \in K^{(n-1)}} V_n(\sigma).$$

Then, it follows that $|K^{(n)}| \cup \operatorname{cl} V_{n-1} \subset V_n$. Thus, we have obtained $V_n(\sigma)$ for every $\sigma \in K^{(n)}$ such that conditions (1), (2), and (3) are satisfied. The proof is completed by induction. □

[12]Note that $|K|$ is paracompact by Corollary 4.7.12, so it is collection-wise normal by Theorem 2.6.1. Observe that $\{\sigma \setminus V_{n-1} \mid \sigma \in K^{(n)},\ \dim \sigma = n\}$ is discrete in $|K|$. Then, we can obtain $\{\operatorname{cl} V_n(\sigma) \mid \sigma \in K^{(n)},\ \dim \sigma = n\}$ without taking V'_{n-1}.

When X is a locally compact paracompact space, each $\mathcal{U} \in \operatorname{cov}(X)$ has a locally finite open refinement $\mathcal{V}$ such that $\operatorname{cl} V$ is compact for each $V \in \mathcal{V}$. Then, as is easily observed, $\mathcal{V}$ is star-finite, so the nerve $N(\mathcal{V})$ is locally finite. Thus, we have the following theorem:

Theorem 4.9.10. *Every open cover of a locally compact paracompact space has a star-finite open refinement whose nerve is locally finite.* □

In addition, we can show that:

Theorem 4.9.11. *Every open cover of a regular Lindelöf space has a countable star-finite open refinement whose nerve is countable and locally finite.*

Proof. Due to Corollary 2.6.4, a regular Lindelöf space X is paracompact. Then, each $\mathcal{U} \in \operatorname{cov}(X)$ has a countable locally finite open refinement $\mathcal{V} = \{V_i \mid i \in \mathbb{N}\}$. Indeed, take open refinements $\mathcal{V}'' \prec \mathcal{V}' \prec \mathcal{U}$ so that $\mathcal{V}'$ is locally finite and $\mathcal{V}''$ is countable. Let $\varphi : \mathcal{V}'' \to \mathcal{V}'$ be a function such that $V \subset \varphi(V)$ for each $V \in \mathcal{V}''$. Then, $\mathcal{V} = \{\varphi(V) \mid V \in \mathcal{V}''\}$ is countable and locally finite. According to Lemma 2.7.1, $\mathcal{V}$ has an open refinement $\{V_{i,1} \mid i \in \mathbb{N}\}$ such that $\operatorname{cl} V_{i,1} \subset V_i$ for each $i \in \mathbb{N}$. We can inductively choose open sets $V_{i,j}$ in X so that $\operatorname{cl} V_{i,j-1} \subset V_{i,j} \subset \operatorname{cl} V_{i,j} \subset V_i$. For each $i, j \in \mathbb{N}$, let

$$W_{i,j} = V_{i,j} \setminus \bigcup_{i'+j'<i+j-1} \operatorname{cl} V_{i',j'}.$$

Then, $\mathcal{W} = \{W_{i,j} \mid i, j \in \mathbb{N}\} \in \operatorname{cov}(X)$ is a star-finite countable open refinement of $\mathcal{U}$. □

4.10 The Inverse Limits of Metric Polyhedra

An **inverse sequence** $(X_i, f_i)_{i\in\mathbb{N}}$ is a sequence of spaces X_i and maps $f_i : X_{i+1} \to X_i$, that is,

$$X_1 \xleftarrow{f_1} X_2 \xleftarrow{f_2} X_3 \xleftarrow{f_3} \cdots,$$

where f_i are called the **bonding maps**. For $i < j$, we denote

$$f_{i,j} = f_i \cdots f_{j-1} : X_j \to X_i.$$

Then, $f_{i,i+1} = f_i$ for each $i \in \mathbb{N}$. For convenience, we denote $f_{i,i} = \mathrm{id}_{X_i}$. The **inverse limit** $\varprojlim(X_i, f_i)$ is defined as the following subspace of the product space $\prod_{i\in\mathbb{N}} X_i$:

$$\varprojlim(X_i, f_i) = \left\{x \in \textstyle\prod_{i\in\mathbb{N}} X_i \;\middle|\; x(i) = f_i(x(i+1)) \text{ for each } i \in \mathbb{N}\right\},$$

which is closed in $\prod_{i\in\mathbb{N}} X_i$. For each $i \in \mathbb{N}$, the restriction $p_i : \varprojlim(X_i, f_i) \to X_i$ of the projection $\mathrm{pr}_i : \prod_{i\in\mathbb{N}} \to X_i$ is called the **(inverse limit) projection** of $\varprojlim(X_i, f_i)$ to X_i. Note that $f_i p_{i+1} = p_i$ for each $i \in \mathbb{N}$. When the bonding maps are evidently known, we simply write $\varprojlim(X_i, f_i) = \varprojlim X_i$. It is possible that $\varprojlim(X_i, f_i) = \emptyset$ even if $X_i \neq \emptyset$ for every $i \in \mathbb{N}$ (cf. Proposition 4.10.9(1)). For example, the following inverse sequence has the empty limit:

$$[1,\infty) \supset [2,\infty) \supset [3,\infty) \supset \cdots,$$

where the bonding maps are the inclusions. A **nested sequence** $X_1 \supset X_2 \supset \cdots$ is an inverse sequence of subspaces such that the inclusions are the bonding maps.

Proposition 4.10.1. *For every nested sequence $X_1 \supset X_2 \supset \cdots$, there exists a homeomorphism $h : \bigcap_{i\in\mathbb{N}} X_i \to \varprojlim X_i$ such that $p_i h : \bigcap_{i\in\mathbb{N}} X_i \to X_i$ is the inclusion.*

Proof. Note that $p_1(\varprojlim X_i) = \bigcap_{i\in\mathbb{N}} X_i$, where $p_1 : \varprojlim X_i \to X_1$ is the projection. Let $h : \bigcap_{i\in\mathbb{N}} X_i \to \varprojlim X_i$ be the diagonal map defined by $h(x) = (x, x, \ldots)$. Then, $p_1 h = \mathrm{id}$ and $h p_1 = \mathrm{id}$. □

Proposition 4.10.2. *For an inverse sequence $(X_i, f_i)_{i\in\mathbb{N}}$, if every bonding map f_i is surjective then the projection $p_n : \varprojlim(X_i, f_i) \to X_n$ is also surjective for each $n \in \mathbb{N}$.*

Proof. For each $x \in X_n$ and each $i \leq n$, define $x_i = f_{i,n}(x)$. For each $i \geq n$, inductively choose $x_{i+1} \in f_i^{-1}(x_i)$. Then, $(x_i)_{i\in\mathbb{N}} \in \varprojlim(X_i, f_i)$ and $p_n((x_i)_{i\in\mathbb{N}}) = x_n = x$. □

Let $(X_i, f_i)_{i\in\mathbb{N}}$ and $(Y_i, g_i)_{i\in\mathbb{N}}$ be inverse sequences. Given maps $h_i : X_i \to Y_i$, $i \in \mathbb{N}$, such that $h_i f_i = g_i h_{i+1}$ for every $i \in \mathbb{N}$, we can define a map $\varprojlim h_i : \varprojlim(X_i, f_i) \to \varprojlim(Y_i, g_i)$ as follows:

$$(\varprojlim h_i)((x_i)_{i\in\mathbb{N}}) = (h_i(x_i))_{i\in\mathbb{N}}.$$

Then, $q_i \varprojlim h_i = h_i p_i$ for every $i \in \mathbb{N}$, where $p_i : \varprojlim(X_i, f_i) \to X_i$ and $q_i : \varprojlim(Y_i, g_i) \to Y_i$ are the projections.

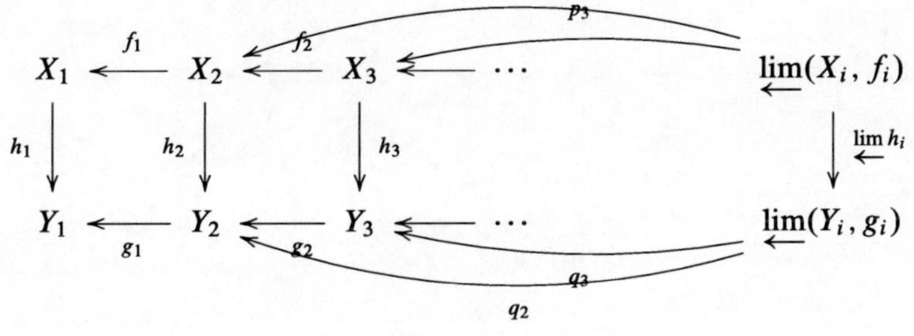

Proposition 4.10.3. *Let $(X_i, f_i)_{i\in\mathbb{N}}$ be an inverse sequence. For any increasing sequence $n(1) < n(2) < \cdots \in \mathbb{N}$, the following $\mathcal{B}$ is an open basis for $\varprojlim(X_i, f_i)$:*

$$\mathcal{B} = \{p_{n(i)}^{-1}(V) \mid i \in \mathbb{N},\ V \text{ is open in } X_{n(i)}\}.$$

Proof. For each open set U in $\varprojlim(X_i, f_i)$ and $x \in U$, we have open sets V_i in X_i, $i = 1, \ldots, n$, such that $x \in \bigcap_{i=1}^{n} p_i^{-1}(V_i) \subset U$. Choose $k \in \mathbb{N}$ so that $n(k) \geq n$. Then, $V = \bigcap_{i=1}^{n} f_{i,n(k)}^{-1}(V_i)$ is an open set in $X_{n(k)}$ and $x \in p_{n(k)}^{-1}(V) = \bigcap_{i=1}^{n} p_i^{-1}(V_i) \subset U$. □

For an inverse sequence $(X_i, f_i)_{i\in\mathbb{N}}$ and an increasing sequence $n(1) < n(2) < \cdots \in \mathbb{N}$, we have the inverse sequence $(X_{n(i)}, f_{n(i),n(i+1)})_{i\in\mathbb{N}}$:

$$X_{n(1)} \xleftarrow{\ f_{n(1),n(2)}\ } X_{n(2)} \xleftarrow{\ f_{n(2),n(3)}\ } X_{n(3)} \longleftarrow \cdots,$$

which is called a **subsequence** of $(X_i, f_i)_{i\in\mathbb{N}}$. Proposition 4.10.3 shows that the map $h : \varprojlim(X_{n(i)}, f_{n(i),n(i+1)}) \to \varprojlim(X_i, f_i)$ obtained as the restriction of the projection of $\prod_{i\in\mathbb{N}} X_i$ onto $\prod_{i\in\mathbb{N}} X_{n(i)}$ is open, hence h is a homeomorphism. Thus, we have the corollary:

Corollary 4.10.4. *Let $(X_i, f_i)_{i\in\mathbb{N}}$ be an inverse sequence. For any increasing sequence $n(1) < n(2) < \cdots \in \mathbb{N}$, there exists a homeomorphism $h : \varprojlim(X_i, f_i) \to \varprojlim(X_{n(i)}, f_{n(i),n(i+1)})$ such that $p'_i h = p_{n(i)}$ for each $i \in \mathbb{N}$, where $p_i : \varprojlim(X_i, f_i) \to X_i$ and $p'_i : \varprojlim(X_{n(i)}, f_{n(i),n(i+1)}) \to X_{n(i)}$ are the projections.* □

The following can be easily proved using Corollary 4.10.4:

Corollary 4.10.5. *Let $(X_i, f_i)_{i\in\mathbb{N}}$ and $(Y_i, g_i)_{i\in\mathbb{N}}$ be inverse sequences. Suppose that there exists an increasing sequence $n(1) < n(2) < \cdots \in \mathbb{N}$ and maps $\varphi_i : X_{n(2i)} \to Y_{n(2i-1)}$ and $\psi_i : Y_{n(2i+1)} \to X_{n(2i)}$ such that $\psi_i\varphi_{i+1} = f_{n(2i),n(2i+2)}$ and $\varphi_i\psi_i = g_{n(2i-1),n(2i+1)}$, that is, the following diagram is commutative:*

$$\begin{array}{ccccccc}
X_{n(2)} & \xleftarrow{\ f_{n(2),n(4)}\ } & X_{n(4)} & \xleftarrow{\ f_{n(4),n(6)}\ } & X_{n(6)} & \longleftarrow & \cdots \\
\downarrow{\scriptstyle \varphi_1} & \nwarrow{\scriptstyle \psi_1} & \downarrow{\scriptstyle \varphi_2} & \nwarrow{\scriptstyle \psi_2} & \downarrow{\scriptstyle \varphi_3} & \nwarrow{\scriptstyle \psi_3} & \\
Y_{n(1)} & \xleftarrow[\ g_{n(1),n(3)}\ ]{} & Y_{n(3)} & \xleftarrow[\ g_{n(3),n(5)}\ ]{} & Y_{n(5)} & \longleftarrow & \cdots
\end{array}$$

Then, $\varprojlim(X_i, f_i)$ is homeomorphic to $\varprojlim(Y_i, g_i)$.

Sketch of Proof. Consider the following inverse sequence:

$$Y_{n(1)} \xleftarrow{\varphi_1} X_{n(2)} \xleftarrow{\psi_1} Y_{n(3)} \xleftarrow{\varphi_2} X_{n(4)} \xleftarrow{\psi_2} \cdots ,$$

whose limit is homeomorphic to the inverse limits of the upper and lower sequences in the above diagram.

Let $(X_i, f_i)_{i\in\mathbb{N}}$ be an inverse sequence and $p_i : X = \varprojlim(X_i, f_i) \to X_i, i \in \mathbb{N}$, be the projections of the inverse limit. Then, the following hold:

(inv-1) $f_i p_{i+1} = p_i$ for each $i \in \mathbb{N}$;
(inv-2) Given maps $g_i : Y \to X_i$, $i \in \mathbb{N}$, such that $f_i g_{i+1} = g_i$, there exists a unique map $g : Y \to X$ such that $p_i g = g_i$ for each $i \in \mathbb{N}$.

The above two conditions characterize the inverse limit, that is, more formally:

Theorem 4.10.6. *For an inverse sequence $(X_i, f_i)_{i\in\mathbb{N}}$, a space X is homeomorphic to $\varprojlim(X_i, f_i)$ if and only if there are maps $q_i : X \to X_i$, $i \in \mathbb{N}$, with conditions* (inv-1) *and* (inv-2). *In this case, there is a unique homeomorphism $h : X \to \varprojlim(X_i, f_i)$ such that $p_i h = q_i$ for each $i \in \mathbb{N}$.*

Proof. The "only if" part: If there is a homeomorphism $h : X \to \varprojlim(X_i, f_i)$, then the maps $q_i = p_i h : X \to X_i, i \in \mathbb{N}$, satisfy conditions (inv-1) and (inv-2), where $p_i : \varprojlim(X_i, f_i) \to X_i$ is the inverse limit projection.

The "if" part: Because $(q_i)_{i\in\mathbb{N}}$ satisfies condition (inv-1), we apply condition (inv-2) for $(p_i)_{i\in\mathbb{N}}$ to obtain a map $h : X \to \varprojlim(X_i, f_i)$ such that $p_i h = q_i$ for each $i \in \mathbb{N}$. Similarly, we apply condition (inv-2) for $(q_i)_{i\in\mathbb{N}}$ to obtain a map $g : \varprojlim(X_i, f_i) \to X$ such that $q_i g = p_i$ for each $i \in \mathbb{N}$. Since $p_i hg = p_i$ and $q_i gh = q_i$ for each $i \in \mathbb{N}$, $hg = \mathrm{id}$ and $gh = \mathrm{id}$ by the uniqueness in condition (inv-2). Therefore, h is a homeomorphism with $h^{-1} = g$. □

Restricting the natural homeomorphism from $\prod_{i\in\mathbb{N}} X_i \times \prod_{i\in\mathbb{N}} Y_i$ onto $\prod_{i\in\mathbb{N}}(X_i \times Y_i)$, we can state the following:

Proposition 4.10.7. *For inverse sequences $(X_i, f_i)_{i\in\mathbb{N}}$ and $(Y_i, g_i)_{i\in\mathbb{N}}$, the product space $\varprojlim(X_i, f_i) \times \varprojlim(Y_i, g_i)$ is homeomorphic to the inverse limit $\varprojlim(X_i \times Y_i, f_i \times g_i)$.* □

Concerning subspaces, we have the following proposition:

Proposition 4.10.8. *Let $X = \varprojlim(X_i, f_i)$ be the inverse limit of an inverse sequence $(X_i, f_i)_{i\in\mathbb{N}}$ with the projections $p_i : X \to X_i$, $i \in \mathbb{N}$.*

(1) *For each $i \in \mathbb{N}$, let A_i be a subspace of X_i such that $f_i(A_{i+1}) \subset A_i$. Then, $A = \varprojlim(A_i, f_i|A_{i+1})$ is a subspace of X and $p_i|A$, $i \in \mathbb{N}$, are the inverse limit projections.*
(2) *For every closed subspace A of X, $A = \varprojlim(p_i(A), f_i|p_{i+1}(A))$ and $p_i|A$, $i \in \mathbb{N}$, are the inverse limit projections.*

Proof. For (1), there is no proof needed. For (2), $A \subset \varprojlim(p_i(A), f_i|p_{i+1}(A))$ is trivial. Conversely, for each $x \in \varprojlim(p_i(A), f_i|p_{i+1}(A))$, we can find $a_i \in A$, $i \in \mathbb{N}$, such that $p_i(a_i) = x(i)$. Then, $(a_i)_{i\in\mathbb{N}}$ converges to x in $\prod_{i\in\mathbb{N}} X_i$, hence $x \in \operatorname{cl} A = A$. □

Remark 15. If $(X_i, f_i)_{i\in\mathbb{N}}$ is a nested sequence, then Proposition 4.10.8(2) is valid for an arbitrary subspace A of X. However, in general, it is necessary to assume the closedness of A. In fact, if every X_i is finite then the inverse limit X is compact (cf. Proposition 4.10.9 below). For $A \subset X$, since each $p_i(A)$ is finite, the inverse limit $\varprojlim(p_i(A), f_i|p_{i+1}(A))$ is also compact. Then, $A \neq \varprojlim(p_i(A), f_i|p_{i+1}(A))$ unless A is closed in X.

Proposition 4.10.9. *Let $X = \varprojlim(X_i, f_i)$ be the inverse limit of an inverse sequence $(X_i, f_i)_{i\in\mathbb{N}}$. Then, the following hold:*

(1) *If every X_i is compact then X is compact, where $X \neq \emptyset$ if $X_i \neq \emptyset$ for every $i \in \mathbb{N}$;*
(2) *If every X_i is (completely) metrizable then X is (completely) metrizable;*
(3) *If every X_i is locally compact and each f_i is proper, then X is locally compact and the inverse limit projections $p_i : X \to X_i$, $i \in \mathbb{N}$, are proper.*

Proof. Since X is a closed subspace of the product space $\prod_{i\in\mathbb{N}} X_i$, (2) and the first half of (1) are trivial. For the second half of (1), we define the nested sequence $X_1^* \supset X_2^* \supset \cdots$ of closed sets in $\prod_{i\in\mathbb{N}} X_i$ as follows:

$$X_n^* = \left\{x \in \textstyle\prod_{i\in\mathbb{N}} X_i \;\middle|\; x(i) = f_{i,n}(x(n)) \text{ for each } i \leq n\right\}.$$

If each X_i is a non-empty compact space, then $\prod_{i\in\mathbb{N}} X_i$ is compact and $X_n^* \neq \emptyset$ for each $i \in \mathbb{N}$, hence $X = \bigcap_{n\in\mathbb{N}} X_n^* \neq \emptyset$.

(3): According to Proposition 4.10.3, it is enough to show that each projection $p_n : X \to X_n$ is proper. For each compact set A in X_n, $p_n^{-1}(A)$ is closed in X, so is closed in $\prod_{i\in\mathbb{N}} X_i$. We define

$$A_i = \begin{cases} f_{i,n}(A) & \text{if } i \leq n, \\ f_{n,i}^{-1}(A) & \text{if } i > n, \end{cases}$$

where $f_{i,j}$ is proper for each $i < j$. Then, $p_n^{-1}(A)$ is contained in the compact set $\prod_{i\in\mathbb{N}} A_i$ in $\prod_{i\in\mathbb{N}} X_i$. Indeed, $p_i p_n^{-1}(A) = f_{i,n} p_n p_n^{-1}(A) \subset f_{i,n}(A) = A_i$ for $i \leq n$ and $p_i p_n^{-1}(A) = p_i p_i^{-1} f_{n,i}^{-1}(A) \subset f_{n,i}^{-1}(A) = A_i$ for $i > n$. Thus, $p_n^{-1}(A)$ is compact. □

For the remainder of this section, we show that every completely metrizable space is represented by the inverse limit of some inverse sequence of metric polyhedra. This can be stated in the following theorem as:

Theorem 4.10.10. *Every completely metrizable space X is homeomorphic to the inverse limit $\varprojlim(|K_i|_{\mathrm{m}}, f_i)$ of an inverse sequence of metric polyhedra and PL* *maps such that each* K_i *is locally finite-dimensional,* card $K_i \le w(X)$, *and each* $f_i : K_{i+1} \to K_i'$ *is simplicial for some admissible subdivision* K_i' *of* K_i, *where admissibility of* K_i' *guarantees the continuity of* $f_i : |K_{i+1}|_{\mathrm{m}} \to |K_i|_{\mathrm{m}}$.

Proof. Because this theorem is obvious if X is finite, we may assume that $w(X) \ge \aleph_0$. Let d be an admissible complete metric for X. By induction, we will construct a refining sequence of open covers of X:

$$\mathcal{U}_1 \succ \mathcal{U}_2^{\mathrm{cl}} \succ \mathcal{U}_2 \succ \mathcal{U}_3^{\mathrm{cl}} \succ \mathcal{U}_3 \succ \cdots$$

with admissible subdivisions K_i' of the nerves $K_i = N(\mathcal{U}_i)$ and simplicial maps $f_i : K_{i+1} \to K_i'$ satisfying the following conditions:

(1) $\mathcal{U}_i$ is locally finite;
(2) $K_i = N(\mathcal{U}_i)$ is locally finite-dimensional;
(3) $\mathrm{mesh}_d\, \mathcal{U}_i < 2^{-i}$;
(4) $\mathrm{cl}\, U \subset \varphi_i^{-1}(O_{K_i'}(f_i(U)))$ for each $U \in K_{i+1}^{(0)} = \mathcal{U}_{i+1}$;
(5) $\mathrm{mesh}_{\rho_j} f_{j,i}(\mathcal{S}_{K_i}) < 2^{-(i-j)}$ for $j < i$,

where φ_i is a canonical map and $\rho_j = \rho_{K_j}$. Note that card $K_i \le \aleph_0 w(X) = w(X)$ by (1) and (2).

First, by Theorem 4.9.9, we have $\mathcal{U}_1 \in \mathrm{cov}(X)$ satisfying (1)–(3). Next, assume that $\mathcal{U}_1, \ldots, \mathcal{U}_i$ and $f_1, \ldots, f_{i-1}$ have been defined so as to satisfy (1)–(5). By Theorem 4.8.8, we can find an admissible subdivision K_i' of K_i such that

$$\mathrm{mesh}_{\rho_i} \mathcal{S}_{K_i'} < 1/2 \text{ and } \mathrm{mesh}_{\rho_j} f_{j,i}(\mathcal{S}_{K_i'}) < 2^{-(i+1-j)} \text{ for all } j < i.$$

Note that $\mathcal{O}_{K_i'}$ is an open cover of $|K_i'|_{\mathrm{m}} = |K_i|_{\mathrm{m}}$ (cf. Lemma 4.8.1). Let $\varphi_i : X \to |K_i|_{\mathrm{m}}$ be a canonical map. Then,

$$\varphi_i^{-1}(\mathcal{O}_{K_i'}) \in \mathrm{cov}(X) \text{ and } \varphi_i^{-1}(\mathcal{O}_{K_i'}) \prec \varphi_i^{-1}(\mathcal{O}_{K_i}) \prec \mathcal{U}_i.$$

By the regularity of X and Theorem 4.9.9, we have $\mathcal{U}_{i+1} \in \mathrm{cov}(X)$ satisfying (1)–(3) and $\mathcal{U}_{i+1}^{\mathrm{cl}} \prec \varphi_i^{-1}(\mathcal{O}_{K_i'})$. Then, there is a simplicial map $f_i : K_{i+1} \to K_i'$ satisfying (4), that is,

$$\mathrm{cl}\, U \subset \varphi_i^{-1}(O_{K_i'}(f_i(U))) \text{ for each } U \in K_{i+1}^{(0)} = \mathcal{U}_{i+1}.$$

Indeed, for each $\langle U_1, \ldots, U_k \rangle \in K_{i+1}$, $\bigcap_{j=1}^k \varphi_i^{-1}(O_{K_i'}(f_i(U_j))) \ne \emptyset$, hence $\bigcap_{j=1}^k O_{K_i'}(f_i(U_j)) \ne \emptyset$, which implies that $\langle f_i(U_1), \ldots, f_i(U_k) \rangle \in K_i'$. Since $f_i(\mathcal{S}_{K_{i+1}}) \prec \mathcal{S}_{K_i'}$, (5) holds for $j < i+1$, that is,

$$\mathrm{mesh}_{\rho_j} f_{j,i+1}(\mathcal{S}_{K_{i+1}}) < 2^{-(i+1-j)} \text{ for each } j < i+1.$$

We now construct a homeomorphism $h : X \to \varprojlim(|K_i|_{\mathrm{m}}, f_i)$. For each $x \in X$ and $i \in \mathbb{N}$, let $\tau_i(x) = \langle \mathcal{U}_i[x] \rangle \in K_i$. We will show that $f_i(\tau_{i+1}(x)) \subset \tau_i(x)$. Let $\sigma \in K_i$ be the smallest simplex containing $f_i(\tau_{i+1}(x)) \in K_i'$. Then,

$$\bigcup_{V \in \mathcal{U}_{i+1}[x]} c_{K_i}(f_i(V))^{(0)} = \sigma^{(0)}.$$

Indeed, for each $V \in \mathcal{U}_{i+1}[x] = \tau_{i+1}(x)^{(0)}$, we have $c_{K_i}(f_i(V)) \leq \sigma$, that is, $c_{K_i}(f_i(V))^{(0)} \subset \sigma^{(0)}$. If $\bigcup_{\mathcal{U}_{i+1}[x]} c_{K_i}(f_i(V))^{(0)} \subsetneq \sigma^{(0)}$ then $\tau_{i+1}(x) = \langle \mathcal{U}_{i+1}[x] \rangle$ is contained in a proper face of σ, which contradicts the minimality of σ. For each $V \in \tau_{i+1}(x)^{(0)} = \mathcal{U}_{i+1}[x]$ and $U \in c_{K_i}(f_i(V))^{(0)}$, we have $O_{K_i'}(f_i(V)) \subset O_{K_i}(U)$ by Proposition 4.2.15. Thus,

$$\bigcap_{V \in \mathcal{U}_{i+1}[x]} O_{K_i'}(f_i(V)) \subset \bigcap_{U \in \sigma^{(0)}} O_{K_i}(U).$$

Then, it follows that

$$\begin{aligned} x \in \bigcap_{V \in \mathcal{U}_{i+1}[x]} V &\subset \bigcap_{V \in \mathcal{U}_{i+1}[x]} \varphi_i^{-1}(O_{K_i'}(f_i(V))) \\ &\subset \bigcap_{U \in \sigma^{(0)}} \varphi_i^{-1}(O_{K_i}(U)) \subset \bigcap_{U \in \sigma^{(0)}} U. \end{aligned}$$

Hence, $\sigma^{(0)} \subset \mathcal{U}_i[x]$, that is, $\sigma \leq \tau_i(x)$, so $f_i(\tau_{i+1}(x)) \subset \tau_i(x)$. Now, we have the following nested sequence of compacta:

$$\tau_i(x) = f_{i,i}(\tau_i(x)) \supset f_{i,i+1}(\tau_{i+1}(x)) \supset f_{i,i+2}(\tau_{i+2}(x)) \supset \cdots.$$

Since $\bigcap_{j \geq i} f_{i,j}(\tau_j(x)) \neq \emptyset$ and $\lim_{k\to\infty} \operatorname{diam}_{\rho_i} f_{i,i+k}(\tau_{i+k}(x)) = 0$ by (5), it follows that $\bigcap_{j \geq i} f_{i,j}(\tau_j(x))$ is a singleton. Thus, we have $h_i : X \to |K_i|$ such that

$$\{h_i(x)\} = \bigcap_{j \geq i} f_{i,j}(\tau_j(x)) \text{ for each } x \in X.$$

Observe that $f_i h_{i+1} = h_i$. Thus, we have $h : X \to \varprojlim(|K_i|_{\mathrm{m}}, f_i)$ such that $p_i h = h_i$ for each $i \in \mathbb{N}$, where $p_i : \varprojlim(|K_i|_{\mathrm{m}}, f_i) \to |K_i|_{\mathrm{m}}$ is the projection.

We now verify the continuity of h. For each $i \in \mathbb{N}$, $x \in X$, and $\varepsilon > 0$, we can choose $j > i$ so that $\operatorname{mesh}_{\rho_i} f_{i,j}(\mathcal{S}_{K_j}) < \varepsilon$ by (5). Then, $\bigcap \mathcal{U}_j[x]$ is an open neighborhood of x in X. Since $y \in \bigcap \mathcal{U}_j[x]$ implies $\tau_j(x) \leq \tau_j(y)$, we have $h_j(y) \in \tau_j(y) \in \operatorname{St}(\tau_j(x), K_j)$, hence

$$h_i(\bigcap \mathcal{U}_j[x]) = f_{i,j} h_j(\bigcap \mathcal{U}_j[x]) \subset f_{i,j}(|\operatorname{St}(\tau_j(x), K_j)|).$$

Then, it follows that

$$\operatorname{diam}_{\rho_i} h_i(\bigcap \mathcal{U}_j[x]) \leq \operatorname{diam}_{\rho_i} f_{i,j}(|\operatorname{St}(\tau_j(x), K_j)|) \leq \operatorname{mesh}_{\rho_i} f_{i,j}(\mathcal{S}_{K_j}) < \varepsilon.$$

Thus, each $h_i : X \to |K_i|_{\mathrm{m}}$ is continuous, so h is continuous.

Next, we show that h is injective. For $x \neq x' \in X$, we can choose $i \in \mathbb{N}$ so that $\operatorname{mesh}_d \mathcal{U}_i < d(x, x')$ by (3). Then, it follows that

$$\tau_i(x)^{(0)} \cap \tau_i(x')^{(0)} = \mathcal{U}_i[x] \cap \mathcal{U}_i[x'] = \emptyset,$$

which means $\tau_i(x) \cap \tau_i(x') = \emptyset$, hence $h_i(x) \neq h_i(x')$, so $h(x) \neq h(x')$. Thus, h is injective.

To prove that h is surjective, let $y \in \varprojlim(|K_i|_{\mathrm{m}}, f_i)$. For each $i \in \mathbb{N}$, let $\tau_i = c_{K_i}(p_i(y)) \in K_i$ be the carrier of $p_i(y)$. Then, we have an open set $W_i = \bigcup_{U \in \tau_i^{(0)}} U \subset X$. Since $\bigcap_{U \in \tau_i^{(0)}} U \neq \emptyset$, it follows that $\operatorname{diam}_d W_i \leq 2 \operatorname{mesh}_d \mathcal{U}_i$, hence $\lim_{i \to \infty} \operatorname{diam}_d W_i = 0$ by (3). For each $U \in \tau_{i+1}^{(0)}$, choose $V \in \tau_i^{(0)}$ so that $f_i(U) \in O_{K_i}(V)$. Since $f_i(U) \in K_i'^{(0)}$ and $K_i' \lhd K_i$, it follows that $O_{K_i'}(f_i(U)) \subset O_{K_i}(V)$, hence by (4),

$$\operatorname{cl} U \subset \varphi_i^{-1}(O_{K_i'}(f_i(U))) \subset \varphi_i^{-1}(O_{K_i}(V)) \subset V.$$

Therefore, $\operatorname{cl} W_{i+1} \subset W_i$. By the completeness of X, we have $z \in X$ such that $\{z\} = \bigcap_{i \in \mathbb{N}} \operatorname{cl} W_i = \bigcap_{i \in \mathbb{N}} W_i$. For each $i \in \mathbb{N}$, z is contained in some $U \in \tau_i^{(0)}$, so $\tau_i^{(0)} \cap \tau_i(z)^{(0)} \neq \emptyset$, that is, $\tau_i \cap \tau_i(z) \neq \emptyset$. It follows from the definition of h_i that $h_i(z) = p_i(y)$ for each $i \in \mathbb{N}$, which means $h(z) = y$.

To conclude that h is a homeomorphism, it remains to show that h is an open map. Let V be a neighborhood of x in X. By (3), $\operatorname{st}(x, \mathcal{U}_i) \subset V$ for some $i \in \mathbb{N}$. Since $h_i(x) \in \tau_i(x)$, we have $U \in \tau_i(x)^{(0)} = \mathcal{U}_i[x]$ such that $h_i(x) \in O_{K_i}(U)$, which means $h(x) \in p_i^{-1}(O_{K_i}(U))$. For each $y \in p_i^{-1} O_{K_i}(U)$, let $z = h^{-1}(y)$. Then, U is a vertex of the carrier $\tau_i \in K_i$ of $p_i(y)$. It follows from the above argument that $z \in \operatorname{st}(U, \mathcal{U}_i) \subset V$, hence $y = h(z) \in h(V)$. Thus, we have $p_i^{-1}(O_{K_i}(U)) \subset h(V)$. Hence, $h(V)$ is a neighborhood of $h(x)$ in $\varprojlim(|K_i|_{\mathrm{m}}, f_i)$. This shows that h is an open map. □

In the above proof, if X is compact then each $\mathcal{U}_i$ can be finite, so each K_i is finite. When X is separable and completely metrizable, each $\mathcal{U}_i$ can be countable and star-finite by virtue of Theorem 4.9.10, which means that each K_i is locally finite. Thus, we have the following:

Corollary 4.10.11. *Every compactum (resp. separable completely metrizable space) is homeomorphic to the inverse limit of an inverse sequence of compact (resp. separable locally compact) polyhedra and PL maps.* □

Note. The compact case of the above Corollary 4.10.11 can be easily proved as follows: By virtue of Corollary 2.3.8, we may assume that X is a closed set in the Hilbert cube $\mathbf{I}^{\mathbb{N}}$. For each $i \in \mathbb{N}$, let $p_i : \mathbf{I}^{\mathbb{N}} \to \mathbf{I}^i$ be the projection defined by $p_i(x) = (x(1), \ldots, x(i))$. Construct polyhedra $P_i \subset \mathbf{I}^i$ so that $p_i(X) \subset \operatorname{int}_{\mathbf{I}^i} P_i$,

$P_{i+1} \subset P_i \times \mathbf{I}$ (hence $p_{i+1}^{-1}(P_{i+1}) \subset p_i^{-1}(P_i)$) and $\bigcap_{i \in \mathbb{N}} p_i^{-1}(P_i) = X$. For each $i \in \mathbb{N}$, let $f_i : P_{i+1} \to P_i$ be the restriction of the projection of $P_i \times \mathbf{I}$ onto P_i. Then, $X = \bigcap_{i \in \mathbb{N}} p_i^{-1}(P_i) \approx \varprojlim(P_i, f_i)$ by the homeomorphism induced by $p_i|X : X \to P_i$.

In the case that X is separable and locally compact metrizable, not only can each K_i be locally finite, but each f_i can also be proper. Indeed, we can take $\mathcal{U}_i$ such that each $U \in \mathcal{U}_i$ has the compact closure, hence $\mathcal{U}_{i+1}[U]$ is finite for each $U \in \mathcal{U}_i$, which means that $K_i = N(\mathcal{U}_i)$ is locally finite, hence $|K_i|_{\mathrm{m}} = |K_i|$ is locally compact. For each $U \in K_{i+1}^{(0)} = \mathcal{U}_{i+1}$ and $V \in K_i^{(0)} = \mathcal{U}_i$, if $f_i(U) \in O_{K_i}(V)$ then

$$\operatorname{cl} U \subset \varphi_i^{-1}(O_{K_i'}(f_i(U))) \subset \varphi_i^{-1}(O_{K_i}(V)) \subset V.$$

Since V contains only finitely many members of $\mathcal{U}_{i+1}$, $f_i^{-1}(f_i(U))$ contains only finitely many vertices of K_{i+1}. Hence, f_i is proper.

Corollary 4.10.12. *Every separable locally compact metrizable space is homeomorphic to the inverse limit of an inverse sequence of separable locally compact polyhedra and proper PL-maps.* □

We now present the following useful remark.

Remark 16. In the proof of Theorem 4.10.10, we needed the following condition that is derived from condition (5):

$$(*) \quad \lim_{k\to\infty} \operatorname{mesh}_{\rho_i} f_{i,i+k}(\mathcal{S}_{K_{i+k}}) = 0.$$

In Theorem 4.10.10, if $\dim K_i < \infty$ for each $i \in \mathbb{N}$, we can take $K_i' = \operatorname{Sd}^{n_i} K_i$ for some $n_i \in \mathbb{N}$. Indeed, choose $n_i \in \mathbb{N}$ in the proof so that

$$\lim_{i\to\infty} \left(\frac{\dim K_i}{\dim K_i + 1}\right)^{n_i} = 0.$$

Then, condition $(*)$ holds because the following inequality is obtained from Lemma 4.7.3:

$$\begin{aligned}
\operatorname{mesh}_{\rho_i} f_{i,i+k}(\operatorname{Sd}^{n_{i+k}} K_{i+k}) &\le \left(\frac{\dim K_{i+k}}{\dim K_{i+k} + 1}\right)^{n_{i+k}} \operatorname{mesh}_{\rho_i} f_{i,i+k}(K_{i+k}) \\
&\le \left(\frac{\dim K_{i+k}}{\dim K_{i+k} + 1}\right)^{n_{i+k}} \operatorname{mesh}_{\rho_i} \operatorname{Sd}^{n_i} K_i \\
&\le 2\left(\frac{\dim K_i}{\dim K_i + 1}\right)^{n_i} \left(\frac{\dim K_{i+k}}{\dim K_{i+k} + 1}\right)^{n_{i+k}}.
\end{aligned}$$

When $\dim K_i \le n$ for every $i \in \mathbb{N}$, we can take $K_i' = \operatorname{Sd} K_i$. Indeed, condition $(*)$ is satisfied because the following inequality follows from Lemma 4.7.3:

$$\begin{aligned}
\operatorname{mesh}_{\rho_i} f_{i,i+k}(\operatorname{Sd} K_{i+k}) &\le \frac{n}{n+1} \operatorname{mesh}_{\rho_i} f_{i,i+k}(K_{i+k}) \\
&\le \frac{n}{n+1} \operatorname{mesh}_{\rho_i} f_{i,i+k-1}(\operatorname{Sd} K_{i+k-1}) \\
&\le \left(\frac{n}{n+1}\right)^2 \operatorname{mesh}_{\rho_i} f_{i,i+k-1}(K_{i+k-1}) \\
&\le \cdots \le \left(\frac{n}{n+1}\right)^k \operatorname{mesh}_{\rho_i} \operatorname{Sd} K_i \\
&\le 2\left(\frac{n}{n+1}\right)^{k+1}.
\end{aligned}$$

Remark 17. By replacing $\mathbb{N}$ with a directed set, we can generalize the inverse sequence to the inverse system. Using an inverse system of polyhedra with the Whitehead topology, Theorem 4.10.10 is valid for every paracompact space, that is, every paracompact space (hence every metrizable space) can be represented as the inverse limit of an inverse system of polyhedra with the Whitehead topology. Refer to Notes for Chap. 4.

4.11 The Mapping Cylinders

The **mapping cylinder** M_f of a map $f : X \to Y$ is defined as the following adjunction space:

$$M_f = Y \cup_{f \circ \mathrm{pr}_X | X \times \{0\}} X \times \mathbf{I},$$

that is, M_f is obtained from the topological sum $Y \oplus (X \times \mathbf{I})$ by identifying the points $(x, 0) \in X \times \{0\}$ with the points $f(x) \in Y$. By $q_f : Y \oplus (X \times \mathbf{I}) \to M_f$, we denote the natural quotient map. The map $c_f : M_f \to Y$, called the **collapsing** of M_f, is defined by $c_f|Y = \mathrm{id}$ and $c_f|X \times (0, 1] = f \circ \mathrm{pr}_X$. By $i_f : X \to M_f$, we denote the natural embedding defined by $i_f(x) = (x, 1)$. Then, $c_f \circ i_f = f$ and $c_f \simeq \mathrm{id}$ rel. Y in M_f which is realized by the homotopy $h^f : M_f \times \mathbf{I} \to M_f$ defined by $h^f_0 = c_f$, $h^f_t|Y = \mathrm{id}$ for each $t \in \mathbf{I}$ and $h^f_t(x, s) = (x, st)$ for each $(x, s) \in X \times (0, 1]$ and $t \in (0, 1]$.

When Y is a singleton, the mapping cylinder M_f is homeomorphic to the quotient space $(X \times \mathbf{I})/(X \times \{0\})$, which is called the **cone** over X. We regard the mapping cylinder M_{id_X} of the identity map id_X as the product space $X \times \mathbf{I}$.

If X is a *closed subspace* of Y, the mapping cylinder M_i of the inclusion $i : X \subset Y$ can be regarded as the subspace $(Y \times \{0\}) \cup (X \times \mathbf{I})$ of the product space $Y \times \mathbf{I}$. However, if Y is perfectly normal and X is not closed in Y then the mapping cylinder M_i of the inclusion $i : X \subset Y$ cannot be regarded as the subspace of

$Y \times \mathbf{I}$. Indeed, let $y \in (\mathrm{cl}_Y X) \setminus X$. Then, we have a map $k : Y \to \mathbf{I}$ such that $\{y\} = k^{-1}(0)$. Observe that the graph $G = \{(x, k(x)) \mid x \in X\}$ of the map k is closed in M_i but is not closed in the subspace $(Y \times \{0\}) \cup (X \times \mathbf{I})$ of $Y \times \mathbf{I}$.

For spaces X and Y with $A \subset X \cap Y$, we denote

$$X \simeq Y \ \text{ rel. } A$$

if there exist homotopy equivalences $f : X \to Y$ and $g : Y \to X$ such that $f|A = g|A = \mathrm{id}$, $gf \simeq \mathrm{id}_X$ rel. A, and $fg \simeq \mathrm{id}_Y$ rel. A. Consider one more space Z with $A \subset Z$. Then, it easily follows that $X \simeq Y$ rel. A and $Y \simeq Z$ rel. A imply $X \simeq Z$ rel. A.

Theorem 4.11.1. *For maps $f, g : X \to Y$, the following are equivalent:*

(a) $f \simeq g$;
(b) $M_f \simeq M_g$ rel. $Y \cup (X \times \{1\})$;
(c) *There is a map $\varphi : M_f \to M_g$ such that $\varphi|Y \cup (X \times \{1\}) = \mathrm{id}$.*

Proof. The implication (b) ⇒ (c) is obvious. Using the above homotopy h^f, we can see (c) ⇒ (a) as follows:

$$f = c_f \circ i_f = c_g \circ \varphi \circ h_0^f \circ i_f \simeq c_g \circ \varphi \circ h_1^f \circ i_f = c_g \circ \varphi \circ i_f = c_g \circ i_g = g.$$

(a) ⇒ (b): Let $h : X \times \mathbf{I} \to Y$ be a homotopy from f to g. We define maps $\varphi : M_f \to M_g$ and $\psi : M_g \to M_f$ as follows: $\varphi|Y = \psi|Y = \mathrm{id}$ and

$$\varphi(x,s) = \begin{cases} (x, 2s-1) & \text{for } (x,s) \in X \times (\frac{1}{2}, 1], \\ h_{2s}(x) & \text{for } (x,s) \in X \times (0, \frac{1}{2}]; \end{cases}$$

$$\psi(x,s) = \begin{cases} (x, 2s-1) & \text{for } (x,s) \in X \times (\frac{1}{2}, 1], \\ h_{1-2s}(x) & \text{for } (x,s) \in X \times (0, \frac{1}{2}]. \end{cases}$$

See Fig. 4.12. Then, it follows that $\psi\varphi|Y = \mathrm{id}$ and

$$\psi\varphi(x,s) = \begin{cases} (x, 4s-3) & \text{for } (x,s) \in X \times (\frac{3}{4}, 1], \\ h_{3-4s}(x) & \text{for } (x,s) \in X \times (\frac{1}{2}, \frac{3}{4}], \\ h_{2s}(x) & \text{for } (x,s) \in X \times (0, \frac{1}{2}]. \end{cases}$$

We define an auxiliary map $\theta : M_f \to M_f$ as follows: $\theta|Y = \mathrm{id}$ and

$$\theta(x,s) = \begin{cases} (x, 4s-3) & \text{for } (x,s) \in X \times (\frac{3}{4}, 1], \\ f(x) & \text{for } (x,s) \in X \times (0, \frac{3}{4}]. \end{cases}$$

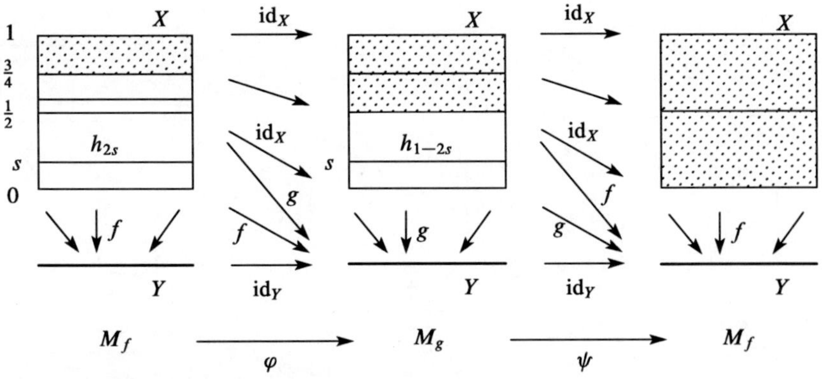

Fig. 4.12 The mapping cylinders of homotopic maps

Then, we have $\psi\varphi \simeq \theta \simeq \mathrm{id}_{M_f}$ rel. $Y \cup (X \times \{1\})$ by the homotopies $\xi : M_f \times \mathbf{I} \to M_f$ and $\zeta : M_f \times \mathbf{I} \to M_f$ defined as follows: $\xi_t|Y = \zeta_t|Y = \mathrm{id}$ and

$$\xi_t(x,s) = \begin{cases} (x, 4s-3) & \text{for } (x,s) \in X \times (\frac{3}{4}, 1], \\ h_{(3-4s)t}(x) & \text{for } (x,s) \in X \times (\frac{1}{2}, \frac{3}{4}], \\ h_{2st}(x) & \text{for } (x,s) \in X \times (0, \frac{1}{2}]; \end{cases}$$

$$\zeta_t(x,s) = \begin{cases} \left(x, \dfrac{4s-3t}{4-3t}\right) & \text{for } (x,s) \in X \times (\frac{3}{4}t, 1], \\ f(x) & \text{for } (x,s) \in X \times (0, \frac{3}{4}t]. \end{cases}$$

Similarly, we can see that $\varphi\psi \simeq \mathrm{id}$ rel. $Y \cup (X \times \{1\})$. □

Theorem 4.11.2. *For maps $f : X \to Y$ and $g : Y \to Z$,*

$$M_g \cup_{i_g} M_f \simeq M_{gf} \quad \text{rel. } Z \cup (X \times \{1\}).$$

Proof. We define maps $\varphi : M_g \cup_{i_g} M_f \to M_{gf}$ and $\psi : M_{gf} \to M_g \cup_{i_g} M_f$ as follows: $\varphi|M_g = c_g$, $\varphi|X \times (0,1] = \mathrm{id}$ and $\psi|Z = \mathrm{id}$,

$$\psi(x,s) = \begin{cases} (x, 2s-1) \in M_f & \text{for } (x,s) \in X \times (\frac{1}{2}, 1], \\ (f(x), 2s) \in M_g & \text{for } (x,s) \in X \times (0, \frac{1}{2}]. \end{cases}$$

Observe that $\psi\varphi|M_g = c_g$, $\psi\varphi|X \times (0,1] = \psi|X \times (0,1]$, and $\varphi\psi|Z = \mathrm{id}$,

$$\varphi\psi(x,s) = \begin{cases} (x, 2s-1) & \text{for } (x,s) \in X \times (\frac{1}{2}, 1], \\ gf(x) & \text{for } (x,s) \in X \times (0, \frac{1}{2}]. \end{cases}$$

See Figs. 4.13 and 4.14. Then, we have

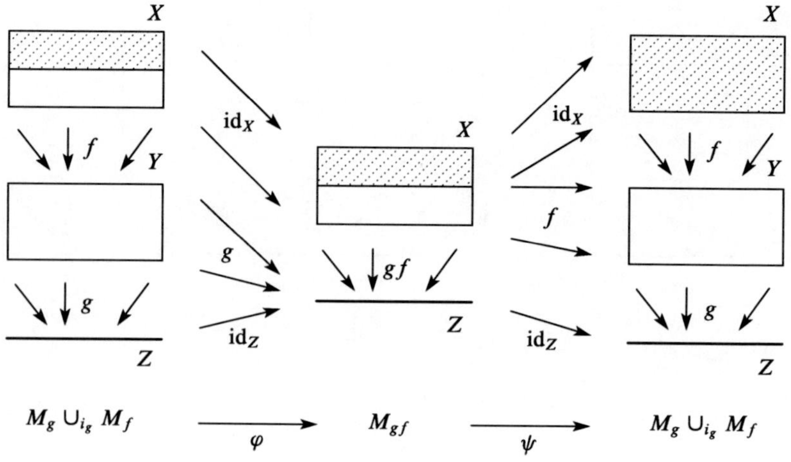

Fig. 4.13 The mapping cylinder of the composition of maps (1)

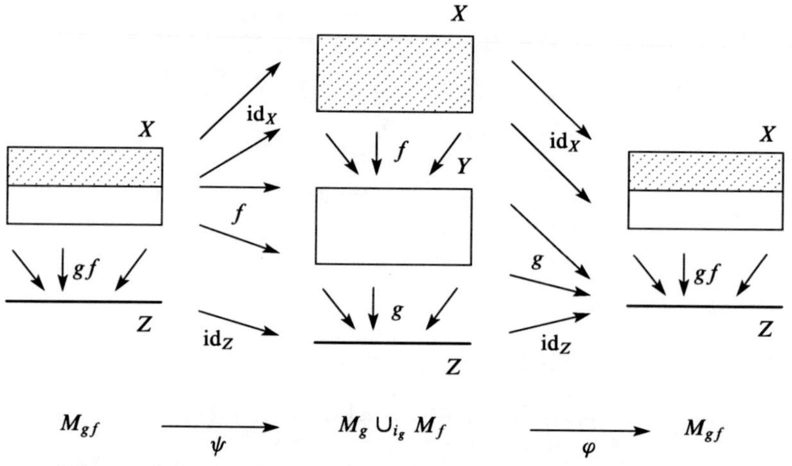

Fig. 4.14 The mapping cylinder of the composition of maps (2)

$$\psi\varphi \simeq \mathrm{id} \ \ \mathrm{rel.}\ Z \cup (X \times \{1\}) \text{ and } \varphi\psi \simeq \mathrm{id} \ \ \mathrm{rel.}\ Z \cup (X \times \{1\}).$$

In fact, these are realized by the homotopies

$$\xi : (M_g \cup_{i_g} M_f) \times \mathbf{I} \to M_g \cup_{i_g} M_f \ \text{ and } \ \zeta : M_{gf} \times \mathbf{I} \to M_{gf}$$

defined as follows: $\xi_t | M_g = h^g_{1-t}$, $\zeta_t | Z = \mathrm{id}$, and

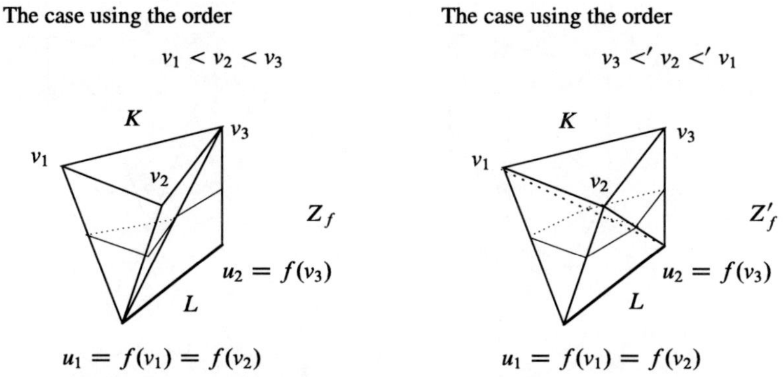

Fig. 4.15 The simplicial mapping cylinders

$$\xi_t(x,s) = \begin{cases} \left(x, \dfrac{2s-t}{2-t}\right) & \text{for } (x,s) \in X \times (\tfrac{1}{2}t, 1], \\ (f(x), 2s+1-t) & \text{for } (x,s) \in X \times (0, \tfrac{1}{2}t]; \end{cases}$$

$$\zeta_t(x,s) = \begin{cases} \left(x, \dfrac{2s-t}{2-t}\right) & \text{for } (x,s) \in X \times (\tfrac{1}{2}t, 1], \\ gf(x) & \text{for } (x,s) \in X \times (0, \tfrac{1}{2}t]. \end{cases} \qquad \square$$

Let K and L be simplicial complexes in the linear spaces E and F, respectively. Consider the linear product space $E \times F \times \mathbb{R}$ and regard K and L as being contained in $E \times \{0\} \times \{1\}$ and $\{0\} \times F \times \{0\}$, respectively. Assume that K is an *ordered simplicial complex*. The **simplicial mapping cylinder** Z_f of a simplicial map $f : K \to L$ is defined as the following simplicial complex:

$$Z_f = L \cup \big\{\langle f(v_1), \cdots, f(v_{m'}), v_m, \cdots, v_n\rangle \mid \\ \langle v_1, \cdots, v_n\rangle \in K,\ v_1 < \cdots < v_n,\ 1 \le m' \le m \le n\big\}.$$

The **collapsing** of Z_f is the simplicial map $\bar{c}_f : Z_f \to L$ defined by $\bar{c}_f(v) = f(v)$ for each $v \in K^{(0)}$ and $\bar{c}_f(u) = u$ for each $u \in L^{(0)}$. Then, $f = \bar{c}_f||K|$. If $L = \{v\}$ is a singleton, Z_f is called the **simplicial cone** over K with v the cone vertex, which is denoted by $v * K$.

Remark 18. Note that the simplicial mapping cylinder changes if the order on $K^{(0)}$ is changed. Moreover, in general, $|Z_f| \not\approx M_f$. For instance, let K and L be the standard triangulations of a 2-simplex $\langle v_1, v_2, v_3\rangle$ and a 1-simplex $\langle u_1, u_2\rangle$, respectively. Consider the simplicial map $f : K \to L$ defined by $f(v_1) = f(v_2) = u_1$ and $f(v_3) = u_2$. Let Z_f be the simplicial mapping cylinder defined by he order $v_1 < v_2 < v_3$ and Z'_f be the one defined by the order $v_3 <' v_2 <' v_1$. Then, we have not only $Z_f \not\equiv Z'_f$ but also $|Z_f| \not\approx |Z'_f|$ (see Fig. 4.15). Evidently, $|Z_f| \not\approx M_f$.

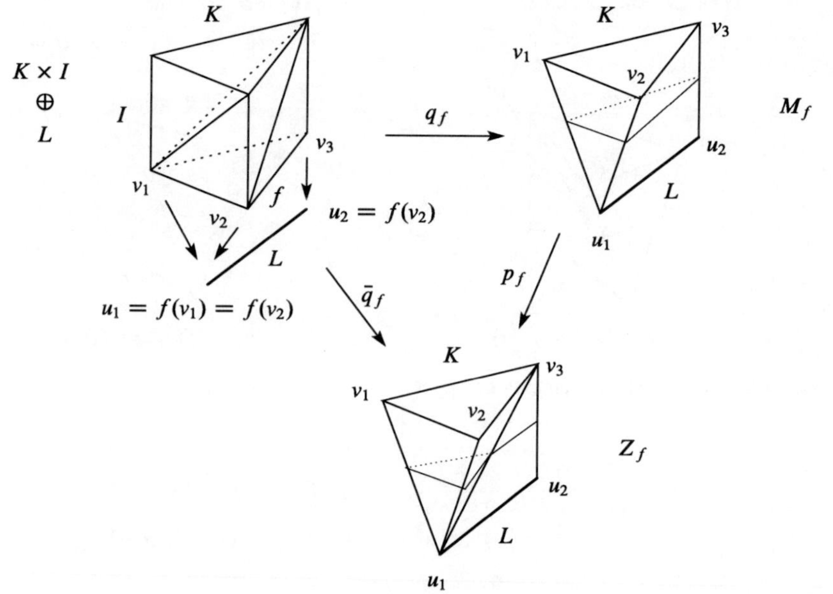

Fig. 4.16 The mapping cylinder and the simplicial mapping cylinder

Because K is assumed to be an ordered simplicial complex, we can consider the product simplicial complex $K \times_s I$ of K and the ordered simplicial complex $I = \{\mathbf{I}, 0, 1\}$ with the natural order, i.e.,

$$K \times_s I = \big\{\langle (v_1, 0), \cdots, (v_{m'}, 0), (v_m, 1), \cdots, (v_n, 1)\rangle \bigm| \langle v_1, \cdots, v_n\rangle \in K,\ v_1 < \cdots < v_n,\ 1 \leq m' \leq m \leq n\big\},$$

where $|K \times_s I| = |K| \times \mathbf{I}$ according to Theorem 4.3.1. We define the simplicial map $\bar{q}_f : L \oplus (K \times_s I) \to Z_f$ by $\bar{q}_f|L^{(0)} = \mathrm{id}$,

$$\bar{q}_f(v, 0) = f(v) \text{ and } \bar{q}_f(v, 1) = v \text{ for each } v \in K^{(0)}.$$

Then, $\bar{q}_f : |L| \oplus (|K| \times \mathbf{I}) \to |Z_f|$ is a quotient map. Indeed, let $A \subset |Z_f|$ and assume that $\bar{q}_f^{-1}(A)$ is closed in $|L| \oplus (|K| \times \mathbf{I})$. Observe that

$$Z_f = L \cup \big\{\bar{q}_f(\tau) \bigm| \tau \in K \times_s I\big\}.$$

For each $\sigma \in L$, $A \cap \sigma = \bar{q}_f^{-1}(A) \cap \sigma$ is closed in σ. For each $\tau \in K \times_s I$, since $\bar{q}_f|\tau$ is a quotient map and $(\bar{q}_f|\tau)^{-1}(A \cap \bar{q}_f(\tau)) = \bar{q}_f^{-1}(A) \cap \tau$ is closed in τ, it follows that $A \cap \bar{q}_f(\tau)$ is closed in $\bar{q}_f(\tau)$. As a consequence, A is closed in $|Z_f|$. Since $\bar{q}_f||L| = \mathrm{id}$ and $\bar{q}_f||K| \times \{0\} = f \circ \mathrm{pr}_{|K|}$, we have the map $p_f : M_f \to |Z_f|$ such that $p_f \circ q_f = \bar{q}_f$. Then, p_f is a quotient map (Fig. 4.16).

For each $(x, s) \in |K| \times \mathbf{I}$, let $c_K(x) = \langle v_1, \cdots, v_n \rangle \in K$ be the carrier of x and write $x = \sum_{i=1}^n t_i v_i$, where $v_1 < \cdots < v_n$ and $t_i = \beta_{v_i}^K(x) > 0$. Choose $m = 1, \cdots, n$ so that

$$\sum_{i=m+1}^{n} t_i \leq s \leq \sum_{i=m}^{n} t_i \quad \left(\text{i.e., } \sum_{i=1}^{m-1} t_i \leq 1 - s \leq \sum_{i=1}^{m} t_i\right),$$

where $\sum_{i=n+1}^n t_i = 0$. Then, it follows that

$$\begin{aligned}(x, s) = &\sum_{i=1}^{m-1} t_i (v_i, 0) + \left(\sum_{i=m}^{n} t_i - s\right)(v_m, 0) \\ &+ \left(s - \sum_{i=m+1}^{n} t_i\right)(v_m, 1) + \sum_{i=m+1}^{n} t_i (v_i, 1) \\ &\in \langle (v_1, 0), \cdots, (v_m, 0), (v_m, 1), \cdots, (v_n, 1) \rangle \in K \times_s I.\end{aligned}$$

By the definition of $\bar{q}_f$, we have

$$\begin{aligned}\bar{q}_f(x, s) = &\sum_{i=1}^{m-1} t_i f(v_i) + \left(\sum_{i=m}^{n} t_i - s\right) f(v_m) \\ &+ \left(s - \sum_{i=m+1}^{n} t_i\right) v_m + \sum_{i=m+1}^{n} t_i v_i \\ &\in \langle f(v_1), \cdots, f(v_m), v_m, \cdots, v_n \rangle \in Z_f.\end{aligned}$$

Suppose $\bar{q}_f(x', s') = \bar{q}_f(x, s)$. If $s = 0$ then $s' = 0$. When $s > 0$, it follows that $v_m, \ldots, v_n$ are vertices of the carrier $c_K(x')$ of x', which are the last $n-m+1$ vertices with respect to the order of $K^{(0)}$. The carrier of $\bar{q}_f(x', s')$ contains no vertices of $K^{(0)}$ except for $v_m, \ldots, v_n$ and $\beta_{v_i}^K(x') = t_i$ for $m < i \leq n$, hence $s' = s$. Then, $\bar{q}_f(x', s') = \bar{q}_f(x, s)$ implies $s' = s$. In addition, if $\bar{q}_f(x, s) = \bar{q}_f(x', s)$, then $\bar{q}_f(x, s') = \bar{q}_f(x', s')$ for every $s' \in [0, s]$, because

$$\sum_{i=m'+1}^{n} t_i \leq s' \leq \sum_{i=m'}^{n} t_i \quad \text{for } m' \geq m.$$

Therefore, we can define a homotopy $\bar{h}^f : |Z_f| \times \mathbf{I} \to |Z_f|$ by $\bar{h}_t^f \| L| = \mathrm{id}$ and $\bar{h}_t^f(\bar{q}_f(x, s)) = \bar{q}_f(x, st)$ on $|Z_f| \setminus |L| \subset \bar{q}_f(|K| \times \mathbf{I})$. Thus, we have

$$\bar{c}_f = \bar{h}_0^f \simeq \bar{h}_1^f = \mathrm{id}_{|Z_f|} \text{ rel. } |L|,$$

where $\bar{h}^f(\bar{c}_f^{-1}(y) \times \mathbf{I}) = \bar{c}_f^{-1}(y)$ for every $y \in |L|$.

$$\begin{array}{ccc} M_f \times \mathbf{I} & \xrightarrow{h^f} & M_f \\ {\scriptstyle p_f \times \mathrm{id}} \downarrow & & \downarrow {\scriptstyle p_f} \\ |Z_f| \times \mathbf{I} & \xrightarrow[\bar{h}^f]{} & |Z_f| \end{array}$$

For each ordered simplicial complex K and its subdivision K', we define the simplicial subdivision $I(K', K)$ of the product simplicial complex $K \times_s I$ as follows:

$$\begin{aligned} I(K', K) &= (K' \times \{0\}) \cup (K \times \{1\}) \\ &\cup \big\{ \langle \sigma \times \{0\} \cup \{v_m, \cdots, v_n\} \times \{1\} \rangle \mid \sigma \in K', \\ &\langle v_1, \cdots, v_n \rangle \in K,\ v_1 < \cdots < v_n,\ \sigma \subset \langle v_1, \cdots, v_m \rangle \big\}, \end{aligned}$$

where $K' \times \{0\} = \{\sigma \times \{0\} \mid \sigma \in K'\}$ and $K \times \{1\} = \{\tau \times \{1\} \mid \tau \in K\}$. For any simplicial map $f : K' \to L$, given an order on $K'^{(0)}$ so that K' is an ordered simplicial complex, we have the simplicial mapping cylinder Z_f. Identifying $K' \subset Z_f$ with $K' \times \{0\} \subset I(K', K)$, we have the simplicial complex

$$Z_f^K = Z_f \cup_{K' = K' \times \{0\}} I(K', K),$$

from which it follows that $|Z_f^K| = |Z_f| \cup_{\mathrm{pr}_{|K|} | |K| \times \{0\}} |K| \times \mathbf{I}$.

Theorem 4.11.3. *Let K and K' be ordered simplicial complexes such that $K' \lhd K$ and let L be a simplicial complex. For any simplicial approximation $f : K' \to L$ of a map $g : |K| \to |L|$,*

$$|Z_f^K| \simeq M_g \quad \text{rel. } |L| \cup (|K| \times \{1\}).$$

Proof. Since $f \simeq g$, we may assume that $f = g$ by virtue of Theorem 4.11.1. We define maps $\varphi : |Z_f^K| \to M_f$ and $\psi : M_f \to |Z_f^K|$ as follows: $\varphi | |Z_f| = \bar{c}_f$, $\varphi | |K| \times (0, 1] = \mathrm{id}$, and $\psi | |L| = \mathrm{id}$,

$$\psi(x, s) = \begin{cases} (x, 2s - 1) & \text{for } (x, s) \in |K| \times (\frac{1}{2}, 1], \\ \bar{q}_f(x, 2s) & \text{for } (x, s) \in |K| \times (0, \frac{1}{2}]. \end{cases}$$

Then, see Figs. 4.17 and 4.18, and observe that

$$\varphi | |L| \cup (|K| \times \{1\}) = \psi | |L| \cup (|K| \times \{1\}) = \mathrm{id},$$

$$\psi\varphi | |Z_f| = \bar{c}_f = \bar{h}_0^f \text{ and } \psi\varphi | |K| \times (0, 1] = \psi | |K| \times (0, 1].$$

We now define the homotopies $\xi : |Z_f^K| \times \mathbf{I} \to |Z_f^K|$ and $\zeta : M_f \times \mathbf{I} \to M_f$ as follows: $\xi_t | |Z_f| = \bar{h}_{1-t}^f$, $\zeta_t | |L| = \mathrm{id}$, and

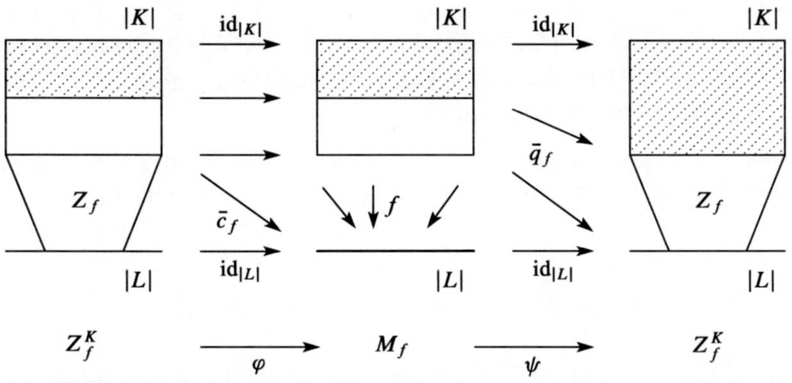

Fig. 4.17 The map $\psi\varphi : Z_f^K \to Z_f^K$

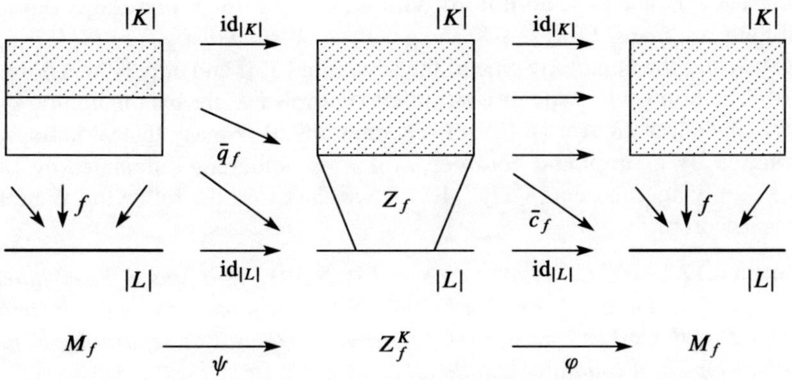

Fig. 4.18 The map $\varphi\psi : M_f \to M_f$

$$\xi_t(x,s) = \begin{cases} \left(x, \dfrac{2s-t}{2-t}\right) & \text{for } (x,s) \in |K| \times (\tfrac{1}{2}t, 1], \\ \bar{q}_f(x, 2s+1-t) & \text{for } (x,s) \in |K| \times (0, \tfrac{1}{2}t]; \end{cases}$$

$$\zeta_t(x,s) = \begin{cases} \left(x, \dfrac{2s-t}{2-t}\right) & \text{for } (x,s) \in |K| \times (\tfrac{1}{2}t, 1], \\ f(x) & \text{for } (x,s) \in |K| \times (0, \tfrac{1}{2}t]. \end{cases}$$

By these homotopies, we have

$$\psi\varphi = \xi_1 \simeq \xi_0 = \mathrm{id} \ \text{ rel. } |L| \cup (|K| \times \{1\}) \text{ and}$$
$$\varphi\psi = \zeta_1 \simeq \zeta_0 = \mathrm{id} \ \text{ rel. } |L| \cup (|K| \times \{1\}).$$

This completes the proof. □

In the above, consider the case that $f = g$ and $f : K \to L$ is simplicial. Then, $|Z_f| \simeq |Z_f^K|$ rel. $|L|$. Indeed, let $r : |Z_f^K| \to |Z_f|$ be the map defined by $r||Z_f| = \mathrm{id}$ and $r||K| \times \mathbf{I} = \mathrm{pr}_{|K|}$. As is easily observed, $r \simeq \mathrm{id}$ rel. $|Z_f|$ in $|Z_f^K|$. Thus, we have the following corollary:

Corollary 4.11.4. *Let K be an ordered simplicial complex and L be a simplicial complex. For every simplicial map $f : K \to L$, $|Z_f| \simeq M_f$ rel. $|L|$.* □

4.12 The Homotopy Type of Simplicial Complexes

In this section, we apply the mapping cylinder and the simplicial mapping cylinder to prove the Whitehead–Milnor Theorem on the homotopy type of simplicial complexes.

A space X is said to be **homotopy dominated by** Y (or Y **homotopy dominates** X) if there are maps $f : X \to Y$ and $g : Y \to X$ such that $gf \simeq \mathrm{id}$. It is easy to see that X has the homotopy type of the singleton $\{0\}$ if and only if X is dominated by $\{0\}$. The **homotopy type of a simplicial complex** K means the homotopy type of the polyhedron $|K|$ (or $|K|_{\mathrm{m}}$) (cf. Theorem 4.9.6). We say that X is **homotopy dominated by a simplicial complex** K if X is homotopy dominated by $|K|$ (or $|K|_{\mathrm{m}}$). Applying Theorems 4.11.1–4.11.3, we can prove the following Whitehead–Milnor Theorem:

Theorem 4.12.1 (J.H.C. Whitehead; Milnor). *If a space X is homotopy dominated by a simplicial complex K, then X has the homotopy type of a simplicial complex L with* $\mathrm{card}\, L^{(0)} = \mathrm{dens}\, |L| \leq \mathrm{dens}\, X$. *When X is separable, X has the homotopy type of a countable simplicial complex.*

Proof. We may assume that X is infinite, so $\mathrm{dens}\, X \geq \aleph_0$. Let $f : X \to |K|$ and $g : |K| \to X$ be maps such that $gf \simeq \mathrm{id}$. Then, we may assume that $\mathrm{card}\, K \leq \mathrm{dens}\, X$. Indeed, take a dense set D in X with $\mathrm{card}\, D = \mathrm{dens}\, X$ and define

$$K_0 = \{\tau \in K \mid \exists x \in D \text{ such that } \tau \leq c_K(f(x))\},$$

where $c_K(f(x))$ is the carrier of $f(x)$. Observe that $\mathrm{card}\, K_0 \leq \mathrm{card}\, D$ and $f(X) = f(\mathrm{cl}\, D) \subset \mathrm{cl}\, f(D) \subset |K_0|$. Because $(g||K_0|)f \simeq \mathrm{id}$, we can replace K with K_0.

By the Simplicial Approximation Theorem 4.7.14, $fg : |K| \to |K|$ has a simplicial approximation $\varphi : K' \to K$, where K' is a subdivision of K. Given orders on $K'^{(0)}$ and $K^{(0)}$ so that K' and K are ordered simplicial complexes, we obtain the simplicial complex Z_φ^K with $\mathrm{card}\, Z_\varphi^K \leq \mathrm{dens}\, X$.

For each $n \in \mathbb{Z}$, let L_n be a copy of Z_φ^K. Identifying K of L_{n+1} with $K \times \{1\}$ of L_n for each $n \in \mathbb{Z}$, we have a simplicial complex $L = \bigcup_{n\in\mathbb{Z}} L_n$, where $\mathrm{card}\, L \leq \mathrm{dens}\, X$. For each $n \in \mathbb{Z}$, let M_{2n-1} and M_{2n} be copies of M_f and M_g, respectively. Identifying $X \times \{1\} \subset M_{2n-1}$ with $X \subset M_{2n}$ and $|K| \times \{1\} \subset M_{2n}$ with $|K| \subset M_{2n+1}$, let $M = \bigcup_{n\in\mathbb{Z}} M_n$.

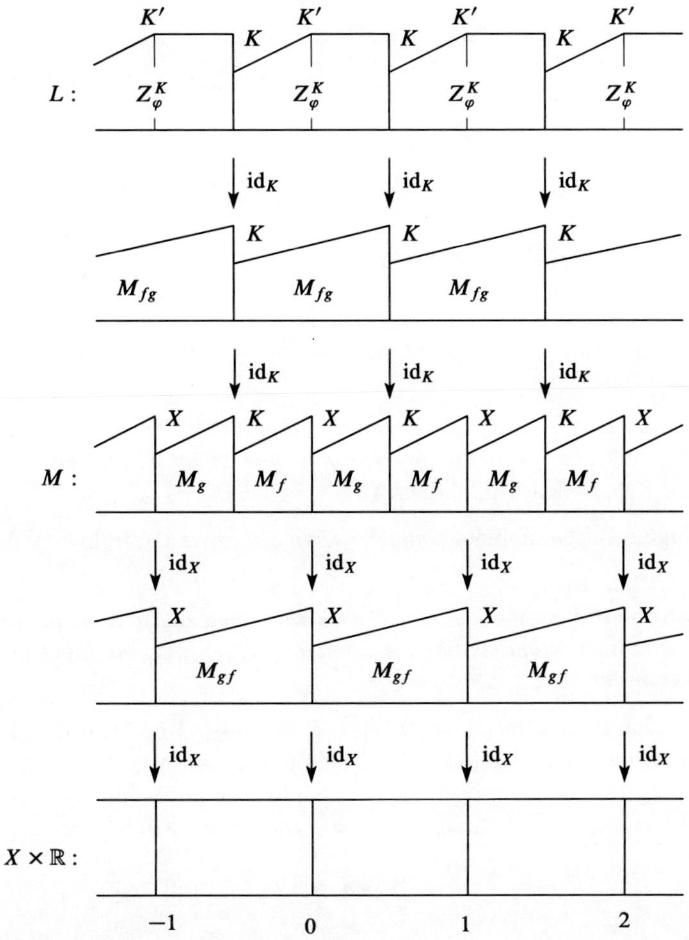

Fig. 4.19 $X \times \mathbb{R} \simeq |L|$

By Theorems 4.11.3 and 4.11.2, we have

$$|L_n| = |Z_\varphi^K| \simeq M_{fg} \simeq M_f \cup_{i_f} M_g = M_{2n-1} \cup M_{2n} \ \text{rel.}\ |K| \cup (|K| \times \{1\}),$$

which implies that $L \simeq M$. By Theorems 4.11.2 and 4.11.1,

$$M_{2n} \cup M_{2n+1} = M_g \cup_{i_g} M_f \simeq M_{gf} \simeq M_{\mathrm{id}_X} \ \text{rel.}\ X \cup (X \times \{1\}).$$

Regarding M_{id_X} as $X \times [n, n+1]$, we have $M \simeq X \times \mathbb{R}$. Thus, we have $X \simeq X \times \mathbb{R} \simeq |L|$ — Fig. 4.19. □

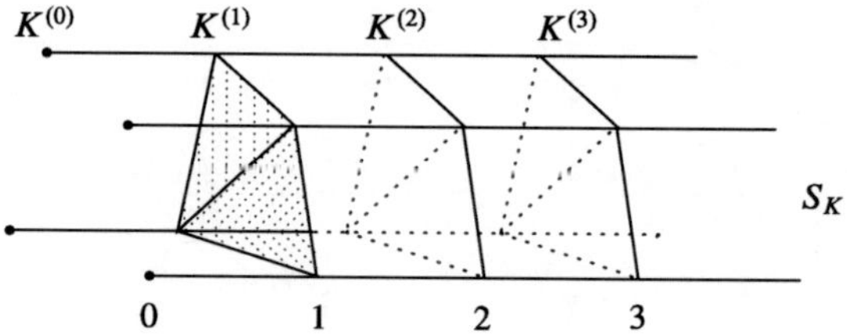

Fig. 4.20 The locally finite simplicial complex S_K

In the above proof, if $\dim K < \infty$ then $\dim L < \infty$. Therefore, we have the following corollary:

Corollary 4.12.2. *A space X homotopy dominated by a finite-dimensional simplicial complex has the homotopy type of a finite-dimensional simplicial complex K with* $\operatorname{card} K^{(0)} = \operatorname{dens}|K| \leq \operatorname{dens} X$. □

Applying product simplicial complexes, we can prove the following theorem:

Theorem 4.12.3. *Every simplicial complex has the homotopy type of a locally finite-dimensional simplicial complex with the same density. In addition, every countable simplicial complex has the homotopy type of a countable locally finite simplicial complex.*

Proof. For each $n \in \omega$, let $L_n = \{i,\ [i, i+1] \mid i \geq n\}$ be the ordered simplicial complex with the natural order. For each simplicial complex K, assuming that K is an ordered simplicial complex, we define a locally finite-dimensional simplicial complex $S_K = \bigcup_{n\in\omega} K^{(n)} \times_s L_n$ (Fig. 4.20). By Theorem 4.3.1, we have

$$|S_K| = \bigcup_{n\in\omega} |K^{(n)}| \times [n, \infty) \subset |K| \times [0, \infty).$$

To prove that $|K| \simeq |S_K|$, it suffices to show that $|K| \times [0, \infty) \simeq |S_K|$ because $|K| \simeq |K| \times [0, \infty)$. For each $n \in \omega$, let

$$T_n = |K^{(n)}| \times [0, n] \cup |S_K|.$$

Then, $|S_K| = T_0 \subset T_1 \subset T_2 \subset \cdots$ and $|K| \times [0, \infty) = \varinjlim T_n$ (cf. Sect. 2.8). For each n-simplex $\tau \in K$, we have a map

$$p_\tau : \tau \times [0, n] \to \partial\tau \times [0, n] \cup \tau \times \{n\} = \tau \times [0, n] \cap T_{n-1}$$

defined as follows: $p_\tau(\hat{\tau}, s) = (\hat{\tau}, n)$ and

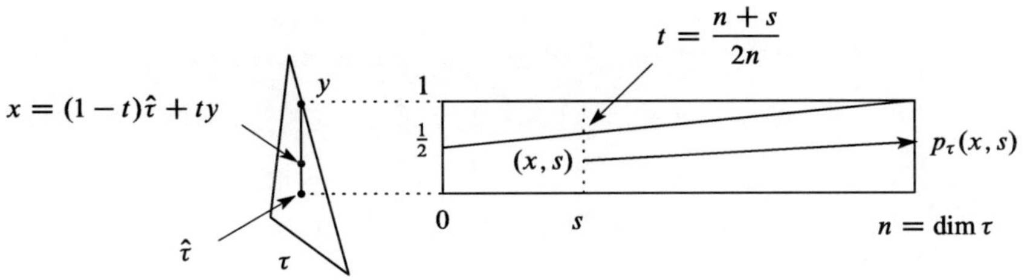

Fig. 4.21 The map p_τ

$$p_\tau(x, s) = \begin{cases} (y, s + 2n(1-t)) & \text{if } \dfrac{n+s}{2n} \leq t \leq 1, \\ \left(\left(1 - \dfrac{2n}{n+s}t\right)\hat{\tau} + \dfrac{2n}{n+s}ty, n\right) & \text{if } 0 < t \leq \dfrac{n+s}{2n}, \end{cases}$$

where $x = (1-t)\hat{\tau} + ty \in \tau$, $y \in \partial\tau$, $0 < t \leq 1$ (Fig. 4.21).

For each $n \in \mathbb{N}$, let $p_n : T_n \to T_{n-1}$ be the map defined by $p_n|T_{n-1} = \mathrm{id}$ and $p_n|\tau \times [0, n] = p_\tau$ for each n-simplex $\tau \in K$. Then, we can define a map $r : |K| \times [0, \infty) \to |S_K|$ by $r|T_n = p_1 \cdots p_n$ for each $n \in \mathbb{N}$. Note that $r||S_K| = \mathrm{id}$ and $r(\tau \times [0, \infty)) \subset \tau \times [0, \infty)$ for each $\tau \in K$. Observe that $r \simeq \mathrm{id}$ rel. $|S_K|$ in $|K| \times [0, \infty)$ by the straight-line homotopy (i.e., $t \mapsto (1-t)r(x, s) + t(x, s)$). Therefore, we have $|K| \times [0, \infty) \simeq |S_K|$.

If K is countable, K has a tower $K_1 \subset K_2 \subset \cdots$ of finite subcomplexes with $K = \bigcup_{n \in \mathbb{N}} K_n$. Then, $T_K = \bigcup_{n \in \omega} K_n \times_s L_n$ is a countable locally finite simplicial complex. As above, we have $|K| \simeq |T_K|$. □

By Theorems 4.12.1 and 4.12.3, we have the following corollary:

Corollary 4.12.4. *A space X homotopy dominated by a simplicial complex has the homotopy type of a locally finite-dimensional simplicial complex K with* $\mathrm{card}\, K^{(0)} = \mathrm{dens}\, |K| \leq \mathrm{dens}\, X$. *If X is separable, X has the homotopy type of a countable locally finite simplicial complex.* □

4.13 Weak Homotopy Equivalences

Let $n \in \omega$. A map $f : X \to Y$ is called an ***n*-equivalence** if it satisfies the following condition $(\pi)_i$ for each $i = 0, \ldots, n$:

$(\pi)_i$ For each map $\alpha : \mathbf{S}^{i-1} \to X$, if $f\alpha$ extends to a map $\beta : \mathbf{B}^i \to Y$, then α extends to a map $\bar{\alpha} : \mathbf{B}^i \to X$ such that $f\bar{\alpha} \simeq \beta$ rel. $\mathbf{S}^{i-1}$,

where $\mathbf{B}^0 = \{0\}$ and $\mathbf{S}^{-1} = \emptyset$.

$$\begin{array}{ccc} X & \xrightarrow{f} & Y \\ \alpha\uparrow & \nwarrow^{\bar{\alpha}} \;\simeq & \uparrow\beta \\ \mathbf{S}^{i-1} & \subset & \mathbf{B}^i \end{array}$$

When $f : X \to Y$ is an n-equivalence for every $n \in \omega$, we call f a **weak homotopy equivalence**. For convenience, a weak homotopy equivalence is sometimes called an **∞-equivalence**.

In the next section, we will give a characterization of n-equivalences in the framework of homotopy groups. In particular, it will be shown that a map is a weak homotopy equivalence if and only if it induces a bijection between the sets of path-components and an isomorphism between their homotopy groups in every dimension. The following proposition is an immediate consequence of this characterization, but we will give a direct proof.

Proposition 4.13.1. *Every homotopy equivalence $f : X \to Y$ is a weak homotopy equivalence.*

Proof. Let $g : Y \to X$ be a homotopy inverse of f. Then, there are homotopies $h : X \times \mathbf{I} \to X$ and $k : Y \times \mathbf{I} \to Y$ such that $h_0 = \mathrm{id}_X$, $h_1 = gf$, $k_0 = \mathrm{id}_Y$, and $k_1 = fg$. For each pair of maps $\alpha : \mathbf{S}^{i-1} \to X$ and $\beta : \mathbf{B}^i \to Y$ with $f\alpha = \beta|\mathbf{S}^{i-1}$, we can extend α to the map $\bar{\alpha} : \mathbf{B}^i \to X$ defined as follows:

$$\bar{\alpha}(x) = \begin{cases} g\beta(8x) & \text{if } 0 \le \|x\| \le 1/8, \\ gfh(\alpha(\|x\|^{-1}x), 8\|x\| - 1) & \text{if } 1/8 \le \|x\| \le 1/4, \\ gk(f\alpha(\|x\|^{-1}x), 2 - 4\|x\|) & \text{if } 1/4 \le \|x\| \le 1/2, \\ h(\alpha(\|x\|^{-1}x), 2 - 2\|x\|) & \text{if } 1/2 \le \|x\| \le 1. \end{cases}$$

It remains to show that $f\bar{\alpha} \simeq \beta$ rel. $\mathbf{S}^{i-1}$.

We define an auxiliary map $\gamma : \mathbf{B}^i \to Y$ as follows:

$$\gamma(x) = \begin{cases} \beta(8x) & \text{if } 0 \le \|x\| \le 1/8, \\ fh(\alpha(\|x\|^{-1}x), 8\|x\| - 1) & \text{if } 1/8 \le \|x\| \le 1/4, \\ k(f\alpha(\|x\|^{-1}x), 2 - 4\|x\|) & \text{if } 1/4 \le \|x\| \le 1/2, \\ k(f\alpha(\|x\|^{-1}x), 4\|x\| - 2) & \text{if } 1/2 \le \|x\| \le 3/4, \\ fh(\alpha(\|x\|^{-1}x), 4 - 4\|x\|) & \text{if } 3/4 \le \|x\| \le 1. \end{cases}$$

Then, $\gamma \simeq f\bar{\alpha}$ rel. $\mathbf{S}^{i-1}$ by the homotopy $\varphi : \mathbf{B}^i \times \mathbf{I} \to Y$ defined as follows:

$$\varphi_t(x) = \begin{cases} k_t(\beta(8x)) & \text{if } 0 \leq \|x\| \leq 1/8, \\ k_t(fh(\alpha(\|x\|^{-1}x), 8\|x\| - 1)) & \text{if } 1/8 \leq \|x\| \leq 1/4, \\ k_t k(f\alpha(\|x\|^{-1}x), 2 - 4\|x\|) & \text{if } 1/4 \leq \|x\| \leq 1/2, \\ k(f\alpha(\|x\|^{-1}x), 4\|x\| - 2 + t) & \text{if } 1/2 \leq \|x\| \leq (3-t)/4, \\ fh(\alpha(\|x\|^{-1}x), (4 - 4\|x\|)/(1+t)) & \text{if } (3-t)/4 \leq \|x\| \leq 1. \end{cases}$$

On the other hand, let $\gamma', \gamma'' : \mathbf{B}^i \to Y$ be the maps defined as follows:

$$\gamma'(x) = \begin{cases} \beta(8x) & \text{if } 0 \leq \|x\| \leq 1/8, \\ fh(\alpha(\|x\|^{-1}x), 8\|x\| - 1) & \text{if } 1/8 \leq \|x\| \leq 1/4, \\ fgf\alpha(\|x\|^{-1}x) & \text{if } 1/4 \leq \|x\| \leq 3/4, \\ fh(\alpha(\|x\|^{-1}x), 4 - 4\|x\|) & \text{if } 3/4 \leq \|x\| \leq 1; \end{cases}$$

$$\gamma''(x) = \begin{cases} \beta(8x) & \text{if } 0 \leq \|x\| \leq 1/8, \\ f\alpha(\|x\|^{-1}x) & \text{if } 1/8 \leq \|x\| \leq 1. \end{cases}$$

It is then easy to obtain the homotopies $\gamma \simeq \gamma' \simeq \gamma'' \simeq \beta$ rel. $\mathbf{S}^{i-1}$. □

In this section, we will show the converse of Proposition 4.13.1 when X and Y are polyhedra. Namely, we will prove that every weak homotopy equivalence between polyhedra is a homotopy equivalence.

A **path-component** of a space X is a maximal path-connected subset of X. The set of path-components of X is denoted by $\pi_0(X)$. Every map $f : X \to Y$ induces the function $f_\sharp : \pi_0(X) \to \pi_0(Y)$, which sends the path-component of $x \in X$ to the path-component of $f(x) \in Y$. The following propositions are easily proved:

Proposition 4.13.2. *(1) A map $f : X \to Y$ is a 0-equivalence if and only if every path-component of Y meets $f(X)$, i.e., $f_\sharp : \pi_0(X) \to \pi_0(Y)$ is a surjection.*
(2) Every 1-equivalence $f : X \to Y$ induces the bijection $f_\sharp : \pi_0(X) \to \pi_0(Y)$ and the surjection $f_\sharp : [(\mathbf{I}, \partial\mathbf{I}), (X, x_0)] \to [(\mathbf{I}, \partial\mathbf{I}), (Y, f(x_0))]$ for every $x_0 \in X$.[13]

□

Proposition 4.13.3. *Let $n \in \omega \cup \{\infty\}$. The composition of n-equivalences is also an n-equivalence.* □

The following proposition, similar to Proposition 4.13.1, is an immediate consequence of the characterization of an n-equivalence in the framework of homotopy groups, which will be discussed in the next section. However, here we will give a direct proof.

---

[13]That is, $f_\sharp : \pi_1(X, x_0) \to \pi_1(Y, f(x_0))$ is an epimorphism for every $x_0 \in X$.

Proposition 4.13.4. *Let $n \in \omega \cup \{\infty\}$. If a map $f : X \to Y$ is homotopic to an n-equivalence, then f is also an n-equivalence.*

Proof. For each $i < n + 1$,[14] let $\alpha : \mathbf{S}^{i-1} \to X$ and $\beta : \mathbf{B}^i \to Y$ be maps such that $f\alpha = \beta|\mathbf{S}^{i-1}$. By the assumption, we have a homotopy $h : X \times \mathbf{I} \to Y$ such that $h_0 = f$ and h_1 is an n-equivalence. According to the Homotopy Extension Theorem 4.3.3, there is a homotopy $\varphi : \mathbf{B}^i \times \mathbf{I} \to Y$ such that $\varphi_0 = \beta$ and $\varphi|\mathbf{S}^{i-1} \times \mathbf{I} = h(\alpha \times \mathrm{id}_\mathbf{I})$. Since $h_1\alpha$ extends to the map $\varphi_1 : \mathbf{B}^i \to Y$, α extends to a map $\bar{\alpha} : \mathbf{B}^i \to X$ such that $h_1\bar{\alpha} \simeq \varphi_1$ rel. $\mathbf{S}^{i-1}$.

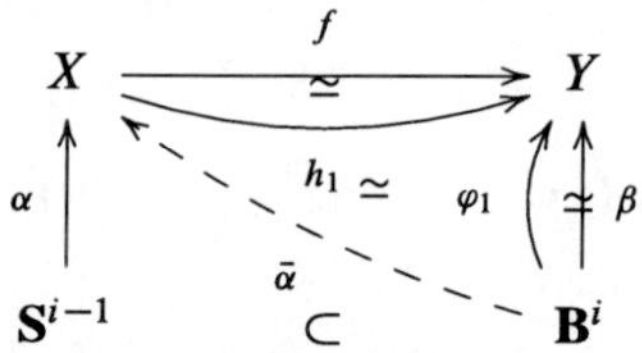

Let $\psi : \mathbf{B}^i \times \mathbf{I} \to Y$ be a homotopy such that $\psi_0 = h_1\bar{\alpha}$, $\psi_1 = \varphi_1$, and $\psi_t|\mathbf{S}^{i-1} = h_1\alpha$ for every $t \in \mathbf{I}$. We define a homotopy $k : \partial(\mathbf{B}^i \times \mathbf{I}) \times \mathbf{I} \to Y$ as follows:

$$k_t(x,s) = \begin{cases} \varphi_{1-t}(x) & \text{if } s = 1, \\ h_{1-t}\bar{\alpha}(x) & \text{if } s = 0, \\ h_{1-t}\alpha(x) & \text{if } x \in \mathbf{S}^{i-1}. \end{cases}$$

Refer to Fig. 4.22. Then, ψ is an extension of k_0. Hence, we can apply the Homotopy Extension Theorem 4.3.3 to extend k_1 to a map $\psi' : \mathbf{B}^i \times \mathbf{I} \to Y$, which is a homotopy from $f\bar{\alpha}$ to β with $\psi'_s|\mathbf{S}^{i-1} = f\alpha$ for every $s \in \mathbf{I}$, i.e., $f\bar{\alpha} \simeq \beta$ rel. $\mathbf{S}^{i-1}$. Therefore, f is an n-equivalence. □

When f is the inclusion, we have the following:

Lemma 4.13.5. *For the inclusion $X \subset Y$, each of the following is equivalent to condition $(\pi)_i$:*

$(\pi)'_i$ *Every map $\alpha : \mathbf{B}^i \to Y$ with $\alpha(\mathbf{S}^{i-1}) \subset X$ is null-homotopic by a homotopy $\varphi : \mathbf{B}^i \times \mathbf{I} \to Y$ with $\varphi(\mathbf{S}^{i-1} \times \mathbf{I}) \subset X$, i.e., $[(\mathbf{B}^i, \mathbf{S}^{i-1}), (Y, X)] = \{0\}$;*[15]

$(\pi)''_i$ *For a homotopy $h : \mathbf{S}^{i-1} \times \mathbf{I} \to Y$ with $h_1(\mathbf{S}^{i-1}) \subset X$, if h_0 extends to a map $\beta : \mathbf{B}^i \to Y$, then h extends to a homotopy $\bar{h} : \mathbf{B}^i \times \mathbf{I} \to Y$ such that $\bar{h}_0 = \beta$ and $\bar{h}_1(\mathbf{B}^i) \subset X$.*

[14] We use the convention that $\infty + 1 = \infty$.

[15] In terms of the homotopy groups, this means that $\pi_i(Y, X, x_0) = 0$ for each $x_0 \in X$. In Proposition 4.14.7, we will give another proof of the equivalence between this condition and $(\pi)_i$.

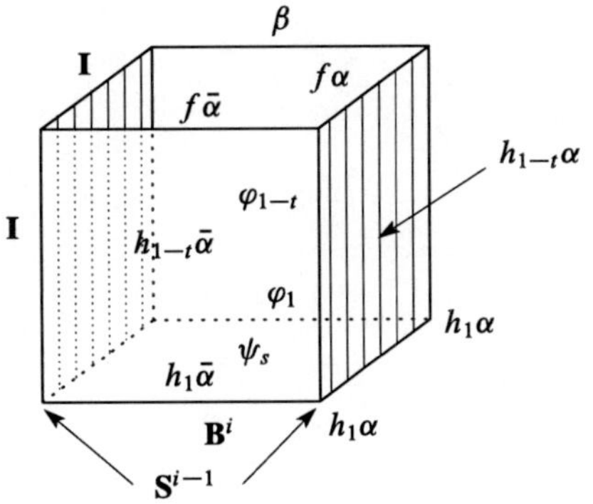

Fig. 4.22 The homotopy k

$(\pi)'_i$

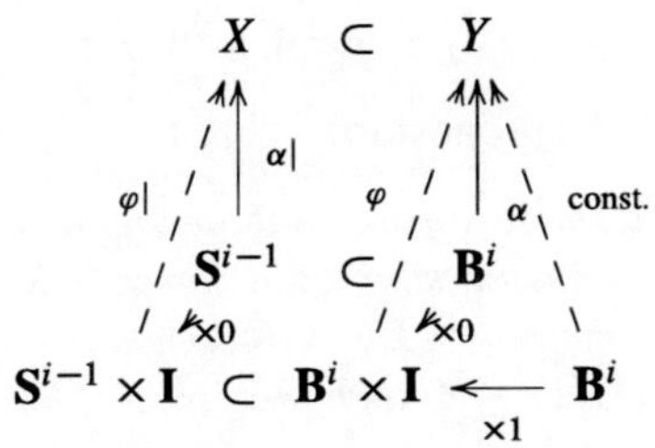

$(\pi)''_i$

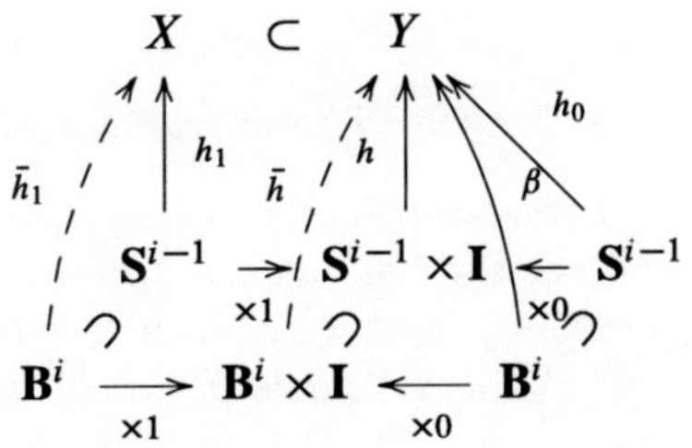

Proof. $(\pi)_i \Rightarrow (\pi)'_i$: From condition $(\pi)_i$, we have a homotopy $h : \mathbf{B}^i \times \mathbf{I} \to Y$ such that $h_0 = \alpha$, $h_1(\mathbf{B}^i) \subset X$ and $h_t|\mathbf{S}^{i-1} = \alpha|\mathbf{S}^{i-1}$ for every $t \in \mathbf{I}$. Because $\mathbf{B}^i$ is contractible, we have $h_1 \simeq 0$ in X. Then, the homotopy φ in $(\pi)'_i$ is easily defined.

$(\pi)'_i \Rightarrow (\pi)''_i$: We have a homeomorphism $\psi : \mathbf{B}^i \times \mathbf{I} \to \mathbf{B}^i \times \mathbf{I}$ such that

$$\psi\big((\mathbf{S}^{i-1} \times \mathbf{I}) \cup (\mathbf{B}^i \times \{0\})\big) = \mathbf{B}^i \times \{0\},$$
$$\psi(\mathbf{S}^{i-1} \times \{1\}) = \mathbf{S}^{i-1} \times \{0\} \text{ and}$$
$$\psi(\mathbf{B}^i \times \{1\}) = (\mathbf{S}^{i-1} \times \mathbf{I}) \cup (\mathbf{B}^i \times \{1\}).$$

Let $\beta : \mathbf{B}^i \to Y$ be an extension of h_0 and define a map $\alpha : \mathbf{B}^i \to Y$ as follows:

$$\alpha(x) = \begin{cases} h\psi^{-1}(x,0) & \text{if } x \in \psi(\mathbf{S}^{i-1} \times \mathbf{I}), \\ \beta \mathrm{pr}_{\mathbf{B}^i}\psi^{-1}(x,0) & \text{if } x \in \psi(\mathbf{B}^i \times \{0\}). \end{cases}$$

Since $\alpha(\mathbf{S}^{i-1}) = h_1(\mathbf{S}^{i-1}) \subset X$, we can apply $(\pi)'_i$ to obtain a homotopy φ : $\mathbf{B}^i \times \mathbf{I} \to Y$ such that $\varphi_0 = \alpha$, $\varphi(\mathbf{S}^{i-1} \times \mathbf{I}) \subset X$, and φ_1 is constant. Then, $\bar{h} = \varphi\psi : \mathbf{B}^i \times \mathbf{I} \to Y$ is the desired extension of h.

$(\pi)''_i \Rightarrow (\pi)_i$: Given a map $\beta : \mathbf{B}^i \to Y$ such that $\alpha = \beta|\mathbf{S}^{i-1} : \mathbf{S}^{i-1} \to X$, let $h : \mathbf{S}^{i-1} \times \mathbf{I} \to Y$ be the constant homotopy defined by $h_t = \alpha$ for each $t \in \mathbf{I}$.

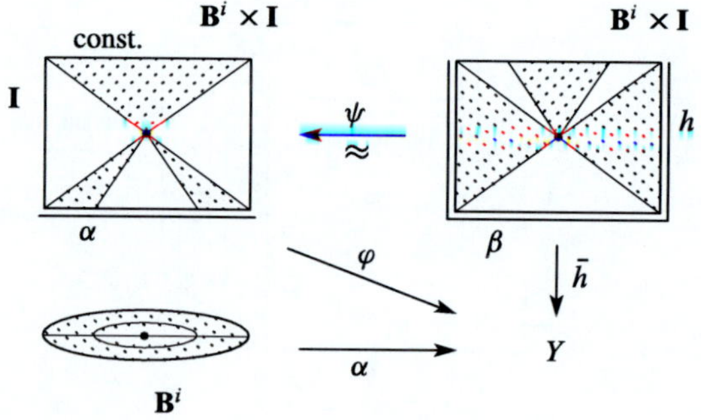

Fig. 4.23 The construction of $\bar{h}$

Due to condition $(\pi)''_i$, h extends to a homotopy $\bar{h} : \mathbf{B}^i \times \mathbf{I} \to Y$ such that $\bar{h}_0 = \beta$ and $\bar{h}_1(\mathbf{B}^i) \subset X$ (Fig. 4.23). Then, $\bar{h}_1|\mathbf{S}^{i-1} = \alpha$ and $\bar{h}_1 \simeq \beta$ rel. $\mathbf{S}^{i-1}$ in Y. □

Using this lemma, we can prove the following proposition:

Proposition 4.13.6. *Let X be a subspace of a space Y such that the inclusion $X \subset Y$ is an n-equivalence, where $n \in \omega \cup \{\infty\}$. Given a simplicial complex K and a subcomplex $L \subset K$ with $K \setminus L \subset K^{(n)}$, if a map $f : |L| \to X$ extends to a map $f' : |K| \to Y$, then f extends to a map $f'' : |K| \to X$ such that $f' \simeq f''$ rel. $|L|$ in Y.*

Proof. According to Theorem 4.3.1,

$$|L \cup K^{(i)}| \times \mathbf{I} = |(L \cup K^{(i)}) \times_c I| \quad \text{for each } i \in \omega,$$

where $I = \{\mathbf{I}, 0, 1\}$. Applying condition $(\pi)''_i$ in Lemma 4.13.5 simplex-wise, we can inductively construct homotopies $h^{(i)} : |L \cup K^{(i)}| \times \mathbf{I} \to Y, i \in \omega$, such that

$$h_0^{(i)} = f'||L \cup K^{(i)}|,\ h_1^{(i)}(|L \cup K^{(i)}|) \subset X \text{ and}$$
$$h^{(i)}||L \cup K^{(i-1)}| \times \mathbf{I} = h^{(i-1)},$$

where $h^{(-1)} : |L| \times \mathbf{I} \to Y$ is the constant homotopy defined by $h_t^{(-1)} = f||L| = f'||L|$ for every $t \in \mathbf{I}$. Because $|K| \times \mathbf{I} = |K \times_c I|$, we can define a homotopy $h : |K| \times \mathbf{I} \to Y$ by $h||L \cup K^{(i)}| \times \mathbf{I} = h^{(i)}$ for each $i \in \omega$. Hence, $f'' = h_1$ is the desired map. □

In the following proposition, we identify X with the subspace $X \times \{1\}$ of the mapping cylinder M_f of $f : X \to Y$.

Proposition 4.13.7. *Let $n \in \omega \cup \{\infty\}$. For a map $f : X \to Y$, the following are equivalent:*

(a) *f is an n-equivalence;*
(b) *The inclusion $i_f : X \subset M_f$ is an n-equivalence*[16]*;*
(c) *Given a simplicial complex K and a subcomplex $L \subset K$ with $K \setminus L \subset K^{(n)}$, for a map $g : |L| \to X$, if fg extends to a map $h : |K| \to Y$ then g extends to a map $\bar{g} : |K| \to X$ such that $f\bar{g} \simeq h$ rel. $|L|$.*

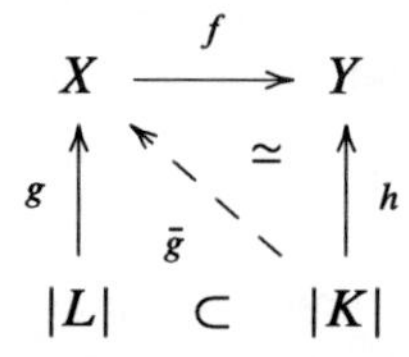

Proof. Since $(\mathbf{B}^i, \mathbf{S}^{i-1}) \approx (|F(\Delta^i)|, |F(\partial\Delta^i)|)$, the implication (c) ⇒ (a) is trivial. Since the collapsing $c_f : M_f \to Y$ is a homotopy equivalence and the inclusion $i_Y : Y \subset M_f$ is its homotopy inverse, the map $i_Y f = i_Y c_f i_f : X \to Y \subset M_f$ is homotopic to $i_f : X \to M_f$. By virtue of Propositions 4.13.1, 4.13.3, and 4.13.4, $f = c_f i_Y f : X \to Y$ is an n-equivalence if and only if i_f is. Thus, we have the equivalence (a) ⇔ (b). It remains to show the implication (b) ⇒ (c).

(b) ⇒ (c): Let K be a simplicial complex and $L \subset K$ be a subcomplex such that $\dim\sigma \leq n$ for every $\sigma \in K \setminus L$. Let $g : |L| \to X$ and $h : |K| \to Y$ be maps such that $fg = h||L|$. The homotopy $h^f : M_f \times \mathbf{I} \to M_f$ from $i_Y c_f$ to id induces a homotopy $\varphi : |L| \times \mathbf{I} \to M_f$ from $i_Y c_f i_f g = i_Y fg = i_Y h||L|$ to $i_f g$, which is defined by $\varphi_t = h^f_t i_f g$. Then, $c_f \varphi_t = c_f h^f_t i_f g = c_f i_f g = fg$ for each $t \in \mathbf{I}$. According to the Homotopy Extension Theorem 4.3.3, φ extends to a homotopy $\bar{\varphi} : |K| \times \mathbf{I} \to M_f$ with $\bar{\varphi}_0 = h$, so $\bar{\varphi}_1$ is an extension of $i_f g$. By (b) and Proposition 4.13.6, g extends to a map $\bar{g} : |K| \to X$ such that $\bar{g} \simeq \bar{\varphi}_1$ rel. $|L|$ in M_f, hence $f\bar{g} = c_f\bar{g} \simeq c_f\bar{\varphi}_1$ rel. $|L|$ in Y. On the other hand, $c_f\bar{\varphi}_1 \simeq c_f\bar{\varphi}_0 = c_f h = h$ rel. $|L|$ in Y because $c_f\bar{\varphi}_t||L| = c_f\varphi_t = fg$ for each $t \in \mathbf{I}$. Thus, we have $f\bar{g} \simeq h$ rel. $|L|$ in Y. □

For polyhedra, we have the following theorem:

Theorem 4.13.8. *Let K and L be simplicial complexes and $n \in \mathbb{N}$. If $\dim K \leq n - 1$ and $\dim L \leq n$, then every n-equivalence $f : |K| \to |L|$ is a homotopy equivalence.*

Proof. According to the Simplicial Approximation Theorem 4.7.14, f has a simplicial approximation $g : K' \to L$ for some $K' \lhd K$. Then, $M_f \simeq |Z^K_g|$ rel. $|L| \cup |K|$ by Theorem 4.11.3, where $|K|$ is identified with $|K| \times \{1\}$ in M_f and $|Z^K_g|$. Since the inclusion $|K| \subset M_f$ is an n-equivalence by (b) in Proposition 4.13.7, it follows that the inclusion $|K| \subset |Z^K_g|$ is also an n-equivalence. Note that

---

[16]According to Lemma 4.13.5 and the previous footnote 15, this condition means that $\pi_i(M_f, X, x_0) = 0$ for each $x_0 \in X$ and $i < n + 1$ (where $\infty + 1 = \infty$).

$\dim Z_g^K \leq n$. We can now apply condition (c) of Proposition 4.13.7 to obtain a map $\varphi : |Z_g^K| \to |K|$ such that $\varphi||K| = \mathrm{id}$ and $\varphi \simeq \mathrm{id}$ rel. $|K|$ in $|Z_g^K|$, which means that the inclusion $|K| \subset |Z_g^K|$ is a homotopy equivalence, hence the inclusion $|K| \subset M_f$ is also a homotopy equivalence. Since the collapsing $c_f : M_f \to |L|$ is a homotopy equivalence, the map $f = c_f||K|$ is a homotopy equivalence. □

In the case that $n = \infty$, $K^{(\infty)} = K$ in Propositions 4.13.6 and 4.13.7, so the above argument is valid without the dimensional assumption. Thus, we have the following theorem:

Theorem 4.13.9. *For simplicial complexes K and L, every weak homotopy equivalence $f : |K| \to |L|$ is a homotopy equivalence.* □

Remark 19. It should be noted that there are connected polyhedra X and Y such that $X \not\simeq Y$ but $\pi_n(X) \cong \pi_n(Y)$ for every $n \in \mathbb{N}$. For example, $X = \mathbf{S}^2 \times \mathbb{R}\mathrm{P}^3$ and $\mathbb{R}\mathrm{P}^2 \times \mathbf{S}^3$ are not homotopy equivalent but they have the isomorphic homotopy groups.[17]

Combining the above with Theorem 4.12.1 and Proposition 4.13.1, we have the following corollary:

Corollary 4.13.10. *Let X and Y be spaces that are homotopy dominated by simplicial complexes. Then, every weak homotopy equivalence $f : X \to Y$ is a homotopy equivalence.* □

4.14 Appendix: Homotopy Groups

In this section, we review several definitions related to the homotopy groups together with their basic results. Some of them are stated without proof. For details, refer to any textbook on Homotopy Theory or Algebraic Topology.

For a pair of paths $\alpha, \beta : \mathbf{I} \to X$ with $\alpha(1) = \beta(0)$, we can define the **join** $\alpha * \beta : \mathbf{I} \to X$ as follows:

$$\alpha * \beta(t) = \begin{cases} \alpha(2t) & \text{if } 0 \leq t \leq 1/2, \\ \beta(2t-1) & \text{if } 1/2 \leq t \leq 1. \end{cases}$$

The **inverse** $\alpha^{\leftarrow} : \mathbf{I} \to X$ of a path α is defined by $\alpha^{\leftarrow}(t) = \alpha(1-t)$ for each $t \in \mathbf{I}$. For another pair of paths α', β', the following holds:

$$\alpha' \simeq \alpha,\ \beta' \simeq \beta \text{ rel. } \partial\mathbf{I} \Rightarrow \alpha * \beta \simeq \alpha' * \beta',\ \alpha^{\leftarrow} \simeq \alpha'^{\leftarrow} \text{ rel. } \partial\mathbf{I}.$$

Moreover, for three paths $\alpha, \beta, \gamma : \mathbf{I} \to X$ with $\alpha(1) = \beta(0)$ and $\beta(1) = \gamma(0)$,

[17] Here, $\mathbb{R}\mathrm{P}^2$ is the real projective plane and $\mathbb{R}\mathrm{P}^3$ is the 3-dimensional real projective space. For example, refer to Hatcher's book "Algebraic Topology."

$$(\alpha * \beta) * \gamma \simeq \alpha * (\beta * \gamma) \text{ rel. } \partial\mathbf{I};$$
$$\alpha * \alpha^{\leftarrow} \simeq c_{\alpha(0)},\ \alpha^{\leftarrow} * \alpha \simeq c_{\alpha(1)} \text{ rel. } \partial\mathbf{I},$$

where $c_x : \mathbf{I} \to X$ is the constant path with $c_x(\mathbf{I}) = \{x\}$. A path $\alpha : \mathbf{I} \to X$ is called a **loop** (at x) if $\alpha(0) = \alpha(1)\ (= x)$.

For each pointed space (X, x_0), we denote $\Omega(X, x_0) = \mathrm{C}((\mathbf{I}, \partial\mathbf{I}), (X, x_0))$, which is the set of all loops in X (at x_0). The space $\Omega(X, x_0)$ admitting the compact-open topology is called the **loop space**. The base point of $\Omega(X, x_0)$ is the constant map $c_{x_0} : \mathbf{I} \to \{x_0\} \subset X$, denoted by $\mathbf{0}$. Each pointed map $f : (X, x_0) \to (Y, y_0)$ induces the pointed map $f_* : \Omega(X, x_0) \to \Omega(Y, y_0)$ defined by $f_*(\alpha) = f\alpha$ (cf. 1.1.3 (1)). Now, we can define the **fundamental group** $\pi_1(X, x_0)$ or the **first homotopy group** of X at x_0 (or (X, x_0)) as follows:

$$\pi_1(X, x_0) = [(\mathbf{I}, \partial\mathbf{I}), (X, x_0)] = \Omega(X, x_0)/\simeq$$

with the operation $[\alpha][\beta] = [\alpha * \beta]$ and the inverse $[\alpha]^{-1} = [\alpha^{\leftarrow}]$, where $[\alpha]$ is the equivalence class of $\alpha \in \Omega(X, x_0)$ with respect to $\simeq$.[18] In general, $\pi_1(X, x_0)$ is not commutative (i.e., non-Abelian). By the way, we have $\pi_1(X, x_0) = \pi_0(\Omega(X, x_0))$ by Proposition 1.1.2.

For each $n > 1$ and each $\alpha, \beta \in \mathrm{C}((\mathbf{I}^n, \partial\mathbf{I}^n), (X, x_0))$, we define $\alpha * \beta$ and $\alpha^{\leftarrow} \in \mathrm{C}((\mathbf{I}^n, \partial\mathbf{I}^n), (X, x_0))$ as follows:

$$\alpha * \beta(t_1, \dots, t_n) = \begin{cases} \alpha(2t_1, t_2, \dots, t_n) & \text{if } 0 \le t_1 \le 1/2, \\ \beta(2t_1 - 1, t_2, \dots, t_n) & \text{if } 1/2 \le t_1 \le 1; \end{cases}$$
$$\alpha^{\leftarrow}(t_1, \dots, t_n) = \alpha(1 - t_1, t_2, \dots, t_n).$$

Then, for each $\alpha, \alpha', \beta, \beta', \gamma \in \mathrm{C}((\mathbf{I}^n, \partial\mathbf{I}^n), (X, x_0))$,

$$\alpha \simeq \alpha',\ \beta \simeq \beta' \Rightarrow \alpha * \beta \simeq \alpha' * \beta',\ \alpha^{\leftarrow} \simeq \alpha'^{\leftarrow};$$
$$(\alpha * \beta) * \gamma \simeq \alpha * (\beta * \gamma)\ ;\ \alpha * \alpha^{\leftarrow} \simeq \alpha^{\leftarrow} * \alpha \simeq 0.$$ [19]

Furthermore, we can see that $\alpha * \beta \simeq \beta * \alpha$. Thus, we have the additive group

$$\pi_n(X, x_0) = [(\mathbf{I}^n, \partial\mathbf{I}^n), (X, x_0)] = \mathrm{C}((\mathbf{I}^n, \partial\mathbf{I}^n), (X, x_0))/\simeq,$$

with the operation of addition $[\alpha] + [\beta] = [\alpha * \beta]$ and the inverse $-[\alpha] = [\alpha^{\leftarrow}]$ of $[\alpha]$. The group $\pi_n(X, x_0)$ is called the ***n*-th homotopy group** of X at x_0 (or (X, x_0)).

Every pointed map $f : (X, x_0) \to (Y, y_0)$ induces the homomorphisms $f_\sharp : \pi_n(X, x_0) \to \pi_n(Y, y_0)$, $n \in \mathbb{N}$, defined by $f_\sharp[\alpha] = [f\alpha]$. For another pointed map

[18]For loops $\alpha, \beta \in \Omega(X, x_0)$, $\alpha \simeq \beta$ means $\alpha \simeq \beta$ rel. $\partial\mathbf{I}$.

[19]For $\alpha, \beta \in \mathrm{C}((\mathbf{I}^n, \partial\mathbf{I}^n), (X, x_0))$, $\alpha \simeq \beta$ means $\alpha \simeq \beta$ rel. $\partial\mathbf{I}^n$.

$f' : (X, x_0) \to (Y, y_0)$, $f'_\sharp = f_\sharp$ if $f' \simeq f$.[20] For each pair of pointed maps $f : (X, x_0) \to (Y, y_0)$ and $g : (Y, y_0) \to (Z, z_0)$, we have $(gf)_\sharp = g_\sharp f_\sharp$ and $(\mathrm{id}_X)_\sharp = \mathrm{id}_{\pi_n(X,x_0)}$. If $f : (X, x_0) \to (Y, y_0)$ is a pointed homotopy equivalence (i.e., there is a pointed map $g : (Y, y_0) \to (X, x_0)$ such that $gf \simeq \mathrm{id}_X$ rel. x_0 and $fg \simeq \mathrm{id}_Y$ rel. y_0) then $f_\sharp$ is an isomorphism (with $(f_\sharp)^{-1} = g_\sharp$).

For a pointed space (X, x_0), $\pi_0(X, x_0)$ is defined as the pointed set $\pi_0(X)$ whose base point is the path-component of X containing x_0. Then, a pointed map $f : (X, x_0) \to (Y, y_0)$ induces the pointed function $f_\sharp : \pi_0(X, x_0) \to \pi_0(Y, y_0)$.

Suppose that $x_0, x_1 \in X$ are contained in the same path-component of X. Then, we have a path $\omega : \mathbf{I} \to X$ from x_0 to x_1. For each $n \in \mathbb{N}$ and each $\alpha \in \mathrm{C}((\mathbf{I}^n, \partial\mathbf{I}^n), (X, x_1))$, we define $\alpha^\omega \in \mathrm{C}((\mathbf{I}^n, \partial\mathbf{I}^n), (X, x_0))$ as follows:

$$\alpha^\omega(z) = \begin{cases} \alpha(2z - \tfrac{1}{2}\mathbf{1}) & \text{if } \|2z - \mathbf{1}\|_\infty \leq \tfrac{1}{2}, \\ \omega(2 - 2\|2z - \mathbf{1}\|_\infty) & \text{if } \|2z - \mathbf{1}\|_\infty \geq \tfrac{1}{2}. \end{cases}$$

where $\|z\|_\infty = \max\{|z(i)| \mid i = 1, \dots, n\}$ and $\mathbf{1} = (1, \dots, 1) \in \mathbb{R}^n$. Then, ω induces the isomorphism $\omega_* : \pi_n(X, x_1) \to \pi_n(X, x_0)$ defined by $\omega_*[\alpha] = [\alpha^\omega]$, where $(\omega_*)^{-1} = \omega^{\leftarrow}_*$. For a path $\omega' : \mathbf{I} \to X$ with $\omega'(0) = x_0$ and $\omega'(1) = x_1$, if $\omega \simeq \omega'$ rel. $\partial\mathbf{I}$ then $\omega_* = \omega'_*$. If $\omega'' : \mathbf{I} \to X$ is another path with $\omega''(0) = x_1$ and $\omega''(1) = x_2$ then $(\omega * \omega'')_* = \omega_* \omega''_* : \pi_n(X, x_2) \to \pi_n(X, x_0)$. When X is path-connected, $\pi_n(X, x_0) \cong \pi_n(X, x_1)$ for any pair of points $x_0, x_1 \in X$.

Let $h : X \times \mathbf{I} \to Y$ be a homotopy with $h_0 = f$ and $h_1 = f'$. For each $x_0 \in X$, the homotopy h gives a path $\omega : \mathbf{I} \to Y$ defined by $\omega(t) = h(x_0, t)$. For each $\alpha \in \mathrm{C}((\mathbf{I}^n, \partial\mathbf{I}^n), (X, x_0))$, $h(\alpha \times \mathrm{id})$ is a homotopy from $f\alpha$ to $f'\alpha$ and $h(\alpha(\partial\mathbf{I}^n) \times \{t\}) = \{\omega(t)\}$. Then, it can be seen that $f\alpha \simeq (f'\alpha)^\omega$ rel. $\partial\mathbf{I}^n$. Thus, it follows that $f_\sharp = \omega_* f'_\sharp$, that is, the following diagram commutes:

$$\begin{array}{ccc} & \overset{f'_\sharp}{\nearrow} & \pi_n(Y, f'(x_0)) \\ \pi_n(X, x_0) & & \downarrow \omega_* \\ & \underset{f_\sharp}{\searrow} & \pi_n(Y, f(x_0)). \end{array}$$

Using this fact, we can show that every homotopy equivalence $f : X \to Y$ induces the isomorphisms $f_\sharp : \pi_n(X, x_0) \to \pi_n(Y, f(x_0))$, $n \in \mathbb{N}$, $x_0 \in X$. In fact, let $g : Y \to X$ be a homotopy inverse of f, that is, $gf \simeq \mathrm{id}_X$ and $fg \simeq \mathrm{id}_Y$. A homotopy from gf to id gives a path $\omega : \mathbf{I} \to X$ from $gf(x_0)$ to x_0 and a homotopy from fg to id_Y gives a path $\omega' : \mathbf{I} \to Y$ from $fgf(x_0)$ to $f(x_0)$. For each $x_0 \in X$ and $n \in \mathbb{N}$, we have the following commutative diagram:

[20]For pointed maps $f, f' \in \mathrm{C}((X, x_0), (Y, y_0))$, $f' \simeq f$ means $f' \simeq f$ rel. x_0.

$$\begin{array}{ccc}
\pi_n(X, x_0) & \xrightarrow{\omega_*} & \pi_n(X, gf(x_0)) \\
{\scriptstyle f_\sharp}\downarrow & \nearrow{\scriptstyle g_\sharp} & \downarrow{\scriptstyle f_\sharp} \\
\pi_n(Y, f(x_0)) & \xrightarrow[\omega'_*]{} & \pi_n(Y, fgf(x_0)),
\end{array}$$

where it should be noted that the left and the right $f_\sharp$ in the diagram are different from each other. Since $g_\sharp f_\sharp = \omega_*$ and $f_\sharp g_\sharp = \omega'_*$ are isomorphisms, $g_\sharp$ is also an isomorphism. Then, it follows that $f_\sharp : \pi_n(X, x_0) \to \pi_n(Y, f(x_0))$ is an isomorphism.

For each $n \geq 2$, we define

$$\mathbf{J}^{n-1} = (\partial\mathbf{I}^{n-1} \times \mathbf{I}) \cup (\mathbf{I}^{n-1} \times \{1\}).$$

Identifying $\mathbf{I}^{n-1} = \mathbf{I}^{n-1} \times \{0\} \subset \mathbf{I}^n$, we have

$$\mathbf{I}^{n-1} \cup \mathbf{J}^{n-1} = \partial\mathbf{I}^n \text{ and } \mathbf{I}^{n-1} \cap \mathbf{J}^{n-1} = \partial\mathbf{I}^{n-1}.$$

For a space X with $x_0 \in A \subset X$, the **relative *n*-th homotopy group** of (X, A) at x_0 is defined as the group

$$\pi_n(X, A, x_0) = [(\mathbf{I}^n, \partial\mathbf{I}^n, \mathbf{J}^{n-1}), (X, A, x_0)]$$

with the operations defined by the analogy of $\pi_n(X, x_0)$. In general, the group $\pi_2(X, A, x_0)$ is non-commutative, so we describe it as a multiplicative group like the fundamental group. On the other hand, for $n \geq 3$, $\pi_n(X, A, x_0)$ is represented as the additive group because it is commutative. It should be noted that $\pi_n(X, \{x_0\}, x_0) = \pi_n(X, x_0)$. We also define

$$\pi_1(X, A, x_0) = \mathrm{C}((\mathbf{I}, \partial\mathbf{I}, 0), (X, A, x_0))/\simeq,$$

which is regarded as the pointed set whose base point is the homotopy class $[c_{x_0}]$. As in the case with pointed spaces, every map $f : (X, A, x_0) \to (Y, B, y_0)$ induces the homomorphisms $f_\sharp : \pi_n(X, A, x_0) \to \pi_n(Y, B, y_0)$, $n \geq 2$, and the pointed function $f_\sharp : \pi_1(X, A, x_0) \to \pi_1(Y, B, y_0)$.

For each $n \geq 2$, let $\partial : \pi_n(X, A, x_0) \to \pi_{n-1}(A, x_0)$ be the homomorphism induced by the restriction operator, i.e., $\partial[\alpha] = [\alpha|\mathbf{I}^{n-1}]$, which is called the **boundary operator**. In addition, we define the pointed function $\partial : \pi_1(X, A, x_0) \to \pi_0(A, x_0)$ as follows: $\partial[\alpha]$ is the path-component of $\alpha(1)$ for each $\alpha \in \mathrm{C}((\mathbf{I}, \partial\mathbf{I}, 0), (X, A, x_0))$. Then, the following diagram commutes:

$$\begin{array}{ccccccc}
\pi_n(A, x_0) & \xrightarrow{i_\sharp} & \pi_n(X, x_0) & \xrightarrow{j_\sharp} & \pi_n(X, A, x_0) & \xrightarrow{\partial} & \pi_{n-1}(A, x_0) \\
\downarrow {\scriptstyle (f|A)_\sharp} & & \downarrow {\scriptstyle f_\sharp} & & \downarrow {\scriptstyle f_\sharp} & & \downarrow {\scriptstyle (f|A)_\sharp} \\
\pi_n(B, y_0) & \xrightarrow[i_\sharp]{} & \pi_n(Y, y_0) & \xrightarrow[j_\sharp]{} & \pi_n(Y, B, y_0) & \xrightarrow[\partial]{} & \pi_{n-1}(B, y_0),
\end{array}$$

where the $i_\sharp$ are the homomorphisms induced by the inclusions $i : (A, x_0) \subset (X, x_0)$ and $i : (B, y_0) \subset (Y, y_0)$ and the $j_\sharp$ are the ones induced by the inclusions $j : (X, \{x_0\}, x_0) \subset (X, A, x_0)$ and $j : (Y, \{y_0\}, y_0) \subset (Y, B, y_0)$. Moreover, $\operatorname{im} i_\sharp = \ker j_\sharp$, $\operatorname{im} j_\sharp = \ker \partial$ and $\operatorname{im} \partial = \ker i_\sharp$. Namely, the following sequence is exact:

$$\cdots \xrightarrow{\partial} \pi_n(A, x_0) \xrightarrow{i_\sharp} \pi_n(X, x_0) \xrightarrow{j_\sharp} \pi_n(X, A, x_0) \xrightarrow{\partial}$$

$$\pi_{n-1}(A, x_0) \xrightarrow{i_\sharp} \cdots \xrightarrow{\partial} \pi_1(A, x_0) \xrightarrow{i_\sharp} \pi_1(X, x_0)$$

$$\xrightarrow{j_\sharp} \pi_1(X, A, x_0) \xrightarrow{\partial} \pi_0(A, x_0) \xrightarrow{i_\sharp} \pi_0(X, x_0).$$

This sequence is called the **homotopy exact sequence** of (X, A, x_0).

Let $\omega : \mathbf{I} \to A$ be a path from x_0 to x_1 and

$$\alpha \in \mathrm{C}((\mathbf{I}^n, \partial\mathbf{I}^n, \mathbf{J}^{n-1}), (X, A, x_1)).$$

We now define α^ω somewhat differently from the case of pointed spaces, that is,

$$\alpha^\omega(z) = \begin{cases} \alpha(2z(1) - \frac{1}{2}, \ldots, 2z(n-1) - \frac{1}{2}, 2z(n)) & \text{if } \|z'\|_\infty \leq \frac{1}{2}, \\ \omega(2 - 2\|z'\|_\infty) & \text{if } \|z'\|_\infty \geq \frac{1}{2}, \end{cases}$$

$$\text{where } z' = (2z(1) - 1, \ldots, 2z(n-1) - 1, z(n)).$$

Now, similar to the case of pointed spaces, we can define the isomorphism $\omega_* : \pi_n(X, A, x_1) \to \pi_n(X, A, x_0)$ by $\omega_*[\alpha] = [\alpha^\omega]$. Then, the following diagram commutes:

$$\begin{array}{ccccccc}
\pi_n(A, x_1) & \xrightarrow{i_\sharp} & \pi_n(X, x_1) & \xrightarrow{j_\sharp} & \pi_n(X, A, x_1) & \xrightarrow{\partial} & \pi_{n-1}(A, x_1) \\
\downarrow {\scriptstyle \omega_*} & & \downarrow {\scriptstyle \omega_*} & & \downarrow {\scriptstyle \omega_*} & & \downarrow {\scriptstyle \omega_*} \\
\pi_n(A, x_0) & \xrightarrow[i_\sharp]{} & \pi_n(X, x_0) & \xrightarrow[j_\sharp]{} & \pi_n(X, A, x_0) & \xrightarrow[\partial]{} & \pi_{n-1}(A, y_0).
\end{array}$$

Thus, if A is path-connected then $\pi_n(X, A, x_0) \cong \pi_n(X, A, x_1)$ for any pair of points $x_0, x_1 \in A$. In this case, it is acceptable to denote $\pi_n(X, A)$ without the base point x_0.

The proof of the following proposition is quite elementary:

Proposition 4.14.1. *For pointed spaces* $(X_1, x_1), \dots, (X_k, x_k)$ *and* $n \in \mathbb{N}$,

$$\pi_n(X_1 \times \cdots \times X_k, (x_1, \dots, x_k)) \cong \pi_n(X_1, x_1) \times \cdots \times \pi_n(X_k, x_k).$$

Proof. We have the following isomorphism

$$\varphi : \pi_n(X_1 \times \cdots \times X_k, (x_1, \dots, x_k)) \to \pi_n(X_1, x_1) \times \cdots \times \pi_n(X_k, x_k);$$
$$\varphi[\alpha] = ((\mathrm{pr}_1)_\sharp[\alpha], \dots, (\mathrm{pr}_k)_\sharp[\alpha]) = ([\mathrm{pr}_1\alpha], \dots, [\mathrm{pr}_k\alpha]).$$

Indeed, for each set of maps $\alpha_i : (\mathbf{I}^n, \partial\mathbf{I}^n) \to (X_i, x_i)$, $i = 1, \dots, k$, we have the map

$$\alpha : (\mathbf{I}^n, \partial\mathbf{I}^n) \to (X_1 \times \cdots \times X_k, (x_1, \dots, x_k))$$

defined by $\alpha(x) = (\alpha_1(x), \dots, \alpha_k(x))$. Then, $\varphi([\alpha]) = ([\alpha_1], \dots, [\alpha_k])$. Hence, φ is an epimorphism. If every $\alpha_1, \dots, \alpha_k$ are null-homotopic, then the map α is null-homotopic, which means that φ is a monomorphism. □

For additive groups $G_1, \dots, G_k$, the direct product $G_1 \times \cdots \times G_k$ is regarded as the direct sum $G_1 \oplus \cdots \oplus G_k$. Thus, when $n \geq 2$ in Theorem 4.14.1,

$$\pi_n(X_1 \times \cdots \times X_k, (x_1, \dots, x_k)) \cong \pi_n(X_1, x_1) \oplus \cdots \oplus \pi_n(X_k, x_k).$$

Proposition 4.14.2. *For every pair of pointed spaces* (X, x_0) *and* (Y, y_0), *there exists a natural bijection*

$$\varphi : \mathrm{C}((X \times \mathbf{I}, H_X), (Y, y_0)) \to \mathrm{C}((X, x_0), (\Omega(Y, y_0), c_{y_0})),$$

where $H_X = (X \times \partial\mathbf{I}) \cup (\{x_0\} \times \mathbf{I})$. *Here,* φ *is natural in the following sense: given pointed maps* $f : (X', x_0') \to (X, x_0)$ *and* $g : (Y, y_0) \to (Y', y_0')$, *the diagrams below are commutative:*

$$\begin{array}{ccc}
\mathrm{C}((X' \times \mathbf{I}, H_{X'}), (Y, y_0)) & \xrightarrow{\varphi} & \mathrm{C}((X', x_0'), (\Omega(Y, y_0), c_{y_0})) \\
\uparrow{\scriptstyle (f \times \mathrm{id})^*} & & \uparrow{\scriptstyle f^*} \\
\mathrm{C}((X \times \mathbf{I}, H_X), (Y, y_0)) & \xrightarrow{\varphi} & \mathrm{C}((X, x_0), (\Omega(Y, y_0), c_{y_0})) \\
\downarrow{\scriptstyle g_*} & & \downarrow{\scriptstyle (g_*)_*} \\
\mathrm{C}((X \times \mathbf{I}, H_X), (Y', y_0')) & \xrightarrow[\varphi]{} & \mathrm{C}((X, x_0), (\Omega(Y', y_0'), c_{y_0'})).
\end{array}$$

Proof. We define the functions

$$\varphi : \mathrm{C}((X \times \mathbf{I}, H_X), (Y, y_0)) \ni k \mapsto \varphi(k) \in \mathrm{C}((X, x_0), (\Omega(Y, y_0), c_{y_0})),$$
$$\psi : \mathrm{C}((X, x_0), (\Omega(Y, y_0), c_{y_0})) \ni k \mapsto \psi(k) \in \mathrm{C}((X \times \mathbf{I}, H_X), (Y, y_0))$$

by $\varphi(k)(x)(s) = k(x, s)$ and $\psi(k)(x, s) = k(x)(s)$, respectively. It is easy to see that $\psi\varphi = \mathrm{id}$ and $\varphi\psi = \mathrm{id}$. Then, φ is a bijection with $\varphi^{-1} = \psi$.

To show the commutativity of the diagram, let $k \in \mathrm{C}((X \times \mathbf{I}, H_X), (Y, y_0))$. For each $x' \in X'$ and $s \in \mathbf{I}$,

$$\begin{aligned}\varphi((f \times \mathrm{id})^*(k))(x')(s) &= \varphi(k(f \times \mathrm{id}))(x')(s) = k(f \times \mathrm{id})(x', s) \\ &= k(f(x'), s) = \varphi(k)(f(x'))(s) = f^*(\varphi(k))(x')(s).\end{aligned}$$

Hence, $\varphi((f \times \mathrm{id})^*(k)) = f^*(\varphi(k))$. On the other hand, for each $x \in X$ and $s \in \mathbf{I}$,

$$\begin{aligned}\varphi(g_*(k))(x)(s) &= \varphi(gk)(x)(s) = gk(x, s) = g(\varphi(k)(x)(s)) \\ &= g_*(\varphi(k)(x))(s) = (g_*)_*(\varphi(k))(x)(s).\end{aligned}$$

Hence, $\varphi(g_*(k)) = (g_*)_*(\varphi(k))$. □

In the above situation, given a homotopy $h : X \times \mathbf{I} \times \mathbf{I} \to Y$ such that $h_t(H_X) = \{y_0\}$ for every $t \in \mathbf{I}$, we have the homotopy $\varphi(h) : X \times \mathbf{I} \to \Omega(Y, y_0)$ defined by $\varphi(h)_t = \varphi(h_t)$, that is, $\varphi(h)(x, t)(s) = \varphi(h_t)(x)(s) = h_t(x, s) = h(x, s, t)$. Then, φ induces the function

$$\Phi : [(X \times \mathbf{I}, H_X), (Y, y_0)] \to [(X, x_0), (\Omega(Y, y_0), c_{y_0})].$$

Given a homotopy $h : X \times \mathbf{I} \to \Omega(Y, y_0)$ such that $h_t(x_0) = c_{y_0}$ for every $t \in \mathbf{I}$, we have the homotopy $\psi(h) : X \times \mathbf{I} \times \mathbf{I} \to Y$ defined by $\psi(h)_t = \psi(h_t)$, that is, $\psi(h)(x, s, t) = \psi(h_t)(x, s) = h_t(x)(s) = h(x, t)(s)$. Then, ψ induces the function

$$\Psi : [(X, x_0), (\Omega(Y, y_0), c_{y_0})] \to [(X \times \mathbf{I}, H_X), (Y, y_0)].$$

Furthermore, given a homotopy $h : X \times \mathbf{I} \times \mathbf{I} \to Y$ such that $h_t(H_X) = \{y_0\}$ for every $t \in \mathbf{I}$, $\psi(\varphi(h))_t = \psi(\varphi(h_t)) = h_t$. Similarly, given a homotopy $h : X \times \mathbf{I} \to \Omega(Y, y_0)$ such that $h_t(x_0) = c_{y_0}$ for every $t \in \mathbf{I}$, $\varphi(\psi(h))_t = \varphi(\psi(h_t)) = h_t$. Then, it follows that Φ is a bijection with $\Phi^{-1} = \Psi$. Thus, we have the following:

Proposition 4.14.3. *For every two pointed spaces (X, x_0) and (Y, y_0), there exists a natural bijection*

$$\Phi : [(X \times \mathbf{I}, H_X), (Y, y_0)] \to [(X, x_0), (\Omega(Y, y_0), c_{y_0})],$$

where $H_X = (X \times \partial\mathbf{I}) \cup (\{x_0\} \times \mathbf{I})$. Here, Φ is natural in a similar sense to Proposition 4.14.2. □

In the above situation, replace (X, x_0) and (Y, y_0) with $(\mathbf{I}^n, \partial\mathbf{I}^n)$ and (X, x_0), respectively. Since $(X \times \mathbf{I}, H_X)$ corresponds to $(\mathbf{I}^{n+1}, \partial\mathbf{I}^{n+1})$, we have a bijection

$$\Phi : [(\mathbf{I}^{n+1}, \partial\mathbf{I}^{n+1}), (X, x_0)] \to [(\mathbf{I}^n, \partial\mathbf{I}^n), (\Omega(X, x_0), c_{x_0})],$$

that is, $\Phi : \pi_{n+1}(X, x_0) \to \pi_n(\Omega(X, x_0), c_{x_0})$. As can be easily seen, $\varphi(\alpha * \alpha') = \varphi(\alpha) * \varphi(\alpha')$ and $\varphi(\alpha^{\leftarrow}) = \varphi(\alpha)^{\leftarrow}$ for each $\alpha, \alpha' \in \mathrm{C}((\mathbf{I}^{n+1}, \partial\mathbf{I}^{n+1}), (X, x_0))$. Hence, $\Phi : \pi_{n+1}(X, x_0) \to \pi_n(\Omega(X, x_0), c_{x_0})$ is an isomorphism. Thus, we have the following theorem:

Theorem 4.14.4. *For every pointed space (X, x_0) and $n \in \mathbb{N}$,*

$$\pi_n(\Omega(X, x_0), c_{x_0}) \cong \pi_{n+1}(X, x_0). \qquad \square$$

Note. Recall that the loop space $\Omega(X, x_0)$ has the operations $\alpha * \beta$ and $\alpha^{\leftarrow}$, which induce the operations of the fundamental group $\pi_1(X, x_0)$. Using these operations of $\Omega(X, x_0)$, we can define the operations $\alpha \hat{*} \beta$ and $\alpha^{\wedge}$ on $\mathrm{C}((\mathbf{I}^n, \partial\mathbf{I}^n), (\Omega(X, x_0), c_{x_0}))$ as follows:

$$(\alpha \hat{*} \beta)(z) = \alpha(z) * \beta(z) \text{ and } \alpha^{\wedge}(z) = \alpha(z)^{\leftarrow} \text{ for each } z \in \mathbf{I}^n.$$

The operations of the homotopy group $\pi_n(\Omega(X, x_0), c_{x_0})$ are also induced by the above operations. In fact, we can define the operations $\alpha \underline{*} \beta$ and $\alpha_{\leftarrow}$ on $\mathrm{C}((\mathbf{I}^{n+1}, \partial\mathbf{I}^{n+1}), (X, x_0))$ as follows:

$$\alpha \underline{*} \beta(t_1, \ldots, t_{n+1}) = \begin{cases} \alpha(t_1, \ldots, t_n, 2t_{n+1}) & \text{if } 0 \le t_{n+1} \le 1/2, \\ \beta(t_1, \ldots, t_n, 2t_{n+1} - 1) & \text{if } 1/2 \le t_{n+1} \le 1, \end{cases}$$

$$\alpha_{\leftarrow}(t_1, \ldots, t_{n+1}) = \alpha(t_1, \ldots, t_n, 1 - t_{n+1}),$$

which induce the operations of the homotopy group $\pi_n(\Omega(X, x_0), c_{x_0})$, i.e., $\alpha \underline{*} \beta \simeq \alpha * \beta$ and $\alpha_{\leftarrow} \simeq \alpha^{\leftarrow}$. Let ψ $(= \varphi^{-1})$ be the natural bijection in the proof of Proposition 4.14.2. For each $\alpha, \beta \in \mathrm{C}((\mathbf{I}^n, \partial\mathbf{I}^n), (\Omega(X, x_0), c_{x_0}))$,

$$\psi(\alpha \hat{*} \beta) = \psi(\alpha) \underline{*} \psi(\beta) \simeq \psi(\alpha) * \psi(\beta) = \psi(\alpha * \beta) \text{ and}$$
$$\psi(\alpha^{\wedge}) = \psi(\alpha)_{\leftarrow} \simeq \psi(\alpha)^{\leftarrow} = \psi(\alpha^{\leftarrow}),$$

hence $\alpha \hat{*} \beta \simeq \alpha * \beta$ and $\alpha^{\wedge} \simeq \alpha^{\leftarrow}$.

Inductively, we can define $\Omega^n(X, x_0) = \Omega(\Omega^n(X, x_0), 0)$. Then, we have

$$\pi_n(X, x_0) = \pi_0(\Omega^n(X, x_0), 0) = \pi_1(\Omega^{n-1}(X, x_0), 0).$$

Let $\mathbf{e}_1 = (1, 0, \ldots, 0) \in \mathbb{R}^{n+1}$. For each $n \in \mathbb{N}$,

$$(\mathbf{I}^n/\partial\mathbf{I}^n, \partial\mathbf{I}^n/\partial\mathbf{I}^n) \approx (\mathbf{S}^n, \mathbf{e}_1) \text{ and}$$
$$(\mathbf{I}^{n+1}/\mathbf{J}^n, \partial\mathbf{I}^{n+1}/\mathbf{J}^n, \mathbf{J}^n/\mathbf{J}^n) \approx (\mathbf{B}^{n+1}, \mathbf{S}^n, \mathbf{e}_1).$$

Then, $\mathrm{C}((\mathbf{I}^n, \partial\mathbf{I}^n), (X, x_0))$ and $\mathrm{C}((\mathbf{I}^n, \partial\mathbf{I}^n, \mathbf{J}^{n-1}), (X, A, x_0))$ can be identified with $\mathrm{C}((\mathbf{S}^n, \mathbf{e}_1), (X, x_0))$ and $\mathrm{C}((\mathbf{B}^n, \mathbf{S}^{n-1}, \mathbf{e}_1), (X, A, x_0))$, respectively. In particular, $\mathrm{C}((\mathbf{S}^1, \mathbf{e}_1), (X, x_0))$ can be regarded as the loop space $\Omega(X, x_0)$. Thus, we can make the following identifications:

$$\pi_n(X, x_0) = [(\mathbf{S}^n, \mathbf{e}_1), (X, x_0)] \text{ and}$$

$$\pi_{n+1}(X, A, x_0) = [(\mathbf{B}^{n+1}, \mathbf{S}^n, \mathbf{e}_1), (X, A, x_0)].$$

In this case, the boundary operator $\partial : \pi_{n+1}(X, A, x_0) \to \pi_n(A, x_0)$ is defined by $\partial[\alpha] = [\alpha|\mathbf{S}^{n-1}]$.

When X (resp. $A \subset X$) is path-connected, as observed before, it is acceptable to denote $\pi_n(X)$ (resp. $\pi_n(X, A)$) without a base point. However, this does not mean that $\pi_n(X)$ (resp. $\pi_n(X, A)$) can be identified with $[\mathbf{S}^n, X]$ (resp. $[(\mathbf{B}^n, \mathbf{S}^{n-1}), (X, A)]$). It is said that X is **simply connected** if it is path-connected and $\pi_1(X, x_0) = \{0\}$ for any/some $x_0 \in X$. The latter condition is equivalent to the condition that every map $\alpha : \mathbf{S}^1 \to X$ is null-homotopic or α extends over $\mathbf{B}^2$. When X (resp. A) is simply connected, $\pi_n(X)$ (resp. $\pi_n(X, A)$) can be identified with $[\mathbf{S}^n, X]$ (resp. $[(\mathbf{B}^n, \mathbf{S}^{n-1}), (X, A)]$). In fact, we have the following proposition:

Proposition 4.14.5. (1) *If X is simply connected, for any $x_0 \in X$ and $n \geq 2$, the inclusion*

$$i : \mathrm{C}((\mathbf{S}^n, \mathbf{e}_1), (X, x_0)) \subset \mathrm{C}(\mathbf{S}^n, X)$$

induces a bijection from $\pi_n(X, x_0)$ onto $[\mathbf{S}^n, X]$.

(2) *If $A \subset X$ is simply connected, for any $x_0 \in A$ and $n \geq 2$, the inclusion*

$$i : \mathrm{C}((\mathbf{B}^n, \mathbf{S}^{n-1}, \mathbf{e}_1), (X, A, x_0)) \subset \mathrm{C}((\mathbf{B}^n, \mathbf{S}^{n-1}), (X, A))$$

induces a bijection from $\pi_n(X, A, x_0)$ onto $[(\mathbf{B}^n, \mathbf{S}^{n-1}), (X, A)]$.

Proof. (1): For each $\alpha \in \mathrm{C}(\mathbf{S}^n, X)$, take a path $\omega : \mathbf{I} \to X$ from $\alpha(\mathbf{e}_1)$ to x_0. Applying the Homotopy Extension Theorem 4.3.3, we can obtain a homotopy $h : \mathbf{S}^n \times \mathbf{I} \to X$ such that $h_0 = \alpha$ and $h(\mathbf{e}_1, t) = \omega(t)$ for each $t \in \mathbf{I}$. Thus, α is homotopic to the map $h_1 \in \mathrm{C}((\mathbf{S}^n, \mathbf{e}_1), (X, x_0))$. This means that i induces the surjection.

Let $\alpha, \beta \in \mathrm{C}((\mathbf{S}^n, \mathbf{e}_1), (X, x_0))$ and assume that there is a homotopy $h : \mathbf{S}^n \times \mathbf{I} \to X$ from α to β. Since X is simply connected, we have a map $k : \mathbf{I}^2 \to X$ such that $k(t, 0) = h(\mathbf{e}_1, t)$ and $k(0, t) = k(1, t) = k(t, 1) = x_0$ for each $t \in \mathbf{I}$. Applying the Homotopy Extension Theorem 4.3.3, we have a homotopy $\tilde{\varphi} : \mathbf{S}^n \times \mathbf{I} \times \mathbf{I} \to X$ that is an extension of the homotopy

$$\varphi : ((\mathbf{S}^n \times \{0, 1\}) \cup (\{\mathbf{e}_1\} \times \mathbf{I})) \times \mathbf{I} \to X$$

defined by $\varphi_t(x, 0) = h(x, 0) = \alpha(x)$ and $\varphi_t(x, 1) = h(x, 1) = \beta(x)$ for each $x \in \mathbf{S}^n$, and $\varphi_t(\mathbf{e}_1, s) = k(s, t)$ for each $s \in \mathbf{I}$. Then, $\tilde{\varphi}_1 : \mathbf{S}^n \times \mathbf{I} \to X$ is a homotopy from α to β with $\tilde{\varphi}_1(\mathbf{e}_1, t) = x_0$ for every $t \in \mathbf{I}$. This means that i induces the injection.

(2): For each $\alpha \in C((\mathbf{B}^n, \mathbf{S}^{n-1}), (X, A))$, take a path $\omega : \mathbf{I} \to A$ from $\alpha(\mathbf{e}_1)$ to x_0. Applying the Homotopy Extension Theorem 4.3.3 twice, we can obtain a homotopy $h : \mathbf{B}^n \times \mathbf{I} \to X$ such that $h_0 = \alpha$, $h(\mathbf{S}^{n-1} \times \mathbf{I}) \subset A$ and $h(\mathbf{e}_1, t) = \omega(t)$ for each $t \in \mathbf{I}$. Thus, α is homotopic to the map $h_1 \in C((\mathbf{B}^n, \mathbf{S}^{n-1}, \mathbf{e}_1), (X, A, x_0))$. This means that i induces the surjection.

Let $\alpha, \beta \in C((\mathbf{B}^n, \mathbf{S}^{n-1}, \mathbf{e}_1), (X, A, x_0))$ and assume that there is a homotopy $h : \mathbf{B}^n \times \mathbf{I} \to X$ from α to β such that $h(\mathbf{S}^{n-1} \times \mathbf{I}) \subset A$. Since A is simply connected, we have a map $k : \mathbf{I}^2 \to A$ such that $k(t, 0) = h(\mathbf{e}_1, t)$ and $k(0, t) = k(1, t) = k(t, 1) = x_0$ for each $t \in \mathbf{I}$. Applying the Homotopy Extension Theorem 4.3.3 twice, we have a homotopy $\tilde{\varphi} : \mathbf{B}^n \times \mathbf{I} \times \mathbf{I} \to X$ such that $\tilde{\varphi}(\mathbf{S}^{n-1} \times \mathbf{I} \times \mathbf{I}) \subset A$ and $\tilde{\varphi}$ is an extension of the homotopy

$$\varphi : ((\mathbf{B}^n \times \{0, 1\}) \cup (\{\mathbf{e}_1\} \times \mathbf{I})) \times \mathbf{I} \to X$$

defined by $\varphi_t(x, 0) = h(x, 0) = \alpha(x)$ and $\varphi_t(x, 1) = h(x, 1) = \beta(x)$ for each $x \in \mathbf{B}^n$, and $\varphi_t(\mathbf{e}_1, s) = k(s, t)$ for each $s \in \mathbf{I}$. Then, $\tilde{\varphi}_1 : \mathbf{B}^n \times \mathbf{I} \to X$ is a homotopy from α to β with $\tilde{\varphi}_1(\mathbf{e}_1, t) = x_0$ for every $t \in \mathbf{I}$. This means that i induces the injection. □

If X is not simply connected, for two maps $\alpha, \alpha' : \mathbf{S}^n \to X$ with $\alpha(\mathbf{e}_1) = \alpha'(\mathbf{e}_1)$, $\alpha \simeq \alpha'$ does not imply $\alpha \simeq \alpha'$ rel. $\mathbf{e}_1$. However, we have the following proposition:

Proposition 4.14.6. *For a map $\alpha : \mathbf{S}^n \to X$, the following are equivalent:*

(a) *α extends over $\mathbf{B}^{n+1}$;*
(b) *α is null-homotopic, i.e., $[\alpha] = 0 \in [\mathbf{S}^n, X]$;*
(c) *$\alpha \simeq c_{\alpha(\mathbf{e}_1)}$ rel. $\mathbf{e}_1$, i.e., $[\alpha] = 0 \in \pi_n(X, \alpha(\mathbf{e}_1))$.*

Proof. The implication (c) ⇒ (b) is obvious. For (b) ⇒ (a), let $h : \mathbf{S}^n \times \mathbf{I} \to X$ be a homotopy from α to a constant map. Then, α extends to the map $\beta : \mathbf{B}^{n+1} \to X$ defined by

$$\beta(x) = \begin{cases} h(\|x\|^{-1}x, 1 - \|x\|) & \text{if } x \neq 0, \\ h_1(\mathbf{e}_1) & \text{if } x = 0. \end{cases}$$

For (a) ⇒ (c), using an extension $\beta : \mathbf{B}^{n+1} \to X$ of α, we can define the homotopy $h : \mathbf{S}^n \times \mathbf{I} \to X$ by $h(x, t) = \beta((1 - t)x + t\mathbf{e}_1)$, which realizes $\alpha \simeq 0$ rel. $\mathbf{e}_1$. □

Even if $A \subset X$ is not simply connected, we have the following:

Proposition 4.14.7. *For a map $\alpha : (\mathbf{B}^{n+1}, \mathbf{S}^n) \to (X, A)$, the following are equivalent:*

(a) *There is a map $\beta : \mathbf{B}^{n+1} \to A$ such that $\beta|\mathbf{S}^n = \alpha|\mathbf{S}^n$ and $\alpha \simeq \beta$ rel. $\mathbf{S}^n$;*
(b) *$\alpha \simeq 0$ by a homotopy $h : \mathbf{B}^{n+1} \times \mathbf{I} \to X$ such that $h(\mathbf{S}^n \times \mathbf{I}) \subset A$, i.e., $[\alpha] = 0 \in [(\mathbf{B}^{n+1}, \mathbf{S}^n), (X, A)]$;*
(c) *$\alpha \simeq 0$ by a homotopy $h : \mathbf{B}^{n+1} \times \mathbf{I} \to X$ such that $h(\mathbf{S}^n \times \mathbf{I}) \subset A$ and $h(\{\mathbf{e}_1\} \times \mathbf{I}) = \{\alpha(\mathbf{e}_1)\}$, i.e., $[\alpha] = 0 \in \pi_{n+1}(X, A, \alpha(\mathbf{e}_1))$.*

Proof. The implications (c) $\Rightarrow$ (b) $\Rightarrow$ (a) are trivial. Given the homotopy in condition (a), we can connect it with a similar homotopy to that in the proof of (a) $\Rightarrow$ (c) in Proposition 4.14.6 to obtain the homotopy in (c). □

Concerning the homomorphism induced by a map, we can state that:

Proposition 4.14.8. *For a map $f : X \to Y$, the following are equivalent:*

(a) *$f_\sharp : \pi_n(X, x_0) \to \pi_n(Y, f(x_0))$ is a monomorphism for every $x_0 \in X$;*
(b) *For each map $\alpha : \mathbf{S}^n \to X$, $f\alpha \simeq 0$ rel. $\mathbf{e}_1$ implies $\alpha \simeq 0$ rel. $\mathbf{e}_1$;*
(c) *Any map $\alpha : \mathbf{S}^n \to X$ is null-homotopic if $f\alpha \simeq 0$;*
(d) *Any map $\alpha : \mathbf{S}^n \to X$ extends over $\mathbf{B}^{n+1}$ if $f\alpha$ extends over $\mathbf{B}^{n+1}$.*

Proof. Identifying $(\mathbf{I}^n/\partial\mathbf{I}^n, \partial\mathbf{I}^n/\partial\mathbf{I}^n) = (\mathbf{S}^n, \mathbf{e}_1)$, we can easily obtain the equivalence (a) $\Leftrightarrow$ (b). By Proposition 4.14.6, conditions (b), (c), and (d) are equivalent to each other. □

The following result can be obtained in the same way as (a) $\Leftrightarrow$ (b) in Proposition 4.14.8:

Proposition 4.14.9. *For a map $f : X \to Y$, the following are equivalent:*

(a) *$f_\sharp : \pi_n(X, x_0) \to \pi_n(Y, f(x_0))$ is an epimorphism for every $x_0 \in X$;*
(b) *For each map $\beta : \mathbf{S}^n \to Y$, there is a map $\alpha : \mathbf{S}^n \to X$ such that $f\alpha(\mathbf{e}_1) = \beta(\mathbf{e}_1)$ and $f\alpha \simeq \beta$ rel. $\mathbf{e}_1$.* □

The following proposition can be obtained by the homotopy exact sequence:

Proposition 4.14.10. *For a pair (X, A) of spaces, let $i : A \subset X$ be the inclusion map. For each $n \in \mathbb{N}$ and $x_0 \in A$, the following hold:*

(1) *$i_\sharp : \pi_n(A, x_0) \to \pi_n(X, x_0)$ is a monomorphism if $\pi_{n+1}(X, A, x_0) = \{0\}$;*
(2) *$i_\sharp : \pi_n(A, x_0) \to \pi_n(X, x_0)$ is an epimorphism if $\pi_n(X, A, x_0) = \{0\}$;*
(3) *$i_\sharp : \pi_n(A, x_0) \to \pi_n(X, x_0)$ is an isomorphism if $\pi_{n+1}(X, A, x_0) = \{0\}$ and $\pi_n(X, A, x_0) = \{0\}$;*
(4) *$\pi_{n+1}(X, A, x_0) = \{0\}$ if $i_\sharp : \pi_n(A, x_0) \to \pi_n(X, x_0)$ is a monomorphism and $i_\sharp : \pi_{n+1}(A, x_0) \to \pi_{n+1}(X, x_0)$ is an epimorphism.*

Direct Proof. First of all, note that (3) is a combination of (1) and (2).

(1): Let $\alpha : \mathbf{S}^n \to A$ be a map such that $\alpha \simeq 0$ in X. According to Proposition 4.14.6, α extends to a map $\beta : \mathbf{B}^{n+1} \to X$. Then, $[\beta] \in \pi_{n+1}(X, A) = \{0\}$, so there is a homotopy $h : \mathbf{B}^{n+1} \times \mathbf{I} \to X$ such that $h(\mathbf{S}^n \times \mathbf{I}) \subset A$, $h_0 = \beta$, and h_1 is a constant map. Hence, $\alpha \simeq 0$ in A by the restriction $h|\mathbf{S}^n \times \mathbf{I}$. Thus, $i_\sharp$ is a monomorphism.

(2): Each map $\alpha : (\mathbf{I}^n, \partial\mathbf{I}^n) \to (X, x_0)$ can be regarded as

$$\alpha \in \mathrm{C}((\mathbf{I}^n, \partial\mathbf{I}^n, \mathbf{J}^{n-1}), (X, A, x_0)),$$

and then $[\alpha] \in \pi_n(X, A, x_0) = \{0\}$. Hence, there is a homotopy $h : \mathbf{I}^n \times \mathbf{I} \to X$ such that $h(\partial\mathbf{I}^n \times \mathbf{I}) \subset A$, $h(\mathbf{J}^{n-1} \times \mathbf{I}) = \{x_0\}$, $h_0 = \alpha$, and $h_1(\mathbf{I}^n) = \{x_0\}$. We define $\alpha' : \mathbf{I}^n \to A$ as follows:

$$\alpha'(z) = h((z(1), \ldots, z(n-1), 0), z(n)) \quad \text{for each } z \in \mathbf{I}^n.$$

Then, $\alpha'(\partial\mathbf{I}^n) = \{x_0\}$, that is, $\alpha' : (\mathbf{I}^n, \partial\mathbf{I}^n) \to (A, x_0)$. It is easy to see that $\alpha' \simeq \alpha$ rel. $\partial\mathbf{I}^n$ in X. Therefore, $i_\sharp$ is an epimorphism.

(4): Let $\alpha \in C((\mathbf{I}^{n+1}, \partial\mathbf{I}^{n+1}, \mathbf{J}^n), (X, A, x_0))$. Note that α itself is a homotopy realizing $\alpha|\mathbf{I}^n \simeq 0$ rel. $\partial\mathbf{I}^n$ in X. Since $i_\sharp : \pi_n(A, x_0) \to \pi_n(X, x_0)$ is a monomorphism, it follows that $\alpha|\mathbf{I}^n \simeq 0$ rel. $\partial\mathbf{I}^n$ in A. By the Homotopy Extension Theorem 4.3.3, we have a homotopy $h : \mathbf{I}^{n+1} \times \mathbf{I} \to X$ such that $h_0 = \alpha$, $h(\mathbf{I}^n \times \mathbf{I}) \subset A$, and $h(\mathbf{J}^n \times \mathbf{I}) = \{x_0\}$. Then, $h_1 \in C((\mathbf{I}^{n+1}, \partial\mathbf{I}^{n+1}), (X, x_0))$. Since $i_\sharp : \pi_{n+1}(A, x_0) \to \pi_{n+1}(X, x_0)$ is an epimorphism, we have $\alpha' \in C((\mathbf{I}^{n+1}, \partial\mathbf{I}^{n+1}), (A, x_0))$ such that $\alpha' \simeq h_1$ rel. $\partial\mathbf{I}^{n+1}$ in X. Now, we define a homotopy $k : \mathbf{I}^{n+1} \times \mathbf{I} \to X$ as follows:

$$k(z, t) = \begin{cases} \alpha'(z(1), \ldots, z(n), z(n+1) + t) & \text{if } z(n+1) + t \leq 1, \\ x_0 & \text{otherwise.} \end{cases}$$

Then, $k_0 = \alpha'$, $k_1 = c_{x_0}$, $k(\mathbf{I}^n \times \mathbf{I}) = \alpha'(\mathbf{I}^{n+1}) \subset A$, and $k(\mathbf{J}^n \times \mathbf{I}) = \{x_0\}$. Thus, we have $[\alpha] = [h_1] = [\alpha'] = 0 \in \pi_{n+1}(X, A, x_0)$. □

Now, in the framework of the homotopy groups, we will give a characterization of n-equivalences. The condition $(\pi)_i$ can be divided into two conditions as follows:

Lemma 4.14.11. *Let $i \in \mathbb{N}$. For a map $f : X \to Y$, condition $(\pi)_i$ is equivalent to the combination of the following two conditions:*

$(\pi_{i-1}^{\text{mono}})$ *$f_\sharp : \pi_{i-1}(X, x_0) \to \pi_{i-1}(Y, f(x_0))$ is a monomorphism for every $x_0 \in X$;*
(π_i^{epi}) *$f_\sharp : \pi_i(X, x_0) \to \pi_i(Y, f(x_0))$ is an epimorphism for every $x_0 \in X$.*

Proof. $(\pi)_i \Rightarrow (\pi_{i-1}^{\text{mono}}) + (\pi_i^{\text{epi}})$: The case $i = 1$ is Proposition 4.13.2. When $i > 1$, since $(\pi)_i$ implies (d) of Proposition 4.14.8, we have $(\pi_{i-1}^{\text{mono}})$.

To see (π_i^{epi}), let $x_0 \in X$ and $\beta : \mathbf{S}^i \to Y$ be a map with $\beta(\mathbf{e}_1) = f(x_0)$. Identifying $(\mathbf{B}^i/\partial\mathbf{B}^i, \partial\mathbf{B}^i/\partial\mathbf{B}^i) = (\mathbf{S}^i, \mathbf{e}_1)$, let $q : (\mathbf{B}^i, \mathbf{S}^{i-1}) \to (\mathbf{S}^i, \mathbf{e}_1)$ be the quotient map. We can apply $(\pi)_i$ to obtain a map $\bar{\alpha} : \mathbf{B}^i \to X$ such that $\bar{\alpha}(\mathbf{S}^{i-1}) = \{x_0\}$ and $f\bar{\alpha} \simeq \beta q$ rel. $\mathbf{S}^{i-1}$ in Y. Since $\bar{\alpha}(\mathbf{S}^{i-1}) = \{x_0\}$ is a singleton, the map $\bar{\alpha}$ induces a map $\tilde{\beta} : \mathbf{S}^i \to X$ such that $\bar{\alpha} = \tilde{\beta} q$. Then, $\tilde{\beta}(\mathbf{e}_1) = x_0$ and $f\tilde{\beta} \simeq \beta$ rel. $\mathbf{e}_1$ in Y because $f\tilde{\beta}q = f\bar{\alpha} \simeq \beta q$ rel. $\mathbf{S}^{i-1}$.

$(\pi_{i-1}^{\text{mono}}) + (\pi_i^{\text{epi}}) \Rightarrow (\pi)_i$: Let $\alpha : \mathbf{S}^{i-1} \to X$ and $\beta : \mathbf{B}^i \to Y$ be maps such that $f\alpha = \beta|\mathbf{S}^{i-1}$. Due to Proposition 4.14.8, $(\pi_{i-1}^{\text{mono}})$ implies that α extends to a map $\alpha' : \mathbf{B}^i \to X$. Let $\gamma : \mathbf{S}^i \to Y$ be the map defined by $\gamma|\mathbf{S}^i_+ = f\alpha'\mathrm{pr}_{\mathbb{R}^i}|\mathbf{S}^i_+$ and $\gamma|\mathbf{S}^i_- = \beta\mathrm{pr}_{\mathbb{R}^i}|\mathbf{S}^i_-$, where

$$\mathbf{S}^i_+ = \mathbf{S}^i \cap (\mathbb{R}^i \times [0, \infty)) \text{ and } \mathbf{S}^i_- = \mathbf{S}^i \cap (\mathbb{R}^i \times (-\infty, 0]).$$

Replacing $\mathbf{e}_1$ with $\mathbf{e}_{i+1} = (0, \ldots, 0, 1) \in \mathbb{R}^{i+1}$, we can apply (π_i^{epi}) to obtain a map $\gamma' : \mathbf{S}^i \to X$ such that $\gamma'(\mathbf{e}_{i+1}) = \alpha'(0)$ and $f\gamma' \simeq \gamma$ rel. $\mathbf{e}_{i+1}$. Similarly, let $q : \mathbf{B}^i \to \mathbf{S}^i$ be the quotient map with $q(\mathbf{S}^{i-1}) = \{\mathbf{e}_{i+1}\}$, where we can assume that

$\mathrm{pr}_{\mathbb{R}^i} q(x) = 2x$ if $\|x\| \leq 1/2$ and $\mathrm{pr}_{\mathbb{R}^i} q(x) = (2 - 2\|x\|)\|x\|^{-1}x$ if $\|x\| \geq 1/2$. Then, we can define a map $\bar{\alpha} : \mathbf{B}^i \to X$ as follows:

$$\bar{\alpha}(x) = \begin{cases} \gamma' q(2x) & \text{if } \|x\| \leq 1/2, \\ \alpha'((2\|x\| - 1)\|x\|^{-1}x) & \text{if } \|x\| \geq 1/2. \end{cases}$$

Since $f\gamma' \simeq \gamma$ rel. $\mathbf{e}_{i+1}$, it follows that $f\bar{\alpha} \simeq \beta'$ rel. $\mathbf{S}^{i-1}$, where $\beta' : \mathbf{B}^i \to Y$ is defined as follows:

$$\beta'(x) = \begin{cases} \gamma q(2x) & \text{if } \|x\| \leq 1/2, \\ f\alpha'((2\|x\| - 1)\|x\|^{-1}x) & \text{if } \|x\| \geq 1/2. \end{cases}$$

Observe that

$$\gamma q(2x) = \beta \mathrm{pr}_{\mathbb{R}^i} q(2x) = \beta(4x) \text{ if } \|x\| \leq 1/4, \text{ and}$$

$$\gamma q(2x) = f\alpha' \mathrm{pr}_{\mathbb{R}^i} q(2x) = f\alpha'((2 - 4\|x\|)\|x\|^{-1}x) \text{ if } 1/4 \leq \|x\| \leq 1/2.$$

Now, it is easy to see that $\beta' \simeq \beta''$ rel. $\mathbf{S}^{i-1}$, where $\beta'' : \mathbf{B}^i \to Y$ is defined by

$$\beta''(x) = \begin{cases} \beta(4x) & \text{if } \|x\| \leq 1/4, \\ f\alpha'(\|x\|^{-1}x) = f\alpha(\|x\|^{-1}x) & \text{if } \|x\| \geq 1/4. \end{cases}$$

Then, it follows that $\beta'' \simeq \beta$ rel. $\mathbf{S}^{i-1}$. Thus, we have $f\bar{\alpha} \simeq \beta$ rel. $\mathbf{S}^{i-1}$. □

The above Lemma 4.14.11 yields the following characterization of n-equivalences:

Theorem 4.14.12. *Let $n \in \mathbb{N} \cup \{\infty\}$. A map $f : X \to Y$ is an n-equivalence if and only if f induces the bijection $f_\sharp : \pi_0(X) \to \pi_0(Y)$ and, for every $x_0 \in X$, $f_\sharp : \pi_i(X, x_0) \to \pi_i(Y, f(x_0))$ is an isomorphism for every $i < n$ and $f_\sharp : \pi_n(X, x_0) \to \pi_n(Y, f(x_0))$ is an epimorphism (if $n < \infty$).*[21] □

Since every homotopy equivalence $f : X \to Y$ induces the isomorphisms $f_\sharp : \pi_n(X, x_0) \to \pi_n(Y, f(x_0))$, $n \in \mathbb{N}$, $x_0 \in X$, it is a weak homotopy equivalence by Theorem 4.14.12. Thus, Proposition 4.13.1 is a corollary of Theorem 4.14.12.

In the case that f is the inclusion, by combining Theorem 4.14.12 and Proposition 4.14.10, we have the following characterization:

Corollary 4.14.13. *For each $n \in \mathbb{N}$, the inclusion $X \subset Y$ is an n-equivalence if and only if each path-component of Y contains exactly one path-component of X, $\pi_i(Y, X, x_0) = \{0\}$ for each $i \leq n$ and $x_0 \in X$.* □

---

[21] This is the definition of an n-equivalence in Homotopy Theory. However, the literature is not consistent on the use of the term "n-equivalence" (some texts require that $f_\sharp : \pi_n(X, x_0) \to \pi_n(Y, f(x_0))$ is an isomorphism).

For a 1-equivalence $f : X \to Y$, conditions (π_i^{mono}) and (π_i^{epi}) can be modified to the conditions without base points in the next two propositions.

Proposition 4.14.14. *For a* 1*-equivalence* $f : X \to Y$*, condition* (π_i^{mono}) *is equivalent to the following condition:*

(π_i^{mono})′ f *induces the injection from* $[\mathbf{S}^i, X]$ *to* $[\mathbf{S}^i, Y]$*, that is, any two maps* $\alpha, \alpha' : \mathbf{S}^i \to X$ *are homotopic if* $f\alpha \simeq f\alpha'$ *in* Y.

Proof. When α' is a constant map, (π_i^{mono})′ is equal to condition (c) of Proposition 4.14.8, so we have the implication (π_i^{mono})′ $\Rightarrow$ (π_i^{mono}).

(π_i^{mono}) $\Rightarrow$ (π_i^{mono})′: For maps $\alpha, \alpha' : \mathbf{S}^i \to X$, assume that $f\alpha \simeq f\alpha'$ in Y. Let $h : \mathbf{S}^i \times \mathbf{I} \to Y$ be a homotopy from $f\alpha'$ to $f\alpha$. Since f is a 1-equivalence, there is a path $\gamma : \mathbf{I} \to X$ from $\alpha'(\mathbf{e}_1)$ to $\alpha(\mathbf{e}_1)$ such that $f\gamma \simeq hj$ rel. $\partial\mathbf{I}$, where $j : \mathbf{I} \to \{\mathbf{e}_1\} \times \mathbf{I} \subset \mathbf{S}^i \times \mathbf{I}$ is the natural injection. We have a map $r : \mathbf{S}^i \to \mathbf{S}^i$ such that $r \simeq \text{id}$ rel. $\mathbf{e}_1$ and $r(\mathbf{S}^i \cap \text{pr}_1^{-1}(\mathbb{R}_+)) = \{\mathbf{e}_1\}$, where $\text{pr}_1 : \mathbb{R}^{i+1} \to \mathbb{R}$ is the projection onto the first factor. We define maps $\alpha'' : \mathbf{S}^i \to X$ and $\beta : \mathbf{S}^i \to Y$ as follows:

$$\alpha''(x) = \begin{cases} \alpha' r(x) & \text{if } x(1) \le 0, \\ \gamma(x(1)) & \text{if } t \ge 1, \end{cases} \quad \text{and } \beta(x) = \begin{cases} f\alpha' r(x) & \text{if } x(1) \le 0, \\ hj(x(1)) & \text{if } t \ge 1. \end{cases}$$

Then, $\alpha'' \simeq \alpha' r \simeq \alpha'$ and $\alpha''(\mathbf{e}_1) = \alpha(\mathbf{e}_1)$. Since $f\gamma \simeq hj$ rel. $\partial\mathbf{I}$, it follows that $f\alpha'' \simeq \beta$ rel. $\mathbf{e}_1$. On the other hand, we have a map $\varphi : \mathbf{S}^i \times \mathbf{I} \to \mathbf{S}^i \times \mathbf{I}$ such that $\varphi \simeq \text{id}$ rel. $\mathbf{S}^i \times \{1\}$, $\varphi(\{\mathbf{e}_1\} \times \mathbf{I}) = \{(\mathbf{e}_1, 1)\}$,

$$\varphi(x, 0) = \begin{cases} j(x(1)) & \text{if } x(1) \le 0, \\ \varphi(x, 0) = r(x) & \text{if } x(1) \le 0. \end{cases}$$

Then, $h\varphi$ is a homotopy from β to $f\alpha$ with $h\varphi(\{\mathbf{e}_1\} \times \mathbf{I}) = \{\beta(\mathbf{e}_1)\}$, which means that $\beta \simeq f\alpha$ rel. $\mathbf{e}_1$. Hence, $f\alpha'' \simeq f\alpha$ rel. $\mathbf{e}_1$. It follows from (π_i^{mono}) that $\alpha'' \simeq \alpha$ rel. $\mathbf{e}_1$. Since $\alpha'' \simeq \alpha'$, we have $\alpha \simeq \alpha'$ in X. □

Proposition 4.14.15. *For a* 1*-equivalence* $f : X \to Y$*, condition* (π_i^{epi}) *is equivalent to the following:*

(π_i^{epi})′ f *induces the surjection from* $[\mathbf{S}^i, X]$ *to* $[\mathbf{S}^i, Y]$*, that is, for each map* $\beta : \mathbf{S}^i \to Y$*, there is a map* $\alpha : \mathbf{S}^i \to X$ *such that* $f\alpha \simeq \beta$ *in* Y.

Proof. (π_i^{epi}) $\Rightarrow$ (π_i^{epi})′: Let $\beta : \mathbf{S}^i \to Y$ be a map. According to Proposition 4.13.2(1), there exists a path $\gamma : \mathbf{I} \to Y$ with $x_0 \in X$ such that $\gamma(0) = \alpha(\mathbf{e}_1)$ and $\gamma(1) = f(x_0)$. Using the map $r : \mathbf{S}^i \to \mathbf{S}^i$ from the proof of (π_i^{mono}) $\Rightarrow$ (π_i^{mono})′ in Proposition 4.14.14, we define a map $\beta' : \mathbf{S}^i \to Y$ by

$$\beta'(x) = \begin{cases} \beta r(x) & \text{if } x(1) \le 0, \\ \gamma(x(1)) & \text{if } x(1) \ge 0. \end{cases}$$

Then, $\beta' \simeq \beta r \simeq \beta$ and $\beta'(\mathbf{e}_1) = f(x_0)$. By ($\pi_i^{\text{epi}}$), we have a map $\alpha : \mathbf{S}^i \to X$ such that $f\alpha \simeq \beta'$ rel. $\mathbf{e}_1$, hence $f\alpha \simeq \beta$.

$(\pi_i^{\text{epi}})' \Rightarrow (\pi_i^{\text{epi}})$: Let $x_0 \in X$ and $\beta : \mathbf{S}^i \to Y$ be a map with $\beta(\mathbf{e}_1) = f(x_0)$. By $(\pi_i^{\text{epi}})'$, we have a map $\alpha' : \mathbf{S}^i \to X$ such that $f\alpha' \simeq \beta$ in Y. Let $h : \mathbf{S}^i \times \mathbf{I} \to Y$ be a homotopy from $f\alpha'$ to β. Since f is a 1-equivalence, we have a path $\gamma : \mathbf{I} \to X$ with $\gamma(0) = \alpha'(\mathbf{e}_1)$, $\gamma(1) = x_0$, and $f\gamma \simeq hj$ rel. $\partial\mathbf{I}$, where $j : \mathbf{I} \to \{\mathbf{e}_1\} \times \mathbf{I} \subset \mathbf{S}^i \times \mathbf{I}$ is the natural injection. Using the above map $r : \mathbf{S}^i \to \mathbf{S}^i$, we now define maps $\alpha : \mathbf{S}^i \to X$ and $\beta' : \mathbf{S}^i \to Y$ as follows:

$$\alpha(x) = \begin{cases} \alpha' r(x) & \text{if } x(1) \leq 0, \\ \gamma(x(1)) & \text{if } x(1) \geq 0, \end{cases} \quad \text{and} \quad \beta'(x) = \begin{cases} f\alpha' r(x) & \text{if } x(1) \leq 0, \\ hj(x(1)) & \text{if } x(1) \geq 0. \end{cases}$$

Since $f\gamma \simeq hj$ rel. $\partial\mathbf{I}$, it follows that $f\alpha \simeq \beta'$ rel. $\mathbf{e}_1$. Moreover, consider the same map $\varphi : \mathbf{S}^i \times \mathbf{I} \to \mathbf{S}^i \times \mathbf{I}$ used in the proof of $(\pi_i^{\text{mono}}) \Rightarrow (\pi_i^{\text{mono}})'$ of Proposition 4.14.14. Then, $h\varphi$ is a homotopy from β' to β with $h\varphi(\{\mathbf{e}_1\} \times \mathbf{I}) = \{\beta(\mathbf{e}_1)\}$, which means that $\beta' \simeq \beta$ rel. $\mathbf{e}_1$. Thus, we have $f\alpha \simeq \beta$ rel. $\mathbf{e}_1$. □

Theorem 4.14.12 can be reformulated using Propositions 4.14.14 and 4.14.15. In particular, we have the following corollary:

Corollary 4.14.16. *A map $f : X \to Y$ is a weak homotopy equivalence if and only if f is a 1-equivalence and f induces the bijection between $[\mathbf{S}^n, X]$ and $[\mathbf{S}^n, Y]$ for every $n \geq 1$.* □

Notes for Chap. 4

There are no good textbooks for studying non-locally finite simplicial complexes or infinite-dimensional simplicial complexes. For the study of Piecewise Linear (PL) Topology, we recommend

- C.P. Rourke and B.J. Sanderson, *Introduction to Piecewise-Linear Topology*, Springer Study Edition, (Springer-Verlag, Berlin, 1982)

The following classical lecture notes are still excellent resources for PL Topology:

- J.F.P. Hudson, *Piecewise Linear Topology* (W.A. Benjamin, Inc., New York, 1969)
- J. Stallings, *Lectures on Polyhedral Topology* (Tata Institute, Bombay, 1967)
- E.C. Zeeman, *Seminar on Combinatorial Topology* (Institute des Hautes Etude Sci., Paris, 1963)

Stellar subdivisions are discussed in Volume I of

- L.C. Glaser, *Geometrical Combinatorial Topology*, Vol.I, Vol.II (Van Nostrand Reinhold Co., New York, 1970, 1972)

There are many good textbooks on homotopy groups. Here we list two of them, a classical one and a recent one:

- S.-T. Hu, *Homotopy Theory* (Academic Press, Inc., New York, 1959)
- A. Hatcher, *Algebraic Topology* (Cambridge Univ. Press, Cambridge, 2002)

The example of Proposition 4.3.2 was given by Dowker [4]. The German terminology "Hauptvermutung" was introduced by Kneser [10], but it was claimed by Poincaré [17, 18] and formulated in 1908 as a conjecture by Steinitz [23] and Tietze [24]. In 1961, Hauptvermutung for polyhedra was disproved by Milnor [14]. In 1969, Kirby and Siebenmann [9] demonstrated that Hauptvermutung does not hold for n-manifolds for $n \geq 5$. Nevertheless, it was discovered in [21] that for any two countable simplicial complexes K and L, $|K| \simeq |L|$ implies $K \times_s \mathrm{F}(\mathbb{N}) \cong L \times_s \mathrm{F}(\mathbb{N})$, where $\mathrm{F}(\mathbb{N})$ is the countable infinite full complex. The completions of the metrics ρ_K and ρ_K^∞ on $|K|$ are discussed in [20]. The proof of Proposition 4.6.5 presented in this book is given by A. Yamashita.

Theorem 4.7.11 was established by Whitehead [25]. A characterization of admissible subdivisions in Lemma 4.8.1 can be found in [7] and the one in Theorem 4.8.4 in [15]. Theorem 4.8.8 was established by Henderson [7]. However, his proof is valid only for locally finite-dimensional complexes, and the complete proof presented here was given in [22].

The nerves of open covers were introduced by Alexandroff [1, 2] and canonical maps by Kuratowski [11]. Theorems 4.9.6 and 4.9.9 were given by Dowker [3, 4]. In Theorem 4.9.6, it can be asserted that $\mathrm{id} : |K| \to |K|_{\mathrm{m}}$ is a **fine homotopy equivalence**, which will be defined in Sect. 6.7. This can be obtained by combining Theorems 4.9.6 and 4.8.8. For an alternative proof, refer to [19].

The compact case of Theorem 4.10.10 was established by Freudenthal [6]. In fact, he proved that every compactum X is homeomorphic to the inverse limit of an inverse sequence of compact polyhedra of $\dim \leq \dim X$ (cf. Corollary 5.2.6). In [8, 7.2], Theorem 4.10.10 was proved under the assumption that every open cover of X has an open refinement whose nerve is finite-dimensional. In fact, this is proved in a more general setting (for a complete uniform space). It follows from [16, Theorem 3.2] that every paracompact space is homeomorphic to the inverse limit of an inverse system of polyhedra (with the Whitehead topology), but this does not imply that every metrizable space is homeomorphic to the inverse limit of an inverse sequence of polyhedra with the metric topology.

The countable case of Theorem 4.12.3 was proved in the proof of [26, Theorem 13]. The simplicial mapping cylinder defined in Sect. 4.12 is different from the mapping cylinder of a simplicial map in [25]. Note that our collapsing is simplicial. Theorem 4.12.1 was first established by Whitehead [27] in the separable case and extended by Milnor [13] to the general case. The mapping cylinder technique used in the proof of Theorem 4.12.1 was essentially invented by Mather [12] (cf. [5]).

References

1. P. Alexandroff, Simpliziale Approximationen in der allgemeinen Topologie. Math. Ann. **96**, 489–511 (1927)
2. P. Alexandroff, Une définition des nombre de Betti pour un ensemble fermé quelconque. C.R. Acad. Sci. Paris **184**, 317–319 (1927)
3. C.H. Dowker, Mapping theorems for non-compact spaces. Am. J. Math. **69**, 200–242 (1947)
4. C.H. Dowker, Topology of metric complexes. Am. J. Math. **74**, 555–577 (1952)
5. S. Ferry, Homotopy, simple homotopy and compacta. Topology **19**, 101–110 (1980)
6. H. Freudenthal, Entwicklungen von Räumen und ihren Gruppen. Compositio Math. **4**, 145–234 (1937)
7. D.W. Henderson, Z-sets in ANR's. Trans. Am. Math. Soc. **213**, 205–216 (1975)
8. J.R. Isbell, Uniform neighborhood retracts. Pac. J. Math. **11**, 609–648 (1961)
9. R.C. Kirby, L.C. Siebenmann, On the triangulation of manifolds and the Hauptvermutung. Bull. Am. Math. Soc. **75**, 742–749 (1969)

10. H. Kneser, Die Topologie der Mannigfaltigkeiten. Jahresbericht DMV **34**, 1–14 (1926)
11. K. Kuratowski, Sur un theorème fondamental concernant le nerf d'un système d'ensambles. Fund. Math. **20**, 191–196 (1933)
12. M. Mather, Counting homotopy types of manifolds. Topology **4**, 93–94 (1965)
13. J. Milnor, On spaces having the homotopy type of a CW-complex. Trans. Am. Math. Soc. **90**, 272–280 (1959)
14. J. Milnor, Two complexes which are homeomorphic but combinatorially distinct. Ann. Math. **74**, 575–590 (1961)
15. K. Mine, K. Sakai, Subdivisions of simplicial complexes preserving the metric topology. Can. Math. Bull. **55**, 157–163 (2012)
16. K. Morita, On expansions of Tychonoff spaces into inverse systems of polyhedra. Sci. Rep. Tokyo Kyoiku Daigaku, Sect.A **13**, 66–74 (1975)
17. H. Poincaré, Analysis situs. J. École Polytech. **1**, 1–123 (1895); *Œuvres de Henri Poincaré*, vol. 6 (Gauthier-Villars, Paris, 1953), pp. 193–288. www.archive.org/details/uvresdehenripoin06poin
18. H. Poincaré, Complément à l'Analysis situs. Rend. Cir. Mat. Palermo **13**, 285–343 (1899); *Œuvres de Henri Poincaré*, vol. 6 (Gauthier-Villars, Paris, 1953), pp. 290–337. www.archive.org/details/uvresdehenripoin06poin
19. K. Sakai, Fine homotopy equivalences of simplicial complexes. Bull. Polish Acad. Sci. Math. **34**, 89–97 (1986)
20. K. Sakai, Completions of metric simplicial complexes by using ℓ_p-norms. Topology Proc. **11**, No. 1., 177–208 (1987)
21. K. Sakai, Combinatorial infinite-dimensional manifolds and $\mathbb{R}^\infty$-manifolds. Topology Appl. **26**, 43–64 (1987)
22. K. Sakai, Small subdivisions of simplicial complexes with the metric topology. J. Math. Soc. Jpn. **63**, 789–800 (2011)
23. E. Steinitz, Beitröge zur Analysis situs. Sitz.-Ber. Berlin Math. Ges. **7**, 29–49 (1908)
24. H. Tietze, Über die topologischen Invarianten mehrdimensionaler Mannigfaltigkeiten. Monat. Math. und Phys. **19**, 1–118 (1908)
25. J.H.C. Whitehead, Simplicial spaces, nuclei, and *m*-groups. Proc. Lond. Math. Soc. (2) **45**, 243–327 (1939)
26. J.H.C. Whitehead, Combinatorial homotopy, I. Bull. Am. Math. Soc. **55**, 213–245 (1949)
27. J.H.C. Whitehead, A certain exact sequence. Ann. Math. **52**, 51–110 (1950)

Chapter 5
Dimensions of Spaces

For an open cover $\mathcal{U}$ of a space X, $\operatorname{ord}\mathcal{U} = \sup\{\operatorname{card}\mathcal{U}[x] \mid x \in X\}$ is called the **order** of $\mathcal{U}$. Note that $\operatorname{ord}\mathcal{U} = \dim N(\mathcal{U}) + 1$, where $N(\mathcal{U})$ is the nerve of $\mathcal{U}$. The **(covering) dimension** of X is defined as follows: $\dim X \leq n$ if each *finite* open cover of X has a *finite* open refinement $\mathcal{U}$ with $\operatorname{ord}\mathcal{U} \leq n + 1$. and then, $\dim X = n$ if $\dim X \leq n$ and $\dim X \not\leq n$. By $\dim X = -1$, we mean that $X = \emptyset$. We say that X is ***n*-dimensional** if $\dim X = n$ and that X is **finite-dimensional (f.d.)** ($\dim X < \infty$) if $\dim X \leq n$ for some $n \in \omega$. Otherwise, X is said to be **infinite-dimensional (i.d.)** ($\dim X = \infty$). The dimension is a topological invariant (i.e., $\dim X = \dim Y$ if $X \approx Y$).

This chapter is devoted to lectures on Dimension Theory. Fundamental theorems are proved and some examples of infinite-dimensional spaces are given. In this context, we discuss the Brouwer Fixed Point Theorem and the characterization of the Cantor set. We also construct finite-dimensional universal spaces such as the Nöbeling spaces and the Menger compacta.

We will use the results in Chaps. 2 and 4. In particular, we will need the combinatorial techniques treated in Chap. 4. Also, the concept of the nerves of open covers is very important in Dimension Theory.

5.1 The Brouwer Fixed Point Theorem

It is said that a space X has the **fixed point property** if any map $f : X \to X$ has a fixed point, i.e., $f(x) = x$ for some $x \in X$. In this section, we prove the following Brouwer Fixed Point Theorem:

Theorem 5.1.1 (Brouwer Fixed Point Theorem). *For every $n \in \mathbb{N}$, the n-cube $\mathbf{I}^n$ has the fixed point property.*

To prove this theorem, we need two lemmas. Let K be a simplicial complex and K' a simplicial subdivision of K. A simplicial map $h : K' \to K$ is called a

K. Sakai, *Geometric Aspects of General Topology*, Springer Monographs in Mathematics, DOI 10.1007/978-4-431-54397-8_5, © Springer Japan 2013

Sperner map if for each $v \in K'^{(0)}$, $h(v)$ is a vertex of the carrier $c_K(v)^{(0)}$ of v in K, equivalently $v \in O_K(h(v))$. In other words, h is a simplicial approximation of $\mathrm{id}_{|K|}$. Indeed, for each $x \in |K'| = |K|$, $c_{K'}(x) \subset c_K(x)$. Since $c_K(v) \leq c_K(x)$ for every $v \in c_{K'}(x)^{(0)}$, it follows that $h(c_{K'}(x)^{(0)}) \subset c_K(x)^{(0)}$, hence $h(x) \in h(c_{K'}(x)) \leq c_K(x)$.

Lemma 5.1.2 (Sperner). *Let τ be an n-simplex, and K' a subdivision of $F(\tau)$, where $F(\tau)$ is the natural triangulation of τ. If $h : K' \to F(\tau)$ is a Sperner map, then the number of n-simplexes $\tau' \in K'$ such that $h(\tau') = \tau$ is odd; hence, there exists such an n-simplex $\tau' \in K'$.*

Proof. We prove the lemma by induction with respect to n. The case $n = 0$ is obvious. Assume the lemma has been established for any $(n-1)$-simplex. Let σ be an $(n-1)$-face of τ. Then, $h(\sigma) \subset \sigma$. The natural triangulation $F(\sigma)$ of σ is a subcomplex of $F(\tau)$. Let L' be the subdivision of $F(\sigma)$ induced by K'. As is easily observed, $h|\sigma : L' \to F(\sigma)$ is also a Sperner map. Let a be the number of $(n-1)$-simplexes $\sigma' \in L'$ such that $h(\sigma') = \sigma$. Then, a is odd by the inductive assumption. Let $\mathcal{S}$ be the set of all $(n-1)$-simplexes $\sigma' \in K'$ such that $h(\sigma') = \sigma$. For each n-simplex $\tau' \in K'$, let $b(\tau')$ denote the number of faces σ' of τ' that belong to $\mathcal{S}$, i.e., $h(\sigma') = \sigma$. Then, it follows that

$$b(\tau') = \begin{cases} 2 & \text{if } h(\tau') = \sigma; \\ 1 & \text{if } h(\tau') = \tau; \\ 0 & \text{otherwise.} \end{cases}$$

Let c be the number of n-simplexes $\tau' \in K'$ such that $h(\tau') = \tau$. Then,

$$\sum_{\tau' \in K' \setminus K'^{(n-1)}} b(\tau') - c \quad \text{is even.}$$

On the other hand, a is equal to the number of $(n-1)$-simplexes σ' of $\mathcal{S}$ such that $\sigma' \subset \sigma$. For each $\sigma' \in \mathcal{S}$, σ' is a common face of exactly two n-simplexes of K' if and only if $\sigma' \not\subset \sigma$. Hence,

$$\sum_{\tau' \in K' \setminus K'^{(n-1)}} b(\tau') - a \quad \text{is even.}$$

Therefore, $a - c$ is also even. Recall that a is odd. Thus, c is also odd. □

Lemma 5.1.3. *Let $\tau = \langle v_1, \ldots, v_{n+1} \rangle$ be an n-simplex and $F_1, \ldots, F_{n+1}$ be closed sets in τ. If $\langle v_{i(1)}, \ldots, v_{i(m)} \rangle \subset F_{i(1)} \cup \cdots \cup F_{i(m)}$ for each $1 \leq i(1) < \cdots < i(m) \leq n+1$, then $F_1 \cap \cdots \cap F_{n+1} \neq \emptyset$.*

Proof. Assume that $F_1 \cap \cdots \cap F_{n+1} = \emptyset$. Then,

$$\mathcal{U} = \{\tau \setminus F_1, \ldots, \tau \setminus F_{n+1}\} \in \operatorname{cov}(\tau).$$

Let K' be a subdivision of $F(\tau)$ that refines $\mathcal{U}$. For each $v \in K'^{(0)}$, choose a vertex v_i of the carrier of v in $F(\tau)$ so that $v \in F_i$, and let $h(v) = v_i$. Then, we have a Sperner map $h : K' \to F(\tau)$. By Lemma 5.1.2, there is a simplex $\tau' \in K'$ such that $h(\tau') = \tau$. Write $\tau' = \langle v'_1, \ldots, v'_{n+1} \rangle$ so that $h(v'_i) = v_i$. By the definition of h, $v'_i \in F_i$ for each $i = 1, \ldots, n+1$. Thus, τ' is not contained in any $\tau \setminus F_i$, which is a contradiction. □

Proof of Theorem 5.1.1. It suffices to show that any map $f : \Delta^n \to \Delta^n$ has a fixed point, where $\Delta^n \subset \mathbb{R}^{n+1}$ is the standard n-simplex. For each $i = 1, \ldots, n+1$, let

$$F_i = \{x \in \Delta^n \mid \operatorname{pr}_i(f(x)) \leq \operatorname{pr}_i(x)\},$$

where $\operatorname{pr}_i : \mathbb{R}^{n+1} \to \mathbb{R}$ is the projection onto the i-th factor. Then, F_i is closed in Δ^n. Moreover, each face $\sigma = \langle \mathbf{e}_{i(1)}, \ldots, \mathbf{e}_{i(m)} \rangle \leq \Delta^n$ is contained in $F_{i(1)} \cup \cdots \cup F_{i(m)}$, where $\{\mathbf{e}_1, \ldots, \mathbf{e}_{n+1}\}$ is the canonical orthonormal basis for $\mathbb{R}^{n+1}$. In fact, if $x \in \sigma$ then

$$\sum_{j=1}^{m} \operatorname{pr}_{i(j)}(f(x)) \leq 1 = \sum_{j=1}^{m} \operatorname{pr}_{i(j)}(x),$$

which implies that $\operatorname{pr}_{i(j)}(f(x)) \leq \operatorname{pr}_{i(j)}(x)$ for some $j = 1, \ldots, m$. By Lemma 5.1.3, we have a point $a \in F_1 \cap \cdots \cap F_{n+1}$. Since $0 \leq \operatorname{pr}_i(f(a)) \leq \operatorname{pr}_i(a)$ for each $i = 1, \ldots, n+1$ and

$$\sum_{i=1}^{n+1} \operatorname{pr}_i(f(a)) = 1 = \sum_{i=1}^{n+1} \operatorname{pr}_i(a),$$

it follows that $\operatorname{pr}_i(f(a)) = \operatorname{pr}_i(a)$ for each $i = 1, \ldots, n+1$, which means that $f(a) = a$. □

The following is the infinite-dimensional version of Theorem 5.1.1:

Corollary 5.1.4. *The Hilbert cube* $\mathbf{I}^{\mathbb{N}}$ *has the fixed point property.*

Proof. For each $n \in \mathbb{N}$, let $p_n : \mathbf{I}^{\mathbb{N}} \to \mathbf{I}^n$ be the projection onto the first n factors and $i_n : \mathbf{I}^n \to \mathbf{I}^{\mathbb{N}}$ the natural injection defined by

$$i_n(x) = (x(1), \ldots, x(n), 0, 0, \ldots).$$

For each map $f : \mathbf{I}^{\mathbb{N}} \to \mathbf{I}^{\mathbb{N}}$, consider the map $f_n = p_n f i_n : \mathbf{I}^n \to \mathbf{I}^n$.

$$\begin{array}{ccc} \mathbf{I}^{\mathbb{N}} & \xrightarrow{f} & \mathbf{I}^{\mathbb{N}} \\ {\scriptstyle p_n}\big\downarrow\big\uparrow{\scriptstyle i_n} & & \big\downarrow{\scriptstyle p_n} \\ \mathbf{I}^n & \xrightarrow[f_n]{} & \mathbf{I}^n \end{array}$$

By the Brouwer Fixed Point Theorem 5.1.1, f_n has a fixed point. We define

$$K_n = \{x \in \mathbf{I}^{\mathbb{N}} \mid p_n f(x) = p_n(x)\},$$

which is closed in $\mathbf{I}^{\mathbb{N}}$ and $K_n \supset K_{n+1}$ for each $n \in \mathbb{N}$. Moreover, $K_n \neq \emptyset$. Indeed, if $y \in \mathbf{I}^n$ is a fixed point of f_n, then $p_n f(i_n(y)) = f_n(y) = y = p_n(i_n(y))$, i.e., $i_n(y) \in K_n$. By compactness, we have $a \in \bigcap_{n \in \mathbb{N}} K_n$. Since $p_n f(a) = p_n(a)$ for every $n \in \mathbb{N}$, we have $f(a) = a$. □

As another corollary of the Brouwer Fixed Point Theorem 5.1.1, we have the following:

Corollary 5.1.5 (No Retraction Theorem). *There does not exist any map* $r : \mathbf{B}^n \to \mathbf{S}^{n-1}$ *with* $r|\mathbf{S}^{n-1} = \mathrm{id}$.[1]

Proof. Suppose that there is a map $r : \mathbf{B}^n \to \mathbf{S}^{n-1}$ with $r|\mathbf{S}^{n-1} = \mathrm{id}$. We define a map $f : \mathbf{B}^n \to \mathbf{B}^n$ by $f(x) = -r(x)$. Then, f has no fixed points, which contradicts the Brouwer Fixed Point Theorem 5.1.1. □

Remark 1. It should be noted that the Brouwer Fixed Point Theorem 5.1.1 can be derived from the No Retraction Theorem 5.1.5. Indeed, if there is a map $f : \mathbf{B}^n \to \mathbf{B}^n$ without fixed points, then we have a map $r : \mathbf{B}^n \to \mathbf{S}^{n-1}$ such that $x \in \langle f(x), r(x) \rangle$ for each $x \in \mathbf{B}^n$, which implies that $r|\mathbf{S}^{n-1} = \mathrm{id}$. In fact, such a map r can be defined as follows:

$$r(x) = (1 + \alpha(x))x - \alpha(x) f(x),$$

where $\alpha(x) \geq 0$ can be obtained by solving the equation

$$\alpha(x)^2 \|x - f(x)\|^2 + 2\alpha(x)\langle x - f(x), x\rangle + \|x\|^2 - 1 = 0,$$

where $\langle y, z\rangle = \sum_{i=1}^n y(i)z(i)$ is the inner product (Fig. 5.1). Therefore, the No Retraction Theorem 5.1.5 implies that $\mathbf{I}^n \approx \mathbf{B}^n$ has the fixed point property. Thus, the Brouwer Fixed Point Theorem 5.1.1 and the No Retraction Theorem 5.1.5 are equivalent.

[1] Such a map r is called a **retraction**, which will be discussed in Chap. 6.

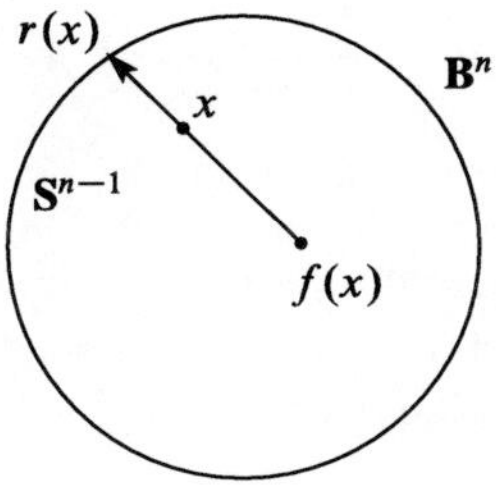

Fig. 5.1 The construction of r

Note. In Algebraic Topology, the homotopy groups or the homology groups are used to prove the No Retraction Theorem 5.1.5, and then the Brouwer Fixed Point Theorem 5.1.1 is proved as the above Remark 1.

Using the Tietze Extension Theorem 2.2.2, we have the following extension theorem:

Theorem 5.1.6. *Let A be a closed set in a normal space X and $n \in \mathbb{N}$.*

(1) *Every map $f : A \to \mathbf{B}^n$ extends over X.*
(2) *Every map $f : A \to \mathbf{S}^{n-1}$ extends over a neighborhood of A in X.*

Proof. By the coordinate-wise application of the Tietze Extension Theorem 2.2.2, each map $f : A \to \mathbf{I}^n$ can be extended over X, which implies (1) because $\mathbf{B}^n \approx \mathbf{I}^n$.

To prove (2), let $f : A \to \mathbf{S}^{n-1}$ be a map. By (1), f extends to a map $\tilde{f} : X \to \mathbf{B}^n$. Then, $W = \tilde{f}^{-1}(\mathbf{B}^n \setminus \{0\})$ is an open neighborhood of A in X. Let $r : \mathbf{B}^n \setminus \{0\} \to \mathbf{S}^{n-1}$ be the radial projection, i.e., $r(x) = x/\|x\|$. Then, $r\tilde{f}|W : W \to \mathbf{S}^{n-1}$ is an extension of f. □

Using the No Retraction Theorem 5.1.5 and Theorem 5.1.6, we can obtain the following characterization of boundary points of a closed set X in Euclidean space $\mathbb{R}^n$:

Theorem 5.1.7. *Let X be a closed subset of Euclidean space $\mathbb{R}^n$. For a point $x \in X$, $x \in \operatorname{bd} X$ if and only if each neighborhood U of x in X contains a neighborhood V of x in X such that every continuous map $f : X \setminus V \to \mathbf{S}^{n-1}$ extends to a continuous map $\tilde{f} : X \to \mathbf{S}^{n-1}$.*

Proof. To show the "only if" part, for each neighborhood U of x in X, choose $\varepsilon > 0$ so that $\overline{\mathrm{B}}(x, \varepsilon) \cap X \subset U$. Then, $V = \mathrm{B}(x, \varepsilon) \cap X$ is the desired neighborhood of x in X. Indeed, every map $f : X \setminus V \to \mathbf{S}^{n-1}$ can be extended to a map $g : X \to \mathbf{B}^n$ by Theorem 5.1.6. Choose $0 < \delta < \varepsilon$ so that $g(X \setminus \mathrm{B}(x, \delta)) \subset \mathbf{B}^n \setminus \{0\}$. Let $r : \mathbf{B}^n \setminus \{0\} \to \mathbf{S}^{n-1}$ be the canonical radial retraction (i.e., $r(y) = \|y\|^{-1}y$). Because $x \in \operatorname{bd} X$, we have $z \in \mathrm{B}(x, \frac{1}{2}(\varepsilon - \delta)) \setminus X$. Let $\lambda = \frac{1}{2}(\varepsilon + \delta) > 0$. Observe that $\mathrm{B}(x, \delta) \subset \mathrm{B}(z, \lambda) \subset \mathrm{B}(x, \varepsilon)$. We define a map $h : X \to X \setminus \mathrm{B}(z, \lambda) \subset X \setminus \mathrm{B}(x, \delta)$ by $h|X \setminus \mathrm{B}(z, \lambda) = \mathrm{id}$ and

$$h(y) = z + \frac{\lambda}{\|y - z\|}(y - z) \quad \text{for } y \in X \cap \mathrm{B}(z, \lambda).$$

Then, $rgh : X \to \mathbf{S}^{n-1}$ is a continuous extension of f.

To prove the "if" part, assume that $x \in \operatorname{int} X$. Then, $\overline{\mathrm{B}}(x, \delta) \subset X$ for some $\delta > 0$. By the condition, $\mathrm{B}(x, \delta)$ contains a neighborhood V of x such that every map $f : X \setminus V \to \mathbf{S}^{n-1}$ extends to a map $\tilde{f} : X \to \mathbf{S}^{n-1}$. It is easy to construct a retraction

$$r : \mathbb{R}^n \setminus \{x\} \to \operatorname{bd} \mathrm{B}(x, \delta) \approx \mathbf{S}^{n-1}.$$

Then, $r|X \setminus V$ extends to a retraction $\tilde{r} : X \to \operatorname{bd} \mathrm{B}(x, \delta)$. Since $\overline{\mathrm{B}}(x, \delta) \subset X$, $\operatorname{bd} \mathrm{B}(x, \delta)$ is a retract of $\overline{\mathrm{B}}(x, \delta)$, which contradicts the No Retraction Theorem 5.1.5 because $(\overline{\mathrm{B}}(x, \delta), \operatorname{bd} \mathrm{B}(x, \delta)) \approx (\mathbf{B}^n, \mathbf{S}^{n-1})$. Thus, we have $x \in \operatorname{bd} X$. □

As a corollary of Theorem 5.1.7, we have the so-called INVARIANCE OF DOMAIN:

Corollary 5.1.8 (INVARIANCE OF DOMAIN). *For each* $X, Y \subset \mathbb{R}^n$, $X \approx Y$ *implies* $\operatorname{int} X \approx \operatorname{int} Y$.

Proof. Let $h : X \to Y$ be a homeomorphism. For each $x \in \operatorname{bd} X$ and each neighborhood U of $h(x)$ in Y, $h^{-1}(U)$ is a neighborhood of x in X that contains a neighborhood V of x such that every map $f : X \setminus V \to \mathbf{S}^{n-1}$ extends to a map $\tilde{f} : X \to \mathbf{S}^{n-1}$. Then, $h(V)$ is a neighborhood of $h(x)$ in Y such that $h(V) \subset U$, and every continuous map $g : Y \setminus h(V) \to \mathbf{S}^{n-1}$ extends to a continuous map $\tilde{g} : Y \to \mathbf{S}^{n-1}$. Indeed, $gh : X \setminus V \to \mathbf{S}^{n-1}$ extends to a continuous map $\tilde{f} : X \to \mathbf{S}^{n-1}$. Then, $\tilde{f}h^{-1} : Y \to \mathbf{S}^{n-1}$ is a continuous extension of g. □

5.2 Characterizations of Dimension

Recall that we define $\dim X \leq n$ if each *finite* open cover of X has a *finite* open refinement $\mathcal{U}$ with $\operatorname{ord} \mathcal{U} \leq n + 1$. The following lemma shows that the refinement $\mathcal{U}$ in this definition need not be finite.

Lemma 5.2.1. *Let* $\mathcal{U}$ *be an open cover of a space* X *and* $\mathcal{V}$ *an open refinement of* $\mathcal{U}$. *Then,* $\mathcal{U}$ *has an open refinement* $\mathcal{W} = \{W_U \mid U \in \mathcal{U}\}$ *such that* $W_U \subset U$ *for each* $U \in \mathcal{U}$ *and* $\operatorname{card} \mathcal{W}[x] \leq \operatorname{card} \mathcal{V}[x]$ *for each* $x \in X$, *which implies that* $\operatorname{ord} \mathcal{W} \leq \operatorname{ord} \mathcal{V}$ *and if* $\mathcal{U}$ *is (locally) finite (or* σ*-discrete) then so is* $\mathcal{W}$.

Proof. Let $\varphi : \mathcal{V} \to \mathcal{U}$ be a function such that $V \subset \varphi(V)$ for each $V \in \mathcal{V}$. For each $U \in \mathcal{U}$, define

$$W_U = \bigcup \varphi^{-1}(U) = \bigcup \{V \in \mathcal{V} \mid \varphi(V) = U\}.$$

Then, $\mathcal{W} = \{W_U \mid U \in \mathcal{U}\}$ is the desired refinement of $\mathcal{U}$. □

The following is a particular case of the Open Cover Shrinking Lemma 2.7.1, which is easily proved directly.

Lemma 5.2.2. *Each finite open cover* $\{U_1, \dots, U_n\}$ *of a normal space* X *has an open refinement* $\{V_1, \dots, V_n\}$ *such that* $\operatorname{cl} V_i \subset U_i$ *for each* $i = 1, \dots, n$.

Proof. Using the normality of X, V_i can be inductively chosen so that

$$\operatorname{cl} V_i \subset U_i \quad \text{and} \quad V_1 \cup \dots \cup V_i \cup U_{i+1} \cup \dots \cup U_n = X. \quad \square$$

We now prove the following characterizations of dimension:

Theorem 5.2.3. *For* $n \in \omega$ *and a normal space* X, *the following are equivalent:*

(a) $\dim X \leq n$;
(b) *Every open cover* $\{U_1, \dots, U_{n+2}\}$ *of* X *has an open refinement* $\mathcal{V}$ *with* $\operatorname{ord} \mathcal{V} \leq n + 1$;
(c) *For each open cover* $\{U_1, \dots, U_{n+2}\}$ *of* X, *there exists an open cover* $\{V_1, \dots, V_{n+2}\}$ *of* X *such that* $V_1 \cap \dots \cap V_{n+2} = \emptyset$ *and* $\operatorname{cl} V_i \subset U_i$ *for each* $i = 1, \dots, n + 2$;
(d) *For every open cover* $\{U_1, \dots, U_{n+2}\}$ *of* X, *there exists a closed cover* $\{A_1, \dots, A_{n+2}\}$ *of* X *such that* $A_1 \cap \dots \cap A_{n+2} = \emptyset$ *and* $A_i \subset U_i$ *for each* $i = 1, \dots, n+2$;
(e) *For every* $k \geq n$, *each map* $f : A \to \mathbf{S}^k$ *of any closed set* A *in* X *extends over* X;
(f) *Each map* $f : A \to \mathbf{S}^n$ *of any closed set* A *in* X *extends over* X.

Proof. Consider the following diagram of implications:

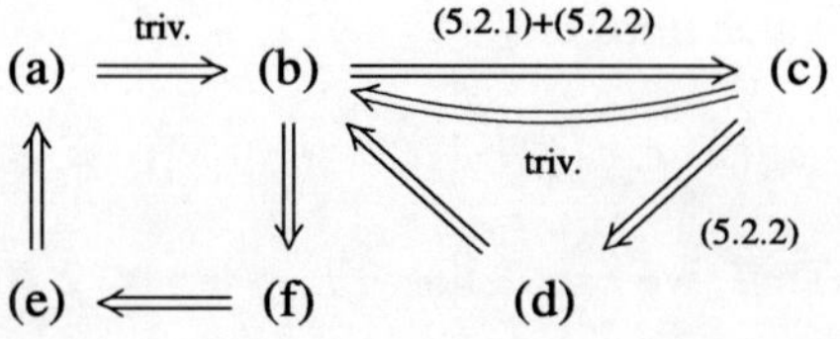

The implications (a) ⇒ (b) and (c) ⇒ (b) are obvious. By Lemmas 5.2.1 and 5.2.2, we have (b) ⇒ (c), hence (b) ⇔ (c). The implication (c) ⇒ (d) follows from Lemma 5.2.2 (or, (d) can be obtained by twice using (c)). Lastly, we prove the implications (d) ⇒ (b) ⇒ (f) ⇒ (e) ⇒ (a).

(d) ⇒ (b): In condition (d), note that

$$\{X \setminus A_1, \dots, X \setminus A_{n+2}\} \in \operatorname{cov}(X).$$

By Lemma 5.2.2, we have a closed cover $\{B_1, \dots, B_{n+2}\}$ of X such that $B_i \subset X \setminus A_i$ for each $i = 1, \dots, n + 2$. Observe

$$(X \setminus B_1) \cap \dots \cap (X \setminus B_{n+2}) = X \setminus (B_1 \cup \dots \cup B_{n+2}) = \emptyset.$$

For each $i = 1, \ldots, n+2$, let $V_i = U_i \setminus B_i \subset U_i$. Since $A_i \subset U_i \cap (X \setminus B_i) = V_i$, we have $\mathcal{V} = \{V_1, \ldots, V_{n+2}\} \in \operatorname{cov}(X)$. Moreover, $V_1 \cap \cdots \cap V_{n+2} = \emptyset$, which means $\operatorname{ord} \mathcal{V} \le n+1$.

(b) $\Rightarrow$ (f): Let Δ^{n+1} be the standard $(n+1)$-simplex and $K = F(\partial \Delta^{n+1})$ (i.e., the simplicial complex consisting of all proper faces of Δ^{n+1}). Then, $|K| = \partial \Delta^{n+1} \approx \mathbf{S}^n$. To extend a given map $f : A \to \mathbf{S}^n$ over X, we consider $\mathbf{S}^n = |K|$. By Theorem 5.1.6(2), $f : A \to |K|$ is extended to a map $\tilde{f} : \operatorname{cl} W \to |K|$, where W is an open neighborhood of A in X. Note that $\operatorname{card} K^{(0)} = n+2$. By (b), X has a finite open cover $\mathcal{V}$ such that $\operatorname{ord} \mathcal{V} \le n+1$ and

$$\mathcal{V} \prec \{\tilde{f}^{-1}(O_K(v)) \cup (X \setminus \operatorname{cl} W) \mid v \in K^{(0)}\}.$$

We have a function $\varphi : \mathcal{V} \to K^{(0)}$ such that

$$V \subset \tilde{f}^{-1}(O_K(\varphi(V))) \cup (X \setminus \operatorname{cl} W) \quad \text{for each } V \in \mathcal{V},$$

which defines a simplicial map $\varphi : N(\mathcal{V}) \to K$ because every $n+1$ many vertices span a simplex of K and each simplex of $N(\mathcal{V})$ has at most $n+1$ many vertices. Since $\mathcal{V}$ is finite, there is a canonical map $g : X \to |N(\mathcal{V})|$ for $N(\mathcal{V})$ by Theorem 4.9.4. For each $x \in W$, $\tilde{f}(x)$ and $\varphi g(x)$ are contained in the same simplex of K. In fact, let $\tau \in K$ be the carrier of $\tilde{f}(x)$, i.e., $\tilde{f}(x) \in \operatorname{rint} \tau$. Then, for each $V \in \mathcal{V}[x]$,

$$x \in V \cap W \subset \tilde{f}^{-1}(O_K(\varphi(V))),$$

hence $\tilde{f}(x) \in O_K(\varphi(V))$. Thus, we have $\tilde{f}(x) \in \bigcap_{V \in \mathcal{V}[x]} O_K(\varphi(V))$, which implies that $\varphi(V) \in \tau^{(0)}$ for each $V \in \mathcal{V}[x]$, i.e., $\langle \varphi(\mathcal{V}[x]) \rangle \le \tau$. On the other hand, $g(x) \in \langle \mathcal{V}[x] \rangle$, which implies

$$\varphi g(x) \in \varphi(\langle \mathcal{V}[x] \rangle) = \langle \varphi(\mathcal{V}[x]) \rangle \le \tau,$$

so $\varphi g(x), \tilde{f}(x) \in \tau$. Thus, we can define a map $h : X \times \{0\} \cup W \times \mathbf{I} \to |K|$ as follows:

$$h(x, 0) = \varphi g(x) \quad \text{for } x \in X \quad \text{and}$$

$$h(x, t) = (1-t)\varphi g(x) + t\tilde{f}(x) \quad \text{for } (x, t) \in W \times \mathbf{I}.$$

Let $k : X \to \mathbf{I}$ be an Urysohn map with $X \setminus W \subset k^{-1}(0)$ and $A \subset k^{-1}(1)$. Then, an extension $f^* : X \to |K|$ of f can be defined by $f^*(x) = h(x, 0)$ $(= \varphi g(x))$ for $x \in X \setminus W$ and $f^*(x) = h(x, k(x))$ for $x \in W$.

(f) $\Rightarrow$ (e): By induction on $k \ge n$, we show that each map $f : A \to \mathbf{S}^{k+1}$ of any closed set A in X extends over X. Let

$$\mathbf{S}^{k+1}_+ = \mathbf{S}^{k+1} \cap (\mathbb{R}^{k+1} \times \mathbb{R}_+) \text{ and } \mathbf{S}^{k+1}_- = -\mathbf{S}^{k+1}_+,$$

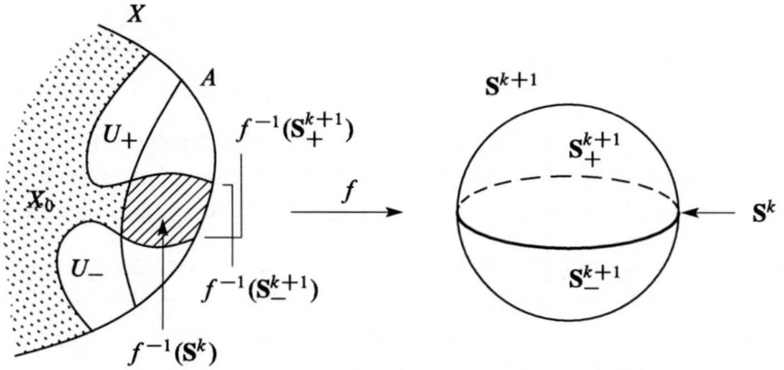

Fig. 5.2 Extending a map $f : A \to \mathbf{S}^{k+1}$

where we identify $\mathbf{S}^k = \mathbf{S}^k \times \{0\} = \mathbf{S}_+^{k+1} \cap \mathbf{S}_-^{k+1} \subset \mathbf{S}^{k+1}$. We have disjoint open sets U_+ and U_- in X such that

$$U_+ \cap A = A \setminus f^{-1}(\mathbf{S}_-^{k+1}) \text{ and } U_- \cap A = A \setminus f^{-1}(\mathbf{S}_+^{k+1}).$$

In fact, by Theorem 5.1.6(2), f extends to a map $f' : U \to \mathbf{S}^{k+1}$ of an open neighborhood of A in X. Then, $U_\pm = f'^{-1}(\mathbf{S}^{k+1} \setminus \mathbf{S}_\mp^{k+1})$ are the desired open sets.

Now, let $X_0 = X \setminus (U_+ \cup U_-)$ and $A_0 = A \cap X_0 = f^{-1}(\mathbf{S}^k)$ (Fig. 5.2). Since $f|A_0 : A_0 \to \mathbf{S}^k$ extends over X by the inductive assumption, $f|A_0$ extends to a map $f_0 : X_0 \to \mathbf{S}^k$. Let

$$X_+ = X_0 \cup U_+ = X \setminus U_- \text{ and } X_- = X_0 \cup U_- = X \setminus U_+,$$

which are closed in X, and hence they are normal. Note that X_0 is closed in both X_+ and X_-. Since $\mathbf{S}_+^{k+1} \approx \mathbf{S}_-^{k+1} \approx \mathbf{B}^{k+1}$, f_0 extends to maps $f_+ : X_+ \to \mathbf{S}_+^{k+1}$ and $f_- : X_- \to \mathbf{S}_-^{k+1}$ by Theorem 5.1.6(1). Then, the desired extension $\tilde{f} : X \to \mathbf{S}^{k+1}$ of f can be defined by $\tilde{f}|X_+ = f_+$ and $\tilde{f}|X_- = f_-$.

(e) $\Rightarrow$ (a): For each finite open cover $\mathcal{U}$ of X, let $K = N(\mathcal{U})$ be the nerve of $\mathcal{U}$ with $f : X \to |K|$ a canonical map (cf. Theorem 4.9.4). If $f(X) \subset |K^{(n)}|$, $f^{-1}(\mathcal{O}_{K^{(n)}}) \in \operatorname{cov}(X)$ is a finite open refinement of $\mathcal{U}$ and

$$\operatorname{ord} f^{-1}(\mathcal{O}_{K^{(n)}}) \leq \operatorname{ord} \mathcal{O}_{K^{(n)}} = \dim K^{(n)} + 1 \leq n + 1.$$

Otherwise, choose $m > n$ so that $f(X) \subset |K^{(m)}|$ but $f(X) \not\subset |K^{(m-1)}|$. Let $\tau_1, \ldots, \tau_k$ be the m-simplexes of K. Since $\partial\tau_i \approx \mathbf{S}^{m-1}$ and $m - 1 \geq n$, we have maps $f_i : X \to \partial\tau_i$ such that $f_i|f^{-1}(\partial\tau_i) = f|f^{-1}(\partial\tau_i)$ by (e). Let $f' : X \to |K|$ be the map defined by

$$f'|f^{-1}(|K^{(m-1)}|) = f|f^{-1}(|K^{(m-1)}|) \text{ and}$$
$$f'|f^{-1}(\tau_i) = f_i|f^{-1}(\tau_i) \text{ for each } i = 1,\dots,k.$$

Then, $f'(X)\subset|K^{(m-1)}|$. Since $c_K(f'(x))\leq c_K(f(x))\leq\langle\mathcal{U}[x]\rangle$ for each $x\in X$, f' is still a canonical map. By the downward induction on $m\geq n$, we can obtain a canonical map $f : X\to|K|$ such that $f(X)\subset|K^{(n)}|$. This completes the proof. □

Remark 2. In the above proof of (e) ⇒ (a), instead of a finite open cover $\mathcal{U}$ of X, let us take a local finite open cover $\mathcal{U}$ whose nerve $K = N(\mathcal{U})$ is *locally finite-dimensional* (l.f.d.). It can be shown that $\mathcal{U}$ has a locally finite open refinement $\mathcal{V}$ with $\operatorname{ord}\mathcal{V} \leq n+1$ (i.e., $\dim N(\mathcal{V}) \leq n$).

Indeed, since K is the nerve of a locally finite open cover, by Theorem 4.9.4, we have a canonical map $f : X \to |K|$ such that each $x \in X$ has a neighborhood V_x in X with $f(V_x) \subset |K_x|$ for some finite subcomplex K_x of K. Note that K might be infinite-dimensional.

Now, consider the following subcomplexes of K:

$$\begin{aligned} K_i &= K \setminus \{\tau \in K \mid \dim\tau > n,\ \tau \text{ is principal in } K_{i-1}\} \\ &= K^{(n)} \cup \{\tau \in K_{i-1} \mid \tau \text{ is not principal in } K_{i-1}\},\ i \in \mathbb{N}, \end{aligned}$$

where $K_0 = K$. Then, $K^{(n)} = \bigcap_{i\in\mathbb{N}} K_i$ because K is l.f.d. We will inductively construct canonical maps $f_i : X \to |K|$, $i \in \mathbb{N}$, such that

$$f_i|f_{i-1}^{-1}(|K_i|) = f_{i-1}|f_{i-1}^{-1}(|K_i|),\ f_i(X) \subset |K_i| \text{ and}$$
$$f_i(V_x) \subset |K_x| \text{ for each } x \in X,$$

where $f_0 = f$. Suppose f_{i-1} have been constructed. For each $\tau \in K_{i-1} \setminus K_i$, since $\dim\tau > n$, we can apply (e) to obtain an extension $f_\tau : X \to \partial\tau$ of $f_{i-1}|f_{i-1}^{-1}(\partial\tau)$. We can define $f_i : X \to |K|$ as follows:

$$f_i|f_{i-1}^{-1}(|K_i|) = f_{i-1}|f_{i-1}^{-1}(|K_i|) \text{ and}$$
$$f_i|f_{i-1}^{-1}(\tau) = f_\tau|f_{i-1}^{-1}(\tau) \text{ for each } \tau \in K_{i-1} \setminus K_i.$$

Then, $f_i(X) \subset |K_i|$. Since $f_i(f_{i-1}^{-1}(\tau)) \subset \partial\tau \subset \tau$ for each $\tau \in K_{i-1}\setminus K_i$, it follows that $f_i(V_x) \subset |K_x|$ for each $x \in X$, so f_i is continuous because each K_x is finite. Moreover, $c_K(f_i(x)) \leq c_K(f_{i-1}(x))$ for each $x \in X$, hence f_i is also a canonical map.

For each $x \in X$, since K_x is finite, $K_x^{(n)} = K_x \cap K_{i(x)}$ for some $i(x) \in \mathbb{N}$. For every $i \geq i(x)$, because $K_x \cap K_i = K_x \cap K_{i(x)}$, we have $f_i|V_x = f_{i(x)}|V_x$. Therefore, we can define a map $\tilde{f} : X \to |K^{(n)}|$ by $\tilde{f}|V_x = f_{i(x)}|V_x$ for each $x \in X$. Then, $\mathcal{V} = \tilde{f}^{-1}(\mathcal{O}_{K^{(n)}}) \in \operatorname{cov}(X)$ is an open refinement of $\mathcal{U}$ with ord $\leq n+1$. By applying Lemma 5.2.1, we can obtain the desired refinement $\mathcal{V}$ of $\mathcal{U}$.

When X is paracompact, since every open cover of X has a locally finite (and σ-discrete) open refinement with the l.f.d. nerve by Theorem 4.9.9, if $\dim X \leq n$, then an arbitrary open cover of X has a (locally finite σ-discrete) open refinement $\mathcal{V}$ with $\operatorname{ord} \mathcal{V} \leq n+1$ by the above remark. Since the converse obviously holds, we have the following characterization:

Theorem 5.2.4. *For $n \in \omega$ and a paracompact space X, $\dim X \leq n$ if and only if an arbitrary open cover of X has a (locally finite σ-discrete) open refinement $\mathcal{V}$ with $\operatorname{ord} \mathcal{V} \leq n+1$.* □

Instead of Theorem 4.9.9, we can use Theorem 4.9.10 to obtain the following corollary:

Corollary 5.2.5. *Let X be regular Lindelöf and $n \in \omega$. Then, $\dim X \leq n$ if and only if an arbitrary open cover of X has a (star-finite and countable) open refinement $\mathcal{V}$ with $\operatorname{ord} \mathcal{V} \leq n+1$.* □

In the proof of Theorem 4.10.10, we can apply Theorem 5.2.4 (Corollary 5.2.5) to obtain $\mathcal{U}_i$ with $\operatorname{ord} \mathcal{U}_i \leq n+1$, namely $\dim K_i \leq n$. By Remark 16 at the end of Sect. 4.10, we have the following version of Theorem 4.10.10 (Corollaries 4.10.11 and 4.10.12).

Corollary 5.2.6. *Every completely metrizable space X with $\dim X \leq n < \infty$ is homeomorphic to the inverse limit of an inverse sequence $(|K_i|_{\mathrm{m}}, f_i)_{i \in \mathbb{N}}$ of metric polyhedra and PL maps such that $\dim K_i \leq n$, $\operatorname{card} K_i \leq \aleph_0 w(X)$, and $f_i : K_{i+1} \to \operatorname{Sd} K_i$ is simplicial. Moreover, if X is compact metrizable (resp. separable and completely metrizable), then each $|K_i|_{\mathrm{m}} = |K_i|$ is compact (resp. locally compact). If X is separable and locally compact metrizable, each $|K_i|_{\mathrm{m}} = |K_i|$ is locally compact and each f_i is proper.* □

Now, we can prove the following theorem:

Theorem 5.2.7. *For each $n \in \mathbb{N}$, $\dim \mathbf{B}^n = n$.*

Proof. For any $\mathcal{U} \in \operatorname{cov}(\Delta^n)$, Δ^n has a triangulation K such that $\mathcal{O}_K \prec \mathcal{S}_K \prec \mathcal{U}$ (Corollary 4.7.7). Since $\operatorname{ord} \mathcal{O}_K = \dim K + 1 = n+1$ and $|K| = \Delta^n \approx \mathbf{B}^n$, it follows that $\dim \mathbf{B}^n \leq n$. If $\dim \mathbf{B}^n \leq n-1$, then we apply Theorem 5.2.3 to obtain a map $r : \mathbf{B}^n \to \mathbf{S}^{n-1}$ such that $r|\mathbf{S}^{n-1} = \mathrm{id}$, which contradicts the No Retraction Theorem 5.1.5. Consequently, we have $\dim \mathbf{B}^n = n$. □

Proposition 5.2.8. *For a normal space X, if there exists a map $f : X \to \mathbf{S}^n$ that is not null-homotopic, then $\dim X \geq n$.*

Proof. Define $\mathbf{S}^n_+$ and $\mathbf{S}^n_-$ as in the proof of Theorem 5.2.3 (f) ⇒ (e) and identify $\mathbf{S}^{n-1} = \mathbf{S}^n_+ \cap \mathbf{S}^n_- \subset \mathbf{S}^n$. If $\dim X \leq n-1$ then $f|f^{-1}(\mathbf{S}^{n-1})$ extends to a map $f' : X \to \mathbf{S}^{n-1}$ by Theorem 5.2.3. We can define a map $g : X \to \mathbf{S}^n$ as follows:

$$g|f^{-1}(\mathbf{S}^n_+) = f'|f^{-1}(\mathbf{S}^n_+) \text{ and } g|f^{-1}(\mathbf{S}^n_-) = f|f^{-1}(\mathbf{S}^n_-).$$

Then, $g \simeq f$ rel. $f^{-1}(\mathbf{S}^n_-)$. Indeed, we have a homeomorphism $\varphi : \mathbf{S}^n_+ \to \mathbf{B}^n$ with $\varphi|\mathbf{S}^{n-1} = \mathrm{id}$. Then, $\varphi f|f^{-1}(\mathbf{S}^n_+) \simeq \varphi f'|f^{-1}(\mathbf{S}^n_+)$ rel. $f^{-1}(\mathbf{S}^{n-1})$ in $\mathbf{B}^n$, which is realized by the straight-line homotopy. Hence, $f|f^{-1}(\mathbf{S}^n_+) \simeq f'|f^{-1}(\mathbf{S}^n_+)$ rel. $f^{-1}(\mathbf{S}^{n-1})$ in $\mathbf{S}^n_+$, which implies $g \simeq f$ rel. $f^{-1}(\mathbf{S}^n_-)$. Since $g(X) \subset \mathbf{S}^n_- \approx \mathbf{B}^n$, it follows that $f \simeq g \simeq 0$. This is a contradiction. □

Remark 3. The converse of Proposition 5.2.8 does not hold. In fact, if X is an n-dimensional contractible space then every map $f : X \to \mathbf{S}^n$ is null-homotopic.

Using simplicial complexes, we can characterize the dimension of paracompact spaces as follows:

Theorem 5.2.9. *Let X be paracompact and $n \in \omega$. Then, $\dim X \leq n$ if and only if, for every simplicial complex K, each map $f : X \to |K|$ (or $f : X \to |K|_{\mathrm{m}}$) is contiguous to a map $g : X \to |K^{(n)}|$ (or $g : X \to |K^{(n)}|_{\mathrm{m}}$). In this case, each $g(x)$ is contained in the carrier $c_K(f(x)) \in K$ of $f(x)$.*

Proof. First, we will show the "if" part. Each (finite) open cover $\mathcal{U}$ of X has an open star-refinement $\mathcal{V}$. Let $K = N(\mathcal{V})$ be the nerve of $\mathcal{V}$. A canonical map $f : X \to |K|$ is contiguous to a map $g : X \to |K^{(n)}|$. Then, $g^{-1}(\mathcal{O}_{K^{(n)}}) \in \mathrm{cov}(X)$ with

$$\operatorname{ord} g^{-1}(\mathcal{O}_{K^{(n)}}) \leq \operatorname{ord} \mathcal{O}_{K^{(n)}} = \dim K^{(n)} + 1 \leq n + 1.$$

Let $V \in \mathcal{V} = K^{(0)}$ and $x \in g^{-1}(O_{K^{(n)}}(V))$. We have $\sigma \in K$ such that $f(x), g(x) \in \sigma$. Then, $c_K(f(x)) \leq \sigma$ and $V \in \sigma^{(0)}$. Since f is canonical, we have $c_K(f(x))^{(0)} \subset \mathcal{V}[x]$ (Proposition 4.9.1). It follows that $V \cap V' \neq \emptyset$ and $x \in V'$ for any $V' \in c_K(f(x))^{(0)}$, which implies $x \in \mathrm{st}(V, \mathcal{V})$. Thus, $g^{-1}(O_{K^{(n)}}(V)) \subset \mathrm{st}(V, \mathcal{V})$, which means $g^{-1}(\mathcal{O}_{K^{(n)}}) \prec \mathcal{U}$. Therefore, $\dim X \leq n$.

To prove the "only if" part, let $f : X \to |K|$ be a map. Because $\dim X \leq n$, X has an open cover $\mathcal{U} \prec f^{-1}(\mathcal{O}_K)$ with $\operatorname{ord}\mathcal{U} \leq n + 1$ by Theorem 5.2.4. Let $L = N(\mathcal{U})$ be the nerve of $\mathcal{U}$ with $\varphi : X \to |L|$ a canonical map. Then, we have a function $\psi : L^{(0)} = \mathcal{U} \to K^{(0)}$ such that $U \subset f^{-1}(O_K(\psi(U)))$, i.e., $f(U) \subset O_K(\psi(U))$. By Proposition 4.4.5, $\psi : L^{(0)} \to K^{(0)}$ induces the simplicial map $\psi : L \to K$. Since $\dim L \leq n$, it follows that $\psi\varphi(X) \subset \psi(|L|) \subset |K^{(n)}|$. Thus, we have a map $g = \psi\varphi : X \to |K^{(n)}|$.

We will show that $g(x) \in c_K(f(x))$ for every $x \in X$. For each $x \in X$, $\varphi(x) \in \langle\mathcal{U}[x]\rangle \in L$ because φ is canonical. Then, $g(x) = \psi\varphi(x) \in \psi(\langle\mathcal{U}[x]\rangle) \in K$. For each $U \in \mathcal{U}[x]$, $f(x) \in f(U) \subset O_K(\psi(U))$, which means $\psi(U) \in c_K(f(x))^{(0)}$. Hence, $\psi(\langle\mathcal{U}[x]\rangle) \leq c_K(f(x))$. Thus, $g(x) \in c_K(f(x))$ for every $x \in X$. □

Remark 4. In the above proof of the "only if" part, when K is *locally finite-dimensional*, we can apply the same argument used in Remark 2 to obtain a map $g : X \to |K^{(n)}|$ contiguous to f.

As a corollary of Theorems 5.2.7 and 5.2.9, we have the following:

Corollary 5.2.10. *For any simplicial complex K,*

$$\dim K = \dim |K| = \dim |K|_{\mathrm{m}}.$$

Proof. An n-simplex $\tau \in K$ is closed in both $|K|$ and $|K|_{\mathrm{m}}$, and $\dim \tau = n$ by Theorem 5.2.7. By the definition of dimension, $\dim |K| \geq \dim \tau$ and $\dim |K|_{\mathrm{m}} \geq \dim \tau$. On the other hand, combining Theorem 5.2.9 with the Simplicial Approximation Theorem 4.7.14, we arrive at $\dim |K| \leq \dim K$ and $\dim |K|_{\mathrm{m}} \leq \dim K$. □

Since the n-dimensional Euclidean space $\mathbb{R}^n$ has an n-dimensional triangulation, we have the following corollary:

Corollary 5.2.11. *For each $n \in \mathbb{N}$, $\dim \mathbb{R}^n = n$.* □

Let A and B be disjoint closed sets in a space X. A closed set C in X is called a **partition** between A and B in X if there exist disjoint open sets U and V in X such that $A \subset U$, $B \subset V$, and $X \setminus C = U \cup V$. A family $(A_\gamma, B_\gamma)_{\gamma \in \Gamma}$ of pairs of disjoint closed sets in X is **inessential** in X if there are partitions L_γ between A_γ and B_γ with $\bigcap_{\gamma \in \Gamma} L_\gamma = \emptyset$. Note that if one of A_γ or B_γ is empty then (A_γ, B_γ) is inessential. If $(A_\gamma, B_\gamma)_{\gamma \in \Gamma}$ is not inessential in X (i.e., $\bigcap_{\gamma \in \Gamma} L_\gamma \neq \emptyset$ for any partitions L_γ between A_γ and B_γ), it is said to be **essential** in X.

A map $f : X \to \mathbf{I}^n$ is said to be **essential** if every map $g : X \to \mathbf{I}^n$ with $g|f^{-1}(\partial \mathbf{I}^n) = f|f^{-1}(\partial \mathbf{I}^n)$ is surjective, where it should be noted that g is also essential. It is said that f is **inessential** if it is not essential, i.e., there is a map $g : X \to \mathbf{I}^n$ such that $g|f^{-1}(\partial \mathbf{I}^n) = f|f^{-1}(\partial \mathbf{I}^n)$ and $g(X) \neq \mathbf{I}^n$. Then, for an inessential map $f : X \to \mathbf{I}^n$, there is a map $g : X \to \partial \mathbf{I}^n$ such that $g|f^{-1}(\partial \mathbf{I}^n) = f|f^{-1}(\partial \mathbf{I}^n)$.

Lemma 5.2.12. *For two maps $f, g : X \to \mathbf{B}^n$, if $f(x) \neq g(x)$ for any $x \in X$, then there is a map $h : X \to \mathbf{S}^{n-1}$ such that $h|f^{-1}(\mathbf{S}^{n-1}) = f|f^{-1}(\mathbf{S}^{n-1})$.*

Proof. In the same way as for the map r in the remark for the No Retraction Theorem 5.1.5, we can obtain a map $h : X \to \mathbf{S}^{n-1}$ such that $f(x) \in \langle h(x), g(x) \rangle$ for each $x \in X$, which implies $h|f^{-1}(\mathbf{S}^{n-1}) = f|f^{-1}(\mathbf{S}^{n-1})$. □

For a map $f : X \to \mathbf{B}^n$ with $f(X) \neq \mathbf{B}^n$, by taking g as a constant map, the following is a special case of Lemma 5.2.12.

Lemma 5.2.13. *If a map $f : X \to \mathbf{B}^n$ is not surjective, then there is a map $h : X \to \mathbf{S}^{n-1}$ such that $h|f^{-1}(\mathbf{S}^{n-1}) = f|f^{-1}(\mathbf{S}^{n-1})$.* □

Proposition 5.2.14. *Let X be a normal space and $h : X \times \mathbf{I} \to \mathbf{I}^n$ be a homotopy such that h_0 is essential and $h(f^{-1}(\partial \mathbf{I}^n) \times \mathbf{I}) \subset \partial \mathbf{I}^n$. Then, h_1 is also essential, hence it is surjective.*

Proof. Let $h_0 = f$ and assume that h_1 is inessential. By Lemma 5.2.13, there is a map $g : X \to \partial \mathbf{I}^n$ such that $g|h_1^{-1}(\partial \mathbf{I}^n) = h_1|h_1^{-1}(\partial \mathbf{I}^n)$. Then, $f^{-1}(\partial \mathbf{I}^n) \subset h_1^{-1}(\partial \mathbf{I}^n)$ and $h_1 \simeq g$ rel. $h_1^{-1}(\partial \mathbf{I}^n)$ by the straight-line homotopy:

$$(1-t)h_1(x) + tg(x) \quad \text{for each } (x,t) \in X \times \mathbf{I}.$$

Connecting this to h, we obtain a homotopy $\varphi : X \times \mathbf{I} \to \mathbf{I}^n$ such that

$$\varphi(f^{-1}(\partial\mathbf{I}^n) \times I) \subset \partial\mathbf{I}^n, \ \varphi_0 = f \ \text{ and } \ \varphi_1 = g.$$

Then, $A = \mathrm{pr}_X(\varphi^{-1}([\frac{1}{3}, \frac{2}{3}]^n))$ is a closed set in X. Observe

$$\varphi^{-1}([\tfrac{1}{3}, \tfrac{2}{3}]^n) \cap (f^{-1}(\partial\mathbf{I}^n) \times \mathbf{I}) = \emptyset,$$

which implies $A \cap f^{-1}(\partial\mathbf{I}^n) = \emptyset$. Taking an Urysohn map $k : X \to \mathbf{I}$ with $k(f^{-1}(\partial\mathbf{I}^n)) = 0$ and $k(A) = 1$, we define a map $g' : X \to \mathbf{I}^n$ as follows:

$$g'(x) = \varphi(x, k(x)) \ \text{ for each } x \in X.$$

Then, $g'|f^{-1}(\partial\mathbf{I}^n) = f|f^{-1}(\partial\mathbf{I}^n)$ but $g'(X) \neq \mathbf{I}^n$. In fact, $g'(A) = g(A) \subset \partial\mathbf{I}^n$ and

$$g'(X \setminus A) \subset \varphi((X \setminus A) \times \mathbf{I}) \subset \varphi\big((X \times \mathbf{I}) \setminus \varphi^{-1}([\tfrac{1}{3}, \tfrac{2}{3}]^n)\big) \subset \mathbf{I}^n \setminus [\tfrac{1}{3}, \tfrac{2}{3}]^n.$$

This is a contradiction because $h_0 = f$ is essential. □

Essential maps can be characterized as follows:

Theorem 5.2.15. *Let X be a normal space. For a map $f : X \to \mathbf{I}^n$, the following are equivalent:*

(a) *f is essential;*
(b) *For each map $g : X \to \mathbf{I}^n$, there is some $x \in X$ such that $f(x) = g(x)$;*
(c) *$(f^{-1}(\mathrm{pr}_i^{-1}(0)), f^{-1}(\mathrm{pr}_i^{-1}(1)))_{i=1}^n$ is essential in X.*

Proof. The implication (a) ⇒ (b) follows from Lemma 5.2.12.

(b) ⇒ (c): Assume that $(f^{-1}(\mathrm{pr}_i^{-1}(0)), f^{-1}(\mathrm{pr}_i^{-1}(1)))_{i=1}^n$ is inessential, that is, there are partitions L_i between $f^{-1}(\mathrm{pr}_i^{-1}(0))$ and $f^{-1}(\mathrm{pr}_i^{-1}(1))$ such that $\bigcap_{i=1}^n L_i = \emptyset$. Then, we have disjoint open sets U_i and V_i in X such that

$$X \setminus L_i = U_i \cup V_i, \ f^{-1}(\mathrm{pr}_i^{-1}(0)) \subset U_i \ \text{ and } \ f^{-1}(\mathrm{pr}_i^{-1}(1)) \subset V_i.$$

Applying Lemma 5.2.2 to the open cover $\{X \setminus L_i \mid i = 1, \ldots, n\}$ of X, we have a closed cover $\{F_i \mid i = 1, \ldots, n\}$ of X such that $F_i \subset X \setminus L_i = U_i \cup V_i$, where we may assume that

$$f^{-1}(\mathrm{pr}_i^{-1}(0)) \cup f^{-1}(\mathrm{pr}_i^{-1}(1)) \subset F_i.$$

Each $U_i \cap F_i = F_i \setminus V_i$ and $V_i \cap F_i = F_i \setminus U_i$ are disjoint closed sets in X. Using Urysohn maps for $U_i \cap F_i$ and $V_i \cap F_i$, we can define a map $g : X \to \mathbf{I}^n$ such that $\mathrm{pr}_i g(U_i \cap F_i) = 1$ and $\mathrm{pr}_i g(V_i \cap F_i) = 0$. Observe

$$\bigcup_{i=1}^{n}(U_i \cap F_i) \cup (V_i \cap F_i) = \bigcup_{i=1}^{n}(U_i \cup V_i) \cap F_i = \bigcup_{i=1}^{n} F_i = X,$$

$$(\mathrm{pr}_i f)^{-1}(1) \subset V_i \subset X \setminus U_i \text{ and } (\mathrm{pr}_i f)^{-1}(0) \subset U_i \subset X \setminus V_i.$$

It follows that $g(x) \neq f(x)$ for any $x \in X$.

(c) $\Rightarrow$ (a): Suppose that f is inessential. Then, there is a map $h : X \to \partial\mathbf{I}^n$ with $h|f^{-1}(\partial\mathbf{I}^n) = f|f^{-1}(\partial\mathbf{I}^n)$ by Lemma 5.2.13. Note that

$$f^{-1}(\mathrm{pr}_i^{-1}(0)) \subset h^{-1}(\mathrm{pr}_i^{-1}(0)) \text{ and } f^{-1}(\mathrm{pr}_i^{-1}(1)) \subset h^{-1}(\mathrm{pr}_i^{-1}(1)).$$

Each $h^{-1}(\mathrm{pr}_i^{-1}(\frac{1}{2}))$ is a partition between $f^{-1}(\mathrm{pr}_i^{-1}(0))$ and $f^{-1}(\mathrm{pr}_i^{-1}(1))$, and then

$$\bigcap_{i=1}^{n} h^{-1}(\mathrm{pr}_i^{-1}(\tfrac{1}{2})) = h^{-1}(\tfrac{1}{2}, \ldots, \tfrac{1}{2}) = \emptyset.$$

Thus, $(f^{-1}(\mathrm{pr}_i^{-1}(0)), f^{-1}(\mathrm{pr}_i^{-1}(1)))_{i=1}^{n}$ is inessential. □

The Brouwer Fixed Point Theorem 5.1.1 means that the identity map of $\mathbf{I}^n$ satisfies condition (b) in Theorem 5.2.15, hence we have the following corollary:

Corollary 5.2.16. *The family* $(\mathrm{pr}_i^{-1}(0), \mathrm{pr}_i^{-1}(1))_{i=1}^{n}$ *is essential in* $\mathbf{I}^n$. □

Remark 5. Due to Theorem 5.2.15, this Corollary 5.2.16 is equivalent to the Brouwer Fixed Point Theorem 5.1.1.

Using essential families and essential maps, we can also characterize dimension as follows:

Theorem 5.2.17 (Eilenberg–Otto; Alexandroff). *Let X be a normal space and $n \in \mathbb{N}$. Then, the following are equivalent:*

(a) $\dim X \geq n$;
(b) *X has an essential map $f : X \to \mathbf{I}^n$;*
(c) *X has an essential family of n pairs of disjoint closed sets.*

Proof. The implication (b) $\Rightarrow$ (c) follows from Theorem 5.2.15. For an essential map $f : X \to \mathbf{I}^n$, $f|f^{-1}(\partial\mathbf{I}^n) : f^{-1}(\partial\mathbf{I}^n) \to \partial\mathbf{I}^n$ cannot extend to any map from X to $\partial\mathbf{I}^n$, which means $\dim X \geq n$ by Theorem 5.2.3. Thus, we have also (b) $\Rightarrow$ (a). The implications (a) $\Rightarrow$ (b) and (c) $\Rightarrow$ (b) remain to be proved.

(a) $\Rightarrow$ (b): By Theorem 5.2.3, there exists a map $f' : A \to \partial\mathbf{I}^n$ of a closed set A in X that cannot extend over X. Nevertheless, f' can be extended to a map $f : X \to \mathbf{I}^n$ by Theorem 5.1.6(1). If there is a map $g : X \to \mathbf{I}^n$ such that $g(x) \neq f(x)$ for any $x \in X$, then we have a map $h : X \to \partial\mathbf{I}^n$ such that $h|f^{-1}(\partial\mathbf{I}^n) = f^{-1}|f^{-1}(\partial\mathbf{I}^n)$ by Lemma 5.2.12. This is a contradiction because h is an extension of f'. Therefore, for each map $g : X \to \mathbf{I}^n$, there is some $x \in X$ such that $f(x) = g(x)$. By Theorem 5.2.15, f is essential.

(c) $\Rightarrow$ (b): Let $(A_i, B_i)_{i=1}^n$ be an essential family of n pairs of disjoint closed sets in X. Using Urysohn maps for A_i and B_i, we can define a map $f : X \to \mathbf{I}^n$ so that $\mathrm{pr}_i f(A_i) = 0$ and $\mathrm{pr}_i f(B_i) = 1$ for each $i = 1, \ldots, n$. Since $A_i \subset f^{-1}(\mathrm{pr}_i^{-1}(0))$ and $B_i \subset f^{-1}(\mathrm{pr}_i^{-1}(1))$, it follows that $(f^{-1}(\mathrm{pr}_i^{-1}(0)), f^{-1}(\mathrm{pr}_i^{-1}(1)))_{i=1}^n$ is essential, which means that f is essential according to Theorem 5.2.15. □

Conditions (b) and (c) are called the ALEXANDROFF CHARACTERIZATION the EILENBERG–OTTO CHARACTERIZATION of dimension. Using Theorem 5.2.17, we can easily show the following corollary:

Corollary 5.2.18. *Every non-degenerate* 0*-dimensional normal space is disconnected. Equivalently, every non-degenerate connected normal space is positive dimensional.* □

5.3 Dimension of Metrizable Spaces

In this section, we will give characterizations of dimension for metrizable spaces. For metric spaces, the following characterization can be established:

Theorem 5.3.1. *Let* $X = (X, d)$ *be a metric space. Then,* $\dim X \leq n$ *if and only if* X *has a sequence* $\mathcal{U}_1 \succ \mathcal{U}_2 \succ \cdots$ *of (locally finite* σ*-discrete) open covers such that* $\mathrm{ord}\,\mathcal{U}_i \leq n+1$ *and* $\lim_{i\to\infty} \mathrm{mesh}\,\mathcal{U}_i = 0$.

Proof. When $\dim X \leq n$, using the "only if" of Theorem 5.2.4, we can inductively construct locally finite σ-discrete open covers $\mathcal{U}_1 \succ \mathcal{U}_2 \succ \cdots$ of X such that $\mathrm{ord}\,\mathcal{U}_i \leq n+1$ and $\lim_{i\to\infty} \mathrm{mesh}\,\mathcal{U}_i = 0$. Thus, the "only if" part holds.

To show the "if" part, let $\mathcal{W}$ be a finite open cover of X. We have a function $\varphi_i : \mathcal{U}_{i+1} \to \mathcal{U}_i$ such that $U \subset \varphi_i(U)$ for each $U \in \mathcal{U}_{i+1}$. For each $j > i$, let $\varphi_{i,j} = \varphi_i \circ \cdots \circ \varphi_{j-1} : \mathcal{U}_j \to \mathcal{U}_i$ and $\varphi_{i,i} = \mathrm{id}_{\mathcal{U}_i}$.

For each $i \in \mathbb{N}$, let

$$X_i = \bigcup \{U \in \mathcal{U}_i \mid \mathrm{st}(U, \mathcal{U}_i) \text{ is contained in some } W \in \mathcal{W}\}.$$

Then, $X_1 \subset X_2 \subset \cdots$ and $X = \bigcup_{i\in\mathbb{N}} X_i$ because $\lim_{i\to\infty} \mathrm{mesh}\,\mathcal{U}_i = 0$. Moreover, let $\mathcal{U}_i' = \mathcal{U}_i[X_i]$ and $\mathcal{U}_i'' = \mathcal{U}_i' \setminus \mathcal{U}_i[X_{i-1}]$, where $X_0 = \emptyset$.

For each $i \in \mathbb{N}$ and $U \in \mathcal{U}_i'$, we define

$$k_i(U) = \min \{k \leq i \mid \varphi_{k,i}(U) \cap X_k \neq \emptyset\}.$$

Observe that $\varphi_{k_i(U),i}(U) \in \mathcal{U}_{k_i(U)}''$ and $k_{k_i(U)}(\varphi_{k_i(U),i}(U)) = k_i(U)$. As is easily seen, $\mathcal{U}_i'' \cap \mathcal{U}_j'' = \emptyset$ if $i \neq j$. For each $U \in \bigcup_{i\in\mathbb{N}} \mathcal{U}_i''$, there is a unique $j(U) \in \mathbb{N}$ such that $U \in \mathcal{U}_{j(U)}''$. Then, we can define

$$U^* = \bigcup \{U' \cap X_i \mid U' \in \mathcal{U}_i',\ i \geq j(U) = k_i(U'),\ \varphi_{j(U),i}(U') = U\} \subset U.$$

Note that if $k_{j(U)}(U) < j(U)$ then $U^* = \emptyset$.

Each $x \in X$ is contained in some X_i, hence $x \in U' \cap X_i$ for some $U' \in \mathcal{U}'_i$. Let $U = \varphi_{k_i(U'),i}(U') \in \mathcal{U}''_{k_i(U')}$. Then, $k_i(U') = j(U)$ and $x \in U' \cap X_i \subset U^*$. Thus, we have

$$\mathcal{V} = \{U^* \mid U \in \bigcup_{i\in\mathbb{N}} \mathcal{U}''_i\} \in \operatorname{cov}(X).$$

Each $U \in \mathcal{U}''_i$ meets X_i, hence it meets some $U' \in \mathcal{U}_i$ such that $\operatorname{st}(U',\mathcal{U}_i)$ is contained in some $W \in \mathcal{W}$. Then, $U^* \subset U \subset \operatorname{st}(U',\mathcal{U}_i) \subset W$. Therefore, $\mathcal{V} \prec \mathcal{W}$.

For each $x \in X$, choose $k \in \mathbb{N}$ so that $x \in X_k \setminus X_{k-1}$. For each $U^* \in \mathcal{V}[x]$, we can find $U' \in \mathcal{U}'_i$ such that $i \geq j(U) = k_i(U')$, $\varphi_{j(U),i}(U') = U$, and $x \in U' \cap X_i$. Then, $k \leq i$ because $x \in X_i$ and $x \notin X_{k-1}$. Thus, we have $\varphi_{k,i}(U') \in \mathcal{U}_k[x]$. On the other hand, $j(U) \leq k$ because $U \cap X_k \neq \emptyset$ and $U \cap X_{j(U)-1} = \emptyset$. Then, $\varphi_{j(U),k}\varphi_{k,i}(U') = \varphi_{j(U),i}(U') = U$. This means that $\mathcal{V}[x] \ni U^* \mapsto \varphi_{k,i}(U') \in \mathcal{U}_k[x]$ is a well-defined injection. Therefore,

$$\operatorname{card}\mathcal{V}[x] \leq \operatorname{card}\mathcal{U}_k[x] \leq \operatorname{ord}\mathcal{U}_k \leq n+1.$$

The proof is complete. □

Applying Theorem 5.3.1, we can show that the inverse limit preserves the dimension.

Theorem 5.3.2. *Let $X = \varprojlim(X_i, f_i)$ be the inverse limit of an inverse sequence $(X_i, f_i)_{i\in\mathbb{N}}$ of metrizable spaces. If $\dim X_i \leq n$ for infinitely many $i \in \mathbb{N}$ then $\dim X \leq n$.*

Proof. By Corollary 4.10.4, we may assume that $\dim X_i \leq n$ for every $i \in \mathbb{N}$. Recall that X is the following subspace of the product space $\prod_{i\in\mathbb{N}} X_i$:

$$X = \{x \in \prod_{i\in\mathbb{N}} X_i \mid x(i) = f_i(x(i+1)) \text{ for every } i \in \mathbb{N}\}.$$

We define $d \in \operatorname{Metr}(X)$ as follows:

$$d(x,y) = \sup_{i\in\mathbb{N}} \min\{d_i(x(i), y(i)), 2^{-i}\},$$

where $d_i \in \operatorname{Metr}(X_i)$. For each $i \in \mathbb{N}$, we can inductively choose $\mathcal{V}_i \in \operatorname{cov}(X_i)$ so that $\operatorname{ord}\mathcal{V}_i \leq n+1$,

$$\mathcal{V}_i \prec (f_j \dots f_{i-1})^{-1}(\mathcal{V}_j) \text{ and } \operatorname{mesh} f_j \dots f_{i-1}(\mathcal{V}_i) < 2^{-i} \text{ for } j < i.$$

Let $\mathcal{U}_i = p_i^{-1}(\mathcal{V}_i) \in \operatorname{cov}(X)$, where $p_i = \operatorname{pr}_i|X : X \to X_i$ is the inverse limit projection. Then, $\mathcal{U}_1 \succ \mathcal{U}_2 \succ \cdots$, $\operatorname{ord}\mathcal{U}_i \leq n+1$, and $\operatorname{mesh}\mathcal{U}_i < 2^{-i}$. Therefore, $\dim X \leq n$ by Theorem 5.3.1. □

The following is obvious by definition:

- If Y is a *closed* set in X then $\dim Y \leq \dim X$.

There exists a 0-dimensional compact space X that contains a subspace Y with $\dim Y > 0$. Such a space will be constructed in Sect. 5.5 (cf. Theorem 5.5.3). However, when X is metrizable, we have the following theorem as a corollary of Theorem 5.3.1.

Theorem 5.3.3 (Subset Theorem). *For every subset Y of a metrizable space X, $\dim Y \le \dim X$.* □

We can apply Theorem 5.3.1 to prove the following completion theorem:

Theorem 5.3.4. *Every n-dimensional metrizable space X can be embedded in an n-dimensional completely metrizable space as a dense set.*

Proof. We can regard X as a dense subset of a complete metric space $Y = (Y, d)$ (Corollary 2.3.10). Applying Theorem 5.3.1, we can obtain a sequence $\mathcal{U}_1 \succ \mathcal{U}_2 \succ \cdots \in \operatorname{cov}(X)$ such that $\operatorname{ord} \mathcal{U}_i \le n+1$ and $\operatorname{mesh}_d \mathcal{U}_i \to 0$ as $i \to \infty$. For each $i \in \mathbb{N}$, there is a collection $\tilde{\mathcal{U}}_i$ of open sets in Y such that $\tilde{\mathcal{U}}_i | X = \mathcal{U}_i$. Then, $\operatorname{ord} \tilde{\mathcal{U}}_i = \operatorname{ord} \mathcal{U}_i \le n+1$ and $\operatorname{mesh}_d \tilde{\mathcal{U}}_i = \operatorname{mesh}_d \mathcal{U}_i$. For each $i \in \mathbb{N}$, $G_i = \bigcup \tilde{\mathcal{U}}_i$ is an open set in Y. Thus, we have a G_δ-set $\tilde{X} = \bigcap_{i \in \mathbb{N}} G_i$ in Y and X is dense in $\tilde{X}$. According to Theorem 2.5.3(2), $\tilde{X}$ is completely metrizable. Moreover, $\dim \tilde{X} \le n$ by Theorem 5.3.1 and $\dim \tilde{X} \ge n$ by Theorem 5.3.3. Consequently, we have $\dim \tilde{X} = n$. □

A subset of a space X is called a **clopen set** in X if it is both closed and open in X. A **clopen basis** for X is an open basis consisting of clopen sets. For metrizable spaces, we characterize the 0-dimensionality as follows:

Theorem 5.3.5. *For a metrizable space X ($\neq \emptyset$), $\dim X = 0$ if and only if X has a σ-locally finite clopen basis.*

Proof. First, assume that $\dim X = 0$ and let $d \in \operatorname{Metr}(X)$. By Theorem 5.3.1, X has a sequence of locally finite open covers $\mathcal{B}_1 \succ \mathcal{B}_2 \succ \cdots$ such that $\operatorname{ord} \mathcal{B}_i = 1$ and $\lim_{i\to\infty} \operatorname{mesh} \mathcal{B}_i = 0$. Note that each $B \in \mathcal{B}_i$ is clopen in X because $B = X \setminus \bigcup\{B' \in \mathcal{B}_i \mid B' \neq B\}$. It is easy to see that $\mathcal{B} = \bigcup_{i \in \mathbb{N}} \mathcal{B}_i$ is a σ-locally finite clopen basis for X.

To show the "if" part, let $\mathcal{B} = \bigcup_{i \in \mathbb{N}} \mathcal{B}_i$ be a σ-locally finite clopen basis for X, where each $\mathcal{B}_i$ is locally finite. Let $\{U_1, U_2\} \in \operatorname{cov}(X)$. For each $i \in \mathbb{N}$, let

$$V_{2i-1} = \bigcup \{B \in \mathcal{B}_i \mid B \subset U_1\} \text{ and } V_{2i} = \bigcup \{B \in \mathcal{B}_i \mid B \subset U_2\}.$$

Because V_i is clopen, we have an open set $W_i = V_i \setminus \bigcup_{j<i} V_j$ in X. Then, $\mathcal{W} = \{W_i \mid i \in \mathbb{N}\}$ is an open refinement of $\{U_1, U_2\}$. Indeed, $W_{2i-1} \subset U_1$, $W_{2i} \subset U_2$, and

$$\bigcup_{i \in \mathbb{N}} W_i = \bigcup_{i \in \mathbb{N}} V_i = \bigcup_{i \in \mathbb{N}} V_{2i-1} \cup \bigcup_{i \in \mathbb{N}} V_{2i} = U_1 \cup U_2 = X.$$

Since $\operatorname{ord} \mathcal{W} \le 1$ by definition, we have $\dim X \le 0$ by Theorem 5.2.3. □

Using the above characterization, we can easily show that $\dim \mathbb{Q} = \dim(\mathbb{R} \setminus \mathbb{Q}) = 0$ and $\dim \mu^0 = 0$, where μ^0 is the Cantor (ternary) set. The following theorem can also be easily proved by applying this characterization.

Theorem 5.3.6. *The countable product of* 0*-dimensional metrizable spaces is* 0*-dimensional.* □

The following simple lemma is very useful in Dimension Theory.

Lemma 5.3.7 (Partition Extension). *Let A, B be closed and U, V be open sets in a metrizable space X such that $A \subset U$ and $B \subset V$ and $\operatorname{cl} U \cap \operatorname{cl} V = \emptyset$. For any subspace Y of X, if Y has a partition S between $Y \cap \operatorname{cl} U$ and $Y \cap \operatorname{cl} V$, then X has a partition L between A and B with $Y \cap L \subset S$.*

Proof. Let U' and V' be disjoint open sets in Y such that $Y \setminus S = U' \cup V'$, $Y \cap \operatorname{cl} U \subset U'$, and $Y \cap \operatorname{cl} V \subset V'$. From $U \cap V' = \emptyset$, it follows that $A \cap \operatorname{cl} V' = \emptyset$. Then,

$$(A \cup U') \cap \operatorname{cl}(B \cup V') = (A \cup U') \cap (B \cup \operatorname{cl} V') = \emptyset.$$

Similarly, we have $(B \cup V') \cap \operatorname{cl}(A \cup U') = \emptyset$. Let $d \in \operatorname{Metr}(X)$ and define

$$U'' = \{x \in X \mid d(x, A \cup U') < d(x, B \cup V')\} \quad \text{and}$$
$$V'' = \{x \in X \mid d(x, B \cup V') < d(x, A \cup U')\}.$$

Then, U'' and V'' are disjoint open sets in X, $A \cup U' \subset U''$, and $B \cup V' \subset V''$. Hence, $L = X \setminus (U'' \cup V'')$ is the desired partition. □

Note that it does not suffice to assume that S is a partition between $A \cap Y$ and $B \cap Y$ in Y. In fact, $A = [-1, 0] \times \{0\}$ and $B = [0, 1] \times \{1\}$ are disjoint closed sets in $X = \mathbb{R}^2$. Let

$$Y = \mathbb{R}^2 \setminus (\mathbb{Q} \times \mathbf{2}) \subset X,$$

where $\mathbf{2} = \{0, 1\}$ is the discrete space of two points. Then, $S = \{0\} \times \mathbb{R}$ is a partition between $A \cap Y$ and $B \cap Y$ in Y but X has no partition L between A and B such that $Y \cap L \subset S$.

Using partitions, we can characterize the dimension for metrizable spaces as in the following theorem:

Theorem 5.3.8. *Let X be metrizable and $n \in \omega$. Then, $\dim X \leq n$ if and only if, for any pair of disjoint closed sets A and B in X, there is a partition L in X between A and B with $\dim L \leq n - 1$.*

Proof. To prove the "if" part, let $(A_i, B_i)_{i=1}^{n+1}$ be a family of pairs of disjoint closed sets in X. Let L_{n+1} be a partition between A_{n+1} and B_{n+1} with $\dim L_{n+1} \leq n - 1$. For each $i = 1, \ldots, n$, let U_i and V_i be open sets in X such that $A_i \subset U_i$, $B_i \subset V_i$ and $\operatorname{cl} U_i \cap \operatorname{cl} V_i = \emptyset$. By Theorem 5.2.17, L_{n+1} has partitions S_i between $L_{n+1} \cap \operatorname{cl} U_i$ and $L_{n+1} \cap \operatorname{cl} V_i$ such that $\bigcap_{i=1}^n S_i = \emptyset$. By the Partition Extension Lemma 5.3.7, X has partitions L_i between A_i and B_i such that $L_i \cap L_{n+1} \subset S_i$.

Then, $\bigcap_{i=1}^{n+1} L_i \subset \bigcap_{i=1}^{n} S_i = \emptyset$. Therefore, $(A_i, B_i)_{i=1}^{n+1}$ is inessential. Thus, we have $\dim X \leq n$ by Theorem 5.2.17.

To show the "only if" part, let $d \in \mathrm{Metr}(X)$ such that $\mathrm{dist}(A, B) > 1$. (Such a metric can be obtained by a metric for X and an Urysohn map for A and B.) Using Theorem 5.2.4 (cf. Theorem 5.3.1), we can construct a sequence $(\mathcal{U}_i)_{i\in\mathbb{N}}$ of locally finite open covers of X such that $\mathrm{ord}\,\mathcal{U}_i \leq n+1$, $\mathrm{mesh}\,\mathcal{U}_i < i^{-1}$, and $\mathcal{U}_{i+1}^{\mathrm{cl}} \prec \mathcal{U}_i$. Let A_0 and B_0 be open neighborhoods of A and B in X, respectively, such that $\mathrm{dist}(A_0, B_0) > 1$. We inductively define sets A_i and B_i ($i \in \mathbb{N}$) as follows:

$$A_i = X \setminus \bigcup\{\mathrm{cl}\,U \mid U \in \mathcal{U}_i[B_{i-1}]\} \quad \text{and}$$

$$B_i = X \setminus \bigcup\{\mathrm{cl}\,U \mid U \in \mathcal{U}_i \setminus \mathcal{U}_i[B_{i-1}]\}.$$

Then, $A_i \cap B_i = \emptyset$. Because of the local finiteness of $\mathcal{U}_i^{\mathrm{cl}}$, A_i and B_i are open in X. Since $B_{i-1} \cap U = \emptyset$ if and only if $B_{i-1} \cap \mathrm{cl}\,U = \emptyset$ for each $U \in \mathcal{U}_i$, it follows that $B_{i-1} \subset B_i$ for each $i \in \mathbb{N}$. Then, $\mathcal{U}_i[B_{i-1}] \subset \mathcal{U}_i[B_i]$. We also have $\mathcal{U}_i[B_i] \subset \mathcal{U}_i[B_{i-1}]$. Indeed, each $U \in \mathcal{U}_i[B_i]$ contains some point of B_i, where that point does not belong to any member of $\mathcal{U}_i \setminus \mathcal{U}_i[B_{i-1}]$. This means $U \in \mathcal{U}_i[B_{i-1}]$. Therefore, $\mathcal{U}_i[B_i] = \mathcal{U}_i[B_{i-1}]$ for each $i \in \mathbb{N}$.

We will show that $\mathrm{cl}\,A_{i-1} \subset A_i$ for each $i \in \mathbb{N}$. This follows from the fact that $\mathrm{cl}\,A_{i-1} \cap \mathrm{cl}\,U = \emptyset$ for each $U \in \mathcal{U}_i[B_{i-1}]$. This fact can be shown as follows: The case $i = 1$ follows from $\mathrm{mesh}\,\mathcal{U}_1 < 1$ and $\mathrm{dist}(A_0, B_0) > 1$. When $i > 1$, for each $U \in \mathcal{U}_i[B_{i-1}]$, $\mathrm{cl}\,U$ is contained in some $V \in \mathcal{U}_{i-1}$. Since $V \in \mathcal{U}_{i-1}[B_{i-1}] = \mathcal{U}_{i-1}[B_{i-2}]$, it follows that $A_{i-1} \cap V = \emptyset$, and hence $\mathrm{cl}\,A_{i-1} \cap V = \emptyset$, which implies $\mathrm{cl}\,A_{i-1} \cap \mathrm{cl}\,U = \emptyset$.

For each $i \in \mathbb{N}$, let $L_i = X \setminus (A_i \cup B_i)$ and let $L = \bigcap_{i\in\mathbb{N}} L_i$. Then, L is a partition between A and B. Indeed, $X \setminus L = \left(\bigcup_{i\in\mathbb{N}} A_i\right) \cup \left(\bigcup_{i\in\mathbb{N}} B_i\right)$, $A \subset \bigcup_{i\in\mathbb{N}} A_i$, $B \subset \bigcup_{i\in\mathbb{N}} B_i$, and

$$\left(\bigcup_{i\in\mathbb{N}} A_i\right) \cap \left(\bigcup_{i\in\mathbb{N}} B_i\right) = \bigcup_{i,j\in\mathbb{N}} (A_i \cap B_j) = \bigcup_{i,j\in\mathbb{N}} (A_{\max\{i,j\}} \cap B_{\max\{i,j\}})$$

$$= \bigcup_{i\in\mathbb{N}} (A_i \cap B_i) = \emptyset.$$

For each $i \in \mathbb{N}$, we have

$$\mathcal{W}_i = \{U \cap L \mid U \in \mathcal{U}_i[B_{i-1}]\} \in \mathrm{cov}(L).$$

Indeed, each $x \in L$ is not contained in A_{i+1}, so $x \in \mathrm{cl}\,V$ for some $V \in \mathcal{U}_{i+1}[B_i]$. Choose $U \in \mathcal{U}_i$ so that $\mathrm{cl}\,V \subset U$. Then, $U \in \mathcal{U}_i[B_i] = \mathcal{U}_i[B_{i-1}]$, hence $x \in U \cap L \in \mathcal{W}_i$. Therefore, $\mathcal{W}_i \in \mathrm{cov}(L)$. Note that $\mathrm{mesh}\,\mathcal{W}_i \leq \mathrm{mesh}\,\mathcal{U}_i < i^{-1}$. Moreover, $\mathcal{W}_{i+1} \prec \mathcal{W}_i$ because each $V \in \mathcal{U}_{i+1}[B_i]$ is contained in some $U \in \mathcal{U}_i$, where $U \in \mathcal{U}_i[B_i] = \mathcal{U}_i[B_{i-1}]$.

We will show that $\operatorname{ord} \mathcal{W}_i \leq n$. Suppose that there are $n+1$ many distinct $U_1, \ldots, U_{n+1} \in \mathcal{U}_i[B_{i-1}]$ that contain a common point $x \in L$. Since $x \notin B_i$, $x \in \operatorname{cl} U_{n+2}$ for some $U_{n+2} \in \mathcal{U}_i \setminus \mathcal{U}_i[B_{i-1}]$. Since $\bigcap_{j=1}^{n+1} U_j$ is a neighborhood of x, it follows that $\bigcap_{j=1}^{n+2} U_j \neq \emptyset$, which is contrary to $\operatorname{ord} \mathcal{U}_i \leq n+1$. Therefore, we have $\dim L \leq n-1$ by Theorem 5.3.1 □

Remark 6. It should be noted that the Partition Extension Lemma 5.3.7 and the "if" part of Theorem 5.3.8 are valid for completely normal (= hereditarily normal) spaces (cf. Sect. 2.2).

5.4 Fundamental Theorems on Dimension

In this section, we prove several fundamental theorems on dimension. We begin with two types of sum theorem.

Theorem 5.4.1 (COUNTABLE SUM THEOREM). *Let $X = \bigcup_{i \in \mathbb{N}} F_i$ be normal and $n \in \omega$, where each F_i is closed in X. If $\dim F_i \leq n$ for every $i \in \mathbb{N}$, then $\dim X \leq n$.*

Proof. It suffices to show the case $n < \infty$. Let $\{U_1, \ldots, U_{n+2}\} \in \operatorname{cov}(X)$. By induction on $i \in \mathbb{N}$, we can define $\mathcal{U}_i = \{U_{i,1}, \ldots, U_{i,n+2}\} \in \operatorname{cov}(X)$ so that

$$\operatorname{cl} U_{i,j} \subset U_{i-1,j} \quad \text{and} \quad U_{i,1} \cap \cdots \cap U_{i,n+2} \cap F_i = \emptyset,$$

where $U_{0,j} = U_j$. Indeed, assume that $\mathcal{U}_{i-1}$ has been defined, where $F_0 = \emptyset$. By Theorem 5.2.3, we have $\{V_{i,1}, \ldots, V_{i,n+2}\} \in \operatorname{cov}(F_i)$ such that

$$V_{i,j} \subset U_{i-1,j} \quad \text{and} \quad V_{i,1} \cap \cdots \cap V_{i,n+2} = \emptyset.$$

Let $W_{i,j} = V_{i,j} \cup (U_{i-1,j} \setminus F_i)$. Then, $\{W_{i,1}, \ldots, W_{i,n+2}\} \in \operatorname{cov}(X)$. By normality, we can find $\mathcal{U}_i = \{U_{i,1}, \ldots, U_{i,n+2}\} \in \operatorname{cov}(X)$ such that $\operatorname{cl} U_{i,j} \subset W_{i,j}$. Observe that $\mathcal{U}_i$ is as desired.

For each $j = 1, \ldots, n+2$, let $A_j = \bigcap_{i \in \mathbb{N}} U_{i,j}$. Observe that $A_j = \bigcap_{i \in \mathbb{N}} \operatorname{cl} U_{i,j}$ is closed in X. Since

$$A_1 \cap \cdots \cap A_{n+2} \cap F_i \subset U_{i,1} \cap \cdots \cap U_{i,n+2} \cap F_i = \emptyset,$$

we have $A_1 \cap \cdots \cap A_{n+2} = \emptyset$. For each $x \in X$, $\{i \in \mathbb{N} \mid x \in U_{i,j}\}$ is infinite for some j. Then, $x \in \bigcap_{i \in \mathbb{N}} U_{i,j} = A_j$. Hence, $X = A_1 \cup \cdots \cup A_{n+2}$. According to Theorem 5.2.3, we have $\dim X \leq n$. □

Theorem 5.4.2 (LOCALLY FINITE SUM THEOREM). *Let X be normal and $n \in \omega$. If X has a locally finite closed cover $\{F_\lambda \mid \lambda \in \Lambda\}$ such that $\dim F_\lambda \leq n$ for each $\lambda \in \Lambda$, then $\dim X \leq n$.*

Proof. We may assume that $n < \infty$, $\Lambda = (\Lambda, \leq)$ is a well-ordered set, and $F_{\min \Lambda} = \emptyset$. Let $\{U_1, \ldots, U_{n+2}\} \in \operatorname{cov}(X)$. Using transfinite induction, we will define $\mathcal{U}_\lambda = \{U_{\lambda,1}, \ldots, U_{\lambda,n+2}\} \in \operatorname{cov}(X)$ so that

$$U_{\min \Lambda, j} = U_j, \; U_{\lambda,1} \cap \cdots \cap U_{\lambda,n+2} \cap F_\lambda = \emptyset, \text{ and}$$

$$\mu < \lambda \Rightarrow U_{\lambda,j} \subset U_{\mu,j}, \; U_{\mu,j} \setminus U_{\lambda,j} \subset \bigcup_{\mu \leq \nu \leq \lambda} F_\nu.$$

Suppose that $\mathcal{U}_\mu$ has been obtained for $\mu < \lambda$. Let $U'_{\lambda,j} = \bigcap_{\mu<\lambda} U_{\mu,j}$. Then, $\{U'_{\lambda,1}, \ldots, U'_{\lambda,n+2}\} \in \operatorname{cov}(X)$. Indeed, if there exists $\mu_0 = \max\{\mu \in \Lambda \mid \mu < \lambda\}$, then $U'_{\lambda,j} = U_{\mu_0,j}$ for each $j = 1, \ldots, n+2$. Otherwise, for each $x \in X$, choose an open neighborhood U of x in X that meets only finitely many F_μ. Then, there exists $\mu_1 < \lambda$ such that $U \cap F_\mu = \emptyset$ for $\mu_1 \leq \mu < \lambda$. If $x \in U_{\mu_1,j} \in \mathcal{U}_{\mu_1}$, then $U \cap U_{\mu_1,j} \subset U_{\mu,j}$ for $\mu_1 \leq \mu < \lambda$ because

$$(U \cap U_{\mu_1,j}) \setminus U_{\mu,j} \subset \bigcup_{\mu_1 \leq \nu \leq \mu} (U \cap F_\nu) = \emptyset.$$

It follows that

$$x \in U \cap U_{\mu_1,j} \subset \bigcap_{\mu_1 \leq \mu < \lambda} U_{\mu,j} = \bigcap_{\mu<\lambda} U_{\mu,j} = U'_{\lambda,j}.$$

We apply Theorem 5.2.3 to obtain $\{V_{\lambda,1}, \ldots, V_{\lambda,n+2}\} \in \operatorname{cov}(F_\lambda)$ such that $V_{\lambda,j} \subset U'_{\lambda,j}$ and $V_{\lambda,1} \cap \cdots \cap V_{\lambda,n+2} = \emptyset$. Now, let $U_{\lambda,j} = V_{\lambda,j} \cup (U'_{\lambda,j} \setminus F_\lambda)$. Then, $\{U_{\lambda,1}, \ldots, U_{\lambda,n+2}\} \in \operatorname{cov}(X)$ is the desired open cover. In fact, if $\mu < \lambda$ then

$$\begin{aligned} U_{\mu,j} \setminus U_{\lambda,j} &\subset F_\lambda \cup ((U_{\mu,j} \setminus F_\lambda) \setminus (U'_{\lambda,j} \setminus F_\lambda)) \subset F_\lambda \cup \Big(U_{\mu,j} \setminus \bigcap_{\nu<\lambda} U_{\nu,j}\Big) \\ &\subset F_\lambda \cup \bigcup_{\mu<\nu<\lambda} (U_{\mu,j} \setminus U_{\nu,j}) = \bigcup_{\mu \leq \nu \leq \lambda} F_\nu. \end{aligned}$$

The proofs of the other properties are easy.

For each $j = 1, \ldots, n+2$, let $U^*_j = \bigcap_{\lambda \in \Lambda} U_{\lambda,j}$. Then, similar to the above, $\{U^*_1, \ldots, U^*_{n+2}\} \in \operatorname{cov}(X)$. Clearly, $U^*_j \subset U_j$ and $U^*_1 \cap \cdots \cap U^*_{n+2} = \emptyset$. Therefore, $\dim X \leq n$ by Theorem 5.2.3. □

The following corollary is a combination of Theorems 5.4.1 and 5.4.2:

Corollary 5.4.3. *Let X be a normal space and $n \in \omega$. If X has a σ-locally finite closed cover $\{F_\lambda \mid \lambda \in \Lambda\}$ such that* $\dim F_\lambda \leq n$ *for each $\lambda \in \Lambda$, then* $\dim X \leq n$. □

The next corollary follows from Theorem 5.4.2:

Corollary 5.4.4. *Let X be a paracompact space and $n \in \omega$. If each point of X has a closed neighborhood with* $\dim \leq n$*, then* $\dim X \leq n$.

Proof. Since X is paracompact and each point of X has a closed neighborhood with $\dim \leq n$, X has a locally finite open cover $\mathcal{U}$ such that $\dim \mathrm{cl}\, U \leq n$ for each $U \in \mathcal{U}$. Then, $\mathcal{U}^{\mathrm{cl}}$ is also locally finite in X, hence $\dim X \leq n$ by Theorem 5.4.2. □

Remark 7. Corollary 5.4.4 can also be proved by applying Michael's Theorem on local properties of closed sets (Corollary 2.6.6). In this case, the proof is reduced to showing that if X is the union of two closed sets X_1 and X_2 with $\dim X_i \leq n$, $i = 1, 2$, then $\dim X \leq n$.

In the remainder of this section, we consider only metrizable spaces. The real line $\mathbb{R}$ is 1-dimensional and we can decompose $\mathbb{R}$ into two 0-dimensional subsets $\mathbb{Q}$ and $\mathbb{R} \setminus \mathbb{Q}$. This can be generalized as follows:

Theorem 5.4.5 (Decomposition Theorem). *Let X be metrizable and $n \in \omega$. Then,* $\dim X \leq n$ *if and only if X is covered by $n+1$ many subsets $X_1, \ldots, X_{n+1}$ with* $\dim X_i \leq 0$.

Proof. To prove the "if" part, let $\mathcal{U}$ be a finite open cover of X. Since $\dim X_i \leq 0$, X_i has a finite open cover $\mathcal{V}_i$ such that $\mathcal{V}_i \prec \mathcal{U}$ and $\mathrm{ord}\, \mathcal{V}_i \leq 1$. For each $V \in \mathcal{V}_i$, choose an open set $W(V)$ in X so that $W(V) \cap X_i = V$ and $W(V)$ is contained in some member of $\mathcal{U}$. Note that $\mathrm{cl}\, W(V) \cap X_i = V$ because V is also closed in X_i. For each $V \in \mathcal{V}_i$, let

$$\tilde{V} = W(V) \setminus \bigcup \{\mathrm{cl}\, W(V') \mid V \neq V' \in \mathcal{V}_i\}.$$

Then, $\tilde{\mathcal{V}}_i = \{\tilde{V} \mid V \in \mathcal{V}_i\}$ is a collection of disjoint open sets in X that covers X_i and refines $\mathcal{U}$. Observe that $\tilde{\mathcal{V}} = \bigcup_{i=1}^{n+1} \tilde{\mathcal{V}}_i$ is an open refinement of $\mathcal{U}$ with $\mathrm{ord}\, \mathcal{V} \leq n+1$. Therefore, $\dim X \leq n$.

The "only if" part can be easily obtained by induction once the following proposition has been proved. □

Proposition 5.4.6. *Let X be metrizable and* $\dim X \leq n < \infty$. *Then, $X = Y \cup Z$ for some $Y, Z \subset X$ with* $\dim Y \leq n-1$ *and* $\dim Z \leq 0$.

Proof. Assume that X is a metric space. For each $i \in \mathbb{N}$, let $\mathcal{U}_i$ be a locally finite open cover of X with $\mathrm{mesh}\, \mathcal{U}_i < i^{-1}$. By paracompactness (Lemma 2.6.2) or normality (Lemma 2.7.1), X has a closed cover $\{F_U \mid U \in \mathcal{U}_i\}$ such that $F_U \subset U$ for all $U \in \mathcal{U}_i$. For each $U \in \mathcal{U}_i$, we apply Theorem 5.3.8 to obtain an open set B_U such that

$$F_U \subset B_U \subset \mathrm{cl}\, B_U \subset U \quad \text{and} \quad \dim \mathrm{bd}\, B_U \leq n-1.$$

It is easy to see that $\mathcal{B} = \{B_U \mid U \in \mathcal{U}_i, i \in \mathbb{N}\}$ is a σ-locally finite open basis for X. Let

$$Y = \bigcup \{\mathrm{bd}\, B \mid B \in \mathcal{B}\} \quad \text{and} \quad Z = X \setminus Y.$$

According to Corollary 5.4.3, $\dim Y \le n-1$. Since $\{B \cap Z \mid B \in \mathcal{B}\}$ is a σ-locally finite clopen basis for Z, we have $\dim Z \le 0$ by Theorem 5.3.5. □

In the above proof of Proposition 5.4.6, the following two facts have been proved:

(1) Each metrizable space X with $\dim X \le n$ has a σ-locally finite open basis $\mathcal{B}$ such that $\dim \operatorname{bd} B \le n-1$ for every $B \in \mathcal{B}$.
(2) If a metrizable space X has such a basis $\mathcal{B}$ then $X = Y \cup Z$ for some $Y, Z \subset X$ with $\dim Y \le n-1$ and $\dim Z \le 0$.

In (2), Y is covered by n many subsets with $\dim \le 0$ by the Decomposition Theorem 5.4.5. Hence, X is covered by $n+1$ many subsets with $\dim \le 0$. By the Decomposition Theorem 5.4.5 again, we have $\dim X \le n$. Thus, (1) implies $\dim X \le n$. Consequently, we have the following characterization of dimension, which is a generalization of Theorem 5.3.5:

Theorem 5.4.7. *Let X be metrizable and $n \in \omega$. Then,* $\dim X \le n$ *if and only if X has a σ-locally finite open basis $\mathcal{B}$ such that* $\dim \operatorname{bd} B \le n-1$ *for each $B \in \mathcal{B}$.* □

The following theorem is obtained as a corollary of the Decomposition Theorem 5.4.5:

Theorem 5.4.8 (ADDITION THEOREM). *For any two subspaces X and Y of a metrizable space,*

$$\dim X \cup Y \le \dim X + \dim Y + 1.$$ □

Regarding the dimension of product spaces, we have the following theorem:

Theorem 5.4.9 (PRODUCT THEOREM). *For any metrizable spaces X and Y,*

$$\dim X \times Y \le \dim X + \dim Y.$$

Proof. If $\dim X = \infty$ or $\dim Y = \infty$, the theorem is obvious.

When $\dim X, \dim Y < \infty$, we prove the theorem by induction on $\dim X + \dim Y$. The case $\dim X = \dim Y = 0$ is a consequence of Theorem 5.3.6. Assume the theorem is true for any two spaces X and Y with $\dim X + \dim Y < k$. Now, let $\dim X = m$, $\dim Y = n$, and $m + n = k$. According to Theorem 5.4.7, X and Y have σ-locally finite open bases $\mathcal{B}_X$ and $\mathcal{B}_Y$ such that $\dim \operatorname{bd} B \le m-1$ for each $B \in \mathcal{B}_X$ and $\dim \operatorname{bd} B \le n-1$ for each $B \in \mathcal{B}_Y$. Then,

$$\mathcal{B} = \left\{ B_1 \times B_2 \;\middle|\; B_1 \in \mathcal{B}_X \text{ and } B_2 \in \mathcal{B}_Y \right\}$$

is a σ-locally finite open basis for $X \times Y$. For each $B_1 \in \mathcal{B}_X$ and $B_2 \in \mathcal{B}_Y$,

$$\operatorname{bd}(B_1 \times B_2) = (\operatorname{bd} B_1 \times \operatorname{cl} B_2) \cup (\operatorname{cl} B_1 \times \operatorname{bd} B_2).$$

Hence, $\dim \mathrm{bd}(B_1 \times B_2) \leq m+n-1$ by the inductive assumption and Theorems 5.4.1 or 5.4.2. Then, we have $\dim X \times Y \leq m+n$ by Theorem 5.4.7. □

Remark 8. In Theorem 5.4.9, the equality $\dim X \times Y = \dim X + \dim Y$ does not hold in general. In fact, there exists a separable metrizable space X such that $\dim X^2 \neq 2 \dim X$. Such a space will be constructed in Theorem 5.12.1. However, if one of X or Y is a locally compact polyhedron or a metric polyhedron (cf. Sect. 4.5), the equality does hold. This will be proved in Theorem 7.9.7.

5.5 Inductive Dimensions

In this section, we introduce two types of dimension defined by induction. First, the **large inductive dimension** $\mathrm{Ind}\, X$ of X can be defined as follows: $\mathrm{Ind}\, \emptyset = -1$ and $\mathrm{Ind}\, X \leq n$ if each closed set $A \subset X$ has an arbitrarily small open neighborhood V with $\mathrm{Ind}\, \mathrm{bd}\, V \leq n-1$. Then, we define $\mathrm{Ind}\, X = n$ if $\mathrm{Ind}\, X \leq n$ and $\mathrm{Ind}\, X \not\leq n-1$. We write $\mathrm{Ind}\, X < \infty$ if $\mathrm{Ind}\, X \leq n$ for some $n \in \mathbb{N}$, and otherwise $\mathrm{Ind}\, X = \infty$. Observe the following:

- If Y is a closed set in X then $\mathrm{Ind}\, Y \leq \mathrm{Ind}\, X$.

For an open set G and a closed set F in X,

$$\mathrm{bd}\,\mathrm{cl}\, G = \mathrm{cl}\, G \setminus \mathrm{int}\,\mathrm{cl}\, G \subset \mathrm{cl}\, G \setminus G = \mathrm{bd}\, G \quad \text{and}$$
$$\mathrm{bd}\,\mathrm{int}\, F = \mathrm{cl}\,\mathrm{int}\, F \setminus \mathrm{int}\, F \subset F \setminus \mathrm{int}\, F = \mathrm{bd}\, F.$$

Then, $\mathrm{Ind}\, X \leq n$ if and only if each closed set A in X has an arbitrarily small *closed* neighborhood V with $\mathrm{Ind}\, \mathrm{bd}\, V \leq n-1$.

As is easily observed, $\mathrm{Ind}\, X \leq n$ if and only if, for any two disjoint closed sets A and B in X, there is a partition L between A and B with $\mathrm{Ind}\, L \leq n-1$. Note that $\mathrm{Ind}\, \emptyset = \dim \emptyset = -1$. The next theorem follows, by induction, from Theorem 5.3.8.

Theorem 5.5.1. *For every metrizable space* X, $\dim X = \mathrm{Ind}\, X$. □

Next, the **small inductive dimension** $\mathrm{ind}\, X$ of X is defined as follows[2]: $\mathrm{ind}\, \emptyset = -1$ and $\mathrm{ind}\, X \leq n$ if each point $x \in X$ has an arbitrarily small open neighborhood V with $\mathrm{ind}\, \mathrm{bd}\, V \leq n-1$; and then $\mathrm{ind}\, X = n$ if $\mathrm{ind}\, X \leq n$ and $\mathrm{ind}\, X \not\leq n-1$. We write $\mathrm{ind}\, X < \infty$ if $\mathrm{ind}\, X \leq n$ for some $n \in \mathbb{N}$, and otherwise $\mathrm{ind}\, X = \infty$. Now,

- $\mathrm{ind}\, Y \leq \mathrm{ind}\, X$ for an *arbitrary* subset $Y \subset X$.

Then, $\mathrm{ind}\, X \leq n$ if and only if each point x of X has an arbitrarily small *closed* neighborhood V with $\mathrm{ind}\, \mathrm{bd}\, V \leq n-1$.

---

[2]In this chapter, spaces are assumed normal, but the small inductive dimension also makes sense for regular spaces.

By definition, $\operatorname{ind} X \leq \operatorname{Ind} X$ and $\operatorname{ind} \emptyset = \operatorname{Ind} \emptyset = \dim \emptyset = -1$. As is easily shown, $\operatorname{ind} X \leq n$ if and only if X has an open basis $\mathcal{B}$ such that $\operatorname{ind} \operatorname{bd} B \leq n-1$ for every $B \in \mathcal{B}$. Comparing this with Theorem 5.4.7, one might expect that the equality $\operatorname{ind} X = \operatorname{Ind} X = \dim X$ holds for an arbitrary metrizable space X. However, there exists a completely metrizable space X such that $\operatorname{ind} X \neq \operatorname{Ind} X$. Before constructing such a space, we first prove the following Coincidence Theorem:

Theorem 5.5.2 (Coincidence Theorem). *For every separable metrizable space X, the equality $\dim X = \operatorname{Ind} X = \operatorname{ind} X$ holds.*

Proof. Because $\operatorname{ind} X \leq \operatorname{Ind} X$ and $\operatorname{Ind} X = \dim X$, it is enough to show that $\dim X \leq \operatorname{ind} X$ when $\operatorname{ind} X < \infty$. We will prove this by induction on $\operatorname{ind} X$. Assume that $\dim X \leq \operatorname{ind} X$ for every separable metrizable space X with $\operatorname{ind} X < n$. Now, let $\operatorname{ind} X = n$. Then, X has an open basis $\mathcal{B}$ such that $\operatorname{ind} \operatorname{bd} B \leq n-1$ for every $B \in \mathcal{B}$. Since X is separable metrizable, X has a countable open basis $\{V_i \mid i \in \mathbb{N}\}$. We define

$$P = \{(i,j) \in \mathbb{N}^2 \mid V_i \subset B \subset V_j \text{ for some } B \in \mathcal{B}\}.$$

For each $p = (i,j) \in P$, choose $B_p \in \mathcal{B}$ so that $V_i \subset B_p \subset V_j$. Then, $\{B_p \mid p \in P\}$ is a countable open basis for X such that $\dim \operatorname{bd} B_p \leq \operatorname{ind} \operatorname{bd} B_p \leq n-1$ for each $p \in P$. By Theorem 5.4.7, we have $\dim X \leq n$. □

In the non-separable case, we have the following theorem:

Theorem 5.5.3. *There exists a completely metrizable space Z such that* $\operatorname{ind} Z = 0$ *but* $\operatorname{Ind} Z = \dim Z = 1$. *Furthermore, Z has a 0-dimensional compactification.*

Example and Proof. Let $\Omega = [0, \omega_1)$ be the space of all countable ordinals with the order topology. Note that the space $\overline{\Omega} = [0, \omega_1]$ is compact and 0-dimensional. In fact, for each open cover $\mathcal{U}$ of $\overline{\Omega}$, we can inductively choose $\omega_1 = \alpha_0 > \alpha_1 > \alpha_2 > \cdots$ so that each $(\alpha_i, \alpha_{i-1}]$ is contained in some member of $\mathcal{U}$. Since Ω is well-ordered, some α_n is equal to 0. Thus, $\mathcal{U}$ has a finite open refinement $\{0\} \cup \{(\alpha_i, \alpha_{i-1}] \mid i = 1, \ldots, n\}$, which is pair-wise disjoint.

Our space is constructed as a subspace of the product $\Omega^{\mathbb{N}}$. Let L be the subset of Ω consisting of infinite limit ordinals and $S = \Omega \setminus L$. For each $k \in \mathbb{N}$, let $S_k = \{\alpha + k \mid \alpha \in L\}$. We define

$$Z = \{z \in \Omega^{\mathbb{N}} \mid z(k) \in L \Rightarrow z(k+1) = z(k) + k, z(k+j) \in S_k \text{ for } j > 1\}.$$

By definition, we have $\operatorname{ind} \Omega^{\mathbb{N}} = 0$, so $\operatorname{ind} Z = 0$. On the other hand, we can write $Z = S^{\mathbb{N}} \cup \bigcup_{k \in \mathbb{N}} Z_k$, where

$$Z_k = \{z \in Z \mid z(k) \in L\} \subset S^{k-1} \times L \times S_k^{\mathbb{N}}.$$

Since S is a discrete space, it follows from Theorem 5.3.6 that $\dim S^{\mathbb{N}} = 0$. As is easily seen, each Z_k is homeomorphic to $S^{k-1} \times S_k^{\mathbb{N}}$ via the following correspondence:

$$Z_k \ni z \mapsto (z(1), \dots, z(k-1), z(k+1), z(k+2), \dots) \in S^{k-1} \times S_k^{\mathbb{N}},$$

where $z(k+1) = z(k) + k$. Then, it follows that $\dim Z_k = 0$ for each $k \in \mathbb{N}$.

(*Neighborhood bases*) For each $\alpha \in L$, choose $\xi_1(\alpha) < \xi_2(\alpha) < \cdots < \alpha$ so that $\sup_{i \in \mathbb{N}} \xi_i(\alpha) = \alpha$. Each $z \in Z$ has the neighborhood basis $\{U_n(z) \mid n \in \mathbb{N}\}$ defined as follows:

$$U_n(z) = \begin{cases} \{x \in Z \mid x(i) = z(i) \ \text{ for } i \leq n\} & \text{if } z \in S^{\mathbb{N}}, \\ \{x \in Z \mid x(i) = z(i) \ \text{ for } k \neq i \leq k+n, & \\ \qquad\qquad \text{and } \xi_n(z(k)) < x(k) \leq z(k)\} & \text{if } z \in Z_k. \end{cases}$$

Note that each $U_n(z)$ is clopen in Z, but $\{U_n(z) \mid z \in S^{\mathbb{N}}\}$ is not locally finite at the point $(\omega, \omega+1, \omega+1, \dots)$ in Z (cf. Theorem 5.3.5). The following statements can be easily proved:

(1) If $z, z' \in S^{\mathbb{N}}$ or $z, z' \in Z_k$, then $U_n(z) \cap U_n(z') \neq \emptyset \Rightarrow U_n(z) = U_n(z')$.
(2) If $z \in Z_k$, $z' \in Z_{k'}$, and $k < k' < n + k$, then $U_n(z) \cap U_{n'}(z') = \emptyset$ for every $n' \in \mathbb{N}$.
(3) If $z \in S^{\mathbb{N}}$, $z' \in Z_k$, and $n < k$, then $U_n(z) \cap U_{n'}(z') \neq \emptyset \Rightarrow U_{n'}(z') \subset U_n(z)$.

Furthermore, we have the following:

(4) For any $z \in S^{\mathbb{N}}$ and $n \in \mathbb{N}$, there exists some $m > n$ such that $U_m(z) \cap U_m(z') = \emptyset$ for every $z' \in \bigcup_{k \leq n} Z_k$.

In fact, if $z(n+1) \notin S_k$ for any $k \leq n$, then $U_{n+1}(z) \cap U_{n+1}(z') = \emptyset$ for every $z' \in \bigcup_{k \leq n} Z_k$. If $z(n+1) \in S_k$ for some $k \leq n$, then $U_m(z) \cap U_m(z') = \emptyset$ for every $m > n$ and $z' \in \bigcup_{k \neq j \leq n} Z_j$. On the other hand, because $z(k+1) \in S$, we can write $z(k+1) = \alpha + r$, where $\alpha \in L \cup \{0\}$ and $r \in \mathbb{N}$. If $\alpha = 0$ or $r \neq k$, then $U_m(z) \cap U_m(z') = \emptyset$ for every $m > k$ and $z' \in Z_k$. When $\alpha \in L$ and $r = k$, choose $m > k$ so that $z(k) \notin (\xi_m(\alpha), \alpha]$. Then, it follows that $U_m(z) \cap U_m(z') = \emptyset$ for every $z' \in Z_k$.

Note that each Z_k is closed in Z by (2) and (4). Then, as mentioned before, we have $\dim Z \leq 1$.

(*Metrizability*) To prove the metrizability, by the Frink Metrization Theorem 2.4.1 it suffices to show that, for each $z \in Z$ and $n \in \mathbb{N}$, there exists $m \in \mathbb{N}$ so that $U_m(z) \cap U_m(z') \neq \emptyset$ implies $U_m(z') \subset U_n(z)$.

When $z \in Z_k$ for some $k \in \mathbb{N}$, if $z' \in \bigcup_{k' < k} Z_{k'}$ or $z' \in \bigcup_{k < k' < n+2k} Z_{k'}$ then $U_{n+k}(z) \cap U_{n+k}(z') = \emptyset$ by (2). Assume $U_{n+k}(z) \cap U_{n+k}(z') \neq \emptyset$. If $z' \in S^{\mathbb{N}} \cup \bigcup_{k' > n+k} Z_{k'}$, then $U_{n+k}(z') \subset U_n(z)$ by definition. If $z' \in Z_k$, then $U_{n+k}(z') = U_{n+k}(z) \subset U_n(z)$ by (1). Thus, we have

$$U_{n+k}(z) \cap U_{n+k}(z') \neq \emptyset \Rightarrow U_{n+k}(z') \subset U_n(z).$$

For $z \in S^{\mathbb{N}}$, we can choose $m > n$ by (4) such that $U_m(z) \cap U_m(z') = \emptyset$ for every $z' \in \bigcup_{k \leq n} Z_k$. Assume $U_m(z) \cap U_m(z') \neq \emptyset$. Then, $z' \in S^{\mathbb{N}}$ or $z' \in Z_k$ for some $k > n$. If $z' \in S^{\mathbb{N}}$, then $U_m(z') = U_m(z) \subset U_n(z)$ by (1). If $z' \in Z_k$ for some $k > n$, then $U_m(z') \subset U_n(z)$ by (3). Thus, we have

$$U_m(z) \cap U_m(z') \neq \emptyset \Rightarrow U_m(z') \subset U_n(z).$$

(*Complete metrizability*) Because of Theorem 2.5.5, to show the complete metrizability of Z, it is enough to prove that Z is a G_δ-set in the compact space $\overline{\Omega}^{\mathbb{N}}$. Extend each $U_n(z)$ to a neighborhood of z in $\overline{\Omega}^{\mathbb{N}}$ as follows:

$$\tilde{U}_n(z) = \begin{cases} \{x \in \overline{\Omega}^{\mathbb{N}} \mid x(i) = z(i) \ \text{ for } i \leq n\} & \text{for } z \in S^{\mathbb{N}}, \\ \{x \in \overline{\Omega}^{\mathbb{N}} \mid x(i) = z(i) \ \text{ for } k \neq i \leq k+n, & \\ \qquad \text{and } \xi_n(z(k)) < x(k) \leq z(k)\} & \text{for } z \in Z_k. \end{cases}$$

Then, each $W_n = \bigcup_{z \in Z} \tilde{U}_n(z)$ is an open neighborhood of Z in $\overline{\Omega}^{\mathbb{N}}$ and $Z = \bigcap_{n \in \mathbb{N}} W_n$. Indeed, if $x \in \bigcap_{n \in \mathbb{N}} W_n \setminus S^{\mathbb{N}}$, then $x(k) \in L$ for some $k \in \mathbb{N}$. For $n > k$, choose $z_n \in Z$ so that $x \in \tilde{U}_n(z_n)$. Since $x(k) \in L$ and $k < n$, it follows that $z_n \notin S^{\mathbb{N}} \cup \bigcup_{k' \neq k} Z_{k'}$, i.e., $z_n \in Z_k$. Then, $x(k+i) = z_n(k+i) \in S_k$ for each $0 < i \leq n$ and $\xi_n(z_n(k)) < x(k) \leq z_n(k)$. Since $x(k+1) = z_n(k+1) = z_n(k)+k$, every $z_n(k)$ is identical, say $z(k)$. Since $z(k) = \sup \xi_n(z(k))$, we have $x(k) = z(k)$. Taking $n \in \mathbb{N}$ arbitrarily large, we can see that $x(i) \in S_k$ for any $i > k$. Hence, $x \in Z_k \subset Z$.

(*1-dimensionality*) It has been shown that Z is metrizable and each Z_k is closed in Z. Then, applying the Countable Sum Theorem (5.4.1) and the Addition Theorem (5.4.8), we have $\dim Z \leq 1$.

To see that $\dim Z > 0$, assume $\dim Z = 0$. Let $\mathcal{W} = \{W_\alpha \mid \alpha \in \Omega\} \in \operatorname{cov}(Z)$, where $W_\alpha = \{z \in Z \mid 0 \leq z(2) \leq \alpha\}$. By the assumption, $\mathcal{W}$ has an open refinement $\mathcal{V}$ with $\operatorname{ord} \mathcal{V} \leq 1$. Then, $\mathcal{V}$ is discrete in Z. Here, we call $s \in S^n$ **regular** if there exist $f : \bigoplus_{i \in \mathbb{N}} S^i \to S$ and $V \in \mathcal{V}$ such that $R(s; f) \subset V$, where

$$R(s; f) = \{x \in S^{\mathbb{N}} \mid x(i) = s(i) \ \text{ for } i \leq n \text{ and}$$
$$x(n+i) \geq f(x(n), \ldots, x(n+i-1)) \ \text{ for } i \in \mathbb{N}\}.$$

Otherwise, s is said to be **irregular**.

First, we verify the following fact:

(5) Every $s \in S$ is irregular.

For each $f : \bigoplus_{i \in \mathbb{N}} S^i \to S$ and $\alpha \in \Omega$, define $s_f^\alpha \in S^{\mathbb{N}}$ as follows: $s_f^\alpha(1) = s$, $s_f^\alpha(2) = \max\{s, f(s), \alpha + 1\}$, and $s_f^\alpha(i+1) = f(s_f^\alpha(1), \ldots, s_f^\alpha(i))$ for $i \geq 2$. Then, $s_f^\alpha \in R(s; f) \setminus W_\alpha$. Hence, $R(s; f)$ is not contained in any $V \in \mathcal{V}$.

Next, we show the following fact:

(6) If $s \in S^n$ is irregular, then $(s,t) \in S^{n+1}$ is irregular for some $t \in S$.

Suppose that (s,t) is regular for every $t \in S$, that is, there are $f_t : \bigoplus_{i\in\mathbb{N}} S^i \to S$ and $V_t \in \mathcal{V}$ such that $R(s,t;f_t) \subset V_t$. When there exist $a \in S$ and $V \in \mathcal{V}$ such that $V_t = V$ for $t \geq \max\{a, f_a(a)\}$, we define $f : \bigoplus_{i\in\mathbb{N}} S^i \to S$ by

$$f(t) = \max\{a, f_a(a)\} \quad \text{for } t \in S \text{ and}$$

$$f(t_1,\ldots,t_i) = f_{t_2}(t_2,\ldots,t_i) \quad \text{for } (t_1,\ldots,t_i) \in S^i, i \geq 2.$$

For $x \in R(s;f)$, let $t = x(n+1)$. Then, $x \in R(s,t;f_t)$ because

$$\begin{aligned} x(n+1+i) &\geq f(x(n),\ldots,x(n+i)) \\ &= f_t(x(n+1),\ldots,x(n+1+(i-1))) \quad \text{for } i \in \mathbb{N}. \end{aligned}$$

Moreover, $t = x(n+1) \geq f(x(n)) = \max\{a, f_a(a)\} \geq a$. Hence, $R(s;f) \subset \bigcup_{t\geq a} R(s,t;f_t) \subset V$, which contradicts the irregularity of s. Therefore, we can obtain an increasing sequence $a_1 < a_2 < \cdots$ in S such that $V_{a_i} \neq V_{a_{i+1}}$ and $a_{i+1} \geq f_{a_i}(a_i)$. Let $\alpha = \sup_{i\in\mathbb{N}} a_i \in L$ and $b_0 = \alpha + n + 1$. For each $j \in \mathbb{N}$, we can inductively choose $b_j \in S_{n+1}$ so that

$$b_j \geq \sup_{i\in\mathbb{N}} f_{a_i}(a_i, b_0, \ldots, b_{j-1}).$$

Then, we have

$$z = (s(1),\ldots,s(n),\alpha,b_0,b_1,b_2,\ldots) \in Z_{n+1} \quad \text{and}$$

$$z_i = (s(1),\ldots,s(n),a_i,b_0,b_1,b_2,\ldots) \in R(s,a_i;f_{a_i}) \subset V_{a_i},$$

where $\lim_{i\to\infty} z_i = z$. This contradicts the discreteness of $\mathcal{V}$ because $V_{a_i} \neq V_{a_{i+1}}$.

By (5) and (6), we obtain $s \in S^{\mathbb{N}}$ such that each $(s(1),\ldots,s(n)) \in S^n$ is irregular. Then, s is contained in some $V \in \mathcal{V}$, from which $U_n(s) \subset V$ for some $n \in \mathbb{N}$, which implies that $(s(1),\ldots,s(n))$ is regular. This is a contradiction.

(*0-dimensional compactification*) Finally, we will show that $\mathrm{cl}_{\overline{\Omega}^{\mathbb{N}}} Z$ is a 0-dimensional compactification of Z. It suffices to show that $\dim \overline{\Omega}^{\mathbb{N}} = 0$. Because $\overline{\Omega}^{\mathbb{N}}$ is compact, each open cover $\mathcal{U}$ of $\overline{\Omega}^{\mathbb{N}}$ has a finite refinement

$$\{p_{m_i}^{-1}(\textstyle\prod_{j=1}^{m_i}[\alpha_{i,j},\beta_{i,j}]) \mid i = 1,\ldots,n\},$$

where $p_k : \overline{\Omega}^{\mathbb{N}} \to \overline{\Omega}^k$ is the projection onto the first k factors. We write

$$\{\alpha_{i,j},\beta_{i,j} \mid i = 1,\ldots,n; j = 1,\ldots,m_i\} = \{\gamma_k \mid k = 1,\ldots,\ell\},$$

where $\gamma_k < \gamma_{k+1}$ for each $k = 1, \ldots, \ell - 1$. Note that $\gamma_1 = 0$ and $\gamma_\ell = \omega_1$. Then, $\mathcal{U}$ has the following pair-wise disjoint open refinement:

$$\{p_m^{-1}\left(\prod_{j=1}^m (\gamma_{k_j - 1}, \gamma_{k_j}]\right) \mid k_j = 1, \ldots, \ell\},$$

where $m = \max\{m_1, \ldots, m_n\}$ and $(\gamma_0, \gamma_1] = \{0\}$. Therefore, $\dim \overline{\Omega}^{\mathbb{N}} = 0$. This completes the proof. □

Remark 9. According to Theorem 5.5.3, there exists a 0-dimensional compact space that contains a 1-dimensional subspace. Thus, in the Subset Theorem 5.3.3, metrizability cannot be replaced by compactness.

Remark 10. The inequality $\dim X \leq \operatorname{Ind} X$ holds for any completely normal (= hereditarily normal) space X because the "if" part of Theorem 5.3.8 is valid for such a space, as was pointed out in Remark 6 (at the end of Sect. 5.3).

5.6 Infinite Dimensions

In this section, several types of infinite dimensions are defined and discussed. According to Theorem 5.2.17, $\dim X = \infty$ if and only if X has an essential family of n pairs of disjoint closed sets for any $n \in \mathbb{N}$. A space X is said to be **strongly infinite-dimensional (s.i.d.)** if X has an infinite essential family of pairs of disjoint closed sets. Obviously, if X is s.i.d. then $\dim X = \infty$. It is said that X is **weakly infinite-dimensional (w.i.d.)** if $\dim X = \infty$ and X is not s.i.d.,[3] that is, for every family $(A_i, B_i)_{i \in \mathbb{N}}$ of pairs of disjoint closed sets in X, there are partitions L_i between A_i and B_i such that $\bigcap_{i \in \mathbb{N}} L_i = \emptyset$.

Theorem 5.6.1. *The Hilbert cube* $\mathbf{I}^{\mathbb{N}}$ *is strongly infinite-dimensional.*

Proof. For each $i \in \mathbb{N}$, let

$$A_i = \{x \in \mathbf{I}^{\mathbb{N}} \mid x(i) = 0\} \text{ and } B_i = \{x \in \mathbf{I}^{\mathbb{N}} \mid x(i) = 1\}.$$

Then, $(A_i, B_i)_{i \in \mathbb{N}}$ is essential in $\mathbf{I}^{\mathbb{N}}$. Indeed, for each $i \in \mathbb{N}$, let L_i be a partition between A_i and B_i. For each $n \in \mathbb{N}$, let $j_n : \mathbf{I}^n \to \mathbf{I}^{\mathbb{N}}$ be the natural injection defined by

$$j_n(x) = (x(1), \ldots, x(n), 0, 0, \ldots).$$

Then, for each $i \leq n$, $j_n^{-1}(L_i)$ is a partition between

$$j_n^{-1}(A_i) = \{x \in \mathbf{I}^n \mid x(i) = 0\} \text{ and } j_n^{-1}(B_i) = \{x \in \mathbf{I}^n \mid x(i) = 1\}.$$

[3]In many articles, the infinite dimensionality is not assumed, i.e., w.i.d. = not s.i.d., so f.d. implies w.i.d. However, here we assume the infinite dimensionality because we discuss the difference among infinite-dimensional spaces.

Since $(j_n^{-1}(A_i), j_n^{-1}(B_i))_{i=1}^n$ is essential in $\mathbf{I}^n$ (Corollary 5.2.16), we have $\bigcap_{i=1}^n j_n^{-1}(L_i) \neq \emptyset$, hence $\bigcap_{i=1}^n L_i \neq \emptyset$. Since $\mathbf{I}^{\mathbb{N}}$ is compact, it follows that $\bigcap_{i\in\mathbb{N}} L_i \neq \emptyset$. □

By definition, a space is strongly infinite-dimensional if it contains an s.i.d. closed subspace. Then, it follows from Theorem 5.6.1 that every space containing a copy of $\mathbf{I}^{\mathbb{N}}$ is strongly infinite-dimensional. For example, ℓ_1, ℓ_2, and $\mathbb{R}^{\mathbb{N}}$ are s.i.d.[4] Moreover, rint $\boldsymbol{Q} = \bigcup_{n\in\mathbb{N}}[-1+2^{-n}, 1-2^{-n}]^{\mathbb{N}}$ and $\mathbf{I}^{\mathbb{N}} \setminus (0,1)^{\mathbb{N}}$ are also s.i.d.[5]

It is said that X is **countable-dimensional (c.d.)** if X is a countable union of f.d. normal subspaces, where it should be noted that subspaces of normal spaces need not be normal (cf. Sect. 2.10). A metrizable space is countable-dimensional if and only if it is a countable union of 0-dimensional subspaces, because an f.d. metrizable space is a finite union of 0-dimensional subspaces by the Decomposition Theorem 5.4.5.

Theorem 5.6.2. *A countable-dimensional metrizable space X with* $\dim X = \infty$ *is weakly infinite-dimensional. In other words, any strongly infinite-dimensional metrizable space is not countable-dimensional.*

Proof. Let $(A_i, B_i)_{i\in\mathbb{N}}$ be a family of pairs of disjoint closed sets in X. We can write $X = \bigcup_{i\in\mathbb{N}} X_i$, where $\dim X_i = 0$. From Theorem 5.2.17 and the Partition Extension Lemma 5.3.7, it follows that for each $i \in \mathbb{N}$, X has a partition L_i between A_i and B_i such that $L_i \cap X_i = \emptyset$. Then, we have

$$\bigcap_{i\in\mathbb{N}} L_i = \Big(\bigcap_{i\in\mathbb{N}} L_i\Big) \cap \Big(\bigcup_{i\in\mathbb{N}} X_i\Big) = \bigcup_{i\in\mathbb{N}}\bigg(\Big(\bigcap_{j\in\mathbb{N}} L_j\Big) \cap X_i\bigg)$$
$$\subset \bigcup_{i\in\mathbb{N}}(L_i \cap X_i) = \emptyset.$$

Therefore, X is w.i.d. □

According to Theorem 5.6.2, the space $\bigoplus_{n\in\mathbb{N}} \mathbf{I}^n$ and its one-point compactification are c.d., hence they are w.i.d. The following space is also c.d. (so w.i.d.):

$$\mathbf{I}^{\mathbb{N}}_f = \{x \in \mathbf{I}^{\mathbb{N}} \mid x(i) = 0 \text{ except for finitely many } i\}.$$

There exists a w.i.d. compactum that is not c.d. As is easily seen, any subspace of a c.d. metrizable space is also c.d. However, a subspace of a w.i.d. metrizable space need not be w.i.d. Such a compactum will be constructed in Theorem 5.13.1.

---

[4]It is known that $\ell_1 \approx \ell_2 \approx \mathbb{R}^{\mathbb{N}}$, where the latter homeomorphy was proved by R.D. Anderson.

[5]Since rint $\boldsymbol{Q}$ and $\mathbf{I}^{\mathbb{N}} \setminus (0,1)^{\mathbb{N}}$ are not completely metrizable, they are not homeomorphic to $\mathbb{R}^{\mathbb{N}}$, but it is known that rint $\boldsymbol{Q} \approx \mathbf{I}^{\mathbb{N}} \setminus (0,1)^{\mathbb{N}}$.

Now, we introduce a strong version of countable dimensionality. We say that X is **strongly countable-dimensional (s.c.d.)** if X is a countable union of f.d. *closed* subspaces. The space $\bigoplus_{n\in\mathbb{N}} \mathbf{I}^n$, its one-point compactification, and the space $\mathbf{I}^{\mathbb{N}}_f$ are s.c.d. Every s.c.d. space is c.d. but the converse does not hold. Let ν_ω be the subspace of the Hilbert cube $\mathbf{I}^{\mathbb{N}}$ defined as follows:

$$\nu_\omega = \{x \in \mathbf{I}^{\mathbb{N}} \mid x(i) \in \mathbf{I} \setminus \mathbb{Q} \text{ except for finitely many } i\}.$$

Theorem 5.6.3. *The space ν_ω is countable-dimensional but not strongly countable-dimensional.*

Proof. Since ν_ω is the countable union of subspaces

$$\{x \in \mathbf{I}^{\mathbb{N}} \mid x(i) \in \mathbf{I} \setminus \mathbb{Q} \text{ for } i \geq n\} \approx \mathbf{I}^n \times (\mathbf{I} \setminus \mathbb{Q})^{\mathbb{N}},$$

it follows that ν_ω is c.d. Moreover, $\dim \nu_\omega = \infty$ because $\mathbf{I}^n \times \{0\} \subset \nu_\omega$ for any $n \in \mathbb{N}$.

Assume that ν_ω is s.c.d., that is, $\nu_\omega = \bigcup_{n\in\mathbb{N}} F_n$, where each F_n is f.d. and closed in ν_ω. Consider the subspace $(\mathbf{I}\setminus\mathbb{Q})^{\mathbb{N}} \subset \nu_\omega$. Since $(\mathbf{I}\setminus\mathbb{Q})^{\mathbb{N}}$ is completely metrizable, at least one $F_n \cap (\mathbf{I} \setminus \mathbb{Q})^{\mathbb{N}}$ has the non-empty interior in $(\mathbf{I} \setminus \mathbb{Q})^{\mathbb{N}}$ by the Baire Category Theorem 2.5.1. Then, we have a non-empty open set U in ν_ω such that $U \cap (\mathbf{I} \setminus \mathbb{Q})^{\mathbb{N}} \subset F_n \cap (\mathbf{I} \setminus \mathbb{Q})^{\mathbb{N}}$. Since U contains a copy of every n-cube $\mathbf{I}^n$, it follows that $\dim U = \infty$, hence $U \setminus F_n \neq \emptyset$ because $\dim F_n < \infty$. Since $(\mathbf{I} \setminus \mathbb{Q})^{\mathbb{N}}$ is dense in ν_ω, we have

$$\big(U \cap (\mathbf{I}\setminus\mathbb{Q})^{\mathbb{N}}\big) \setminus \big(F_n \cap (\mathbf{I}\setminus\mathbb{Q})^{\mathbb{N}}\big) = (U \setminus F_n) \cap (\mathbf{I}\setminus\mathbb{Q})^{\mathbb{N}} \neq \emptyset,$$

which is a contradiction. Therefore, ν_ω is not s.c.d. □

A collection $\mathcal{A}$ of subsets of X is **locally countable** if each $x \in X$ has a neighborhood U that meets only countably many members of $\mathcal{A}$, i.e., $\operatorname{card} \mathcal{A}[U] \leq \aleph_0$.

Basic Properties of (Strong) Countable-Dimension 5.6.4.

(1) If X is a countable union of countable-dimensional subspaces, then X is countable-dimensional.
(2) If X is a countable union of strongly countable-dimensional closed subspaces, then X is strongly countable-dimensional.
(3) Every *closed* subspace of a (strongly) countable-dimensional space is (strongly) countable-dimensional. For a metrizable space, this is valid for a non-closed subspace, that is, every subspace of a (strongly) countable-dimensional *metrizable* space is (strongly) countable-dimensional.

The proofs of the above three items are trivial by definition.

(4) A paracompact space X is (strongly) countable-dimensional if each point $x \in X$ has a (strongly) countable-dimensional neighborhood.

Sketch of Proof. Let $\mathcal{P}$ be the property of closed sets in X being c.d. (or s.c.d.). Apply Michael's Theorem on local properties (Corollary 2.6.6). To show (F-3), use the Locally Finite Sum Theorem 5.4.2.

(5) If a paracompact space X has a locally countable union of countable-dimensional subspaces then X is countable-dimensional.
(6) If a paracompact space X has a locally countable union of strongly countable-dimensional closed subspaces then X is strongly countable-dimensional.

Sketch of Proof of (5) *(and* (6)*).* Let $\mathcal{A}$ be a locally countable (closed) cover of X such that each $A \in \mathcal{A}$ is c.d. (s.c.d.). Each $x \in X$ has an open neighborhood V_x in X such that $\mathcal{A}[V_x]$ is countable. Then, $\mathrm{st}(V_x, \mathcal{A}) = \bigcup \mathcal{A}[V_x]$ is a c.d. (s.c.d.) neighborhood of x in X.

From Theorem 5.3.8, it follows that any finite-dimensional metrizable space X contains an n-dimensional closed set for every $n \leq \dim X$. However, this is not true for an infinite-dimensional space. Namely, there exists an infinite-dimensional compactum such that every subset with $\dim \not\leq 0$ is infinite-dimensional. Such a space is called a **hereditarily infinite-dimensional (h.i.d.)** space. We will construct an h.i.d. compactum in Theorem 5.13.4.

Next, we introduce infinite-dimensional versions of inductive dimensions. By transfinite induction on ordinals $\alpha \geq \omega$, the **large transfinite inductive dimension** $\operatorname{trInd} X$ and the **small transfinite inductive dimension** $\operatorname{trind} X$ are defined as follows: $\operatorname{trInd} X < \omega$ means that $\operatorname{Ind} X < \infty$ and $\operatorname{trInd} X \leq \alpha$ if each closed set $A \subset X$ has an arbitrarily small open neighborhood V with $\operatorname{trInd} \operatorname{bd} V < \alpha$. Similarly, $\operatorname{trind} X < \omega$ means that $\operatorname{ind} X < \infty$ and $\operatorname{trind} X \leq \alpha$ if each $x \in X$ has an arbitrarily small open neighborhood V with $\operatorname{trind} \operatorname{bd} V < \alpha$. Then, we define $\operatorname{trInd} X = \alpha$ (resp. $\operatorname{trind} X = \alpha$) if $\operatorname{trInd} X \leq \alpha$ (resp. $\operatorname{trind} X \leq \alpha$) and $\operatorname{trInd} X \not\leq \beta$ (resp. $\operatorname{trind} X \not\leq \beta$) for any $\beta < \alpha$. It should be noted that $\operatorname{trInd} X < \omega$ (resp. $\operatorname{trind} X < \omega$) implies $\operatorname{trInd} X = \operatorname{Ind} X < \infty$ (resp. $\operatorname{trind} X = \operatorname{ind} X < \infty$). Using transfinite induction, we can show that if $\operatorname{trInd} X = \alpha$ (resp. $\operatorname{trind} X = \alpha$) and $\beta < \alpha$, then X contains a closed set A with $\operatorname{trInd} A = \beta$ (resp. $\operatorname{trind} A = \beta$).

Lemma 5.6.5. *If* $\operatorname{trInd} X = \alpha$ *(resp.* $\operatorname{trind} X = \alpha$*) and* $\beta < \alpha$*, then* X *has a closed set* Y *such that* $\operatorname{trInd} Y = \beta$ *(resp.* $\operatorname{trind} Y = \beta$*).*

Proof. Because of the similarity, we prove the lemma only for trInd. Assume that the lemma holds for any ordinal $< \alpha$. Since $\operatorname{trInd} X \not\leq \beta$, X has disjoint closed sets A and B such that $\operatorname{trInd} L \not< \beta$ for any partition L between A and B. On the other hand, since $\operatorname{trInd} X \leq \alpha$, there is a partition L between A and B such that $\operatorname{trInd} L < \alpha$. If $\beta = \operatorname{trInd} L$, then L is the desired Y. When $\beta < \operatorname{trInd} L$, by the inductive assumption, L has a closed set Y with $\operatorname{trInd} Y = \beta$. □

It is said that a space X has **large** (or **small**) **transfinite inductive dimension** (abbrev. trInd (or trind)) if $\operatorname{trInd} X \leq \alpha$ (or $\operatorname{trind} X \leq \alpha$) for some ordinal α.

Proposition 5.6.6. *For a space* X*, the following statements hold:*

(1) *If* X *has* trInd*, then* X *has* trind *and* $\operatorname{trind} X \leq \operatorname{trInd} X$.

(2) *If* X *has* trind*, then every subspace* A *of* X *also has* trind*, where* $\operatorname{trind} A \leq \operatorname{trind} X$.
(3) *If* X *has* trInd*, then every* closed *subspace* A *of* X *has* trInd*, where* $\operatorname{trInd} A \leq \operatorname{trInd} X$.
(4) *If* X *has no* trInd*, then* X *has a closed set* A *with an open neighborhood* U *such that the boundary of each neighborhood of* A *contained in* U *has no* trInd.
(5) *If* X *has no* trind*, then* X *has a point* $x \in X$ *with an open neighborhood* U *such that the boundary of each neighborhood of* x *contained in* U *has no* trind.

Proof. Statements (1)–(3) are easily proved by the definitions.

(4): Let $\mathcal{P}$ be the collection of pairs (A, U) of closed sets A in X and open sets U in X with $A \subset U$. Suppose that for each $(A, U) \in \mathcal{P}$, A has a neighborhood $V_{(A,U)}$ in X such that $\operatorname{cl} V_{(A,U)} \subset U$ and $\operatorname{bd} V_{(A,U)}$ has trInd. Take an ordinal α so that $\alpha > \operatorname{trInd} \operatorname{bd} V_{(A,U)}$ for every $(A, U) \in \mathcal{P}$. Then, $\operatorname{Ind} X \leq \alpha$, so X has trInd.

(5): In the proof of (4), replace the closed sets A in X with points $x \in X$. □

We now prove that the converse of Proposition 5.6.6(1) does not hold.

Theorem 5.6.7. *The strongly countable-dimensional space* $\bigoplus_{n\in\mathbb{N}} \mathbf{I}^n$ *has no* trInd *but* $\operatorname{trind} \bigoplus_{n\in\mathbb{N}} \mathbf{I}^n = \omega$.

Proof. Each point of $\bigoplus_{n\in\mathbb{N}} \mathbf{I}^n$ is contained in some $\mathbf{I}^n$, hence $\operatorname{trind} \bigoplus_{n\in\mathbb{N}} \mathbf{I}^n \leq \omega$. Because $\operatorname{ind} \bigoplus_{n\in\mathbb{N}} \mathbf{I}^n = \infty$, we have $\operatorname{trind} \bigoplus_{n\in\mathbb{N}} \mathbf{I}^n = \omega$.

On the other hand, assume that $\bigoplus_{n\in\mathbb{N}} \mathbf{I}^n$ has trInd, i.e., $\operatorname{trInd} \bigoplus_{n\in\mathbb{N}} \mathbf{I}^n = \alpha$ for some ordinal α. Then, $\alpha \geq \omega$ because $\dim \bigoplus_{n\in\mathbb{N}} \mathbf{I}^n = \infty$. By Lemma 5.6.5, $\bigoplus_{n\in\mathbb{N}} \mathbf{I}^n$ contains a closed set X with $\operatorname{trInd} X = \omega$. For each $n \in \mathbb{N}$, let $X_n = X \cap \mathbf{I}^n$. Then, each X_n is finite-dimensional, but $\sup_{n\in\mathbb{N}} \dim X_n = \infty$ because $X = \bigoplus_{n\in\mathbb{N}} X_n$. By Theorem 5.3.8, we have disjoint closed sets A_n and B_n in X_n such that $\dim L \geq \dim X_n - 1$ for any partition L between A_n and B_n in X_n. Then, $A = \bigoplus_{n\in\mathbb{N}} A_n$ and $B = \bigoplus_{n\in\mathbb{N}} B_n$ are disjoint closed sets in X. Since $\operatorname{trInd} X = \omega$, we have a partition L in X between A and B such that $\operatorname{trInd} L < \omega$, i.e., $\dim L < \infty$. Choose $n \in \mathbb{N}$ so that $\dim X_n > \dim L + 1$. Then, $X_n \cap L$ is a partition in X_n between A_n and B_n and $\dim X_n \cap L \leq \dim L < \dim X_n - 1$. This is a contradiction. Therefore, $\bigoplus_{n\in\mathbb{N}} \mathbf{I}^n$ has no trInd. □

The above Theorem 5.6.7 also shows that the converse of the following theorem does not hold.

Theorem 5.6.8. *A metrizable space is countable-dimensional if it has* trInd.

Proof. This can be proved by transfinite induction. Assume that all metrizable spaces with $\operatorname{trInd} < \alpha$ are c.d. and let X be a metrizable space with $\operatorname{trInd} X = \alpha$. By the analogy of Proposition 5.4.6, we can construct a σ-locally finite basis $\mathcal{B}$ for X such that $\operatorname{trInd} \operatorname{bd} B < \alpha$ for each $B \in \mathcal{B}$. Let

$$Y = \bigcup \{ \operatorname{bd} B \mid B \in \mathcal{B} \} \quad \text{and} \quad Z = X \setminus Y.$$

Then, $\dim Z \leq 0$ by Theorem 5.3.5. On the other hand, by the assumption, $\operatorname{bd} B$ is c.d. for all $B \in \mathcal{B}$. Then, Y is also c.d. by 5.6.4(5) and (1). Therefore, X is c.d. □

The following theorem can be proved in a similar manner (cf. the proof of Theorem 5.5.2).

Theorem 5.6.9. *A separable metrizable space is countable-dimensional if it has* trind. □

Remark 11. In Theorem 5.6.9, it is unknown whether the separability is necessary or not, that is, the existence of a metrizable space that has trind but is not c.d. is unknown.

As we saw in Theorem 5.6.7, the converse of Theorem 5.6.8 is not true in general, but it is true for compacta. Namely, the following theorem holds:

Theorem 5.6.10. *A compactum has* trInd *if and only if it is countable-dimensional.*

Proof. It is enough to prove the "if" part. Let X be compact and $X = \bigcup_{n \in \mathbb{N}} A_n$, where $\dim A_n \leq 0$ for each $n \in \mathbb{N}$. Suppose that X has no trInd. Then, by Proposition 5.6.6(4), X has a closed set A with an open neighborhood U such that the boundary of each neighborhood of A contained in U has no trInd. Since $\dim A_1 \leq 0$, we can use the Partition Extension Lemma 5.3.7 to find a closed neighborhood V_1 of A contained in U such that $\operatorname{bd} V_1 \cap A_1 = \emptyset$. Then, $X_1 = \operatorname{bd} V_1$ has no trInd and $X_1 \cap A_1 = \emptyset$. By the same argument, we have a closed set $X_2 \subset X_1$ that misses A_2 and has no trInd. Thus, by induction, we can obtain closed sets $X_1 \supset X_2 \supset \cdots$ such that each X_n has no trInd and $X_n \cap A_n = \emptyset$. Then,

$$\bigcap_{n \in \mathbb{N}} X_n = \bigcap_{n \in \mathbb{N}} X_n \cap \bigcup_{n \in \mathbb{N}} A_n \subset \bigcup_{n \in \mathbb{N}} (X_n \cap A_n) = \emptyset,$$

which contradicts the compactness of X. □

Although $\bigoplus_{n \in \mathbb{N}} \mathbf{I}^n$ has no trInd (Theorem 5.6.7), the one-point compactification of $\bigoplus_{n \in \mathbb{N}} \mathbf{I}^n$ has trInd by Theorem 5.6.10. Thus, even if a space X has trInd, it does not imply that a subspace A of X has trInd, that is, Theorem 5.6.6(3) does not hold without the closedness of A.

Theorem 5.6.11. *A completely metrizable space has* trind *if it is countable-dimensional.*

Proof. Let $X = (X, d)$ be a complete metric space and $X = \bigcup_{n \in \mathbb{N}} A_n$, where $\dim A_n \leq 0$ for each $n \in \mathbb{N}$. Suppose that X has no trind. Then, by Proposition 5.6.6(5), X has a point a with an open neighborhood U such that the boundary of each neighborhood of a contained in U has no trind, where we may assume that $\operatorname{diam} U < 2^{-1}$. In the same way as for Theorem 5.6.10, we can inductively obtain non-empty closed sets $X_1 \supset X_2 \supset \cdots$ such that $X_n \cap A_n = \emptyset$ and $\operatorname{diam} X_n < 2^{-n}$ for each $n \in \mathbb{N}$. Then,

$$\bigcap_{n \in \mathbb{N}} X_n = \bigcap_{n \in \mathbb{N}} X_n \cap \bigcup_{n \in \mathbb{N}} A_n \subset \bigcup_{n \in \mathbb{N}} (X_n \cap A_n) = \emptyset.$$

However, the completeness of X implies $\bigcap_{n\in\mathbb{N}} X_n \neq \emptyset$. This is a contradiction. □

Combining Theorems 5.6.9 and 5.6.11, we have the following corollary:

Corollary 5.6.12. *A* separable *completely metrizable space has* trind *if and only if it is countable-dimensional.* □

The next theorem shows that the "if" part of Theorem 5.6.11 does not hold without the completeness.

Theorem 5.6.13. *The strongly countable-dimensional space* $\mathbf{I}_f^{\mathbb{N}}$ *has no* trind.

To prove this theorem, we need the following two lemmas:

Lemma 5.6.14. *Let X be a subspace of a metrizable space M. Then, every open set U in M contains an open set U' in M such that $X \cap U' = X \cap U$ and $X \cap \mathrm{cl}_M U' = \mathrm{cl}_X(X \cap U')$, hence $X \cap \mathrm{bd}_M U' = \mathrm{bd}_X(X \cap U')$.*

Proof. Take $d \in \mathrm{Metr}\, M$ and define

$$U' = \{x \in U \mid d(x, X \cap U) < d(x, X \setminus U)\}.$$

Then, $X \cap U' = X \cap U$. Evidently, $\mathrm{cl}_X(X \cap U') \subset X \cap \mathrm{cl}_M U'$. Assume that $\mathrm{cl}_X(X \cap U') = \mathrm{cl}_X(X \cap U) \neq X \cap \mathrm{cl}_M U'$, that is, we have $x \in X \cap \mathrm{cl}_M U' \setminus \mathrm{cl}_X(X \cap U)$. For each $\varepsilon > 0$, we have $y \in U'$ so that $d(x, y) < \frac{1}{2}\min\{\varepsilon, d(x, X \cap U)\}$. Since $d(y, X \cap U) < d(y, X \setminus U)$, it follows that

$$\begin{aligned} d(x, X \setminus U) &\geq d(y, X \setminus U) - d(x, y) \\ &> d(y, X \cap U) - \tfrac{1}{2}d(x, X \cap U) = \tfrac{1}{2}d(x, X \cap U) > 0. \end{aligned}$$

On the other hand, $x \notin X \cap U$, i.e., $x \in X \setminus U$, which is a contradiction. □

Lemma 5.6.15. *Let M be a separable metrizable space and $X \subset M$ with* $\mathrm{trind}\, X \leq \alpha$. *Then, X is contained in some G_δ-set X^* in M with* $\mathrm{trind}\, X^* \leq \alpha$.

Proof. Assuming that the lemma is true for any ordinal $< \alpha$, we will show the lemma for α. For each $i \in \mathbb{N}$, applying Lemma 5.6.14, we can find a countable open collection $\mathcal{U}_i$ in M such that $X \subset X_i = \bigcup \mathcal{U}_i$, $\mathrm{mesh}\,\mathcal{U}_i < 1/i$, and $\mathrm{trind}\, X \cap \mathrm{bd}_M U < \alpha$ for each $U \in \mathcal{U}_i$, where $X \cap \mathrm{bd}_M U = \mathrm{bd}_X(X \cap U)$ for each $U \in \mathcal{U}_i$. By the inductive assumption, for each $U \in \mathcal{U}_i$, there is a G_δ-set G_U in M such that $X \cap \mathrm{bd}_M U \subset G_U$ and $\mathrm{trind}\, G_U < \alpha$. Then,

$$\begin{aligned} X^* &= \bigcap_{i\in\mathbb{N}} X_i \cap \bigcap_{i\in\mathbb{N}} \bigcap_{U\in\mathcal{U}_i} (G_U \cup (M \setminus \mathrm{bd}_M U)) \\ &= \bigcap_{i\in\mathbb{N}} X_i \setminus \bigcup_{i\in\mathbb{N}} \bigcup_{U\in\mathcal{U}_i} (\mathrm{bd}_M U \setminus G_U) \end{aligned}$$

is a G_δ-set in M and $X \subset X^*$. For any $i \in \mathbb{N}$, every $x \in X^*$ is contained in some $U \in \mathcal{U}_i$. Then, $\operatorname{diam} X^* \cap U < 1/i$ and

$$\begin{aligned}\operatorname{bd}_{X^*}(X^* \cap U) &= \operatorname{cl}_{X^*}(X^* \cap U) \setminus U \\ &\subset (X^* \cap \operatorname{cl}_M U) \setminus U = X^* \cap \operatorname{bd}_M U \subset G_U,\end{aligned}$$

which implies $\operatorname{trind} X^* \cap \operatorname{bd}_M U < \alpha$. Thus, each point $x \in X^*$ has an arbitrarily small neighborhood V with $\operatorname{trind} \operatorname{bd}_{X^*} V < \alpha$. Hence, $\operatorname{trind} X^* \leq \alpha$. □

Proof of Theorem 5.6.13. Assume that $\mathbf{I}_f^{\mathbb{N}}$ has trind. According to Lemma 5.6.15, $\mathbf{I}_f^{\mathbb{N}}$ is contained in some G_δ-set G in $\mathbf{I}^{\mathbb{N}}$ that also has trind. Then, G is c.d. by Theorem 5.6.9. We show that G contains a copy of $\mathbf{I}^{\mathbb{N}}$, hence G is s.i.d., which contradicts Theorem 5.6.2. Thus, we obtain the desired result.

Let $G = \bigcap_{k \in \mathbb{N}} U_k$, where U_k is open in $\mathbf{I}^{\mathbb{N}}$. Note that $0 = (0, 0, \dots) \in \mathbf{I}_f^{\mathbb{N}} \subset G \subset U_1$. Choose $n_1 \in \mathbb{N}$ and $a_1, \dots, a_{n_1} \in (0, 1)$ so that

$$\{x \in \mathbf{I}^{\mathbb{N}} \mid x(i) \leq a_i \text{ for } i = 1, \dots, n_1\} \subset U_1.$$

Note that $\prod_{i=1}^{n_1} [0, a_i] \times \{0\} \subset \mathbf{I}_f^{\mathbb{N}} \subset U_2$. According to the Wallace Theorem 2.1.2, we can choose $n_2 \in \mathbb{N}$ and $a_{n_1+1}, \dots, a_{n_2} \in (0, 1)$ so that $n_2 > n_1$ and

$$\{x \in \mathbf{I}^{\mathbb{N}} \mid x(i) \leq a_i \text{ for } i = 1, \dots, n_2\} \subset U_2.$$

By induction, we can obtain an increasing sequence n_i of natural numbers and a sequence $a_i \in (0, 1)$ such that

$$\{x \in \mathbf{I}^{\mathbb{N}} \mid x(i) \leq a_i \text{ for } i = 1, \dots, n_k\} \subset U_k \quad \text{for each } k \in \mathbb{N}.$$

Then, $G = \bigcap_{k \in \mathbb{N}} U_k$ contains $\prod_{i \in \mathbb{N}} [0, a_i] \approx \mathbf{I}^{\mathbb{N}}$. □

Remark 12. There exists a slightly stronger version of the weak infinite dimension. We say that X is **weakly infinite-dimensional in the sense of Smirnov (S-w.i.d.)** if $\dim X = \infty$, and for every family $(A_i, B_i)_{i \in \mathbb{N}}$ of pairs of disjoint closed sets in X, there are partitions L_i between A_i and B_i such that $\bigcap_{i=1}^n L_i = \emptyset$ for some $n \in \mathbb{N}$. To distinguish w.i.d. from S-w.i.d. the term "**weakly infinite-dimensional in the sense of Alexandroff (A-w.i.d.)**" is used. Obviously, every S-w.i.d. space is (A-)w.i.d. For compact spaces, the converse is also true, that is, the two notions of weak infinite dimension are equivalent. It was shown in [32] that the Stone–Čech compactification of a normal space X is w.i.d. if and only if X is S-w.i.d.[6]

[6]Refer to Engelking's book "Theory of Dimensions, Finite and Infinite," Problem 6.1.E.

5.7 Compactification Theorems

Note that every separable metrizable space X has a metrizable compactification. Indeed, embedding X into the Hilbert cube $\mathbf{I}^{\mathbb{N}}$ (Corollary 2.3.8), the closure of X in $\mathbf{I}^{\mathbb{N}}$ is a metrizable compactification of X. On the other hand, every n-dimensional metrizable space can be embedded in an n-dimensional completely metrizable space as a dense set (Theorem 5.3.4). In this section, we show that every n-dimensional separable metrizable space has an n-dimensional metrizable compactification and that every c.d. (resp. s.c.d.) separable completely metrizable space has a c.d. (resp. s.c.d.) metrizable compactification.

Note. Here is an alternative proof of Corollary 2.3.8. Let $X = (X, d)$ be a separable metric space with $\{a_i \mid i \in \mathbb{N}\}$ a countable dense set. For each $i \in \mathbb{N}$, we define a map $f_i : X \to \mathbf{I}$ by $f_i(x) = \min\{1, d(x, a_i)\}$ for each $x \in X$. Then, the map $f : X \to \mathbf{I}^{\mathbb{N}}$ defined by $f(x) = (f_i(x))_{i\in\mathbb{N}}$ is an embedding. Indeed, for $x \neq y \in X$, choose a_i so that $d(x, a_i) < \min\{1, \frac{1}{2}d(x, y)\}$. Then, $f_i(x) < f_i(y)$ because

$$f_i(x) = d(x, a_i) < \tfrac{1}{2}d(x, y) < d(x, y) - d(x, a_i) \leq d(y, a_i).$$

Thus, f is injective. If f is not an embedding, then there are $x, x_n \in X$, $n \in \mathbb{N}$, and $0 < \delta < 1$ such that $\lim_{n\to\infty} f(x_n) = f(x)$ but $d(x_n, x) \geq \delta$ for all $n \in \mathbb{N}$. Choose a_i so that $d(x, a_i) < \frac{1}{3}\delta$. Then, we have $f_i(x) < 1$. For sufficiently large $n \in \mathbb{N}$,

$$\begin{aligned} f_i(x_n) - f_i(x) &= d(x_n, a_i) - d(x, a_i) \\ &\geq d(x_n, x) - 2d(x, a_i) > \delta - \tfrac{2}{3}\delta = \tfrac{1}{3}\delta, \end{aligned}$$

which contradicts $\lim_{n\to\infty} f_i(x_n) = f_i(x)$. Therefore, f is an embedding.

Recall that a metric space $X = (X, d)$ or a metric d is said to be **totally bounded** provided that, for any $\varepsilon > 0$, there is a finite set $A \subset X$ such that $d(x, A) < \varepsilon$ for every $x \in X$, i.e., $X = \bigcup_{a\in A} \mathrm{B}_d(a, \varepsilon)$. It is now easy to show that X is totally bounded if and only if, for any $\varepsilon > 0$, X has a finite open cover $\mathcal{U}$ with mesh $\mathcal{U} < \varepsilon$. Then, every compact metric space is totally bounded. As is easily seen, any subspace of a totally bounded metric space X is also totally bounded with respect to the metric inherited from X.

Theorem 5.7.1. *A metrizable space is separable if and only if it has an admissible totally bounded metric.*

Proof. If a metrizable space X is separable, then X can be embedded in the Hilbert cube $\mathbf{I}^{\mathbb{N}}$. Restricting a metric for $\mathbf{I}^{\mathbb{N}}$, we can obtain an admissible totally bounded metric on X.

Conversely, if X has an admissible totally bounded metric d, then X has finite subsets A_i, $i \in \mathbb{N}$, so that $d(x, A_i) < 2^{-i}$ for every $x \in X$. Then, $A = \bigcup_{i\in\mathbb{N}} A_i$ is a countable dense subset of X. Hence, X is separable. □

Theorem 5.7.2 (Compactification Theorem). *Every n-dimensional separable metrizable space has an n-dimensional metrizable compactification.*

Proof. Let X be a separable metrizable space with $\dim X = n$. By Theorem 5.7.1, X has an admissible totally bounded metric d. For each $i \in \mathbb{N}$, X has a finite open cover $\mathcal{U}_i = \{U_{i,j} \mid j = 1, \ldots, m_i\}$ such that $\operatorname{ord} \mathcal{U}_i \leq n + 1$, $\operatorname{mesh}_d \mathcal{U}_i < 2^{-i}$, and $\operatorname{mesh} f_{i',j'}(\mathcal{U}_i) < 2^{-i}$ for $i' < i$ and $j' \leq m_{i'}$, where $f_{i,j} : X \to \mathbf{I}$ is the map defined by

$$f_{i,j}(x) = \frac{d(x, X \setminus U_{i,j})}{\sum_{k=1}^{m_i} d(x, X \setminus U_{i,k})}.$$

For each $i \in \mathbb{N}$, we define a map $f_i : X \to \mathbf{I}^{m_i}$ by

$$f_i(x) = (f_{i,1}(x), \ldots, f_{i,m_i}(x)).$$

Then, the map $f : X \to \prod_{i \in \mathbb{N}} \mathbf{I}^{m_i}$ defined by $f(x) = (f_i(x))_{i \in \mathbb{N}}$ is an embedding. Indeed, $\bigcup_{i \in \mathbb{N}} \mathcal{U}_i = \{U_{i,j} \mid i \in \mathbb{N},\ j \leq m_i\}$ is an open basis for X. Since $x \in U_{i,j}$ if and only if $f_{i,j}(x) > 0$, it follows that f is injective, and

$$f(U_{i,j}) = f(X) \cap \left\{z \in \textstyle\prod_{i \in \mathbb{N}} \mathbf{I}^{m_i} \mid z(i)(j) > 0\right\}.$$

The closure $\tilde{X}$ of $f(X)$ in $\prod_{i \in \mathbb{N}} \mathbf{I}^{m_i}$ is a metrizable compactification of X. Let ρ be the admissible metric for $\prod_{i \in \mathbb{N}} \mathbf{I}^{m_i}$ defined by $\rho(z, z') = \sup_{i \in \mathbb{N}} 2^{-i} \rho_i(z(i), z'(i))$, where ρ_i is the metric for $\mathbf{I}^{m_i}$ defined by

$$\rho_i(x, y) = \max\left\{|x(j) - y(j)| \;\middle|\; j = 1, \ldots, m_i\right\} \text{ for } x, y \in \mathbf{I}^{m_i}.$$

For each $i \in \mathbb{N}$ and $j \leq m_i$, let $W_{i,j} = \{z \in \tilde{X} \mid z(i)(j) > 0\}$. Then, $W_{i,j} \cap f(X) = f(U_{i,j})$ is dense in $W_{i,j}$. For $i' < i$

$$\operatorname{diam}_{\rho_{i'}} f_{i'}(U_{i,j}) = \max\left\{\operatorname{diam} f_{i',j'}(U_{i,j}) \;\middle|\; j' \leq m_{i'}\right\} < 2^{-i}.$$

Thus, it follows that $\operatorname{diam}_\rho W_{i,j} = \operatorname{diam}_\rho f(U_{i,j}) \leq 2^{-i}$. For each $z \in \tilde{X}$, we have $x_n \in X$, $n \in \mathbb{N}$, such that $f(x_n) \to z$ $(n \to \infty)$. Note that $\sum_{j=1}^{m_i} f_{i,j}(x_n) = 1$. For each $i \in \mathbb{N}$, we can find $j \leq m_i$ such that $f_{i,j}(x_n) \geq 1/m_i$ for infinitely many $n \in \mathbb{N}$. Because $f_{i,j}(x_n) \to z(i)(j)$ $(n \to \infty)$, we have $z(i)(j) \geq 1/m_i$, i.e., $z \in W_{i,j}$. Therefore, $\mathcal{W}_i = \{W_{i,j} \mid j = 1, \ldots, m_i\} \in \operatorname{cov}(\tilde{X})$ with $\operatorname{mesh}_\rho \mathcal{W}_i \leq 2^{-i}$. Since $W_{i,j} \cap f(X) = f(U_{i,j})$ and $f(X)$ is dense in $\tilde{X}$, it follows that $\operatorname{ord} \mathcal{W}_i = \operatorname{ord} f(\mathcal{U}_i) = \operatorname{ord} \mathcal{U}_i \leq n + 1$. Since $\tilde{X}$ is compact, we can find $i_1 < i_2 < \cdots$ in $\mathbb{N}$ so that $\mathcal{W}_{i_1} \succ \mathcal{W}_{i_2} \succ \cdots$. Then, $\dim \tilde{X} \leq n$ by Theorem 5.3.1. On the other hand, $\dim X \leq \dim \tilde{X}$ by the Subset Theorem 5.3.3. Thus, we have $\dim \tilde{X} = n$. □

In the above proof, suppose now that X is a closed subset of a separable metrizable space Y and d is an admissible totally bounded metric for Y. Then, Y has open covers $\mathcal{V}_i = \{V_{i,j} \mid j = 1, \ldots, m_i\}$ such that $\operatorname{ord} \mathcal{V}_i[X] \leq n + 1$,

$\operatorname{mesh}_d \mathcal{V}_i < 2^{-i}$, and $\operatorname{mesh} g_{i',j'}(\mathcal{V}_i) < 2^{-i}$ for $i' < i$ and $j' \leq m_{i'}$, where $g_{i,j} : Y \to \mathbf{I}$ is the map defined by

$$g_{i,j}(y) = \frac{d(y, Y \setminus V_{i,j})}{\sum_{k=1}^{m_i} d(y, Y \setminus V_{i,k})}.$$

As for f in the above proof, using maps $g_{i,j}$, we can define an embedding $g : Y \to \prod_{i \in \mathbb{N}} \mathbf{I}^{m_i}$. The closure $\tilde{Y}$ of $g(Y)$ in $\prod_{i \in \mathbb{N}} \mathbf{I}^{m_i}$ is a metrizable compactification of Y such that $\dim \operatorname{cl}_{\tilde{Y}} X = \dim X$. Furthermore, we can strengthen this as follows:

Theorem 5.7.3. *Let X be a separable metrizable space and $X_1, X_2, \ldots$ be closed sets in X. Then, there exists a metrizable compactification $\tilde{X}$ of X such that $\dim \operatorname{cl}_{\tilde{X}} X_i = \dim X_i$.*

Sketch of Proof. Assume that $\dim X_i = n_i < \infty$. Let d be an admissible totally bounded metric for X. Construct open covers $\mathcal{U}_{i,j} = \{U_{i,j,k} \mid k = 1, \ldots, m(i,j)\}$ of X so that $\operatorname{ord} \mathcal{U}_{i,j}[X_i] \leq n_i + 1$, $\operatorname{mesh}_d \mathcal{U}_{i,j} < 2^{-i-j}$, and $\operatorname{mesh} f_{i',j',k'}(\mathcal{U}_{i,j}) < 2^{-i-j}$ for $i' + j' < i + j$ and $k' \leq m(i', j')$, where $f_{i,j,k} : X \to \mathbf{I}$ is the map defined by

$$f_{i,j,k}(x) = \frac{d(x, X \setminus U_{i,j,k})}{\sum_{l=1}^{m(i,j)} d(x, X \setminus U_{i,j,l})}.$$

As above, we can now use these maps $f_{i,j,k}$ to define an embedding

$$f : X \to \prod_{n \in \mathbb{N}} \prod_{i+j=n+1} \mathbf{I}^{m(i,j)}.$$

The desired compactification of X is obtained as the closure $\tilde{X}$ of $f(X)$ in the compactum $\prod_{n \in \mathbb{N}} \prod_{i+j=n+1} \mathbf{I}^{m(i,j)}$.

Next, we show the following theorem:

Theorem 5.7.4. *Every separable completely metrizable space X has a metrizable compactification γX such that the remainder $\gamma X \setminus X$ is a countable union of finite-dimensional compact sets, hence it is strongly countable-dimensional.*

Proof. We may assume that X is a subset of a compact metric space $Z = (Z, d)$ with $\operatorname{diam} Z \leq 1$. Since X is completely metrizable, we can write $X = \bigcap_{i \in \mathbb{N}} G_i$, where $G_1 \supset G_2 \supset \cdots$ are open in Z. Since each G_i is totally bounded, G_i has a finite open cover $\mathcal{U}_i$ with $\operatorname{mesh} \mathcal{U}_i < 2^{-i}$. We can write $\bigcup_{i \in \mathbb{N}} \mathcal{U}_i = \{U_n \mid n \in \mathbb{N}\}$. Let $f : Z \to \mathbf{I}^{\mathbb{N}}$ be a map defined by

$$f(z)(n) = d(z, X \setminus U_n), \ n \in \mathbb{N}.$$

Then, $f|X$ is an embedding. In fact, if $x \neq y \in X$, there exists some U_n such that $x \in U_n$ but $y \notin U_n$. Then, $f(x)(n) \neq 0 = f(y)(n)$, which implies that $f(x) \neq f(y)$. For each $x \in X$ and each neighborhood U of x in X, choose $n \in \mathbb{N}$ so that $x \in U_n \cap X \subset U$. Since

$$f(U_n \cap X) = f(X) \cap \{x \in \mathbf{I}^{\mathbb{N}} \mid x(n) > 0\},$$

$f(U)$ is a neighborhood of $f(x)$ in $f(X)$.

Let γX be the closure of $f(X)$ in $\mathbf{I}^{\mathbb{N}}$. Identifying X with $f(X)$, γX is a compactification of X. Note that $\gamma X \subset f(Z)$. If $f(z)(n) > 0$ for infinitely many $n \in \mathbb{N}$, then z is contained in infinitely many G_i, which implies that $z \in \bigcap_{i \in \mathbb{N}} G_i = X$. Thus, we have $f(Z) \setminus f(X) \subset \mathbf{I}^{\mathbb{N}}_f$, hence $\gamma X \setminus X \subset \mathbf{I}^{\mathbb{N}}_f$. Since X is completely metrizable, X is G_δ in γX, hence $\gamma X \setminus X$ is F_σ in γX. Consequently, $\gamma X \setminus X$ is a countable union of f.d. compact sets. □

Now, we prove a compactification theorem for (strongly) countable-dimensional spaces:

Theorem 5.7.5. *Every (strongly) countable-dimensional separable completely metrizable space has a (strongly) countable-dimensional metrizable compactification.*

Proof. The c.d. case is a direct consequence of Theorem 5.7.4. To prove the s.c.d. case, let X be a separable completely metrizable space with $X = \bigcup_{i \in \mathbb{N}} X_i$, where each X_i is closed in X, $\dim X_i < \infty$, and $X_1 \subset X_2 \subset \cdots$. By Theorem 5.7.3, X has a metrizable compactification Y such that $\dim \mathrm{cl}_Y X_i = \dim X_i$. By the complete metrizability of X, we can write $X = \bigcap_{i \in \mathbb{N}} U_i$, where each U_i is open in Y and $Y = U_1 \supset U_2 \supset \cdots$. Let $Z = \bigcup_{i \in \mathbb{N}} U_i \cap \mathrm{cl}_Y X_i$. Then, $X = \bigcup_{i \in \mathbb{N}} X_i \subset Z$. Since each $U_i \cap \mathrm{cl}_Y X_i$ is an F_σ-set in Y, Z is a countable union of f.d. compact sets.

We show that $Y \setminus Z = \bigcup_{i \in \mathbb{N}}((Y \setminus \mathrm{cl}_Y X_i) \setminus U_{i+1})$, which is an F_σ-set in Y. For each $y \in Y \setminus Z$, let $i_0 = \max\{i \in \mathbb{N} \mid y \in U_i\}$. Then, $y \in U_{i_0} \setminus U_{i_0+1}$, which implies $y \notin \mathrm{cl}_Y X_{i_0}$ because $y \notin Z$. Hence, $y \in (Y \setminus \mathrm{cl}_Y X_{i_0}) \setminus U_{i_0+1}$. On the other hand, for each $z \in Z$, we have i_1 such that $z \in U_{i_1} \cap \mathrm{cl}_Y X_{i_1}$. For $i \geq i_1$, $z \notin (Y \setminus \mathrm{cl}_Y X_i) \setminus U_{i+1}$ because $z \in \mathrm{cl}_Y X_{i_1} \subset \mathrm{cl}_Y X_i$. For $i < i_1$, $z \notin (Y \setminus \mathrm{cl}_Y X_i) \setminus U_{i+1}$ because $z \in U_{i_1} \subset U_{i+1}$. Thus, Z is a G_δ-set in a compactum Y, hence it is completely metrizable.

Now, applying Theorem 5.7.4, we have a metrizable compactification $\tilde{Z}$ of Z such that $\tilde{Z} \setminus Z$ is a countable union of f.d. compact sets. Then, $\tilde{Z}$ is a compactification of X and it is a countable union of f.d. compact sets, hence it is s.c.d. □

5.8 Embedding Theorem

Recall that every separable metrizable space X can be embedded into the Hilbert cube $\mathbf{I}^{\mathbb{N}}$ (Corollary 2.3.8). As a finite-dimensional version of this result, we prove the following theorem:

Theorem 5.8.1 (Embedding Theorem). *Every separable metrizable space with* dim $\leq n$ *can be embedded in* $\mathbf{I}^{2n+1}$*, and can hence be embedded in the Euclidean space* $\mathbb{R}^{2n+1}$.

Remark 13. In Theorem 5.8.1, the cube $\mathbf{I}^{2n+1}$ cannot be replaced by a smaller dimensional cube. In fact, there exist n-dimensional compact polyhedra that cannot be embedded into $\mathbf{I}^{2n}$. See Fig. 5.3.

$K(\Delta^4)^{(1)} \approx$ 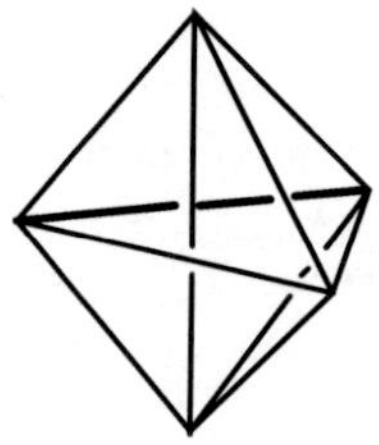 $\not\hookrightarrow \mathbf{I}^2$

Fig. 5.3 A 1-dimensional polyhedron that cannot be embedded in $\mathbf{I}^2$

To prove Theorem 5.8.1, we introduce a new notion. Now, let $X = (X, d)$ be a *compact* metric space. Given $\varepsilon > 0$, a map $f : X \to Y$ is called an ***ε*-map** if $\operatorname{diam} f^{-1}(y) < \varepsilon$ for each $y \in Y$. Then, a map $f : X \to Y$ is an embedding if and only if $f : X \to Y$ is an ε-map for every $\varepsilon > 0$.

Lemma 5.8.2. *Let $f : X \to Y$ be an ε-map from a compact metric space $X = (X, d)$ to a metric space $Y = (Y, \rho)$. Then, there is some $\delta > 0$ such that any map $g : X \to Y$ with $\rho(f, g) < \delta$ is an ε-map.*

Proof. Since f is a closed map, each $y \in Y$ has an open neighborhood V_y in Y such that $\operatorname{diam} f^{-1}(V_y) < \varepsilon$. Since X is compact, we can choose $\delta > 0$ so that each $\mathrm{B}_\rho(f(x), 2\delta)$ is contained in some V_y, hence $\operatorname{diam} f^{-1}(\mathrm{B}_\rho(f(x), 2\delta)) < \varepsilon$. Let $g : X \to Y$ be a map with $\rho(f, g) < \delta$. For $y \in Y$ and $x, x' \in g^{-1}(y)$,

$$\rho(f(x), f(x')) \leq \rho(f(x), g(x)) + \rho(f(x'), g(x')) < 2\delta,$$

which implies that $g^{-1}(y) \subset f^{-1}(\mathrm{B}_\rho(f(x), 2\delta))$. Therefore, $\operatorname{diam} g^{-1}(y) < \varepsilon$, that is, g is an ε-map. □

For spaces X and Y, let $\mathrm{Emb}(X, Y)$ denote the subspace of $\mathrm{C}(X, Y)$ consisting of all closed embeddings.

Theorem 5.8.3. *Let $X = (X, d)$ be a compact metric space and $Y = (Y, \rho)$ a complete metric space. Assume that for each $\varepsilon > 0$ and $\delta > 0$, every map $f : X \to Y$ is δ-close to an ε-map. Then, every map $f : X \to Y$ can be approximated by an embedding, that is,* $\mathrm{Emb}(X, Y)$ *is dense in the space* $\mathrm{C}(X, Y)$ *with the sup-metric.*

Proof. For each $n \in \mathbb{N}$, let G_n be the set of all 2^{-n}-maps from X to Y. Then, G_n is open and dense in the space $\mathrm{C}(X, Y)$ by Lemma 5.8.2 and the assumption. By the Baire Category Theorem 2.5.1, $\mathrm{Emb}(X, Y) = \bigcap_{n\in\mathbb{N}} G_n$ is also dense in $\mathrm{C}(X, Y)$, hence so is the set of embeddings of X into Y. □

The following is called the GENERAL POSITION LEMMA:

Lemma 5.8.4 (GENERAL POSITION). *Let $\{U_i \mid i \in \mathbb{N}\}$ be a countable open collection in $\mathbb{R}^n$ and $A \subset \mathbb{R}^n$ with* $\operatorname{card} A \leq \aleph_0$ *such that each $n + 1$ many points of*

A are affinely independent. Then, there exists $B = \{v_i \mid i \in \mathbb{N}\}$ *such that* $v_i \in U_i \setminus A$ *for each* $i \in \mathbb{N}$ *and each* $n+1$ *many points of* $A \cup B$ *are affinely independent.*

Proof. Assume that $v_1 \in U_1, \ldots, v_k \in U_k$ have been chosen so that each $n+1$ many points of $A \cup \{v_1, \ldots, v_k\}$ are affinely independent. Using the Baire Category Theorem 2.5.1 and the fact that every $(n-1)$-dimensional flat (= hyperplane) in $\mathbb{R}^n$ is nowhere dense in $\mathbb{R}^n$, we can find a point

$$v_{k+1} \in U_{k+1} \setminus \bigcup \big\{ \operatorname{fl}\{x_1, \ldots, x_k\} \bigm| x_i \in A \cup \{v_1, \ldots, v_k\} \big\}.$$

Then, each $n+1$ many points of $A \cup \{v_1, \ldots, v_{k+1}\}$ are affinely independent. By induction, we can obtain the desired set $B = \{v_i \mid i \in \mathbb{N}\} \subset \mathbb{R}^n$. □

Because every separable metrizable space has a metrizable compactification with the same dimension by the Compactification Theorem 5.7.2, the Embedding Theorem 5.8.1 can be obtained as a corollary of the next theorem:

Theorem 5.8.5 (EMBEDDING APPROXIMATION). *Let X be a compact metric space with* $\dim X \leq n$. *Then, every map* $f : X \to \mathbf{I}^{2n+1}$ *can be approximated by embeddings, that is, for each* $\varepsilon > 0$, *there is an embedding* $h : X \to \mathbf{I}^{2n+1}$ *that is* ε*-close to* f. *In particular, every compact metrizable space with* $\dim \leq n$ *can be embedded in* $\mathbf{I}^{2n+1}$.

Proof. Because of Theorem 5.8.3, it is enough to show that for each $\varepsilon > 0$ and $\delta > 0$, every map $f : X \to \mathbf{I}^{2n+1}$ is δ-close to an ε-map. We have a finite open cover $\mathcal{U}$ of X such that $\operatorname{ord}\mathcal{U} \leq n+1$, $\operatorname{mesh}\mathcal{U} < \varepsilon$, and $\operatorname{mesh} f(\mathcal{U}) < \delta/2$. Let $K = N(\mathcal{U})$ be the nerve of $\mathcal{U}$. A canonical map $\varphi : X \to |K|$ is an ε-map because $\varphi^{-1}(\mathcal{O}_K) \prec \mathcal{U}$. By the General Position Lemma 5.8.4, we have points $v_U \in \mathbf{I}^{2n+1}$, $U \in \mathcal{U}$, such that $d(v_U, f(U)) < \delta/2$ and every $2n+2$ many points $v_{U_1}, \ldots, v_{U_{2n+2}}$ are affinely independent. We can define a map $g : |K| \to \mathbf{I}^{2n+1}$ as follows: $g(U) = v_U$ for each $U \in \mathcal{U} = K^{(0)}$ and g is linear on each simplex of K_1. Then, g is injective. Hence, $h = g\varphi : X \to \mathbf{I}^{2n+1}$ is an ε-map. For each $x \in X$, let $\mathcal{U}[x] = \{U_1, \ldots, U_k\}$. Then,

$$\|v_{U_i} - f(x)\| \leq d(v_{U_i}, f(U_i)) + \operatorname{diam} f(U_i) < \delta.$$

Since $\mathrm{B}(f(x), \delta)$ is convex, it follows that

$$g\varphi(x) \in g(\langle U_1, \ldots, U_k \rangle) = \langle v_{U_1}, \ldots, v_{U_k} \rangle \subset \mathrm{B}(f(x), \delta).$$

Therefore, $h = g\varphi$ is δ-close to f. □

We generalize a non-compact version of the Embedding Approximation Theorem 5.8.5. Given $\mathcal{U} \in \operatorname{cov}(X)$, a map $f : X \to Y$ is called a $\boldsymbol{\mathcal{U}}$**-map** if $f^{-1}(\mathcal{V}) \prec \mathcal{U}$ for some $\mathcal{V} \in \operatorname{cov}(Y)$. By $\mathrm{C}_{\mathcal{U}}(X, Y)$, we denote the subspace of $\mathrm{C}(X, Y)$ consisting of all $\mathcal{U}$-maps. In the case that X is a compact metric space, let $\varepsilon > 0$ be a Lebesgue number for $\mathcal{U} \in \operatorname{cov}(X)$. Then, every $\mathcal{U}$-map is an ε-map.

Conversely, if $\mathcal{U} = \{B(x, \varepsilon) \mid x \in X\}$, then every ε-map $f : X \to Y$ is a $\mathcal{U}$-map. Indeed, f is closed because of the compactness of X. For each $y \in f(X)$, take $x_y \in f^{-1}(y)$. Since $f^{-1}(y) \subset B(x_y, \varepsilon)$, y has an open neighborhood V_y in Y such that $f^{-1}(V_y) \subset B(x_y, \varepsilon)$. Then,

$$\mathcal{V} = \{V_y \mid y \in f(X)\} \cup \{Y \setminus f(X)\} \in \operatorname{cov}(Y) \text{ and } f^{-1}(\mathcal{V}) \prec \mathcal{U}.$$

Recall that if Y is completely metrizable then the space $C(X, Y)$ with the limitation topology is a Baire space (Theorem 2.9.4). The limitation topology is the topology in which $\{\mathcal{V}(f) \mid \mathcal{V} \in \operatorname{cov}(Y)\}$ is a neighborhood basis of each $f \in C(X, Y)$,[7] where

$$\mathcal{V}(f) = \{g \in C(X, Y) \mid g \text{ is } \mathcal{V}\text{-close to } f\}.$$

In the following two lemmas, let Y be an arbitrary paracompact space.

Lemma 5.8.6. *For each $\mathcal{U} \in \operatorname{cov}(X)$, $C_{\mathcal{U}}(X, Y)$ is open in the space $C(X, Y)$ with the limitation topology.*

Proof. For each $f \in C_{\mathcal{U}}(X, Y)$, $f^{-1}(\mathcal{V}) \prec \mathcal{U}$ for some $\mathcal{V} \in \operatorname{cov}(Y)$. Let $\mathcal{W} \in \operatorname{cov}(Y)$ such that $\operatorname{st}\mathcal{W} \prec \mathcal{V}$. For each $g \in \mathcal{W}(f)$, $f(g^{-1}(\mathcal{W})) \prec \operatorname{st}\mathcal{W} \prec \mathcal{V}$, so $g^{-1}(\mathcal{W}) \prec f^{-1}(\mathcal{V}) \prec \mathcal{U}$, which implies $g \in C_{\mathcal{U}}(X, Y)$. □

Lemma 5.8.7. *For each complete metric space $X = (X, d)$, $\operatorname{Emb}(X, Y) = \bigcap_{n \in \mathbb{N}} C_{\mathcal{U}_n}(X, Y)$, where $\mathcal{U}_n \in \operatorname{cov}(X)$ with $\operatorname{mesh}\mathcal{U}_n < 2^{-n}$. Thus, when X is a completely metrizable space, $\operatorname{Emb}(X, Y)$ is a G_δ-set in the space $C(X, Y)$ with the limitation topology.*

Proof. Obviously, $\operatorname{Emb}(X, Y) \subset \bigcap_{n \in \mathbb{N}} C_{\mathcal{U}_n}(X, Y)$. Every $f \in \bigcap_{n \in \mathbb{N}} C_{\mathcal{U}_n}(X, Y)$ is injective. For $x_n \in X, n \in \mathbb{N}$, if $(f(x_n))_{n \in \mathbb{N}}$ is convergent, then $(x_n)_{n \in \mathbb{N}}$ is Cauchy, so it is convergent. This means that f is closed, hence $f \in \operatorname{Emb}(X, Y)$. Thus, we have $\operatorname{Emb}(X, Y) = \bigcap_{n \in \mathbb{N}} C_{\mathcal{U}_n}(X, Y)$. □

When Y is completely metrizable, the space $C(X, Y)$ with the limitation topology is a Baire space by Theorem 2.9.4. Then, by Lemmas 5.8.6 and 5.8.7, Theorem 5.8.3 can be generalized as follows:

Theorem 5.8.8. *Let X and Y be completely metrizable spaces. Suppose that, for each $\mathcal{U} \in \operatorname{cov}(X)$, $C_{\mathcal{U}}(X, Y)$ is dense in the space $C(X, Y)$ with the limitation topology. Then, $\operatorname{Emb}(X, Y)$ is also dense in $C(X, Y)$. In other words, if every map $f : X \to Y$ is approximated by $\mathcal{U}$-maps for each $\mathcal{U} \in \operatorname{cov}(X)$, then every map $f : X \to Y$ is approximated by closed embeddings.* □

[7]When Y is paracompact, $\{\mathcal{V}(f) \mid \mathcal{V} \in \operatorname{cov}(Y)\}$ is a neighborhood basis of each $f \in C(X, Y)$ and the topology is Hausdorff.

We consider the case that X and Y are locally compact metrizable. Let $\mathrm{C}^P(X,Y)$ be the subspace of the space $\mathrm{C}(X,Y)$ with the limitation topology consisting of all proper maps.[8] Then, the space $\mathrm{C}^P(X,Y)$ is a Baire space by Theorem 2.9.8. It should be noted that $\mathrm{Emb}(X,Y) \subset \mathrm{C}^P(X,Y)$. Moreover, if X is non-compact, then any constant map of X to Y is not proper, which implies that $\mathrm{Emb}(X,Y)$ is not dense in the space $\mathrm{C}(X,Y)$ with the limitation topology because $\mathrm{C}^P(X,Y)$ is clopen in $\mathrm{C}(X,Y)$ due to Corollary 2.9.7. For an open cover $\mathcal{U} \in \mathrm{cov}(X)$ consisting of open sets with the compact closures, we have $\mathrm{C}_{\mathcal{U}}(X,Y) \subset \mathrm{C}^P(X,Y)$.

Indeed, for each $f \in \mathrm{C}_{\mathcal{U}}(X,Y)$, let $\mathcal{V}$ be a locally finite open cover of Y such that $f^{-1}(\mathcal{V}) \prec \mathcal{U}$. Each compact set A in Y meets only finitely many $V_1,\dots,V_n \in \mathcal{V}$, where each $f^{-1}(V_i)$ is contained in some $U_i \in \mathcal{U}$. Then, $f^{-1}(A) \subset \bigcup_{i=1}^n \mathrm{cl}\, U_i$. Since $\bigcup_{i=1}^n \mathrm{cl}\, U_i$ is compact, $f^{-1}(A)$ is also compact.

The following theorem is the locally compact version of Theorem 5.8.8:

Theorem 5.8.9. *Let X and Y be locally compact metrizable spaces. If $\mathrm{C}_{\mathcal{U}}(X,Y)$ is dense in the space $\mathrm{C}^P(X,Y)$ with the limitation topology for each open cover $\mathcal{U}$ of X consisting of open sets with the compact closures, then* $\mathrm{Emb}(X,Y)$ *is also dense in* $\mathrm{C}^P(X,Y)$. □

Now, we show the following locally compact version of the Embedding Approximation Theorem 5.8.5:

Theorem 5.8.10 (Embedding Approximation). *Let X be a locally compact separable metrizable space with* $\dim X \le n$. *Then,* $\mathrm{Emb}(X,\mathbb{R}^{2n+1})$ *is dense in the space* $\mathrm{C}^P(X,\mathbb{R}^{2n+1})$ *with the limitation topology, that is, for each open cover $\mathcal{U}$ of $\mathbb{R}^{2n+1}$, every proper map $f : X \to \mathbb{R}^{2n+1}$ is $\mathcal{U}$-close to a closed embedding $h : X \to \mathbb{R}^{2n+1}$.*

Proof. Because of Theorem 5.8.9, it suffices to show that $\mathrm{C}_{\mathcal{U}}(X,Y)$ is dense in $\mathrm{C}^P(X,\mathbb{R}^{2n+1})$ for each $\mathcal{U} \in \mathrm{cov}(X)$, that is, for any $\mathcal{V} \in \mathrm{cov}(\mathbb{R}^{2n+1})$, every proper map $f : X \to \mathbb{R}^{2n+1}$ is $\mathcal{V}$-close to some $\mathcal{U}$-map $h : X \to \mathbb{R}^{2n+1}$.

We can find $\mathcal{W} \in \mathrm{cov}(\mathbb{R}^{2n+1})$ such that $\mathcal{W}$ is star-finite ($\mathrm{ord}\, \mathcal{W} \le 2n+2$), $\mathrm{cl}\, W$ is compact for each $W \in \mathcal{W}$, and $\{\langle \mathrm{st}(x,\mathcal{W})\rangle \mid x \in X\} \prec \mathcal{V}$. By replacing a refinement with $\mathcal{U}$, we may assume that $\mathcal{U} \prec f^{-1}(\mathcal{W})$ (i.e., $f(\mathcal{U}) \prec \mathcal{W}$), $\mathcal{U}$ is countable, and $\mathrm{ord}\mathcal{U} \le n+1$ (cf. Corollary 5.2.5). Write $\mathcal{U} = \{U_i \mid i \in \mathbb{N}\}$ and choose $W_i \in \mathcal{W}$, $i \in \mathbb{N}$, so that $f(U_i) \subset W_i$. Let $K = N(\mathcal{U})$ be the nerve of $\mathcal{U}$ with $\varphi : X \to |K|$ a canonical map. Then, $\dim K \le n$ and φ is a $\mathcal{U}$-map because $\varphi^{-1}(O_K(U)) \subset U$ for each $U \in \mathcal{U} = K^{(0)}$.

By the General Position Lemma 5.8.4, we have points $v_i \in \mathbb{R}^{2n+1}$, $i \in \mathbb{N}$, such that $v_i \in W_i$ and every $2n+2$ many points $v_{i_1},\dots,v_{i_{2n+2}}$ are affinely independent. Then, we have a PL-map $g : |K| \to \mathbb{R}^{2n+1}$ such that $g(U_i) = v_i \in W_i$ for each $U_i \in K^{(0)} = \mathcal{U}$ and $g|\sigma$ is affine on each simplex $\sigma \in K$. For each pair of simplexes $\sigma,\tau \in K$, $g(\sigma^{(0)} \cup \tau^{(0)})$ is affinely independent, which implies that $g|\sigma \cup \tau$ is an embedding. Hence, g is injective.

---

[8]In this case, a proper map coincides with a perfect map (Proposition 2.1.5).

To prove that g is a closed embedding, let A be a closed set in $|K|$. Each $y \in \operatorname{cl} g(A)$ is contained in some $W \in \mathcal{W}$. By the star-finiteness of $\mathcal{W}$, $\mathcal{W}[W]$ is finite, hence $g(K^{(0)}) \cap W$ is finite. Since K is star-finite, $W \cap g(\sigma) \neq \emptyset$ for only finitely many simplexes $\sigma \in K$. Let

$$\{\sigma \in K \mid W \cap g(\sigma) \neq \emptyset\} = \{\sigma_1, \ldots, \sigma_m\}.$$

Since g is injective, it follows that $W \cap g(A) = \bigcup_{i=1}^m W \cap g(A \cap \sigma_i)$, which is closed in W, and hence $y \in g(A)$. Therefore, $g(|K|)$ is closed in $\mathbb{R}^{2n+1}$.

It remains to be shown that the $\mathcal{U}$-map $g\varphi : X \to Y$ is $\mathcal{V}$-close to f. For each $x \in X$, take the carrier $\sigma \in K$ of $\varphi(x)$ and let $\sigma^{(0)} = \{U_{i_1}, \ldots, U_{i_k}\}$. Then,

$$g(\varphi(x)) \in g(\sigma) = \langle g(\sigma^{(0)}) \rangle = \langle v_{i_1}, \ldots, v_{i_k} \rangle.$$

On the other hand, since $x \in U_{i_1} \cap \cdots \cap U_{i_k}$, we have $f(x) \in W_{i_1} \cap \cdots \cap W_{i_k}$. Then, it follows that $v_{i_1}, \ldots, v_{i_k} \in \operatorname{st}(f(x), \mathcal{W})$. Recall that $\langle \operatorname{st}(f(x), \mathcal{W}) \rangle$ is contained in some $V \in \mathcal{V}$. Then, we have $g(\varphi(x)), f(x) \in V$. Thus, $g\varphi$ is $\mathcal{V}$-close to f. □

Remark 14. In the Embedding Approximation Theorem 5.8.10, a map $f : X \to \mathbb{R}^{2n+1}$ cannot be approximated by closed embeddings if f is not proper. Indeed, $A = f^{-1}(a)$ is not compact for some $a \in \mathbb{R}^{2n+1}$. If $h : X \to \mathbb{R}^{2n+1}$ is a closed embedding then $h(A)$ is closed in $\mathbb{R}^{2n+1}$. Because $h(A)$ is non-compact, it is unbounded, hence $\sup_{x \in A} \|h(x) - f(x)\| = \infty$.

We now show the following proposition:

Proposition 5.8.11. *Let X be a paracompact space and $n \in \omega$. Suppose that for each $\mathcal{U} \in \operatorname{cov}(X)$, there exist a paracompact space Y with* $\dim Y \leq n$ *and a $\mathcal{U}$-map $f : X \to Y$. Then,* $\dim X \leq n$.

Proof. For each $\mathcal{U} \in \operatorname{cov}(X)$, we have a $\mathcal{U}$-map $f : X \to Y$ such that $\dim Y \leq n$. Then, by Theorem 5.2.4, we have $\mathcal{V} \in \operatorname{cov}(Y)$ such that $f^{-1}(\mathcal{V}) \prec \mathcal{U}$ and $\operatorname{ord} \mathcal{V} \leq n+1$. Note that $\operatorname{ord} f^{-1}(\mathcal{V}) \leq n+1$. Therefore, $\dim X \leq n$ by Theorem 5.2.4. □

When X is a metric space, using Theorem 5.3.1 instead of Theorem 5.2.4, we have the following:

Proposition 5.8.12. *Let X be a metric space and $n \in \omega$. Suppose that for each $\varepsilon > 0$, there exist a paracompact space Y with* $\dim Y \leq n$ *and a closed ε-map $f : X \to Y$. Then,* $\dim X \leq n$. □

5.9 Universal Spaces

Given a class $\mathcal{C}$ of spaces, a space $Y \in \mathcal{C}$ is called a **universal space** for $\mathcal{C}$ if every space $X \in \mathcal{C}$ can be embedded into Y. The Hilbert cube $\mathbf{I}^{\mathbb{N}}$ and $\mathbb{R}^{\mathbb{N}}$ are universal spaces for separable metrizable spaces (Corollary 2.3.8) and the countable power

$J(\Gamma)^{\mathbb{N}}$ of the hedgehog is the universal space for metrizable spaces with weight $\leq \operatorname{card} \Gamma$ (Corollary 2.3.7).[9]

In this section, we show the existence of universal spaces for metrizable spaces with $\dim \leq n$, and for countable-dimensional and strongly countable-dimensional metrizable spaces.

First, we will show that the space $\mathbf{I}^{\mathbb{N}}_f$ is also a universal space for strongly countable-dimensional separable metrizable spaces.

Lemma 5.9.1. *Let X be a separable metrizable space and $X_0 \subset X_1$ be closed sets in X with $\dim X_1 \leq n$. Then, there exists a map $f : X \to \mathbf{I}^{2n+2}$ such that $X_0 = f^{-1}(0)$ and $f|X_1 \setminus X_0$ is an embedding.*

Proof. Applying the Tietze Extension Theorem 2.2.2 coordinate-wise, we can extend an embedding of X_1 into $\mathbf{I}^{2n+1}$ obtained by Theorem 5.8.1 to a map $h : X \to \mathbf{I}^{2n+1}$. Let $g : X \to \mathbf{I}$ be a map with $g^{-1}(0) = X_0$. We define a map $f : X \to \mathbf{I}^{2n+2} = \mathbf{I}^{2n+1} \times \mathbf{I}$ by $f(x) = (g(x)h(x), g(x))$. Then, $f^{-1}(0) = X_0$. It is easy to prove that $f|X_1 \setminus X_0$ is injective. To see that $f|X_1 \setminus X_0$ is an embedding, let $x, x_i \in X_1 \setminus X_0$, $i \in \mathbb{N}$, and assume that $f(x) = \lim_{i\to\infty} f(x_i)$. Since $g(x_i) \to g(x)$ and $g(x_i), g(x) > 0$, we have $g(x_i)^{-1} \to g(x)^{-1}$, which implies that $h(x_i) \to h(x)$ in $\mathbf{I}^{2n+1}$, hence $x_i \to x$ in X. Therefore, $f|X_1 \setminus X_0$ is an embedding. □

Theorem 5.9.2. *The space $\mathbf{I}^{\mathbb{N}}_f$ is a universal space for strongly countable-dimensional separable metrizable spaces.*

Proof. Let X be an s.c.d. separable metric space. We can write $X = \bigcup_{k\in\mathbb{N}} X_k$, where $X_1 \subsetneqq X_2 \subsetneqq \cdots$ are closed in X and $\dim X_k = n_k < \infty$. By Theorem 5.7.3, X has a metrizable compactification Y such that $\dim \operatorname{cl}_Y X_k = \dim X_k = n_k$. By Lemma 5.9.1, we have maps $f_k : Y \to \mathbf{I}^{2n_k+2}$ ($k \in \mathbb{N}$) such that $f_k^{-1}(0) = \operatorname{cl}_Y X_{k-1}$ and $f_k|\operatorname{cl}_Y X_k \setminus \operatorname{cl}_Y X_{k-1}$ is an embedding for each $k \in \mathbb{N}$, where $X_0 = \emptyset$. We define a map $f : Y \to \prod_{k\in\mathbb{N}} \mathbf{I}^{2n_k+2} = \mathbf{I}^{\mathbb{N}}$ by $f(x) = (f_k(x))_{k\in\mathbb{N}}$. Then, $f|\bigcup_{k\in\mathbb{N}} \operatorname{cl}_Y X_k$ is injective. By definition, $f(\bigcup_{k\in\mathbb{N}} \operatorname{cl}_Y X_k) \subset \mathbf{I}^{\mathbb{N}}_f$. For $y \in Y$, if $f(y) \in \mathbf{I}^{\mathbb{N}}_f$ then $f_{k+1}(y) = 0$ for some $k \in \mathbb{N}$, which means that $y \in \operatorname{cl}_Y X_k$. Then, it follows that

$$f(A) \cap \mathbf{I}^{\mathbb{N}}_f = f\big(A \cap \textstyle\bigcup_{k\in\mathbb{N}} \operatorname{cl}_Y X_k\big) \quad \text{for each } A \subset Y.$$

Since f is a closed map, the restriction $f|\bigcup_{k\in\mathbb{N}} \operatorname{cl}_Y X_k : \bigcup_{k\in\mathbb{N}} \operatorname{cl}_Y X_k \to \mathbf{I}^{\mathbb{N}}_f$ is also a closed map. Therefore, $f|\bigcup_{k\in\mathbb{N}} \operatorname{cl}_Y X_k$ is an embedding, hence so is $f|X$. This completes the proof. □

For each $n \in \omega$, let

$$\nu_n = \big\{x \in \mathbf{I}^{\mathbb{N}} \bigm| x(i) \in \mathbf{I} \setminus \mathbb{Q} \text{ except for } n \text{ many } i\big\}.$$[10]

[9] Usually, the phrase "the class of" is omitted.

[10] Recall that ν^0 denotes the space $\mathbb{R} \setminus \mathbb{Q}$. Then, $\nu_0 \subsetneqq \nu^0$ but $\nu_0 \approx ((-1,1) \setminus \mathbb{Q})^{\mathbb{N}} \approx \nu^0$.

Then, $\nu_0 = (\mathbf{I} \setminus \mathbb{Q})^{\mathbb{N}} \subset \nu_1 \subset \nu_2 \subset \cdots \subset \nu_\omega = \bigcup_{n\in\omega} \nu_n$. Recall that ν_ω is c.d. but not s.c.d. (Theorem 5.6.3). We will show that ν_n is a universal space for separable metrizable spaces with $\dim \leq n$ and that ν_ω is the universal space for c.d. separable metrizable spaces. To avoid restricting ourselves to separable spaces, we construct non-separable analogues to ν_n and ν_ω.

Let Γ be an infinite set. Recall that the hedgehog $J(\Gamma)$ is the closed subspace of $\ell_1(\Gamma)$ defined as

$$\begin{aligned} J(\Gamma) &= \{x \in \ell_1(\Gamma) \mid x(\gamma) \in \mathbf{I} \text{ for all } \gamma \in \Gamma \text{ and} \\ &\qquad\qquad x(\gamma) \neq 0 \text{ for at most one } \gamma \in \Gamma\} \\ &= \bigcup_{\gamma\in\Gamma} \langle 0, \mathbf{e}_\gamma \rangle = \bigcup_{\gamma\in\Gamma} \mathbf{I}\mathbf{e}_\gamma \subset \ell_1(\Gamma). \end{aligned}$$

Then, $\dim J(\Gamma) = 1$. Let

$$P(\Gamma) = \{x \in J(\Gamma) \mid x(\gamma) \in (\mathbf{I} \setminus \mathbb{Q}) \cup \{0\}\} = \{0\} \cup \bigcup_{\gamma\in\Gamma} (\mathbf{I} \setminus \mathbb{Q})\mathbf{e}_\gamma.$$

Observe $P(\Gamma) = \{0\} \cup \bigcup_{i\in\mathbb{N}} P_i$, where $P_i = P(\Gamma) \setminus \mathrm{B}(0, 1/i)$. Each P_i is the discrete union of 0-dimensional closed sets in $P(\Gamma)$ that are homeomorphic to $\mathbf{I}\setminus\mathbb{Q}$, hence $\dim P_i = 0$ by the Locally Finite Sum Theorem 5.4.2. Then, $\dim P(\Gamma) = 0$ by the Countable Sum Theorem 5.4.1. Now, we define

$$\nu_\omega(\Gamma) = \{z \in J(\Gamma)^{\mathbb{N}} \mid z(i) \in P(\Gamma) \text{ except for finitely many } i\}.$$

Observe that $\nu_\omega(\Gamma)$ is the countable union of subspaces that are homeomorphic to $J(\Gamma)^n \times P(\Gamma)^{\mathbb{N}}$. Since $\dim J(\Gamma)^n \times P(\Gamma)^{\mathbb{N}} \leq n$ (Product Theorem 5.4.9 and Theorem 5.3.6) and $J(\Gamma)^n$ contains a copy of $\mathbf{I}^n$, we have $\dim J(\Gamma)^n \times P(\Gamma)^{\mathbb{N}} = n$. Therefore, it follows that $\nu_\omega(\Gamma)$ is c.d. For each $n \in \omega$, we define

$$\nu_n(\Gamma) = \{z \in J(\Gamma)^{\mathbb{N}} \mid z(i) \in P(\Gamma) \text{ except for } n \text{ many } i\}.$$

Then, $\nu_0(\Gamma) = P(\Gamma)^{\mathbb{N}} \subset \nu_1(\Gamma) \subset \nu_2(\Gamma) \subset \cdots \subset \nu_\omega(\Gamma) = \bigcup_{n\in\omega} \nu_n(\Gamma)$.

Theorem 5.9.3. *For each $n \in \omega$, $\dim \nu_n(\Gamma) = \dim \nu_n = n$.*

Proof. We only give a proof of $\dim \nu_n(\Gamma) = n$ because $\dim \nu_n = n$ is similar and simpler.

We already proved that $\dim \nu_0(\Gamma) = \dim P(\Gamma)^{\mathbb{N}} = 0$. Assuming that $\dim \nu_{n-1}(\Gamma) = n - 1$ and $n > 0$, we now prove that $\dim \nu_n(\Gamma) = n$. We can write

$$\nu_n(\Gamma) = \nu_0(\Gamma) \cup \bigcup_{i\in\mathbb{N}} \bigcup_{q\in((0,1]\cap\mathbb{Q})} \bigcup_{\gamma\in\Gamma} \nu_n(i, q, \gamma),$$

where $\nu_n(i,q,\gamma)$ is a closed set in $\nu_n(\Gamma)$ defined as follows:

$$\nu_n(i,q,\gamma) = \{z \in \nu_n(\Gamma) \mid z(i)(\gamma) = q\}.$$

Since $\{\nu_n(i,q,\gamma) \mid \gamma \in \Gamma\}$ is discrete in $\nu_n(\Gamma)$ and $\nu_n(i,q,\gamma) \approx \nu_{n-1}(\Gamma)$, $\bigcup_{\gamma\in\Gamma} \nu_n(i,q,\gamma)$ is an $(n-1)$-dimensional closed set in $\nu_n(\Gamma)$ by the Locally Finite Sum Theorem 5.4.2. Then, $\dim \nu_n(\Gamma) \leq n$ by the Countable Sum Theorem 5.4.1 and the Addition Theorem 5.4.8. Since $\nu_n(\Gamma)$ contains an n-dimensional subspace $J(\Gamma)^n \times P(\Gamma)^{\mathbb{N}}$, we have $\dim \nu_n(\Gamma) \geq n$, hence $\dim \nu_n(\Gamma) = n$. The result follows by induction. □

We will show that the space $\nu_n(\Gamma)$ is a universal space for metrizable spaces with $\dim \leq n$ and weight $\leq \operatorname{card}\Gamma$, and that the space $\nu_\omega(\Gamma)$ is a universal space for c.d. metrizable spaces with weight $\leq \operatorname{card}\Gamma$.

Lemma 5.9.4. *Let X be a metrizable space and $X_0, X_1, \dots \subset X$ with $\dim X_n \leq 0$. Suppose that $L_0 = \emptyset, L_1, \dots, L_{m-1}$ are closed sets in X satisfying the following condition:*

(∗) *No $x \in X_n$ are contained in $n+1$ many sets L_i.*

Then, for each pair (A, B) of disjoint closed sets in X, there exists a partition L_m in X between A and B that does not violate the condition (∗).

Proof. Let $C_0 = X_0$. For $n < m$, define

$$C_n = \bigcup\{X_n \cap \bigcap_{j=1}^n L_{i_j} \mid 0 \leq i_1 < i_2 < \dots < i_n < m\}.$$

Then, $C_i \cap C_j = \emptyset$ for $i \neq j$ by (∗). Let $D = \bigcup_{i=0}^{m-1} C_i$. For each $n < m-1$, $\bigcup_{i=n+1}^{m-1} C_i$ is contained in the closed set

$$F = \bigcup\{\bigcap_{j=1}^{n+1} L_{i_j} \mid 0 \leq i_1 < i_2 < \dots < i_{n+1} < m\}.$$

Note that $F \cap \bigcup_{i=0}^n X_i = \emptyset$ by (∗). For this reason, $\bigcup_{i=0}^n C_i = D \setminus F$ is open in D. Therefore, each $C_n = \bigcup_{i=0}^n C_i \setminus \bigcup_{i=0}^{n-1} C_i$ is an F_σ-set in D. It follows from the Subset Theorem 5.3.3 and the Countable Sum Theorem 5.4.1 that $\dim D \leq 0$.

Using Theorem 5.2.17 and the Partition Extension Lemma 5.3.7, we obtain a partition L_m between A and B such that $L_m \cap D = \emptyset$. Condition (∗) is trivial for $n \geq m$. For $n < m$, if $x \in X_n$ is contained in n many sets L_i $(i < m)$, then $x \in C_n \subset D$, which implies $x \notin L_m$. Therefore, condition (∗) is satisfied. □

Lemma 5.9.5. *Let X be a metrizable space and $X_0, X_1, \dots \subset X$ with $\dim X_n \leq 0$ and let $a < b \in \mathbb{R}$. Then, for any sequence $(A_i, B_i)_{i\in\mathbb{N}}$ of pairs of disjoint closed sets in X, there exist maps $f_i : X \to [a,b]$, $i \in \mathbb{N}$, such that $A_i = f_i^{-1}(a)$, $B_i = f_i^{-1}(b)$, and*

$$\operatorname{card}\{i \in \mathbb{N} \mid f_i(x) \in (a,b) \cap \mathbb{Q}\} \leq n \quad \text{for } x \in X_n.$$

Proof. Let $\{q_j \mid j \in \mathbb{N}\} = (a,b) \cap \mathbb{Q}$, where $q_i \neq q_j$ if $i \neq j$. For each $j \in \mathbb{N}$, let

$$\delta_j = \min \{q_i - a, b - q_i, |q_i - q_{i'}| \mid i, i' \leq j,\ i \neq i'\},$$

and define $a_j = q_j - 2^{-j-1}\delta_j$ and $b_j = q_j + 2^{-j-1}\delta_j$. For each $i \in \mathbb{N}$, let $f_{i,0} : X \to [a,b]$ be a map with $A_i = f_{i,0}^{-1}(a)$ and $B_i = f_{i,0}^{-1}(b)$. We construct maps $f_{i,j} : X \to [a,b]$, $i, j \in \mathbb{N}$, so as to satisfy the following conditions:

(1) $A_i = f_{i,j}^{-1}(a)$ and $B_i = f_{i,j}^{-1}(b)$;
(2) $f_{i,j}(x) \neq f_{i,j-1}(x) \Rightarrow f_{i,j-1}(x), f_{i,j}(x) \in (a_j, b_j)$
(i.e., $f_{i,j}|f_{i,j-1}^{-1}([a,a_j] \cup [b_j,b]) = f_{i,j-1}|f_{i,j-1}^{-1}([a,a_j] \cup [b_j,b])$);
(3) No $x \in X_n$ are contained in $n+1$ many $f_{i,j}^{-1}(q_j)$.

For each $(i,j) \in \mathbb{N}^2$, let $k(i,j) = \frac{1}{2}(i+j-2)(i+j-1)+j \in \mathbb{N}$. Then, (i,j) is the $k(i,j)$-th element of $\mathbb{N}^2$ in the ordering

$$(1,1), (2,1), (1,2), (3,1), (2,2), (1,3), \ldots.$$

By induction on $k(i,j)$, we construct maps $f_{i,j}$ satisfying conditions (1), (2), and (3) above. Assume that $f_{i',j'}$ have been defined for $k(i',j') < m$. We will define $f_{i,j}$ for $k(i,j) = m$. Applying Lemma 5.9.4 to $L_0 = \emptyset$, $L_{k(i',j')} = f_{i',j'}^{-1}(q_{j'})$, $k(i',j') < m$, $A = f_{i,j-1}^{-1}([a,a_j])$, and $B = f_{i,j-1}^{-1}([b_j,b])$, we obtain a partition L_m in X between A and B such that

$(*)$ No $x \in X_n$ are contained in $n+1$ many sets L_i.

Then, we can easily obtain a map $f_{i,j} : X \to [a,b]$ such that $L_m = f_{i,j}^{-1}(q_j)$ and $f_{i,j}|A \cup B = f_{i,j-1}|A \cup B$, for which conditions (1), (2), and (3) are satisfied.

Since $b_j - a_j = 2^{-j}\delta_j$, it follows from (2) that $|f_{i,j}(x) - f_{i,j-1}(x)| < 2^{-j}\delta_j$ for each $x \in X$. Then, $(f_{i,j})_{j\in\mathbb{N}}$ uniformly converges to a map $f_i : X \to [a,b]$ and $|f_{i,j}(x) - f_i(x)| \leq 2^{-j}\delta_j$. For each $x \in A_i$, $f_i(x) = \lim_{j\to\infty} f_{i,j}(x) = a$ by (1). For each $x \in X \setminus A_i$, we have $k = \min\{j \in \mathbb{N} \mid f_{i,0}(x) > a_j\}$ because $f_{i,0}(x) > a = \inf_{j\in\mathbb{N}} a_j$. Then, $f_{i,0}(x) = f_{i,1}(x) = \cdots = f_{i,k-1}(x) > a_k$ and $f_{i,k}(x) > a_k = q_k - 2^{-k-1}\delta_k$, hence

$$f_i(x) \geq f_{i,k}(x) - 2^{-k}\delta_k > q_k - \delta_k \geq q_k - (q_k - a) = a.$$

Therefore, $A_i = f_i^{-1}(a)$. Similarly, we have $B_i = f_i^{-1}(b)$.

For each $x \in X_n$, let

$$M = \{i \in \mathbb{N} \mid f_{i,j}(x) = q_j \ \text{ for some } j \in \mathbb{N}\}.$$

Then, M has at most n many elements by (3). For $i \in \mathbb{N} \setminus M$ and $j \in \mathbb{N}$, let $K = \{k > j \mid f_{i,k}(x) \neq f_{i,j}(x)\}$. If $K = \emptyset$, then $f_i(x) = f_{i,j}(x) \neq q_j$ because $i \notin M$. Otherwise, let $k = \min K > j \geq 1$. Since $f_{i,k-1}(x) = f_{i,j}(x) \neq f_{i,k}(x)$, we have $a_k < f_{i,k}(x) < b_k$ by (2). Then, $|f_{i,k}(x) - q_j| \geq \delta_k - 2^{-k-1}\delta_k$. On the other hand, $|f_i(x) - f_{i,k}(x)| < 2^{-k+1}\delta_k$. Therefore,

$$\begin{aligned}|f_i(x) - q_j| &\geq |f_{i,k}(x) - q_j| - |f_i(x) - f_{i,k}(x)| \\ &> \delta_k - 2^{-k-1}\delta_k - 2^{-k+1}\delta_k > \tfrac{1}{4}\delta_k > 0.\end{aligned}$$

Thus, $\operatorname{card}\{i \mid f_i(x) \in (a,b) \cap \mathbb{Q}\} \leq n$ for $x \in X_n$. □

Proposition 5.9.6. *Let X be a metrizable space and Γ be an infinite set with $w(X) \leq \operatorname{card}\Gamma$. For each sequence $X_0, X_1, \dots \subset X$ with $\dim X_n \leq 0$, there exists an embedding $h : X \to J(\Gamma)^{\mathbb{N}}$ such that $h(X_n) \subset \nu_n(\Gamma)$.*

Proof. By Corollary 2.3.2, X has an open basis $\mathcal{B} = \bigcup_{i\in\mathbb{N}} \mathcal{B}_i$, where each $\mathcal{B}_i$ is discrete in X. Then, as is easily observed, $\operatorname{card}\mathcal{B}_i \leq w(X) \leq \operatorname{card}\Gamma$, hence we have $\Gamma_i \subset \Gamma$, $i \in \mathbb{N}$, such that $\operatorname{card}\mathcal{B}_i = \operatorname{card}\Gamma_i$ and $\Gamma_i \cap \Gamma_j = \emptyset$ if $i \neq j$. For each $i \in \mathbb{N}$, we write $\mathcal{B}_i = \{B_\gamma \mid \gamma \in \Gamma_i\}$, where $B_\gamma \neq B_{\gamma'}$ if $\gamma \neq \gamma'$. Let $A_i = X \setminus \bigcup_{\gamma\in\Gamma_i} B_\gamma$. We apply Lemma 5.9.5 to obtain maps $f_i : X \to [0,1]$, $i \in \mathbb{N}$, such that $A_i = f_i^{-1}(0)$ and $\operatorname{card}\{i \in \mathbb{N} \mid f_i(x) \in (0,1) \cap \mathbb{Q}\} \leq n$ for $x \in X_n$. We define $h_i : X \to J(\Gamma)$ by

$$h_i(x)(\gamma) = \begin{cases} f_i(x) & \text{if } x \in B_\gamma, \gamma \in \Gamma_i, \\ 0 & \text{otherwise.} \end{cases}$$

In other words, $h_i(x) = f_i(x)\mathbf{e}_\gamma$ for $x \in B_\gamma$, $\gamma \in \Gamma_i$, and $h_i(x) = 0$ for $x \in A_i$. The desired embedding $h : X \to J(\Gamma)^{\mathbb{N}}$ can be defined by $h(x) = (h_i(x))_{i\in\mathbb{N}}$. Indeed, if $x \neq y \in X$, then $x \in B_\gamma$ and $y \notin B_\gamma$ for some $\gamma \in \Gamma_i$. Then, $h_i(x)(\gamma) = f_i(x) > 0 = h_i(y)(\gamma)$. Thus, h is injective. For each $\gamma \in \Gamma_i$, $U_\gamma = \{z \in J(\Gamma)^{\mathbb{N}} \mid z(i)(\gamma) > 0\}$ is open in $J(\Gamma)^{\mathbb{N}}$. Observe that $h(B_\gamma) = U_\gamma \cap h(X)$. Therefore, h is an embedding of X into $J(\Gamma)^{\mathbb{N}}$. For $x \in X_n$,

$$\operatorname{card}\{i \in \mathbb{N} \mid h_i(x) \notin P(\Gamma)\} = \operatorname{card}\{i \in \mathbb{N} \mid f_i(x) \in \mathbb{Q} \setminus \{0\}\} \leq n.$$

Then, it follows that $h(X_n) \subset \nu_n(\Gamma)$. □

Theorem 5.9.7. *Let Γ be an infinite set. The space $\nu_n(\Gamma)$ is a universal space for metrizable spaces X with $w(X) \leq \operatorname{card}\Gamma$ and $\dim X \leq n$, and the space $\nu_\omega(\Gamma)$ is a universal space for countable-dimensional metrizable spaces X with $w(X) \leq \operatorname{card}\Gamma$.*

Proof. We can write $X = \bigcup_{i\in\omega} X_i$, where $\dim X_i \leq 0$ and $X_i = \emptyset$ for $i > n$ if $\dim X = n$. The theorem follows from Proposition 5.9.6. □

Let X be a separable metrizable space with $\dim X \leq n$. In the proof of Proposition 5.9.6, we can take a $\mathcal{B}_i$ with only one element. Then, replacing $\mathbf{I}$ with $[a,1]$ where $a \in \mathbf{I}\setminus\mathbb{Q}$, the map $h : X \to [a,1]^{\mathbb{N}} \subset \mathbf{I}^{\mathbb{N}}$ defined by $h(x) = (f_i(x))_{i\in\mathbb{N}}$ is an embedding such that $h(X_n) \subset \nu_n$. Similar to Theorem 5.9.7, we have the following separable version:

Theorem 5.9.8. *The space ν_n is a universal space for separable metrizable spaces with* $\dim \leq n$ *and the space ν_ω is a universal space for countable-dimensional separable metrizable spaces.* □

Next, recall that $\mathbf{I}_f^{\mathbb{N}}$ is s.c.d. We now define

$$K_\omega = \bigcup_{n\in\mathbb{N}} \nu_n \times \big((0,1]^n \times \{0\}\big) \subset \nu_\omega \times \mathbf{I}_f^{\mathbb{N}}.$$[11]

For an infinite set Γ, we define

$$K_\omega(\Gamma) = \bigcup_{n\in\mathbb{N}} \nu_n(\Gamma) \times \big((0,1]^n \times \{0\}\big) \subset \nu_\omega(\Gamma) \times \mathbf{I}_f^{\mathbb{N}} \subset J(\Gamma)^{\mathbb{N}} \times \mathbf{I}^{\mathbb{N}}.$$

Then, K_ω is separable and $w(K_\omega(\Gamma)) = \operatorname{card}\Gamma$. Moreover, $K_\omega(\Gamma)$ and K_ω are s.c.d. Indeed, for $(x,y) \in K_\omega(\Gamma)$,

$$(x,y) \in \bigcup_{i=1}^{n} \nu_i(\Gamma) \times \big((0,1]^i \times \{0\}\big) \Leftrightarrow y(n+1) = 0.$$

Hence, $\bigcup_{i=1}^n \nu_i(\Gamma) \times ((0,1]^i \times \{0\})$ is a closed set in $K_\omega(\Gamma)$, which is finite-dimensional by the Product Theorem 5.4.9 and the Addition Theorem 5.4.8.

Theorem 5.9.9. *Let Γ be an infinite set. The space $K_\omega(\Gamma)$ is a universal space for strongly countable-dimensional metrizable spaces X with $w(X) \leq \operatorname{card}\Gamma$.*

Proof. We can write $X = \bigcup_{i\in\mathbb{N}} F_i$, where each F_i is closed in X, $\dim F_i \leq i-1$, and $F_i \subset F_{i+1}$ for each $i \in \mathbb{N}$. By the Decomposition Theorem 5.4.5, we have a sequence $X_1, X_2, \dots \subset X$ such that $\dim X_n \leq 0$ and

$$F_1 = X_1,\ F_2 \setminus F_1 = X_2 \cup X_3,\ F_3 \setminus F_2 = X_4 \cup X_5 \cup X_6,\ \dots,$$

i.e., $F_i \setminus F_{i-1} = \bigcup_{n=k(i-1)+1}^{k(i)} X_n$, where $F_0 = \emptyset$ and $k(i) = \frac{1}{2}i(i+1)$. We apply Proposition 5.9.6 to obtain an embedding $h : X \to J(\Gamma)^{\mathbb{N}}$ such that $h(X_n) \subset \nu_n(\Gamma)$ for each $n \in \mathbb{N}$. For each $i \in \mathbb{N}$, let $f_i : X \to \mathbf{I}$ be a map with $f_i^{-1}(0) = F_{i-1}$, and define a map $f : X \to \mathbf{I}^{\mathbb{N}}$ as follows:

$$f(x) = (f_1(x), f_2(x), f_2(x), f_3(x), f_3(x), f_3(x), \dots),$$

where each $f_i(x)$ appears i times, i.e., $\mathrm{pr}_n f = f_i$ for $k(i-1)+1 \leq n \leq k(i)$. Now, we can define the embedding $g : X \to J(\Gamma)^{\mathbb{N}} \times \mathbf{I}^{\mathbb{N}}$ by $g(x) = (h(x), f(x))$. For each $x \in X$, choose $i \in \mathbb{N}$ and $k(i-1)+1 \leq n \leq k(i)$ so that $x \in X_n \subset F_i \setminus F_{i-1}$. Then, $h(x) \in h(X_n) \subset \nu_n(\Gamma) \subset \nu_{k(i)}(\Gamma)$. Since $x \in F_i \setminus F_{i-1}$, it follows that

[11] This is different from the usual notation. In the literature for Dimension Theory, this space is represented by $K_\omega(\aleph_0)$ and K_ω stands for $\mathbf{I}_f^{\mathbb{N}}$.

$f_j(x) > 0$ for $j \le i$ and $f_j(x) = 0$ for $j \ge i + 1$, i.e., $\mathrm{pr}_j f(x) > 0$ for $j \le k(i)$ and $\mathrm{pr}_j f(x) = 0$ for $j \ge k(i) + 1$. Therefore, $f(x) \in (0, 1]^{k(i)} \times \{0\} \subset \mathbf{I}^{\mathbb{N}}$. Thus, we have

$$g(x) = (h(x), f(x)) \in \nu_{k(i)}(\Gamma) \times \big((0, 1]^{k(i)} \times \{0\}\big) \subset K_\omega(\Gamma).$$

Consequently, X can be embedded into $K_\omega(\Gamma)$. □

Similarly, we can obtain the following separable version:

Theorem 5.9.10. *The space K_ω is a universal space for strongly countable-dimensional separable metrizable spaces.* □

5.10 Nöbeling Spaces and Menger Compacta

In this section, we shall construct two universal spaces for separable metrizable spaces with $\dim \le n$, which are named the n-dimensional Nöbeling space and the n-dimensional Menger compactum.

In the previous section, we defined the universal space ν_n. In the definition of ν_n, we replace $\mathbf{I}^{\mathbb{N}}$ with $\mathbb{R}^{2n+1}$ to define the ***n*-dimensional Nöbeling space** ν^n, that is,

$$\begin{aligned}\nu^n &= \big\{x \in \mathbb{R}^{2n+1} \bigm| x(i) \in \mathbb{R} \setminus \mathbb{Q} \ \text{ except for } n \text{ many } i\big\}\\ &= \big\{x \in \mathbb{R}^{2n+1} \bigm| x(i) \in \mathbb{Q} \ \text{ at most } n \text{ many } i\big\},\end{aligned}$$

which is the n-dimensional version of the space of irrationals $\nu^0 = \mathbb{R} \setminus \mathbb{Q}$. Similar to Theorem 5.9.3, we can see $\dim \nu^n = n$. Observe

$$\mathbb{R}^{2n+1} \setminus \nu^n = \big\{x \in \mathbb{R}^{2n+1} \bigm| x(i) \in \mathbb{Q} \ \text{ at least } n + 1 \text{ many } i\big\},$$

which is a countable union of n-dimensional flats that are closed in $\mathbb{R}^{2n+1}$. Then, ν^n is a G_δ-set in $\mathbb{R}^{2n+1}$, hence it is completely metrizable. Thus, we have the following proposition:

Proposition 5.10.1. *The space ν^n is a separable completely metrizable space with* $\dim \nu^n = n$. □

Moreover, ν^n has the additional property:

Proposition 5.10.2. *Each point of ν^n has an arbitrarily small neighborhood that is homeomorphic to ν^n. In fact, $\nu^n \cap \prod_{i=1}^{2n+1}(a_i, b_i) \approx \nu^n$ for each $a_i < b_i \in \mathbb{Q}$, $i = 1, \dots, 2n + 1$.*

Proof. Let $\varphi : \mathbb{R} \to (-1, 1)$ be the homeomorphism defined by

$$\varphi(t) = \frac{t}{1 + |t|} \quad \left(\varphi_i^{-1}(s) = \frac{s}{1 - |s|}\right).$$

We define a homeomorphism $h : \mathbb{R}^{2n+1} \to \prod_{i=1}^{2n+1}(a_i, b_i)$ as follows:

$$h(x) = (h_1(x(1)), \dots, h_{2n+1}(x(2n+1))),$$

where $h_i : \mathbb{R} \to (a_i, b_i)$ is the homeomorphism defined by

$$h_i(t) = \frac{b_i - a_i}{2}(\varphi(t) + 1) + a_i.$$

Since $h_i(\mathbb{Q}) = \mathbb{Q} \cap (a_i, b_i)$, we have $h(\nu^n) = \nu^n \cap \prod_{i=1}^{2n+1}(a_i, b_i)$. □

We will show the universality of ν^n.

Theorem 5.10.3. *The n-dimensional Nöbeling space ν^n is a universal space for separable metrizable spaces with* $\dim \leq n$.

According to the Compactification Theorem 5.7.2, every n-dimensional separable metrizable space X has an n-dimensional metrizable compactification. Theorem 5.10.3 comes from the following proposition:

Proposition 5.10.4. *For each locally compact separable metrizable space X with $\dim X \leq n$ and $\mathcal{U} \in \operatorname{cov}(\mathbb{R}^{2n+1})$, every proper map $f : X \to \mathbb{R}^{2n+1}$ is $\mathcal{U}$-close to a closed embedding $g : X \to \nu^n$. If X is compact, then $\mathbb{R}^{2n+1}$ can be replaced by $\mathbf{I}^{2n+1}$.*

This can be shown by a modification of the proof of the Embedding Approximation Theorem 5.8.10 (or 5.8.5). To this end, we need the following generalization of Theorem 5.8.9:

Lemma 5.10.5. *Let X and Y be locally compact metrizable spaces and $Y_0 = \bigcap_{n\in\mathbb{N}} G_n \subset Y$, where each G_n is open in Y (hence Y_0 is a G_δ-set in Y). Suppose that for each $n \in \mathbb{N}$ and each open cover $\mathcal{U}$ of X consisting of open sets with the compact closures, $\mathrm{C}_{\mathcal{U}}(X, G_n)$ is dense in the space $\mathrm{C}^P(X, Y)$ with the limitation topology. Then,* $\mathrm{Emb}(X, Y_0)$ *is dense in* $\mathrm{C}^P(X, Y)$.

Proof. Observe that

$$\begin{aligned}\mathrm{Emb}(X, Y_0) &= \mathrm{Emb}(X, Y) \cap \mathrm{C}(X, Y_0) \\ &= \bigcap_{n\in\mathbb{N}} \mathrm{C}_{\mathcal{U}_n}(X, Y) \cap \bigcap_{n\in\mathbb{N}} \mathrm{C}(X, G_n) = \bigcap_{n\in\mathbb{N}} \mathrm{C}_{\mathcal{U}_n}(X, G_n),\end{aligned}$$

where $\mathcal{U}_n \in \operatorname{cov}(X)$ consists of open sets with the compact closures and $\operatorname{mesh}\mathcal{U}_n < 2^{-n}$. By the assumption, each $\mathrm{C}_{\mathcal{U}}(X, G_n)$ is open and dense in $\mathrm{C}^P(X, Y)$. Since $\mathrm{C}^P(X, Y)$ is a Baire space by Theorem 2.9.8, we have the desired result. □

Proof of Proposition 5.10.4. According to the definition of ν^n, we can write

$$\nu^n = \mathbb{R}^{2n+1} \setminus \bigcup_{i\in\mathbb{N}} H_i = \bigcap_{i\in\mathbb{N}}(\mathbb{R}^{2n+1} \setminus H_i),$$

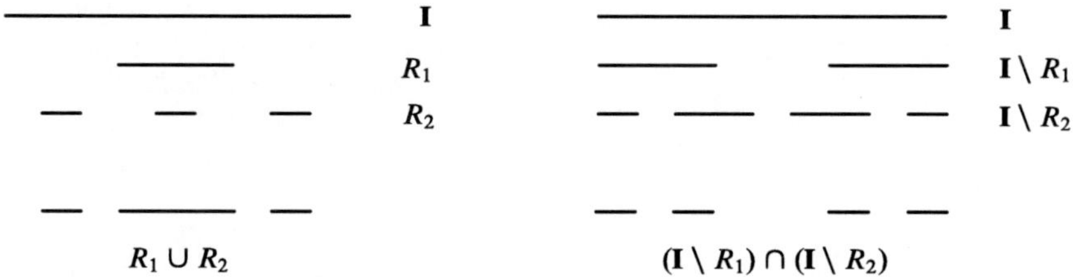

Fig. 5.4 R_k and $\mathbf{I} \setminus R_k$

where each H_i is an n-dimensional flat. Because of Lemma 5.10.5, it suffices to show that, for each n-dimensional flat H in $\mathbb{R}^{2n+1}$ and each $\mathcal{U} \in \operatorname{cov}(X)$ consisting of open sets with the compact closure, $\mathrm{C}_{\mathcal{U}}(X, \mathbb{R}^{2n+1} \setminus H)$ is dense in the space $\mathrm{C}^P(X, \mathbb{R}^{2n+1})$ with the limitation topology.

In the proof of 5.8.10, we can choose $v_i \in \mathbb{R}^{2n+1}$, $i \in \mathbb{N}$, to satisfy the additional condition that the flat hull of every $n+1$ many points $v_{i_1}, \ldots, v_{i_{n+1}}$ misses H (i.e., $\mathrm{fl}\{v_{i_1}, \ldots, v_{i_{n+1}}\} \cap H = \emptyset$). Thus, we can obtain the PL embedding $g : |K| \to \mathbb{R}^{2n+1}$ such that $g(|K|) \cap H = \emptyset$. The map $g\varphi : X \to \mathbb{R}^{2n+1} \setminus H$ is a $\mathcal{U}$-map that is $\mathcal{V}$-close to f.

If X is compact, we can replace $\mathbb{R}^{2n+1}$ by $\mathbf{I}^{2n+1}$ to obtain the additional statement. □

Remark 15. It is known that if X is a separable completely metrizable space with $\dim X \leq n$, then every map $f : X \to \nu^n$ can be approximated by closed embeddings $h : X \to \nu^n$. Refer to Remark 14.

Before defining the n-dimensional Menger compactum, let us recall the construction of the Cantor (ternary) set μ^0. We can geometrically describe $\mu^0 \subset \mathbf{I}$ as follows: For each $k \in \mathbb{N}$, let

$$R_k = \bigcup_{m=0}^{3^{k-1}-1} (m/3^{k-1} + 1/3^k, m/3^{k-1} + 2/3^k) \subset \mathbf{I}.$$

Then, $\mu^0 = \bigcap_{k\in\mathbb{N}}(\mathbf{I} \setminus R_k) = \mathbf{I} \setminus \bigcup_{k\in\mathbb{N}} R_k$ (Fig. 5.4). Observe that

$$\bigcap_{i=1}^{k}(\mathbf{I} \setminus R_i) = [0, 3^{-k}] + V_k^0, \text{ where } V_k^0 = \left\{ \sum_{i=1}^{k} \frac{2x(i)}{3^i} \,\middle|\, x \in \mathbf{2}^k \right\}.$$

Moreover, $\{3^{-k}\mu^0 + v \mid v \in V_k^0\}$ is an open cover of μ^0 with ord $= 1$, where

$$\begin{aligned} \mu^0 \approx 3^{-k}\mu^0 + v &= \mu^0 \cap ([0, 3^{-k}] + v) \\ &= \mu^0 \cap ((-3^{-k-1}, 3^{-k} + 3^{-k-1}) + v). \end{aligned}$$

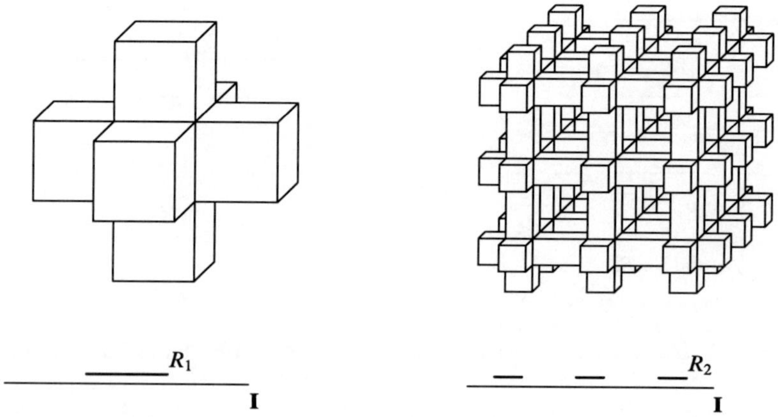

Fig. 5.5 $\mathbf{I}^3 \setminus M_1^3$ and $\mathbf{I}^3 \setminus M_2^3$

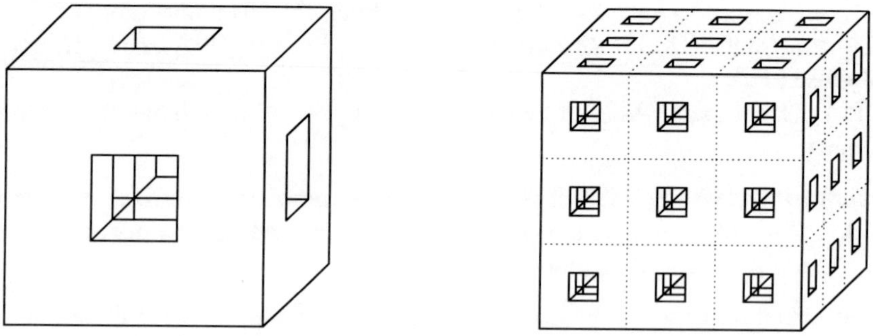

Fig. 5.6 M_1^3 and M_2^3

As the n-dimensional version of μ^0, the ***n*-dimensional Menger compactum** μ^n is defined as follows: For each $k \in \mathbb{N}$, let

$$M_k^{2n+1} = \{x \in \mathbf{I}^{2n+1} \mid x(i) \in \mathbf{I} \setminus R_k \text{ except for } n \text{ many } i\}$$
$$= \{x \in \mathbf{I}^{2n+1} \mid x(i) \in R_k \text{ at most } n \text{ many } i\},$$

where it should be noted that

$$\mathbf{I} \setminus M_k^{2n+1} = \{x \in \mathbf{I}^{2n+1} \mid x(i) \in R_k \text{ at least } n+1 \text{ many } i\}.$$

Now, we define $\mu^n = \bigcap_{k \in \mathbb{N}} M_k^{2n+1}$. Since each M_k^{2n+1} is compact, μ^n is also compact. See Figs. 5.5–5.7.

Proposition 5.10.6. *For each* $n \in \mathbb{N}$, $\dim \mu^n = n$.

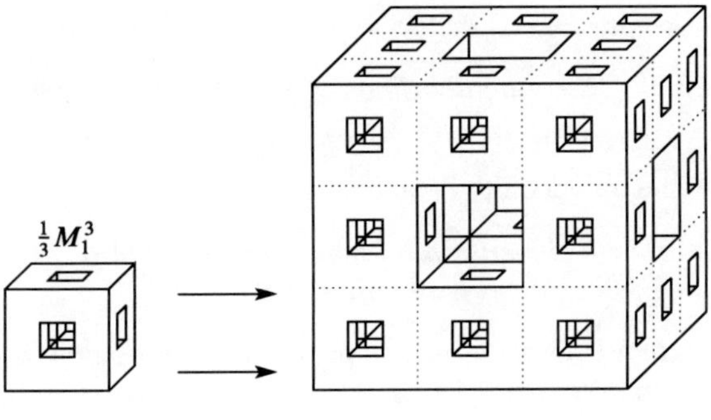

Fig. 5.7 $M_1^3 \cap M_2^3$

Proof. Since μ^n contains every n-face of $\mathbf{I}^{2n+1}$, it follows that $\dim \mu^n \geq n$. We can apply Proposition 5.8.12 to see $\dim \mu^n \leq n$. We use the metric $d \in \mathrm{Metr}(\mathbf{I}^{2n+1})$ defined by

$$d(x, y) = \max\{|x(i) - y(i)| \mid i = 1, \ldots, 2n + 1\}.$$

For each $\varepsilon > 0$, choose $k \in \mathbb{N}$ so large that $2/3^k < \varepsilon$. Let K be the cell complex consisting of all faces of $(2n + 1)$-cubes

$$\prod_{i=1}^{2n+1} \left[\frac{m_i - 1}{3^{k-1}}, \frac{m_i}{3^{k-1}}\right] \subset \mathbf{I}^{2n+1}, \ m_i = 1, \ldots, 3^k.$$

Since $\mu^n \subset M_k^{2n+1}$, it suffices to construct an ε-map of M_k^{2n+1} to $|K^{(n)}|$, where $K^{(n)}$ is the n-skeleton of K.

For each $C \in K$ with $\dim C > n$, let $r_C : C \setminus \{\hat{C}\} \to \partial C$ be the radial retraction, where $\hat{C}$ is the barycenter of C and ∂C is the radial boundary of C. Observe that $M_k^{2n+1} \cap C \subset C \setminus \{\hat{C}\}$ and $r_C(M_k^{2n+1} \cap C) \subset M_k^{2n+1} \cap \partial C$. For each $m \geq n$, we can define a retraction

$$r_m : M_k^{2n+1} \cap |K^{(m+1)}| \to M_k^{2n+1} \cap |K^{(m)}|$$

by $r_m|C = r_C$ for each $(m + 1)$-cell $C \in K$. Since $|K^{(n)}| \subset M_k^{2n+1}$, we have a retraction

$$r = r_n \cdots r_{2n} : M_k^{2n+1} \to |K^{(n)}|.$$

By construction, $r^{-1}(x) \subset \mathrm{st}(x, K)$ for each $x \in |K^{(n)}|$. Since $\mathrm{mesh}\, K = 1/3^{k-1} < \varepsilon/2$, it follows that r is an ε-map. □

For each $k \in \mathbb{N}$, $\mu^n \approx 3^{-k}\mu^n \subset [0, 3^{-k}]$. Let

$$V_k^n = \{v \in 3^{-k}\mathbb{Z}^{2n+1} \mid [0, 3^{-k}]^{2n+1} + v \subset M_k^{2n+1}\}.$$

Then, $M_k^{2n+1} = \bigcup_{v \in V_k^n}([0, 3^{-k}]^{2n+1} + v)$ and $\mu^n = 3^{-k}\mu^n + V_k^n$. Thus, we have the following proposition:

Proposition 5.10.7. *Every neighborhood of each point of μ^n contains a copy of μ^n.* □

We will show the universality of μ^n.

Theorem 5.10.8. *The n-dimensional Menger compactum μ^n is a universal space for separable metrizable spaces with* dim $\leq n$.

Proof. By Theorem 5.10.3, it suffices to prove that every compact set X in $\mathbf{I}^{2n+1} \cap \nu^n$ can be embedded in μ^n.

First, note that

$$X \subset \{x \in \mathbf{I}^{2n+1} \mid x(i) \neq 1/2 \ \text{ except for } n \text{ many } i\}.$$

Then, we have a rational $q_1 > 0$ such that

$$X \subset \{x \in \mathbf{I}^{2n+1} \mid x(i) \in \mathbf{I} \setminus R_1^X \ \text{ except for } n \text{ many } i\},$$

where $R_1^X = (1/2 - q_1, 1/2 + q_1) \subset (0,1)$. Let $A_1^X = \{1/2 - q_1, 1/2 + q_1\}$ be the set of end-points of R_1^X and let $g_1 : \mathbf{I} \to \mathbf{I}$ be the PL homeomorphism defined by $g_1(0) = 0$, $g_1(1) = 1$, and $g_1(1/2 \pm q_1) = 1/2 \pm 1/6$, i.e., $g_1(A_1^X) = \{1/3, 2/3\}$. Observe that $|g_1(s) - s| < 3^{-1}$ for every $s \in \mathbf{I}$.

Let B_1^X be the set of mid-points of components of $\mathbf{I} \setminus A_1^X$, i.e., $B_1^X = \{1/2, 1/2^2 - q_1/2, 3/2^2 + q_1/2\} \subset \mathbb{Q}$. Note that

$$X \subset \{x \in \mathbf{I}^{2n+1} \mid x(i) \in \mathbf{I} \setminus B_1^X \ \text{ except for } n \text{ many } i\}.$$

Then, we have a rational $q_2 > 0$ such that $2q_2$ is smaller than the diameter of each component of $\mathbf{I} \setminus A_1^X$, and

$$X \subset \{x \in \mathbf{I}^{2n+1} \mid x(i) \in \mathbf{I} \setminus R_2^X \ \text{ except for } n \text{ many } i\},$$

where $R_2^X = \bigcup_{b \in B_1^X}(b - q_2, b + q_2)$. Let A_2^X be the set of end-points of components of R_2^X and let $g_2 : \mathbf{I} \to \mathbf{I}$ be the PL homeomorphism defined by $g_2(0) = 0$, $g_2(1) = 1$, $g_2(A_1^X \cup A_2^X) = \{m/3^2 \mid m = 1, \ldots, 3^2 - 1\}$. Then, $g_2|A_1^X = g_1|A_1^X$ and $|g_2(s) - g_1(s)| < 3^{-2}$ for every $s \in \mathbf{I}$.

Let B_2^X be the set of mid-points of components of $\mathbf{I} \setminus (A_1^X \cup A_2^X)$. Then, $B_2^X \subset \mathbb{Q}$. Since

$$X \subset \{x \in \mathbf{I}^{2n+1} \mid x(i) \in \mathbf{I} \setminus B_2^X \ \text{ except for } n \text{ many } i\},$$

we have a rational $q_3 > 0$ such that $2q_3$ is smaller than the diameter of each component of $\mathbf{I} \setminus (A_1^X \cup A_2^X)$, and

$$X \subset \{x \in \mathbf{I}^{2n+1} \mid x(i) \in \mathbf{I} \setminus R_3^X \ \text{ except for } n \text{ many } i\},$$

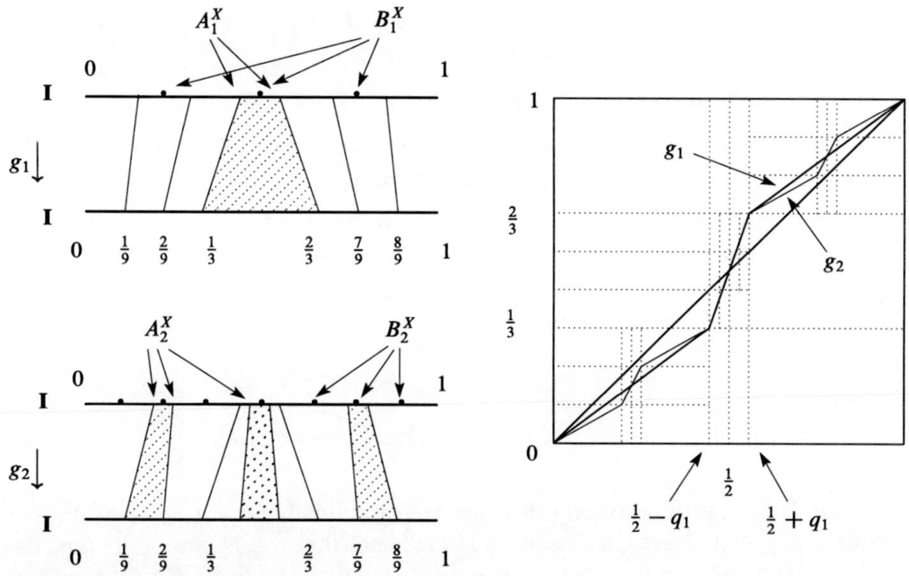

Fig. 5.8 Homeomorphisms $g_1, g_2, \dots$

where $R_3^X = \bigcup_{b \in B_2^X} (b - q_2, b + q_2)$. Let A_3^X be the set of end-points of components of R_3^X and let $g_3 : \mathbf{I} \to \mathbf{I}$ be the PL homeomorphism defined by $g_3(0) = 0$, $g_3(1) = 1$, $g_3(A_1^X \cup A_2^X \cup A_3^X) = \{m/3^3 \mid m = 1, \dots, 3^3 - 1\}$. Then, $g_3|A_1^X \cup A_2^X = g_2|A_1^X \cup A_2^X$ and $|g_3(s) - g_2(s)| < 3^{-3}$ for every $s \in \mathbf{I}$ — (Fig. 5.8).

By induction, we obtain $R_k^X, A_k^X \subset \mathbf{I}$ ($k \in \mathbb{N}$) such that R_k^X is the union of 3^{k-1} many disjoint open intervals, A_k^X is the set of all end-points of components of R_k^X, each component of R_k^X is contained in some component of $\mathbf{I} \setminus A_{k-1}^X$, and

$$X \subset \{x \in \mathbf{I}^{2n+1} \mid x(i) \in \mathbf{I} \setminus R_k^X \text{ except for } n \text{ many } i\}.$$

Hence, X is contained in

$$\mu_X^n = \bigcap_{k \in \mathbb{N}} \{x \in \mathbf{I}^{2n+1} \mid x(i) \in \mathbf{I} \setminus R_k^X \text{ except for } n \text{ many } i\}.$$

At the same time, we have the PL homeomorphisms $g_k : \mathbf{I} \to \mathbf{I}$, $k \in \mathbb{N}$, such that

$$g_k(0) = 0,\ g_k(1) = 1,\ g_k\big(\textstyle\bigcup_{i=1}^{k} A_i^X\big) = \{m/3^k \mid m = 1, \dots, 3^k - 1\},$$

$$g_k \Big| \bigcup_{i=1}^{k-1} A_i^X = g_{k-1} \Big| \bigcup_{i=1}^{k-1} A_i^X \text{ and } \big|g_k(s) - g_{k-1}(s)\big| < 3^{-k} \text{ for every } s \in \mathbf{I}.$$

Then, $(g_k)_{k \in \mathbb{N}}$ uniformly converges to a map $g : \mathbf{I} \to \mathbf{I}$. Since $A = \bigcup_{i=1}^{\infty} A_i^X$ is dense in $\mathbf{I}$ and g maps A onto $\{m/3^k \mid k \in \mathbb{N},\ m = 1, \dots, 3^k - 1\}$ in the same order, it follows that g is bijective, hence g is a homeomorphism. Let $h : \mathbf{I}^{2n+1} \to \mathbf{I}^{2n+1}$

be the homeomorphism defined by $h(x) = (g(x(1)), \ldots, g(x(2n+1)))$. As is easily observed, $h(\mu^n_X) = \mu^n$, hence $h(X) \subset \mu^n$. □

We also have the following theorem:

Theorem 5.10.9. *Let X be a compactum with* $\dim X \leq n$. *Then, every map $f : X \to \mu^n$ can be approximated by embeddings into μ^n.*

Proof. By Proposition 5.10.4, f can be approximated by embeddings f' into $M_k^{2n+1} \cap \nu^n$ for an arbitrarily large $k \in \mathbb{N}$. Replacing X by $f'(X)$ in the proof of Theorem 5.10.8, we can take $R_i^X = R_i$ and $g_i = \mathrm{id}$ for $i \leq k$. Therefore, f can be approximated by embeddings like hf'. □

5.11 Total Disconnectedness and the Cantor Set

A space X is said to be **totally disconnected** provided that, for any two distinct points $x \neq y \in X$, there is a clopen set H in X such that $x \in H$ but $y \notin H$ (i.e., the empty set is a partition between any two distinct points). Equivalently, for each $x \in X$ the intersection of all clopen sets containing x is the singleton $\{x\}$. According to Theorem 5.3.8, the 0-dimensionality implies the total disconnectedness. We say that X is **hereditarily disconnected** if every non-degenerate subset of X is disconnected (i.e., every component of X is a singleton). Clearly, the total disconnectedness implies the hereditary disconnectedness. Therefore, we have the following fact:

Fact. *Every 0-dimensional space is totally disconnected, and every totally disconnected space is hereditarily disconnected.*

The converse assertions are true for compact spaces. To see this, we prove the following lemma:

Lemma 5.11.1. *Let X be compact, $x \in X$, and C be the intersection of all clopen sets in X containing x.*

(1) *For each open neighborhood U of C in X, there is a clopen set H in X such that $C \subset H \subset U$.*
(2) *C is the component of X containing x.*

Proof. (1): Let $\mathcal{H}$ be all the clopen sets in X containing x. Since $X \setminus U$ is compact and $\{X \setminus H \mid H \in \mathcal{H}\}$ is its open cover in X, there are $H_1, \ldots, H_n \in \mathcal{H}$ such that

$$X \setminus U \subset \bigcup_{i=1}^{n} (X \setminus H_i) = X \setminus \bigcap_{i=1}^{n} H_i .$$

Thus, we have $H = \bigcap_{i=1}^{n} H_i \in \mathcal{H}$ and $C \subset H \subset U$.

(2): Since C clearly contains the component of X containing x, it suffices to show that C is connected. Now assume that $C = A \cup B$, where A and B are disjoint closed sets in C and $x \in A$. From the normality, it follows that there are disjoint open sets U and V in X such that $A \subset U$ and $B \subset V$. By (1), we have a clopen set H in X such that $C \subset H \subset U \cup V$. Since $H \cap U$ is open in X and $H \setminus V$ is closed in X, $H \cap U = H \setminus V$ is clopen in X. Then, $C \subset H \cap U \subset U$, which implies that $B \subset C \cap V = \emptyset$. Thus, C is connected. □

Theorem 5.11.2. *For every non-empty compact space X, the following are equivalent:*

(a) $\dim X = 0$;
(b) *X is totally disconnected;*
(c) *X is hereditarily disconnected.*

Proof. The implications (a) $\Rightarrow$ (b) $\Rightarrow$ (c) follow from the above Fact. Here, we will prove the converse implications.

(c) $\Rightarrow$ (b): For each $x \in X$, the intersection of all clopen sets in X containing x is a component of X by Lemma 5.11.1(2). It is, in fact, the singleton $\{x\}$, which means that X is totally disconnected.

(b) $\Rightarrow$ (a): Let $\mathcal{U}$ be a finite open cover of X. Each $x \in X$ belongs to some $U \in \mathcal{U}$. Because of the total disconnectedness of X, the singleton $\{x\}$ is the intersection of all clopen sets in X containing x. By Lemma 5.11.1(1), we have a clopen set H_x in X such that $x \in H_x \subset U$. From the compactness, it follows that $X = \bigcup_{i=1}^n H_{x_i}$ for some $x_1, \ldots, x_n \in X$. Let

$$V_1 = H_{x_1}, V_2 = H_{x_2} \setminus H_{x_1}, \ldots, V_n = H_{x_n} \setminus (H_{x_1} \cup \cdots \cup H_{x_{n-1}}).$$

Then, $\mathcal{V} = \{V_1, \ldots, V_n\}$ is an open refinement of $\mathcal{U}$ and $\operatorname{ord} \mathcal{V} = 1$. Hence, we have $\dim X = 0$. □

The implications (c) $\Rightarrow$ (b) $\Rightarrow$ (a) in Theorem 5.11.2 do not hold in general. In the next section, we will show the existence of nonzero-dimensional totally disconnected spaces, i.e., counter-examples for (b) $\Rightarrow$ (a). Here, we give a counter-example for (c) $\Rightarrow$ (b) via the following theorem:

Theorem 5.11.3. *There exists a separable metrizable space that is hereditarily disconnected but not totally disconnected.*

Example and Proof. Take a countable dense set D in the Cantor set μ^0 and define

$$X = D \times \mathbb{Q} \cup (\mu^0 \setminus D) \times (\mathbb{R} \setminus \mathbb{Q}) \subset \mu^0 \times \mathbb{R}.$$

Let $p : X \to \mu^0$ be the restriction of the projection of $\mu^0 \times \mathbb{R}$ onto μ^0.

First, we show that X is hereditarily disconnected. Let $A \subset X$ be a non-degenerate subset. When $\operatorname{card} p(A) > 1$, since μ^0 is hereditarily disconnected, $p(A)$ is disconnected, which implies that A is disconnected. When $\operatorname{card} p(A) = 1$, $A \subset p(A) \times \mathbb{Q} \approx \mathbb{Q}$ or $A \subset p(A) \times (\mathbb{R} \setminus \mathbb{Q}) \approx \mathbb{R} \setminus \mathbb{Q}$. Since both $\mathbb{Q}$ and $\mathbb{R} \setminus \mathbb{Q}$ are hereditarily disconnected, A is disconnected.

Next, we prove that X is not totally disconnected. Assume that X is totally disconnected and let $x_0 \in D \subset \mu^0$. Because $(x_0, 0), (x_0, 1) \in X$, we have closed sets F_0 and F_1 in $\mu^0 \times \mathbb{R}$ such that $X \subset F_0 \cup F_1$, $F_0 \cap F_1 \cap X = \emptyset$, and $(x_0, i) \in F_i$. Then, $F_0 \cap X$ and $F_1 \cap X$ are clopen in X. Choose an open neighborhood U_0 of x_0 in μ^0 so that

$$(U_0 \cap D) \times \{i\} = U_0 \times \{i\} \cap X \subset F_i \quad \text{for } i = 0, 1.$$

Since F_i is closed in $\mu^0 \times \mathbb{R}$ and D is dense in μ^0, it follows that $U_0 \times \{i\} \subset F_i$. For each $r \in \mathbb{Q}$, let

$$C_r = \{x \in U_0 \mid (x, r) \in F_0 \cap F_1\}.$$

Then, each C_r is closed and nowhere dense in U_0. Indeed, for each $x \in U_0 \setminus C_r$, because $(x, r) \notin F_0 \cap F_1$ and $F_0 \cap F_1$ is closed in $\mu^0 \times \mathbb{R}$, x has a neighborhood U in U_0 such that $U \times \{r\} \cap F_0 \cap F_1 = \emptyset$. Then, $(y, r) \notin F_0 \cap F_1$ for all $y \in U$, i.e., $U \cap C_r = \emptyset$, so C_r is closed in U_0. Since $F_0 \cap F_1 \cap X = \emptyset$ and $(x, r) \in X$ for $x \in D$, we have $C_r \subset U_0 \setminus D$, which implies that C_r is nowhere dense in U_0.

We will show that $U_0 \setminus D = \bigcup_{r \in \mathbb{Q}} C_r$. Then,

$$U_0 = \bigcup_{r \in \mathbb{Q}} C_r \cup \bigcup_{x \in D \cap U_0} \{x\},$$

which is contrary to the Baire Category Theorem 2.5.1. Thus, it would follow that X is not totally disconnected. For each $x \in U_0 \setminus D$,

$$\{x\} \times \mathbb{R} = \mathrm{cl}_{\mu^0 \times \mathbb{R}} \{x\} \times (\mathbb{R} \setminus \mathbb{Q}) \subset \mathrm{cl}_{\mu^0 \times \mathbb{R}} X \subset F_0 \cup F_1.$$

If $x \notin \bigcup_{r \in \mathbb{Q}} C_r$, then $F_0 \cap F_1 \cap \{x\} \times \mathbb{Q} = \emptyset$ because $x \notin C_r$ for all $r \in \mathbb{Q}$. Therefore,

$$F_0 \cap F_1 \cap \{x\} \times \mathbb{R} = F_0 \cap F_1 \cap \{x\} \times (\mathbb{R} \setminus \mathbb{Q}) \subset F_0 \cap F_1 \cap X = \emptyset.$$

Because $(x, i) \in F_i \cap \{x\} \times \mathbb{R}$, this contradicts the connectedness of $\mathbb{R}$. Therefore, $x \in \bigcup_{r \in \mathbb{Q}} C_r$ and the proof is complete. □

In the remainder of this section, we give a characterization of the Cantor set μ^0 and show that every compactum is a continuous image of μ^0. Recall that $\mu^0 \approx \mathbf{2}^{\mathbb{N}}$, where $\mathbf{2} = \{0, 1\}$ is the discrete space of two points. In the following, μ^0 can be replaced by $\mathbf{2}^{\mathbb{N}}$ (cf. Sect. 1.1).

Theorem 5.11.4 (Characterization of the Cantor Set). *A space X is homeomorphic to the Cantor set μ^0 if and only if X is a totally disconnected compactum with no isolated points.*

Proof. It suffices to show the "if" part. Since $\mu^0 \approx \mathbf{2}^{\mathbb{N}}$, we will construct a homeomorphism $h : \mathbf{2}^{\mathbb{N}} \to X$. Let $d \in \mathrm{Metr}(X)$ with $\mathrm{diam}\, X < 1$. First, note that

(∗) Each non-empty open set in X can be written as the disjoint union of an arbitrary finite number of non-empty open sets.

In fact, because X has no isolated points, each non-empty open set U in X is non-degenerate and $\dim U = 0$ by Theorem 5.11.2 and the Subset Theorem 5.3.3. We apply Theorem 5.2.3 iteratively to obtain the fact $(*)$.

Using the fact $(*)$, we will construct a sequence $1 = n_0 < n_1 < \cdots$ in $\mathbb{N}$ and

$$\mathcal{E}_n = \{E(x) \mid x \in \mathbf{2}^n\} \in \operatorname{cov}(X), \quad n \in \mathbb{N},$$

so that

(1) Each $E(x) \in \mathcal{E}_n$ is non-empty, so non-degenerate;
(2) $\operatorname{mesh} \mathcal{E}_{n_i} < 2^{-i}$;
(3) $E(x) \cap E(y) = \emptyset$ if $x \neq y \in \mathbf{2}^n$; and
(4) $E(x) = E(x(1), \ldots, x(n), 0) \cup E(x(1), \ldots, x(n), 1)$ for all $x \in \mathbf{2}^n$.

By $(*)$, we have $\mathcal{E}_1 = \{E(0), E(1)\} \in \operatorname{cov}(X)$ such that $E(0)$ and $E(1)$ are non-empty, $E(0) \cap E(1) = \emptyset$, and $\operatorname{mesh} \mathcal{E}_1 \leq \operatorname{diam} X < 1 = 2^0$. Assume that $1 = n_0 < \cdots < n_{i-1}$ and $\mathcal{E}_1, \ldots, \mathcal{E}_{n_{i-1}}$ have been defined. For each $x \in \mathbf{2}^{n_{i-1}}$, $E(x) \in \mathcal{E}_{n_{i-1}}$ is a compactum as a clopen set in X. Since $\dim E(x) = 0$, $E(x)$ has a finite open cover $\mathcal{U}_x$ with $\operatorname{ord} \mathcal{U}_x = 1$ and $\operatorname{mesh} \mathcal{U}_x < 2^{-i}$ (Theorem 5.3.1). Choose $m \in \mathbb{N}$ so that $\operatorname{card} \mathcal{U}_x \leq 2^m$ for each $x \in \mathbf{2}^{n_{i-1}}$. Using the fact $(*)$, as a refinement of $\mathcal{U}_x$, we can obtain

$$\mathcal{E}_x = \{E(x, y) \mid y \in \mathbf{2}^m\} \in \operatorname{cov}(E(x)),$$

where $E(x, y) \neq \emptyset$ for every $y \in \mathbf{2}^m$. Then, $\operatorname{mesh} \mathcal{E}_x < 2^{-i}$. We define $n_i = m + n_{i-1} > n_{i-1}$ and

$$\mathcal{E}_{n_i} = \bigcup_{x \in \mathbf{2}^{n_{i-1}}} \mathcal{E}_x = \{E(x, y) \mid (x, y) \in \mathbf{2}^{n_{i-1}} \times \mathbf{2}^m = \mathbf{2}^{n_i}\}.$$

Thus, we have $\mathcal{E}_{n_i} \in \operatorname{cov}(X)$ with $\operatorname{mesh} \mathcal{E}_{n_i} < 2^{-i}$. By the downward induction using formula (4), we can define $\mathcal{E}_{n_i - 1}, \ldots, \mathcal{E}_{n_{i-1}+1} \in \operatorname{cov}(X)$. Therefore, we obtain $\mathcal{E}_1, \ldots, \mathcal{E}_{n_i} \in \operatorname{cov}(X)$.

For each $x \in \mathbf{2}^{\mathbb{N}}$, $\bigcap_{n \in \mathbb{N}} E(x(1), \ldots, x(n)) \neq \emptyset$ because of the compactness of X. Since

$$\lim_{n \to \infty} \operatorname{diam} E(x(1), \ldots, x(n)) = 0,$$

we can define $h : \mathbf{2}^{\mathbb{N}} \to X$ by

$$\{h(x)\} = \bigcap_{n \in \mathbb{N}} E(x(1), \ldots, x(n)).$$

To show that h is a homeomorphism, it suffices to prove that h is a continuous bijection because $\mathbf{2}^{\mathbb{N}}$ is compact. For each $\varepsilon > 0$, choose $i \in \mathbb{N}$ so that $2^{-i} < \varepsilon$. Then, $\operatorname{mesh} \mathcal{E}_{n_i} < \varepsilon$ by (2). For each $x, y \in \mathbf{2}^{\mathbb{N}}$, $x(1) = y(1), \ldots, x(n_i) = y(n_i)$ imply $h(x), h(y) \in E(x(1), \ldots, x(n_i)) \in \mathcal{E}_{n_i}$, so $d(h(x), h(y)) < \varepsilon$. Hence, h is continuous. It easily follows from (3) that h is injective. By (4), for each $y \in X$, we can inductively choose $x(n) \in \mathbf{2}$, $n \in \mathbb{N}$, so that $y \in E(x(1), \ldots, x(n))$. Then, we have $x \in \mathbf{2}^{\mathbb{N}}$ such that $y \in \bigcap_{n \in \mathbb{N}} E(x(1), \ldots, x(n))$, i.e., $y = h(x)$. Hence, h is surjective. This completes the proof. □

The Cantor set is very important because of the following theorem:

Theorem 5.11.5. *Every compactum X is a continuous image of the Cantor set, that is, there exists a continuous surjection $f : \mu^0 \to X$.*

The proof consists of a combination of the following two propositions.

Proposition 5.11.6. *Every separable metrizable space X is a continuous image of a subspace of the Cantor set.*

Proof. We have a natural continuous surjection $\varphi : \mu^0 \to \mathbf{I}$ defined by $\varphi(\sum_{i \in \mathbb{N}} 2x_i/3^i) = \sum_{i \in \mathbb{N}} x_i/2^i$, where $x_i \in \mathbf{2} = \{0, 1\}$. Since $(\mu^0)^{\mathbb{N}} \approx \mu^0$, the Hilbert cube $\mathbf{I}^{\mathbb{N}}$ is a continuous image of the Cantor set. Therefore, the result follows from the fact that every separable metrizable space can be embedded in $\mathbf{I}^{\mathbb{N}}$ (Corollary 2.3.8). □

Proposition 5.11.7. *Any non-empty closed set A in μ^0 is a retract of μ^0, that is, there is a map $r : \mu^0 \to A$ with $r|A = \mathrm{id}$.*

Proof. Since $\mu^0 \approx \mathbf{2}^{\mathbb{N}}$, we may replace μ^0 by $\mathbf{2}^{\mathbb{N}}$. For each $x \in \mathbf{2}^{\mathbb{N}}$ and $n \in \mathbb{N}$, we inductively define $x^A(n) \in \mathbf{2}$ as follows:

$$x^A(n) = \begin{cases} x(n) & \text{if } (x^A(1), \ldots, x^A(n-1), x(n)) \in p_n(A), \\ 1 - x(n) & \text{otherwise,} \end{cases}$$

where $p_n : \mathbf{2}^{\mathbb{N}} \to \mathbf{2}^n$ is the projection onto the first n factors. Since $A \neq \emptyset$, $(x^A(1), \ldots, x^A(n)) \in p_n(A)$ for each $n \in \mathbb{N}$. Since A is closed in $\mathbf{2}^{\mathbb{N}}$, it follows that $x^A = (x^A(n))_{n \in \mathbb{N}} \in A$. It is obvious that $x^A = x$ for $x \in A$. We can define a retraction $r : \mathbf{2}^{\mathbb{N}} \to A$ by $r(x) = x^A$. For each $x, y \in \mathbf{2}^{\mathbb{N}}$,

$$p_n(x) = p_n(y) \Rightarrow p_n(r(x)) = p_n(x^A) = p_n(y^A) = p_n(r(y)),$$

hence r is continuous. □

5.12 Totally Disconnected Spaces with dim ≠ 0

In this section, we will construct totally disconnected separable metrizable spaces X with $\dim X \neq 0$. The first example called the **Erdös space** is constructed in the proof of the following theorem. This space is also an example of spaces X such that $\dim X^2 \neq 2 \dim X$.

Theorem 5.12.1. *There exists a 1-dimensional totally disconnected separable metrizable space X that is homeomorphic to $X^2 = X \times X$.*

Example and Proof. The desired space X is a subspace of the Hilbert space ℓ_2 defined as follows:

$$X = \{x \in \ell_2 \mid x(i) \in \mathbb{Q} \text{ for all } i \in \mathbb{N}\}.$$

The space $\ell_2 \times \ell_2$ has the norm $\|(x,y)\| = (\|x\|^2 + \|y\|^2)^{1/2}$. Then, the map $h : \ell_2 \times \ell_2 \to \ell_2$ defined by $h(x,y)(2i-1) = x(i)$ and $h(x,y)(2i) = y(i)$ is an isometry, hence it is a homeomorphism. Since $h(X \times X) = X$, we have $X \times X \approx X$.

To prove the total disconnectedness of X, let $x \neq y \in X$. Then, $x(i_0) \neq y(i_0)$ for some $i_0 \in \mathbb{N}$. Without loss of generality, we may assume that $x(i_0) < y(i_0)$. Choose $t \in \mathbb{R} \setminus \mathbb{Q}$ so that $x(i_0) < t < y(i_0)$. Then, $H = \{z \in X \mid z(i_0) < t\}$ is clopen in X and $x \in H$ but $y \notin H$. Hence, X is totally disconnected.

Note that $\dim X = \operatorname{ind} X$ by the Coincidence Theorem 5.5.2. Next, we show that $\operatorname{ind} X > 0$ and $\operatorname{ind} X \leq 1$. If so, we would have $\dim X = \operatorname{ind} X = 1$.

To show that $\operatorname{ind} X > 0$, it suffices to prove that $\operatorname{bd} U \neq \emptyset$ for every open neighborhood U of 0 contained in $\mathrm{B}(0,1) = \{x \in X \mid \|x\| < 1\}$. We can inductively choose $a_1, a_2, \dots \in \mathbb{Q}$ so that

$$x_n = (a_1, \dots, a_n, 0, 0, \dots) \in U \quad \text{and} \quad d(x_n, X \setminus U) < 1/n.$$

In fact, when $a_1, \dots, a_n$ have been chosen, let

$$k_0 = \min\{k \in \mathbb{N} \mid (a_1, \dots, a_n, k/(n+2), 0, 0, \dots) \notin U\}.$$

Then, $(k_0 - 1)/(n+2) \in \mathbb{Q}$ is the desired a_{n+1}. Since $\sum_{i=1}^{n} a_i^2 < 1$ for each n, it follows that $\sum_{i=1}^{\infty} a_i^2 \leq 1 < \infty$, hence $x_0 = (a_i)_{i \in \mathbb{N}} \in X$. Since $x_n \to x_0$ $(n \to \infty)$, it follows that $x_0 \in \operatorname{cl} U$. On the other hand, since $d(x_n, X \setminus U) < 1/n$, we have $x_0 \in \operatorname{cl}(X \setminus U)$. Therefore, $x_0 \in \operatorname{bd} U$.

To show that $\operatorname{ind} X \leq 1$, it suffices to prove that each $F_n = \{x \in X \mid \|x\| = 1/n\}$ is 0-dimensional. Note that $F_n \subset \mathbb{Q}^{\mathbb{N}}$ as sets. Furthermore, the topology on F_n coincides with the product inherited from the product space $\mathbb{Q}^{\mathbb{N}}$ (Proposition 1.2.4). Since $\dim \mathbb{Q}^{\mathbb{N}} = 0$, we have $\dim F_n = 0$ by the Subset Theorem 5.3.3. The proof is complete. □

To construct totally disconnected metrizable spaces X of arbitrarily large dimensions, we need the following lemmas:

Lemma 5.12.2. *Let $(A_\gamma, B_\gamma)_{\gamma \in \Gamma}$ be an essential family of pairs of disjoint closed sets in a compact space X and $\gamma_0 \in \Gamma$. For each $\gamma \in \Gamma \setminus \{\gamma_0\}$, let L_γ be a partition between A_γ and B_γ in X and $L = \bigcap_{\gamma \in \Gamma \setminus \{\gamma_0\}} L_\gamma$. Then, L has a component that meets both A_{γ_0} and B_{γ_0}.*

Proof. Assume that L has no components that meet both A_{γ_0} and B_{γ_0}. Let D be the union of all components of L that meet A_{γ_0}, where we allow the case $D = \emptyset$ or $D = L$. For each $x \in L \setminus D$, the component C_x of L containing x misses A_{γ_0}. By Lemma 5.11.1(1), we have a clopen set E_x in L such that $C_x \subset E_x \subset L \setminus A_{\gamma_0}$. For each $y \in E_x$, the component C_y of L with $y \in C_y$ is contained in E_x, hence $C_y \cap A_{\gamma_0} = \emptyset$. Then, it follows that $E_x \subset L \setminus D$. Therefore, $L \setminus D$ is open in L, that is, D is closed in L, so it is compact.

For each $x \in D$, the component of L containing x misses B_{γ_0} by the assumption. As above, we have a clopen set E_x in L such that $x \in E_x \subset L \setminus B_{\gamma_0}$. Since D is compact, $D \subset \bigcup_{i=1}^n E_{x_i}$ for some $x_1, \dots, x_n \in D$. Then, $E = \bigcup_{i=1}^n E_{x_i}$ is clopen in L and $A_{\gamma_0} \cap L \subset D \subset E \subset L \setminus B_{\gamma_0}$.

By the normality of X, we have disjoint open sets U and V in X such that $A_{\gamma_0} \cup E \subset U$ and $B_{\gamma_0} \cup (L \setminus E) \subset V$. Then, $L_{\gamma_0} = X \setminus (U \cup V)$ is a partition between A_{γ_0} and B_{γ_0} in X and $\bigcap_{\gamma \in \Gamma} L_\gamma = L \cap L_{\gamma_0} = \emptyset$. This is contrary to the essentiality of $(A_\gamma, B_\gamma)_{\gamma \in \Gamma}$. Therefore, L has a component that meets both A_{γ_0} and B_{γ_0}. □

Lemma 5.12.3. *Let X be a compactum and $f : X \to Y$ be a continuous surjection. Then, X has a G_δ-subset S that meets each fiber of f at precisely one point, that is,*

$$\operatorname{card}(f^{-1}(y) \cap S) = 1 \quad \textit{for each } y \in Y.$$

Proof. We may assume that $X \subset \mathbf{I}^{\mathbb{N}}$. For each $y \in Y$, since $f^{-1}(y)$ is non-empty and compact, we can define $g(y) \in X$ as follows:

$$g(y)(1) = \min \operatorname{pr}_1(f^{-1}(y)) \quad \text{and}$$

$$g(y)(n) = \min \operatorname{pr}_n\big(f^{-1}(y) \cap \textstyle\bigcap_{i=1}^{n-1} \operatorname{pr}_i^{-1}(g(y)(i))\big) \quad \text{for } n > 1.$$

Then, $\emptyset \neq f^{-1}(y) \cap \bigcap_{i=1}^n \operatorname{pr}_i^{-1}(g(y)(i)) \subset \operatorname{pr}_n^{-1}(g(y)(n))$. By the compactness of $f^{-1}(y)$, we have

$$\emptyset \neq f^{-1}(y) \cap \bigcap_{i \in \mathbb{N}} \operatorname{pr}_i^{-1}(g(y)(i)) \subset \bigcap_{n \in \mathbb{N}} \operatorname{pr}_n^{-1}(g(y)(n)) = \{g(y)\},$$

which means $g(y) \in f^{-1}(y)$. Thus, the set $S = \{g(y) \mid y \in Y\}$ meets each fiber of f at precisely one point.

For each $n, m \in \mathbb{N}$, let

$$F_{n,m} = \big\{x \in X \;\big|\; \exists z \in X \text{ such that } z(i) = x(i) \text{ for } i < n,$$
$$z(n) \leq x(n) - \tfrac{1}{m} \text{ and } f(z) = f(x)\big\}.$$

Since X is a compactum, it is easy to see that $F_{n,m}$ is closed in X, hence $U_{n,m} = X \setminus F_{n,m}$ is open in X. We show that $S = \bigcap_{n,m \in \mathbb{N}} U_{n,m}$, which is a G_δ-set in X. For each $y \in Y$, if $z \in X$, $z(i) = g(y)(i)$ for all $i < n$ and $z(n) \leq g(y)(n) - \frac{1}{m}$, then $f(z) \neq y = f(g(y))$; otherwise $g(y)(n) \leq z(n)$ $(< g(y)(n))$ by the definition of $g(y)$. Thus, $g(y) \in U_{n,m}$ for all $n, m \in \mathbb{N}$, i.e., $S \subset \bigcap_{n,m \in \mathbb{N}} U_{n,m}$. Conversely, for each $x \in \bigcap_{n,m \in \mathbb{N}} U_{n,m}$, let $y = f(x)$ (i.e., $x \in f^{-1}(y)$). Then, $x = g(y) \in S$. Otherwise, let $n = \min\{i \in \mathbb{N} \mid x(i) \neq g(y)(i)\}$. Since $g(y)(n) < x(n)$ by the definition of $g(y)$, it follows that $g(y)(n) \leq x(n) - \frac{1}{m}$ for some $m \in \mathbb{N}$, i.e., $x \in F_{n,m} = X \setminus U_{n,m}$, which is a contradiction. □

For a metric space $X = (X, d)$, let $\mathrm{Comp}(X)$ be the space of all non-empty compact sets in X that admits the **Hausdorff metric** d_H defined as follows:

$$\begin{aligned} d_H(A, B) &= \inf\{r > 0 \mid A \subset \mathrm{N}_d(B, r),\ B \subset \mathrm{N}_d(A, r)\} \\ &= \max\{\sup_{a\in A} d(a, B),\ \sup_{b\in B} d(b, A)\}. \end{aligned}$$

According to the following proposition, the topology of $\mathrm{Comp}(X)$ induced by the Hausdorff metric d_H coincides with the Vietoris topology defined in Sect. 3.8.

Proposition 5.12.4. *For a metric space $Y = (Y, d)$, the Vietoris topology on $\mathrm{Comp}(Y)$ is induced by the Hausdorff metric d_H. Consequently, the space $\mathrm{Comp}(X)$ with the Vietoris topology is metrizable if Y is metrizable.*

Proof. For each $A \in \mathrm{Comp}(Y)$ and $r > 0$, we can choose $a_1, \dots, a_n \in A$ so that $A \subset \bigcup_{i=1}^n \mathrm{B}(a_i, r/2)$. Then,

$$A \in \left(\bigcup_{i=1}^n \mathrm{B}(a_i, r/2)\right)^+ \cap \bigcap_{i=1}^n \mathrm{B}(a_i, r/2)^- \cap \mathrm{Comp}(Y) \subset \mathrm{B}_{d_H}(A, r),$$

which means that $\mathrm{B}_{d_H}(A, r)$ is a neighborhood of A in the Vietoris topology.[12]

Let $A \in \mathrm{Comp}(Y)$. For each open set U in Y with $A \in U^-$, taking $a \in A \cap U$, we have $\mathrm{B}_{d_H}(A, d(a, Y \setminus U)) \subset U^-$. On the other hand, for each open set U in Y with $A \in U^+$, we have $\mathrm{B}_{d_H}(A, \delta) \subset U^+$, where $\delta = \mathrm{dist}(A, Y \setminus U) > 0$. Thus, $\{\mathrm{B}_{d_H}(A, r) \mid r > 0\}$ is a neighborhood basis at $A \in \mathrm{Comp}(Y)$. □

Note. When $Y = (Y, d)$ is a bounded metric space, the Hausdorff metric d_H is defined on the set $\mathrm{Cld}(Y)$ consisting of all non-empty closed sets in Y, which induces a topology different from the Vietoris topology if Y is non-compact. If Y is unbounded, then $d_H(A, B) = \infty$ for some $A, B \in \mathrm{Cld}(Y)$. But, even in this case, d_H induces the topology on $\mathrm{Cld}(Y)$. We should note that this topology is dependent on the metric d. For example, $\mathrm{Cld}(\mathbb{R})$ is non-separable with respect to the Hausdorff metric induced by the usual metric. In fact, it has no countable open basis because $\mathfrak{P}_0(\mathbb{N})$ is an uncountable discrete set of $\mathrm{Cld}(\mathbb{R})$. On the other hand, $\mathbb{R}$ is homeomorphic to the unit open interval $(0, 1)$ and $\mathrm{Cld}((0, 1))$ is separable with respect to the Hausdorff metric induced by the usual metric because $\mathrm{Fin}((0, 1))$ is dense in $\mathrm{Cld}((0, 1))$.

As observed in Sect. 3.8, the space $\mathrm{Cld}(Y)$ with the Vietoris topology is Hausdorff if and only if Y is regular. Here, it is remarked that $\mathrm{Cld}(Y)$ is metrizable if and only if Y is compact and metrizable. Indeed, if Y is compact metrizable then $\mathrm{Cld}(Y) = \mathrm{Comp}(Y)$ is metrizable by Proposition 5.12.4. Conversely, if Y is non-compact then Y contains a countable discrete set. Then, $\mathfrak{P}_0(\mathbb{N}) = \mathrm{Cld}(\mathbb{N})$ can be embedded into $\mathrm{Cld}(Y)$ as a subspace, which implies that $\mathfrak{P}_0(\mathbb{N})$ is metrizable. Note that $\mathfrak{P}_0(\mathbb{N})$ is separable because $\mathrm{Fin}(\mathbb{N})$ is dense in $\mathfrak{P}_0(\mathbb{N})$. Thus, $\mathfrak{P}_0(\mathbb{N})$ is second countable. Let $\mathcal{B}$ be a countable open base for $\mathfrak{P}_0(\mathbb{N})$. For each $A \in \mathfrak{P}_0(\mathbb{N})$, choose $B_A \in \mathcal{B}$ so that $A \in B_A \subset A^+$. When $A \neq A' \in \mathfrak{P}_0(\mathbb{N})$, we may assume $A \setminus A' \neq \emptyset$. Then, $A \notin B_{A'}$. Hence, we have $B_A \neq B_{A'}$. Consequently, $\mathrm{card}\,\mathcal{B} \geq \mathrm{card}\,\mathfrak{P}_0(\mathbb{N}) = 2^{\aleph_0}$, which is a contradiction.

[12] Recall $U^- = \{A \subset Y \mid A \cap U \neq \emptyset\}$ and $U^+ = \{A \subset Y \mid A \subset U\}$.

Theorem 5.12.5. *Let $X = (X, d)$ be a metric space.*

(1) *If X is totally bounded then so is* $\mathrm{Comp}(X)$ *with respect to d_H.*
(2) *If X is complete then so is* $\mathrm{Comp}(X)$ *with respect to d_H.*
(3) *If X is compact then so is* $\mathrm{Comp}(X)$.

Proof. (1): For each $\varepsilon > 0$, we have $F \in \mathrm{Fin}(X)$ such that $d(x, F) < \varepsilon$ for every $x \in X$. Then, $\mathrm{Fin}(F)$ is a finite subset of $\mathrm{Comp}(X)$. For each $A \in \mathrm{Comp}(X)$, let $F_A = \{z \in F \mid d(z, A) < \varepsilon\}$. For each $x \in A$, we have $z \in F$ such that $d(x, z) < \varepsilon$, which implies that $z \in F_A$. Then, $F_A \neq \emptyset$ (i.e., $F_A \in \mathrm{Fin}(F)$) and $d_H(A, F_A) < \varepsilon$. Hence, $\mathrm{Comp}(X)$ is totally bounded.

(2): Let $(A_n)_{n\in\mathbb{N}}$ be a Cauchy sequence in $\mathrm{Comp}(X)$. If $(A_n)_{n\in\mathbb{N}}$ has a convergent subsequence, then $(A_n)_{n\in\mathbb{N}}$ itself is convergent. Hence, it can be assumed that $d_H(A_n, A_i) < 2^{-n-1}$ for each $n < i$. Then, we prove that $(A_n)_{n\in\mathbb{N}}$ converges to

$$A_0 = \bigcap_{n\in\mathbb{N}} \mathrm{cl}\, \mathrm{N}(A_n, 2^{-n}) \in \mathrm{Comp}(X).$$

To this end, since A_0 is closed in X and $A_0 \subset \mathrm{N}(A_n, 2^{-n+1})$ for each $n \in \mathbb{N}$, it suffices to show that A_0 is totally bounded and $A_n \subset \mathrm{N}(A_0, 2^{-n})$ for each $n \in \mathbb{N}$.

First, we show that $A_n \subset \mathrm{N}(A_0, 2^{-n})$. For each $x \in A_n$, inductively choose $x_i \in A_i$, $i > n$, so that $d(x_i, x_{i-1}) < 2^{-i}$, where $x = x_n$. Since $(x_i)_{i\geq n}$ is a Cauchy sequence in X, it converges to some $x_0 \in X$. For each $i \geq n$,

$$d(x_i, x_0) \leq \sum_{j=i}^{\infty} d(x_j, x_{j+1}) < \sum_{j=i}^{\infty} 2^{-j-1} = 2^{-i},$$

hence $d(x_0, A_i) < 2^{-i}$ and $d(x_0, x) < 2^{-n}$. Moreover, for each $i < n$,

$$d(x_0, A_i) \leq d(x_0, x) + d(x, A_i) < 2^{-n} + 2^{-i-1} \leq 2^{-i}.$$

Therefore, $x_0 \in \bigcap_{i\in\mathbb{N}} \mathrm{N}(A_i, 2^{-i}) \subset A_0$, so $A_0 \neq \emptyset$ and $x \in \mathrm{N}(A_0, 2^{-n})$.

To see the total boundedness of A_0, let $\varepsilon > 0$. Choose $n \in \mathbb{N}$ so that $2^{-n+1} < \varepsilon/3$, and take a finite $\varepsilon/3$-dense subset $\{u_1, \ldots, u_k\}$ of A_n.[13] For each $i = 1, \ldots, k$, choose $v_i \in A_0$ so that $d(u_i, v_i) < 2^{-n}$. Then, $\{v_1, \ldots, v_k\}$ is an ε-dense subset of A_0. Indeed, for each $x \in A_0$, we have $y \in A_n$ such that $d(x, y) < 2^{-n+1}$. Then, $d(y, u_i) < \varepsilon/2$ for some $i = 1, \ldots, k$. Hence,

$$d(x, v_i) \leq d(x, y) + d(y, u_i) + d(u_i, v_i) < 2^{-n+1} + \varepsilon/3 + 2^{-n} < \varepsilon.$$

(3): This is a combination of (1) and (2). □

Theorem 5.12.6. *For each $n \in \mathbb{N}$, there exists an n-dimensional totally disconnected separable completely metrizable space. In addition, there exists a strongly infinite-dimensional totally disconnected separable completely metrizable space.*

---

[13] In a metric space $X = (X, d)$, $A \subset X$ is said to be **ε-dense** if $d(x, A) < \varepsilon$ for each $x \in X$.

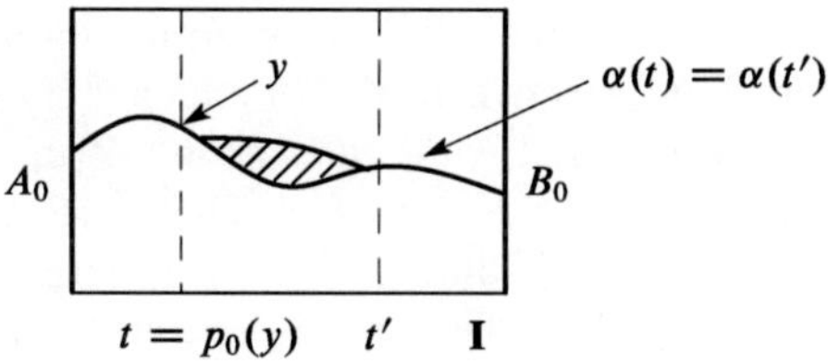

Fig. 5.9 $\alpha(t), t \in \mu^0$

Proof (Example and Proof). To construct the examples simultaneously, let $X = \mathbf{I} \times \mathbf{I}^\Gamma$ and $d \in \operatorname{Metr}(X)$ where $\Gamma = \{1, \ldots, n\}$ in the n-dimensional case and $\Gamma = \mathbb{N}$ in the infinite-dimensional case. Let $p_0 : X \to \mathbf{I}$ be the projection onto the first factor. Put $A_0 = p_0^{-1}(0)$ and $B_0 = p_0^{-1}(1)$ and define

$$\mathcal{E} = \{E \in \operatorname{Comp}(X) \mid E \text{ is connected, } E \cap A_0 \neq \emptyset,\ E \cap B_0 \neq \emptyset\}.$$

Then, $\mathcal{E}$ is closed in $\operatorname{Comp}(X)$. Indeed, let $D \in \operatorname{Comp}(X) \setminus \mathcal{E}$. When D is not connected, it can be written as the disjoint union of two non-empty closed subsets D_1 and D_2. Let $\varepsilon = \frac{1}{2}\operatorname{dist}_d(D_1, D_2) > 0$. Then, every $E \in \mathrm{B}_{d_H}(D, \varepsilon)$ is not connected because E is contained in $\mathrm{N}_d(D_1, \varepsilon) \cup \mathrm{N}_d(D_2, \varepsilon)$ and meets both $\mathrm{N}_d(D_1, \varepsilon)$ and $\mathrm{N}_d(D_2, \varepsilon)$. Hence, $\mathrm{B}_{d_H}(D, \varepsilon) \cap \mathcal{E} = \emptyset$. If $D \cap A_0 = \emptyset$, then $\mathrm{N}_d(D, \delta) \cap A_0 = \emptyset$, where $\delta = \operatorname{dist}_d(A_0, D) > 0$. Every $E \in \mathrm{B}_{d_H}(D, \delta)$ also misses A_0, which implies $\mathrm{B}_{d_H}(D, \delta) \cap \mathcal{E} = \emptyset$. The case $D \cap B_0 = \emptyset$ is identical.

Since $\operatorname{Comp}(X)$ is compact by Theorem 5.12.5(3), $\mathcal{E}$ is also compact. Then, we have a map $\alpha : \mu^0 \to \mathcal{E}$ of the Cantor set μ^0 onto $\mathcal{E}$ by Theorem 5.11.5. We define

$$Y = \{y \in p_0^{-1}(\mu^0) \mid y \in \alpha p_0(y)\} \subset X.$$

Obviously, $p_0(Y) \subset \mu^0$. For each $t \in \mu^0$, since $\alpha(t)$ is a continuum that meets both A_0 and B_0, it follows that $p_0\alpha(t) = \mathbf{I}$, so $t = p_0(y)$ for some $y \in \alpha(t)$, where $y \in Y$ (Fig. 5.9). Thus, we have $p_0(Y) = \mu^0$. Moreover, Y is closed in X, so is compact. Indeed, let $(y_i)_{i\in\mathbb{N}}$ be a sequence in Y converging to $y \in X$. Since $p_0(y_i) \in \mu^0$ for every $i \in \mathbb{N}$ and $(p_0(y_i))_{i\in\mathbb{N}}$ converges to $p_0(y)$, we have $p_0(y) \in \mu^0$. Since $y_i \in \alpha p_0(y_i)$ for every $i \in \mathbb{N}$ and $(\alpha p_0(y_i))_{i\in\mathbb{N}}$ converges to $\alpha p_0(y)$ in $\mathcal{E}$, it easily follows that $y \in \alpha p_0(y)$, hence $y \in Y$.

By Lemma 5.12.3, Y has a G_δ-subset S such that

$$\operatorname{card}(p_0^{-1}(t) \cap S) = 1 \text{ for each } t \in \mu^0.$$

Since Y is compact, S is completely metrizable. Since $p_0|S : S \to \mu^0$ is a continuous bijection and μ^0 is totally disconnected, it follows that S is also totally disconnected. Moreover, $S \cap E \neq \emptyset$ for every $E \in \mathcal{E}$. Indeed, because $\mathcal{E} = \alpha p_0(S)$, we can find $y \in S \subset Y$ such that $E = \alpha p_0(y)$, where $y \in \alpha p_0(y) = E$.

Now, for each $i \in \Gamma$, let $p_i : X \to \mathbf{I}$ be the projection onto the i-th coordinates of the second factor $\mathbf{I}^\Gamma$. Since $p_i^{-1}(0), p_i^{-1}(1) \in \mathcal{E}$, it follows that $A_i = S \cap p_i^{-1}(0) \neq \emptyset$ and $B_i = S \cap p_i^{-1}(1) \neq \emptyset$. Then, $(A_i, B_i)_{i\in\Gamma}$ is essential in S. In fact, by the

Partition Extension Lemma 5.3.7, for each partition L_i between A_i and B_i in S, we have a partition $\tilde{L}_i$ between $p_i^{-1}(0)$ and $p_i^{-1}(1)$ in X such that $\tilde{L}_i \cap S \subset L_i$. According to Lemma 5.12.2, the intersection of the partions $\tilde{L}_i$ has a component $E \in \mathcal{E}$. Then, $\bigcap_{i \in \Gamma} L_i \supset E \cap S \neq \emptyset$. Therefore, S is s.i.d. when $\Gamma = \mathbb{N}$. In the case that $\Gamma = \{1, \ldots, n\}$, $\dim S \geq n$ by Theorem 5.2.17. Since $S \subset \mu^0 \times \mathbf{I}^\Gamma$, $\dim S \leq n$ by the Subset Theorem 5.3.3 and the Product Theorem 5.4.9, hence $\dim S = n$. □

5.13 Examples of Infinite-Dimensional Spaces

In this section, we construct two infinite-dimensional compacta. One is weakly infinite-dimensional but not countable-dimensional. The other is hereditarily infinite-dimensional. First, we present the following theorem:

Theorem 5.13.1. *There exists a weakly infinite-dimensional compact metrizable space that contains a strongly infinite-dimensional subspace, and hence it is not countable-dimensional.*

Example and Proof. Let S be an s.i.d. totally disconnected separable completely metrizable space (Theorem 5.12.6) and let $X = \gamma S$ be a compactification of S with the c.d. remainder (Theorem 5.7.4). Then, we show that X is the required example.

First, X contains the s.i.d. subset S, so X is not c.d. (Theorem 5.6.2). To see that X is w.i.d., let $(A_i, B_i)_{i \in \omega}$ be a family of pairs of disjoint closed sets in X. Since $X \setminus S$ is c.d., $X \setminus S = \bigcup_{i \in \mathbb{N}} X_i$, where $\dim X_i = 0$ for each $i \in \mathbb{N}$. For each $i \in \mathbb{N}$, by Theorem 5.3.8 and the Partition Extension Lemma 5.3.7, X has a partition L_i between A_i and B_i such that $L_i \cap X_i = \emptyset$. Then,

$$L = \bigcap_{i \in \mathbb{N}} L_i \subset \bigcap_{i \in \mathbb{N}} X \setminus X_i = X \setminus \bigcup_{i \in \mathbb{N}} X_i = S.$$

If $L \neq \emptyset$, then L is compact and totally disconnected, which implies $\dim L = 0$ by Theorem 5.11.2. Again by Theorem 5.3.8 and the Partition Extension Lemma 5.3.7, X has a partition L_0 between A_0 and B_0 such that $L_0 \cap L = \emptyset$, which means $\bigcap_{i \in \omega} L_i = \emptyset$. □

For a compact space X and a metric space $Y = (Y, d)$, let $\mathrm{C}(X, Y)$ be the space of all maps from X to Y admitting the topology induced by the sup-metric $d(f, g) = \sup_{x \in X} d(f(x), g(x))$, which is identical to the compact-open topology because X is compact (cf. 1.1.3(6)). Then, from 1.1.3(5), we have the following lemma:

Lemma 5.13.2. *Let X be a compactum and $Y = (Y, d)$ be a separable metric space. The space $\mathrm{C}(X, Y)$ is separable.*

Note. This lemma can be proved directly as follows:

Sketch of Direct Proof. Let $\{U_i \mid i \in \mathbb{N}\}$ and $\{V_j \mid j \in \mathbb{N}\}$ be open bases for X and Y, respectively. For each $i, j \in \mathbb{N}$, let

$$W_{i,j} = \{f \in \mathrm{C}(X,Y) \mid f(\mathrm{cl}\,U_i) \subset V_j\}.$$

It is easy to prove that each $W_{i,j}$ is open in $\mathrm{C}(X,Y)$.

To see that $\{W_{i,j} \mid i, j \in \mathbb{N}\}$ is an open subbase for $\mathrm{C}(X,Y)$, let $f \in \mathrm{C}(X,Y)$ and $\varepsilon > 0$. For each $x \in X$, choose $i(x), j(x) \in \mathbb{N}$ so that $x \in U_{i(x)}$, $\mathrm{diam}\,V_{j(x)} < \varepsilon$ and $f(\mathrm{cl}\,U_{i(x)}) \subset V_{j(x)}$, i.e., $f \in W_{i(x),j(x)}$. Because of the compactness of X, we have $x_1, \ldots, x_n \in X$ such that $X = U_{i(x_1)} \cup \cdots U_{i(x_n)}$. Then, observe $f \in \bigcap_{k=1}^{n} W_{i(x_k),j(x_k)} \subset \mathrm{B}_d(f,\varepsilon)$.

To construct a hereditarily infinite-dimensional space, we need the following key lemma:

Lemma 5.13.3. *Let $C \subset \mathbf{I}$ be homeomorphic to the Cantor set, $n \in \mathbb{N}$, and $\Gamma \subset \mathbb{N} \setminus \{n\}$ such that Γ and $\mathbb{N} \setminus \Gamma$ are infinite. Then, there exists a collection $\{S_i \mid i \in \Gamma\}$ of partitions S_i between $A_i = \mathrm{pr}_i^{-1}(0)$ and $B_i = \mathrm{pr}_i^{-1}(1)$ in $\mathbf{I}^{\mathbb{N}}$ such that every subset $X \subset \bigcap_{i\in\Gamma} S_i$ is strongly infinite-dimensional if $C \subset \mathrm{pr}_n(X)$, where $\mathrm{pr}_i : \mathbf{I}^{\mathbb{N}} \to \mathbf{I}$ is the projection of $\mathbf{I}^{\mathbb{N}}$ onto the i-th factor.*

Proof. Without loss of generality, we may assume that $n = 1$ and $\Gamma = \{2i \mid i \in \mathbb{N}\}$. For each $i \in \mathbb{N}$, let $C_i = \mathrm{pr}_i^{-1}([0, \frac{1}{4}])$ and $D_i = \mathrm{pr}_i^{-1}([\frac{3}{4}, 1])$. We define

$$\Omega = \{f \in \mathrm{C}(\mathbf{I}^{\mathbb{N}}, \mathbf{I}^{\mathbb{N}}) \mid \forall i \in \mathbb{N},\ f^{-1}(A_i) = C_{2i},\ f^{-1}(B_i) = D_{2i}\}.$$

Since Ω is separable by Lemma 5.13.2, there exist $T \subset C$ and a continuous surjection $\psi : T \to \Omega$ by Proposition 5.11.6. Let $E = \mathrm{pr}_1^{-1}(T) \subset \mathbf{I}^{\mathbb{N}}$ and define a map $\varphi : E \to \mathbf{I}^{\mathbb{N}}$ by $\varphi(x) = (\psi\mathrm{pr}_1(x))(x)$. For each $i \in \mathbb{N}$,

$$\begin{aligned}\varphi^{-1}(A_i) &= \{x \in E \mid \varphi(x) = (\psi\mathrm{pr}_1(x))(x) \in A_i\} \\ &= \{x \in E \mid x \in (\psi\mathrm{pr}_1(x))^{-1}(A_i) = C_{2i}\} = E \cap C_{2i}\end{aligned}$$

and similarly $\varphi^{-1}(B_i) = E \cap D_{2i}$. Since $\mathrm{pr}_i^{-1}(\frac{1}{2})$ is a partition between A_i and B_i in $\mathbf{I}^{\mathbb{N}}$, $\varphi^{-1}(\mathrm{pr}_i^{-1}(\frac{1}{2}))$ is a partition between $C_{2i} \cap E$ and $D_{2i} \cap E$ in E. By the Partition Extension Lemma 5.3.7, we have a partition S_{2i} between A_{2i} and B_{2i} in $\mathbf{I}^{\mathbb{N}}$ such that $S_{2i} \cap E \subset \varphi^{-1}(\mathrm{pr}_i^{-1}(\frac{1}{2}))$. It should be noted that $(A_{2i} \cap \mathrm{pr}_1^{-1}(x), B_{2i} \cap \mathrm{pr}_1^{-1}(x))_{i\in\mathbb{N}}$ is essential in $\mathrm{pr}_1^{-1}(x)$ for every $x \in C$. Then, $\mathrm{pr}_1^{-1}(x) \cap \bigcap_{i\in\mathbb{N}} S_{2i} \neq \emptyset$ for every $x \in C$, hence $C \subset \mathrm{pr}_1\big(\bigcap_{i\in\mathbb{N}} S_{2i}\big)$.

Take $X \subset \bigcap_{i\in\mathbb{N}} S_{2i}$ such that $C \subset \mathrm{pr}_1(X)$. We will show that X is s.i.d., that is, X has an infinite essential family of pairs of disjoint closed sets. For each $i \in \mathbb{N}$, let $C'_i = \mathrm{pr}_i^{-1}([0, \frac{1}{3}]) \cap X$ and $D'_i = \mathrm{pr}_i^{-1}([\frac{2}{3}, 1]) \cap X$. To see that $(C'_{2i}, D'_{2i})_{i\in\mathbb{N}}$ is essential, let L_i be a partition between C'_{2i} and D'_{2i} in X. By the Partition Extension Lemma 5.3.7, we have a partition H_i between C_{2i} and D_{2i} in $\mathbf{I}^{\mathbb{N}}$ such that $H_i \cap X \subset L_i$. There is a map $f_i : \mathbf{I}^{\mathbb{N}} \to \mathbf{I}$ such that $f_i^{-1}(0) = C_{2i}$, $f_i^{-1}(1) = D_{2i}$, and

$f_i^{-1}(\frac{1}{2}) = H_i$.[14] Indeed, let U_i and V_i be disjoint open sets in $\mathbf{I}^{\mathbb{N}}$ such that $C_{2i} \subset U_i$, $D_{2i} \subset V_i$, and $X \setminus H_i = U_i \cup V_i$. We can take maps $g_i : X \setminus V_i \to \mathbf{I}$ and $h_i : X \setminus U_i \to \mathbf{I}$ such that $g_i^{-1}(0) = C_{2i}$, $g_i^{-1}(1) = H_i$, $h_i^{-1}(0) = H_i$, and $h_i^{-1}(1) = D_{2i}$ (cf. Theorem 2.2.6). The desired f_i can be defined by

$$f_i(x) = \begin{cases} \frac{1}{2}g_i(x) & \text{if } x \in X \setminus V_i, \\ \frac{1}{2} + \frac{1}{2}h_i(x) & \text{if } x \in X \setminus U_i. \end{cases}$$

Now, we define a map $f : \mathbf{I}^{\mathbb{N}} \to \mathbf{I}^{\mathbb{N}}$ by $f(x) = (f_i(x))_{i \in \mathbb{N}}$. For each $i \in \mathbb{N}$, $f^{-1}(A_i) = f^{-1}(\mathrm{pr}_i^{-1}(0)) = f_i^{-1}(0) = C_{2i}$ and similarly $f^{-1}(B_i) = D_{2i}$, which implies that $f \in \Omega = \psi(T)$, hence $f = \psi(t)$ for some $t \in T$. Since $T \subset C \subset \mathrm{pr}_1(X)$, we have $x \in X$ such that $t = \mathrm{pr}_1(x)$. Then, $\varphi(x) = (\psi \mathrm{pr}_1(x))(x) = f(x)$. On the other hand, since $x \in \mathrm{pr}_1^{-1}(T) = E$, we have

$$x \in X \cap E \subset \bigcap_{i \in \mathbb{N}} S_{2i} \cap E \subset \bigcap_{i \in \mathbb{N}} \varphi^{-1}(\mathrm{pr}_i^{-1}(\tfrac{1}{2})) = \varphi^{-1}(\tfrac{1}{2}, \tfrac{1}{2}, \ldots).$$

Then, $f(x) = \varphi(x) = (\frac{1}{2}, \frac{1}{2}, \ldots)$, i.e., $f_i(x) = \frac{1}{2}$ for each $i \in \mathbb{N}$, hence $x \in \bigcap_{i \in \mathbb{N}} H_i \cap X \subset \bigcap_{i \in \mathbb{N}} L_i$. Therefore, $(C'_{2i}, D'_{2i})_{i \in \mathbb{N}}$ is essential. □

Theorem 5.13.4. *There exists a hereditarily infinite-dimensional compact metrizable space.*

Example and Proof. Let $\{C_n \mid n \in \mathbb{N}\}$ be a collection of Cantor sets in $\mathbf{I}$ such that every non-degenerate subinterval of $\mathbf{I}$ contains some C_n. Let $\Gamma_{i,n}$ $(i, n \in \mathbb{N})$ be disjoint infinite subsets of $\mathbb{N} \setminus \{1\}$ such that $i \notin \Gamma_{i,n}$. For each $i, n \in \mathbb{N}$, by Lemma 5.13.3, we have a compact set $S_{i,n} \subset \mathbf{I}^{\mathbb{N}}$ that is the intersection of partitions between $A_j = \mathrm{pr}_j^{-1}(0)$ and $B_j = \mathrm{pr}_j^{-1}(1)$ $(j \in \Gamma_{i,n})$ and has the property that $X \subset S_{i,n}$ is s.i.d. if $C_n \subset \mathrm{pr}_i(X)$.

We will show that $S = \bigcap_{i,n \in \mathbb{N}} S_{i,n}$ is h.i.d. Since S is the intersection of partitions between A_j and B_j $(j \in \bigcup_{i,n \in \mathbb{N}} \Gamma_{i,n})$ and $(A_j, B_j)_{j \in \mathbb{N}}$ is essential, S meets every partition between A_1 and B_1, which implies that $\dim S \neq -1, 0$. Now, let $\emptyset \neq X \subset S$. In the case that $\dim \mathrm{pr}_i(X) = 0$ for every $i \in \mathbb{N}$, since $\dim \prod_{i \in \mathbb{N}} \mathrm{pr}_i(X) = 0$ by Theorem 5.3.6 and $X \subset \prod_{i \in \mathbb{N}} \mathrm{pr}_i(X)$, we have $\dim X = 0$ by the Subset Theorem 5.3.3. When $\dim \mathrm{pr}_i(X) \neq 0$ for some $i \in \mathbb{N}$, $\mathrm{pr}_i(X)$ contains a non-degenerate subinterval of $\mathbf{I}$, hence it contains some C_n. Then, it follows that X is s.i.d. □

[14] Refer to the last Remark of Sect. 2.2.

5.14 Appendix: The Hahn–Mazurkiewicz Theorem

The content of this section is not part of Dimension Theory but is related to the content of Sect. 5.11. According to Theorem 5.11.5, every compact metrizable space is the continuous image of the Cantor (ternary) set μ^0. In this section, we will prove the following characterization of the continuous image of the interval $\mathbf{I}$:

Theorem 5.14.1 (Hahn–Mazurkiewicz). *A space X is the continuous image of the interval $\mathbf{I}$ if and only if X is a locally connected continuum.*[15]

Here, X is **locally connected** if each point $x \in X$ has a neighborhood basis consisting of connected neighborhoods. Because of Theorem 5.14.1, a locally connected continuum is called a **Peano continuum** in honor of the first mathematician who showed that the square $\mathbf{I}^2$ is the continuous image of the interval $\mathbf{I}$.

The continuous image of a continuum is also a continuum, where the metrizability follows from 2.4.5(1). Since every closed map is a quotient map, the "only if" part of Theorem 5.14.1 comes from the following proposition:

Proposition 5.14.2. *Let $f : X \to Y$ be a quotient map. If X is locally connected, then so is Y. Namely, the quotient space of a locally connected space is also locally connected.*

Proof. Let $y \in Y$. For each open neighborhood U of y in Y, let C be the connected component of U with $y \in C$. Since X is locally connected, each $x \in f^{-1}(C)$ has a connected neighborhood $V_x \subset f^{-1}(U)$. Note that $f(V_x)$ is connected, $f(V_x) \subset U$, and $f(V_x) \cap C \neq \emptyset$. Since C is a connected component of U, it follows that $f(V_x) \subset C$, hence $V_x \subset f^{-1}(C)$. Therefore, $f^{-1}(C)$ is open in X, which means that C is open in Y. Thus, C is a connected neighborhood of y in Y with $C \subset U$. □

To prove the "if" part of Theorem 5.14.1, we introduce a simple chain in a metric space $X = (X, d)$. A finite sequence $(U_1, \dots, U_n)$ of *connected open* sets[16] in X is called a **chain** (an **ε-chain**) if

$$U_i \cap U_{i+1} \neq \emptyset \text{ for each } i = 1, \dots, n-1$$

(and $\operatorname{diam} U_i < \varepsilon$ for every $i = 1, \dots, n$), where n is called the **length** of this chain. A chain is said to be **simple** provided that

$$\operatorname{cl} U_i \cap \operatorname{cl} U_j = \emptyset \text{ if } |i - j| > 1.\text{[17]}$$

[15] Recall that a continuum is a compact connected metrizable space.

[16] In general, each link U_i is not assumed to be connected and open.

[17] This condition is stronger than usual, and is adopted to simplify our argument. Usually, it is said that $(U_1, \dots, U_n)$ is a simple chain if $U_i \cap U_j \neq \emptyset \Leftrightarrow |i - j| \leq 1$. However, in our definition, $U_i \cap U_j \neq \emptyset \Leftrightarrow \operatorname{cl} U_i \cap \operatorname{cl} U_j \neq \emptyset \Leftrightarrow |i - j| \leq 1$.

It is said that two distinct points $a, b \in X$ are connected by a simple (ε-)chain $(U_1, \ldots, U_n)$ if $a \in U_1 \setminus \operatorname{cl} U_2$ and $b \in U_n \setminus \operatorname{cl} U_{n-1}$ (when $n = 1$, this means $a, b \in U_1$), where $(U_1, \ldots, U_n)$ is called a simple (ε-)chain from a to b. Given open sets U and V in X with $\operatorname{dist}(\operatorname{cl} U, \operatorname{cl} V) > 0$, it is said that U and V are connected by a simple (ε-)chain $(U_1, \ldots, U_n)$ if

$$U \cap U_1 \neq \emptyset, \ \operatorname{cl} U \cap \operatorname{cl}(U_2 \cup \cdots \cup U_n) = \emptyset,$$
$$V \cap U_n \neq \emptyset, \text{ and } \operatorname{cl} V \cap \operatorname{cl}(U_1 \cup \cdots \cup U_{n-1}) = \emptyset,$$

where $(U_1, \ldots, U_n)$ is called a simple (ε-)chain from U to V. When U is connected (and $\operatorname{diam} U < \varepsilon$), $(U, U_1, \ldots, U_n, V)$ is a simple (ε-)chain.

Lemma 5.14.3. *Let $X = (X, d)$ be a connected, locally connected metric space, and $a \neq b \in X$. Then, the following hold:*

(1) *Each pair of distinct points are connected by a simple ε-chain for any $\varepsilon > 0$.*
(2) *Each pair of open sets U and V in X with $\operatorname{cl} U \cap \operatorname{cl} V = \emptyset$ are connected by a simple ε-chain for any $\varepsilon > 0$.*
(3) *Each pair of open sets U and V in X with $\operatorname{dist}(U, V) > 0$ are connected by a simple chain of length n for any $n \in \mathbb{N}$.*

Proof. (1): Let W be the subset of X consisting of all points $x \in X$ satisfying the following condition:

- a and x are connected by a simple ε-chain.

Then, W is open in X by the definition. Using the local connectedness of X, we can easily show that $a \in W$ and $X \setminus W$ is open in X. Since X is connected, it follows that $W = X$. Then, we have $b \in W$. This gives (1).

(2): Take points $a \in U$ to $b \in V$ and apply (1) to them, we have a simple ε-chain $(W_1, \ldots, W_m)$ from a to b. Let

$$k_0 = \max\{i \mid \operatorname{cl} W_i \cap \operatorname{cl} U \neq \emptyset\} > 1 \text{ and}$$
$$k_1 = \min\{i \geq k_0 \mid \operatorname{cl} W_i \cap \operatorname{cl} V \neq \emptyset\} \geq k_0.$$

If $W_{k_0} \cap U \neq \emptyset$, then $(W_{k_0}, \ldots, W_{k_1})$ is a simple ε-chain from U to V. When $W_{k_0} \cap U = \emptyset$ or $W_{k_1} \cap V = \emptyset$ (except for the case that $k_0 = k_1$ and $W_{k_0} \cap U = W_{k_0} \cap V = \emptyset$), we take a connected open neighborhood U' of some $x \in \operatorname{cl} W_{k_0} \cap \operatorname{cl} U$ with $\operatorname{diam} U' < \varepsilon - \operatorname{diam} W_{k_0}$ or a connected open neighborhood V' of some $y \in \operatorname{cl} W_{k_1} \cap \operatorname{cl} V$ with $\operatorname{diam} V' < \varepsilon - \operatorname{diam} W_{k_1}$ (in the except case, $\operatorname{diam} U'$, $\operatorname{diam} V' < \frac{1}{2}(\varepsilon - \operatorname{diam} W_{k_0})$). Then, replacing W_{k_0} by $U' \cup W_{k_0}$ or W_{k_1} by $V' \cup W_{k_1}$ (in the except case, replacing $W_{k_0} = W_{k_1}$ by $U' \cup V' \cup W_{k_0}$), we can obtain a simple ε-chain from U to V.

(3): For each $n \in \mathbb{N}$, let $\varepsilon = n^{-1} \operatorname{dist}(U, V) > 0$. By (2), we have a simple ε-chain $(W_1, \ldots, W_k)$ from U to V. Then, $n < k$ because

$$\operatorname{dist}(U, V) \leq \operatorname{diam} W_1 + \cdots + \operatorname{diam} W_k < k\varepsilon = n^{-1}k \operatorname{dist}(U, V).$$

Hence, U and V are connected by a simple chain $(W_1, \ldots, \bigcup_{i=n}^{k} W_i)$ of length n.

□

Recall that X is **path-connected** if every pair of points $x, y \in X$ can be connected by a path, i.e., there is a path $f : \mathbf{I} \to X$ with $f(0) = x$ and $f(1) = y$. It is said that X is **arcwise connected** if every two distinct points $x, y \in X$ can be connected by an arc, i.e., there is an arc $f : \mathbf{I} \to X$ with $f(0) = x$ and $f(1) = y$.[18] A space X is **locally path-connected** (or **locally arcwise connected**) if each neighborhood U of each point $x \in X$ contains a neighborhood V of x such that every two (distinct) points $y, y' \in V$ can be connected by a path (or an arc) in U. According to the following lemma, the local path-connectedness and the local arcwise connectedness can be defined in the same manner as the local connectedness.

Lemma 5.14.4. *For a locally path-connected (or locally arcwise connected) space X, the following hold:*

(1) *Every component of X is open and path-connected (or arcwise connected).*
(2) *Each point of a locally path-connected (or locally arcwise connected) space X has a neighborhood basis consisting of path-connected (or arcwise connected) open neighborhoods.*

Proof. (1): For each $x \in X$, let W be a subset of X consisting of all points connected with x by a path (or an arc) in X (and x itself). Then, it is easy to see that W is a connected clopen set in X, and hence it is a component of X.

(2): Every open neighborhood U of each $x \in X$ is also locally path-connected (or locally arcwise connected). It follows from (1) that the component of U containing x is a path-connected (or arcwise connected) open neighborhood of x. □

Obviously, every arcwise connected (resp. locally arcwise connected) space is path-connected (resp. locally path-connected), and every path-connected (resp. locally path-connected) space is connected (resp. locally connected). However, according to the following theorem, for connected locally compact metrizable spaces, the local connectedness implies the local arcwise connectedness.

Theorem 5.14.5. *Every connected, locally connected, locally compact metrizable space X is arcwise connected and locally arcwise connected.*

Proof. Because of the local compactness of X and 2.7.7(1), it can be assumed that $X = (X, d)$ is a metric space such that $\overline{\mathrm{B}}(x, 1)$ is compact for each $x \in X$, so $X = (X, d)$ is complete. Let $a, b \in X$ be two distinct points. By induction on

[18] Recall that an arc is an injective path, i.e., an embedding of $\mathbf{I}$.

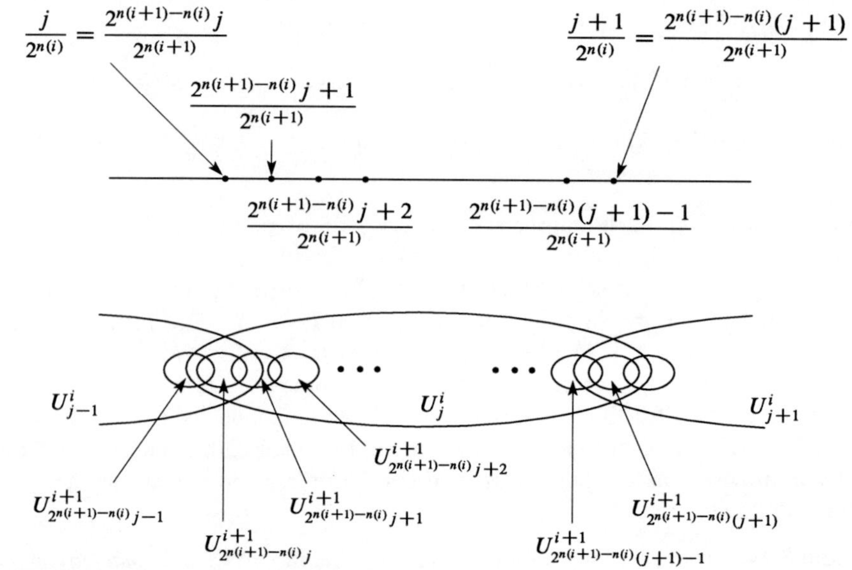

Fig. 5.10 Illustration of condition (2)

$i \in \mathbb{N}$, we will construct a simple 2^{-i}-chain $(U^i_0, U^i_1, \ldots, U^i_{2^{n(i)}-1})$ from a to b so that

(1) $n(1) < n(2) < \cdots$; and
(2) $U^{i+1}_k \subset U^i_j$ for $2^{n(i+1)-n(i)}j \le k < 2^{n(i+1)-n(i)}(j+1)$ (Fig. 5.10).

Since X is locally connected, a and b have connected open neighborhoods U and V, respectively, such that $\operatorname{diam} U$, $\operatorname{diam} V < 2^{-1}$, and $\operatorname{cl} U \cap \operatorname{cl} V = \emptyset$. Using Lemma 5.14.3(2), we can obtain $n(1) \ge 2$ and a simple 2^{-1}-chain $(U^1_1, \ldots, U^1_{2^{n(1)}-2})$ in X from U to V. Let $U^1_0 = U$ and $U^1_{2^{n(1)}-1} = V$. Thus, we have a simple 2^{-1}-chain $(U^1_0, \ldots, U^1_{2^{n(1)}-1})$ from a to b.

Next, suppose that a simple 2^{-i}-chain $(U^i_0, U^i_1, \ldots, U^i_{2^{n(i)}-1})$ from a to b has been obtained. Let U and V be connected open neighborhoods of a and b in X, respectively, such that $\operatorname{cl} U \subset U^i_0$ and $\operatorname{cl} V \subset U^i_{2^{n(i)}-1}$. Since each U^i_j is connected and locally connected, we can apply inductively Lemma 5.14.3(2) to obatin a simple $2^{-(i+1)}$-chain $(V^j_0, \ldots, V^j_{k(j)})$ in U^i_j from $U^i_j \cap V^{j-1}_{k(j-1)}$ to $U^i_j \cap U^i_{j+1}$, where $V^{-1}_{k(-1)} = U$ and $U^i_{2^{n(i)}} = V$. Choose $n(i+1) > n(i)$ so that

$$2^{n(i+1)-n(i)} > \max\{k(j) \mid j = 0, 1, \ldots, 2^{n(i)} - 1\}.$$

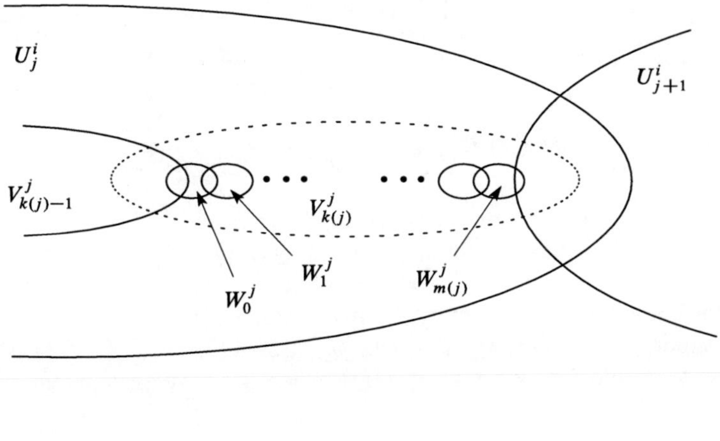

Fig. 5.11 A simple chain $(W^j_0, W^j_1, \dots, W^j_{m(j)})$ in $V^j_{k(j)}$

For each $j = 0, 1, \dots, 2^{n(i)} - 1$, let $m(j) = 2^{n(i+1)-n(i)} - k(j) - 1$ (i.e., $k(j) + m(j) = 2^{n(i+1)-n(i)} - 1$). By Lemma 5.14.3(3), we have a simple chain $(W^j_0, \dots, W^j_{m(j)})$ in $V^j_{k(j)}$ from $V^j_{k(j)} \cap V^j_{k(j)-1}$ and $V^j_{k(j)} \cap V^{j+1}_1$ (Fig. 5.11). Now, we define

$$U^{i+1}_{2^{n(i+1)-n(i)}j} = V^j_0, \dots, U^{i+1}_{2^{n(i+1)-n(i)}j+k(j)-1} = V^j_{k(j)-1},$$

$$U^{i+1}_{2^{n(i+1)-n(i)}j+k(j)} = W^j_0, \cdots, U^{i+1}_{2^{n(i+1)-n(i)}j+2^{n(i+1)-n(i)}-1} = W^j_{m(j)},$$

which are contained in U^i_j. Let $U^{i+1}_0 = U$ and $U^{i+1}_{2^{n(i+1)}-1} = V$. Then, $(U^{i+1}_0, U^{i+1}_1, \dots, U^{i+1}_{2^{n(i+1)}-1})$ is the desired simple $2^{-(i+1)}$-chain.

For each $x \in \mathbf{2}^{\mathbb{N}} = \{0, 1\}^{\mathbb{N}}$, observe $0 \le \sum_{j=1}^{n(i)} 2^{n(i)-j} x(j) \le 2^{n(i)} - 1$ and

$$\sum_{j=1}^{n(i)} 2^{n(i)-j} x(j) = 2^{n(i)-n(i-1)} \sum_{j=1}^{n(i-1)} 2^{n(i-1)-j} x(j) + \sum_{j=n(i-1)+1}^{n(i)} 2^{n(i)-j} x(j),$$

where $0 \le \sum_{j=n(i-1)+1}^{n(i)} 2^{n(i)-j} x(j) < 2^{n(i)-n(i-1)}$. Then, it follows from (4) that

$$U^i_{\sum_{j=1}^{n(i)} 2^{n(i)-j} x(j)} \subset U^{i-1}_{\sum_{j=1}^{n(i-1)} 2^{n(i-1)-j} x(j)}.$$

By (3) and the completeness of X, the following is a singleton:

$$\bigcap_{i \in \mathbb{N}} \operatorname{cl} U^i_{\sum_{j=1}^{n(i)} 2^{n(i)-j} x(j)} \neq \emptyset.$$

Then, we have a map $f : \mathbf{2}^{\mathbb{N}} \to X$ such that

$$\{f(x)\} = \bigcap_{i \in \mathbb{N}} \operatorname{cl} U^i_{\sum_{j=1}^{n(i)} 2^{n(i)-j} x(j)},$$

where $f(0) = a$ and $f(1) = b$. For $x, y \in \mathbf{2}^{\mathbb{N}}$, if $x(j) = y(j)$ for $j < 2^{n(i)}$, then

$$f(x), f(y) \in U^i_{\sum_{j=1}^{n(i)} 2^{n(i)-j} x(j)} = U^i_{\sum_{j=1}^{n(i)} 2^{n(i)-j} y(j)},$$

hence $d(f(x), f(y)) < 2^{-i}$ by (2), which implies that f is continuous.

Let $\varphi : \mathbf{2}^{\mathbb{N}} \to \mathbf{I}$ be the quotient map defined by $\varphi(x) = \sum_{i=1}^{\infty} 2^{-i} x(i)$. For each $x, y \in \mathbf{2}^{\mathbb{N}}$, we will show that $\varphi(x) = \varphi(y)$ if and only if $f(x) = f(y)$, hence f induces the embedding $h : \mathbf{I} \to X$ with $h(0) = a$ and $h(1) = b$.

First, suppose that $\varphi(x) = \varphi(y)$, i.e., $\sum_{i=1}^{\infty} 2^{-i} x(i) = \sum_{i=1}^{\infty} 2^{-i} y(i)$. When $x \neq y$, let $k = \min\{i \in \mathbb{N} \mid x(i) \neq y(i)\}$, where we may assume that $x(k) = 1$ and $y(k) = 0$. Then,

$$\begin{aligned}\sum_{i=1}^{\infty} 2^{-i} x(i) \geq \sum_{i=1}^{k} 2^{-i} x(i) &= \sum_{i=1}^{k-1} 2^{-i} x(i) + 2^{-k} \\ &= \sum_{i=1}^{k} 2^{-i} y(i) + \sum_{j=k+1}^{\infty} 2^{-j} \geq \sum_{i=1}^{k} 2^{-i} y(i),\end{aligned}$$

which implies that $x(i) = 0$ and $y(i) = 1$ for every $i > k$. Thus, we have

$$\sum_{j=1}^{k-1} 2^{k-1-j} x(j) = \sum_{j=1}^{k-1} 2^{k-1-j} y(j) \ \text{ and}$$

$$\sum_{j=1}^{m} 2^{m-j} x(j) = \sum_{j=1}^{m} 2^{m-j} y(j) + 1 \ \text{ for every } m \geq k.$$

Then, it follows that

$$U^i_{\sum_{j=1}^{n(i)} 2^{n(i)-j} x(j)} \cap U^i_{\sum_{j=1}^{n(i)} 2^{n(i)-j} y(j)} \neq \emptyset \ \text{ for every } i \in \mathbb{N},$$

which implies that $d(f(x), f(y)) = 0$ by (3), hence $f(x) = f(y)$.

Conversely, suppose that $f(x) = f(y)$. For every $i \in \mathbb{N}$,

$$U^i_{\sum_{j=1}^{n(i)-1} 2^{n(i)-j} x(j)} \cap U^i_{\sum_{j=1}^{n(i)-1} 2^{n(i)-j} y(j)} \neq \emptyset,$$

which means $\left|\sum_{j=1}^{n(i)} 2^{n(i)-j} x(j) - \sum_{j=1}^{n(i)} 2^{n(i)-j} y(j)\right| \leq 1$. Therefore,

$$|\varphi(x) - \varphi(y)| = \left|\sum_{j=1}^{\infty} 2^{-j} x(j) - \sum_{j=1}^{\infty} 2^{-j} y(j)\right|$$

$$
\begin{aligned}
&= \lim_{i\to\infty} \left| \sum_{j=1}^{n(i)-1} 2^{-j} x(j) - \sum_{j=1}^{n(i)-1} 2^{-j} y(j) \right| \\
&= \lim_{i\to\infty} 2^{-n(i)} \left| \sum_{j=1}^{n(i)-1} 2^{n(i)-j} x(j) - \sum_{j=1}^{n(i)-1} 2^{n(i)-j} y(j) \right| \\
&\leq \lim_{i\to\infty} 2^{-n(i)} = 0,
\end{aligned}
$$

that is, $\varphi(x) = \varphi(y)$. Thus, we have proved that X is arcwise connected.

Finally, note that every neighborhood of each point $x \in X$ contains a connected open neighborhood U in X. Since U is also completely metrizable, it follows that U is also arcwise connected. This means that X is locally arcwise connected. □

By the "only if" part of Theorems 5.14.1 and 5.14.5, we have the following corollary:

Corollary 5.14.6. *Let X be an arbitrary space. Then, each pair of distinct points $x \neq y \in X$ are connected by a path in X if and only if they are connected by an arc in X. In this case, the image of the arc is contained in the image of the path.*

Proof. The "if" part is obvious. To see the "only if" part, let $f : \mathbf{I} \to X$ be a path with $f(0) = x$ and $f(1) = y$. Since the image $f(\mathbf{I})$ is a locally connected continuum (i.e., a Peano continuum) by the "only if" part of Theorem 5.14.1, we have an arc from x to y in $f(\mathbf{I})$ ($\subset X$) by Theorem 5.14.5. □

Thus, we know that there is no difference between the (local) path-connectedness and the (local) arcwise connectedness of an arbitrary space. This allows us to sate the following:

Corollary 5.14.7. *An arbitrary space X is path-connected if and only if X is arcwise connected. Moreover, X is locally path-connected if and only if X is locally arcwise connected.* □

A metric space $X = (X, d)$ is said to be **uniformly locally path-connected** provided that, for every $\varepsilon > 0$, there is $\delta > 0$ such that each pair of points $x, y \in X$ with $d(x, y) < \delta$ can be connected by a path with diam $< \varepsilon$.

Proposition 5.14.8. *A compact metric space X is uniformly locally path-connected if it is locally path-connected.*

Proof. For each $\varepsilon > 0$, we apply Lemma 5.14.4(2) to obtain $\mathcal{U} \in \operatorname{cov}(X)$ consisting of path-connected open sets with mesh $\mathcal{U} < \varepsilon$. Let $\delta > 0$ be a Lebesgue number for $\mathcal{U}$. Then, each pair of points $x, y \in X$ with $d(x, y) < \delta$ can be connected by a path with diam $< \varepsilon$. □

We are now ready to prove the "if" part of the Hahn–Mazurkiewicz Theorem 5.14.1.

Proof of the "if" part of Theorem 5.14.1. We may assume that $X = (X, d)$ is a compact connected metric space. Let μ^0 be the Cantor (ternary) set in $\mathbf{I}$. By Theorem 5.11.5, there exists a continuous surjection $f : \mu^0 \to X$. By Theorem 5.14.5, X is path-connected and locally path-connected (arcwise connected and locally arcwise connected). According to Proposition 5.14.8, we have $\delta_1 > \delta_2 > \cdots > 0$ such that every two distinct points within δ_n can be connected by a path with diam $< 1/n$, where we may assume that $\delta_n \leq 1/n$.

Because of the construction of μ^0, the complement $\mathbf{I} \setminus \mu^0$ has only finitely many components $C_i = (a_i, b_i)$, $i = 1, \ldots, m$, such that $d(f(a_i), f(b_i)) \geq \delta_1$. Indeed, there is some $k \in \mathbb{N}$ such that

$$a, b \in \mu^0, \ |a - b| < 3^{-k} \Rightarrow d(f(a), f(b)) < \delta_1,$$
$$\big(\text{i.e., } d(f(a), f(b)) \geq \delta_1 \Rightarrow |a - b| \geq 3^{-k}\big),$$

which implies that $m \leq \sum_{i=1}^{k} 2^{i-1}$. For each $i = 1, \ldots, m$, let $f_i : \operatorname{cl} C_i = [a_i, b_i] \to X$ be a path with $f_i(a_i) = f(a_i)$ and $f_i(b_i) = f(b_i)$. Then, we can extend f to the map

$$f' : M = \mu^0 \cup \bigcup_{i=1}^{m} \operatorname{cl} C_i \to X$$

that is defined by $f'|\operatorname{cl} C_i = f_i$ for each $i = 1, \ldots, m$.

For each component $C = (a, b)$ of $\mathbf{I} \setminus M$ (which is a component of $\mathbf{I} \setminus \mu^0$), $f(a) = f(b)$ or $0 < d(f(a), f(b)) < \delta_1$. In the former case, let $f_C : \operatorname{cl} C = [a, b] \to X$ be the constant path with $f_C([a, b]) = \{f(a)\}$ $(= \{f(b)\})$. In the latter case, choose $n \in \mathbb{N}$ so that $\delta_{n+1} \leq d(f(a), f(b)) < \delta_n$ and take a path $f_C : \operatorname{cl} C = [a, b] \to X$ such that $f_C(a) = f(a)$, $f_C(b) = f(b)$, and diam $f_C([a, b]) < 1/n$. Then, f' can be extended to the map $f^* : \mathbf{I} \to X$ by $f^*|\operatorname{cl} C = f_C$ for every component C of $\mathbf{I} \setminus M$.

It remains to verify the continuity of f^*. Since each component C of $\mathbf{I} \setminus M$ is an open interval, the continuity of f^* at a point of $\mathbf{I} \setminus M$ follows from the continuity of f_C. The continuity of f^* at a point of $\operatorname{int} M$ comes from the continuity of f'. We will show the continuity of f^* at a point $x \in \operatorname{bd} M$ $(= \mu^0)$. For each $\varepsilon > 0$, choose $n \in \mathbb{N}$ so that $1/n < \varepsilon/2$. Since f' is continuous at x, we have a neighborhood U of x in $\mathbf{I}$ such that $f'(U \cap M) \subset \mathrm{B}(f'(x), \delta_n/2)$ $(\subset \mathrm{B}(f^*(x), \varepsilon/2)$ because $\delta_n \leq 1/n < \varepsilon/2)$. In the case that $x \notin \operatorname{bd} C$ for any component $C = (a, b)$ of $\mathbf{I} \setminus M$ with $d(f(a), f(b)) \geq \delta_n$, U can be chosen so that $U \cap \operatorname{cl} C = \emptyset$ for any component $C = (a, b)$ of $\mathbf{I} \setminus M$ with $d(f(a), f(b)) \geq \delta_n$. In the case $x \in \operatorname{bd} C_0$ for some component $C_0 = (a_0, b_0)$ of $\mathbf{I} \setminus M$ with $d(f(a_0), f(b_0)) \geq \delta_n$ (such a component C_0 is unique if it exists), U can be chosen so that $f_{C_0}(U \cap C_0) \subset \mathrm{B}(f'(x), \varepsilon/2)$. Now, let $C = (a, b)$ be a component of $\mathbf{I} \setminus M$ with $\operatorname{cl} C \cap U \neq \emptyset$. Then, $a \in U \cap M$ or $b \in U \cap M$, and so $d(f'(a), f'(x)) < \varepsilon/2$ or $d(f'(b), f'(x)) < \varepsilon/2$, respectively. If $f'(a) = f'(b)$, then $f^*(C) = f_C(C) = \{f(a)\} \subset \mathrm{B}(f'(x), \varepsilon/2)$. If $0 < d(f(a), f(b)) < \delta_n$, then diam $f_C([a, b]) < 1/n < \varepsilon/2$, which implies that $f^*(C) = f_C([a, b]) \subset$

$B(f'(x), \varepsilon)$. When $d(f(a), f(b)) \geq \delta_n$, it follows that $x \in \operatorname{bd} C$, which means that $C = C_0$. Then, $f^*(U \cap C) = f_{C_0}(U \cap C_0) \subset B(f'(x), \varepsilon/2)$. Consequently, we have $f^*(U) \subset B(f^*(x), \varepsilon)$. This completes the proof. □

Notes for Chap. 5

Below, we list only three among textbooks on Dimension Theory:

- R. Engelking, *Theory of Dimensions, Finite and Infinite*, Sigma Ser. in Pure Math. **10** (Heldermann Verlag, Lembo, 1995)
- W. Hurewicz and H. Wallman, *Dimension Theory* (Princeton University Press, Princeton, 1941)
- K. Nagami, *Dimension Theory* (Academic Press, Inc., New York, 1970)

For a more comprehensive study of Dimension Theory, we refer to Engelking's book, which also contains excellent historical notes. Nagami's book is quite readable and contains an appendix titled "Cohomological Dimension Theory" by Kodama. The classical book by Hurewicz and Wallman is still a worthwhile read. Nothing fundamental has yet changed in the framework of Dimension Theory since its publication. In this book, Hurewicz and Wallman discuss the Hausdorff dimension, which is useful in the field of Fractal Geometry. However, we do not discuss this here. In the following textbook of van Mill, Chap. 5 is devoted to Dimension Theory, and was used to prepare the last two sections of this chapter.

- J. van Mill, *Infinite-Dimensional Topology, Prerequisites and Introduction*, North-Holland Math. Library **43** (Elsevier Sci. Publ. B.V., Amsterdam, 1989)

The definition of dim, which is due to Čech [11], is based on a property of covers of $\mathbf{I}^n$ discovered by Lebesgue [28]. The Brouwer Fixed Point Theorem 5.1.1 was established in [8]. The proof using Sperner's Lemma 5.1.2 in [53] is due to Knaster et al. [26].

The equivalence between (a), (b), and (d) in Theorem 5.2.3 was established by Hemmingsen [20] and the equivalence between (a) and (d) was proved independently by Alexandroff [2] and Dowker [12]. The equivalence between (a) and (f) was first established for compact metrizable spaces by Hurewicz [23] and for normal spaces by Alexandroff [2], Hemmingsen [20], and Dowker [12], independently.

The compact case of Corollary 5.2.6 was established by Freudenthal [17], and was generalized to compact Hausdorff spaces by Mardešić [33].

In [22], a map $f : X \to \mathbf{I}^n$ is called a **universal map** if it satisfies condition (b) in Theorem 5.2.15. The equivalence between (b) and (c) in Theorem 5.2.15 is due to Holszyński [22]. The equivalence between (a) and (b) in Theorem 5.2.17 was established by Alexandroff [1]. The equivalence between (a) and (c) in Theorem 5.2.17 was first established by Eilenberg and Otto [14] in the separable metrizable case and extended to normal spaces by Hemmingsen [20].

Theorem 5.3.1 was established by Vopěnka [55] and Theorem 5.3.2 was proved by Nagami [40]. The Subset Theorem was proved by Dowker [13]. The Countable Sum Theorem (5.4.1) was established by Čech [11] and the Locally Finite Sum Theorem (5.4.2) was proved independently by Morita [Mo] and Katětov [24]. The Addition Theorem (5.4.8) was proved by Smirnov [52]. The Decomposition and Product Theorems (5.4.5, 5.4.9) were proved independently by Katětov [24] and Morita [39].

An inductive definition of dimension was outlined by Poincaré [44]. The first precise definition of a dimension function was introduced by Brouwer [9]. His function coincides with Ind in the class of locally connected compact metrizable spaces. The definition of Ind was formulated by Čech [10]. On the other hand, the definition of ind was formulated by Urysohn [54] and Menger [37]. The first example in Theorem 5.5.3 was constructed by Roy [47,48] but the example presented here was constructed by Kulesza [27] and the proof of dim > 0 was simplified by Levin [31].

The weak infinite dimension was first introduced by Alexandroff in [3]. In Remark 12, we mentioned the weak infinite dimension in the sense of Smirnov, which was first studied in [32] and [51].

Theorem 5.7.4 is due to Lelek [30] and the simple proof presented here is taken from Engelking and Pol [15].

In [42], Nöbeling introduced the spaces ν^n and showed their universality. The spaces μ^n were introduced by Menger [38], who showed that the universality μ^1 is a universal space for compacta with dim ≤ 1. Theorem 5.10.8 is due to Bothe [6]. In [29], Lefschetz constructed a universal space for compacta with dim $\leq n$. In [5], Bestvina gave the topological characterization of μ^n. Using Bestvina's characterization, we can see that Lefschetz' universal space is homeomorphic to μ^n; the result for $n = 1$ had been obtained by Anderson [4]. Recently, in [41], Nagórko established the topological characterization of ν^n.

The total disconnectedness and the hereditary disconnectedness were respectively introduced by Sierpiński [50] and Hausdorff [19]. The example of Theorem 5.11.3 is due to Knaster and Kuratowski [25] (their example is the one in the Remark).

The example of Theorem 5.12.1 was described by Erdös [16]. Lemma 5.12.3 is due to Bourbaki [7, Chap. 9] and the proof presented here is due to van Mill (Chap. 5 in his book listed above). The first completely metrizable nonzero-dimensional totally disconnected space was constructed by Sierpiński [50] (his example is 1-dimensional). Theorem 5.12.6 was established by Mazurkiewicz [36] but the example and proof presented here is due to Rubin et al. [49] with some help from [45].

The example of Theorem 5.13.1 is presented by Pol [45]. Theorem 5.13.4 is due to Walsh [56] but the example given here is due to Pol [46]. The earlier example of a compact metrizable space, whose compact subsets are all either 0-dimensional or infinite-dimensional, was constructed by Henderson [21].

In 1890, Peano [43] showed that the square $\mathbf{I}^2$ is the continuous image of $\mathbf{I}$. The Hahn–Mazurkiewicz Theorem 5.14.1 was independently proved by Hahn [18] for planar sets and by Mazurkiewicz [34] for subspaces of Euclidean space. In [35], Mazurkiewicz gave a systematic exposition.

For more details, consult the historical and bibliographical notes at the end of each section of Engelking's book.

References

1. P. Alexandroff, Dimensionstheorie. Ein Beitrag zur Geometrie der abgeschlossenen Mengen. Math. Ann. **106**, 161–238 (1932)
2. P. Alexandroff, On the dimension of normal spaces. Proc. Roy. Soc. Lond. **189**, 11–39 (1947)
3. P. Alexandroff, *Preface to the Russian translation* (Russian). Russian translation of W. Hurewicz, H. Wallman, *Dimension Theory*, Izdat. Inostr. Literatura, Moscow (1948)
4. R.D. Anderson, A characterization of the universal curve and a proof of its homogeneity. Ann. Math. **67**, 313–324 (1958)
5. M. Bestvina, Characterizing k-dimensional universal Menger compacta. Mem. Am. Math. Soc. **71**(380) (1988)
6. H.G. Bothe, Eine Einbettung m-dimensionaler Mengen in einen $(m + 1)$-dimensionalen absoluten Retrakt. Fund. Math. **52**, 209–224 (1963)
7. N. Bourbaki, *Topologie Générale*, 2nd edition. (Hermann, Paris, 1958)
8. L.E.J. Brouwer, Über Abbildung von Mannigfaltigkeiten. Math. Ann. **71**, 97–115 (1912)
9. L.E.J. Brouwer, Über den natürlichen Dimensionsbegriff. J. für die reine und angrew Math. **142**, 146–152 (1913)
10. E. Čech, Sur la théorie de la dimension. C.R. Acad. Paris **193**, 976–977 (1931)

11. E. Čech, Přispěvek k theorii dimense, (Contribution to dimension theory). Časopis Pěst. Mat. Fys. **62**, 277–291 (1933); French translation in Topological Papers, Prague (1968), pp. 129–142
12. C.H. Dowker, Mapping theorems for non-compact spaces. Am. J. Math. **69**, 200–242 (1947)
13. C.H. Dowker, Local dimension of normal spaces. Q. J. Math. Oxf. **6**, 101–120 (1955)
14. S. Eilenberg, E. Otto, Quelques propiétés caractéristique de la dimension. Fund. Math. **31**, 149–153 (1938)
15. R. Engelking, R. Pol, Compactifications of countable-dimensional and strongly countable-dimensional spaces. Proc. Am. Math. Soc. **104**, 985–987 (1988)
16. P. Erdös, The dimension of the rational points in Hilbert space. Ann. Math. **41**, 734–736 (1940)
17. H. Freudenthal, Entwicklungen von Räumen und ihren Gruppen. Compositio Math. **4**, 145–234 (1937)
18. H. Hahn, Mengentheoretische Charakterisierung der stetigen Kurve. Sitzungsberichte Akad. Wiss. Wien Abt. IIa **123**, 2433–2489 (1914)
19. F. Hausdorff, *Grundzüge der Mengenlehre* (Teubner, Leipzig, 1914)
20. E. Hemmingsen, Some theorems in dimension theory for normal Hausdorff spaces. Duke Math. J. **13**, 495–504 (1946)
21. D.W. Henderson, An infinite-dimensional compactum with no positive-dimensional compact subsets – a simpler construction. Am. J. Math. **89**, 105–121 (1967)
22. W. Holszyński, Une généralisation de théorème de Brouwer sur les points invariant. Bull. Acad. Polon. Sci., Sér. Sci. Math. Astr. et Phys. **12**, 603–606 (1964)
23. W. Hurewicz, Über Abbildungen topologischer Räume auf die n-dimensionale Sphäre. Fund. Math. **24**, 144–150 (1935)
24. M. Katětov, On the dimension of non-separable spaces, I (Russian). Czech. Math. J. **2**, 333–368 (1952)
25. B. Knaster, K. Kuratowski, Sur les ensembles connexes. Fund. Math. **2**, 206–255 (1921)
26. B. Knaster, K. Kuratowski, S. Mazurkiewicz, Ein Beweis des Fixpunktsatzes für n-dimensionale Simplexe. Fund. Math. **14**, 132–137 (1929)
27. J. Kulesza, An example in the dimension theory of metrizable spaces. Topology Appl. **35**, 109–120 (1990)
28. H. Lebesgue, Sur la non-applicabilité de deux domaines appartenant respectivement à des espaces à n et $n + p$ dimensions. Math. Ann. **70**, 166–168 (1911)
29. S. Lefschetz, On compact spaces. Ann. Math. **32**, 521–538 (1931)
30. A. Lelek, On the dimensionality of remainders in compactifications (Russian). Dokl. Akad. Nauk SSSR **160**, 534–537 (1965); English translation in Sov. Math. Dokl. **6**, 136–140 (1965)
31. M. Levin, A remark on Kulesza's example. Proc. Am. Math. Soc. **128**, 623–624 (2000)
32. B.T. Levšenko, On strongly infinite-dimensional spaces (Russian). Vestnik Moskov. Univ., Ser. Mat. 219–228 (1959)
33. S. Mardešić, On covering dimension and inverse limits of compact spaces. Illinois J. Math. **4**, 278–291 (1960)
34. S. Mazurkiewicz, O arytmetyzacji contnuów. C.R. Varsovie **6**, 305–311 (1913); O arytmetyzacji contnuów, II. C.R. Varsovie **6**, 941–945 (1913); French translation Travaux de topologie, Warszawa, pp. 37–41, 42–45 (1969)
35. S. Mazurkiewicz, Sur les lignes de Jordan. Fund. Math. **1**, 166–209 (1920)
36. S. Mazurkiewicz, Sur les problème κ et λ de Urysohn. Fund. Math. **10**, 311–319 (1927)
37. K. Menger, Über die Dimensionalität von Punktmengen I. Monatsh. für Math. und Phys. **33**, 148–160 (1923)
38. K. Menger, Allgemeine Räume und Cartesische Räume, Zweite Mitteilung: "Über umfassendste n-dimensionale Mengen". Proc. Akad. Amterdam **29**, 1125–1128 (1926)
39. K. Morita, On the dimension of normal spaces, II. J. Math. Soc. Jpn. **2**, 16–33 (1950)
40. K. Nagami, Finite-to-one closed mappings and dimension, II. Proc. Jpn. Acad. **35**, 437–439 (1959)
41. A. Nagórko, Characterization and topological rigidity of Nöbeling manifolds. Ph.D. Thesis, University of Warsaw (2006)

42. G. Nöbeling, Über eine n-dimensionale Universalmenge in R^{2n+1}. Math. Ann. **104**, 71–80 (1931)
43. G. Peano, Sur une courbe, qui remplit toute une aire plane. Math. Ann. **36**, 157–160 (1890)
44. H. Poincaré, Pourquoi l'espace a trois dimensions. Revue de Metaph. et de Morale **20**, 483–504 (1912)
45. R. Pol, A weakly infinite-dimensional compactum which is not countable-dimensional. Proc. Am. Math. Soc. **82**, 634–636 (1981)
46. R. Pol, Countable-dimensional universal sets. Trans. Am. Math. Soc. **297**, 255–268 (1986)
47. P. Roy, Failure of equivalence of dimension concepts for metric spaces. Bull. Am. Math. Soc. **68**, 609–613 (1962)
48. P. Roy, Nonequality of dimensions for metric spaces. Trans. Am. Math. Soc. **134**, 117–132 (1968)
49. L. Rubin, R.M. Schori, J.J. Walsh, New dimension-theory techniques for constructing infinite-dimensional examples. Gen. Topology Appl. **10**, 93–102 (1979)
50. W. Sierpiński, Sur les ensembles connexes et non connexes. Fund. Math. **2**, 81–95 (1921)
51. E.G. Skljarenko, On dimensional properties of infinite-dimensional spaces (Russian). Isv. Akad. Nauk. SSSR Ser. Mat. **23**, 197–212 (1959); English translation in Am. Math. Soc. Transl. Ser. 2 **21**, 35–50 (1962)
52. Ju.M. Smirnov, Some relations in the theory of dimensions (Russian). Mat. Sb. **29**, 151–172 (1951)
53. E. Sperner, Neuer Beweis für die Invarianz des Dimensionszahl und des Gebietes. Abh. Math. Semin. Hamburg Univ. **6**, 265–272 (1928)
54. P. Urysohn, Kes multiplicités Cantoriennes. C.R. Acad. Paris **175**, 440–442 (1922)
55. P. Vopěnka, Remark on the dimension of metric spaces (Russian). Czech. Math. J. **9**, 519–521 (1959)
56. J.J. Walsh, Infinite-dimensional compacta containing no n-dimensional ($n \geq 1$) subsets. Topology **18**, 91–95 (1979)

Chapter 6
Retracts and Extensors

A subset A of a space X is called a **retract** of X if there is a map $r : X \to A$ such that $r|A = \mathrm{id}$, which is called a **retraction**. As is easily observed, every retract of a space X is *closed* in X. A **neighborhood retract** of X is a *closed* set in X that is a retract of some neighborhood in X. A *metrizable* space X is called an **absolute neighborhood retract** (**ANR**) (resp. an **absolute retract** (**AR**)) if X is a neighborhood retract (or a retract) of an arbitrary metrizable space that contains X as a closed subspace. A space Y is called an **absolute neighborhood extensor for metrizable spaces** (**ANE**) if each map $f : A \to Y$ from any closed set A in an arbitrary metrizable space X extends over some neighborhood U of A in X. When f can always be extended over X (i.e., $U = X$ in the above), we call Y an **absolute extensor for metrizable spaces** (**AE**). As is easily observed, every *metrizable* ANE (resp. a *metrizable* AE) is an ANR (resp. an AR). As will be shown, the converse is also true. Thus, a *metrizable* space is an ANE (resp. an AE) if and only if it is an ANR (resp. an AR).

This chapter is devoted to lectures on ANR Theory (Theory of Retracts). We will prove the basic properties, fundamental theorems, and various characterizations of ANEs and ANRs.

The results in Chaps. 2 and 4 are used frequently. For topological linear spaces, refer to Chap. 3. To characterize countable-dimensional ANRs, finite-dimensional ANEs and ANRs, we need some theorems from Chap. 5.

6.1 The Dugundji Extension Theorem and ANEs

Recall that a topological linear space E is locally convex if $0 \in E$ has a neighborhood basis consisting of convex sets. In this section, we prove the following extension theorem:

K. Sakai, *Geometric Aspects of General Topology*, Springer Monographs in Mathematics,
DOI 10.1007/978-4-431-54397-8_6, © Springer Japan 2013

Theorem 6.1.1 (DUGUNDJI EXTENSION THEOREM). *Let E be a locally convex topological linear space, X be a metrizable space, and A be closed in X. Then, each map $f : A \to E$ can be extended to a map $\tilde{f} : X \to E$ such that the image $\tilde{f}(X)$ is contained in the convex hull $\langle f(A) \rangle$ of $f(A)$.*

Due to the above Theorem 6.1.1, every locally convex topological linear space and its convex set are AEs, and, if they are metrizable, then they are ARs.

Recall that every metrizable space X is paracompact, that is, any open cover of X has a locally finite open refinement (Theorem 2.3.1), and that, for each locally finite open cover $\mathcal{U} \in \operatorname{cov}(X)$, X has a partition of unity $(\lambda_U)_{U \in \mathcal{U}}$ such that $\operatorname{supp} \lambda_U \subset U$ for each $U \in \mathcal{U}$ (Theorem 2.7.2), where $\operatorname{supp} \lambda_U = \operatorname{cl} \lambda_U^{-1}((0,1])$.

Note. For a locally finite open cover $\mathcal{U}$ of a metric space $X = (X, d)$, a partition of unity can be directly defined as follows: For each $U \in \mathcal{U}$, define a map $\lambda_U : X \to \mathbf{I}$ by

$$\lambda_U(x) = \frac{d(x, X \setminus U)}{\sum_{V \in \mathcal{U}} d(x, X \setminus V)}.$$

Then, $(\lambda_U)_{U \in \mathcal{U}}$ is a partition of unity on X such that $\lambda_U^{-1}((0,1]) = U$ (hence $\operatorname{supp} \lambda_U = \operatorname{cl} U$) for each $U \in \mathcal{U}$.

By shrinking $\mathcal{U}$ (the Open Cover Shrinking Lemma 2.7.1), we can require $\operatorname{supp} \lambda_U \subset U$.

The following is a key to the proof of Theorem 6.1.1:

Lemma 6.1.2. *Let $X = (X, d)$ be a metric space and $A \neq \emptyset$ be a proper closed set in X. Then, there exists a locally finite open cover $\mathcal{U}$ of $X \setminus A$ with $a_U \in A$, $U \in \mathcal{U}$, such that $x \in U \in \mathcal{U}$ implies $d(x, a_U) \leq 2d(x, A)$.*

Proof. Let $\mathcal{U}$ be a locally finite open cover of $X \setminus A$ such that

$$\mathcal{U} \prec \left\{ \mathrm{B}(x, \tfrac{1}{4} d(x, A)) \mid x \in X \setminus A \right\}.$$

For each $U \in \mathcal{U}$, choose $x_U \in X \setminus A$ so that $U \subset \mathrm{B}(x_U, \frac{1}{4} d(x_U, A))$, and then choose $a_U \in A$ so that $d(x_U, a_U) < \frac{5}{4} d(x_U, A)$. If $x \in U \in \mathcal{U}$ then

$$d(x, a_U) \leq d(x, x_U) + d(x_U, a_U) < \tfrac{3}{2} d(x_U, A) < 2d(x, A)$$

because $d(x, A) \geq d(x_U, A) - d(x, x_U) > \frac{3}{4} d(x_U, A)$. □

We call the above $(a_U)_{U \in \mathcal{U}}$ a **Dugundji system** for $A \subset X$.

Proof of Theorem 6.1.1. By the above lemma, we have a Dugundji system $(a_U)_{U \in \mathcal{U}}$ for $A \subset X$. Let $(\lambda_U)_{U \in \mathcal{U}}$ be a partition of unity on $X \setminus A$ such that $\operatorname{supp} \lambda_U \subset U$ for each $U \in \mathcal{U}$. (Here, it is enough to require that $\lambda_U^{-1}((0,1]) \subset U$ for each $U \in \mathcal{U}$.) We define $\tilde{f} : X \to E$ as follows:

$$\tilde{f}(x) = \begin{cases} f(x) & \text{if } x \in A, \\ \sum_{U \in \mathcal{U}} \lambda_U(x) f(a_U) & \text{if } x \in X \setminus A. \end{cases}$$

By the local finiteness of $\mathcal{U}$, $\tilde{f}$ is continuous at $x \in X \setminus A$. To prove the continuity of $\tilde{f}$ at $a \in A$, let V be any convex neighborhood of $f(a)$ in E. From the continuity of f, we have $\delta > 0$ such that $f(\mathrm{B}(a, \delta) \cap A) \subset V$. For each $x \in \mathrm{B}(a, \frac{1}{3}\delta) \setminus A$, if $\lambda_U(x) \neq 0$ then $x \in U$, hence

$$d(a, a_U) \leq d(a, x) + d(x, a_U) \leq d(a, x) + 2d(x, A) \leq 3d(x, a) < \delta,$$

which implies $f(a_U) \in V$. From the convexity of V, it follows that $\tilde{f}(x) \in V$. Thus, $\tilde{f}(x) \in V$ for all $x \in \mathrm{B}(a, \frac{1}{3}\delta)$. □

Remark 1. It is easy to show that $\mathrm{C}(X, E)$ and $\mathrm{C}(A, E)$ are topological linear spaces with respect to the compact-open topology. If E is locally convex, so is $\mathrm{C}(X, E)$.

In fact, let $\mathcal{U}$ be an open neighborhood basis at $0 \in E$. The set of all non-empty compact sets in X is denoted by $\mathrm{Comp}(X)$. For each $K \in \mathrm{Comp}(X)$ and $U \in \mathcal{U}$, let

$$\mathrm{W}(K, U) = \{f \in \mathrm{C}(X, E) \mid f(K) \subset U\}.$$

Then, the following $\mathcal{W}$ satisfies all conditions of Proposition 3.4.1:

$$\mathcal{W} = \{\mathrm{W}(K, U) \mid K \in \mathrm{Comp}(X),\ U \in \mathcal{U}\}.$$

Let $R : \mathrm{C}(X, E) \to \mathrm{C}(A, E)$ be the restriction operator, i.e., $R(f) = f|A$ for each $f \in \mathrm{C}(X, E)$. Then, R is linear and continuous. Fixing a Dugundji system $(a_U)_{U \in \mathcal{U}}$ for $A \subset X$ and a partition of unity $(\lambda_U)_{U \in \mathcal{U}}$, the extension operator $L : \mathrm{C}(A, E) \to \mathrm{C}(X, E)$ is defined by $L(f) = \tilde{f}$ as above. Then, $R \circ L = \mathrm{id}$ and L is linear by definition. Moreover, L is continuous.

To prove the continuity of L at $0 \in \mathrm{C}(A, E)$, let $\mathfrak{V}$ be a neighborhood of $0 \in \mathrm{C}(X, E)$. Then, we have a compact set K in X and an open convex neighborhood W of $0 \in E$ such that

$$\{g \in \mathrm{C}(X, E) \mid g(K) \subset W\} \subset \mathfrak{V}.$$

It suffices to find a compact set $K_A \subset A$ such that $f(K_A) \subset W$ implies $\tilde{f}(K) \subset W$ for $f \in \mathrm{C}(A, E)$. We define

$$K_A = \{a_U \mid U \in \mathcal{U}[K]\} \cup (A \cap K).$$

For $f \in \mathrm{C}(A, E)$, $f(K_A) \subset W$ implies $\tilde{f}(K) \subset W$ by the convexity of W and the definition of $\tilde{f}$. To prove the compactness of K_A, it suffices to show that every sequence $(a_k)_{k \in \mathbb{N}}$ in K_A has a convergent subsequence. If $a_k \in A \cap K$ for infinitely many $k \in \mathbb{N}$, the compactness of K_A follows from the compactness of $A \cap K$. Otherwise, we may assume that $a_k \in A \setminus K$ for all $k \in \mathbb{N}$, i.e., $a_k = a_{U_k}$ for some $U_k \in \mathcal{U}[K]$, where we may also assume that $U_k \neq U_{k'}$ if $k \neq k'$. For each $k \in \mathbb{N}$, we have $x_k \in U_k \cap K$. By the compactness of K, $(x_k)_{k \in \mathbb{N}}$ has a subsequence $(x_{k_j})_{j \in \mathbb{N}}$ converging to some $x_0 \in K$. Since $\{U_{k_j} \mid j \in \mathbb{N}\}$ is locally finite in $X \setminus A$, it follows that $x_0 \in K \cap A \subset K_A$. Hence,

$$d(x_{k_j}, a_{k_j}) = d(x_{k_j}, a_{U_{k_j}}) \leq 2d(x_{k_j}, A) \leq 2d(x_{k_j}, x_0) \to 0,$$

which implies that $(a_{k_j})_{j \in \mathbb{N}}$ also converges to x_0. Therefore, K_A is compact.

Remark 2. When $E = (E, \|\cdot\|)$ is a normed linear space, let $\mathrm{C}^B(X, E)$ be the normed linear space of all bounded maps with the sup-norm $\|f\| = \sup_{x\in X}\|f(x)\|$. For a closed set A in X, given a Dugundji system $(a_U)_{U\in\mathcal{U}}$ and a partition of unity $(\lambda_U)_{U\in\mathcal{U}}$, Theorem 6.1.1 gives the extension operator $L : \mathrm{C}^B(A, E) \to \mathrm{C}^B(X, E)$, which is an isometry (i.e., $\|L(f)\| = \|f\|$ for every $f \in \mathrm{C}^B(A, E)$). In this case, the restriction operator $R : \mathrm{C}^B(X, E) \to \mathrm{C}^B(A, E)$ satisfies $\|R(f)\| \leq \|f\|$ for every $f \in \mathrm{C}^B(X, E)$.

Remark 3. In Theorem 6.1.1, the local convexity is essential. Actually, there exists a metric linear space that is not an AE (i.e., an AR). Such an example will be constructed in Sect. 7.12.

Remark 4. Theorem 6.1.1 is valid even if E is a Fréchet space and X is a paracompact space. Namely, every Fréchet space is an AE for paracompact spaces. Indeed, applying the Michael Selection Theorem 3.8.8 to the set-valued function $\varphi : X \ni x \mapsto \mathrm{cl}\langle f(A)\rangle \in \mathrm{Conv}(E)$, we can extend f to a map $\tilde{f} : X \to \mathrm{cl}\langle f(A)\rangle \subset E$. Here, the complete metrizability is required for $\mathrm{cl}\langle f(A)\rangle$ rather than E itself. As a result, every complete convex set in a locally convex metric linear space is an AE for paracompact spaces.

A polyhedron (with the Whitehead topology) is another example of an ANE. To prove this, we need the following lemma:

Lemma 6.1.3. *Let X be metrizable and $A \subset X$. Then, each open set U in A can be extended to an open set $E(U)$ in X so that*

(1) $E(\emptyset) = \emptyset$, $E(A) = X$;
(2) $E(U) \cap A = U$;
(3) $U \subset V \Rightarrow E(U) \subset E(V)$;
(4) $E(U \cap V) = E(U) \cap E(V)$.

Proof. Using $d \in \mathrm{Metr}(X)$, for each open set U in A, we define

$$E(U) = \{x \in X \mid d(x, U) < d(x, A \setminus U)\},$$

where $d(x, \emptyset) = \infty$. Then, $E(U)$ is an open set in X. Conditions (1), (2), and (3) are obvious. We show $E(U) \cap E(V) \subset E(U \cap V)$, which implies (4) by combining (3). For each $x \in E(U) \cap E(V)$, it suffices to prove

$(*)$ $d(x, U \cap V) < d(x, A \setminus (U \cap V)) = \min\{d(x, A \setminus U),\ d(x, A \setminus V)\}$.

Then, without loss of generality, we may assume that

$$d(x, A \setminus (U \cap V)) = d(x, A \setminus U) \leq d(x, A \setminus V).$$

Since $x \in E(U)$, there exists a $y \in U$ such that $d(x, y) < d(x, A \setminus U)$. Then, it follows that $d(x, y) < d(x, A \setminus V)$, which implies that $y \in V$. Hence, $y \in U \cap V$, so we have the inequality $(*)$. □

Theorem 6.1.4. *The polyhedron $|K|$ of any simplicial complex K is an ANE.*

Proof. Let X be a metrizable space and $f : A \to |K|$ be a map from a closed set A in X to $|K|$. Let $(\beta_v(x))_{v\in K^{(0)}}$ be the barycentric coordinate of $x \in |K|$. Since $\beta_v : |K| \to \mathbf{I}$ is continuous, each $U_v = f^{-1}(O_K(v))$ is open in A, where $O_K(v) = \beta_v^{-1}((0,1])$ is the open star at v with respect to K. We show that $\{U_v \mid v \in K^{(0)}\}$ is locally finite in A. Assume that it is not locally finite at $x_0 \in A$. Since $\{v \in K^{(0)} \mid x_0 \in U_v\} = c_K(f(x_0))^{(0)}$ is finite, we can choose $x_1, x_2, \cdots \in A$ and $v_1, v_2, \cdots \in K^{(0)}$ so that $x_0 \notin U_{v_n}$, $x_n \in U_{v_n}$, $\lim_{n\to\infty} x_n = x_0$, and $v_n \neq v_m$ if $n \neq m$. Then, $f(x_0) = \lim_{n\to\infty} f(x_n)$. On the other hand, each $\sigma \in K$ does not meet $O_K(v)$ except for $v \in \sigma^{(0)}$, so it contains only finitely many $f(x_n)$, which implies that $\{f(x_n) \mid n \in \mathbb{N}\}$ is closed in $|K|$. Hence, $f(x_0) = f(x_n)$ for some $n \in \mathbb{N}$, which is contrary to $x_0 \notin U_{v_n}$.

Now, for each $x \in A$, choose an open neighborhood V_x of x in A so that $V_x \cap U_v = \emptyset$ except for finitely many $v \in K^{(0)}$. Let V be an open neighborhood of A in X such that $\operatorname{cl} V \subset \bigcup_{x\in A} E(V_x)$, where $E(U)$ is defined in Lemma 6.1.3. For each $v \in K^{(0)}$, we define $\tilde{U}_v = E(U_v) \cap V$. Then, $\{\tilde{U}_v \mid v \in K^{(0)}\}$ is locally finite in X. We can apply the Tietze Extension Theorem 2.2.2 to construct a map $g_v : X \to \mathbf{I}$ such that $g_v|A = \beta_v \circ f$ and $g_v(X \setminus \tilde{U}_v) = 0$. Then, $g = \sum_{v\in K^{(0)}} g_v : X \to \mathbb{R}_+$ is well-defined and continuous. Since $g(x) = \sum_{v\in K^{(0)}} \beta_v f(x) = 1$ for each $x \in A$, we have an open neighborhood $W = g^{-1}((0,\infty))$ of A in X. We can extend f to $\tilde{f} : W \to |K|$ by

$$\tilde{f}(x) = \sum_{v\in K^{(0)}} \frac{g_v(x)}{g(x)} v \quad \left(\text{i.e., } \beta_v(\tilde{f}(x)) = \frac{g_v(x)}{g(x)}\right).$$

Then, $\tilde{f}$ is well-defined and continuous. Indeed, for each $x \in W$, only finitely many $g_v(x)$ are nonzero, say

$$\{v \in K^{(0)} \mid g_v(x) \neq 0\} = \{v_1, \cdots, v_n\}.$$

Then, it follows that

$$E\left(\textstyle\bigcap_{i=1}^n U_{v_i}\right) = E(U_{v_1}) \cap \cdots \cap E(U_{v_n}) \supset \tilde{U}_{v_1} \cap \cdots \cap \tilde{U}_{v_n} \neq \emptyset,$$

which implies that $\bigcap_{i=1}^n U_{v_i} \neq \emptyset$, i.e., $\bigcap_{i=1}^n O_K(v_i) \neq \emptyset$. Therefore, $\tilde{f}(x) \in \langle v_1, \cdots, v_n\rangle \in K$. Thus, we have $\tilde{f}(x) \in |K|$. Each $x \in W$ has a neighborhood W_x in W that meets only finitely many $g_v^{-1}((0,\infty))$, $v \in K^{(0)}$, i.e., $\tilde{f}(W_x) \subset |L|$ for a finite subcomplex L of K. Since $\tilde{f}|V_x : V_x \to |L|_{\mathrm{m}} = |L|$ is continuous, $\tilde{f} : W \to |K|$ is also continuous. □

Remark 5. Since every metrizable ANE is an ANR, it follows that $|K|$ $(= |K|_{\mathrm{m}})$ is an ANR for every locally finite simplicial complex K. If K is not locally finite, then there are many maps from a compactum to $|K|_{\mathrm{m}}$ that are not continuous with respect to the Whitehead topology. Indeed, K contains a subcomplex L such that $|L|_{\mathrm{m}}$ is homeomorphic to the hedgehog

$$J(\mathbb{N}) = \big\{x \in \ell_1 \bigm| x(i) \in \mathbf{I} \text{ for all } i \in \mathbb{N} \text{ and}$$
$$x(i) \neq 0 \text{ at most one } i \in \mathbb{N}\big\} \subset \ell_1.$$

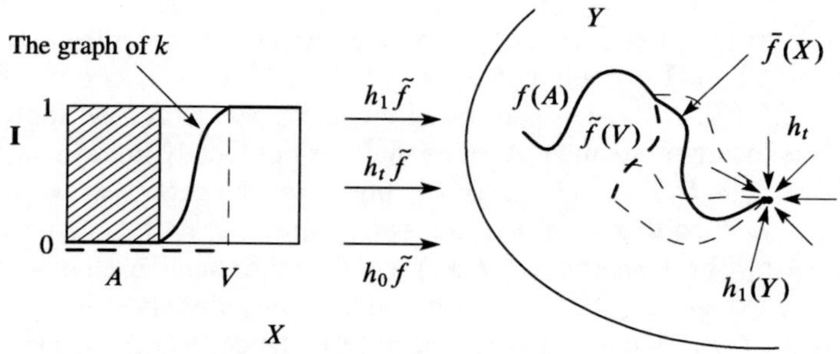

Fig. 6.1 Extending a map using a contraction

It is easy to construct a map $f : \mathbf{I} \to |L|_{\mathrm{m}} = J(\mathbb{N})$ such that $f(1/i) = 2^{-i}\mathbf{e}_i$ for each $i \in \mathbb{N}$. Then, $f : \mathbf{I} \to |L|$ $(\subset |K|)$ is not continuous because $f(\mathbf{I})$ is not contained in the polyhedron of any finite subcomplex of L. In the next section, we will show that $|K|_{\mathrm{m}}$ is an ANR for an arbitrary simplicial complex K (Theorem 6.2.6).

Let X be a space and $A \subset X$. It is said that A is **contractible** in X if the inclusion map $A \subset X$ is null-homotopic, i.e., there is a homotopy $h : A \times \mathbf{I} \to X$ such that $h_0 = \mathrm{id}$ and h_1 is constant. Such a homotopy h is called a **contraction** of A in X. If $A = X$, we simply say that X is **contractible** and h is called a **contraction** of X. A space X is contractible if and only if X has the homotopy type of a singleton.

Theorem 6.1.5. *A contractible ANE is an AE.*

Proof. Let X be a metrizable space, A be a closed set in X, and Y be an ANE with a contraction h. Each map $f : A \to Y$ extends to a map $\tilde{f} : U \to Y$ from a neighborhood U of A in X. Choose an open neighborhood V of A in X so that $\operatorname{cl} V \subset U$, and let $k : X \to \mathbf{I}$ be an Urysohn map with $k(A) = 0$ and $k(X \setminus V) = 1$. We can extend f to a map $\bar{f} : X \to Y$ by $\bar{f}(X \setminus V) = h_1(Y)$ and

$$\bar{f}(x) = h(\tilde{f}(x), k(x)) \quad \text{for each } x \in \operatorname{cl} V.$$

See Fig. 6.1. □

By Theorem 5.1.6(2) and the No Retraction Theorem 5.1.5, the unit sphere $\mathbf{S}^n$ is an ANE but not an AE. Then, we have the following corollary:

Corollary 6.1.6. *For every $n \in \omega$, the unit sphere $\mathbf{S}^n$ is not contractible.*

For any full complex K, $|K|$ is contractible. Indeed, fixing $v_0 \in K^{(0)}$, we can define a contraction $h : |K| \times \mathbf{I} \to |K|$ by

$$\beta_v(h(x,t)) = (1-t)\beta_v(x) + t\beta_v(v_0), \ v \in K^{(0)}.$$

Thus, we have the following:

Corollary 6.1.7. *The polyhedron* $|K|$ *of a full complex* K *is an AE.* □

Recall that a tower $X_1 \subset X_2 \subset \cdots$ of spaces is said to be **closed** if each X_n is closed in X_{n+1}.

Theorem 6.1.8. *For a closed tower* $Y_1 \subset Y_2 \subset \cdots$ *of ANEs, the direct limit* $Y = \varinjlim Y_n$ *is also an ANE.*

Proof. Let $f : A \to Y$ be a map from a closed set A in a metric space $X = (X, d)$. For each $n \in \mathbb{N}$, let $A_n = f^{-1}(Y_n)$ and define

$$X_n = \{x \in X \mid d(x, A_n) \leq d(x, A \setminus A_n)\}.$$

Then, each X_n is closed in X and $X_n \cap A = A_n$. By induction, we can obtain $N_1 \subset N_2 \subset \cdots \subset X$ and maps $f_n : A \cup N_n \to Y$ such that N_n is a closed neighborhood of A_n in X_n, $f_n(N_n) \subset Y_n$, and $f_n|A \cup N_{n-1} = f_{n-1}$, where $N_0 = \emptyset$, $f_0 = f$. Indeed, assume that f_n has been obtained. Since Y_{n+1} is an ANE and $A_{n+1} \cup N_n$ is a closed subset of X_{n+1}, $A_{n+1} \cup N_n$ has a closed neighborhood N_{n+1} in X_{n+1} and $f_n|A_{n+1} \cup N_n : A_{n+1} \cup N_n \to Y_{n+1}$ extends to a map $f' : N_{n+1} \to Y_{n+1}$. The map f_{n+1} is defined by $f_{n+1}|A = f$ and $f_{n+1}|N_{n+1} = f'$. Let $N = \bigcup_{n\in\mathbb{N}} \operatorname{int}_X N_n$ and define $\tilde{f} : N \to Y$ by $\tilde{f}|\operatorname{int}_X N_n = f_n|\operatorname{int}_X N_n$. If $A \subset \bigcup_{n\in\mathbb{N}} \operatorname{int}_X N_n$, then N is a neighborhood of A in X and the continuity of $\tilde{f}$ follows from that of each f_n.

Since each N_n is a neighborhood of A_n in X_n, it suffices to show that $A \subset \bigcup_{n\in\mathbb{N}} \operatorname{int}_X X_n$. On the contrary, assume that there is an $a \in A \setminus \bigcup_{n\in\mathbb{N}} \operatorname{int}_X X_n$. Then, a is contained in some A_k. For each $i \in \mathbb{N}$, we can choose $x_i \in X \setminus X_{k+i}$ so that $\lim_{n\to\infty} x_i = a$. By the definition of X_i, $d(x_i, a_i) < d(x_i, A_{k+i})$ for some $a_i \in A \setminus A_{k+i}$. Since $d(x_i, a_i) < d(x_i, a)$, it follows that $\lim_{n\to\infty} a_i = a$, hence $f(a) = \lim_{i\to\infty} f(a_i)$. On the other hand, $\{f(a_i) \mid i \in \mathbb{N}\}$ is closed in Y because each Y_{k+i} contains only finitely many points $f(a_j)$. Therefore, $f(a) = f(a_i)$ for some $i \in \mathbb{N}$, hence $f(a) \notin Y_{k+i}$, which is a contradiction. Thus, the proof is complete. □

Now, we list the basic properties of ANEs, which can be easily proved.

Basic Properties of ANEs 6.1.9.

(1) An arbitrary product of AEs is an AE and a finite product of ANEs is an ANE.

Sketch of Proof. Extend coordinate-wise.

(2) A retract of an AE is an AE and a neighborhood retract of an ANE is an ANE.

Sketch of Proof. Compose an extension with a retraction.

(3) Any open set in an ANE is also an ANE.

Sketch of Proof. Restrict an extension.

(4) A topological sum of ANEs is an ANE.

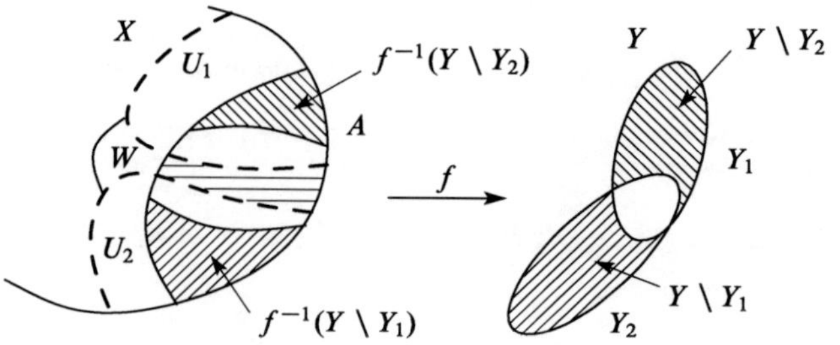

Fig. 6.2 The union of two open ANEs

Sketch of Proof. Let $f : A \to \bigoplus_{\lambda \in \Lambda} Y_\lambda$ be a map from a closed set A in a metrizable space X, where each Y_λ is an ANE. Extend each $f|f^{-1}(Y_\lambda)$ over a neighborhood in $E(f^{-1}(Y_\lambda))$, where E is the extension operator in Lemma 6.1.3.

(5) Let $Y = Y_1 \cup Y_2$, where Y_i is open in Y. If Y_1 and Y_2 are ANEs, then so is Y.

Sketch of Proof. Let $f : A \to Y$ be a map of a closed set $A \subset X$. Separate $f^{-1}(Y_1 \setminus Y_2)$ and $f^{-1}(Y_2 \setminus Y_1)$ by open sets U_1 and U_2 in X. By extending $f| : A \setminus (U_1 \cup U_2) \to Y_1 \cap Y_2$ over a closed neighborhood W in $X \setminus (U_1 \cup U_2)$, we first extend f to $f' : A \cup W \to Y$. Then, extending $f'| : (A \cap U_i) \cup W \to Y_i$ over a neighborhood in $U_i \cup W$ ($i = 1, 2$), we can obtain an extension of f. Refer to Fig. 6.2.

(6) (Hanner's Theorem) A paracompact space is an ANE if it is locally an ANE, i.e., each point has an ANE neighborhood.

Sketch of Proof. Applying Michael's Theorem 2.6.5 on local properties, the proof follows from (3), (4), and (5).

In the above, (4) and (5) are special cases of (6). Combining (6) with Theorem 6.1.1, every n-manifold (with boundary) is an ANE.[1] Due to 2.6.7(4), it is metrizable, hence it is an ANR.

(7) Let $Y = Y_1 \cup Y_2$, where Y_i is closed in Y. If Y_1, Y_2, and $Y_1 \cap Y_2$ are ANEs (AEs) then so is Y. If Y and $Y_1 \cap Y_2$ are ANEs (AEs), then so are Y_1 and Y_2.

Sketch of Proof. The first assertion is similar to (5). Now, U_1 and U_2 are disjoint open sets such that $U_1 \cap A = f^{-1}(Y_1 \setminus Y_2)$ and $U_2 \cap A = f^{-1}(Y_2 \setminus Y_1)$ (Fig. 6.3).

For the second assertion, let $f : A \to Y_1$ be a map of a closed set $A \subset X$. First, extend f to a map $\tilde{f} : U \to Y$ of a neighborhood U of A in X. Then, extending $\tilde{f}| : \tilde{f}^{-1}(Y_1 \cap Y_2) \to Y_1 \cap Y_2$ over a neighborhood U_0 of $\tilde{f}^{-1}(Y_1 \cap Y_2)$ in $\tilde{f}^{-1}(Y_2)$, we can extend $\tilde{f}|\tilde{f}^{-1}(Y_1)$ over $U_1 = \tilde{f}^{-1}(Y_1) \cup U_0$. See Fig. 6.4.

[1] A paracompact space M is called an ***n*-manifold (possibly with boundary)** if each point has a neighborhood that is homeomorphic to an open set in $\mathbf{I}^n$. The **boundary** ∂M of M is the subset of M consisting of all points with no neighborhood homeomorphic to an open set in $\mathbb{R}^n$. The **interior** $\mathrm{Int} M = M \setminus \partial M$ is an n-manifold without boundary. It is known that the boundary ∂M is an $(n-1)$-manifold without boundary.

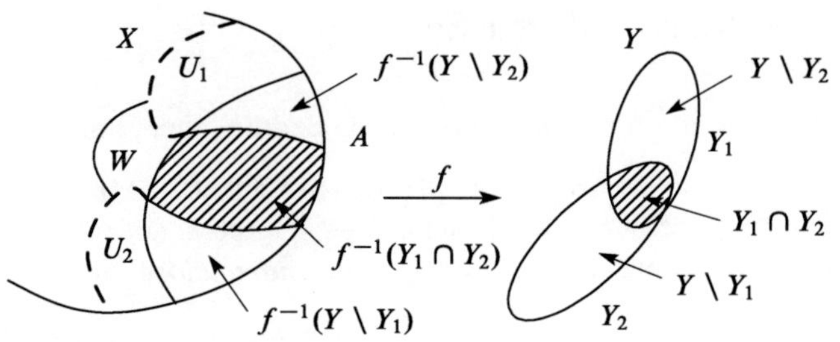

Fig. 6.3 The union of two closed ANEs

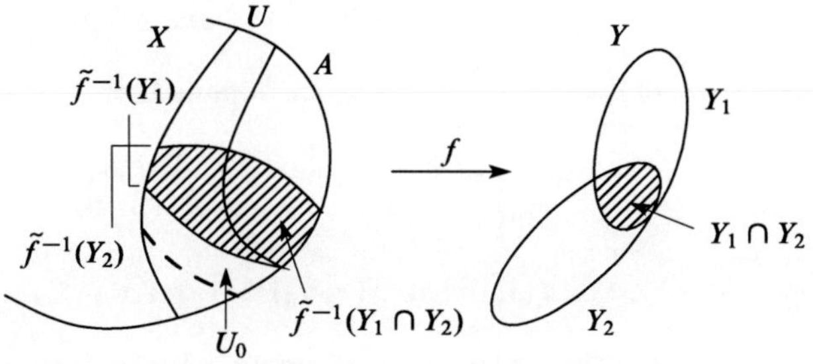

Fig. 6.4 Two closed sets whose union and intersection are ANEs

(7) For a compactum X and an ANE (resp. AE) Y, the space $\mathrm{C}(X, Y)$ of all maps from X to Y with the compact-open topology is an ANE (resp. AE).

Sketch of Proof. Let Z be a metrizable space with C a closed set. A map $f : C \to \mathrm{C}(X, Y)$ induces a map $g : C \times X \to Y$ by 1.1.3(4). Then, g extends over a neighborhood W of $C \times X$ in $Z \times X$. By the compactness of X, $U \times X \subset W$ for some neighborhood U of C in Z. The extension of g induces a map from U to $\mathrm{C}(X, Y)$ by Proposition 1.1.1.

(8) For a pair (X, A) of compacta and a pair (Y, B) of ANEs, the space $\mathrm{C}((X, A), (Y, B))$ of all maps from (X, A) to (Y, B) with the compact-open topology is an ANE, where B is not necessarily closed in Y.

Sketch of Proof. Let Z be a metrizable space with C a closed set. A map $f : C \to \mathrm{C}((X, A), (Y, B))$ induces a map $g : (C \times X, C \times A) \to (Y, B)$ (cf. 1.1.3(4)). As in (8), extend $g|C \times A$ to a map $g' : V \times A \to B$, where V is a closed neighborhood of $C \times A$ in $Z \times A$. Define $g'' : (C \times X) \cup (V \times A) \to Y$ by $g''|C \times X = g$ and $g''|V \times A = g'$. Next, extend g'' over a neighborhood W of $(C \times X) \cup (V \times A)$ in $Z \times X$ and find a neighborhood U of C in Z so that $U \subset V$ and $U \times X \subset W$. The restriction of this extension of g'' to $U \times X$ induces an extension $\tilde{f} : U \to \mathrm{C}((X, A), (Y, B))$ of f (cf. Proposition 1.1.1).

6.2 Embeddings of Metric Spaces and ANRs

In this section, we first prove the following embedding theorem:

Theorem 6.2.1 (ARENS–EELLS EMBEDDING THEOREM). *Every (complete) metric space X can be isometrically embedded in a (complete) normed linear space E with* $\operatorname{dens} E = \aleph_0 \operatorname{dens} X$ *as a linearly independent closed set.*

Proof. When X is finite, let $X' = X \cup \{y_i \mid i \in \mathbb{N}\}$, where $y_i \notin X$ and $y_i \neq y_j$ if $i \neq j$. The metric d of X can be extended to a metric on X' by defining

$$d(y_i, y_j) = 1 \ \text{ for } i \neq j \text{ and } \ d(x, y_i) = \operatorname{diam} X \ \text{ for } x \in X.$$

The linear span of X in the normed linear space obtained for X' is the desired one. Thus, we may assume that X is infinite.

Let $X^* = X \cup \{y_0\}$, where $y_0 \notin X$. Fix a point $x_0 \in X$ and extend the metric d of X to a metric on X^* by defining

$$d(x, y_0) = d(y_0, x) = d(x, x_0) + 1 \ \text{ for } x \in X.$$

By $\operatorname{Fin}(X^*)$, we denote the set of all non-empty finite subsets of X^*, and by $\ell_\infty(\operatorname{Fin}(X^*))$ the Banach space of all bounded real functions on $\operatorname{Fin}(X^*)$ with the sup-norm $\|\cdot\|$. For each $x \in X^*$, we define $\varphi(x) \in \ell_\infty(\operatorname{Fin}(X^*))$ by

$$\varphi(x)(F) = d(x, F) - d(y_0, F) \ \text{ for } \ F \in \operatorname{Fin}(X^*),$$

where $|\varphi(x)(F)| \leq d(x, y_0) = d(x, x_0) + 1$. Then, $\|\varphi(x) - \varphi(y)\| = d(x, y)$ for each $x, y \in X^*$. Indeed, for each $F \in \operatorname{Fin}(X^*)$,

$$\begin{aligned}|\varphi(x)(F) - \varphi(y)(F)| &= |d(x, F) - d(y, F)| \\ &\leq d(x, y) = |\varphi(x)(\{y\}) - \varphi(y)(\{y\})|.\end{aligned}$$

Thus, we have an isometry $\varphi : X^* \to \ell_\infty(\operatorname{Fin}(X^*))$, where it should be noted that $\varphi(y_0) = 0$. Let E be the linear subspace of $\ell_\infty(\operatorname{Fin}(X^*))$ spanned by $\varphi(X)$, i.e.,

$$E = \left\{ \textstyle\sum_{i=1}^n \lambda_i \varphi(x_i) \;\middle|\; n \in \mathbb{N},\ \lambda_i \in \mathbb{R} \ \text{ and } \ x_i \in X \right\}.$$

We show that $\operatorname{dens} E = \operatorname{dens} X$ when X is infinite. Take a dense set D in X with $\operatorname{card} D = \operatorname{dens} X$, and define

$$E_D = \left\{ \textstyle\sum_{i=1}^n \lambda_i \varphi(x_i) \;\middle|\; n \in \mathbb{N},\ \lambda_i \in \mathbb{Q} \ \text{ and } \ x_i \in D \right\}.$$

Then, E_D is dense in E and $\operatorname{card} E_D = \aleph_0 \operatorname{card} D = \operatorname{dens} X$.

We show that $\varphi(X)$ is closed in E. Let $f = \sum_{i=1}^m \lambda_i \varphi(y_i) \in E$ and $x_n \in X$, $n \in \mathbb{N}$, such that $\varphi(x_n)$ converges to f. Let $F = \{y_0, y_1, \cdots, y_m\} \in \operatorname{Fin}(X^*)$. Then, $d(x_n, F) = \varphi(x_n)(F)$ converges to

$$f(F) = \sum_{i=1}^{m} \lambda_i \varphi(y_i)(F) = \sum_{i=1}^{m} \lambda_i d(y_i, F) = 0.$$

Since F is finite and $d(y_0, X) = 1$, $(x_n)_{n \in \mathbb{N}}$ has a subsequence that converges to some $y \in F \setminus \{y_0\} \subset X$, so $f = \varphi(y) \in \varphi(X)$. Therefore, $\varphi(X)$ is closed in E. If X is complete, $\varphi(X)$ is closed in $\ell_\infty(\mathrm{Fin}(X^*))$, so is closed in the closure $\mathrm{cl}\, E$ of E in $\ell_\infty(\mathrm{Fin}(X^*))$.

Finally, we show that $\varphi(X)$ is linearly independent. Let $\sum_{i=1}^{m} \lambda_i \varphi(x_i) = 0$, where $\lambda_i \in \mathbb{R}$, $x_i \in X$, and $x_i \neq x_j$ if $i \neq j$. Then,

$$\sum_{i=1}^{m} \lambda_i (d(x_i, F) - d(y_0, F)) = 0 \quad \text{for each } F \in \mathrm{Fin}(X^*).$$

For each $i = 1, \ldots, m$, applying this to

$$F_i = \{y_0, x_1, \cdots, x_m\} \setminus \{x_i\} \in \mathrm{Fin}(X^*),$$

we have $\lambda_i d(x_i, F_i) = 0$, hence $\lambda_i = 0$ because $d(x_i, F_i) \neq 0$. □

Remark 6. Due to Theorem 2.3.9, every metric space X can be embedded into the Banach space $\mathrm{C}^B(X)$ by the isometry $\varphi : X \to \mathrm{C}^B(X)$ defined as above, i.e., $\varphi(x)(z) = d(x, z) - d(x_0, z)$, where $x_0 \in X$ is fixed. In this case, $\varphi(X)$ is closed in the convex hull $\langle \varphi(X) \rangle$.

Indeed, for each $f \in \langle \varphi(X) \rangle \setminus \varphi(X)$, write $f = \sum_{i=1}^{n} t_i \varphi(x_i)$, where $x_i \in X$, $t_i \in \mathbf{I}$ with $\sum_{i=1}^{n} t_i = 1$. Choose $\delta > 0$ so that $\|f - \varphi(x_i)\| > \delta$ for every $i = 1, \ldots, n$. To prove $\mathrm{B}(f, \delta) \cap \varphi(X) = \emptyset$, assume the contrary, i.e., $\|f - \varphi(x)\| \leq \delta$ for some $x \in X$. Then, $\sum_{i=1}^{n} t_i d(x_i, x) = |f(x) - \varphi(x)(x)| \leq \delta$, which implies that $d(x, x_i) \leq \delta$ for some $i = 1, \ldots, n$. On the other hand, $|f(x)(z) - \varphi(x_i)(z)| > \delta$ for some $z \in X$. However, $|f(x)(z) - \varphi(x_i)(z)| \leq \left| \sum_{i=1}^{n} t_i (d(x_i, z) - d(x, z)) \right| \leq d(x, x_i) \leq \delta$, which is a contradiction.

The following is a very useful procedure to extend homeomorphisms.

Theorem 6.2.2 (KLEE'S TRICK). *Let E and F be metrizable topological linear spaces that are AEs and let A and B be closed sets in E and F, respectively. Then, each homeomorphism $f : A \times \{0\} \to \{0\} \times B$ extends to a homeomorphism $\tilde{f} : E \times F \to E \times F$.*

Proof. Let $i_E : E \to E \times \{0\} \subset E \times F$ and $i_F : F \to \{0\} \times F \subset E \times F$ be the natural injections. Then, $\mathrm{pr}_F \circ f \circ i_E | A : A \to F$ and $\mathrm{pr}_E \circ f^{-1} \circ i_F | B : B \to E$ extend to maps $g_1 : E \to F$ and $g_2 : F \to E$, respectively. We define homeomorphisms $f_1, f_2 : E \times F \to E \times F$ as follows:

$$f_1(x, y) = (x, y + g_1(x)) \quad \text{and} \quad f_2(x, y) = (x + g_2(y), y).$$

Then, the homeomorphism $\tilde{f} = f_2^{-1} \circ f_1 : E \times F \to E \times F$ is an extension of f (cf. Fig. 6.5). Indeed, for each $x \in A$, $f(x, 0) = (0, \mathrm{pr}_F f(x, 0))$ and then

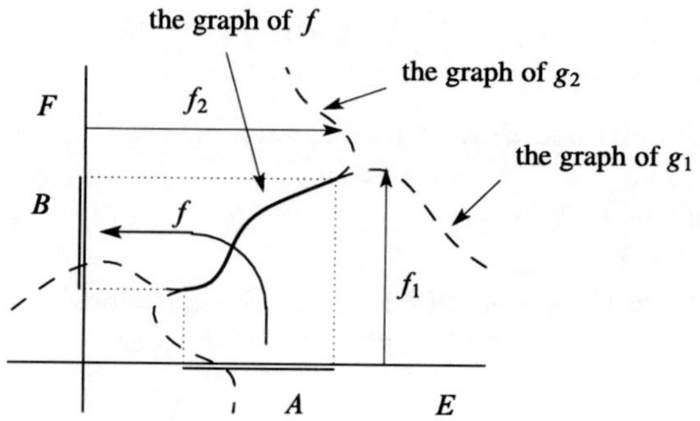

Fig. 6.5 Klee's trick

$$
\begin{aligned}
\tilde{f}(x,0) &= f_2^{-1}(x, g_1(x)) = (x - g_2(g_1(x)), g_1(x)) \\
&= (x - \mathrm{pr}_E \circ f^{-1}(0, \mathrm{pr}_F \circ f(x,0)), \mathrm{pr}_F \circ f(x,0)) \\
&= (x - \mathrm{pr}_E \circ f^{-1}(f(x,0)), \mathrm{pr}_F \circ f(x,0)) \\
&= (0, \mathrm{pr}_F \circ f(x,0)) = f(x,0).
\end{aligned}
$$

□

As a corollary, we prove the following metric extension theorem:

Theorem 6.2.3 (HAUSDORFF'S METRIC EXTENSION THEOREM). *Let A be a closed set in a (completely) metrizable space X. Every admissible (complete) metric on A extends to an admissible (complete) metric on X.*

Proof. By Theorem 6.2.1, we have a closed embedding $g : X \to E$ of X into a (complete) normed linear space $E = (E, \|\cdot\|_E)$. Let d be an admissible (complete) metric on A. Again, by Theorem 6.2.1, we have a closed isometry $h : A \to F$ of $A = (A, d)$ into a (complete) normed linear space $F = (F, \|\cdot\|_F)$. Since E and F are AEs by the Dugundji Extension Theorem 6.1.1, we can apply Theorem 6.2.2 to obtain a homeomorphism $f : E \times F \to E \times F$ such that

$$f(g(x), 0) = (0, h(x)) \quad \text{for all } x \in A.$$

Let $i : E \to E \times \{0\} \subset E \times F$ be the natural injection. Then, $f \circ i \circ g : X \to E \times F$ is a closed embedding of X into the product normed linear space $E \times F$ with the norm

$$\|(x, y)\| = \|x\|_E + \|y\|_F.$$

Since $f \circ i \circ g | A$ is an isometry with respect to d and $\|\cdot\|$, we can extend d to a (complete) metric $\tilde{d}$ on X as follows:

$$\tilde{d}(x, y) = \|f \circ i \circ g(x) - f \circ i \circ g(y)\|.$$

□

In Theorem 6.2.1, if an embedding is not required to be an isometry, we have the following:

Theorem 6.2.4. *Every completely metrizable space can be embedded in a Hilbert space with the same density as a closed set. Moreover, every metrizable space can be embedded in a pre-Hilbert space (that is, a linear subspace of a Hilbert space) with the same density as a closed set.*

Proof. Let $X = (X, d)$ be a metric space. We may assume that X is infinite. For each $n \in \mathbb{N}$, X has a locally finite partition of unity $(f_\gamma)_{\gamma \in \Gamma(n)}$ such that $\operatorname{diam} \operatorname{supp} f_\gamma < 2^{-n}$ for each $\gamma \in \Gamma(n)$, which implies that $f_\gamma(x) f_\gamma(y) = 0$ if $d(x, y) \geq 2^{-n}$. Let $\Gamma = \bigcup_{n \in \mathbb{N}} \Gamma(n)$, where it can be assumed that $\Gamma(n) \cap \Gamma(m) = \emptyset$ if $n \neq m$. For each $\gamma \in \Gamma$, let $g_\gamma = 2^{-n} f_\gamma$, where $\gamma \in \Gamma(n)$. We define $h : X \to \ell_2(\Gamma)$ by $h(x) = (g_\gamma(x)^{\frac{1}{2}})_{\gamma \in \Gamma}$. Observe that $h(X)$ is contained in the unit sphere of $\ell_2(\Gamma)$. Then, the continuity of h follows from the continuity of g_γ, $\gamma \in \Gamma$ (Proposition 1.2.4). It is easy to show that

$$2^{-n} \leq d(x, y) \ (< 2^{-n+1}) \Rightarrow \|h(x) - h(y)\|^2 \geq 2^{-n+1} \ (> d(x, y)).$$

Hence, the inverse h^{-1} of h is uniformly continuous. Thus, h is an embedding. Since $(f_\gamma)_{\gamma \in \Gamma(n)}$ is locally finite, we have $\operatorname{card} \Gamma(n) \leq w(X) = \operatorname{dens} X$, hence $\operatorname{card} \Gamma \leq \operatorname{dens} X$. Observe

$$\operatorname{dens} X \leq \operatorname{dens} \ell_2(\Gamma) = \operatorname{card} \Gamma \leq \operatorname{dens} X,$$

so we have $\operatorname{dens} \ell_2(\Gamma) = \operatorname{dens} X$. When X is complete, $h(X)$ is also complete, so closed in $\ell_2(\Gamma)$.

When X is not complete, let E be the linear subspace of $\ell_2(\Gamma)$ generated by $h(X)$. We now show that $h(X)$ is closed in E. Let $y \in \operatorname{cl}_E h(X)$. Then, $y = \sum_{i=1}^m t_i h(x_i)$ for some $x_i \in X$ and $t_i \in \mathbb{R}$. We may assume that $x_i \neq x_j$ if $i \neq j$ and $t_i \neq 0$ for all i. Suppose that $m > 1$. Choose $n \in \mathbb{N}$ so that $d(x_i, x_j) \geq 2^{-n+1}$ if $i \neq j$. For each $i = 1, \ldots, m$, we have $\gamma_i \in \Gamma(n)$ such that $f_{\gamma_i}(x_i) > 0$, i.e., $h(x_i)(\gamma_i) > 0$. If $i \neq j$ then $f_{\gamma_i}(x_j) = 0$, i.e., $h(x_j)(\gamma_i) = 0$. Hence, $y(\gamma_i) > 0$. Since $y \in \operatorname{cl} h(X)$, we have $x \in X$ such that $h(x)(\gamma_i) > 0$, i.e., $f_{\gamma_i}(x) > 0$ for all i. This implies that $d(x, x_i) < 2^{-n}$ for each i, hence $d(x_1, x_2) < 2^{-n+1}$, which is a contradiction. Therefore, $y = t_1 h(x_1)$. Observe that $y(\gamma) \geq 0$ for every $\gamma \in \Gamma$ and $\|y\| = 1$. Thus, we have $y = h(x_1)$. Hence, $h(X)$ is closed in E. □

Note. By Corollary 3.8.12 and the Arens–Eells Embedding Theorem 6.2.1, every completely metrizable space X can be embedded in $\ell_1(\Gamma)$ as a closed set, where $\operatorname{card} \Gamma = \operatorname{dens} X$. It is known that $\ell_1(\Gamma) \approx \ell_2(\Gamma)$. Thus, we have an alternative proof for Theorem 6.2.4.

As is easily observed, a metrizable AE (resp. ANE) is an AR (resp. ANR). The converse is also true by the Arens–Eells Embedding Theorem 6.2.1 (or Theorem 6.2.4), the Dugundji Extension Theorem 6.1.1, and 6.1.9(2). Thus, we have the following theorem:

Theorem 6.2.5. *For a metrizable space X, the following hold:*

(1) *X is an AR if and only if X is an AE.*
(2) *X is an ANR if and only if X is an ANE.* □

The following theorem corresponds to Theorem 6.1.4:

Theorem 6.2.6. *For any simplicial complex K, $|K|_{\mathrm{m}}$ is an ANR.*

Proof. We can regard K as a subcomplex of the full simplicial complex F with $F^{(0)} = K^{(0)}$. Note that the canonical representation $\beta^F : |F| \to \ell_1(F^{(0)})$ is an isometry with respect to the metric ρ_F (cf. Sect. 4.5) and $\beta^F(|F|)$ is a convex set in $\ell_1(F^{(0)})$. Then, $|F|_{\mathrm{m}}$ is an AE by Theorem 6.1.1. Combining 6.1.9(2) with Theorem 6.2.5(2), it suffices to show that $|K|_{\mathrm{m}}$ is a neighborhood retract of $|F|_{\mathrm{m}}$ (refer to 6.2.10(2)).

By Theorem 4.6.7, $|K|_{\mathrm{m}} = |\operatorname{Sd}^2 K|_{\mathrm{m}}$ and $|F|_{\mathrm{m}} = |\operatorname{Sd}^2 F|_{\mathrm{m}}$ as spaces. Let

$$N = \bigcup_{v \in (\operatorname{Sd}^2 K)^{(0)}} \operatorname{St}(v, \operatorname{Sd}^2 F).$$

Then, N is a subcomplex of $\operatorname{Sd}^2 F$ and $|N|$ is a neighborhood of $|K|$ in $|F|_{\mathrm{m}}$. Each vertex $v \in N^{(0)}$ is the barycenter of $\sigma \in \operatorname{Sd} F$ with $\sigma \cap |K| \neq \emptyset$, where it should be noted that $\sigma \cap |K| = \langle \sigma^{(0)} \cap |K| \rangle \in \operatorname{Sd} K$. Let $r_0(v)$ be the barycenter of $\sigma \cap |K|$. Then, for $v_1, \ldots, v_n \in N^{(0)}$,

$$\langle v_1, \ldots, v_n \rangle \in N \Rightarrow \langle r_0(v_1), \ldots, r_0(v_n) \rangle \in \operatorname{Sd}^2 K.$$

Hence, the function $r_0 : N^{(0)} \to (\operatorname{Sd}^2 K)^{(0)}$ defines the simplicial map $r : N \to \operatorname{Sd}^2 K$. By definition, $r|\operatorname{Sd}^2 K = \mathrm{id}$, which means that r is a retraction. Thus, $|K|_{\mathrm{m}}$ is a neighborhood retract of $|F|_{\mathrm{m}}$. This completes the proof. □

Let X be a subspace of Y. A homotopy $h : X \times \mathbf{I} \to Y$ with $h_0 = \mathrm{id}$ is called a **deformation** of X in Y. When $X = Y$, h is called a **deformation** of X. A subset A of X is said to be a **deformation retract** of X if there is a homotopy $h : X \times \mathbf{I} \to X$ such that $h_0 = \mathrm{id}$ and h_1 is a retraction of X onto A, where h_1 is called a **deformation retraction** of X onto A. When $h_t|A = \mathrm{id}$ for all $t \in \mathbf{I}$, we call A a **strong deformation retract** of X and h_1 a **strong deformation retraction** of X onto A. A deformation retraction (resp. a strong deformation retraction) $r : X \to A$ ($\subset X$) is a retraction with $r \simeq \mathrm{id}_X$ (resp. $r \simeq \mathrm{id}_X$ rel. A). A *closed* set A in X is called a **neighborhood deformation retract** of X if A has a neighborhood U in X with a homotopy $h : U \times \mathbf{I} \to X$ such that $h_0 = \mathrm{id}$ and h_1 is a retraction of U onto A, where h_1 is called a **deformation retraction** of U onto A in X. When $h_t|A = \mathrm{id}$ for all $t \in \mathbf{I}$, we call A a **strong neighborhood deformation retract** of X and h_1 a **strong deformation retraction** of U onto A in X.

Proposition 6.2.7. *A retract of an AR is a strong deformation retract and a neighborhood retract of an ANR is a strong neighborhood deformation retract.*

Proof. Let X be an ANR and A be a closed set in X with $r : U \to A$ a retraction of an open neighborhood U of A in X. Since X is an ANE by Theorem 6.2.5, we have a neighborhood W of $U \times \{0,1\} \cup A \times \mathbf{I}$ in $U \times \mathbf{I}$ and a map $h : W \to X$ such that $h(x,0) = x$, $h(x,1) = r(x)$ for $x \in U$, and $h(x,t) = x$ for $x \in A$. Choose a neighborhood V of A in U so that $V \times \mathbf{I} \subset W$. Then, $h|V \times \mathbf{I}$ is the desired homotopy. When X is an AR, we can take $U = X$ and $W = X \times \mathbf{I}$ in the above. □

Remark 7. By Proposition 6.2.7, a deformation retract of an AR is a strong deformation retract. In Sect. 6.4, it will be proved that this is valid for any ANR, that is, a deformation retract of an ANR is a strong deformation retract (Theorem 6.4.4).

A space X is **locally contractible** if each neighborhood U of any point $x \in X$ contains a neighborhood of x that is contractible in U. The following proposition can be proved by letting $A = \{x\}$ in the proof of Proposition 6.2.7:

Proposition 6.2.8. *Every ANR is locally contractible and every AR is contractible.* □

By Theorems 6.1.5 and 6.2.5, a contractible ANR is an AR. Thus, we have the following characterization of ARs:

Corollary 6.2.9. *A metrizable space is an AR if and only if it is a contractible ANR.* □

By Theorem 6.2.5, we can translate 6.1.9 as follows:

Basic Properties of ANRs 6.2.10.

(1) A countable product of ARs is an AR and a finite product of ANRs is an ANR.

The metrizability requires the countable product.

(2) A retract of an AR is an AR and a neighborhood retract of an ANR is an ANR.
(3) Any open set in an ANR is also an ANR.
(4) (HANNER'S THEOREM) A paracompact space is an ANR if it is locally an ANR, that is, each point has an ANR neighborhood.

See the remark on 6.1.9(6). The metrizability of X follows from 2.6.7(4). Every n-manifold is an ANR.

(5) Let $X = X_1 \cup X_2$, where X_i is closed in X, $i = 1, 2$. If X_1, X_2, and $X_1 \cap X_2$ are ANRs (ARs) then so is X. If X and $X_1 \cap X_2$ are ANRs (ARs), then so are X_1 and X_2.

Sketch of Proof. In the first assertion, the metrizability of X follows from 2.4.5(2). The second assertion can also be proved by showing that X_1 and X_2 are neighborhood retracts of X (cf. (2)).

(6) The space $\mathrm{C}(X, Y)$ of all maps from a compactum X to an ANR (resp. an AR) Y with the compact-open topology is an ANR (resp. an AR).

For the metrizability of $\mathrm{C}(X, Y)$, refer to 1.1.3(6).

(7) For a locally compact separable metrizable space X and an AR Y, the space $\mathrm{C}(X, Y)$ of all maps from X to Y with the compact-open topology is an AR.

Sketch of Proof. Regard Y as a retract of a normed linear space E by the Arens–Eells Embedding Theorem 6.2.1. Then, $\mathrm{C}(X, Y)$ can be regarded as a retract of $\mathrm{C}(X, E)$ by 1.1.3(1), where $\mathrm{C}(X, E)$ is a locally convex topological linear space by Remark 1 in Sect. 6.1, and hence is an AE by the Dugundji Extension Theorem 6.1.1. Since $\mathrm{C}(X, E)$ is metrizable by 1.1.3(7), it is an AR, so $\mathrm{C}(X, Y)$ is an AR.

(8) There exists a locally compact separable metrizable space X such that the space $\mathrm{C}(X, \mathbf{S}^1)$ with the compact-open topology is not locally path-connected, so it is not an ANR (cf. 6.1.9(8)).

Example. Let $X = \bigcup_{i\in\mathbb{N}}(\mathbf{S}^1 + 2i\mathbf{e}_1)$ be a subspace of $\mathbb{R}^2$ and define a map $f : X \to \mathbf{S}^1$ by

$$f(x + 2i, y) = ((-1)^i x, y) \text{ for every } (x, y) \in \mathbf{S}^1.$$

For each $n \in \mathbb{N}$, let $X_n = \bigcup_{i=1}^n(\mathbf{S}^1 + 2i\mathbf{e}_1)$ and $r_n : X \to X_n$ be the retraction defined by $r_n(X \setminus X_n) = \{(2n+1)\mathbf{e}_1\}$. Each neighborhood $\mathcal{U}$ of f in $\mathrm{C}(X, \mathbf{S}^1)$ contains some fr_n. But $fr_n \not\simeq f$ in $\mathbf{S}^1$, which means that fr_n and f cannot be connected by any path in $\mathrm{C}(X, \mathbf{S}^1)$.

6.3 Small Homotopies and LEC Spaces

For a space X, let Δ_X denote the diagonal of X^2, that is,

$$\Delta_X = \{(x, x) \mid x \in X\} \subset X^2.$$

For each $A \subset X^2$ and $x \in X$, we define $A(x) = \{y \in X \mid (x, y) \in A\}$. Each neighborhood U of Δ_X in X^2 gives every $x \in X$ its neighborhood $U(x)$ simultaneously. Given an open cover $\mathcal{U}$ of X, we have an open neighborhood $W = \bigcup_{U\in\mathcal{U}} U^2$ of Δ_X in X^2, where $W(x) = \mathrm{st}(x, \mathcal{U})$ for each $x \in X$. Such open neighborhoods of Δ_X in X^2 form a neighborhood basis of Δ_X. Indeed, let U be an open neighborhood of Δ_X in X^2. Each $x \in X$ has an open neighborhood V_x in X such that $V_x^2 \subset U$. Thus, we have an open cover $\mathcal{V} = \{V_x \mid x \in X\}$ such that $\bigcup_{x\in X} V_x^2 \subset U$

A space X said to be **locally equi-connected** (**LEC**) if the diagonal Δ_X of X^2 has a neighborhood U and there is a map $\lambda : U \times \mathbf{I} \to X$ such that

$$\lambda(x, y, 0) = x \text{ and } \lambda(x, y, 1) = y \text{ for each } (x, y) \in U \text{ and}$$

$$\lambda(x, x, t) = x \text{ for each } x \in X \text{ and } t \in \mathbf{I},$$

where $\mathbf{I} \ni t \mapsto \lambda(x, y, t) \in X$ is a path from x to y in X. Such a map λ is called an **equi-connecting map** for X. If $U = X^2$, X is said to be **equi-connected** (**EC**). For example, every convex set X in a topological linear space is EC, where a natural EC map λ is defined by

$$\lambda(x, y, t) = (1 - t) \cdot x + t \cdot y.$$

In particular, every topological linear space is EC. More generally, a contractible topological group X is also EC and a semi-locally contractible topological group X is LEC, where X is said to be **semi-locally contractible** if each point of X has a neighborhood that is contractible in X. In fact, let $\varphi : V \times \mathbf{I} \to X$ be a contraction of a neighborhood of the unit $1 \in X$ ($V = X$ when X is contractible). Then, $U = \{(x, y) \mid x \cdot y^{-1} \in V\}$ is a neighborhood of Δ_X in X^2 ($U = X^2$ if $V = X$). We can define an equi-connecting map $\lambda : U \times \mathbf{I} \to X$ by

$$\lambda(x, y, t) = \varphi(1, t)^{-1} \cdot \varphi(x \cdot y^{-1}, t) \cdot y.$$

The following proposition is easily proved:

Proposition 6.3.1. *An AR is EC and an ANR is LEC.*

Sketch of Proof. An equi-connecting map λ for an AR (or an ANR) X can be obtained as an extension of the map of $(X^2 \times \{0, 1\}) \cup (\Delta_X \times \mathbf{I})$ to X defined by the conditions of an equi-connecting map.

The converse of Proposition 6.3.1 does not hold (cf. Remark 3 after Proof of Theorem 6.1.1; Sect. 7.12).

Now, we will characterize LEC spaces via the following theorem.

Theorem 6.3.2. *For an arbitrary space X,*

(1) *X is EC if and only if Δ_X is a strong deformation retract of X^2, and*
(2) *X is LEC if and only if Δ_X is a strong neighborhood deformation retract of X^2.*

Proof. To prove the "only if" part of both (1) and (2), let $\lambda : U \times \mathbf{I} \to X$ be an equi-connecting map for X, where $U = X^2$ for (1) or U is a neighborhood of Δ_X in X^2 for (2). Let $h : U \times \mathbf{I} \to X^2$ be the homotopy defined by $h(x, y, t) = (x, \lambda(x, y, 1 - t))$. Then, $h_0 = \mathrm{id}$, $h_t|\Delta_X = \mathrm{id}$ for each $t \in \mathbf{I}$, and h_1 is a retraction of U onto Δ_X.

To show the "if" part of (1) and (2), let $h : U \times \mathbf{I} \to X^2$ be a homotopy such that $h_0 = \mathrm{id}$, $h_t|\Delta_X = \mathrm{id}$ for each $t \in \mathbf{I}$ and h_1 is a retraction of U onto Δ_X, where $U = X^2$ or U is a neighborhood of Δ_X in responce to (1) or (2). An equi-connecting map $\lambda : U \times \mathbf{I} \to X$ for X can be defined as follows:

$$\lambda(x, y, t) = \begin{cases} \mathrm{pr}_1 h(x, y, 2t) & \text{for } 0 \le t \le \frac{1}{2}, \\ \mathrm{pr}_2 h(x, y, 2 - 2t) & \text{for } \frac{1}{2} \le t \le 1, \end{cases}$$

where $\mathrm{pr}_1, \mathrm{pr}_2 : X^2 \to X$ are the projections onto the first and the second coordinates, respectively. □

By the proof of Theorem 6.3.2, for every LEC space X, the diagonal Δ_X has a neighborhood U in X^2 with a pr_1-preserving homotopy $h : U \times \mathbf{I} \to X^2$ such that $h_0 = \mathrm{id}$, $h_1 : U \to \Delta_X$ is a retraction and $h_t|\Delta_X = \mathrm{id}$ for all $t \in \mathbf{I}$, where h is **pr_1-preserving** if $\mathrm{pr}_1 h_t = \mathrm{pr}_1|U$ for each $t \in \mathbf{I}$. Thus, every LEC space X has the following property:

(∗) Each neighborhood U of Δ_X in X^2 contains a neighborhood V of Δ_X with a pr_1-preserving homotopy $h : V \times \mathbf{I} \to U$ such that $h_0 = \mathrm{id}$ and $h_1(V) = \Delta_X$.

In the above (∗), for each $x \in X$, $U(x) = \{y \in X \mid (x, y) \in U\}$ is a neighborhood of x in X. We say that X is **unified locally contractible (ULC)**[2] when X has the above property (∗). A ULC space is locally contractible, but, as will be seen, the converse does not hold. As seen in the above, an LEC space is ULC but the converse is unknown.

Question If X is ULC, is X LEC?

Theorem 6.3.3. *Let X be a space such that X^2 is normal.*[3] *Then, X is EC if and only if X is contractible and LEC.*

Proof. An equi-connecting map $\lambda : X^2 \times \mathbf{I} \to X$ induces the contraction $h : X \times \mathbf{I} \to X$ defined by $h(x, t) = \lambda(x, x_0, t)$, where $x_0 \in X$ is fixed. Thus, we have proved the "only if" part.

To show the "if" part, let X be contractible and LEC. By Theorem 6.3.2, Δ_X has an open neighborhood U in X^2 and there is a homotopy $h : U \times \mathbf{I} \to X^2$ such that $h_0 = \mathrm{id}$, $h_t|\Delta_X = \mathrm{id}$ for each $t \in \mathbf{I}$, and $h_1(U) = \Delta_X$. On the other hand, X has a contraction $\varphi : X \times \mathbf{I} \to X$. Note that $(\varphi(x, t), \varphi(x, t)) \in \Delta_X$ for each $x \in X$ and $t \in \mathbf{I}$. Hence, there exists an open neighborhood V of Δ_X in X^2 such that $(\varphi(x, t), \varphi(y, t)) \in U$ for $(x, y) \in V$ and $t \in \mathbf{I}$. Let $k : X^2 \to \mathbf{I}$ be an Urysohn map with $k(\Delta_X) = 0$ and $k(X^2 \setminus V) = 1$. Observe that $(\varphi(x, k(x, y)), \varphi(y, k(x, y))) \in U$ for every $(x, y) \in X^2$. Then, we can define a homotopy $\tilde{h} : X^2 \times \mathbf{I} \to X^2$ as follows:

$$\tilde{h}(x, y, t) = \begin{cases} (\varphi(x, 2k(x, y)t), \varphi(y, 2k(x, y)t)) & \text{for } 0 \le t \le \frac{1}{2}, \\ h(\varphi(x, k(x, y)), \varphi(y, k(x, y)), 2t - 1) & \text{for } \frac{1}{2} \le t \le 1. \end{cases}$$

Observe that $\tilde{h}_0 = \mathrm{id}$, $\tilde{h}_t|\Delta_X = \mathrm{id}$ for each $t \in \mathbf{I}$, and $\tilde{h}_1(X^2) = \Delta_X$. Therefore, Δ_X is a strong deformation retract of X^2. Consequently, it follows from Theorem 6.3.2 that X is EC by Theorem 6.3.2. □

---

[2]This concept was introduced by F.D. Ancel. He adopted the term "uniformly locally contractible" but here we replace "uniformly" by "unified" because we say that a metric space X is **uniformly locally contractible** if, for each $\varepsilon > 0$, there is some $\delta > 0$ such that $\mathrm{B}(x, \delta)$ is contractible in $\mathrm{B}(x, \varepsilon)$ for each $x \in X$.

[3]As we saw in 2.10.2, the Sorgenfrey line S is (perfectly) normal but S^2 is not normal.

Let $\mathcal{V}$ be an open refinement of an open cover $\mathcal{U}$ of a space X. We call $\mathcal{V}$ an **h-refinement** (resp. **$\hbar$-refinement**) of $\mathcal{U}$ if any two $\mathcal{V}$-close maps $f, g : Y \to X$ defined on an arbitrary space Y are $\mathcal{U}$-homotopic (resp. $\mathcal{U}$-homotopic rel. $\{y \in Y \mid f(y) = g(y)\}$), where we write

$$\mathcal{V} \underset{h}{\prec} \mathcal{U} \quad \text{or} \quad \mathcal{U} \underset{h}{\succ} \mathcal{V} \quad \left(\text{resp. } \mathcal{V} \underset{\hbar}{\prec} \mathcal{U} \quad \text{or} \quad \mathcal{U} \underset{\hbar}{\succ} \mathcal{V}\right).$$

Using $\hbar$-refinements, we can characterize LEC spaces as follows:

Theorem 6.3.4. *A space X is LEC if and only if each open cover of X has an $\hbar$-refinement.*

Proof. To prove the "only if" part, let $\lambda : U \times \mathbf{I} \to X$ be an equi-connecting map for X and $\mathcal{U} \in \operatorname{cov}(X)$. For each $x \in X$, choose $U_x \in \mathcal{U}$ so that $x \in U_x$. From the continuity of λ, each $x \in X$ has an open neighborhood V_x such that $V_x^2 \subset U$ and $\lambda(V_x^2 \times \mathbf{I}) \subset U_x$. Then, $\mathcal{V} = \{V_x \mid x \in X\}$ is the desired refinement of $\mathcal{U}$. Indeed, any two $\mathcal{V}$-close maps $f, g : Y \to X$ of an arbitrary space Y are $\mathcal{U}$-homotopic by a homotopy $h : Y \times \mathbf{I} \to X$ defined by $h(y, t) = \lambda(f(y), g(y), t)$.

Now, we show the "if" part. By the condition, the open cover $\mathcal{U} = \{X\}$ has an $\hbar$-refinement $\mathcal{V}$. Then, $U = \bigcup_{V \in \mathcal{V}} V^2$ is a neighborhood of Δ_X in X^2. Since $\operatorname{pr}_1|U$ and $\operatorname{pr}_2|U$ are $\mathcal{V}$-close and $\operatorname{pr}_1|\Delta_X = \operatorname{pr}_2|\Delta_X$, we have a homotopy $\lambda : U \times \mathbf{I} \to X$ such that $\lambda_0 = \operatorname{pr}_1|U$, $\lambda_1 = \operatorname{pr}_2|U$ and $\lambda_t|\Delta_X = \operatorname{pr}_1|\Delta_X = \operatorname{pr}_2|\Delta_X$ for each $t \in \mathbf{I}$. This homotopy λ is an equi-connecting map for X. □

By Proposition 6.3.1, we have the following corollary:

Corollary 6.3.5. *Every open cover of an ANR has an $\hbar$-refinement (hence, it has an h-refinement).* □

Theorem 6.3.6. *A paracompact space X is ULC if and only if each open cover of X has an h-refinement.*

Proof. To prove the "only if" part, for each $\mathcal{U} \in \operatorname{cov}(X)$, let $\mathcal{U}'$ be a star-refinement of $\mathcal{U}$. Then, $W = \bigcup_{U \in \mathcal{U}'} U^2$ contains a neighborhood W_0 of Δ_X in X^2 with a pr_1-preserving homotopy $h : W_0 \times \mathbf{I} \to W$ such that $h_0 = \mathrm{id}$ and $h_1(W_0) = \Delta_X$. Choose $\mathcal{V} \in \operatorname{cov}(X)$ so that $\bigcup_{V \in \mathcal{V}} V^2 \subset W_0$. Observe that $V \subset \operatorname{st}(x, \mathcal{U}')$ for each $V \in \mathcal{V}$ and $x \in V$. Then, it follows that $\mathcal{V} \prec \mathcal{U}$. Moreover, $\mathcal{V}$ is an h-refinement of $\mathcal{U}$. Indeed, let $f, g : Y \to X$ be $\mathcal{V}$-close maps. Since $(f(y), g(y)) \in W_0$ for each $y \in Y$, we can define a homotopy $h^* : Y \times \mathbf{I} \to X$ by $h^*(y, t) = \operatorname{pr}_2 h(f(y), g(y), t)$. Then, $h^*_0 = g$ and $h^*_1 = f$. For each $y \in Y$ and $t \in \mathbf{I}$, since $h(f(y), g(y), t) \in W$, we have $U \in \mathcal{U}'$ such that $(f(y), h^*(y, t)) = h(f(y), g(y), t) \in U^2$, resulting in $U \in \mathcal{U}'[f(y)]$ and $h^*(y, t) \in U$. Therefore, $h^*(\{y\} \times \mathbf{I}) \subset \operatorname{st}(f(y), \mathcal{U}')$. Thus, h^* is a $\mathcal{U}$-homotopy.

To show the "if" part, let W be a neighborhood of Δ_X in X^2. Choose $\mathcal{U} \in \operatorname{cov}(X)$ so that $\bigcup_{U \in \mathcal{U}} U^2 \subset W$. Taking an h-refinement $\mathcal{V}$ of $\mathcal{U}$, define $W_0 = \bigcup_{V \in \mathcal{V}} V^2$. Then, W_0 is a neighborhood of Δ_X in W. Since $\operatorname{pr}_1|W_0$ and $\operatorname{pr}_2|W_0$ are $\mathcal{V}$-close, there is a $\mathcal{U}$-homotopy $h : W_0 \times \mathbf{I} \to X$ such that $h_0 = \operatorname{pr}_1|W_0$ and $h_1 = \operatorname{pr}_2|W_0$. We define a pr_1-preserving homotopy $\tilde{h} : W_0 \times \mathbf{I} \to X^2$ by $\tilde{h}(x, y, t) = (x, h(x, y, 1 - t))$.

Then, $\tilde{h}_0 = \mathrm{id}$ and $\tilde{h}_1(W_0) = \Delta_X$. For each $(x, y) \in W_0$, $h(\{(x, y)\} \times \mathbf{I})$ is contained in some $U \in \mathcal{U}$, which implies that $\tilde{h}(\{(x, y)\} \times \mathbf{I}) \subset U^2$. Therefore, we have $\tilde{h}(W_0 \times \mathbf{I}) \subset W$. □

Next, consider the following related theorem:

Theorem 6.3.7. *A space X has an open cover $\mathcal{U}$ such that any two $\mathcal{U}$-close maps $f, g : Y \to X$ defined on an arbitrary space Y are homotopic (i.e., $\mathcal{U}$ is an h-refinement of $\{X\}$) if and only if Δ_X is a neighborhood deformation retract of X^2.*

Proof. To show the "only if" part, given an open cover $\mathcal{U}$ in the condition, $W = \bigcup_{U \in \mathcal{U}} U^2$ is a neighborhood of Δ_X in X^2. Since $\mathrm{pr}_1|W$ and $\mathrm{pr}_2|W$ are $\mathcal{U}$-close, there is a homotopy $h : W \times I \to X$ such that $h_0 = \mathrm{pr}_1|W$ and $h_1 = \mathrm{pr}_2|W$. Then, we can define $\tilde{h} : W \times I \to X^2$ by $\tilde{h}(x, y, t) = (x, h(x, y, 1 - t))$. Observe that $\tilde{h}_0 = \mathrm{id}$ and $\tilde{h}_1 : W \to \Delta_X$ is a retraction. Thus, Δ_X is a neighborhood deformation retract of X^2.

To prove the "if" part, let W be a neighborhood of Δ_X in X^2 and $\varphi : W \times \mathbf{I} \to X^2$ a homotopy such that $\varphi_0 = \mathrm{id}$ and $\varphi_1 : W \to \Delta_X$ is a retraction. Then, we have $\mathcal{U} \in \mathrm{cov}(X)$ such that $U^2 \subset W$ for each $U \in \mathcal{U}$. Any two $\mathcal{U}$-close maps $f, g : Y \to X$ of an arbitrary space Y are homotopic by a homotopy $h : Y \times \mathbf{I} \to X$ defined as follows:

$$h(y, t) = \begin{cases} \mathrm{pr}_1\varphi(f(y), g(y), 2t) & \text{for } 0 \le t \le \frac{1}{2}, \\ \mathrm{pr}_2\varphi(f(y), g(y), 2 - 2t) & \text{for } \frac{1}{2} \le t \le 1. \end{cases}$$ □

In general, a locally contractible space is not ULC, as shown in the following theorem:

Theorem 6.3.8 (Borsuk). *There exists a continuum X such that each point has a neighborhood basis consisting of contractible neighborhoods but the cover $\{X\}$ has no h-refinements (hence, X is not ULC).*

Example and Proof. Let $X_0 = \{x \in \mathbf{I}^{\mathbb{N}} \mid x(1) = 0\}$ and, for each $n \in \mathbb{N}$, let $X_n = \partial C_n$ be the boundary n-sphere of the $(n + 1)$-cube

$$C_n = \left[(n + 1)^{-1}, n^{-1}\right] \times \mathbf{I}^n \times \{0\} \times \{0\} \times \cdots \subset \mathbf{I}^{\mathbb{N}}.$$

Then, we prove that $X = \bigcup_{n \in \omega} X_n$ $(\subset \mathbf{I}^{\mathbb{N}})$ is the desired continuum.

First, we show that each point of X has a neighborhood basis consisting of contractible neighborhoods. Since $\bigcup_{n \in \mathbb{N}} X_n = X \setminus X_0$ is a polyhedron that is open in X, each $x \in \bigcup_{n \in \mathbb{N}} X_n$ has such a neighborhood basis. When $x \in X_0$, for each neighborhood U of x in X, we can find $m \ge 2$ and a convex neighborhood W of $(x(2), \ldots, x(m))$ in $\mathbf{I}^{m-1}$ such that $p_m^{-1}([0, m^{-1}] \times W) \subset U$, where $p_m : X \to \mathbf{I}^m$ is the restriction of the projection onto the first m coordinates. When $x(m + 1) \le \frac{1}{2}$ (resp. $\ge \frac{1}{2}$), we define a neighborhood V of x in X as follows:

$$V = p_{m+1}^{-1}([0, m^{-1}] \times W \times [0, 1)) \quad \left(\text{resp. } p_{m+1}^{-1}([0, m^{-1}] \times W \times (0, 1])\right).$$

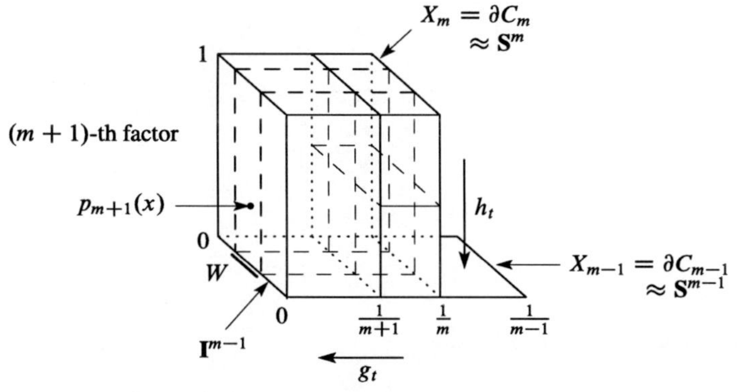

Fig. 6.6 Borsuk's example

Then, $V \subset U$ and V is contractible. Indeed, let

$$V_0 = p_{m+1}^{-1}(\{0\} \times W \times \{0\}) \subset X_0 \quad (\text{resp. } p_{m+1}^{-1}(\{0\} \times W \times \{1\})) \text{ and}$$
$$V_1 = p_{m+1}^{-1}([0, m^{-1}] \times W \times \{0\}) \quad (\text{resp. } p_{m+1}^{-1}([0, m^{-1}] \times W \times \{1\})).$$

Then, V_1 is a strong deformation retract of V by a deformation $h : V \times \mathbf{I} \to V$ sliding along the $(m+1)$-th coordinate, and V_0 is a strong deformation retract of V_1 by a deformation $g : V_1 \times \mathbf{I} \to V_1$ sliding along the first coordinate (Fig. 6.6). Since V_0 is contractible, V is also contractible.

Next, we show that $\{X\}$ has no h-refinements. On the contrary, assume that $\{X\}$ has an h-refinement $\mathcal{U} \in \operatorname{cov}(X)$. By the compactness of X, we can take $m \in \mathbb{N}$ such that $\{X \cap p_m^{-1}(x) \mid x \in \mathbf{I}^m\} \prec \mathcal{U}$. Then, the map $f : X \to X$ defined by $f(x) = (x(1), \ldots, x(m), 0, 0, \ldots)$ is $\mathcal{U}$-close to id, hence $f \simeq \mathrm{id}$. Let $r : X \to X_m$ be a retraction defined as follows:

$$r(x) = \begin{cases} ((m+1)^{-1}, x(2), \ldots, x(m+1), 0, 0, \ldots) & \text{if } x(1) \le (m+1)^{-1}, \\ (m^{-1}, x(2), x(3), \ldots) & \text{if } x(1) \ge m^{-1}, \\ x & \text{otherwise.} \end{cases}$$

Then, $\mathrm{id} = r|X_m \simeq rf|X_m : X_m \to X_m$. Moreover,

$$rf(X_m) = f(X_m) = [(m+1)^{-1}, m^{-1}] \times \mathbf{I}^{m-1} \times \{0\} \times \cdots,$$

hence $rf|X_m \simeq 0$. Thus, $\mathbf{S}^m \approx X_m$ is contractible, which contradicts Corollary 6.1.6. □

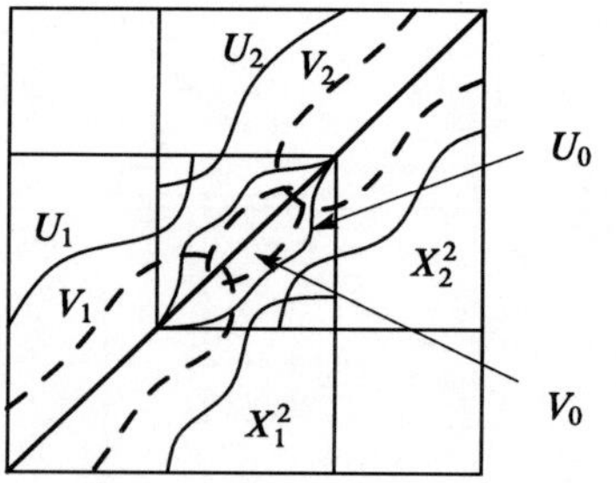

Fig. 6.7 The union of two open LEC spaces

Basic Properties of LEC Spaces 6.3.9.

(1) An arbitrary product of EC spaces is EC and a finite product of LEC spaces is LEC.

Sketch of Proof. Define an equi-connecting map coordinate-wise.

(2) A retract of an EC space is EC and a neighborhood retract of an LEC space is LEC.

Sketch of Proof. Compose a retraction with the restriction of an equi-connecting map.

(3) Any open set in an LEC space is LEC.

Sketch of Proof. Restrict an equi-connecting map.

(4) Let $X = X_1 \cup X_2$, where X_i is open in X. Suppose that X^2 is normal. If X_1 and X_2 are LEC, then so is X.

Sketch of Proof. Apply Theorem 6.3.2. Note that $X_0 = X_1 \cap X_2$ is also LEC by (3). For $i = 0, 1, 2$, we have a neighborhood U_i of Δ_{X_i} in X_i^2 and a homotopy $h^i : U_i \times \mathbf{I} \to X_i^2$ such that $h^i_0 = \mathrm{id}$, h^i_1 is a retraction onto Δ_{X_i} and $h^i_t|\Delta_{X_i} = \mathrm{id}$ for all $t \in \mathbf{I}$. Choose open sets V_0, V'_0, V_1, V_2 in X^2 so that

$$\mathrm{cl}\, V_1 \cap \mathrm{cl}\, V_2 = \emptyset,\ \Delta_X \setminus X_2^2 \subset V_1 \subset U_1,\ \Delta_X \setminus X_1^2 \subset V_2 \subset U_2,$$
$$\Delta_X \setminus (V_1 \cup V_2) \subset V_0 \subset \mathrm{cl}\, V_0 \subset V'_0 \subset \mathrm{cl}\, V'_0 \subset U_0$$

and $h^0(\mathrm{cl}\, V'_0 \times \mathbf{I}) \subset U_1 \cap U_2$. Then, $V = V_0 \cup V_1 \cup V_2$ is a neighborhood of Δ_X in X^2. Using h^i $(i = 0, 1, 2)$ and an Urysohn map $k : X^2 \to \mathbf{I}$ with $k(X^2 \setminus V'_0) = 0$ and $k(\mathrm{cl}\, V_0) = 1$, define a homotopy $h : V \times \mathbf{I} \to X^2$ such that $h_0 = \mathrm{id}$, h_1 is a retraction onto Δ_X, and $h_t|\Delta_X = \mathrm{id}$ for all $t \in \mathbf{I}$. See Fig. 6.7.

(5) A metrizable space X is LEC if each point of X has an LEC neighborhood in X.

Sketch of Proof. Apply Michael's Theorem 2.6.5 on local properties.

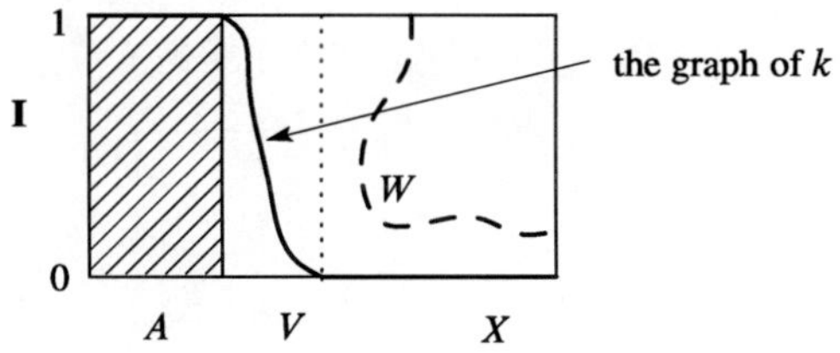

Fig. 6.8 An extension of a homotopy

6.4 The Homotopy Extension Property

We say that a space Y has the **homotopy extension property (HEP)** for metrizable spaces provided any map $f : A \to Y$ of a closed set A in an arbitrary metrizable space X extends over X if f is homotopic to a map extending over X. Due to the following theorem, every ANE has the HEP for metrizable spaces.

Theorem 6.4.1 (Homotopy Extension Theorem). *Let Y be an ANE, $\mathcal{U}$ be an open cover of Y, and $h : A \times \mathbf{I} \to Y$ be a $\mathcal{U}$-homotopy of a closed set A in a metrizable space X. If h_0 extends to a map $f : X \to Y$, then h extends to a $\mathcal{U}$-homotopy $\tilde{h} : X \times \mathbf{I} \to Y$ with $\tilde{h}_0 = f$.*

Proof. Since Y is an ANE, h extends to a map $h' : W \to Y$ from a neighborhood W of $X \times \{0\} \cup A \times \mathbf{I}$ in $X \times \mathbf{I}$ such that $h'(x, 0) = f(x)$ for each $x \in X$. For each $a \in A$, choose $U_a \in \mathcal{U}$ so that $h'(\{a\} \times \mathbf{I}) = h(\{a\} \times \mathbf{I}) \subset U_a$. Then, each $a \in A$ has an open neighborhood V_a in X such that $V_a \times \mathbf{I} \subset W$ and $h'(V_a \times \mathbf{I}) \subset U_a$. Let $V = \bigcup_{a \in A} V_a$ and let $k : X \to \mathbf{I}$ be an Urysohn map with $k(X \setminus V) = 0$ and $k(A) = 1$. Then, the desired homotopy $\tilde{h} : X \times \mathbf{I} \to Y$ can be defined by $\tilde{h}(x, t) = h'(x, tk(x))$. See Fig. 6.8. □

Using Corollary 6.3.5, we can prove another Homotopy Extension Theorem:

Theorem 6.4.2. *Any open cover $\mathcal{U}$ of an ANR Y has an open refinement $\mathcal{V}$ satisfying the following condition:*

($*$) *For any $\mathcal{V}$-homotopy $h : A \times \mathbf{I} \to Y$ of a closed set A in a metrizable space X, if h_0 and h_1 extend to $\mathcal{V}$-close maps $f, g : X \to Y$, respectively, then h extends to a $\mathcal{U}$-homotopy $\tilde{h} : X \times \mathbf{I} \to Y$ such that $h_0 = f$ and $h_1 = g$.*

Proof. By Corollary 6.3.5, we can take open refinements of $\mathcal{U}$ as follows:

$$\mathcal{V} \underset{h}{\prec} \mathcal{W}' \overset{*}{\prec} \mathcal{W} \underset{h}{\prec} \mathcal{U}' \overset{*}{\prec} \mathcal{U}.$$

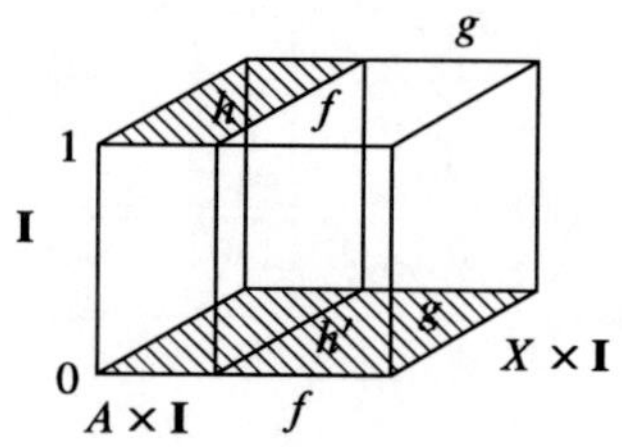

Fig. 6.9 An extension of a homotopy

We will show that $\mathcal{V}$ is the desired refinement. Let $h : A \times \mathbf{I} \to Y$ be a $\mathcal{V}$-homotopy of a closed set A in a metrizable space X such that h_0 and h_1 extend to $\mathcal{V}$-close maps $f, g : X \to Y$, respectively. Then, we have a $\mathcal{W}'$-homotopy $h' : X \times \mathbf{I} \to Y$ such that $h'_0 = f$ and $h'_1 = g$. On the other hand, h extends to a map $h'' : A \times \mathbf{I} \cup X \times \{0,1\} \to Y$ defined by $h''(x,0) = f(x)$ and $h''(x,1) = g(x)$ for all $x \in X$. Since $h'|A \times \mathbf{I} \cup X \times \{0,1\}$ and h'' are $\mathcal{W}$-close, there is a $\mathcal{U}'$-homotopy

$$\varphi : (A \times \mathbf{I} \cup X \times \{0,1\}) \times \mathbf{I} \to Y$$

such that $\varphi_0 = h'|A \times \mathbf{I} \cup X \times \{0,1\}$ and $\varphi_1 = h''$ (cf. Fig. 6.9). By Theorem 6.4.1, φ extends to a $\mathcal{U}'$-homotopy $\tilde{\varphi} : (X \times \mathbf{I}) \times \mathbf{I} \to Y$ such that $\tilde{\varphi}_0 = h'$ and $\tilde{\varphi}_1|A \times \mathbf{I} \cup X \times \{0,1\} = h''$. Then, $\tilde{h} = \tilde{\varphi}_1 : X \times \mathbf{I} \to Y$ is a $\mathcal{U}$-homotopy such that $\tilde{h}_0 = f$, $\tilde{h}_1 = g$, and $\tilde{h}|A \times \mathbf{I} = h$. □

Using the HEP, we can characterize ANRs as follows:

Theorem 6.4.3. *For a metrizable space X, the following are equivalent:*

(a) *X is an ANR;*
(b) *X is semi-locally contractible and has the HEP for metrizable spaces;*
(c) *X has an open cover $\mathcal{V}$ such that for any $\mathcal{V}$-homotopy $h : A \times \mathbf{I} \to X$ of a closed set A in a metrizable space Y, if h_0 and h_1 respectively extend to $\mathcal{V}$-close maps $f, g : Y \to X$, then h extends to a homotopy $\tilde{h} : Y \times \mathbf{I} \to X$ such that $h_0 = f$ and $h_1 = g$;*
(d) *X is locally contractible and has an open cover $\mathcal{V}$ such that for any $\mathcal{V}$-homotopy $h : A \times \mathbf{I} \to X$ of a closed set A in a metrizable space Y, if h_0 extends to a map $f : Y \to X$ then h extends to a homotopy $\tilde{h} : Y \times \mathbf{I} \to X$ such that $\tilde{h}_0 = f$;*
(e) *Each $x \in X$ has a neighborhood V in X such that any map $f : A \to V$ of a closed set A in a metrizable space Y extends to a map $\tilde{f} : Y \to X$;*
(f) *Each $x \in X$ has a neighborhood V in X such that any map $f : A \to V$ of a closed set A in a metrizable space Y extends to a map $\tilde{f} : V \to X$ of a neighborhood V of A in Y to X.*

Sketch of Proof. Observe the following implications:

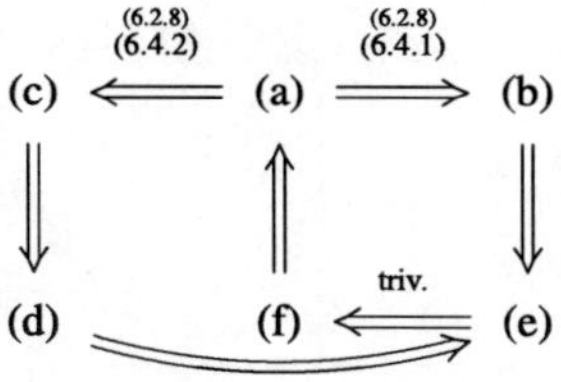

(b) (or (d)) $\Rightarrow$ (e): Choose a neighborhood V of $x \in X$ so as to be contractible in X (or in some member of $\mathcal{V}$ in (d)).

(c) $\Rightarrow$ (d): Take $\mathcal{V}$ such as in (c). Each $x \in X$ is contained in some $V \in \mathcal{V}$. Using (c), construct a contraction $g : V \times \mathbf{I} \to X$ so that $g_t(x) = x$ for each $t \in \mathbf{I}$. Restricting g to a small neighborhood of x, we can show the local contractibility. For f and h of (d), applying (c) to the $\mathcal{V}$-homotopy $h^* : A \times \mathbf{I} \to X$ defined by $h^*_t = h_{2t}$ for $t \leq \frac{1}{2}$ and $h^*_t = h_{2-2t}$ for $t \geq \frac{1}{2}$, we can obtain an extension of h.

(f) $\Rightarrow$ (a): In condition (f), $\operatorname{int} V$ is an ANE, hence an ANR. Then, Hanner's Theorem 6.2.10(4) can be applied.

Using the Homotopy Extension Theorem 6.4.1, we can prove the following theorem, which was announced in the remark after Proposition 6.2.7:

Theorem 6.4.4. *A deformation retract of an ANR is a strong deformation retract.*

Proof. Let A be a deformation retract of an ANR X. Then, there is a homotopy $h : X \times \mathbf{I} \to X$ such that $h_0 = \mathrm{id}$ and $h_1 : X \to A$ is a retraction. Since $h_1 h_1 = h_1$ and $h_1|A = \mathrm{id}$, we can define a homotopy

$$\varphi : (A \times \mathbf{I} \cup X \times \{0, 1\}) \times \mathbf{I} \to X$$

as follows:

$$\varphi(x, t, s) = \begin{cases} h_{t(1-s)}(x) & \text{if } x \in A, \\ h_{1-s}h_1(x) & \text{if } t = 1, \\ x & \text{if } t = 0, \\ h_t(x) & \text{if } s = 0. \end{cases}$$

Refer to Fig. 6.10. Since φ_0 extends to h, φ extends a homotopy $\tilde{\varphi} : (X \times \mathbf{I}) \times \mathbf{I} \to X$ by the Homotopy Extension Theorem 6.4.1. Then, $h' = \tilde{\varphi}_1 : X \times \mathbf{I} \to X$ is a homotopy such that $h'_t|A = h_0|A = \mathrm{id}$ for every $t \in \mathbf{I}$, $h'_1 = h_0 h_1 = h_1$, and $h'_0 = \mathrm{id}$. Thus, h'_1 is a strong deformation retraction of X onto A, so A is a strong deformation retract of X. □

6.5 Complementary Pairs of ANRs

First, we prove the following lemma, which is often used in extending maps.

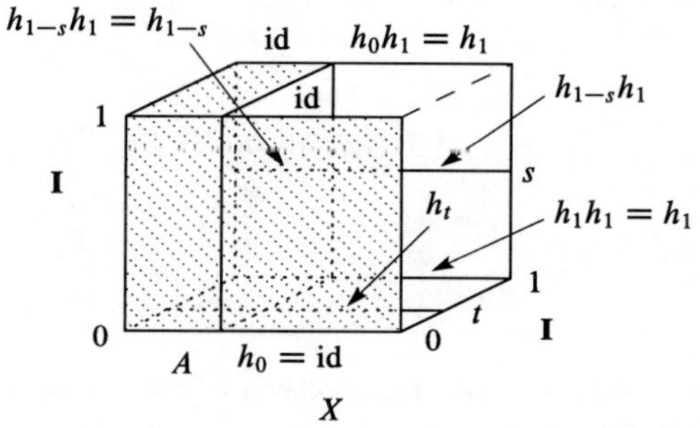

Fig. 6.10 The homotopy φ

Lemma 6.5.1 (C.H. Dowker). *Let X and Y be metrizable spaces and $f : A \to Y$ be a map of a closed set A in X. Suppose there exists a homotopy $h : A \times \mathbf{I} \to Y$ such that $h_0 = f$ and $h|A \times (0,1]$ extends over a neighborhood W of $A \times (0,1]$ in $X \times (0,1]$. Then, f extends over a neighborhood of A in X. If $X \times \{1\} \subset W$, then f extends over X.*

Proof. For simplicity, let d stand for admissible metrics of X and Y as well as the metric on $X \times \mathbf{I}$ defined as follows:

$$d((x,t),(x',t')) = \max\{d(x,x'),\ |t-t'|\}.$$

We may assume that h is defined on the set $W \cup A \times \{0\}$ and that $h|A \times \mathbf{I}$ and $h|W$ are continuous. However, it is not assumed that h is continuous. First, we find a neighborhood W^* of $A \times (0,1]$ in $X \times (0,1]$ such that $W^* \subset W$ and $h|W^* \cup A \times \{0\}$ is continuous. To this end, for each $(a,t) \in A \times (0,1]$, choose a neighborhood $W(a,t)$ of (a,t) in W so that $\operatorname{diam} W(a,t) < \frac{1}{2}t$ and $\operatorname{diam} h(W(a,t)) < \frac{1}{2}t$. Then, $W^* = \bigcup_{(a,t)\in A\times(0,1]} W(a,t)$ is a neighborhood of $A \times (0,1]$ in $X \times (0,1]$. To verify the continuity of $h|W^* \cup A \times \{0\}$ at $(a,0) \in A \times \{0\}$, let $\varepsilon > 0$. By the continuity of $h|A \times \mathbf{I}$ at $(a,0)$, we can choose $\delta > 0$ so that, if $(a',t') \in A \times \mathbf{I}$ and $d((a,0),(a',t')) < \delta$, then $d(h(a,0),h(a',t')) < \frac{1}{2}\varepsilon$. Let $(x,t) \in W^*$ with $d((a,0),(x,t)) < \frac{1}{2}\min\{\delta,\varepsilon\}$. Then, we have $(a',t') \in A \times (0,1]$ such that $(x,t) \in W(a',t')$, resulting in $d((x,t),(a',t')) < \frac{1}{2}t'$ and $d(h(x,t),h(a',t')) < \frac{1}{2}t'$. Since $|t-t'| < \frac{1}{2}t'$, it follows that $\frac{1}{2}t' < t < \frac{1}{2}\min\{\delta,\varepsilon\}$. Thus, $d((x,t),(a',t')) < \frac{1}{2}\delta$ and $d(h(x,t),h(a',t')) < \frac{1}{2}\varepsilon$. Since

$$d((a,0),(a',t')) \le d((a,0),(x,t)) + d((x,t),(a',t')) < \delta,$$

it follows that $d(h(a,0),h(a',t')) < \frac{1}{2}\varepsilon$. Then,

$$d(h(a,0),h(x,t)) \le d(h(a,0),h(a',t')) + d(h(a',t'),h(x,t)) < \varepsilon.$$

Therefore, $h|W^* \cup A \times \{0\}$ is continuous at $(a,0)$.

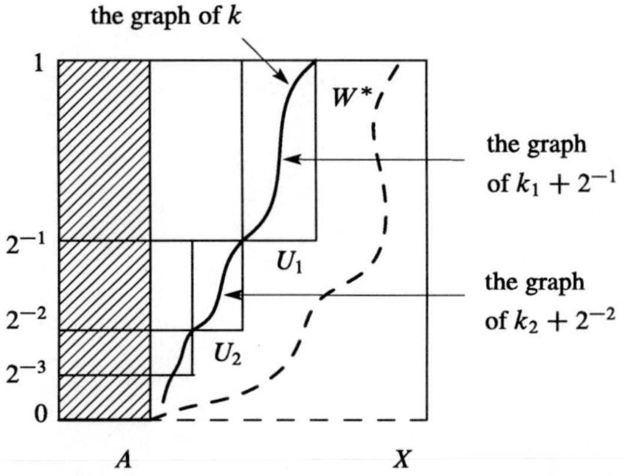

Fig. 6.11 Extending a map

Choose open neighborhoods U_n of A in X, $n \in \mathbb{N}$, so that $U_n \times [2^{-n}, 1] \subset W^*$, $\operatorname{cl} U_{n+1} \subset U_n$, and $\bigcap_{n \in \mathbb{N}} U_n = A$. Take maps $k_n : X \to [0, 2^{-n}]$, $n \in \mathbb{N}$, so that $k_n(\operatorname{cl} U_{n+1}) = 0$ and $k_n(X \setminus U_n) = 2^{-n}$, and define a map $k : X \to \mathbf{I}$ by $k(x) = \sum_{n \in \mathbb{N}} k_n(x)$. Observe that $k^{-1}(0) = A$ and $k(x) \in [2^{-n}, 2^{-n+1}]$ for each $x \in U_n \setminus U_{n+1}$, which implies that $(x, k(x)) \in A \times \{0\} \cup W^*$ for all $x \in U_1$. Then, f extends to a map $\tilde{f} : U_1 \to Y$ defined by $\tilde{f}(x) = h(x, k(x))$. See Fig. 6.11.

If $X \times \{1\} \subset W$, we define $W^* = \bigcup_{(a,t) \in A \times (0,1]} W(a, t) \cup X \times \{1\}$. Then, the map $\tilde{f}$ above can be defined over X. □

Remark 8. In the above proof, it is not enough to take a map k with $k^{-1}(0) = A$. Indeed, we define a map

$$h : \mathbf{I} \times (0, 1] \cup \{0\} \times \{0\} \to \mathbb{R}$$

by $h(x, t) = x/t$ if $t \neq 0$ and $h(0, 0) = 0$, where $(X, A) = (\mathbf{I}, \{0\})$, $Y = \mathbb{R}$ and $W = \mathbf{I} \times (0, 1]$. Moreover, let $k = \mathrm{id} : \mathbf{I} \to \mathbf{I}$. Then, $k^{-1}(0) = \{0\}$. Using this k, we define $\tilde{f} : \mathbf{I} \to \mathbb{R}$ as in the above proof, i.e., $\tilde{f} = h(x, k(x))$ for each $x \in \mathbf{I}$. Then, $\tilde{f}$ is not continuous at $0 \in \mathbf{I}$ because $\tilde{f}(0) = h(0, 0) = 0$, but $\tilde{f}(x) = h(x, k(x)) = x/k(x) = 1$ if $x \neq 0$.

Applying Lemma 6.5.1, we prove the following theorem:

Theorem 6.5.2 (Kruse–Liebnitz). *Let X be metrizable and A be a strong neighborhood deformation retract of X. If A and $X \setminus A$ are ANRs, then so is X.*

Proof. From the assumption, we have a homotopy $h : \operatorname{cl} U \times \mathbf{I} \to X$ of an open neighborhood U of A in X such that $h_0 = \mathrm{id}$, h_1 is a retraction of $\operatorname{cl} U$ onto A, and $h_t|A = \mathrm{id}$ for every $t \in \mathbf{I}$. Given an admissible metric d for X, we may assume that $\operatorname{diam} h(\{x\} \times \mathbf{I}) < 1$ for every $x \in U$. Let $f : B \to X$ be a map from a closed set B in an arbitrary metrizable space Y. We apply Lemma 6.5.1 to extend f over a neighborhood of B in Y.

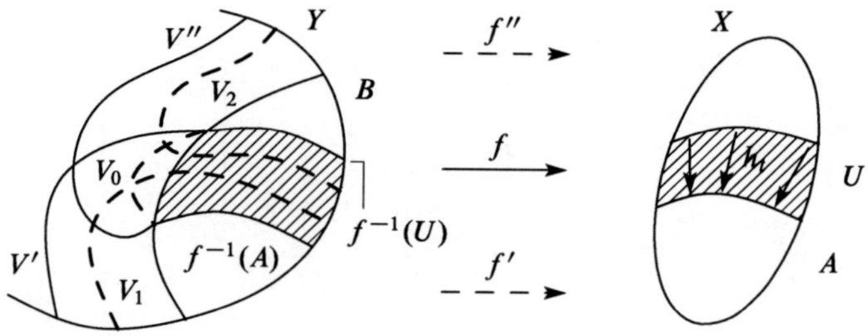

Fig. 6.12 Extensions of $h_1 f|f^{-1}(U)$ and $f|f^{-1}(X \setminus A)$

Note that $f^{-1}(U)$ and $f^{-1}(X \setminus A)$ are closed in $Y \setminus f^{-1}(X \setminus U)$ and $Y \setminus f^{-1}(A)$, respectively. Since A and $X \setminus A$ are ANRs, maps $h_1 f|f^{-1}(U)$ and $f|f^{-1}(X \setminus A)$ extend to maps $f' : V' \to A$ and $f'' : V'' \to X \setminus A$, respectively, where V' and V'' are open sets in $Y \setminus f^{-1}(X \setminus U)$ and $Y \setminus f^{-1}(A)$, respectively, such that $V' \cap B = f^{-1}(U)$ and $V'' \cap B = f^{-1}(X \setminus A)$. Observe that

$$V' \cap V'' \cap B = f^{-1}(U \setminus A) \text{ and } f'|V' \cap V'' \cap B = h_1 f''|V' \cap V'' \cap B.$$

Note that every open cover of an ANR has an $\hbar$-refinement by Corollary 6.3.5. Then, we can find an open set V_0 in $V' \cap V''$ such that $V_0 \cap B = V' \cap V'' \cap B$ and $f'|V_0 \simeq h_1 f''|V_0$ rel. $V_0 \cap B$. Let $\varphi : V_0 \times \mathbf{I} \to A$ be a homotopy such that

$$\varphi_0 = f'|V_0,\ \varphi_1 = h_1 f''|V_0 \text{ and } \varphi_t|V_0 \cap B = f'|V_0 \cap B \text{ for all } t \in \mathbf{I}.$$

Choose disjoint open sets V_1 and V_2 in Y so that $f^{-1}(A) \subset V_1 \subset V'$ and $B \setminus f^{-1}(U) \subset V_2 \subset V''$. Then, $V = V_0 \cup V_1 \cup V_2$ is an open neighborhood of B in Y. Refer to Fig. 6.12.

The function $U \ni x \mapsto \operatorname{diam} h(\{x\} \times \mathbf{I}) \in \mathbf{I}$ is continuous. Indeed, for each $x \in U$ and $\varepsilon > 0$, using the compactness of $\mathbf{I}$, we can find $\delta > 0$ such that $d(x, x') < \delta$ implies $d(h(x,t), h(x',t)) < \varepsilon/2$ for every $t \in \mathbf{I}$. Let $x' \in U$ with $d(x, x') < \delta$. For each $t_0, t_1 \in \mathbf{I}$,

$$\begin{aligned} d(h(x', t_0), h(x', t_1)) &\leq d(h(x, t_0), h(x, t_1)) \\ &\quad + d(h(x, t_0), h(x', t_0)) + d(h(x, t_1), h(x', t_1)) \\ &< \operatorname{diam} h(\{x\} \times \mathbf{I}) + \varepsilon, \end{aligned}$$

which implies that

$$\operatorname{diam} h(\{x'\} \times \mathbf{I}) \leq \operatorname{diam} h(\{x\} \times \mathbf{I}) + \varepsilon.$$

Replacing x' with x in the above, we have

$$\operatorname{diam} h(\{x\} \times \mathbf{I}) \leq \operatorname{diam} h(\{x'\} \times \mathbf{I}) + \varepsilon.$$

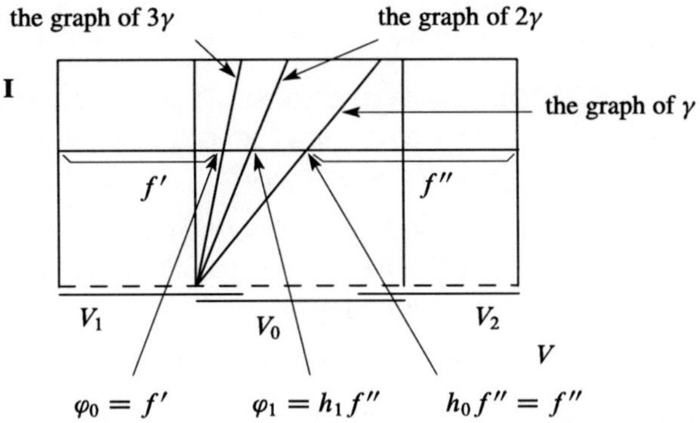

Fig. 6.13 The map g

Let $\alpha : (V \setminus V_0) \cup B \to \mathbf{I}$ be an Urysohn map with $\alpha(V_1 \setminus V_0) = 0$ and $\alpha(V_2 \setminus V_0) = 1$. We define the map $\beta : (V \setminus V_0) \cup B \to \mathbf{I}$ by $\beta(V_1 \setminus V_0) = 0$, $\beta(V_2 \setminus V_0) = 1$, and

$$\beta(x) = \min\left\{1,\ \operatorname{diam} h(\{f(x)\} \times \mathbf{I}) + \alpha(x)\right\} \quad \text{for } x \in f^{-1}(\operatorname{cl} U).$$

Note that $\operatorname{diam} h(\{f(x)\} \times \mathbf{I}) \leq \beta(x)$ for every $x \in f^{-1}(\operatorname{cl} U)$ and $\beta^{-1}(0) \cap B = f^{-1}(A)$. By the Tietze Extension Theorem 2.2.2, β extends to a map $\gamma : V \to \mathbf{I}$. Observe that $\gamma^{-1}((0,1)) \subset V_0$, $\gamma^{-1}([0,1)) \subset V'$, and $\gamma^{-1}((0,1]) \subset V''$. Then, we can define a map $g : V \times (0,1] \to X$ as follows:

$$g(x,t) = \begin{cases} f''(x) & \text{if } t \leq \gamma(x), \\ h(f''(x), \gamma(x)^{-1}t - 1) & \text{if } \gamma(x) \leq t \leq 2\gamma(x), \\ \varphi(x, 3 - \gamma(x)^{-1}t) & \text{if } 2\gamma(x) \leq t \leq 3\gamma(x), \\ f'(x) & \text{if } t \geq 3\gamma(x). \end{cases}$$

Refer to Fig. 6.13. For each $(x,t) \in B \times (0,1]$, we have $d(g(x,t), f(x)) \leq t$. Indeed, recall $\beta|B = \gamma|B$ and $\varphi_t|V_0 \cap B = h_1 f|V_0 \cap B$ for every $t \in \mathbf{I}$. Observe the following:

$$t \leq \gamma(x) \Rightarrow g(x,t) = f(x);$$
$$\gamma(x) \leq t \leq 2\gamma(x) \Rightarrow d(g(x,t), f(x)) \leq \operatorname{diam} h(\{f(x)\} \times \mathbf{I}) \leq \gamma(x) \leq t;$$
$$t \geq 2\gamma(x) \Rightarrow d(g(x,t), f(x)) = d(h_1 f(x), f(x)) \leq \gamma(x) \leq t.$$

Therefore, $g|B \times (0,1]$ can be extended to the homotopy $\tilde{g} : B \times \mathbf{I} \to X$ with $\tilde{g}_0 = f$. By Lemma 6.5.1, f extends over a neighborhood of B in Y. □

Recall that a perfect map $f : X \to Y$ is a closed map such that $f^{-1}(y)$ is compact for every $y \in Y$. The perfect image of a metrizable space is also metrizable

(2.4.5(1)). As a corollary of Theorem 6.5.2, we have the following theorem on the adjunction spaces:

Theorem 6.5.3 (Borsuk–Whitehead–Hanner). *Let X and Y be ANRs and $f : A \to Y$ be a map from a closed set A in X. Then, the adjunction space $Y \cup_f X$ is an ANR if f is perfect and A is an ANR.*

Proof. Since A is a strong neighborhood deformation retract of X by Proposition 6.2.7, it follows that Y is a strong neighborhood deformation retract of $Y \cup_f X$. The result follows from Theorem 6.5.2. □

As special cases of Theorem 6.5.3, we have the following:

Corollary 6.5.4. *For any perfect map $f : X \to Y$ between ANRs, the mapping cylinder M_f is an ANR.* □

When Y is a singleton in the above, the mapping cylinder M_f is regarded as the cone $(X \times \mathbf{I})/(X \times \{0\})$. Since a contractible ANR is an AR (Corollary 6.2.9), we have the following corollary:

Corollary 6.5.5. *The cone $(X \times \mathbf{I})/(X \times \{0\})$ over a compact ANR X is an AR.* □

Mapping cylinders are very useful, as seen in Sect. 4.12. However, they are not, in general, metrizable. We introduce the mapping cylinder that is different from the quotient space. For a map $f : X \to Y$, let

$$M(f) = Y \cup \big(X \times (0, 1]\big) \quad \text{(the disjoint union)},$$

whose topology is generated by open sets in $X \times (0, 1]$ and sets $U \cup (f^{-1}(U) \times (0, \varepsilon))$, where U is open in Y and $0 < \varepsilon < 1$. Then, we have the natural continuous bijection $\phi_f : M_f \to M(f)$, where $\phi_f|Y = \mathrm{id}$ and $\phi_f|X \times (0, 1] = \mathrm{id}$.

When X and Y are bounded closed sets in normed linear spaces E and F, respectively, we can define a closed embedding $\varphi : M(f) \to E \times F \times \mathbb{R}$ as follows: $\varphi(y) = (0, y, 0)$ for each $y \in Y$ and

$$\varphi(x, t) = (tx, (1 - t) f(x), t) \quad \text{for each } (x, t) \in X \times (0, 1].$$

To verify the continuity at $y \in \mathrm{cl}\, f(X)$, let $\varepsilon > 0$. Choose $\delta_1 > 0$ so that $(1 - t)\mathrm{B}(y, \delta_1) \subset \mathrm{B}(y, \varepsilon)$ if $0 < t < \delta_1$. Since X is bounded in E, we have $0 < \delta < \delta_1$ $(< \varepsilon)$ such that $\delta X \subset \mathrm{B}(0, \varepsilon)$. Let $U = Y \cap \mathrm{B}(y, \delta)$. Then, $U \cup f^{-1}(U) \times (0, \delta)$ is an open neighborhood of y in $M(f)$ and

$$\varphi\big(U \cup (f^{-1}(U) \times (0, \delta))\big) \subset \mathrm{B}(0, \varepsilon) \times \mathrm{B}(y, \varepsilon) \times (-\varepsilon, \varepsilon).$$

To verify the closedness of φ, let $(z_i)_{i\in\mathbb{N}}$ be a sequence in $M(f)$ such that $(\varphi(z_i))_{i\in\mathbb{N}}$ converges to $(x, y, t) \in E \times F \times \mathbb{R}$. We will show that $(z_i)_{i\in\mathbb{N}}$ has a convergent subsequence. Then, we may assume that $z_i \in Y$ for every $i \in \mathbb{N}$ or $z_i \in X \times (0, 1]$ for every $i \in \mathbb{N}$. In the first case, since $\varphi(z_i) = (0, z_i, 0)$, $(z_i)_{i\in\mathbb{N}}$ converges to y. In the second case, let $z_i = (x_i, t_i)$ for each $i \in \mathbb{N}$. Then, $t = \lim_{i\to\infty} t_i \in \mathbf{I}$. Hence, $x = t \lim_{i\to\infty} x_i$ and $y = (1 - t) \lim_{i\to\infty} f(x_i)$. If $t > 0$ then $\lim_{i\to\infty} x_i = t^{-1}x$, where $t^{-1}x \in X$ because X is closed in E. Hence, $(z_i)_{i\in\mathbb{N}}$ is convergent. When $t = 0$, we have $\lim_{i\to\infty} f(x_i) = y$, where $y \in Y$ because Y is closed in F. For each open neighborhood U of y in Y and

$0 < \varepsilon < 1$, we have $i_0 \in \mathbb{N}$ such that $i \geq i_0$ implies $t_i < \varepsilon$ and $f(x_i) \in U$, i.e., $z_i = (x_i, t_i) \in U \cup f^{-1}(U) \times (0, \varepsilon)$, which means that $(z_i)_{i \in \mathbb{N}}$ is convergent.

Due to the Arens–Eells Embedding Theorem 6.2.1, every metrizable (resp. completely metrizable) space can be embedded into a normed linear space (resp. a Banach space). Then, $M(f)$ is (completely) metrizable if X and Y are. The metrizability of $M(f)$ can also be shown by applying the Bing Metrization Theorem 2.3.4.

For any map $f : X \to Y$, we can identify $M(f)$ with M_f as sets (by the natural map ϕ_f), but the topology of M_f is finer than the topology of $M(f)$. Let $q_f : Y \oplus (X \times \mathbf{I}) \to M_f$ be the quotient map. If $f : X \to Y$ is *perfect*, $M(f) = M_f$ as spaces, where the natural map ϕ_f is a homeomorphism.

Indeed, for each neighborhood U of $y \in Y \setminus f(X)$ in M_f, since $f^{-1}(y) \times \{0\} \subset q_f(U) \cap (X \times \mathbf{I})$ and $f^{-1}(y)$ is compact, we have a neighborhood W of $f^{-1}(y)$ in X with $\delta \in (0, 1)$ such that $W \times [0, \delta) \subset q_f(U) \cap (X \times \mathbf{I})$. Since f is closed, there is a neighborhood V of y in Y such that $f^{-1}(V) \subset W$ and $V \subset q_f(U) \cap Y$. Then, $V \cup (f^{-1}(V) \times (0, \delta)) \subset \phi_f(U)$. Thus, $\phi_f(U)$ is a neighborhood of y in $M(f)$.

We also call $M(f)$ the **mapping cylinder** of f (or the **metrizable mapping cylinder** of f when we need to distinguish it from M_f). As with M_f, we define the **collapsing** $c_f : M(f) \to Y$, which is continuous. Then, $c_f \simeq \mathrm{id}$ rel. Y in $M(f)$ by the homotopy $h^f : M(f) \times \mathbf{I} \to M(f)$ defined in the same manner as $h^f : M_f \times \mathbf{I} \to M_f$. Hence, Y is a strong deformation retract of $M(f)$. The natural map $\phi_f q_f : Y \oplus (X \times \mathbf{I}) \to M(f)$ is abbreviated by q_f, the same notation as the natural quotient map.

From Theorem 6.5.2 we deduce:

Corollary 6.5.6. *For any map $f : X \to Y$ between ANRs, the mapping cylinder $M(f)$ is an ANR.* □

The mapping cylinder $M(\mathrm{id}_X) = M_{\mathrm{id}_X}$ of the identity map id_X of X is regarded as the product space $X \times \mathbf{I}$. When X is a subspace of Y, the mapping cylinder $M(i)$ of the inclusion map $i : X \subset Y$ can be regarded as a subspace $(Y \times \{0\}) \cup (X \times \mathbf{I})$ of the product space $Y \times \mathbf{I}$, but M_i cannot be regarded thus unless X is *closed* in Y (cf. Sect. 4.11). If $Y = \{0\}$, we denote $C(X) = M(f)$, i.e.,

$$C(X) = \{0\} \cup (X \times (0, 1]),$$

which has the topology generated by open sets in the product space $X \times (0, 1]$ and sets $\{0\} \cup (X \times (0, \varepsilon))$, where $0 < \varepsilon \leq 1$. We call $C(X)$ the **(metrizable) cone** over X. The following subspace of $C(X)$ is called the **(metrizable) open cone** over X:

$$C^o(X) = \{0\} \cup (X \times (0, 1)).$$

Then, $C(X)$ and $C^o(X)$ are contractible.

Corollary 6.5.7. *The cone $C(X)$ over any ANR X is an AR. Hence, so is the open cone $C^o(X)$.* □

For a map to a paracompact space, we have the following lemma:

Lemma 6.5.8. *Let Y be a paracompact space and $f : X \to Y$ be a map. For each open cover $\mathcal{U}$ of $M(f)$, Y has a locally finite open cover $\mathcal{V}$ with a map $\alpha : Y \to (0, 1)$ such that, for each $V \in \mathcal{V}$, $V \cup (f^{-1}(V) \times (0, \sup\alpha(V)])$ is contained in some member of $\mathcal{U}$.*

Proof. For each $y \in Y$, let

$$\gamma(y) = \sup\big\{s \in (0,1) \bigm| \exists U \in \mathcal{U},\ \exists V : \text{open in } Y \text{ such that} \\ y \in V \cup (f^{-1}(V) \times (0,s)) \subset U\big\} > 0$$

Then, $\gamma : Y \to (0,1)$ is lower semi-continuous (l.s.c.). By Theorem 2.7.6, we have a map $\alpha : Y \to (0,1)$ such that $\alpha(y) < \gamma(y)$ for every $y \in Y$. For each $y \in Y$, we have $s_y > \alpha(y)$ and an open neighborhood V_y of y in Y such that $V_y \cup (f^{-1}(V_y) \times (0, s_y))$ is contained in some $U \in \mathcal{U}$ and $\alpha(y') < s_y$ for every $y' \in V_y$. Let $\mathcal{V} \in \operatorname{cov}(Y)$ be a locally finite open refinement of the open cover $\{V_y \mid y \in Y\}$. Then, α and $\mathcal{V}$ are as required. □

As seen above, the topology of $M(f)$ is different from that of M_f, but we have the following theorem:

Theorem 6.5.9. *For each map $f : X \to Y$, the natural bijection $\phi_f : M_f \to M(f)$ is a homotopy equivalence with a homotopy inverse $\psi : M(f) \to M_f$ such that*

$$\psi\phi_f \simeq \mathrm{id} \ \textit{rel.}\ Y \cup (X \times \{1\}) \ \textit{and}\ \phi_f\psi \simeq \mathrm{id}\ \textit{rel.}\ Y \cup (X \times \{1\}),$$

hence $M(f) \simeq M_f$ rel. $Y \cup X \times \{1\}$. If Y is paracompact, for each open cover $\mathcal{U}$ of $M(f)$, the homotopy inverse ψ can be chosen such that

$$\psi\phi_f \simeq_{\phi_f^{-1}(\mathcal{U})} \mathrm{id}\ \textit{rel.}\ Y \cup (X \times \{1\}) \ \textit{and}\ \phi_f\psi \simeq_{\mathcal{U}} \mathrm{id}\ \textit{rel.}\ Y \cup (X \times \{1\}).$$

Proof. We first prove the case where Y is paracompact. By Lemma 6.5.8, for each open cover $\mathcal{U}$ of $M(f)$, we have a map $\alpha : Y \to (0,1)$ such that

$$\big\{\{y\} \cup (f^{-1}(y) \times (0, \alpha(y)]) \bigm| y \in Y\big\} \prec \mathcal{U}.$$

Let $q_f : Y \oplus (X \times \mathbf{I}) \to M_f$ be the quotient map. Then, we can define a map $\psi : M(f) \to M_f$ by $\psi|Y = q_f|Y$ and

$$\psi(x,s) = \begin{cases} q_f(x,s) = (x,s) & \text{if } x \in X,\ s \geq \alpha(f(x)), \\ q_f(x, 2s - \alpha(f(x))) & \text{if } x \in X,\ \alpha(f(x))/2 \leq s \leq \alpha(f(x)), \\ q_f(x,0) = f(x) & \text{if } x \in X,\ 0 < s \leq \alpha(f(x))/2. \end{cases}$$

The continuity of ψ follows from that of $\psi|X \times (0,1]$ and $\psi|W$, where

$$W = Y \cup \{(x,s) \in X \times (0,1] \mid s < \alpha(f(x))/2\}.$$

Now, we define a homotopy $h : (Y \oplus (X \times \mathbf{I})) \times \mathbf{I} \to Y \oplus (X \times \mathbf{I})$ as follows: $h_0 = \mathrm{id}$, $h_t|Y = \mathrm{id}$ for each $t \in \mathbf{I}$, and

$$h_t(x,s) = \begin{cases} (x,s) & \text{if } x \in X,\ s \geq t\alpha(f(x)), \\ (x,\ 2s - t\alpha(f(x))) & \text{if } x \in X,\ t\alpha(f(x))/2 \leq s \leq t\alpha(f(x)), \\ (x,0) & \text{if } x \in X,\ s \leq t\alpha(f(x))/2. \end{cases}$$

Then, as is easily shown, h induces a $\mathcal{U}$-homotopy $h' : M(f) \times \mathbf{I} \to M(f)$ and the $\phi_f^{-1}(\mathcal{U})$-homotopy $h'' : M_f \times \mathbf{I} \to M_f$ such that

$$h'_0 = \mathrm{id},\ h'_1 = \phi_f\psi, \text{ and } h'_t|Y \cup (X \times \{1\}) = \mathrm{id} \text{ for each } t \in \mathbf{I};$$

$$h''_0 = \mathrm{id},\ h''_1 = \psi\phi_f, \text{ and } h''_t|Y \cup (X \times \{1\}) = \mathrm{id} \text{ for each } t \in \mathbf{I}.$$

The continuity of h' can be verified as follows: Since $\phi_f : M_f \to M(f)$ is continuous, the continuity of $h'|M(f) \times (0,1]$ can be shown in a manner similar to that of ψ. Observe that $X \times (0,1] \times \{0\}$ has the following open neighborhood in $M(f) \times \mathbf{I}$:

$$W' = \{(x,s,t) \in X \times (0,1] \times \mathbf{I} \mid s > t\alpha(f(x))\}.$$

The continuity of $h'|W'$ is obvious. For each neighborhood V of y in Y and $\varepsilon \in (0,1)$, it is easy to show that

$$h'((V \cup (f^{-1}(V) \times (0,\varepsilon))) \times \mathbf{I}) \subset V \cup (f^{-1}(V) \times (0,\varepsilon)).$$

Thus, h' is also continuous at $(y,0)$.

In the general case, since there are no covering estimations, we can take a constant map as α (e.g., $\alpha(y) = \frac{1}{2}$) in the above proof. □

For maps $f_i : X_i \to X_{i-1}$, $i = 1, \ldots, n$, let $M(f_1, \ldots, f_n) = \bigcup_{i=1}^n M(f_i)$ be the adjunction space, where each $X_i \times \{1\} \subset M(f_i)$ is identified with $X_i \subset M(f_{i+1})$. We call $M(f_1, \ldots, f_n)$ the **mapping telescope** of $f_1, \ldots, f_n$. By Theorem 6.5.9, Theorems 4.11.1 and 4.11.2 are also valid for the metrizable mapping cylinders.

Corollary 6.5.10. *For maps $f, g : X \to Y$, the following conditions are equivalent:*

(a) $f \simeq g$;
(b) $M(f) \simeq M(g)$ *rel.* $Y \cup X \times \{1\}$;
(c) *There is a map* $\varphi : M(f) \to M(g)$ *with* $\varphi|Y \cup X \times \{1\} = \mathrm{id}$. □

Corollary 6.5.11. *For maps $f : X \to Y$ and $g : Y \to Z$, $M(gf) \simeq M(g,f)$ rel.* $Z \cup X \times \{1\}$. □

These corollaries can also be proved directly in the same manner as Theorems 4.11.1 and 4.11.2.

6.6 Realizations of Simplicial Complexes

Let X be a space and $\mathcal{U} \in \operatorname{cov}(X)$. Let K be a simplicial complex and L be a subcomplex of K with $K^{(0)} \subset L$. Then, a map $f : |L| \to X$ is called a **partial $\mathcal{U}$-realization** of K in X if $\{f(\sigma \cap |L|) \mid \sigma \in K\} \prec \mathcal{U}$. A **full $\mathcal{U}$-realization** of K in X is a map $f : |K| \to X$ such that $\{f(\sigma) \mid \sigma \in K\} \prec \mathcal{U}$. We call $\mathcal{V} \in \operatorname{cov}(X)$ a **Lefschetz refinement** of $\mathcal{U}$ and denote

$$\mathcal{V} \underset{L}{\prec} \mathcal{U} \quad \text{or} \quad \mathcal{U} \underset{L}{\succ} \mathcal{V}$$

if $\mathcal{V} \prec \mathcal{U}$ and any partial $\mathcal{V}$-realization of an arbitrary simplicial complex in X extends to a full $\mathcal{U}$-realization in X. The following is called LEFSCHETZ'S CHARACTERIZATION of ANRs.

Theorem 6.6.1 (LEFSCHETZ). *A metrizable space X is an ANR if and only if any open cover of X admits a Lefschetz refinement.*

Proof. To prove the "only if" part, by the Arens–Eells Embedding Theorem 6.2.1, we may assume that X is a closed set in a normed linear space E. Then, we have an open neighborhood W of X in E and a retraction $r : W \to X$. For each open cover $\mathcal{U}$ of X, $r^{-1}(\mathcal{U})$ is an open cover of W, which has a refinement $\mathcal{V}$ consisting of open convex sets in E. We show that $\mathcal{V}|X$ is a Lefschetz refinement of $\mathcal{U}$.

Let K be a simplicial complex, L be a subcomplex of K with $K^{(0)} \subset L$, and let $f_0 : |L| \to X$ be a partial $(\mathcal{V}|X)$-realization of K. By induction, we can obtain maps $f_n : |L \cup K^{(n)}| \to W$, $n \in \mathbb{N}$, so that

$$f_n||L \cup K^{(n-1)}| = f_{n-1} \quad \text{and}$$

$$f_n(\sigma \cap |L \cup K^{(n)}|) \subset \langle f_0(\sigma \cap |L|)\rangle \quad \text{for each } \sigma \in K.$$

Indeed, given f_{n-1}, then for each $\sigma \in K^{(n)} \setminus (L \cup K^{(n-1)})$, $f_{n-1}|\partial\sigma$ extends to a map $f_\sigma : \sigma \to \langle f_0(\sigma \cap |L|)\rangle$ by the Dugundji Extension Theorem 6.1.1. Thus, f_n can be defined by $f_n|\sigma = f_\sigma$ for $\sigma \in K^{(n)} \setminus (L \cup K^{(n-1)})$. For each $\sigma \in K$, we write $\sigma \cap |L \cup K^{(n)}| = \bigcup_{i=1}^m \sigma_i$, where $\sigma_i \in L \cup K^{(n)}$ for each $i = 1, \cdots, m$. Then,

$$f_n(\sigma \cap |L \cup K^{(n)}|) = \bigcup_{i=1}^m f_{\sigma_i}(\sigma_i) \subset \bigcup_{i=1}^m \langle f_0(\sigma_i \cap |L|)\rangle \subset \langle f_0(\sigma \cap |L|)\rangle.$$

Let $f : |K| \to W$ be the map defined by $f||L \cup K^{(n)}| = f_n$. For each $\sigma \in K$, $f_0(\sigma \cap |L|)$ is contained in some $V \in \mathcal{V}$, which implies that

$$f(\sigma) \subset \langle f_0(\sigma \cap |L|)\rangle \subset V.$$

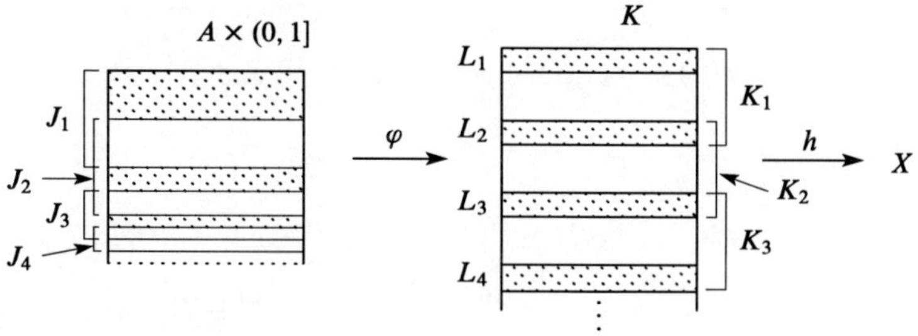

Fig. 6.14 The nerves K_n and L_n

Thus, f is a full $\mathcal{V}$-realization of K in W. Since $\mathcal{V} \prec r^{-1}(\mathcal{U})$, $r \circ f : |K| \to X$ is a full $\mathcal{U}$-realization of K in X, which is an extension of f_0.

To prove the "if" part, it suffices to show that X is an ANE. Let $f : A \to X$ be a map from a closed set A in a metrizable space Y. Given $d \in \mathrm{Metr}(X)$, take open covers of X as follows:

$$\mathcal{U}_1 \underset{L}{\succ} \mathcal{V}_1 \overset{*}{\succ} \mathcal{U}_2 \underset{L}{\succ} \mathcal{V}_2 \overset{*}{\succ} \mathcal{U}_3 \underset{L}{\succ} \mathcal{V}_3 \overset{*}{\succ} \cdots,$$

where $\operatorname{mesh} \mathcal{U}_n < 2^{-n}$ for each $n \in \mathbb{N}$. For each $n \in \mathbb{N}$, let

$$\mathcal{W}_n = \{ f^{-1}(U) \times J_n \mid U \in \mathcal{U}_{n+2} \},$$

where $J_1 = (2^{-1}, 1]$ and $J_n = (2^{-n}, 2^{-n}3)$ for $n > 1$. Then, $\mathcal{W} = \bigcup_{n \in \mathbb{N}} \mathcal{W}_n$ is an open cover of $A \times (0, 1]$. Let K be the nerve of $\mathcal{W}$ with $\varphi : A \times (0, 1] \to |K|$ a canonical map. For each $n \in \mathbb{N}$, let K_n and L_n be the nerves of $\mathcal{W}_n \cup \mathcal{W}_{n+1}$ and $\mathcal{W}_n$, respectively, which are naturally regarded as subcomplexes of K (Fig 6.14). Then, we have

$$K = \bigcup_{n \in \mathbb{N}} K_n, \ K_n \cap K_{n+1} = L_{n+1} \ \text{ and } \ L_n \cap L_{n+1} = \emptyset.$$

For each $n \in \mathbb{N}$, let $g'_n : |L_n^{(0)}| \to X$ be a map such that $g'_n(W) \in f \mathrm{pr}_A(W)$ for each $W \in \mathcal{W}_n$. Observe that g'_n is a partial $\mathcal{V}_{n+1}$-realization of L_n in X. Then, g'_n extends to a full $\mathcal{U}_{n+1}$-realization $g_n : |L_n| \to X$. We define $h'_n : |L_n \cup L_{n+1}| \to X$ by $h'_n||L_n| = g_n$ and $h'_n||L_{n+1}| = g_{n+1}$. Observe that h'_n is a partial $\mathcal{V}_n$-realization of K_n in X. Then, h'_n extends to a full $\mathcal{U}_n$-realization $h_n : |K_n| \to X$. We can define a map $h : |K| \to X$ by $h||K_n| = h_n$. Thus, we have the map $h\varphi : A \times (0, 1] \to X$.

For each $(a, t) \in A \times (0, 1]$, let $\sigma \in K$ be the carrier of $\varphi(a, t)$ and $W \in \sigma^{(0)}$. Since φ is a canonical map, we have $(a, t) \in W$ (Proposition 4.9.1). When $\sigma \in K_n$, $h(\sigma) = h_n(\sigma)$ is contained in some $U \in \mathcal{U}_n$, hence $h\varphi(a, t) \in U$. Moreover, since $W \in \mathcal{W}_n \cup \mathcal{W}_{n+1}$, we have

$$h(W) = g_i'(W) \in f\,\mathrm{pr}_A(W) \cap U \quad (\text{where } i = n \text{ or } n+1),$$

$$f(a) \in f\,\mathrm{pr}_A(W) \in \mathcal{U}_{n+2} \cup \mathcal{U}_{n+3} \text{ and } t > 2^{-n-1}.$$

Then, it follows that

$$\begin{aligned} d(h\varphi(a,t), f(a)) &\leq d(h\varphi(a,t), h(W)) + d(h(W), f(a)) \\ &\leq \operatorname{diam} U + \operatorname{diam} f\,\mathrm{pr}_A(W) \\ &< 2^{-n} + 2^{-n-2} < 2^{-n-1}3 < 3t. \end{aligned}$$

Therefore, $h\varphi$ can be extended to a homotopy $\tilde{h} : A \times \mathbf{I} \to X$ by $\tilde{h}_0 = f$.

On the other hand, since $|K|$ is an ANE, φ extends over a neighborhood of $A \times (0,1]$ in $Y \times (0,1]$, hence so does $h\varphi$. Then, we can apply Lemma 6.5.1 to extend f over a neighborhood of A in Y. □

Recall that a space X is **homotopy dominated** by a space Y if there are maps $f : X \to Y$ and $g : Y \to X$ such that $gf \simeq \mathrm{id}_X$. If $gf \simeq_{\mathcal{U}} \mathrm{id}_X$ for an open cover $\mathcal{U}$ of X (resp. $gf \simeq_\varepsilon \mathrm{id}_X$ for $\varepsilon > 0$), we say that X is **$\mathcal{U}$-homotopy dominated** (resp. **ε-homotopy dominated**) by Y. The following theorem is HANNER'S CHARACTERIZATION of ANRs.

Theorem 6.6.2 (HANNER). *For a metric space $X = (X,d)$, the following are equivalent:*

(a) *X is an ANR;*
(b) *For each open cover $\mathcal{U}$ of X, there is a simplicial complex K such that X is $\mathcal{U}$-homotopy dominated by $|K|$;*
(c) *For any $\varepsilon > 0$, X is ε-homotopy dominated by an ANE.*

Proof. The implication (b) ⇒ (c) follows from the fact that $|K|$ is an ANE for any simplicial complex K. We will prove the implications (a) ⇒ (b) and (c) ⇒ (a).

(a) ⇒ (b): For each open cover $\mathcal{U}$ of X, applying Theorems 6.6.1 and 6.3.6 with Proposition 6.3.1, we can take open refinements as follows:

$$\mathcal{U} \underset{h}{\succ} \mathcal{V} \overset{*}{\succ} \mathcal{V}' \underset{L}{\succ} \mathcal{W} \overset{*}{\succ} \mathcal{W}'.$$

Let K be the nerve of $\mathcal{W}'$ with a canonical map $f : X \to |K|$. For each $W \in \mathcal{W}' = K^{(0)}$, choosing $g'(W) \in W$, we can obtain a partial $\mathcal{W}$-realization $g' : K^{(0)} \to X$ of K, which extends to a full $\mathcal{V}'$-realization $g : |K| \to X$ of K. Observe that gf is $\mathcal{V}$-close to id_X, which means $gf \simeq_{\mathcal{U}} \mathrm{id}_X$. Thus, X is $\mathcal{U}$-homotopy dominated by K.

(c) ⇒ (a): We show that X is an ANE. Let $f : A \to X$ be a map of a closed set A in a metrizable space Y. For each $n \in \mathbb{N}$, we have an ANE X_n, maps $j_n : X \to X_n$ and $k_n : X_n \to X$, and a 3^{-n}-homotopy $h^n : X \times \mathbf{I} \to X$ with $h_0^n = \mathrm{id}_X$ and $h_1^n = k_n j_n$. For each $n \in \mathbb{N}$, let $I_n' = [3^{-n}2, 3^{-(n-1)}]$ and $I_n'' = [3^{-n}, 3^{-n}2]$. Since each X_n is an ANE, $j_n f$ extends to a map $f_n : U_n \to X_n$ from an open neighborhood U_n of A in Y. We may assume that $U_{n+1} \subset U_n$ for each $n \in \mathbb{N}$. For each $n \in \mathbb{N}$, we define a map $g'_n : U_n \times I'_n \to X$ by

$$g'_n(y,t) = h^{n+1}(k_n f_n(y), 3 - 3^n t).$$

Then, for each $y \in U_n$,

$$g'_n(y, 3^{-(n-1)}) = k_n f_n(y) \quad \text{and} \quad g'_n(y, 3^{-n}2) = k_{n+1} j_{n+1} k_n f_n(y).$$

Also, for each $(a,t) \in A \times I'_n$,

$$\begin{aligned} d(g'_n(a,t), f(a)) &= d(h^{n+1}(k_n j_n f(a), 3 - 3^n t), f(a)) \\ &\leq 3^{-(n+1)} + 3^{-n} < 3^{-n}2 \leq t. \end{aligned}$$

For each $n \in \mathbb{N}$, let

$$A_n = A \times I''_n \cup U_{n+1} \times \partial I''_n \subset U_{n+1} \times I''_n.$$

We define a map $h'_n : A_n \to X_{n+1}$ as follows:

$$h'_n(y,t) = \begin{cases} j_{n+1} k_n f_n(y) & \text{if } (y,t) \in U_{n+1} \times \{3^{-n}2\}, \\ f_{n+1}(y) & \text{if } (y,t) \in U_{n+1} \times \{3^{-n}\}, \\ j_{n+1} h^n(f(y), 3^n t - 1) & \text{if } (y,t) \in A \times I''_n. \end{cases}$$

Since X_{n+1} is an ANE, h'_n extends to a map $h''_n : V_n \to X_{n+1}$ of a neighborhood V_n of A_n in $U_{n+1} \times I''_n$. Thus, we obtain the map $g''_n = k_{n+1} h''_n : V_n \to X$. Observe, for each $y \in U_{n+1}$,

$$g''_n(y, 3^{-n}2) = k_{n+1} j_{n+1} k_n f_n(y) \quad \text{and} \quad g''_n(y, 3^{-n}) = k_{n+1} f_{n+1}(y)$$

and for each $(a,t) \in A \times I''_n$,

$$\begin{aligned} d(g''_n(a,t), f(a)) &= d(k_{n+1} j_{n+1} h^n(f(a), 3^n t - 1), f(a)) \\ &< 3^{-(n+1)} + 3^{-n} < 3^{-n}2 \leq 2t. \end{aligned}$$

Then, $W = \bigcup_{n \in \mathbb{N}} ((U_n \times I'_n) \cup V_n)$ is a neighborhood of $A \times (0,1]$ in $Y \times (0,1]$. We define a map $g : W \to X$ by

$$g|U_n \times I'_n = g'_n \quad \text{and} \quad g|V_n = g''_n.$$

Since $d(g(a,t), f(a)) < 2t$ for each $(a,t) \in A \times (0,1]$, we can apply Lemma 6.5.1 to extend f over a neighborhood of A in X. — Fig. 6.15. □

Remark 9. In condition (b) of Theorem 6.6.2, we can take K as a locally finite-dimensional simplicial complex with card $K^{(0)} \leq w(X)$. Indeed, if X is finite,

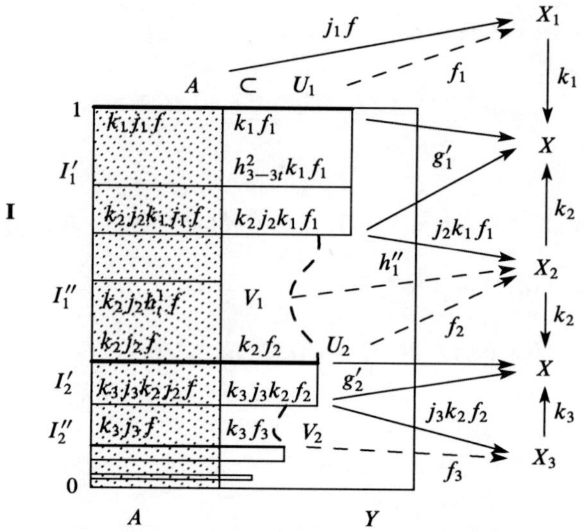

Fig. 6.15 Extension using Lemma 6.5.1

then X is itself a 0-dimensional simplicial complex. When X is infinite, by Theorem 4.9.9, each open cover of X has a locally finite σ-discrete open refinement with the locally finite-dimensional nerve. Then, in the proof of (a) $\Rightarrow$ (b), we can take $\mathcal{W}'$ such that $\mathcal{W}'$ is σ-discrete and the nerve K of $\mathcal{W}'$ is locally finite-dimensional. It follows that

$$\operatorname{card} K \leq \aleph_0 \operatorname{card} \mathcal{W}' \leq \aleph_0 w(X) = w(X).$$

Corollary 6.6.3. *Let $f : X \to Y$ be a map from a paracompact space X to an ANR Y and $\mathcal{U}$ an open cover of Y. Then, each open cover $\mathcal{V}$ of X has an open refinement $\mathcal{W}$ with a map $\psi : |N(\mathcal{W})| \to Y$ such that $\psi\varphi \simeq_{\mathcal{U}} f$ for any canonical map $\varphi : X \to |N(\mathcal{W})|$.*

Proof. Let $\mathcal{U}' \in \operatorname{cov}(Y)$ be such that $\operatorname{st}\mathcal{U}'$-close maps are $\mathcal{U}$-homotopic. By Theorem 6.6.2, we have a simplicial complex K with maps $\varphi' : Y \to |K|$ and $\psi' : |K| \to Y$ such that $\psi'\varphi' \simeq_{\mathcal{U}'} \mathrm{id}_Y$, hence $f \simeq_{\mathcal{U}'} \psi'\varphi' f$. Replacing K with a small subdivision, we may assume that $\mathcal{O}_K \prec \psi'^{-1}(\mathcal{U}')$ (Theorem 4.7.11). Let $\mathcal{W} \in \operatorname{cov}(X)$ be a common refinement of $(\varphi' f)^{-1}(\mathcal{O}_K)$ and $\mathcal{V}$. Assigning to each $W \in \mathcal{W} = N(\mathcal{W})^{(0)}$ a $g(W) \in K^{(0)}$ such that $W \subset \varphi'^{-1}(O_K(g(W)))$, we can obtain a simplicial map $g : N(\mathcal{W}) \to K$. Thus, we have a map $\psi = \psi' g : |N(\mathcal{W})| \to Y$. Let $\varphi : X \to |N(\mathcal{W})|$ be a canonical map, that is, $\varphi^{-1}(O_{N(\mathcal{W})}(W)) \subset W$ for each $W \in \mathcal{W}$. For each $x \in X$, choose $W \in \mathcal{W}$ so that $x \in \varphi^{-1}(O_{N(\mathcal{W})}(W))$. Then, $g\varphi(x) \in g(O_{N(\mathcal{W})}(W)) \subset O_K(g(W))$ and $\varphi' f(x) \in \varphi' f(W) \subset O_K(g(W))$. Since $\psi'(\mathcal{O}_K) \prec \mathcal{U}'$, it follows that $\psi\varphi = \psi' g\varphi$

is $\mathcal{U}'$-close to $\psi'\varphi' f$. Consequently, $\psi\varphi$ is $\mathrm{st}\mathcal{U}'$-close to f, which implies that $\psi\varphi \simeq_{\mathcal{U}} f$. □

Remark 10. In Corollary 6.6.3, $\mathcal{W}$ can be taken to be locally finite and σ-discrete in X and to have a locally finite-dimensional nerve $N(\mathcal{W})$ (cf. Theorem 4.9.9). If X is separable, we can take a star-finite countable open refinement $\mathcal{W}$, which has the locally finite nerve $N(\mathcal{W})$ (cf. Theorem 4.9.10). When X is compact, we can take a finite open refinement $\mathcal{W}$.

By Theorem 6.6.2 and Corollary 4.12.4, we have the following corollary:

Corollary 6.6.4. *Every ANR X has the homotopy type of a locally finite-dimensional simplicial complex K with* $\operatorname{card} K^{(0)} \leq w(X)$. *In particular, every separable ANR has the homotopy type of a countable locally finite simplicial complex.* □

For every simplicial complex K, we have $|K| \simeq |K|_{\mathrm{m}}$ by Theorem 4.9.6 and $|K|_{\mathrm{m}}$ is an ANR by Theorem 6.2.6. Then, we have the following corollary:

Corollary 6.6.5. *A space X has the homotopy type of a simplicial complex if and only if it has the homotopy type of an ANR.* □

Moreover, by Corollary 4.13.10, we can obtain:

Corollary 6.6.6. *Let X and Y be ANRs. Then, every weak homotopy equivalence $f : X \to Y$ is a homotopy equivalence.* □

A subset $A \subset X$ is said to be **homotopy dense** in X if there exists a homotopy $h : X \times \mathbf{I} \to X$ such that $h_0 = \mathrm{id}$ and $h(X \times (0,1]) \subset A$. When X is paracompact, for each open cover $\mathcal{U} \in \mathrm{cov}(X)$, X is $\mathcal{U}$-homotopy dominated by A and A is $\mathcal{U}|A$-homotopy dominated by X. In fact, we have a lower semi-continuous function $\gamma : X \to (0,1]$ defined by

$$\gamma(x) = \sup\{t \in \mathbf{I} \mid h(\{x\} \times [0,t]) \subset U \text{ for some } U \in \mathcal{U}\}.$$

By Theorem 2.7.6, there is a map $\alpha : X \to (0,1]$ such that $\alpha(x) < \gamma(x)$ for each $x \in X$. We can define a $\mathcal{U}$-homotopy $h^\alpha : X \times \mathbf{I} \to X$ by $h^\alpha(x,t) = h(x, t\alpha(x))$. Then, $h^\alpha_1 : X \to A$, $h^\alpha_1 \simeq_{\mathcal{U}} \mathrm{id}_X$ in X, and $h^\alpha_1|A \simeq_{\mathcal{U}} \mathrm{id}_A$ in A. Thus, from Theorem 6.6.2, we have the following corollary:

Corollary 6.6.7. *Let X be a metrizable space and A be a homotopy dense subset of X. Then, X is an ANR if and only if A is an ANR.* □

Evidently, if $A \subset B \subset X$ and A is homotopy dense in X, then B is also homotopy dense in X. For any $A \subset \partial\mathbf{I}^n$, $\mathbf{I}^n \setminus A$ is homotopy dense in $\mathbf{I}^n$. As is easily observed, the radial interior $\mathrm{rint}\, \boldsymbol{Q} = \bigcup_{n\in\mathbb{N}}[-1+2^{-n}, 1-2^{-n}]^{\mathbb{N}}$ and the pseudo-interior $(-1,1)^{\mathbb{N}}$ of the Hilbert cube $\boldsymbol{Q} = [-1,1]^{\mathbb{N}}$ are homotopy dense in $\boldsymbol{Q}$. By the following proposition, $\boldsymbol{Q}_f = [-1,1]^{\mathbb{N}}_f$ and $(-1,1)^{\mathbb{N}}_f$ are also homotopy dense in $\boldsymbol{Q}$.

Proposition 6.6.8. *Let $x_0 \in A \subset X$. If A is homotopy dense in X and X is contractible then the following set $A^{\mathbb{N}}_f$ is homotopy dense in $X^{\mathbb{N}}$:*

$$A^{\mathbb{N}}_f = \{x \in A^{\mathbb{N}} \mid x(n) = x_0 \text{ except for finitely many } n \in \mathbb{N}\}.$$

Proof. We have a homotopy $h : X \times \mathbf{I} \to X$ such that $h_0 = \mathrm{id}$, $h_t(X) \subset A$ for every $t > 0$, and $h_1(X) = \{x_0\}$. Indeed, let $f : X \times \mathbf{I} \to X$ be a contraction and $g : X \times \mathbf{I} \to X$ be a homotopy such that $g_0 = \mathrm{id}$ and $g_t(X) \subset A$ for every $t > 0$. Then, h can be defined by

$$h_t(x) = \begin{cases} g_{2t} f_{2t}(x) & \text{for } 0 \le t \le 1/2, \\ g_{2-2t} f_{2-2t}(x_0) & \text{for } 1/2 \le t \le 1. \end{cases}$$

Now, we define the homotopy $\varphi : X^{\mathbb{N}} \times \mathbf{I} \to X^{\mathbb{N}}$ as follows: $\varphi_0 = \mathrm{id}$,

$$\begin{aligned} \varphi_1(x) &= (x_0, x_0, x_0, \dots) \\ \varphi_{2^{-1}}(x) &= (h_{2^{-1}}(x(1)), x_0, x_0, \dots) \\ \varphi_{2^{-2}}(x) &= (h_{2^{-2}}(x(1)), h_{2^{-1}}(x(2)), x_0, x_0, \dots) \\ &\vdots \\ \varphi_{2^{-n}}(x) &= (h_{2^{-n}}(x(1)), h_{2^{-n+1}}(x(2)), \dots, h_{2^{-1}}(x(n)), x_0, x_0, \dots) \\ &\vdots \end{aligned}$$

and for $2^{-n} < t < 2^{-n+1}$,

$$\varphi_t(x) = (h_t(x(1)), \dots, h_{2^{n-2}t} x(n-1), h_{2^{n-1}t}(x(n)), x_0, x_0, \dots).$$

Then, $\varphi_t(X^{\mathbb{N}}) \subset A^{\mathbb{N}}_f$ for every $t > 0$. □

A space X is said to be **homotopically trivial** if, for each $n \in \mathbb{N}$, every map $f : \mathbf{S}^{n-1} \to X$ is null-homotopic, that is, f extends over $\mathbf{B}^n$. As is easily observed, X is homotopically trivial if and only if the map from X to the singleton $\{0\}$ is a weak homotopy equivalence.

Proposition 6.6.9. *Let K be a simplicial (or cell) complex and L be a subcomplex of K. If X is homotopically trivial then any map $f : |L| \to X$ extends to a map $\tilde{f} : |K| \to X$.*

Proof. For each $n \in \omega$, let $K_n = K^{(n)} \cup L$. Then, $K = \bigcup_{n \in \omega} K_n$. Since X is homotopically trivial, we can inductively construct maps $f_n : |K_n| \to X$, $n \in \omega$, so that $f_n||K_{n-1}| = f_{n-1}$, where $K_{-1} = L$ and $f_{-1} = f$. The desired extension $\tilde{f}$ is defined by $\tilde{f}||K_n| = f_n$ for each $n \in \omega$. □

Every contractible space is homotopically trivial. For an ANR X, the converse is also true. In fact, if X is homotopically trivial, it follows from Corollary 6.6.6 that X is homotopy equivalent to the singleton $\{0\}$, which means that X is contractible. As a consequence, the following characterization follows from Corollary 6.2.9:

Theorem 6.6.10. *A metrizable space is an AR if and only if it is a homotopically trivial ANR.* □

6.7 Fine Homotopy Equivalences

For an open cover $\mathcal{U}$ of a space Y, a map $f : X \to Y$ is called a **$\mathcal{U}$-homotopy equivalence** if there is a map $g : Y \to X$ such that $gf \simeq_{f^{-1}(\mathcal{U})} \mathrm{id}_X$ and $fg \simeq_{\mathcal{U}} \mathrm{id}_Y$, where g is called a **$\mathcal{U}$-homotopy inverse** of f. For every simplicial complex K, $\mathrm{id} : |K| \to |K|_{\mathrm{m}}$ is an $\mathcal{O}_K$-homotopy equivalence (cf. Theorem 4.9.6). We call $f : X \to Y$ a **fine homotopy equivalence** if f is a $\mathcal{U}$-homotopy equivalence for each open cover $\mathcal{U}$ of Y. For example, if $X = Y \times Z$ and Z is contractible, then the projection $\mathrm{pr}_Y : X \to Y$ is a fine homotopy equivalence. Note that the image of a fine homotopy equivalence is dense in its range (or codomain).

Proposition 6.7.1. *A subset X of a metrizable space Y is homotopy dense in Y if and only if the inclusion map $X \subset Y$ is a fine homotopy equivalence.*

Proof. The "only if" part was shown in the arguments before Corollary 6.6.7.

To prove the "if" part, assume that the inclusion map $X \subset Y$ is a fine homotopy equivalence and let $d \in \mathrm{Metr}(Y)$. For each $n \in \mathbb{N}$, we have a map $f_n : Y \to X$ with 3^{-n}-homotopies $\varphi^{(n)} : Y \times \mathbf{I} \to Y$ and $\psi^{(n)} : X \times \mathbf{I} \to X$ such that $\varphi^{(n)}_0 = \mathrm{id}_Y$, $\varphi^{(n)}_1 = f_n$, $\psi^{(n)}_0 = \mathrm{id}_X$, and $\psi^{(n)}_1 = f_n|X$. We can define a homotopy $h : Y \times \mathbf{I} \to Y$ as follows: $h_0 = \mathrm{id}$ and for $t > 0$,

$$h_t = \begin{cases} f_i \varphi^{(i+1)}_{3-3^i t} & \text{if } 3^{-i}2 \le t \le 3^{-i+1}, \\ \psi^{(i)}_{3^i t-1} f_{i+1} & \text{if } 3^{-i} \le t \le 3^{-i}2. \end{cases}$$

Then, $d(h_t, \mathrm{id}) < 2t$ because

$$d(h_t, \mathrm{id}) < 3^{-i} + 3^{-i-1} < 3^{-i}2 \text{ if } 3^{-i} \le t \le 3^{-i+1}.$$

Hence, h is continuous at points of $Y \times \{0\}$. Since $h(Y \times (0,1]) \subset X$, it follows that X is homotopy dense in Y. □

Proposition 6.7.2. *For each map $f : X \to Y$, both of the collapsings $c_f : M(f) \to Y$ and $c_f : M_f \to Y$ are fine homotopy equivalences.*

Proof. Let $j : Y \to M(f)$ be the inclusion map. Then, $c_f j = \mathrm{id}_Y$. The homotopy $h^f : M(f) \times \mathbf{I} \to M(f)$ is a $c_f^{-1}(\mathcal{U})$-homotopy from $c_f = jc_f$ to $\mathrm{id}_{M(f)}$ for every $\mathcal{U} \in \mathrm{cov}(Y)$. Thus, $c_f : M(f) \to Y$ is a fine homotopy equivalence. The same proof is valid for $c_f : M_f \to Y$. □

The following has been shown in Theorem 6.5.9 as an additional statement:

Proposition 6.7.3. *Let Y be a paracompact space. For each map $f : X \to Y$, the natural bijection $\phi_f : M_f \to M(f)$ is a fine homotopy equivalence.* □

Here, we note the following:

Proposition 6.7.4. *Let $f : X \to Y$ be a fine homotopy equivalence and Z be a paracompact space. Then, a map $g : Y \to Z$ is a fine homotopy equivalence if and only if gf is a fine homotopy equivalence.*

$$\begin{array}{ccc} X & \xrightarrow{f} & Y \\ & {\scriptstyle gf}\searrow & \downarrow {\scriptstyle g} \\ & & Z \end{array}$$

Proof. First, assume that g is a fine homotopy equivalence. For each $\mathcal{U} \in \operatorname{cov}(Z)$, we have $\mathcal{V} \in \operatorname{cov}(Z)$ such that $\operatorname{st}^2 \mathcal{V} \prec \mathcal{U}$. Then, g has a $\mathcal{V}$-homotopy inverse $g' : Z \to Y$ and f has a $g^{-1}(\mathcal{V})$-homotopy inverse $f' : Y \to X$. We show that $f'g'$ is a $\mathcal{U}$-homotopy inverse of gf. Since $g(g^{-1}(\mathcal{V})) \prec \mathcal{V}$, we have $gff'g' \simeq_{\mathcal{V}} gg' \simeq_{\mathcal{V}} \mathrm{id}_Z$, hence $gff'g' \simeq_{\mathcal{U}} \mathrm{id}_Z$. On the other hand,

$$f'g'gf \simeq_{f'(g^{-1}(\mathcal{V}))} f'f \simeq_{f^{-1}(g^{-1}(\mathcal{V}))} \mathrm{id}_X,$$

where $f'(g^{-1}(\mathcal{V})) \prec f^{-1}(g^{-1}(\operatorname{st}\mathcal{V}))$. Hence, $f'g'gf \simeq_{f^{-1}(g^{-1}(\mathcal{U}))} \mathrm{id}_X$. Thus, gf is a fine homotopy equivalence.

Next, assume that gf is a fine homotopy equivalence. For each $\mathcal{U} \in \operatorname{cov}(Z)$, we have $\mathcal{V} \in \operatorname{cov}(Z)$ such that $\operatorname{st}^3 \mathcal{V} \prec \mathcal{U}$. Then, f has a $g^{-1}(\mathcal{V})$-homotopy inverse $f' : Y \to X$ and gf has a $\mathcal{V}$-homotopy inverse $h : Z \to X$. Since $fhg(g^{-1}(\mathcal{V})) \prec fh(\mathcal{V})$, we have

$$fhg \simeq_{fh(\mathcal{V})} fhgff' \simeq_{g^{-1}(\mathcal{V})} ff' \simeq_{g^{-1}(\mathcal{V})} \mathrm{id}_Y,$$

where $fh(\mathcal{V}) \prec g^{-1}(\operatorname{st}\mathcal{V})$ because $gfh(\mathcal{V}) \prec \operatorname{st}\mathcal{V}$. Hence, $fhg \simeq_{g^{-1}(\mathcal{U})} \mathrm{id}_Y$. On the other hand, $gfh \simeq_{\mathcal{V}} \mathrm{id}_Z$. Thus, g is a fine homotopy equivalence. □

From Theorem 6.6.2, it follows that the range of a fine homotopy equivalence of an ANR is an ANR if it is metrizable. This extends as follows:

Theorem 6.7.5 (G. Kozlowski). *Let $f : X \to Y$ be a map from an ANR X to a metrizable space Y such that $f(X)$ is dense in Y and, for each open cover $\mathcal{U}$ of Y, there is a map $g : Y \to X$ with $gf \simeq_{f^{-1}(\mathcal{U})} \mathrm{id}_X$. Then, f is a fine homotopy equivalence and Y is an ANR.*

Proof. First, we prove the following two claims:

Claim (1). For a map $g : Y \to X$, if gf is $f^{-1}(\mathcal{U})$-close to id_X then fg is $\mathrm{st}\mathcal{U}$-close to id_Y.

For each $y \in Y$, choose $U_1, U_2 \in \mathcal{U}$ so that $y \in U_1$ and $fg(y) \in U_2$, i.e., $y \in U_1 \cap (fg)^{-1}(U_2)$. Since $f(X)$ is dense in Y, we have an $x \in X$ such that $f(x) \in U_1 \cap (fg)^{-1}(U_2)$. Since gf is $f^{-1}(\mathcal{U})$-close to id_X, we have a $U_3 \in \mathcal{U}$ such that $x, gf(x) \in f^{-1}(U_3)$, i.e., $f(x), fgf(x) \in U_3$. Then, $U_1 \cap U_3 \neq \emptyset$ and $U_2 \cap U_3 \neq \emptyset$. Hence, it follows that $y, fg(y) \in \mathrm{st}(U_3, \mathcal{U})$.

Claim (2). Each open cover $\mathcal{U}$ of Y has an open refinement $\mathcal{V}$ such that $f^{-1}(\mathcal{V})$ is an h-refinement of $f^{-1}(\mathcal{U})$.

For an open star-refinement $\mathcal{W}$ of $\mathcal{U}$, we have a map $g : Y \to X$ such that $gf \simeq_{f^{-1}(\mathcal{W})} \mathrm{id}_X$. By Corollary 6.3.5, $f^{-1}(\mathcal{W})$ has an h-refinement $\mathcal{W}'$. Let $\mathcal{V} \in \mathrm{cov}(Y)$ be a common refinement of $g^{-1}(\mathcal{W}')$ and $\mathcal{U}$. Then, $\mathcal{V}$ is the desired refinement. Indeed, let $k, k' : Z \to X$ be $f^{-1}(\mathcal{V})$-close maps. Then, gfk and gfk' are $\mathcal{W}'$-close, so $gfk \simeq_{f^{-1}(\mathcal{W})} gfk'$. Since $gf \simeq_{f^{-1}(\mathcal{W})} \mathrm{id}_X$, it follows that $k \simeq_{\mathrm{st}\, f^{-1}(\mathcal{W})} k'$, so $k \simeq_{f^{-1}(\mathcal{U})} k'$.

Now, we will prove the theorem. Because of Theorem 6.6.2, it suffices to show that f is a fine homotopy equivalence. For each open cover $\mathcal{U}$ of Y, Y has an admissible metric ρ such that $\{\overline{\mathrm{B}}_\rho(y, 1) \mid y \in Y\} \prec \mathcal{U}$ (2.7.7(1)). Using Claim (2) inductively, we can take $\mathcal{U}_i \in \mathrm{cov}(Y)$, $i \in \omega$, such that

$$\mathcal{U}_0 \overset{*}{\succ} \mathcal{U}_1 \overset{*}{\succ} \cdots, \ \mathrm{mesh}\,\mathcal{U}_n < 2^{-n-1} \text{ and } f^{-1}(\mathcal{U}_n) \underset{h}{\prec} f^{-1}(\mathcal{U}_{n-1}).$$

By the condition on f, we have maps $g_n : Y \to X$, $n \in \mathbb{N}$, such that $g_n f \simeq_{f^{-1}(\mathcal{U}_{n+2})} \mathrm{id}_X$. Then, fg_n is $\mathcal{U}_{n+1}$-close to id_Y by Claim (1). Since fg_n and fg_{n+1} are $\mathcal{U}_n$-close, that is, g_n and g_{n+1} are $f^{-1}(\mathcal{U}_n)$-close, we have an $f^{-1}(\mathcal{U}_{n-1})$-homotopy $h_n : Y \times \mathbf{I} \to X$ such that $h_{n,0} = g_n$ and $h_{n,1} = g_{n+1}$. Let $h : Y \times \mathbf{I} \to Y$ be the homotopy defined as follows:

$$h(y,t) = \begin{cases} fh_n(y, 2 - 2^n t) & \text{if } 2^{-n} \le t \le 2^{-n+1},\ n \in \mathbb{N} \\ y & \text{if } t = 0. \end{cases}$$

Then, h is a $\mathcal{U}$-homotopy with $h_0 = \mathrm{id}_Y$ and $h_1 = fg_1$. Indeed, for each $y \in Y$,

$$\begin{aligned} \mathrm{diam}\, h(\{y\} \times \mathbf{I}) &= \mathrm{diam}\, h(\{y\} \times (0,1]) = \mathrm{diam} \bigcup_{n \in \mathbb{N}} fh_n(\{y\} \times \mathbf{I}) \\ &\le \sum_{n \in \mathbb{N}} \mathrm{mesh}\,\mathcal{U}_{n-1} < \sum_{n \in \mathbb{N}} 2^{-n} = 1. \end{aligned}$$

Thus, we have $fg_1 \simeq_{\mathcal{U}} \mathrm{id}_Y$. Recall $g_1 f \simeq_{f^{-1}(\mathcal{U}_3)} \mathrm{id}_X$. Then, it follows that g_1 is a $\mathcal{U}$-homotopy inverse of f. Consequently, f is a fine homotopy equivalence. □

In view of Proposition 6.7.1, Theorem 6.7.5 yields the following:

Corollary 6.7.6. *Let X be an ANR that is a dense set in a metrizable space Y. Suppose that, for each open cover $\mathcal{U}$ of Y, there is a map $g : Y \to X$ with $g|X \simeq_{\mathcal{U}} \mathrm{id}_X$ in X. Then, X is homotopy dense in Y and Y is an ANR.* □

Using the mapping cylinders, we can characterize fine homotopy equivalences between ANRs as follows:

Theorem 6.7.7. *For a map $f : X \to Y$ between ANRs, the following are equivalent:*

(a) *f is a fine homotopy equivalence;*
(b) *$X \times (0, 1]$ is homotopy dense in $M(f)$;*
(c) *The natural map $q_f : X \times \mathbf{I} \to M(f)$ is a fine homotopy equivalence.*

Proof. Because of Proposition 6.7.1, condition (b) means that the inclusion $X \times (0, 1] \subset M(f)$ is a fine homotopy equivalence. The equivalence (b) ⇔ (c) follows from Proposition 6.7.4.

$$\begin{array}{ccc} X \times (0,1] & \subset & X \times \mathbf{I} \\ \cap & \swarrow{\scriptstyle q_f} & \\ M(f) & & \end{array}$$

(c) ⇒ (a): Since the projection $\mathrm{pr}_X : X \times \mathbf{I} \to X$ and the collapsing $c_f : M(f) \to Y$ are fine homotopy equivalences (Proposition 6.7.2), we can apply Proposition 6.7.4 to see that if the natural map $q_f : X \times \mathbf{I} \to M(f)$ is a fine homotopy equivalence, then $f : X \to Y$ is a fine homotopy equivalence.

$$\begin{array}{ccc} X \times \mathbf{I} & \xrightarrow{q_f} & M(f) \\ {\scriptstyle \mathrm{pr}_X}\downarrow & \searrow & \downarrow{\scriptstyle c_f} \\ X & \xrightarrow[f]{} & Y \end{array}$$

(a) ⇒ (b): Due to Corollary 6.7.6, for each open cover $\mathcal{U}$ of $M(f)$, it suffices to construct a map $k : M(f) \to X \times (0, 1]$ such that $k|X \times (0, 1]$ is $\mathcal{U}$-homotopic to id in $X \times (0, 1]$. By Lemma 6.5.8, we have an open cover $\mathcal{V}$ with a map $\alpha : Y \to (0, 1)$ such that, for each $V \in \mathcal{V}$,

$$V \cup (f^{-1}(V) \times (0, \sup \alpha(V)])$$

is contained in some member of $\mathcal{U}$. Then, there exist a map $g : Y \to X$ and an $f^{-1}(\mathcal{V})$-homotopy $h : X \times \mathbf{I} \to X$ such that $h_0 = \mathrm{id}_X$ and $h_1 = gf$. We define a map $k : M(f) \to X \times (0, 1]$ as follows:

$$k(x,s) = \begin{cases} (x,s) & \text{if } x \in X \text{ and } s \geq \alpha f(x), \\ (h(x, (2\alpha f(x) - 2s)/\alpha f(x)),\ s) & \\ & \text{if } x \in X \text{ and } \alpha f(x)/2 \leq s \leq \alpha f(x), \\ (gf(x), \alpha f(x)/2) & \text{if } x \in X \text{ and } s \leq \alpha f(x)/2, \end{cases}$$

and $k(y) = (g(y), \alpha fg(y)/2)$ for $y \in Y$. Then, $k|X \times (0, 1]$ is $\mathcal{U}$-homotopic to id in $X \times (0, 1]$ by the homotopy $\varphi : X \times (0, 1] \times \mathbf{I} \to X \times (0, 1]$ defined as follows[4]:

$$\varphi_t(x,s) = \begin{cases} (x,s) & \text{if } x \in X \text{ and } s \geq \alpha f(x), \\ (h(x, t(2\alpha f(x) - 2s)/\alpha f(x)),\ s) & \\ & \text{if } x \in X \text{ and } \alpha f(x)/2 \leq s \leq \alpha f(x), \\ (h_t(x), (1-t)s + t\alpha f(x)/2) & \\ & \text{if } x \in X \text{ and } s \leq \alpha f(x)/2. \end{cases}$$

This completes the proof. □

Remark 11. In Theorem 6.7.7, the implications (b) ⇔ (c) ⇒ (a) hold for any metrizable spaces X and Y. In fact, for a map $f : X \to Y$ between metrizable spaces, the following are equivalent:

(a) For each open cover $\mathcal{U}$ of Y, there is a map $g : Y \to X$ such that $gf \simeq_{f^{-1}(\mathcal{U})} \mathrm{id}_X$;
(b) For each open cover $\mathcal{U}$ of $M(f)$, there is a map $g : M(f) \to X \times (0, 1]$ such that $g|X \times (0, 1] \simeq_{\mathcal{U}} \mathrm{id}$ in $X \times (0, 1]$;
(c) For each open cover $\mathcal{U}$ of $M(f)$, there is a map $g : M(f) \to X \times \mathbf{I}$ such that $gq_f \simeq_{q_f^{-1}(\mathcal{U})} \mathrm{id}_{X \times \mathbf{I}}$.

It should be remarked that if X is an ANR then condition (a) above implies that f is a fine homotopy equivalence and that Y is also an ANR by Theorem 6.7.5.

Regarding the inverse limit of an inverse sequence of ANRs, we have the following theorem:

Theorem 6.7.8. *Let $X = \varprojlim(X_i, \varphi_i)$ be the inverse limit of an inverse sequence $(X_i, \varphi_i)_{i \in \mathbb{N}}$ of completely metrizable ANRs such that each bonding map $\varphi_i : X_{i+1} \to X_i$ is a fine homotopy equivalence. Then, X is an ANR. Moreover, if X_1 is an AR (so every X_i is an AR), then X is an AR.*

---

[4]It is not required that $k \simeq_{\mathcal{U}} \mathrm{id}$ in $M(f)$.

Proof. Let $f : A \to X$ be a map from a closed set A in a metrizable space Y. For each $i \in \mathbb{N}$, let $p_i : X \to X_i$ be the inverse limit projection. Take an admissible complete metric d_i for X_i and choose $\mathcal{U}_i \in \operatorname{cov}(X_i)$ so that $\operatorname{mesh}_{d_j} \varphi_{j,i}(\mathcal{U}_i) < 2^{-i-1}$ for each $j \leq i$, where $\varphi_{i,i} = \operatorname{id}$ and $\varphi_{j,i} = \varphi_j \varphi_{j+1} \cdots \varphi_{i-1} : X_i \to X_j$ for $j < i$. Since φ_i is a fine homotopy inverse, it has a $\mathcal{U}_i$-homotopy inverse $\psi_i : X_i \to X_{i+1}$.

Now, since X_1 is an ANR, the map $p_1 f$ extends to a map $f_1 : U \to X_1$ of a neighborhood U of A in Y. By induction, we can obtain maps $f_i : U \to X_i$, $i \in \mathbb{N}$, such that $f_i|A = p_i f$ and $d_j(\varphi_{j,i} f_i, \varphi_{j,i+1} f_{i+1}) < 2^{-i}$ for each $j \leq i$. Indeed, suppose that f_i has been obtained. Then,

$$p_{i+1} f \simeq_{\varphi_i^{-1}(\mathcal{U}_i)} \psi_i \varphi_i p_{i+1} f = \psi_i p_i f = \psi_i f_i|A.$$

By the Homotopy Extension Theorem 6.4.1, $p_{i+1} f$ extends to a map $f_{i+1} : U \to X_{i+1}$ that is $\varphi_i^{-1}(\mathcal{U}_i)$-homotopic to $\psi_i f_i$, hence $\varphi_{i+1} f_{i+1}$ is $\operatorname{st}\mathcal{U}_i$-close to f_i, which implies that $d_j(\varphi_{j,i+1} f_{i+1}, \varphi_{j,i} f_i) < 2^{-i}$.

For each $j \in \mathbb{N}$, since d_j is complete, we have a map $\tilde{f}_j = \lim_{i\to\infty} \varphi_{j,i} f_i : U \to X_j$. Then, it follows that

$$\varphi_j \tilde{f}_{j+1} = \varphi_j \big(\lim_{i\to\infty} \varphi_{j+1,i} f_i\big) = \lim_{i\to\infty} \varphi_{j,i} f_i = \tilde{f}_j.$$

Therefore, we have a map $\tilde{f} : U \to X$ such that $p_j \tilde{f} = \tilde{f}_j$ for every $j \in \mathbb{N}$. Since $p_j \tilde{f}|A = \tilde{f}_j|A = p_j f$ for every $j \in \mathbb{N}$, it follows that $\tilde{f}|A = f$.

If X_1 is an AR, we can take $U = X_1$ in the above. Then, f extends over Y. □

Remark 12. In Theorem 6.7.8, it suffices to require of each bonding map φ_i the condition that for any $\mathcal{U}_i \in \operatorname{cov}(X_i)$, there is a map $\psi_i : X_i \to X_{i+1}$ such that $\psi_i \varphi_i \simeq_{\varphi_i^{-1}(\mathcal{U}_i)} \operatorname{id}_{X_{i+1}}$ and $\varphi_i \psi_i$ is $\mathcal{U}_i$-close to id_{X_i}. In this case, since $\varphi_i(X_{i+1})$ is dense in X_i, the map $\varphi_i : X_{i+1} \to X_i$ is a fine homotopy equivalence by Theorem 6.7.5.

6.8 Completions of ANRs and Uniform ANRs

For a metric space $X = (X, d)$, a sequence $\mathcal{U} = (\mathcal{U}_n)_{n\in\mathbb{N}}$ of open covers of X is called a **zero-sequence** if $\lim_{n\to\infty} \operatorname{mesh} \mathcal{U}_n = 0$. For a zero-sequence $\mathcal{U}$, we define the **telescope** $\operatorname{Tel}(\mathcal{U})$ as follows:

$$\operatorname{Tel}(\mathcal{U}) = \bigcup_{n\in\mathbb{N}} N(\mathcal{U}_n \cup \mathcal{U}_{n+1}),$$

where $N(\mathcal{U}_n \cup \mathcal{U}_{n+1})$ is the nerve of $\mathcal{U}_n \cup \mathcal{U}_{n+1}$ and we regard $\mathcal{U}_n \cap \mathcal{U}_m = \emptyset$ for $n \neq m$ as the sets of vertices of the nerves. The following characterization of ANRs is due to Nguyen To Nhu:

Theorem 6.8.1 (NGUYEN TO NHU). *A metric space $X = (X, d)$ is an ANR if and only if X has a zero-sequence $\mathcal{U} = (\mathcal{U}_n)_{n\in\mathbb{N}}$ of open covers with a map $f : |\operatorname{Tel}(\mathcal{U})| \to X$ satisfying the following conditions:*

(i) *$f(U) \in U$ for each $U \in \bigcup_{n\in\mathbb{N}} \mathcal{U}_n = \operatorname{Tel}(\mathcal{U})^{(0)}$;*
(ii) $\lim_{n\to\infty} \operatorname{mesh}\{f(\sigma) \mid \sigma \in N(\mathcal{U}_n \cup \mathcal{U}_{n+1})\} = 0$.

Proof. To prove the "only if" part, by the Arens–Eells Embedding Theorem 6.2.1, we may assume that X is a neighborhood retract of a normed linear space E. Let $r : W \to X$ be a retraction of an open neighborhood W of X in E onto X. For each $n \in \mathbb{N}$, let $\mathcal{V}_n$ be a collection of open convex sets in E such that $X \subset \bigcup \mathcal{V}_n \subset W$ and $\operatorname{mesh} r(\mathcal{V}_n) < 2^{-n}$. Then, we have a zero-sequence $\mathcal{U} = (\mathcal{U}_n)_{n\in\mathbb{N}}$ of open covers of X such that $\operatorname{st}\mathcal{U}_n \prec \mathcal{V}_n$ and $\mathcal{U}_{n+1} \prec \mathcal{U}_n$. Choosing $f_0(U) \in U$ for each $U \in \bigcup_{n\in\mathbb{N}} \mathcal{U}_n = \operatorname{Tel}(\mathcal{U})^{(0)}$, we have a map $f_0 : \operatorname{Tel}(\mathcal{U})^{(0)} \to X \subset E$, which extends to a map $f : |\operatorname{Tel}(\mathcal{U})| \to E$ such that $f|\sigma$ is affine on each simplex $\sigma \in \operatorname{Tel}(\mathcal{U})$. For each $\sigma \in \operatorname{Tel}(\mathcal{U})$, we have $n(\sigma) \in \mathbb{N}$ and $U_\sigma \in \sigma^{(0)}$ such that $\sigma \in N(\mathcal{U}_{n(\sigma)} \cup \mathcal{U}_{n(\sigma)+1})$ and $U_\sigma \in \mathcal{U}_{n(\sigma)}$. Since $\mathcal{U}_{n(\sigma)+1} \prec \mathcal{U}_{n(\sigma)}$, it follows that $f_0(\sigma^{(0)}) \subset \operatorname{st}(U_\sigma, \mathcal{U}_{n(\sigma)})$, which is contained in some $V_\sigma \in \mathcal{V}_{n(\sigma)}$. By the convexity of V_σ, $f(\sigma) \subset V_\sigma \in \mathcal{V}_{n(\sigma)}$. Hence, $f(|\operatorname{Tel}(\mathcal{U})|) \subset W$. Thus, we have the map $rf : |\operatorname{Tel}(\mathcal{U})| \to X$, which satisfies condition (i) because rf is an extension of f_0. For each $\sigma \in \operatorname{Tel}(\mathcal{U})$, $\operatorname{diam} rf(\sigma) \leq \operatorname{mesh} r(\mathcal{V}_{n(\sigma)}) < 2^{-n(\sigma)}$, which means that rf satisfies condition (ii).

To prove the "if" part, let $\mathcal{U} = (\mathcal{U}_n)_{n\in\mathbb{N}}$ be a zero-sequence of open covers of X and $f : |\operatorname{Tel}(\mathcal{U})| \to X$ be a map satisfying conditions (i) and (ii). For each $n \in \mathbb{N}$, let $\varphi_n : X \to |N(\mathcal{U}_n)|$ be a canonical map. For each $x \in X$, let $\sigma_{n,x}$ be the simplex of $N(\mathcal{U}_n \cup \mathcal{U}_{n+1})$ spanned by vertices of the carrier of $\varphi_n(x)$ in $N(\mathcal{U}_n)$ and the carrier of $\varphi_{n+1}(x)$ in $N(\mathcal{U}_{n+1})$. Then, $\varphi_n(x), \varphi_{n+1}(x) \in \sigma_{n,x}$. Thus, we have a homotopy $g^n : X \times \mathbf{I} \to |N(\mathcal{U}_n \cup \mathcal{U}_{n+1})|$ such that

$$g^n_0 = \varphi_n,\ g^n_1 = \varphi_{n+1} \ \text{ and } \ g^n(\{x\} \times \mathbf{I}) \subset \sigma_{n,x} \ \text{ for each } x \in X.$$

Then, each fg^n_t is α_n-close to id, where

$$\alpha_n = \sup_{i\geq n} \operatorname{mesh}\mathcal{U}_i + \operatorname{mesh}\big\{f(\sigma) \bigm| \sigma \in N(\mathcal{U}_i \cup \mathcal{U}_{i+1})\big\}.$$

Indeed, each $x \in X$ is contained in some $U \in \mathcal{U}_n \cap \sigma^{(0)}_{n,x}$ and $f(U) \in U$, hence

$$\begin{aligned} d(x, fg^n(x,t)) &\leq d(x, f(U)) + d(f(U), fg^n(x,t)) \\ &\leq \operatorname{diam} U + \operatorname{diam} f(\sigma_{n,x}) \\ &\leq \operatorname{mesh}\mathcal{U}_n + \operatorname{diam} f(\sigma_{n,x}) \leq \alpha_n. \end{aligned}$$

Thus, we can define a homotopy $h : X \times \mathbf{I} \to X$ as follows:

$$h(x,t) = \begin{cases} fg^n(x, 2 - 2^n t) & \text{if } 2^{-n} \le t \le 2^{-n+1},\ n \in \mathbb{N}, \\ x & \text{if } t = 0. \end{cases}$$

Observe that $h_0 = \mathrm{id}$, $h_{2^{-n+1}} = f\varphi_n$, and $\operatorname{diam} h(\{x\} \times [0, 2^{-n+1}]) \le 2\alpha_n$ for each $x \in X$. Hence, $f\varphi_n$ is $2\alpha_n$-homotopic to id, which implies that X is $2\alpha_n$-homotopy dominated by $N(\mathcal{U}_n)$. Since $\lim_{n\to\infty} \alpha_n = 0$, X is an ANR by Theorem 6.6.2. □

Remark 13. In Theorem 6.8.1, if $f(|\operatorname{Tel}(\mathcal{U})|)$ is contained in some $A \subset X$, then the homotopy h in the proof above satisfies $h(X \times (0,1]) \subset A$, hence A is homotopy dense in X.

Corollary 6.8.2. *Let X be a subset of a metric space $M = (M, d)$. If X is an ANR, then X is homotopy dense in some G_δ-set Y in M, hence Y is also an ANR.*

Proof. By Theorem 6.8.1, X has a zero-sequence $\mathcal{U} = (\mathcal{U}_n)_{n\in\mathbb{N}}$ of open covers with a map $f : |\operatorname{Tel}(\mathcal{U})| \to X$ satisfying conditions (i) and (ii) in Theorem 6.8.1. Using the operator E as in Lemma 6.1.3 to extend open sets in X to open sets in M, we define

$$Y = \mathrm{cl}_M X \cap \bigcap_{n\in\mathbb{N}} \bigcup_{U\in\mathcal{U}_n} E(U).$$

Then, Y is a G_δ-set in M and $X \subset Y \subset \mathrm{cl}_M X$. For each $n \in \mathbb{N}$, let

$$\widetilde{\mathcal{U}}_n = \{Y \cap E(U) \mid U \in \mathcal{U}_n\} \in \operatorname{cov}(Y).$$

Then, $\widetilde{\mathcal{U}} = (\widetilde{\mathcal{U}}_n)_{n\in\mathbb{N}}$ is a zero-sequence because $\operatorname{mesh} \widetilde{\mathcal{U}}_n = \operatorname{mesh} \mathcal{U}_n \to 0$ (as $n \to \infty$). Observe $\operatorname{Tel}(\widetilde{\mathcal{U}}) \equiv \operatorname{Tel}(\mathcal{U})$ by the correspondence $Y \cap E(U) \leftrightarrow U$. Then, the map $f : |\operatorname{Tel}(\widetilde{\mathcal{U}})| \to X \subset Y$ satisfies conditions (i) and (ii) in Theorem 6.8.1. By the above remark, we have the result. □

Since every metrizable space X can be embedded into a completely metrizable space M (e.g., the metric completion of X for some admissible metric), Corollary 6.8.2 yields:

Corollary 6.8.3. *Every ANR is contained in a completely metrizable ANR as a homotopy dense subset.* □

Let $X = (X, d)$ and $Y = (Y, \rho)$ be metric spaces and A be a closed set in X. A map $f : X \to Y$ is said to be **uniformly continuous at** A if, for each $\varepsilon > 0$, there is a $\delta > 0$ such that for each $x \in X$ and $a \in A$, $d(x, a) < \delta$ implies $\rho(f(x), f(a)) < \varepsilon$.[5] We call A a **uniform retract** of X if there is a retraction $r : X \to A$ that is uniformly continuous at A. A **uniform neighborhood** of A in X is a subset U of X with $\operatorname{dist}(A, X \setminus U) > 0$, that is, there is a $\delta > 0$ such that $\mathrm{B}(a, \delta) \subset U$ for all $a \in A$. We call A a **uniform neighborhood retract** of X if A

[5] We do not use the preposition "on" but "at" here. When we say that $f : X \to Y$ is uniformly continuous on A, it means that $f|A$ is uniformly continuous.

is a uniform retract of a uniform neighborhood of A in X. It is easy to see that every compact (neighborhood) retract of X is a uniform (neighborhood) retract. Observe that the following closed set in the Euclidean plane $\mathbb{R}^2$ is homeomorphic to the real line $\mathbb{R}$ but is not a uniform (neighborhood) retract of $\mathbb{R}^2$:

$$\{(x, y) \in \mathbb{R}^2 \mid y = x^{-1},\ x \geq 1\} \cup (\{1\} \times \mathbf{I}) \cup ([1, \infty) \times \{0\}).$$

A metric space X is called a **uniform AR** (a **uniform ANR**) if X is a uniform retract (a uniform neighborhood retract) of an arbitrary metric space that contains X as a closed metric subspace (i.e., the inclusion of X is an isometrically closed embedding). A metric space Y is called a **uniform AE** (a **uniform ANE**) if for any closed set A in an arbitrary metric space X, any uniformly continuous map $f : A \to Y$ extends to a map $\tilde{f} : X \to Y$ (a map $\tilde{f} : U \to Y$ of a uniform neighborhood of A in X) that is uniformly continuous at A. It should be noted that a metric space uniformly homeomorphic to a uniform AE (or a uniform ANE) is also a uniform AE (or a uniform ANE). Like ANRs and ANEs, we have the following theorem:

Theorem 6.8.4. *A metric space X is a uniform AR (resp. a uniform ANR) if and only if X is a uniform AE (resp. a uniform ANE).*

Proof. The "if" part is trivial. We prove the "only if" part for the case where X is a uniform ANR. By the Arens–Eells Embedding Theorem 6.2.1, we may assume that X is a closed set in a normed linear space $E = (E, \|\cdot\|)$. Then, there is a uniform retraction $r : G \to X$ of a uniform neighborhood G of X in E. It is easy to prove the following:

$(*)$ For each $\varepsilon > 0$, there exists a $\gamma(\varepsilon) > 0$ such that if $S \subset X$ and $\operatorname{diam} S \leq \gamma(\varepsilon)$ then $\langle S \rangle \subset G$ and $\operatorname{diam} r(\langle S \rangle) < \varepsilon$.

Let $Y = (Y, d)$ be a metric space and $f : A \to X$ be a uniformly continuous map from a closed set A in Y. For each $\varepsilon > 0$, we have $\beta(\varepsilon) > 0$ such that for any $a, a' \in A$, $d(a, a') < \beta(\varepsilon)$ implies $\|f(a) - f(a')\| < \varepsilon$. For each $a \in A$, let

$$V_a = \{y \in Y \setminus A \mid d(y, a) < 2d(y, A)\}.$$

Then, $Y \setminus A$ has a locally finite partition of unity $(\lambda_a)_{a \in A}$ subordinated to the open cover $\{V_a \mid a \in A\}$. For each $\varepsilon > 0$, we define $\delta(\varepsilon) = \frac{1}{3}\beta(\frac{1}{2}\gamma(\varepsilon)) > 0$ and

$$U = \{y \in Y \mid d(y, A) < \delta(1)\}.$$

Then, U is a uniform neighborhood of A in Y. We extend f to $\tilde{f} : U \to X$ by

$$\tilde{f}(y) = \begin{cases} f(y) & \text{if } y \in A, \\ r\left(\sum_{a \in A} \lambda_a(y) f(a)\right) & \text{if } y \in U \setminus A. \end{cases}$$

Then, $\tilde{f}$ is well-defined and uniformly continuous at A. Indeed, let $y \in U \setminus A$, $a \in A$, and $d(y,a) < \delta(\varepsilon)$. If $\lambda_{a'}(y) \neq 0$ then $d(y,a') < 2d(y,A) < 2\delta(\varepsilon)$, so $d(a,a') < \beta(\frac{1}{2}\gamma(\varepsilon))$, which implies $\|f(a) - f(a')\| < \frac{1}{2}\gamma(\varepsilon)$. Thus,

$$\operatorname{diam}\left(\{f(a') \mid \lambda_{a'}(y) \neq 0\} \cup \{f(a)\}\right) \leq \gamma(\varepsilon).$$

When $\varepsilon = 1$, $\sum_{a' \in A} \lambda_{a'}(y) f(a') \in G$ by $(*)$, which shows the well-definedness of $\tilde{f}$. Since $\tilde{f}(y) = r\left(\sum_{a' \in A} \lambda_{a'}(y) f(a')\right)$, we have $\|\tilde{f}(y) - f(a)\| < \varepsilon$ by $(*)$, hence $\tilde{f}$ is uniformly continuous at A.

If X is a uniform AR, we can take $G = E$ in the above. Then, $\tilde{f}$ can be defined over Y, which means that X is a uniform AE. □

Remark 14. The above proof is also valid even if X is a closed set in a convex set C in E and G is a neighborhood of X in C. Moreover, in the case where E is not a normed linear space, if condition $(*)$ is satisfied, then the proof can be carried out.

If X itself is a convex set in a locally convex metric linear space E, let $G = C = X$ and $r = \mathrm{id}$ in the above remark. Condition $(*)$ is satisfied with respect to the linear metric for E.[6] Indeed, by the local convexity of E, each $\frac{1}{2}\varepsilon$-neighborhood of $0 \in E$ contains a convex neighborhood of 0, which contains some $\gamma(\varepsilon)$-neighborhood of 0. If $S \subset X$ with $\operatorname{diam} S < \gamma(\varepsilon)$, then $\langle S \rangle \subset C = X$ and $\operatorname{diam}\langle S \rangle < \varepsilon$. Thus, we have the following theorem:

Corollary 6.8.5. *Every convex set in a locally convex metric linear space is a uniform AE, and is hence a uniform AR.* □

According to Corollary 6.3.5, every open cover $\mathcal{U}$ of an ANR X has an h-refinement $\mathcal{V}$, that is, any two $\mathcal{V}$-close maps into X are $\mathcal{U}$-homotopic. The following is a uniform version of this:

Theorem 6.8.6. *Let X be a uniform ANR. Then, for each $\varepsilon > 0$, there is $\delta > 0$ such that any two δ-close maps of an arbitrary space to X are ε-homotopic.*

Proof. By the Arens–Eells Embedding Theorem 6.2.1, we can consider X as a closed set in a normed linear space $E = (E, \|\cdot\|)$. Since X is a uniform ANR, X has a uniform neighborhood U with a retraction $r : U \to X$ that is uniformly continuous at X. For each $\varepsilon > 0$, we have $\delta > 0$ such that $\delta < \operatorname{dist}(X, E \setminus U)$ and, for each $x \in X$ and $y \in U$, $d(x,y) < \delta$ implies $\|x - r(y)\| < \varepsilon/2$. Then, any two δ-close maps $f, g : Y \to X$ are ε-homotopic by the homotopy $h : Y \times \mathbf{I} \to X$ defined as follows:

$$h(y,t) = r((1-t)f(y) + tg(y)).$$ □

The following is a variant of Theorem 6.6.10.

---

[6]The linear metric can be replaced by any admissible invariant metric.

Theorem 6.8.7. *A uniform ANR is a uniform AR if and only if it is homotopically trivial.*

Proof. If a uniform ANR X is homotopically trivial, then X is an AR by Theorem 6.6.10. Let $Y = (Y, d)$ be a metric space and $f : A \to X$ be a uniformly continuous map from a closed set A in Y. Since X is a uniform ANE by Theorem 6.8.4, f extends to a map $\tilde{f} : U \to X$ of a uniform open neighborhood U in Y that is uniformly continuous at A. Choose a uniform neighborhood V of A in Y so that $\operatorname{cl} V \subset U$. Since X is an AE, $\tilde{f}|\operatorname{cl} V$ extends to a map $\bar{f} : Y \to X$. Obviously, $\bar{f}$ is uniformly continuous at A. Therefore, X is a uniform AE, hence it is a uniform AR by Theorem 6.8.4. □

When several metric spaces with common points appear, we denote by $\mathrm{B}_X(x, r)$ the open ball in the space X (centered at x with radius r).

Lemma 6.8.8. *A uniform ANR X is homotopy dense in every metric space Z that contains X isometrically as a dense subset.*

Proof. In the proof of the "only if" part of Theorem 6.8.1, since the retraction $r : W \to X$ can be assumed to be a retraction of a uniform open neighborhood W of X in a normed linear space E that is uniformly continuous at X, we can take $\delta_1 > \delta_2 > \cdots > 0$ so that $\lim_{n\to\infty} \delta_n = 0$, $\mathrm{B}_E(x, \delta_n) \subset W$, and $\operatorname{diam} r(\mathrm{B}_E(x, \delta_n)) < 2^{-n}$ for every $x \in X$ and $n \in \mathbb{N}$. Let $\mathcal{V}_n = \{\mathrm{B}_E(x, \delta_n) \mid x \in X\}$ and $\mathcal{U}_n = \{\mathrm{B}_X(x, \delta_n/3) \mid x \in X\}$. Then, $\mathcal{U} = (\mathcal{U}_n)_{n\in\mathbb{N}}$ is a zero-sequence of open covers of X. Observe that $\operatorname{st}\mathcal{U}_n \prec \mathcal{V}_n$ and $\mathcal{U}_{n+1} \prec \mathcal{U}_n$. Take any map $f_0 : \operatorname{Tel}(\mathcal{U})^{(0)} \to X$ such that $f_0(U) \in U$. Similar to the proof of Theorem 6.8.1, we can extend f_0 to a map f satisfying conditions (i) and (ii) in Theorem 6.8.1.

For each $n \in \mathbb{N}$, let $\tilde{\mathcal{U}}_n = \{\mathrm{B}_Z(x, \delta_n/3) \mid x \in X\}$. Then, $\tilde{\mathcal{U}} = (\tilde{\mathcal{U}}_n)_{n\in\mathbb{N}}$ is a zero-sequence of open covers of Z. By the same argument as in Corollary 6.8.2, we can show that X is homotopy dense in Z. □

Theorem 6.8.9. *For a metric space X, the following are equivalent:*

(a) *X is a uniform ANR;*
(b) *Every metric space Z containing X isometrically as a dense subset is a uniform ANR and X is homotopy dense in Z;*
(c) *X is isometrically embedded in some uniform ANR Z as a homotopy dense subset.*

Proof. The implications (a) ⇒ (c) and (b) ⇒ (a) are obvious by considering $Z = X$.

(a) ⇒ (b): Let Z be a metric space containing X isometrically as a dense subset. Then, X is homotopy dense in Z by Lemma 6.8.8. To prove that Z is a uniform ANR, assume that Z is a closed set in a metric space Y. Let $Y' = Y \setminus (Z \setminus X)$. Since X is a closed set in Y', there is a retraction $r : U \to X$ of a uniform neighborhood U of X in Y' that is uniformly continuous at X. Then, $V = U \cup Z$ is a uniform neighborhood of Z in Y. We extend r to $\tilde{r} : V \to Z$ by $\tilde{r}|Z \setminus X = \mathrm{id}$. It is easy to see that $\tilde{r}$ is uniformly continuous at Z. Hence, Z is a uniform ANR.

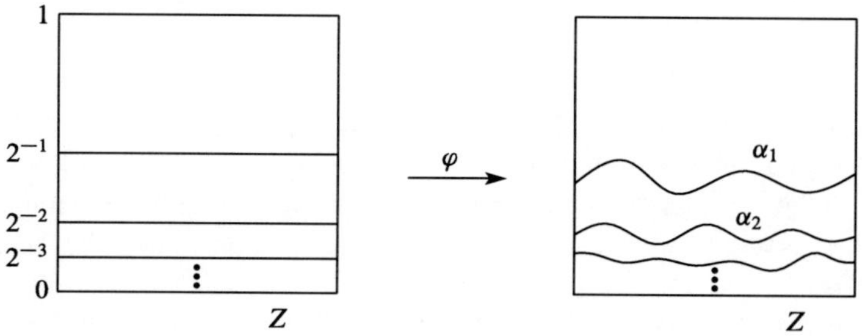

Fig. 6.16 The homeomorphism φ

(c) $\Rightarrow$ (a): Let $Y = (Y, d)$ be a metric space and $f : A \to X$ be a uniformly continuous map from a closed set A in Y. Since Z is a uniform ANE by Theorem 6.8.4, f extends to a map $\tilde{f} : U \to Z$ of a uniform open neighborhood U of A in Y that is uniformly continuous at A, where we may assume that $d(y, A) \leq 1$ for all $y \in U$. On the other hand, we have a homotopy $h : Z \times \mathbf{I} \to Z$ such that $h_0 = \mathrm{id}$ and $h_t(Z) \subset X$ for all $t > 0$. Pushing the image of $\tilde{f}$ into X by this homotopy, we can define a map $U \ni y \mapsto h(\tilde{f}(y), d(y, A)) \in X$ extending f. Unfortunately, in general, this is not uniformly continuous at A, so we need to adjust the homotopy h.

It is easy to construct maps $\alpha_n : Z \to (0, 1)$, $n \in \mathbb{N}$, such that $\alpha_{n+1}(z) < \alpha_n(z)$ ($\leq 2^{-n}$) and $\operatorname{diam} h(\{z\} \times [0, \alpha_n(z)]) < 2^{-n}$. There exists a homeomorphism $\varphi : Z \times \mathbf{I} \to Z \times \mathbf{I}$ such that $\mathrm{pr}_Z \varphi = \mathrm{pr}_Z$, $\varphi | Z \times \{0, 1\} = \mathrm{id}$, and

$$\varphi(z, 2^{-n}) = (z, \alpha_n(z)) \text{ for each } z \in Z \text{ and } n \in \mathbb{N}.$$

See Fig. 6.16. Then, it follows that $d(z, h\varphi(z, t)) < 2^{-n}$ if $t < 2^{-n}$. We define an extension $f' : U \to X$ of f by $f'(y) = h\varphi(\tilde{f}(y), d(y, A))$ for each $y \in U$. For each $\varepsilon > 0$, choose $n \in \mathbb{N}$ so that $2^{-n+1} < \varepsilon$. Since $\tilde{f}$ is uniformly continuous at A, we have $0 < \delta < 2^{-n}$ such that if $y \in U$, $a \in A$, and $d(y, a) < \delta$ ($< 2^{-n}$), then $d(\tilde{f}(y), f(a)) < 2^{-n}$, resulting in

$$\begin{aligned} d(f'(y), f(a)) &\leq d(h\varphi(\tilde{f}(y), d(y, A)), \tilde{f}(y)) + d(\tilde{f}(y), f(a)) \\ &< 2^{-n} + 2^{-n} = 2^{-n+1} < \varepsilon. \end{aligned}$$

Therefore, f' is also uniformly continuous at A. Thus, X is a uniform ANE, hence a uniform ANR by Theorem 6.8.4. □

By Theorem 6.8.9 (combined with Theorem 6.8.7), we have the following corollary, which shows an advantage of the concepts of uniform ARs and uniform ANRs.

Corollary 6.8.10. *A metric space X is a uniform ANR (resp. a uniform AR) if and only if the metric completion $\tilde{X}$ of X is a uniform ANR (resp. a uniform AR) and X is homotopy dense in $\tilde{X}$.* □

Regarding admissible metrics on ANRs, we prove the following theorem:

Theorem 6.8.11. *For any admissible metric d on an AR (resp. ANR) X, X has an admissible metric $\rho \geq d$ such that (X, ρ) is a uniform AR (resp. uniform ANR), where if d is bounded then so is ρ.*

Proof. Since an AR is a contractible ANR (Corollary 6.2.9), by Theorem 6.8.7, it is enough to prove the case where X is an ANR. By the Arens–Eells Embedding Theorem 6.2.1, we may assume that X is a neighborhood retract of a normed linear space E, i.e., there is a retraction $r : G \to X$ of a neighborhood G of X in E. Then, as noted in Remark 14 on the proof of Theorem 6.8.4, it suffices to construct an admissible metric $\rho \geq d$ on X that satisfies condition $(*)$ in the proof of Theorem 6.8.4.

By induction, we will construct admissible metrics ρ_i, $i \in \mathbb{N}$, for X such that, for each $n \in \mathbb{N}$ and $A \subset X$,

$$\operatorname{diam}_{\rho_i} A \leq 2^{-n} \Rightarrow \langle A \rangle \subset G,\ \operatorname{diam}_{\rho_{i-1}} r(\langle A \rangle) \leq n^{-1}2^{-n},$$

where $\rho_0 = d$. Assume ρ_{i-1} has been defined. By the continuity of r, X has open covers $\mathcal{U}_n^i$, $n \in \mathbb{N}$, such that

$$\langle U \rangle \subset G \text{ and } \operatorname{diam}_{\rho_{i-1}} r(\langle U \rangle) \leq n^{-1}2^{-n} \text{ for each } U \in \mathcal{U}_n^i.$$

Then, X has admissible metrics d_n^i, $n \in \mathbb{N}$, such that $\{\overline{\mathrm{B}}_{d_n^i}(x, 1) \mid x \in X\} \prec \mathcal{U}_n^i$ (cf. 2.7.7(1)). We now define ρ_i as follows:

$$\rho_i(x, y) = \sum_{n=1}^{\infty} \min\left\{d_n^i(x, y),\ 2^{-n}\right\}.$$

It is easy to prove that ρ_i is an admissible metric for X. Since $\rho_i(x, y) \leq 2^{-n}$ implies $d_n^i(x, y) \leq 2^{-n}$, it follows that $A \subset X$ with $\operatorname{diam}_{\rho_i} A \leq 2^{-n}$ is contained in some $U \in \mathcal{U}_n^i$. Then, ρ_i satisfies the condition.

Now, we define a metric on X as follows:

$$\rho(x, y) = \rho_0(x, y) + \sum_{i=1}^{\infty} \min\left\{\rho_i(x, y), 2^{-i+1}\right\}.$$

It is easy to see that ρ is an admissible metric for X and, if $d = \rho_0$ is bounded, then so is ρ. Let $A \subset X$ such that $\operatorname{diam}_\rho A \leq \frac{1}{2}$. Since $\operatorname{diam}_{\rho_1} A \leq \frac{1}{2}$, $\langle A \rangle \subset G$. Choose $n \in \mathbb{N}$ so that $2^{-n-1} < \operatorname{diam}_\rho A \leq 2^{-n}$. Since $\operatorname{diam}_{\rho_i} A \leq 2^{-n}$ for each $i \leq n$, $\operatorname{diam}_{\rho_i} r(\langle A \rangle) \leq n^{-1}2^{-n}$ for each $i < n$. Hence,

$$\operatorname{diam}_\rho r(\langle A \rangle) \leq 2^{-n} + \sum_{i \geq n} 2^{-i+1} = 10 \cdot 2^{-n-1} < 10 \operatorname{diam}_\rho A.$$

Then, ρ satisfies condition $(*)$. □

6.9 Homotopy Types of Open Sets in ANRs

In this section, we will prove the following characterization of ANRs:

Theorem 6.9.1 (Cauty). *A metrizable space X is an ANR if and only if every open set in X has the homotopy type of an ANR, i.e., the homotopy type of a simplicial complex (cf. Corollary 6.6.5).*

To prove the "if" part of this theorem, we use the mapping cylinders and the mapping telescopes discussed in Sect. 6.5. For maps $f_i : X_i \to X_{i-1}, i = 1, \ldots, n$, we regard

$$M(f_1, \ldots, f_n) = X_0 \cup X_1 \times (0, 1] \cup \cdots \cup X_n \times (n-1, n],$$

where $M(f_i)$ is identified with $X_{i-1} \times \{i-1\} \cup X_i \times (i-1, i]$ by reparameterizing. The following fact states in detail Corollary 6.5.11.

Fact 1. *For maps $f : X \to Y$ and $g : Y \to Z$, let $\varphi : M(g, f) \to M(gf)$ and $\psi : M(gf) \to M(g, f)$ be the maps defined as follows: $\varphi|Z = \psi|Z = \mathrm{id}_Z$,*

$$\varphi(x, s) = \begin{cases} (x, s-1) & \text{for } (x, s) \in X \times (1, 2], \\ g(x) & \text{for } (x, s) \in Y \times (0, 1]; \end{cases}$$

$$\psi(x, s) = \begin{cases} (x, 2s) & \text{for } (x, s) \in X \times (\tfrac{1}{2}, 1], \\ (f(x), 2s) & \text{for } (x, s) \in X \times (0, \tfrac{1}{2}]. \end{cases}$$

Then, $\psi\varphi \simeq \mathrm{id}$ rel. $Z \oplus X \times \{2\}$ and $\varphi\psi \simeq \mathrm{id}$ rel. $Z \oplus X \times \{1\}$ by the homotopies $\xi : M(g, f) \times \mathbf{I} \to M(g, f)$ and $\zeta : M(gf) \times \mathbf{I} \to M(gf)$ defined as follows: $\xi_t|Z = \zeta_t|Z = \mathrm{id}$ and

$$\xi_t(x, s) = \begin{cases} \left(x, \dfrac{2(s-t)}{2-t}\right) & \text{for } (x, s) \in X \times (1 + \tfrac{1}{2}t, 2], \\ (f(x), 2s - 1 - t) & \text{for } (x, s) \in X \times (1, 1 + \tfrac{1}{2}t], \\ (x, s-t) & \text{for } (x, s) \in Y \times (t, 1], \\ g(x) & \text{for } (x, s) \in Y \times (0, t]; \end{cases}$$

$$\zeta_t(x, s) = \begin{cases} \left(x, \dfrac{2s-t}{2-t}\right) & \text{for } (x, s) \in X \times (\tfrac{1}{2}t, 1], \\ gf(x) & \text{for } (x, s) \in X \times (0, \tfrac{1}{2}t]. \end{cases}$$

Then, we can state

(1) $\xi_t(\{x\} \times (1, 2]) \subset \{f(x)\} \times (0, 1] \cup \{x\} \times (1, 2]$ *for $x \in X$ and $t \in \mathbf{I}$;*
(2) $\xi_t(\{y\} \times (0, 1]) \subset \{g(y)\} \cup \{y\} \times (0, 1]$ *for $y \in Y$ and $t \in \mathbf{I}$;*
(3) $\zeta_t(\{x\} \times (0, 1]) \subset \{gf(x)\} \cup \{x\} \times (0, 1]$ *for $x \in X$ and $t \in \mathbf{I}$.*

We can now also state the following fact:

Fact 2. *Let $f : X \to Y$, $g : Y \to X$ be maps and $h : X \times \mathbf{I} \to X$ be a homotopy such that $h_0 = \mathrm{id}$ and $h_1 = gf$. Define maps $\tilde{f} : X \times \mathbf{I} \to M(f)$, $\tilde{g} : M(f) \to X \times \mathbf{I}$, $\bar{f} : X \times \mathbf{I} \to M(g)$, and $\bar{g} : M(g) \to X \times \mathbf{I}$ as follows:*

$$\tilde{f}(x,t) = \begin{cases} (x,t) & \text{for } t > 0, \\ f(x) & \text{for } t = 0, \end{cases} \qquad (\text{i.e., } \tilde{f} = q_f | X \times \mathbf{I});$$

$$\begin{cases} \tilde{g}(x,s) = \begin{cases} (h_{2-2s}(x), s) & \text{for } (x,s) \in X \times (\frac{1}{2}, 1], \\ (gf(x), s) & \text{for } (x,s) \in X \times (0, \frac{1}{2}], \end{cases} \\ \tilde{g}(y) = (g(y), 0) \qquad\qquad \text{for } y \in Y; \end{cases}$$

$$\bar{f}(x,s) = \begin{cases} (f(x), 2s-1) & \text{for } (x,s) \in X \times (\frac{1}{2}, 1], \\ h_{2s}(x) & \text{for } (x,s) \in X \times [0, \frac{1}{2}]; \end{cases}$$

$$\begin{cases} \bar{g}(y,s) = (g(y), s) & \text{for } (y,s) \in Y \times (0,1], \\ \bar{g}(x) = (x, 0) & \text{for } x \in X. \end{cases}$$

Furthermore, define the homotopies $\tilde{h}, \bar{h} : X \times \mathbf{I} \times \mathbf{I} \to X \times \mathbf{I}$ as follows:

$$\tilde{h}_t(x,s) = \begin{cases} (h_{(2-2s)t}(x), s) & \text{for } (x,s,t) \in X \times (\frac{1}{2}, 1] \times \mathbf{I}, \\ (h_t(x), s) & \text{for } (x,s,t) \in X \times [0, \frac{1}{2}] \times \mathbf{I}; \end{cases}$$

$$\bar{h}_t(x,s) = \begin{cases} (h_t(x), (1+t)s - t) & \text{for } (x,s,t) \in X \times (\frac{1}{2}, 1] \times \mathbf{I}, \\ (h_{2st}(x), (1-t)s) & \text{for } (x,s,t) \in X \times [0, \frac{1}{2}] \times \mathbf{I}. \end{cases}$$

Then,

(1) $\tilde{h}_0 = \mathrm{id}$ *and* $\tilde{h}_1 = \tilde{g}\tilde{f}$;
(2) $\tilde{f}|X \times \{1\} = \tilde{g}|X \times \{1\} = \mathrm{id}$ *and* $\tilde{h}_t|X \times \{1\} = \mathrm{id}$ *for every* $t \in \mathbf{I}$;
(3) $\tilde{h}_t(x,0) = (h_t(x), 0)$ *for each* $x \in X$ *and* $t \in \mathbf{I}$;
(4) $\tilde{h}(\{(x,s)\} \times \mathbf{I}) \subset h(\{x\} \times \mathbf{I}) \times \{s\}$ *for each* $(x,s) \in X \times \mathbf{I}$;
(5) $\tilde{g}(\{f(x)\} \cup \{x\} \times (0,1]) \subset h(\{x\} \times \mathbf{I}) \times \mathbf{I}$ *for each* $x \in X$;
(6) $\bar{h}_0 = \mathrm{id}$ *and* $\bar{h}_1 = \bar{g}\bar{f}$;
(7) $\bar{h}_t|X \times \{0\} = \mathrm{id}$ *for every* $t \in \mathbf{I}$;
(8) $\bar{h}_t(x,1) = (h_t(x), 1)$ *for each* $x \in X$ *and* $t \in \mathbf{I}$;
(9) $\bar{h}(\{x\} \times \mathbf{I} \times \mathbf{I}) \subset h(\{x\} \times \mathbf{I}) \times \mathbf{I}$ *for each* $x \in X$;
(10) $\bar{g}(\{g(y)\} \cup \{y\} \times (0,1]) \subset \{g(y)\} \times \mathbf{I}$ *for each* $y \in Y$.

Remark 15. We can define $\tilde{g} : M(f) \to X \times \mathbf{I}$ as follows:

$$\begin{cases} \tilde{g}(x,s) = (h_{1-s}(x), s) & \text{for } (x,s) \in X \times (0,1], \\ \tilde{g}(y) = (g(y), 0) & \text{for } y \in Y. \end{cases}$$

This definition is natural, but the continuity of $\tilde{g}$ is not guaranteed because of the topology of $M(f)$.

The following is the key lemma for Theorem 6.9.1.

Lemma 6.9.2. *Suppose that X is a metrizable space such that each open set in X has the homotopy type of some ANR. Let Y_0 be an ANR, X_0 be a closed set in X, W_0 be an open set in X with $X_0 \subset W_0$, $\mathcal{V}_0$ be an open cover of W_0, and let $f_0 : W_0 \to Y_0$, $g_0 : Y_0 \to W_0$ be maps and $h^0 : W_0 \times \mathbf{I} \to W_0$ be a $\mathcal{V}_0$-homotopy such that $h^0_0 = \mathrm{id}$ and $h^0_1 = g_0 f_0$. Given a discrete open collection $\mathcal{W}$ in X, let $W_1 = \bigcup \mathcal{W}$ and*

$$\mathcal{V} = \mathcal{V}_0 \cup \mathcal{W} \cup \{V \cup W \mid V \in \mathcal{V}_0,\ W \in \mathcal{W}[V]\}.$$

Then, there exists an ANR Y such that Y contains Y_0 as a closed set, $f_0|X_0$ and g_0 extend to maps $\tilde{f} : W_0 \cup W_1 \to Y$ and $\tilde{g} : Y \to W_0 \cup W_1$, respectively, and there is a $\mathcal{V}$-homotopy $\tilde{h} : (W_0 \cup W_1) \times \mathbf{I} \to W_0 \cup W_1$ such that $\tilde{h}_0 = \mathrm{id}$, $\tilde{h}_1 = \tilde{g}\tilde{f}$, and $\tilde{h}|X_0 \times \mathbf{I} = h^0|X_0 \times \mathbf{I}$.

Proof. Choose a closed neighborhood N of X_0 in X so that $N \subset W_0$, and let $\alpha : W_0 \cup W_1 \to [0,1]$ be an Urysohn map with $\alpha(N \cup (W_0 \setminus W_1)) = 0$ and $\alpha(W_1 \setminus W_0) = 1$. We define

$$S = W_0 \times [0, \tfrac{1}{6}] \cup (W_0 \cap W_1) \times [\tfrac{1}{6}, \tfrac{5}{6}] \cup W_1 \times [\tfrac{5}{6}, 1] \subset X \times \mathbf{I}$$

and an embedding $j : W_0 \cup W_1 \to S$ by $j(x) = (x, \alpha(x))$. Let $r = \mathrm{pr}_X|S : S \to W_0 \cup W_1$. Then, $rj = \mathrm{id}$ and $j(N \cup (W_0 \setminus W_1)) \subset W_0 \times \{0\}$.

By the hypothesis, there are ANRs Y_1, Y_* and homotopy equivalences $f_1 : W_1 \to Y_1$, $f_* : W_0 \cap W_1 \to Y_*$ with homotopy inverses g_1 and g_*, respectively. Since $\mathcal{W}$ is discrete, Y_1 and Y_* can be written as $Y_1 = \bigoplus_{W \in \mathcal{W}} Y_1^W$, $Y_* = \bigoplus_{W \in \mathcal{W}} Y_*^W$, and $f_1^W = f_1|W : W \to Y_1^W$, $f_*^W = f_*|W_0 \cap W : W_0 \cap W \to Y_*^W$ are homotopy equivalences with $g_1^W = g_1|Y_1^W : Y_1^W \to W$, $g_*^W = g_*|Y_*^W : Y_*^W \to W_0 \cap W$ homotopy inverses, respectively. Let

$$Y = M(f_0 g_*) \cup_{Y_* \times \{1\}} M(f_1 g_*).$$

Then, Y is an ANR by Corollary 6.5.6 with 6.2.10(5). By identifying $Y_0 = Y_0 \times \{0\} \subset M(f_0 g_*)$, Y contains Y_0 as a closed set. We will construct maps $f : S \to Y$, $g : Y \to S$ and an $r^{-1}(\mathcal{V})$-homotopy $h : S \times \mathbf{I} \to S$ such that $h_0 = \mathrm{id}$, $h_1 = gf$, $f(x,0) = f_0(x)$ for each $x \in W_0$, $g(y) = (g_0(y), 0)$ for each $y \in Y_0$, and

$$h_t(x,0) = \begin{cases} (h^0_1(x), 0) & \text{for } x \in W_0 \text{ and } \tfrac{1}{2} \le t \le 1, \\ (h^0_{2t}(x), 0) & \text{for } x \in W_0 \text{ and } 0 \le t \le \tfrac{1}{2}. \end{cases}$$

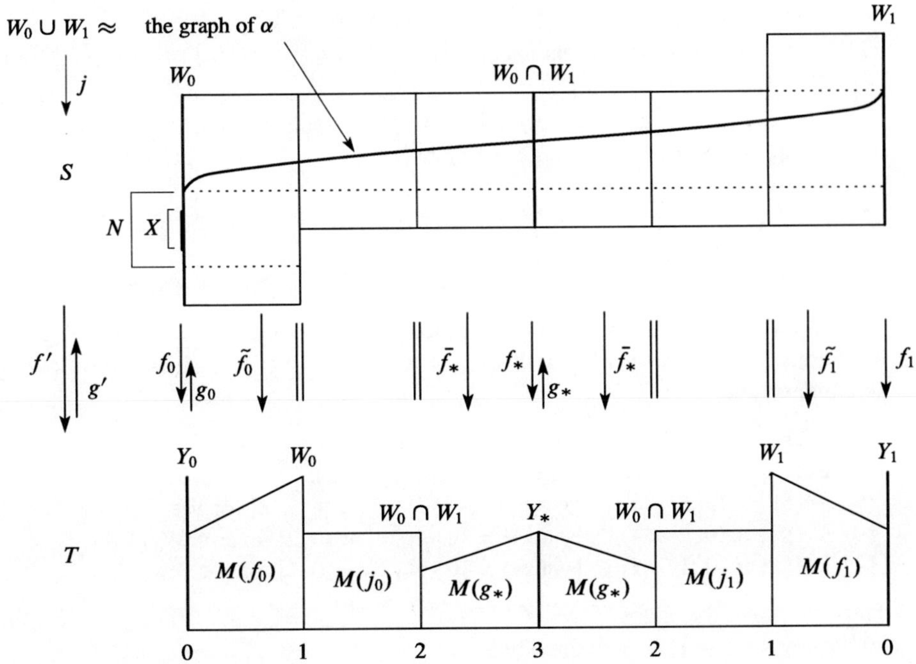

Fig. 6.17 The maps f' and g'

Then, the maps $\tilde{f} = fj$ and $\tilde{g} = rg$ are the desired ones and the $\mathcal{V}$-homotopy $\tilde{h}$ is defined as follows:

$$\tilde{h}_t(x) = rh_{(1+\beta(x))^{-1}t}\, j(x) = rh_{(1+\beta(x))^{-1}t}(x, \alpha(x)),$$

where $\beta : W_0 \cup W_1 \to \mathbf{I}$ is an Urysohn map with $\beta((W_0 \cup W_1) \setminus N) = 0$ and $\beta(X_0) = 1$. Indeed, $\tilde{f}(x) = f(x,0) = f_0(x)$ for $x \in X_0$, $\tilde{g}(y) = rg(y) = g_0(y)$ for $y \in Y_0$ and $\tilde{h}_t(x) = rh_{2^{-1}t}(x,0) = h^0_t(x)$ for $(x,t) \in X_0 \times \mathbf{I}$. For every $x \in W_0 \cup W_1$, $\tilde{h}_0(x) = rh_0(x, \alpha(x)) = x$. If $x \notin N$, then $\tilde{h}_1(x) = rh_1 j(x) = rgfj(x) = \tilde{g}\tilde{f}(x)$. When $x \in N$, since $(1 + \beta(x))^{-1} \geq \frac{1}{2}$, it follows that

$$\tilde{h}_1(x) = rh_{(1+\beta(x))^{-1}}(x, 0) = h^0_1(x) = g_0 f_0(x) = \tilde{g}\tilde{f}(x).$$

For $i = 0, 1$, let $j_i : W_0 \cap W_1 \to W_i$ be the inclusion. We define

$$T = M(f_0, j_0, g_*) \cup_{Y_* \times \{3\}} M(f_1, j_1, g_*).$$

Using Fact 2, we can obtain maps $f' : S \to T$, $g' : T \to S$ and a homotopy $h' : S \times \mathbf{I} \to S$ such that $h'_0 = \mathrm{id}$, $h'_1 = g'f'$, $h'_t(x, 0) = (h^0_t(x), 0)$ for each $x \in W_0$ and $t \in \mathbf{I}$ (cf. Fig. 6.17). The following are consequences of the construction:

(1) $h'_t|(W_0 \cap W_1) \times \left([\frac{1}{6}, \frac{1}{3}] \cup [\frac{2}{3}, \frac{5}{6}]\right) = \mathrm{id}$ for each $t \in \mathbf{I}$;
(2) Each h'_t preserves components of $(W_0 \cap W_1) \times [\frac{1}{3}, \frac{1}{2}]$, $(W_0 \cap W_1) \times [\frac{1}{2}, \frac{2}{3}]$, and $W_1 \times [\frac{5}{6}, 1]$;
(3) $h'(\{x\} \times [0, \frac{1}{6}] \times \mathbf{I}) \subset h^0(\{x\} \times \mathbf{I}) \times [0, \frac{1}{6}]$ for each $x \in W_0$;
(4) $g'(\{g_*(y)\} \times (0, 2] \cup \{y\} \times (2, 3]) \subset h^0(\{g_*(y)\} \times \mathbf{I} \times (0, 1] \cup \{g_*(y)\} \times (1, 3])$ for each $y \in Y_*$;
(5) $g'(\{f_0(x)\} \times \{0\} \cup \{x\} \times (0, 2]) \subset h^0(\{x\} \times \mathbf{I}) \times \mathbf{I} \cup \{x\} \times (1, 2]$ for each $x \in W_0 \cap W_1$;
(6) $g'(\{f_0(x)\} \times \{0\} \cup \{x\} \times (0, 1]) \subset h^0(\{x\} \times \mathbf{I}) \times \mathbf{I}$ for each $x \in W_0$.

Now, we can use Fact 1 to obtain maps $f'' : T \to Y$, $g'' : Y \to T$ and a homotopy $h'' : T \times \mathbf{I} \to T$ such that $h''_0 = \mathrm{id}$ and $h''_1 = g'' f''$. Then, the following statements hold:

(7) $h''_t|Y_0 \times \{0\} \oplus Y_* \times \{3\} \oplus Y_1 \times \{0\} = \mathrm{id}$ for each $t \in \mathbf{I}$;
(8) Each h''_t preserves components of $M(f_1, j_1, g_*)$;
(9) $h''(\{y\} \times (2, 3] \times \mathbf{I}) \subset \{g_*(y)\} \times (0, 2] \cup \{y\} \times (2, 3]$ for $y \in Y_*$;
(10) $h''(\{x\} \times (0, 2] \times \mathbf{I}) \subset \{f_0(x)\} \times \{0\} \cup \{x\} \times (0, 2]$ for $x \in W_0 \cap W_1$;
(11) $h''(\{x\} \times (0, 1] \times \mathbf{I}) \subset \{f_0(x)\} \times \{0\} \cup \{x\} \times (0, 1]$ for $x \in W_0$.

Thus, we have the maps $f = f'' f' : S \to Y$, $g = g' g'' : Y \to S$ and the homotopy $h : S \times \mathbf{I} \to S$ defined by

$$h_t = \begin{cases} g' h''_{2t-1} f' & \text{for } t \geq \frac{1}{2}, \\ h'_{2t} & \text{for } t \leq \frac{1}{2}. \end{cases}$$

Then, $h_0 = \mathrm{id}$ and $h_1 = gf$. Since h^0 is a $\mathcal{V}_0$-homotopy, it is easy to prove that h is a $r^{-1}(\mathcal{V})$-homotopy. Thus, the proof is complete. □

Now, we can prove Theorem 6.9.1:

Proof of Theorem 6.9.1. We must prove the "if" part. For any open cover $\mathcal{U}$ of X, we will show that X is $\mathcal{U}$-homotopy dominated by an ANE. Then, it will follow from Theorem 6.6.2 that X is an ANR.

Let K be the nerve of $\mathcal{U}$ with $\varphi : X \to |K|$ a canonical map. For each simplex $\sigma \in K$, let $X_\sigma = \varphi^{-1}(|\operatorname{St}(\hat{\sigma}, \operatorname{Sd}^2 K)|)$, where $\hat{\sigma}$ is the barycenter of σ. Then, $X = \bigcup_{\sigma \in K} X_\sigma$. For each $n \in \omega$, choose an open set W_n in X so that

$$\bigcup_{\sigma \in K^{(n)} \setminus K^{(n-1)}} X_\sigma \subset W_n \subset \operatorname{cl} W_n \subset \bigcup_{\sigma \in K^{(n)} \setminus K^{(n-1)}} \varphi^{-1}(O_{\operatorname{Sd} K}(\hat{\sigma})),$$

where $\{\varphi^{-1}(O_{\operatorname{Sd} K}(\hat{\sigma})) \mid \sigma \in K^{(n)} \setminus K^{(n-1)}\}$ is pair-wise disjoint. For each $\sigma \in K^{(n)} \setminus K^{(n-1)}$, let $W_\sigma = W_n \cap \varphi^{-1}(O_{\operatorname{Sd} K}(\hat{\sigma}))$. Then,

$$X_\sigma \subset W_\sigma \subset \operatorname{cl} W_\sigma \subset \varphi^{-1}(O_{\operatorname{Sd} K}(\hat{\sigma}))$$

and $\mathcal{W}_n = \{W_\sigma \mid \sigma \in K^{(n)} \setminus K^{(n-1)}\}$ is a discrete open collection in X with $\bigcup \mathcal{W}_n = W_n$. For each $n \in \omega$, we define

$$V_n = \bigcup_{i=0}^{n} W_i = \bigcup_{\sigma \in K^{(n)}} W_\sigma \text{ and } X_n = \bigcup_{\sigma \in K^{(n)}} X_\sigma \subset V_n.$$

Note that $V_n = V_{n-1} \cup W_n$. For each $\sigma \in K$,

$$\begin{aligned}\sigma \subset & \bigcup_{v \in \sigma \cap (\mathrm{Sd}^2 K)^{(0)}} O(v, \mathrm{Sd}^2 K) \\ \subset & \bigcup_{v \in \sigma \cap (\mathrm{Sd}^2 K)^{(0)}} |\mathrm{St}(v, \mathrm{Sd}^2 K)| = \bigcup_{\tau \leq \sigma} |\mathrm{St}(\hat{\tau}, \mathrm{Sd}^2 K)|.\end{aligned}$$

Hence, we have $X = \bigcup_{n \in \omega} \operatorname{int} X_n$. We inductively define

$$\mathcal{V}_n = \mathcal{V}_{n-1} \cup \mathcal{W}_n \cup \{V \cup W \mid V \in \mathcal{V}_{n-1},\ W \in \mathcal{W}_n[V]\},$$

where $\mathcal{V}_0 = \mathcal{W}_0$. Then, each $\mathcal{V}_n$ is an open cover of V_n. By induction, it can be seen that for each $V \in \mathcal{V}_n$, there are some $\sigma_0, \sigma_1, \ldots, \sigma_k \in K^{(n)}$ such that $\dim \sigma_0 < \cdots < \dim \sigma_k$ (hence $k \leq n$), $\sigma_0 \cap \cdots \cap \sigma_k \neq \emptyset$, and $V = \bigcap_{i=0}^{k} W_{\sigma_i}$. Take $v \in \bigcap_{i=0}^{k} \sigma_i^{(0)}$. Then, we have

$$V \subset \varphi^{-1}\big(\bigcup_{i=0}^{k} O_{\mathrm{Sd}\,K}(\hat{\sigma}_i)\big) \subset \varphi^{-1}(O_K(v)),$$

which means that $\mathcal{V}_n \prec \mathcal{U}$.

For each $v \in K^{(0)}$, W_v has the homotopy type of an ANR Y_v. Let $Y_0 = \bigoplus_{v \in K^{(0)}} Y_v$. Since $\mathcal{V}_0 = \mathcal{W}_0$ is discrete in X and $V_0 = W_0$, there exist two maps $f_0 : V_0 \to Y_0$, $g_0 : Y_0 \to V_0$ and a $\mathcal{V}_0$-homotopy $h^0 : V_0 \times \mathbf{I} \to V_0$ with $h^0_0 = g_0 f_0$ and $h^0_1 = \mathrm{id}$. By Lemma 6.9.2, we have an ANR Y_1, maps $f_1 : V_1 \to Y_1$, $g_1 : Y_1 \to V_1$ and a $\mathcal{V}_1$-homotopy $h^1 : V_1 \times \mathbf{I} \to V_1$ such that $h^1_0 = g_1 f_1$, $h^1_1 = \mathrm{id}$, $Y_0 \subset Y_1$, $f_1|X_0 = f_0|X_0$, $g_1|Y_0 = g_0$, and $h^1|X_0 \times \mathbf{I} = h^0|X_0 \times \mathbf{I}$. Again, using Lemma 6.9.2, we obtain an ANR Y_2, maps $f_2 : V_2 \to Y_2$ and $g_2 : Y_2 \to V_2$, and a $\mathcal{V}_2$-homotopy $h^2 : V_2 \times \mathbf{I} \to V_2$ such that $h^2_0 = g_2 f_2$, $h^2_1 = \mathrm{id}$, $Y_1 \subset Y_2$, $f_2|X_1 = f_1|X_1$, $g_2|Y_1 = g_1$, and $h^2|X_1 \times \mathbf{I} = h^1|X_1 \times \mathbf{I}$. Thus, we apply Lemma 6.9.2 inductively to obtain a tower $Y_0 \subset Y_1 \subset Y_2 \subset \cdots$ of ANRs with maps $f_n : V_n \to Y_n$, $g_n : Y_n \to V_n$ and $\mathcal{V}_n$-homotopies $h^n : V_n \times \mathbf{I} \to V_n$ such that $h^n_0 = g_n f_n$, $h^n_1 = \mathrm{id}$, $f_n|X_{n-1} = f_{n-1}|X_{n-1}$, $g_n|Y_{n-1} = g_{n-1}$, and $h^n|X_{n-1} \times \mathbf{I} = h^{n-1}|X_{n-1} \times \mathbf{I}$. Let $Y = \varinjlim Y_n$. We can define maps $f : X \to Y$, $g : Y \to X$, and a $\mathcal{U}$-homotopy $h : X \times \mathbf{I} \to X$ as follows: $f|X_n = f_n|X_n$, $g|Y_n = g_n$, and $h|X_n \times \mathbf{I} = h^n|X_n \times \mathbf{I}$ for $n \geq 0$. Then, $h_0 = gf$ and $h_1 = \mathrm{id}$, hence X is $\mathcal{U}$-homotopy dominated by Y. Moreover, Y is an ANE by Theorem 6.1.8. This completes the proof. □

6.10 Countable-Dimensional ANRs

Recall that X is countable-dimensional if $X = \bigcup_{i\in\mathbb{N}} A_i$ for some countably many subsets $A_i \subset X$ with $\dim A_i \leq 0$ (cf. Sect. 5.6). By the Decomposition Theorem 5.4.5 in Dimension Theory, every n-dimensional metrizable space is the union of at most $n+1$ many 0-dimensional subspaces, hence it is countable-dimensional. In this section, we prove the following theorem:

Theorem 6.10.1. *Every countable-dimensional locally contractible metrizable space is an ANR.*

First, we introduce a covering property related to countable-dimensionality. A space X has **Property C** provided for any open covers $\mathcal{U}_n$ of X, $n \in \mathbb{N}$, there exists an open cover $\mathcal{V} = \bigcup_{n\in\mathbb{N}} \mathcal{V}_n$ of X such that each $\mathcal{V}_n$ is pair-wise disjoint and $\mathcal{V}_n \prec \mathcal{U}_n$.

Lemma 6.10.2. *Let X be a metrizable space. If $X = \bigcup_{i\in\mathbb{N}} A_i$ and each A_i has Property C, then X also has Property C.*

Proof. Let $\mathcal{U}_n \in \operatorname{cov}(X)$, $n \in \mathbb{N}$. Take a bijection $k : \mathbb{N}\times\mathbb{N} \to \mathbb{N}$. Having Property C, each A_i has an open cover $\mathcal{V}'_i = \bigcup_{j\in\mathbb{N}} \mathcal{V}'_{(i,j)}$ such that each $\mathcal{V}'_{(i,j)}$ is pair-wise disjoint and $\mathcal{V}'_{(i,j)} \prec \mathcal{U}_{k(i,j)}$. For each $V \in \mathcal{V}'_{(i,j)}$, choose $U(V) \in \mathcal{U}_{k(i,j)}$ so that $V \subset U(V)$, and define

$$\tilde{V} = \{x \in U(V) \mid d(x,V) < d(x, A_i \setminus V)\} = U(V) \cap E(V),$$

where $E(V)$ is the open set in X defined in the proof of Lemma 6.1.3. Then, $\tilde{V}$ is open in X and $\tilde{V} \cap A_i = V$. For each $n \in \mathbb{N}$, let $\mathcal{V}_n = \{\tilde{V} \mid V \in \mathcal{V}'_{k^{-1}(n)}\}$. Then, each $\mathcal{V}_n$ is pair-wise disjoint and $\mathcal{V}_n \prec \mathcal{U}_n$. Since $\bigcup_{j\in\mathbb{N}} \mathcal{V}_{k(i,j)}$ is an open cover of A_i in X, $\mathcal{V} = \bigcup_{n\in\mathbb{N}} \mathcal{V}_n$ is an open cover of $X = \bigcup_{i\in\mathbb{N}} A_i$. □

Lemma 6.10.3. *Let X be a metrizable space that has Property C. Then, every F_σ set in X also has Property C.*

Proof. It is easy to see that any closed set in X has Property C. Then, we can apply Lemma 6.10.2 to complete the proof. □

Theorem 6.10.4. *Every countable-dimensional metrizable space has Property C.*

Proof. Since any open cover of a 0-dimensional space has a discrete open refinement, any 0-dimensional space has Property C. Then, applying Lemma 6.10.2, we complete the proof. □

Lemma 6.10.5. *If X has Property C, then $X \times \mathbf{I}$ also has Property C.*

Proof. Let $\mathcal{U}_n \in \operatorname{cov}(X \times \mathbf{I})$, $n \in \mathbb{N}$. Using the compactness of $\mathbf{I}$, we can easily find $\mathcal{W}_n \in \operatorname{cov}(X)$, $n \in \mathbb{N}$, and partitions $0 = t_0^W < t_1^W < \cdots < t_{k(W)}^W = 1$, $W \in \mathcal{W}_n$, so that

$$\mathcal{W}_n^* = \{W \times J_i^W \mid W \in \mathcal{W}_n,\ i = 0, 1, \cdots, k(W)\} \prec \mathcal{U}_{2n} \wedge \mathcal{U}_{2n-1},$$

where $J_0^W = [0, t_1^W)$, $J_{k(W)}^W = (t_{k(W)-1}^W, 1]$ and $J_i^W = (t_{i-1}^W, t_{i+1}^W)$ for $0 < i < k(W)$. Since X has Property C, X has an open cover $\mathcal{V} = \bigcup_{n \in \mathbb{N}} \mathcal{V}_n$ such that each $\mathcal{V}_n$ is pair-wise disjoint and $\mathcal{V}_n \prec \mathcal{W}_n$. For each $V \in \mathcal{V}_n$, choose $W(V) \in \mathcal{W}_n$ so that $V \subset W(V)$. For each $n \in \mathbb{N}$, let

$$\mathcal{V}_{2n}^* = \left\{V \times J_{2i}^{W(V)} \mid V \in \mathcal{V}_n, 0 \le i \le \tfrac{1}{2}k(W(V))\right\} \text{ and}$$
$$\mathcal{V}_{2n-1}^* = \left\{V \times J_{2i+1}^{W(V)} \mid V \in \mathcal{V}_n, 0 \le i \le \tfrac{1}{2}(k(W(V)) - 1)\right\}.$$

We have $\mathcal{V}^* = \bigcup_{n \in \mathbb{N}} \mathcal{V}_n^* \in \operatorname{cov}(X \times \mathbf{I})$, where each $\mathcal{V}_n^*$ is pair-wise disjoint. Since

$$\mathcal{V}_{2n-1}^*, \mathcal{V}_{2n}^* \prec \mathcal{W}_n^* \prec \mathcal{U}_{2n-1} \wedge \mathcal{U}_{2n},$$

it follows that $\mathcal{V}_n^* \prec \mathcal{U}_n$ for each $n \in \mathbb{N}$. Hence, $X \times \mathbf{I}$ has Property C. □

By Lemmas 6.10.3 and 6.10.5, we have the following proposition:

Proposition 6.10.6. *If X is metrizable and has Property C, then $X \times (0, 1]$ also has Property C.* □

By Theorem 6.10.4, we can obtain Theorem 6.10.1 as a corollary of the following Extension Theorem:

Theorem 6.10.7. *Let A be a closed set in a metrizable space X such that A has Property C, and let Y be a locally contractible metrizable space. Then, any map $f : A \to Y$ extends over a neighborhood of A in X.*

Proof. Inductively take open covers of Y:

$$\mathcal{U}_1 \succ \mathcal{V}_1 \overset{*}{\succ} \mathcal{U}_2 \succ \mathcal{V}_2 \overset{*}{\succ} \mathcal{U}_3 \succ \cdots,$$

so that $\operatorname{mesh} \mathcal{U}_n < 2^{-n}$ and each member of $\mathcal{V}_n$ is contractible in some member of $\mathcal{U}_n$. For each $n \in \mathbb{N}$, let

$$\mathcal{U}_n^* = \left\{f^{-1}(U) \times J_m \mid m \in \mathbb{N},\ U \in \mathcal{U}_{2n+m}\right\} \in \operatorname{cov}(A \times (0, 1]),$$

where $J_1 = (1/2, 1]$ and $J_m = (2^{-m}, 2^{-m}3)$ for $m > 1$. Since $A \times (0, 1]$ has Property C, it has an open cover $\mathcal{W} = \bigcup_{n \in \mathbb{N}} \mathcal{W}_n$ such that each $\mathcal{W}_n$ is pair-wise disjoint and $\mathcal{W}_n \prec \mathcal{U}_n^*$.

For each $W \in \mathcal{W}$, choose $n(W), m(W), k(W) \in \mathbb{N}$, $U(W) \in \mathcal{U}_{k(W)}$, $V(W) \in \mathcal{V}_{k(W)-1}$, and $\tilde{U}(W) \in \mathcal{U}_{k(W)-1}$ as follows:

(1) $W \in \mathcal{W}_{n(W)}$,
(2) $W \subset f^{-1}(U(W)) \times J_{m(W)} \in \mathcal{U}_{n(W)}^*$,
(3) $U(W) \in \mathcal{U}_{k(W)}$, $k(W) = 2n(W) + m(W)$,
(4) $\operatorname{st}(U(W), \mathcal{U}_{k(W)}) \subset V(W) \in \mathcal{V}_{k(W)-1}$,
(5) $V(W)$ is contractible in $\tilde{U}(W) \in \mathcal{U}_{k(W)-1}$,

and let $\psi^W : V(W) \times \mathbf{I} \to \tilde{U}(W)$ be a contraction, that is, $\psi_0^W = \mathrm{id}$ and $\psi_1^W(V(W)) = \{x_W\}$, where $x_W \in \tilde{U}(W)$.

Let K be the nerve of $\mathcal{W}$ with $\varphi : A \times (0,1] \to |K|$ a canonical map. Since $|K|$ is an ANE, φ extends to a map $\tilde{\varphi} : N \to |K|$ over a neighborhood N of $A \times (0,1]$ in $X \times (0,1]$. We will construct a map $g : |K| \to Y$ so that $d(g\varphi(a,t), f(a)) < t$ for each $(a,t) \in A \times (0,1]$. Then, $g\varphi : A \times (0,1] \to Y$ would be extended over $A \times \mathbf{I}$ by f. Hence, we could apply Lemma 6.5.1 to extend f over a neighborhood of A in X.

For each $\sigma \in K$, take a vertex $W(\sigma) \in \sigma^{(0)}$ $(\subset \mathcal{W})$ such that $n(W(\sigma)) = \min\{n(W) \mid W \in \sigma^{(0)}\}$. Then, such a vertex $W(\sigma) \in \sigma^{(0)}$ is unique. Indeed, $n(W) < n(W')$ or $n(W') < n(W)$ for $W \neq W' \in \sigma^{(0)}$ because each $\mathcal{W}_n$ is pair-wise disjoint. Next, let σ_0 be the $(n-1)$-face of σ such that $W(\sigma) \notin \sigma_0$, i.e., $\sigma = \langle \sigma_0 \cup \{W(\sigma)\}\rangle$. Then, each point of σ can be represented as

$$(1-t)z + tW(\sigma), \; t \in \mathbf{I}, \; z \in \sigma_0.$$

Here, we should notice that, if $\tau < \sigma$ and $W(\sigma) \in \tau^{(0)}$, then $W(\tau) = W(\sigma)$ and $\tau_0 < \sigma_0$.

We will show

(6) $\tilde{U}(W) \subset V(W(\sigma))$ for every $W \in \sigma^{(0)}$.

Since $J_{m(W(\sigma))} \cap J_{m(W)} \neq \emptyset$, we have $|m(W(\sigma)) - m(W)| \leq 1$, which implies $k(W(\sigma)) < k(W)$ by (3). Since $\mathcal{U}_{k(W)-1} \prec \mathcal{U}_{k(W(\sigma))}$, $\tilde{U}(W)$ is contained in some member of $\mathcal{U}_{k(W(\sigma))}$. On the other hand, $U(W) \cap U(W(\sigma)) \neq \emptyset$ by (2). Since $U(W) \subset \tilde{U}(W)$, we have $\tilde{U}(W) \cap U(W(\sigma)) \neq \emptyset$. Thus, it follows from (4) that

$$\tilde{U}(W) \subset \mathrm{st}(U(W(\sigma)), \mathcal{U}_{k(W(\sigma))}) \subset V(W(\sigma)).$$

Now, let $g_0 : |K^{(0)}| \to Y$ be a map such that $g_0(W) = x_W \in \tilde{U}(W)$ for each $W \in K^{(0)}$. Assume that we have maps $g_i : |K^{(i)}| \to Y$, $i < n$, such that $g_i||K^{(i-1)}| = g_{i-1}$ and, if $\sigma \in K^{(i)}$, $t \in \mathbf{I}$ and $z \in \sigma_0$, then $g_{i-1}(z) \in \tilde{U}(W(\sigma_0)) \subset V(W(\sigma))$ and

$$g_i((1-t)z + tW(\sigma)) = \psi^{W(\sigma)}(g_{i-1}(z), t) \in \tilde{U}(W(\sigma)).$$

For each n-simplex $\sigma \in K$, we have $g_{n-1}(\sigma_0) \subset \tilde{U}(W(\sigma_0)) \subset V(W(\sigma))$ by the above assumption and (6). Then, we can define a map $g_\sigma : \sigma \to \tilde{U}(W(\sigma)) \subset Y$ as follows:

$$g_\sigma((1-t)z + tW(\sigma)) = \psi^{W(\sigma)}(g_{n-1}(z), t).$$

It is easy to prove that $g_\sigma|\partial\sigma = g_{n-1}|\partial\sigma$. Hence, g_{n-1} extends to the map $g_n : |K^{(n)}| \to Y$ defined by $g_n|\sigma = g_\sigma$ for each n-simplex $\sigma \in K$, where if $\sigma \in K^{(n)}$, $t \in \mathbf{I}$, and $z \in \sigma_0$, then

$$g_n((1-t)z + tW(\sigma)) = \psi^{W(\sigma)}(g_n(z), t) \in \tilde{U}(W(\sigma)).$$

By induction, we have maps g_n, $n \in \mathbb{N}$, satisfying the above condition. Let $g : |K| \to Y$ be the map defined by $g||K^{(n)}| = g_n$. Then, $g(\sigma) \subset \tilde{U}(W(\sigma))$ for each $\sigma \in K$.

It remains to show that $d(g\varphi(a,t), f(a)) < t$ for each $(a,t) \in A\times(0,1]$. Let $\sigma \in K$ be the carrier of $\varphi(a,t)$. Then, $(a,t) \in W(\sigma)$, which implies that $t \in J_{m(W(\sigma))}$, $f(a) \in U(W(\sigma)) \subset \tilde{U}(W(\sigma))$ and $g\varphi(a,t) \in \tilde{U}(W(\sigma))$. Hence, $t > 2^{-m(W(\sigma))} \geq 2^{-k(W(\sigma))+2}$ and

$$d(g\varphi(a,t), f(a)) \leq \operatorname{diam} \tilde{U}(W(\sigma)) \leq \operatorname{mesh} \mathcal{U}_{k(W(\sigma))-1} < 2^{-k(W(\sigma))+1} < t.$$

This completes the proof. □

6.11 The Local n-Connectedness

In this section, we need a few results from Chap. 5: Theorem 5.2.3, the No Retraction Theorem 5.1.5, the General Position Lemma 5.8.4, and Sect. 5.10.

A space X is said to be ***n*-connected** ($\boldsymbol{C^n}$) if each map $f : \mathbf{S}^i \to X$, $i \leq n$, extends over $\mathbf{B}^{i+1}$.[7] Being 0-connected means being path-connected. We also say that X is **simply connected** instead of 1-connected. Note that $\mathbf{S}^n$ is $(n-1)$-connected but not n-connected by Theorem 5.2.3 and the No Retraction Theorem 5.1.5. We also say that X is **locally *n*-connected** ($\boldsymbol{LC^n}$) if for each $x \in X$, each neighborhood U of x contains a neighborhood V of x such that any map $f : \mathbf{S}^i \to V$, $i \leq n$, extends to a map $\tilde{f} : \mathbf{B}^{i+1} \to U$. To be LC^0 means to be locally path-connected. A space X is said to be **locally simply connected** instead of locally 1-connected. We say that X is $\boldsymbol{C^\infty}$ (or $\boldsymbol{LC^\infty}$) if X is C^n (or LC^n) for all $n \in \omega$. Being C^∞ means being homotopically trivial. Every locally contractible space X is LC^∞ but, in general, the converse does not hold. In fact, being LC^∞ does not imply that every point has an arbitrarily small homotopically trivial (C^∞) neighborhood.

*Example 6.11.1 (*BORSUK*).* There exists an LC^∞ continuum that has a point without homotopically trivial neighborhoods, so it is not locally contractible.

Example and Proof. For each $i \in \mathbb{N}$, let $u_i = 2^{-i}\mathbf{e}_1, v_i = 2^{-i-1}3\mathbf{e}_1 \in \ell_2$ and let S_i be the i-dimensional sphere in ℓ_2 centered at v_i with radius 2^{-i-1}, i.e., $S_i = v_i + 2^{-i-1}\mathbf{S}^i$, where we identify $\mathbf{S}^i = \mathbf{S}^i \times \{0\} \subset \ell_2$. Note that $S_i \cap S_{i+1} = \{u_i\}$ and $S_i \cap S_j = \emptyset$ if $|i-j| > 1$. For each $n \in \mathbb{N}$, let $X_n = \{\mathbf{0}\} \cup \bigcup_{\mathbf{i}=\mathbf{n}}^{\infty} \mathbf{S}_\mathbf{i} \subset \ell_2$. Then, X_1 is compact and connected. Each X_n is the closed 2^{-n+1}-neighborhood of the point $\mathbf{0}$ in X_1. As is easily observed, every S_n is a retract of X_1. Since S_n is not contractible, it follows that S_n is not contractible in any set containing S_n. Then, any neighborhood of $\mathbf{0}$ in X_1 is not homotopically trivial because it contains S_n for sufficiently large n. On the other hand, X_1 is locally contractible at any point of $X_1 \setminus \{\mathbf{0}\}$. It remains to show that X_1 is LC^∞ at the point $\mathbf{0}$.

---

[7] In terms of homotopy groups, this means that $\pi_i(X, x_0) = \{0\}$ for every $x_0 \in X$ and $i \leq n$.

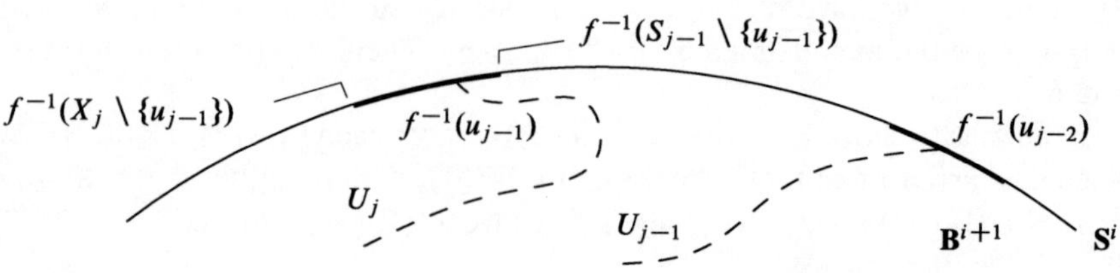

Fig. 6.18 Open sets U_J

To prove that X_1 is LC^∞ at $\mathbf{0}$, it suffices to show that X_m is n-connected for each $m > n$, that is, each map $f : \mathbf{S}^i \to X_m$, $i \leq n$, extends over $\mathbf{B}^{i+1}$. First, we take open sets U_j in $\mathbf{B}^{i+1}$, $j \geq m$, such that

$$f^{-1}(X_j \setminus \{u_{j-1}\}) \subset U_j \subset \operatorname{cl} U_j \subset U_{j-1} \setminus f^{-1}(S_{j-1} \setminus \{u_{j-1}\}).$$

In fact, let $U_m = \mathbf{B}^{i+1} \setminus f^{-1}(u_{m-1})$ and suppose that U_{j-1} has been obtained. Since $\operatorname{cl} f^{-1}(X_j \setminus \{u_{j-1}\}) \subset f^{-1}(X_j)$ and

$$\operatorname{cl}\left((\mathbf{B}^{i+1} \setminus U_{j-1}) \cup f^{-1}(S_{j-1} \setminus \{u_{j-1}\})\right) \subset (\mathbf{B}^{i+1} \setminus U_{j-1}) \cup f^{-1}(S_{j-1}),$$

$f^{-1}(X_j \setminus \{u_{j-1}\})$ and $(\mathbf{B}^{i+1} \setminus U_{j-1}) \cup f^{-1}(S_{j-1} \setminus \{u_{j-1}\})$ are separated in $\mathbf{B}^{i+1}$. By complete normality,[8] $\mathbf{B}^{i+1}$ has the desired open set U_j. — Fig. 6.18.

Next, for each $j \geq m$, observe that $f(\mathbf{S}^i \cap \operatorname{bd} U_j) = \{u_{j-1}\}$ and

$$f(\mathbf{S}^i \cap (\operatorname{cl} U_j \setminus U_{j+1})) \subset X_j \setminus (X_{j+1} \setminus \{u_j\}) = S_j.$$

Since $\dim(\operatorname{cl} U_j \setminus U_{j+1}) = i + 1 \leq j$, we can apply Theorem 5.2.3 to obtain a map $f_j : \operatorname{cl} U_j \setminus U_{j+1} \to S_j$ ($\approx \mathbf{S}^j$) such that

$$f_j(\operatorname{bd} U_j) = \{u_{j-1}\},\ f_j(\operatorname{bd} U_{j+1}) = \{u_j\},\ \text{and}$$
$$f_j|\mathbf{S}^i \cap (\operatorname{cl} U_j \setminus U_{j+1}) = f|\mathbf{S}^i \cap (\operatorname{cl} U_j \setminus U_{j+1}).$$

Now, we can define a map $\tilde{f} : \mathbf{B}^{i+1} \to X_m$ as follows:

$$\tilde{f}|\operatorname{cl} U_j \setminus U_{j+1} = f_j \text{ for } j \geq m,$$
$$\tilde{f}(\mathbf{B}^{i+1} \setminus U_m) = \{u_{m-1}\},\ \text{and}\ \tilde{f}\left(\textstyle\bigcap_{j \geq m} U_j\right) = \{0\}.$$

Then, $\tilde{f}$ is an extension of f. Hence, X_m is LC^n. □

[8] Due to Theorem 2.2.5, X is completely normal if and only if X is hereditarily normal.

Remark 16. In the above example, for a fixed $n \in \mathbb{N}$, let S_i be the n-dimensional sphere in ℓ_2 centered at v_i with radius 2^{-i-1}, i.e., $S_i = v_i + 2^{-i-1}\mathbf{S}^n$. Then, a similar proof shows that $\{0\} \cup \bigcup_{i\in\mathbb{N}} S_i$ is LC^{n-1} but not LC^n. Thus, we have an n-dimensional continuum that is LC^{n-1} but not LC^n.

The universal spaces ν^n and μ^n are also LC^{n-1} but not LC^n as we now explain (cf. Sect. 5.10). The n-dimensional universal Nöbeling space ν^n is defined as follows:

$$\begin{aligned}\nu^n &= \{x \in \mathbb{R}^{2n+1} \mid x(i) \in \mathbb{R} \setminus \mathbb{Q} \text{ except for } n \text{ many } i\} \\ &= \{x \in \mathbb{R}^{2n+1} \mid x(i) \in \mathbb{Q} \text{ at most } n \text{ many } i\}.\end{aligned}$$

For each $k \in \mathbb{N}$, let

$$R_k = \bigcup_{m=0}^{3^{k-1}-1} (m/3^{k-1} + 1/3^k, m/3^{k-1} + 2/3^k) \subset \mathbf{I}.$$

The intersection $\bigcap_{k\in\mathbb{N}}(\mathbf{I} \setminus R_k)$ $(= \mathbf{I} \setminus \bigcup_{k\in\mathbb{N}} R_k)$ is the Cantor (ternary) set μ^0. The n-dimensional universal Menger compactum μ^n is defined as the intersection $\mu^n = \bigcap_{k\in\mathbb{N}} M_k^{2n+1}$, where

$$\begin{aligned}M_k^{2n+1} &= \{x \in \mathbf{I}^{2n+1} \mid x(i) \in \mathbf{I} \setminus R_k \text{ except for } n \text{ many } i\} \\ &= \{x \in \mathbf{I}^{2n+1} \mid x(i) \in R_k \text{ at most } n \text{ many } i\}.\end{aligned}$$

Theorem 6.11.2. *The n-dimensional universal Nöbeling space ν^n and the n-dimensional universal Menger compactum μ^n are C^{n-1} and LC^{n-1}, but they are neither C^n nor LC^n at any point.*

Proof. To prove that ν^n is not C^n or LC^n at any point, it suffices to show that for any neighborhood V of $x \in \nu^n$, there is a map $f : \mathbf{S}^n \to V$ that is not null-homotopic in ν^n. Choose $a_i < b_i \in \mathbb{R}\setminus\mathbb{Q}$, $i = 1, \ldots, 2n+1$, so that $\nu^n \cap \prod_{i=1}^{2n+1}[a_i, b_i] \subset V$, and define $C = \prod_{i=1}^{n+1}[a_i, b_i] \subset \mathbb{R}^{n+1}$. Take a point $v \in \mathbb{Q}^{n+1} \cap \operatorname{int} C \subset \mathbb{R}^{n+1}$. Then, we have a retraction $r : \mathbb{R}^{n+1}\setminus\{v\} \to \partial C$. Since $\{v\}\times\mathbb{R}^n \subset \mathbb{R}^{2n+1}\setminus\nu^n$, we have the map $rp : \nu^n \to \partial C$, where $p : \nu^n \to \mathbb{R}^{n+1}\setminus\{v\}$ is the restriction of the projection of $\mathbb{R}^{2n+1}$ onto the first $n+1$ factors. Now, let $u = (a_{n+2}, \ldots, a_{2n+1}) \in (\mathbb{R}\setminus\mathbb{Q})^n$. Then, $\partial C\times\{u\} \subset \nu^n$ by the definition of ν^n. Let $f : \mathbf{S}^n \to \partial C\times\{u\}$ be a homeomorphism. Assume that f is null-homotopic in ν, that is, f extends to a map $\tilde{f} : \mathbf{B}^{n+1} \to \nu^n$. The map $f^{-1}rp\tilde{f} : \mathbf{B}^{n+1} \to \mathbf{S}^n$ is a retraction, which contradicts the No Retraction Theorem 5.1.5. Hence, f is not null-homotopic in ν.

By analogy, we can show that for any neighborhood V of $x \in \mu^n$, there is a map $f : \mathbf{S}^n \to V$ that is not null-homotopic in μ^n. In the above, replacing ν^n and $[a_i, b_i]$ with μ^n and $[(m_i - 1)/3^k, m_i/3^k]$, $m_i \in \mathbb{N}$, respectively, and defining the points v, u by

$$v = ((m_1 - \tfrac{1}{2})/3^k, \ldots, (m_{n+1} - \tfrac{1}{2})/3^k) \text{ and}$$
$$u = (m_{n+2}/3^k, \ldots, m_{2n+1}/3^k),$$

the same arguments apply.

Suppose that ν^n and μ^n inherit the metric of $\mathbb{R}^{2n+1}$ induced by the norm

$$\|x\|_\infty = \max\{|x(i)| \mid i = 1, \ldots, 2n+1\}.$$

Let $m \leq n$ and denote $K = F(\Delta^m)$ and $L = F(\partial\Delta^m)$. To show that ν^n and μ^n are C^{n-1} and LC^{n-1}, it suffices to show that every map $f : |L| \to \nu^n$ (resp. $f : |L| \to \mu^n$), $m \leq n$, extends to a map $\tilde{f} : |K| \to \nu^n$ (resp. $\tilde{f} : |K| \to \mu^n$) with $\operatorname{diam} \tilde{f}(|K|) \leq 2 \operatorname{diam} f(|L|)$ (resp. $\operatorname{diam} \tilde{f}(|K|) \leq 12 \operatorname{diam} f(|L|)$). Since $C = \prod_{i=1}^{2n+1} \operatorname{pr}_i f(|L|)$ is an AE, f extends to a map $f_0 : |K| \to C$, where

$$\operatorname{diam} f(|L|) \leq \operatorname{diam} f_0(|K|) \leq \operatorname{diam} C = \operatorname{diam} f(|L|).$$

Then, we will deform f_0 to a map $\tilde{f} : |K| \to \nu^n$ (resp. $\tilde{f} : |K| \to \mu^n$) such that $\tilde{f}||L| = f$ and $d(\tilde{f}, f_0) < \frac{1}{2} \operatorname{diam} C$ (resp. $d(\tilde{f}, f_0) < 6 \operatorname{diam} C$).

The case of ν^n: We write $\nu^n = \mathbb{R}^{2n+1} \setminus \bigcup_{i\in\mathbb{N}} H_i$, where each H_i is an n-dimensional flat. By induction, we will construct maps $f_k : |K| \to \mathbb{R}^{2n+1} \setminus \bigcup_{i=1}^k H_i$, $k \in \mathbb{N}$, such that $f_k||L| = f$ and $d(f_k, f_{k-1}) < \varepsilon_k/2$, where $\varepsilon_k > 0$ is defined as follows: $\varepsilon_0 = \operatorname{diam} C$ and

$$\varepsilon_k = \min\{\varepsilon_{k-1}/2,\ \operatorname{dist}(f(|L|), H_k),\ \tfrac{1}{2}\operatorname{dist}(f_{k-1}(|K|), H_{k-1})\} > 0.$$

Observe $\sum_{i=k}^\infty \varepsilon_i/2 \leq \varepsilon_k \leq 2^{-k} \operatorname{diam} C$. Then, $(f_i)_{i\in\mathbb{N}}$ is Cauchy, so it converges to a map $\tilde{f} : |K| \to \mathbb{R}^{2n+1}$ by the completeness of $\mathrm{C}(|K|, \mathbb{R}^{2n+1})$, where $\tilde{f}||L| = f$ and $d(\tilde{f}, f_0) \leq \varepsilon_1 \leq \frac{1}{2} \operatorname{diam} C$. Since $d(\tilde{f}, f_i) \leq \varepsilon_{i+1}$ for each $i \in \mathbb{N}$, it follows that

$$\operatorname{dist}(\tilde{f}(|K|), H_i) \geq \operatorname{dist}(f_i(|K|), H_i) - \varepsilon_{i+1} \geq \varepsilon_{i+1} > 0,$$

hence $\tilde{f}(|K|) \subset \mathbb{R}^{2n+1} \setminus \bigcup_{i\in\mathbb{N}} H_i = \nu^n$.

Now, assume that $f_0, \ldots, f_{k-1}$ have been obtained. Then, $\varepsilon_k > 0$ is defined as above. Choose $K' \lhd K$ so that $\operatorname{diam}\langle f_{k-1}(\sigma)\rangle < \varepsilon_k/2$ for every $\sigma \in K'$, and let L' be the subcomplex of K' with $L' \lhd L$. Then, for each $\sigma \in K'$ with $\sigma \cap |L| \neq \emptyset$, we have

$$\operatorname{dist}(\langle f_{k-1}(\sigma)\rangle, H_k) \geq \operatorname{dist}(\mathrm{N}(f(|L|), \varepsilon_k/2), H_k) \geq \varepsilon_k/2.$$

Since H_k is an n-dimensional flat in $\mathbb{R}^{2n+1}$ and $\dim K = m \leq n$, we can apply the General Position Lemma 5.8.4 to find points $p_v \in \mathbb{R}^{2n+1}$, $v \in K'^{(0)} \setminus L'^{(0)}$, such that $\|p_v - f_{k-1}(v)\| < \varepsilon_k/2$ and

$$\operatorname{fl}\{p_{v_1}, \ldots, p_{v_{m+1}}\} \cap H_k = \emptyset \text{ for every } v_1, \ldots, v_{m+1} \in K'^{(0)} \setminus L'^{(0)}.$$

Using the barycentric coordinates, we can define a map $f_k : |K| \to \mathbb{R}^{2n+1} \setminus H_k$ as follows:

$$f_k(x) = \begin{cases} \sum_{v \in K'^{(0)} \setminus L'^{(0)}} \beta_v^{K'}(x) p_v & \text{if } \alpha(x) = 0, \\ \alpha(x) f\left(\sum_{v \in L'^{(0)}} \alpha(x)^{-1} \beta_v^{K'}(x) v\right) + \sum_{v \in K'^{(0)} \setminus L'^{(0)}} \beta_v^{K'}(x) p_v & \\ & \text{if } \alpha(x) > 0, \end{cases}$$

where $\alpha(x) = \sum_{v \in L'^{(0)}} \beta_v^{K'}(x)$. Then, $f_k||L| = f$. In the above definition, if $\alpha(x) \neq 1$ then

$$\sum_{v \in K'^{(0)} \setminus L'^{(0)}} \beta_v^{K'}(x) p_v = (1 - \alpha(x)) \sum_{v \in K'^{(0)} \setminus L'^{(0)}} \frac{\beta_v^{K'}(x)}{1 - \alpha(x)} p_v,$$

where $1 - \alpha(x) = \sum_{v \in K'^{(0)} \setminus L'^{(0)}} \beta_v^{K'}(x)$. Since $\| p_v - f_{k-1}(v) \| < \varepsilon_k/2$ for each $v \in K'^{(0)} \setminus L'^{(0)}$, it follows that $\| f_k(x) - f_{k-1}(x) \| < \varepsilon_k/2$ for each $x \in |K|$. This completes the induction.

The case of μ^n: Recall $\mu^n = \bigcap_{k \in \mathbb{N}} M_k^{2n+1} \subset \mathbf{I}^{2n+1}$. Take the largest number $k_0 \in \omega$ such that $2 \operatorname{diam} C \leq 1/3^{k_0}$. Then, $1/3^{k_0} < 6 \operatorname{diam} C$. For each $i = 1, \ldots, 2n+1$, let $m_i \in \mathbb{N}$ be the smallest number such that $\max \operatorname{pr}_i(C) \leq m_i/3^{k_0}$. Then, $f_0(|K|)$ is contained in the following cube:

$$D = \prod_{i=1}^{2n+1} \left[\frac{m_i - 1}{3^{k_0}}, \frac{m_i}{3^{k_0}} \right].$$

Observe $D \subset \bigcap_{i=1}^{k_0 - 1} M_i^{2n+1}$. We have a homeomorphism $\varphi : (D, D \cap \mu^n) \to (\mathbf{I}^{2n+1}, \mu^n)$ defined by

$$\varphi(x) = 3^{k_0} x - (m_1 - 1, \ldots, m_{2n+1} - 1).$$

Refer to Fig. 5.7. If the map $\varphi f_0 : |K| \to \mathbf{I}^{2n+1}$ can be deformed to a map $f' : |K| \to \mu^n$, then $\widetilde{f_0} = \varphi^{-1} f' : |K| \to D \cap \mu^n$ is the desired one. Thus, we may assume that $D = \mathbf{I}^{2n+1}$.

By induction, we will construct maps $f_k : |K| \to \bigcap_{i=1}^{k} M_i^{2n+1}$, $k \in \mathbb{N}$, such that $f_k||L| = f$ and f_k is $2/3^k$-close to f_{k-1}. Since $\sum_{i=k}^{\infty} 2/3^i = 1/3^{k-1}$, it follows that $(f_i)_{i \in \mathbb{N}}$ is Cauchy and thus converges to a map $\tilde{f} : |K| \to \mathbf{I}^{2n+1}$ with $\tilde{f}||L| = f$. Since $\tilde{f}$ is the limit of $f_j : |K| \to \bigcap_{i=1}^{k} M_i^{2n+1}$, $j \geq k$, it follows that $\tilde{f}(|K|) \subset \mu^n = \bigcap_{i \in \mathbb{N}} M_i^{2n+1}$.

Assume that $f_0, \dots, f_{k-1}$ have been obtained. Let

$$S = \{m/3^{k-1} + 3/(3^k 2) \mid m = 0, \dots, 3^{k-1} - 1\} \text{ and}$$
$$T = \{x \in \mathbb{R}^{2n+1} \mid x(i) \in S \text{ at least } n+1 \text{ many } i\}.$$

Then, S consists of mid-points of components of R_{k-1}, T is a finite union of n-dimensional flats in $\mathbb{R}^{2n+1}$ and

$$M_k^{2n+1} = \mathbf{I}^{2n+1} \setminus \mathrm{N}(T, 1/(3^k 2)).$$

Note that M_k^{2n+1} and $\bigcap_{i=1}^{k-1} M_i^{2n+1}$ are the polyhedra of subcomplexes of the cell complex consisting of all faces of the following cubes:

$$\prod_{i=1}^{2n+1} \left[\frac{m_i - 1}{3^k}, \frac{m_i}{3^k}\right] \subset \mathbf{I}^{2n+1}, \ m_i \in \{1, \dots, 3^k\}.$$

Choose a simplicial subdivision $K' \lhd K$ so that $\operatorname{diam} f_{k-1}(\sigma) < 1/(3^k 4)$ for each $\sigma \in K'$. As in the case of ν^n, we can apply the General Position Lemma 5.8.4 to find points $p_v \in \mathbf{I}^{2n+1}$, $v \in K'^{(0)} \setminus f_{k-1}^{-1}(M_k^{2n+1})$, so that p_v and $f_{k-1}(v)$ are contained in the same cube in the above and

$$\mathrm{fl}\{p_{v_1}, \dots, p_{v_{m+1}}\} \cap T = \emptyset$$

for every $m+1$ many $v_1, \dots, v_{m+1} \in K'^{(0)} \setminus f_{k-1}^{-1}(M_k^{2n+1})$. Then, in the same way as f_k in the case of ν^n, we can define a map $g : |K| \to \bigcap_{i=1}^{k-1} M_i^{2n+1} \setminus T$ such that $g|f_{k-1}^{-1}(M_k^{2n+1}) = f_{k-1}$ and g is 3^{-k}-close to f_{k-1}. On the other hand, we have a retraction $h : \mathbf{I}^{2n+1} \setminus T \to M_k^{2n+1}$ such that $h(\bigcap_{i=1}^{k-1} M_i^{2n+1} \setminus T) = \bigcap_{i=1}^{k} M_i^{2n+1}$ and h is 3^{-k}-close to id. Then, hg is the desired map f_k. This completes the induction. □

6.12 Finite-Dimensional ANRs

A space Y is called an **absolute neighborhood extensor for metrizable spaces of dimension $\leq n$** (or an **ANE(n)**) if each map $f : A \to Y$ from any closed set A in a metrizable space X of $\dim X \leq n$ extends over some neighborhood U of A in X. When f can always be extended over X (i.e., $U = X$ in the above), we call Y an **absolute extensor for metrizable spaces of dimension $\leq n$** (or an **AE(n)**). In this section, we show that the local n-connectedness characterizes ANE($n+1$)s and n-dimensional ANRs.

For a metrizable space X, $\dim X \leq n$ if and only if each open cover $\mathcal{U}$ of X has an open refinement $\mathcal{V}$ with $\operatorname{ord} \mathcal{V} \leq n+1$.[9] We use the following facts of Dimension Theory:

(i) If $\dim X \leq n$ then $\dim X \times \mathbf{I} \leq n+1$ — the Product Theorem 5.4.9.
(ii) For any subset A of a metrizable space X, $\dim A \leq \dim X$—the Subset Theorem 5.3.3.

In Lemma 6.1.2, given a proper closed set $A \neq \emptyset$ in a metric space $X = (X, d)$, we obtained a Dugundji system $(a_U)_{U \in \mathcal{U}}$, i.e., $\mathcal{U}$ is a locally finite open cover of $X \setminus A$ with $a_U \in A$, $U \in \mathcal{U}$, such that $x \in U \in \mathcal{U}$ implies $d(x, a_U) \leq 2d(x, A)$. If $\dim X \setminus A \leq n+1$, we can take $\mathcal{U}$ with the additional condition that $\operatorname{ord} \mathcal{U} \leq n+2$, i.e., $\dim N(\mathcal{U}) \leq n+1$, where $N(\mathcal{U})$ is the nerve of $\mathcal{U}$. Thus, we have the following variant:

Lemma 6.12.1. *Let $X = (X, d)$ be a metric space and $A \neq \emptyset$ be a proper closed set in X such that* $\dim X \setminus A \leq n+1$. *Then, there exists a Dugundji system $(a_U)_{U \in \mathcal{U}}$ for $A \subset X$ such that* $\dim N(\mathcal{U}) \leq n+1$. □

An open refinement $\mathcal{V}$ of $\mathcal{U} \in \operatorname{cov}(X)$ is called a **C^n-refinement** of $\mathcal{U}$ if each $V \in \mathcal{V}$ is contained in some $U \in \mathcal{U}$ such that any map $f : \mathbf{S}^i \to V$, $i \leq n$, extends to a map $\tilde{f} : \mathbf{B}^{i+1} \to U$. We call $\mathcal{V}$ an **n-Lefschetz refinement** of $\mathcal{U}$ if any partial $\mathcal{V}$-realization of an arbitrary simplicial complex with $\dim \leq n$ in X extends to full $\mathcal{U}$-realization in X. We denote

$$\mathcal{V} \underset{C^n}{\prec} \mathcal{U} \text{ or } \mathcal{U} \underset{C^n}{\succ} \mathcal{V} \quad \left(\text{resp. } \mathcal{V} \underset{L^n}{\prec} \mathcal{U} \text{ or } \mathcal{U} \underset{L^n}{\succ} \mathcal{V}\right)$$

when $\mathcal{V}$ is a C^n-refinement (resp. an n-Lefschetz refinement) of $\mathcal{U}$.

Lemma 6.12.2. *For an LC^n paracompact space Y, each open cover $\mathcal{U}$ of Y has an $(n+1)$-Lefschetz refinement.*

Proof. Since Y is LC^n, $\mathcal{U}$ has the following open refinements:

$$\mathcal{U} = \mathcal{V}_{n+1} \overset{*}{\succ} \mathcal{U}_n \underset{C^n}{\succ} \mathcal{V}_n \overset{*}{\succ} \mathcal{U}_{n-1} \underset{C^{n-1}}{\succ} \mathcal{V}_{n-1} \overset{*}{\succ} \cdots \overset{*}{\succ} \mathcal{U}_0 \underset{C^0}{\succ} \mathcal{V}_0.$$

Let K be a simplicial complex with $\dim K \leq n+1$, L be a subcomplex of K with $K^{(0)} \subset L$, and $f : |L| \to Y$ be a partial $\mathcal{V}_0$-realization of K. By induction, we can define partial $\mathcal{V}_i$-realizations $f_i : |L \cup K^{(i)}| \to Y$ of K such that $f_i||L \cup K^{(i-1)}| = f_{i-1}$, where $f_0 = f$. Indeed, assume that f_{i-1} has been defined. For each $\sigma \in K^{(i)} \setminus (L \cup K^{(i-1)})$, choose $V_\sigma \in \mathcal{V}_{i-1}$ so that $f_{i-1}(\partial\sigma) \subset V_\sigma$. Since $\mathcal{V}_{i-1}$ is a C^{i-1}-refinement of $\mathcal{U}_{i-1}$, $f_{i-1}|\partial\sigma$ extends to a map $f_\sigma : \sigma \to U_\sigma$ for some $U_\sigma \in \mathcal{U}_{i-1}$. Thus, f_i can be defined by $f_i|\sigma = f_\sigma$ for $\sigma \in K^{(i)} \setminus (L \cup K^{(i-1)})$.

---

[9]In the definition of dimension, $\mathcal{U}$ and $\mathcal{V}$ are required to be *finite*. However, by Theorem 5.2.4, this requirement is not necessary for paracompact spaces (so metrizable spaces).

For each $\sigma \in K$, we can write $\sigma \cap |L \cup K^{(i)}| = \bigcup_{j=1}^{k} \sigma_j$, where $\sigma_j \in L \cup K^{(i)}$. Since each $f_i(\sigma_j)$ is contained in some $U_j \in \mathcal{U}_{i-1}$ and $f(\sigma^{(0)})$ is contained in some $U_0 \in \mathcal{U}_{i-1}$, it follows that $f_i(|F(\sigma)^{(i)}|) \subset \mathrm{st}(U_0, \mathcal{U}_{i-1})$. Since $\mathcal{U}_i$ is a star-refinement of $\mathcal{V}_i$, f_i is a $\mathcal{V}_i$-realization of K. Note that f_{n+1} is a full $\mathcal{U}$-realization of K. Therefore, $\mathcal{V}_0$ is an $(n+1)$-Lefschetz refinement of $\mathcal{U}$. □

Now, we prove the following characterization of ANE(n)s:

Theorem 6.12.3. *Let $n \in \omega$. For a metrizable space Y, the following are equivalent:*

(a) *Y is LC^n;*
(b) *Y is an ANE$(n+1)$;*
(c) *If A is a closed set in a metrizable space X with $\dim X \setminus A \le n+1$, then every map $f : A \to Y$ extends over a neighborhood of A in X;*
(d) *If A is a closed set in a metrizable space X with $\dim A \le n$, then every map $f : A \to Y$ extends over a neighborhood of A in X;*
(e) *Each open cover $\mathcal{U}$ of Y has an $(n+1)$-Lefschetz refinement;*
(f) *Each neighborhood U of any $y \in Y$ contains a neighborhood V of y in Y such that every map $f : X \to V$ of a metrizable space X with $\dim X \le n$ is null-homotopic in U.*

Proof. The implication (a) ⇒ (e) is Lemma 6.12.2. The implication (f) ⇒ (a) is trivial and (c) ⇒ (b) follows from (ii). We show the implications (a) ⇒ (c), (b) ⇒ (f), and (e) ⇒ (d) ⇒ (f).

$$\begin{array}{ccccc}
\text{(e)} & \overset{(6.12.2)}{\Longleftarrow} & \text{(a)} & \Longrightarrow & \text{(c)} \\
\Big\Downarrow & & \Big\Uparrow {\scriptstyle \text{triv.}} & & \Big\Downarrow {\scriptstyle \text{(ii)}} \\
\text{(d)} & \Longrightarrow & \text{(f)} & \Longleftarrow & \text{(b)}
\end{array}$$

(a) ⇒ (c): We may assume that X and Y are metric spaces. For simplicity, let d stand for both metrics of X and Y. By (a), we can take open covers of Y as follows:

$$\mathcal{V}_0 = \{Y\} \overset{*}{\succ} \mathcal{U}_1 \underset{C^n}{\succ} \mathcal{V}_1 \overset{*}{\succ} \mathcal{U}_2 \underset{C^n}{\succ} \mathcal{V}_2 \overset{*}{\succ} \cdots$$

and $\mathrm{mesh}\, \mathcal{V}_i < 2^{-i}$ for each $i \in \mathbb{N}$. By Lemma 6.12.1, we have a Dugundji system $(a_W)_{W \in \mathcal{W}}$ for $A \subset X$ such that $\dim N(\mathcal{W}) \le n+1$. Let $\varphi : X \setminus A \to |N(\mathcal{W})|$ be a canonical map, i.e., $\varphi(x) \in \langle \mathcal{W}[x] \rangle$ for each $x \in X \setminus A$ (cf. Proposition 4.9.1). For each $\sigma \in N(\mathcal{W})$, let $A(\sigma) = \{a_W \mid W \in \sigma^{(0)}\} \subset A$. Let K_0 be the subcomplex of $N(\mathcal{W})$ consisting of all simplexes $\sigma \in N(\mathcal{W})$ such that $f(A(\sigma))$ is a singleton. Obviously, $K_0^{(0)} = N(\mathcal{W})^{(0)} = \mathcal{W}$. For each $\sigma \in N(\mathcal{W}) \setminus K_0$, let

$$k(\sigma) = \max \left\{ i \in \omega \;\middle|\; f(A(\sigma)) \subset V \text{ for some } V \in \mathcal{V}_i \right\},$$

and choose $V_0(\sigma) \in \mathcal{V}_{k(\sigma)}$ so that $f(A(\sigma)) \subset V_0(\sigma)$. Observe that if $\sigma \in N(\mathcal{W}) \setminus K_0$ and $\sigma < \tau$, then $\tau \in N(\mathcal{W}) \setminus K_0$ and $k(\sigma) \geq k(\tau)$. We define a subcomplex $K \subset N(\mathcal{W})$ as follows:

$$K = K_0 \cup \{\sigma \in N(\mathcal{W}) \mid k(\sigma) \geq n+1\}.$$

For each $i \in \mathbb{N}$, let $K_i = K_0 \cup K^{(i)}$. Note that $K = K_{n+1}$. Let $g_0 : |K_0| \to Y$ be the map defined by $g_0(\sigma) = f(A(\sigma))$ for each $\sigma \in K_0$. Then, observe

$$g_0(\sigma \cap |K_0|) = f(A(\sigma)) \subset V_0(\sigma) \in \mathcal{V}_{k(\sigma)} \text{ for each } \sigma \in K \setminus K_0.$$

For each 1-simplex $\sigma \in K_1 \setminus K_0$, we have $U_0(\sigma) \in \mathcal{U}_{k(\sigma)}$ such that $g_0|\partial\sigma = g_0|\sigma^{(0)}$ extends to a map $g_\sigma : \sigma \to U_0(\sigma)$. Then, we can extend g_0 to a map $g_1 : |K_1| \to Y$ defined by $g_1|\sigma = g_\sigma$ for $\sigma \in K_1 \setminus K_0$. For each $\sigma \in K \setminus K_0$,

$$g_1(\sigma \cap |K_1|) \subset \mathrm{St}(V_0(\sigma), \mathcal{U}_{k(\sigma)}) \in \mathrm{St}(\mathcal{V}_{k(\sigma)}, \mathcal{U}_{k(\sigma)}) \prec \mathrm{St}\,\mathcal{U}_{k(\sigma)} \prec \mathcal{V}_{k(\sigma)-1},$$

hence $g_1(\sigma \cap |K_1|)$ is contained in some $V_1(\sigma) \in \mathcal{V}_{k(\sigma)-1}$. Next, for each 2-simplex $\sigma \in K_2 \setminus K_1$, we have $U_1(\sigma) \in \mathcal{U}_{k(\sigma)-1}$ such that $g_1|\partial\sigma$ extends to a map $g_\sigma : \sigma \to U_1(\sigma)$. Then, as above, we can extend g_1 to a map $g_2 : |K_2| \to Y$ such that, for each $\sigma \in K \setminus K_0$, $g_2(\sigma \cap |K_2|)$ is contained in some $V_2(\sigma) \in \mathcal{V}_{k(\sigma)-2}$. We continue this process $n+1$ times to obtain a map $g = g_{n+1} : |K| = |K_{n+1}| \to Y$ such that, for each $\sigma \in K = K_{n+1}$, $g(\sigma)$ is contained in some $V_{n+1}(\sigma) \in \mathcal{V}_{k(\sigma)-n-1}$.

Now, we define $\tilde{f} : A \cup \varphi^{-1}(|K|) \to Y$ by $\tilde{f}|A = f$ and $\tilde{f}|\varphi^{-1}(|K|) = g\varphi|\varphi^{-1}(|K|)$. It remains to be shown that $A \cup \varphi^{-1}(|K|)$ is a neighborhood of A in X and $\tilde{f}$ is continuous. To this end, it suffices to prove that $A \cup \varphi^{-1}(|K|)$ is a neighborhood of each $a \in \mathrm{bd}_X A$ in X and $\tilde{f}$ is continuous at each $a \in \mathrm{bd}_X A$. We will prove these claims at the same time. For each $a \in \mathrm{bd}_X A$ and $\varepsilon > 0$, choose $i \geq n+1$ and $V \in \mathcal{V}_i$ so that $2^{-i}(1 + 2^{n+1}) < \varepsilon$ and $f(a) \in V$. Then, we have $\delta > 0$ such that $f(A \cap \mathrm{B}(a, \delta)) \subset V$. For $x \in X \setminus A$ with $d(x, a) < \delta/3$, let $\sigma_x \in N(\mathcal{W})$ be the carrier of $\varphi(x)$. For every $W \in \sigma_x^{(0)}$, since $x \in W \in \mathcal{W}$, we have $d(x, a_W) \leq 2d(x, A) \leq 2d(x, a)$, hence

$$d(a, a_W) \leq d(x, a) + d(x, a_W) \leq 3d(x, a) < \delta.$$

Then, $f(A(\sigma_x)) \subset V \in \mathcal{V}_i$, which implies that $k(\sigma_x) \geq i \geq n+1$, hence $\sigma_x \in K$. Therefore, $\varphi(x) \in |K|$. Thus, $A \cup \varphi^{-1}(|K|)$ is a neighborhood of a in X. On the other hand, note that

$$f(a) \in V,\ g(\sigma_x^{(0)}) = f(A(\sigma_x)) \subset V \text{ and}$$
$$g(\sigma_x) \subset V_{n+1}(\sigma_x) \in \mathcal{V}_{k(\sigma_x)-n-1}.$$

Then, it follows that

$$d(f(a), g\varphi(x)) \leq \operatorname{diam} V + \operatorname{diam} V_{n+1}(\sigma_x)$$
$$< 2^{-i} + 2^{-k(\sigma_x)+n+1} \leq 2^{-i}(1 + 2^{n+1}) < \varepsilon.$$

Therefore, $\tilde{f}$ is continuous at $a \in \mathrm{bd}_X A$. The continuity of $\tilde{f}$ on $\mathrm{int}_X A \cup \varphi^{-1}(|K|)$ is obvious. Thus, f is extended over a neighborhood of A in X.

(e) $\Rightarrow$ (d): Given $d \in \mathrm{Metr}(Y)$, we take open covers of Y as follows:

$$\mathcal{U}_1 \underset{L^{n+1}}{\succ} \mathcal{V}_1 \overset{*}{\succ} \mathcal{U}_2 \underset{L^{n+1}}{\succ} \mathcal{V}_2 \overset{*}{\succ} \mathcal{U}_3 \underset{L^{n+1}}{\succ} \mathcal{V}_3 \overset{*}{\succ} \cdots$$

and $\operatorname{mesh} \mathcal{U}_i < 2^{-i-1}$ for each $i \in \mathbb{N}$. Since $\dim A \leq n$, A has open covers $\mathcal{W}_i$, $i \in \mathbb{N}$, such that $\operatorname{ord} \mathcal{W}_i \leq n+1$ and $\mathcal{W}_i \prec f^{-1}(\mathcal{U}_{i+2}) \wedge \mathcal{W}_{i-1}$, where $\mathcal{W}_0 = \{A\}$. For each $i \in \mathbb{N}$, let K_i be the nerve of $\mathcal{W}_i$ with $\varphi_i : A \to |K_i|$ a canonical map. Then, $\dim K_i \leq n$. Extending a partial $\mathcal{V}_{i+1}$-realization of K_i defined on $K_i^{(0)}$, we can obtain a full $\mathcal{U}_{i+1}$-realization $p_i : |K_i| \to Y$ such that

$$p_i(W) \in f(W) \text{ for each } W \in K_i^{(0)} = \mathcal{W}_i.$$

Since $\mathcal{W}_{i+1} \prec \mathcal{W}_i$, there is a simplicial map $\psi_i : K_{i+1} \to K_i$ such that

$$W \subset \psi_i(W) \in K_i^{(0)} = \mathcal{W}_i \text{ for each } W \in K_{i+1}^{(0)} = \mathcal{W}_{i+1}.$$

Choosing an order on $K_i^{(0)}$ so that K_i is an ordered simplicial complex, we have the simplicial mapping cylinder Z_{ψ_i} of ψ_i. Then, $\dim Z_{\psi_i} \leq n+1$ and $Z_{\psi_i}^{(0)} \subset K_i \oplus K_{i+1}$. Observe that $p_i \oplus p_{i+1} : |K_i| \oplus |K_{i+1}| \to Y$ is a partial $\mathcal{V}_i$-realization of Z_{ψ_i}. Then, we have a full $\mathcal{U}_i$-realization $g_i : |Z_{\psi_i}| \to Y$ such that $g_i||K_i| = p_i$ and $g_i||K_{i+1}| = p_{i+1}$. On the other hand, we also have a map $h_i : A \times \mathbf{I} \to |Z_{\psi_i}|$ such that $h_i(a, 0) = \varphi_i(a)$ and $h_i(a, 1) = \varphi_{i+1}(a)$. Indeed, φ_i and $\psi_i\varphi_{i+1}$ are canonical maps for the cover $\mathcal{W}_i$, hence they are contiguous (Corollary 4.9.2) and h_i can be defined as follows:

$$h_i(a,t) = \begin{cases} \bar{q}_{\psi_i}(\varphi_{i+1}(a), 2t-1) & \text{if } t \geq \frac{1}{2}, \\ (1-2t)\varphi_i(a) + 2t\psi_i\varphi_{i+1}(a) & \text{if } t \leq \frac{1}{2}, \end{cases}$$

where $\bar{q}_{\psi_i} : |K_i| \oplus |K_{i+1} \times I| \to |Z_{\psi_i}|$ is the quotient map (Fig. 6.19). For each $(a, t) \in A \times [0, 1)$, the carrier of $h_i(a, t)$ in Z_{ψ_i} has a vertex $W \in \mathcal{W}_i[a]$. Then,

$$g_i(W) = p_i(W) \in f(W) \subset \operatorname{st}(f(a), \mathcal{U}_{i+2}).$$

Since g_i is a full $\mathcal{U}_i$-realization of Z_{ψ_i}, it follows that $g_i h_i(a, t) \in \operatorname{st}(f(a), \mathcal{V}_{i-1})$.

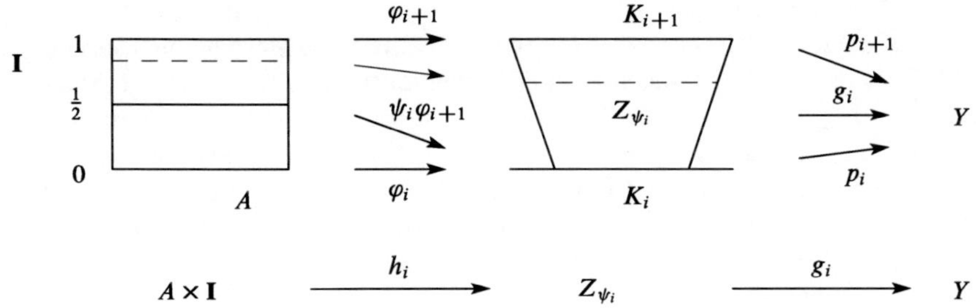

Fig. 6.19 The maps h_i and g_i

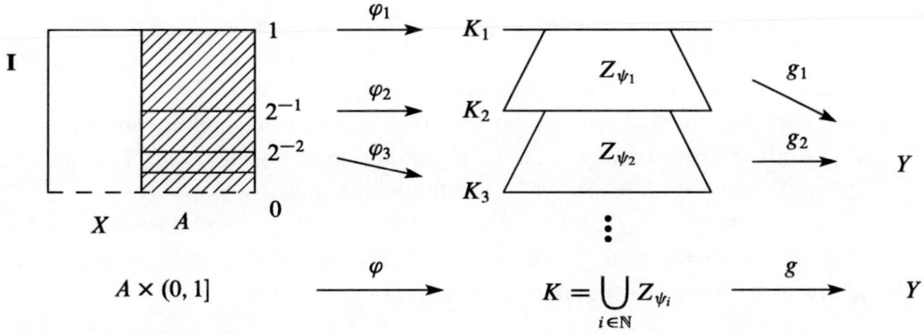

Fig. 6.20 The maps φ and g

Now, we have a simplicial complex $K = \bigcup_{i\in\mathbb{N}} Z_{\psi_i}$ and a map $g : |K| \to Y$ defined by $g|Z_{\psi_i} = g_i$ for each $i \in \mathbb{N}$. We also define a map $\varphi : A \times (0, 1] \to |K|$ as follows:

$$\varphi(a, t) = h_i(a, 2 - 2^i t) \text{ for } 2^{-i} \leq t \leq 2^{-i+1}.$$

See Fig. 6.20. Since $|K|$ is an ANE, the map $g\varphi : A \times (0, 1] \to Y$ extends over a neighborhood of $A \times (0, 1]$ in $X \times (0, 1]$. On the other hand, for each $a \in A$ and $2^{-i} < t \leq 2^{-i+1}$,

$$g\varphi(a, t) = g_i h_i(a, 2 - 2^i t) \in \operatorname{st}(f(a), \mathcal{U}_{i-1}),$$

which implies that $d(g\varphi(a, t), f(a)) < 2^{-i} < t$. Hence, $g\varphi$ can be extended over $A \times \mathbf{I}$ by f. Thus, we can apply Lemma 6.5.1 to extend f over a neighborhood of A in X.

(b) (or (d)) $\Rightarrow$ (f): Let $d \in \operatorname{Metr}(Y)$ and assume that condition (f) does not hold. Then, we have a point $y_0 \in Y$, an open neighborhood U of y_0 in Y and maps $f_i : X_i \to U$, $i \in \mathbb{N}$, such that each X_i is a metrizable space, $\dim X_i \leq n$, $f_i(X_i) \subset \mathrm{B}(y_0, 2^{-i})$, and each f_i is not null-homotopic. Take $v_0 \notin \bigoplus_{i\in\mathbb{N}} X_i \times \mathbf{I}$ and let

$$X = \{v_0\} \cup \bigoplus_{i\in\mathbb{N}} X_i \times \mathbf{I} \text{ and } A = \{v_0\} \cup \bigoplus_{i\in\mathbb{N}} X_i \times \{0, 1\},$$

where X admits the topology generated by the open sets in $X_i \times \mathbf{I}$ and $\{v_0\} \cup \bigoplus_{j \geq i} X_j \times \mathbf{I}$, $i \in \mathbb{N}$. Then, X is metrizable by the Bing Metrization Theorem 2.3.4 and A is closed in X. By facts (i) and (ii) mentioned at the beginning of this section, $\dim X_i \times \mathbf{I} \leq n+1$ and $\dim X_i \times \{0,1\} \leq n$ for every $i \in \mathbb{N}$. By definition, it is easy to show that $\dim X \leq n+1$ and $\dim A \leq n$. We define a map $f : A \to U$ by $f(v_0) = y_0$, $f(x,0) = f_i(x)$, and $f(x,1) = y_0$ for $x \in X_i$. Then, f extends over a neighborhood W of A in X by (b) (or (d)). Note that W contains some $X_i \times \mathbf{I}$ that maps into U by the extension of f. This means that f_i is null-homotopic in U, which is a contradiction. □

Remark 17. The following conditions can also be added to the list of Theorem 6.12.3, as variants of (a) (or (f)) corresponding to (b), (c), and (d).

(b') Each neighborhood U of any point y in Y contains a neighborhood V of y in Y such that every map $f : A \to V$ defined in a closed set A in a metrizable space X with $\dim X \leq n+1$ extends to a map $\tilde{f} : X \to U$;
(c') Each neighborhood U of any point y in Y contains a neighborhood V of y in Y such that every map $f : A \to V$ defined in a closed set A in a metrizable space X with $\dim X \setminus A \leq n+1$ extends to a map $\tilde{f} : X \to U$;
(d') Each neighborhood U of any point y in Y contains a neighborhood V of y in Y such that every map $f : A \to V$ defined in a closed set A in a metrizable space X with $\dim A \leq n$ extends to a map $\tilde{f} : X \to U$.

Sketch of Proof. For $y \in V \subset U$ in each of these conditions, given a map $f : X \to V$ of a metrizable space X with $\dim X \leq n$, let $h : X \times \{0,1\} \to V$ be the map defined by $h(x,0) = f(x)$ and $h(x,1) = y$ for $x \in X$. Then, h can be extended to a homotopy giving $f \simeq 0$ in U. Thus, each of (b'), (c'), and (d') implies condition (f) in Theorem 6.12.3.

The proofs of the implications (b) ⇒ (b'), (c) ⇒ (c'), and (d) ⇒ (d') are similar to the proof of (b) (or (d)) ⇒ (f) in Theorem 6.12.3.

As in Proposition 6.6.9, we can prove the following lemma:

Lemma 6.12.4. *Let K be an $(n+1)$-dimensional simplicial complex and L be a subcomplex of K. If X is n-connected, then any map $f : |L| \to X$ extends over $|K|$.* □

Using this lemma and Theorem 6.12.3, we can obtain the following characterization of AE(n)s:

Theorem 6.12.5. *Let $n \in \omega$. For a metrizable space Y, the following are equivalent:*

(a) *Y is C^n and LC^n;*
(b) *Y is an AE$(n+1)$;*
(c) *If A is a closed set in a metrizable space X with* $\dim X \setminus A \leq n+1$, *then every map $f : A \to Y$ extends over X;*
(d) *If A is a closed set in a metrizable space X with* $\dim A \leq n$, *then every map $f : A \to Y$ extends over X.*

Proof. The implication (c) $\Rightarrow$ (b) follows from (ii). Since conditions (b) and (d) imply that Y is n-connected, the implications (b) $\Rightarrow$ (a) and (d) $\Rightarrow$ (a) can be obtained by Theorem 6.12.3.

In the proof of Theorem 6.12.3 (a) $\Rightarrow$ (c), if Y is n-connected, the map $g : |K| \to Y$ can be extended over $|N(\mathcal{W})|$ by Lemma 6.12.4, hence $\tilde{f}$ can be defined over X. Thus, the implication (a) $\Rightarrow$ (c) of this theorem also holds.

In the proof of Theorem 6.12.3 (e) $\Rightarrow$ (d), we may assume that $\operatorname{diam} Y < 2^{-3}$. If Y is n-connected, we can take $\mathcal{U}_i = \mathcal{V}_i = \{Y\}$ for $i = 1, 2, 3$, so we can also take $\mathcal{W}_1 = \{A\}$. Now, since $\varphi|A \times \{1\}$ is a constant map and $|K|$ is an ANE, $g\varphi$ extends over a neighborhood of $A \times (0, 1] \cup X \times \{1\}$ in $X \times (0, 1]$. Therefore, f extends over X by Lemma 6.5.1. Since condition (a) of this theorem implies condition (e) of Theorem 6.12.3, we have the implication (a) $\Rightarrow$ (d). □

By Theorems 6.12.3 and 6.12.5, we have the following characterization of ARs and ANRs of dimension $\leq n$:

Corollary 6.12.6. *Let X be a metrizable space with* $\dim X \leq n$.

(1) *X is an AR if and only if X is C^n and LC^n (i.e., X is an AE$(n+1)$).*
(2) *X is an ANR if and only if X is LC^n (i.e., X is an ANE$(n+1)$).* □

Theorem 6.12.7. *Let $f : X \to Y$ be a map from a paracompact space X with* $\dim X \leq n$ *to an LC^{n-1} paracompact (resp. an LC^n metrizable) space Y and let $\mathcal{U}$ be an open cover of Y. Then, each open cover $\mathcal{V}$ of X has an open refinement $\mathcal{W}$ with* $\operatorname{ord} \mathcal{W} \leq n+1$ *(i.e.,* $\dim N(\mathcal{W}) \leq n$*) and a full $\mathcal{U}$-realization $\psi : |N(\mathcal{W})| \to Y$ such that $\psi\varphi$ is $\mathcal{U}$-close (resp. $\mathcal{U}$-homotopic) to f for any canonical map $\varphi : X \to |N(\mathcal{W})|$.*

Proof. By Lemma 6.12.2 and the paracompactness of Y, we can take the following open refinements of $\mathcal{U} \in \operatorname{cov}(Y)$:

$$\mathcal{U} \overset{*}{\succ} \mathcal{U}_1 \underset{L^n}{\succ} \mathcal{U}_0 \overset{*}{\succ} \mathcal{U}'.$$

By Theorem 5.2.4, $f^{-1}(\mathcal{U}')$ and $\mathcal{V}$ have a common open refinement $\mathcal{W} \in \operatorname{cov}(X)$ with $\operatorname{ord} \mathcal{W} \leq n + 1$ (i.e., $\dim N(\mathcal{W}) \leq n$). For simplicity, we denote $K = N(\mathcal{W})$. For each $W \in \mathcal{W} = K^{(0)}$, choosing $U_W \in \mathcal{U}'$ so that $f(W) \subset U_W$ and taking a point $\psi_0(W) \in U_W$, we have a partial $\mathcal{U}_0$-realization $\psi_0 : K^{(0)} \to Y$ of K, which extends to a full $\mathcal{U}_1$-realization $\psi : |K| \to Y$ of K. It is easy to prove that $\psi\varphi$ is $\mathcal{U}$-close to f for any canonical map $\varphi : X \to |N(\mathcal{W})|$.

If Y is a metrizable ANE$(n+1)$, we can take $d \in \operatorname{Metr}(Y)$ so that $\{\overline{\mathrm{B}}_d(x, 1) \mid x \in Y\} \prec \mathcal{U}$ by 2.7.7(1). For each $i \in \mathbb{N}$, take $\mathcal{U}_i, \mathcal{V}_i \in \operatorname{cov}(Y)$ such that

$$\mathcal{V}_{i+1} \overset{*}{\prec} \mathcal{U}_i \overset{*}{\prec} \mathcal{V}_i \underset{L^{n+1}}{\prec} \mathcal{B}_i = \{\mathrm{B}_d(x, 2^{-i-3}) \mid x \in Y\}.$$

By the above argument, we can obtain $\mathcal{W}_i \in \operatorname{cov}(X)$ and a full $\mathcal{U}_i$-realization $\psi_i : |K_i| \to Y$ of the nerve K_i of $\mathcal{W}_i$ such that $\mathcal{W}_1 \prec \mathcal{V}$, $\mathcal{W}_{i+1} \prec \mathcal{W}_i$, $\operatorname{ord} \mathcal{W}_i \leq n + 1$, and $\psi_i\varphi_i$ is $\mathcal{V}_i$-close to f, where $\varphi_i : X \to |K_i|$ is any canonical map for $\mathcal{W}_i$. Since $\mathcal{W}_{i+1} \prec \mathcal{W}_i$, there is a simplicial map $k_i : K_{i+1} \to K_i$ such that $k_i\varphi_{i+1}$ is also a canonical map for $\mathcal{W}_i$ (cf. Sect. 4.9). Then, $\psi_i k_i \varphi_{i+1}$ is also $\mathcal{V}_i$-close to f. Since φ_i and $k_i\varphi_{i+1}$ are contiguous, we can define a map $\varphi_i^* : X \times \mathbf{I} \to |Z_{k_i}|$ by

$$\varphi_i^*(x,t) = \begin{cases} \bar{q}_{k_i}(\varphi_{i+1}(x), 2t-1) & \text{if } \frac{1}{2} \le t \le 1, \\ (1-2t)\varphi_i(x) + 2tk_i\varphi_{i+1}(x) & \text{if } 0 \le t \le \frac{1}{2}, \end{cases}$$

where $\bar{q}_{k_i} : |K_i| \oplus |K_{i+1} \times I| \to Z_{k_i}$ is the natural simplicial map from the product simplicial complex $K_{i+1} \times I$ of K_{i+1} and $I = \{\mathbf{I}, 0, 1\}$ to the simplicial mapping cylinder Z_{k_i} of k_i (cf. Sect. 4.12). Since $\psi_i k_i$ is a full $\mathcal{U}_i$-realization of K_{i+1} and ψ_{i+1} is a full $\mathcal{U}_{i+1}$-realization of K_{i+1}, it follows that ψ_i and ψ_{i+1} induce a partial $\mathcal{V}_i$-realization of Z_{k_i}. Since $\dim Z_{k_i} \le n+1$, we have a full $\mathcal{B}_i$-realization $\psi_i^* : |Z_{k_i}| \to Y$ such that $\psi_i^*||K_i| = \psi_i$ and $\psi_i^*||K_{i+1}| = \psi_{i+1}$. Observe that

$$\psi_i^*\varphi_i^*(\{x\} \times \mathbf{I}) \subset \mathrm{st}^2(\psi_i\varphi_i(x), \mathcal{B}_i),$$

hence $\operatorname{diam} \psi_i^*\varphi_i^*(\{x\} \times \mathbf{I}) < 2^{-i}$ for each $x \in X$. Then, we can define a homotopy $h : X \times \mathbf{I} \to Y$ as follows:

$$h_0 = f \ \text{ and } \ h(x,t) = \psi_i^*\varphi_i^*(x, 2 - 2^i t) \ \text{ for } 2^{-i} \le t \le 2^{-i+1}.$$

Since $\operatorname{diam} h(\{x\} \times \mathbf{I}) < 1$ for each $x \in X$, we have $f \simeq_{\mathcal{U}} h_1 = \psi_1\varphi_1$. □

Remark 18. In Theorem 6.12.7, $\mathcal{W}$ can be taken as in Corollary 6.6.3, so as to be locally finite and σ-discrete in X (cf. Theorem 5.2.4). When X is separable, a star-finite countable open refinement $\mathcal{W}$ can be taken, hence $|N(\mathcal{W})|$ is separable and locally compact (cf. Corollary 5.2.5). If X is compact then $\mathcal{W}$ is finite, hence $|N(\mathcal{W})|$ is compact.

The following is easily seen by the same argument as in the above proof:

Proposition 6.12.8. *Let* $\mathcal{U}, \mathcal{V}, \mathcal{W} \in \mathrm{cov}(Y)$ *such that*

$$\mathcal{W} \overset{*}{\prec} \mathcal{V} \underset{L^{n+1}}{\prec} \mathcal{U}.$$

Let P be a polyhedron with $\dim P \le n$ *and Q be a subpolyhedron of P. If two maps $f, g : P \to Y$ are $\mathcal{W}$-close and $f|Q = g|Q$, then $f \simeq_{\mathcal{U}} g$ rel. Q.*

Proof. Given maps $f, g : P \to Y$ that are $\mathcal{W}$-close and $f|Q = g|Q$, we define a map

$$h' : (Q \times \mathbf{I}) \cup (P \times \{0,1\}) \to Y$$

by $h'(x,0) = f(x)$ and $h'(x,1) = g(x)$ for each $x \in P$ and $h'|Q \times \mathbf{I} = f\mathrm{pr}_Q$. Let K be a triangulation of P such that Q is triangulated by a subcomplex of K and $K \prec f^{-1}(\mathcal{W}) \wedge g^{-1}(\mathcal{W})$. Give an order on $K^{(0)}$ so that K is an ordered simplicial complex. Then, the product simplicial complex $K \times_s I$ is a triangulation of $P \times \mathbf{I}$, where $I = \{\mathbf{I}, 0, 1\}$ is the ordered simplicial complex with $|I| = \mathbf{I}$. Moreover,

$$(K \times_s I)^{(0)} \subset (Q \times \mathbf{I}) \cup (P \times \{0, 1\}).$$

Since h' is a partial $\mathcal{V}$-realization of $K \times_s I$, it extends to a full $\mathcal{U}$-realization $h : P \times \mathbf{I} \to Y$ of $K \times_s I$. Then, h is a $\mathcal{U}$-homotopy realizing $f \simeq_{\mathcal{U}} g$ rel. Q. Thus, the proof is complete. □

An open refinement $\mathcal{V}$ of $\mathcal{U} \in \operatorname{cov}(Y)$ is called an ***h*^{n}-refinement** of $\mathcal{U}$ if any two $\mathcal{V}$-close maps $f, g : X \to Y$ defined on an arbitrary *metrizable* space X with $\dim X \leq n$ are $\mathcal{U}$-homotopic, where we denote

$$\mathcal{V} \underset{h^n}{\prec} \mathcal{U} \quad \text{or} \quad \mathcal{U} \underset{h^n}{\succ} \mathcal{V}.$$

Theorem 6.12.9. *A metrizable space Y is LC^n (i.e., an ANE$(n+1)$) if and only if every open cover of Y has an h^n-refinement.*

Proof. To prove the "if" part, it suffices to show that Y is LC^n by Theorem 6.12.3. For each $y \in Y$ and each open neighborhood U of y in Y, the open cover $\mathcal{U} = \{U, Y \setminus \{y\}\}$ of Y has an h^n-refinement $\mathcal{V}$. Let $V \in \mathcal{V}$ such that $y \in V$. Then, $V \subset U$. For each $i \leq n$, every map $f : \mathbf{S}^i \to V$ is $\mathcal{V}$-close to the constant map c_y with $c_y(\mathbf{S}^i) = \{y\}$, hence $f \simeq_{\mathcal{U}} c_y$, which means $f \simeq c_y$ in U. Therefore, Y is LC^n.

Now, we will show the "only if" part. Each $\mathcal{U} \in \operatorname{cov}(Y)$ has the following open refinements:

$$\mathcal{U} \overset{*}{\succ} \mathcal{U}_1 \overset{*}{\succ} \mathcal{U}_0 \underset{L^{n+1}}{\succ} \mathcal{V}_0 \succ \operatorname{st}^2 \mathcal{V} \succ \mathcal{V}.$$

Let X be a metrizable space with $\dim X \leq n$ and $f, g : X \to Y$ be maps that are $\mathcal{V}$-close. By Theorem 6.12.7, X has open covers $\mathcal{W}_i$ and full $\mathcal{V}$-realizations $\psi_i : |N(\mathcal{W}_i)| \to Y$, $i = 1, 2$, such that $\dim N(\mathcal{W}_i) \leq n$, $\psi_1\varphi_1 \simeq_{\mathcal{V}} f$, and $\psi_2\varphi_2 \simeq_{\mathcal{V}} g$ for any canonical maps $\varphi_i : X \to |N(\mathcal{W}_i)|$. Take a common refinement $\mathcal{W} \in \operatorname{cov}(X)$ of $\mathcal{W}_1$ and $\mathcal{W}_2$ with $\dim N(\mathcal{W}) \leq n$ and let $\varphi : X \to |N(\mathcal{W})|$ be a canonical map. We have refining simplicial maps $k_i : N(\mathcal{W}) \to N(\mathcal{W}_i)$, $i = 1, 2$, i.e., $W \subset k_i(W)$ for each $W \in \mathcal{W} = N(\mathcal{W})^{(0)}$. Then, $k_i\varphi$ is also a canonical map for $N(\mathcal{W}_i)$ (Corollary 4.9.3).

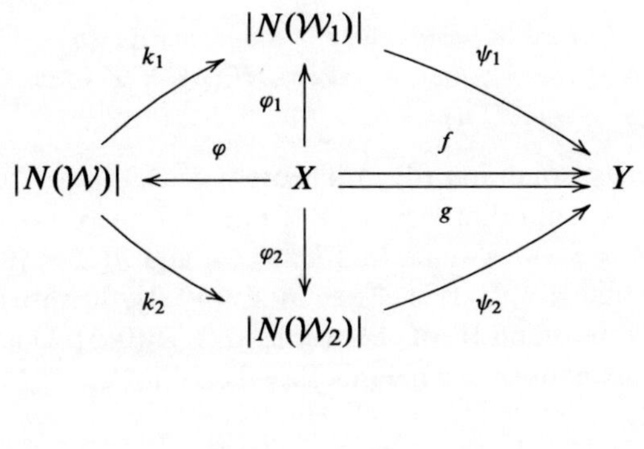

For each principal simplex $\sigma \in N(\mathcal{W})$, take $x \in \bigcap_{W \in \sigma^{(0)}} W$. Since φ is a canonical map, it follows that $\varphi(x) \in \sigma$. On the other hand, we have $V \in \mathcal{V}$ such that $f(x), g(x) \in V$. Then, $\psi_1 k_1 \varphi(x), \psi_2 k_2 \varphi(x) \in \mathrm{st}(V, \mathcal{V})$. Since ψ_i is a full $\mathcal{V}$-realization of $N(\mathcal{W}_i)$, $\psi_i k_i$ is a full $\mathcal{V}$-realization of $N(\mathcal{W})$, so it follows that

$$\psi_1 k_1(\sigma) \cup \psi_2 k_2(\sigma) \subset \mathrm{st}^2(V, \mathcal{V}) \in \mathrm{st}^2 \mathcal{V} \prec \mathcal{V}_0.$$

Thus, $\psi_1 k_1$ and $\psi_2 k_2$ induce a partial $\mathcal{V}_0$-realization of the product simplicial complex $N(\mathcal{W}) \times_s I$, where we regard $N(\mathcal{W})$ as an ordered complex by giving an order on $\mathcal{W} = N(\mathcal{W})^{(0)}$. Then, we have a full $\mathcal{U}_0$-realization

$$\psi : |N(\mathcal{W}) \times_s I| = |N(\mathcal{W})| \times \mathbf{I} \to Y,$$

which is a $\mathcal{U}_1$-homotopy with $\psi_0 = \psi_1 k_1$ and $\psi_1 = \psi_2 k_2$. Thus, we have

$$f \simeq_{\mathcal{V}} \psi_1 k_1 \varphi \simeq_{\mathcal{U}_1} \psi_2 k_2 \varphi \simeq_{\mathcal{V}} g.$$

Therefore, $f \simeq_{\mathcal{U}} g$. □

Remark 19. The above proof is valid even if X is paracompact, but we need the metrizability of Y (cf. Theorem 6.12.7).

A subset Z in a space Y is said to be ***n*-homotopy dense** provided that, for each map $f : X \to Y$ from an arbitrary *metrizable* space X with $\dim X \leq n$, there exists a homotopy $h : X \times \mathbf{I} \to Y$ such that $h_0 = f$ and $h(X \times (0,1]) \subset Z$. The following two theorems are the ANE(n) versions of Corollary 6.6.7:

Theorem 6.12.10. *Every n-homotopy dense set Z in an ANE(n) Y is also an ANE(n). If Y is an AE(n) then so is Z.*

Proof. Let A be a closed set in a metrizable space X with $\dim X \leq n$. Each map $f : A \to Z$ extends to a map $g : U \to Y$, where U is a neighborhood U of A in X. Since $\dim U \leq n$, there exists a homotopy $h : U \times \mathbf{I} \to Y$ such that $h_0 = g$ and $h(A \times (0,1]) \subset Z$. Take $d \in \mathrm{Metr}(X)$ such that $\mathrm{diam}_d(X) \leq 1$ and define a map $\tilde{f} : U \to Z$ by $\tilde{f}(x) = h(x, d(x, A))$. Then, $\tilde{f}$ is an extension of f. Thus, Z is an ANE(n). When Y is an AE(n), we can take $U = X$ in the above, hence Z is an AE(n). □

Theorem 6.12.11. *If a metrizable space Y contains an $(n-1)$-homotopy dense set Z that is an ANE($n+1$), then Y is an ANE(n). If Z is an $(n-1)$-connected ANE($n+1$), then Y is an AE(n).*

Proof. We will verify condition (d) of Theorem 6.12.3. Let A be a closed set in a metrizable space X with $\dim A \leq n-1$. For each map $f : A \to Y$, there is a homotopy $h : A \times \mathbf{I} \to Y$ such that $h_0 = f$ and $h(A \times (0,1]) \subset Z$. Since $\dim A \times (0,1] \leq n$ and $A \times (0,1]$ is closed in $X \times (0,1]$, the restriction $h|A \times (0,1]$ extends over a neighborhood W of $A \times (0,1]$ in $X \times (0,1]$. Due to Lemma 6.5.1, this implies that f extends over a neighborhood of A in X.

Since Z is $(n-1)$-homotopy dense in Y, every map $g : \mathbf{S}^{n-1} \to Y$ is homotopic to a map $g' : \mathbf{S}^{n-1} \to Z$. When Z is $(n-1)$-connected, since g' is null-homotopic in Z, g is null-homotopic in Y, hence Y is also $(n-1)$-connected. Since an $(n-1)$-connected ANE(n) is AE(n) by Theorem 6.12.5, we have the additional statement. □

6.13 Embeddings into Finite-Dimensional ARs

Since a normed linear space is an AR by Theorems 6.1.1 and 6.2.5, the Arens–Eells Embedding Theorem 6.2.1 means that every metrizable space can be embedded in an AR as a closed set. In this section, we consider the finite-dimensional version of this fact. Here, we need more theorems from Dimension Theory, e.g., Theorems 5.3.4, 5.3.2, Corollary 5.2.6, etc.

The n-dimensional Menger compactum μ^n and the n-dimensional Nöbeling space ν^n are AE(n)s by Theorems 6.11.2 and 6.12.5. Then, Theorems 5.10.3 and 5.10.8 imply that every compact metrizable space with dim $\leq n$ can be embedded in an n-dimensional separable metrizable AE(n) as a closed set.[10]

First, we prove the following theorem:

Theorem 6.13.1. *Every n-dimensional metrizable space X can be embedded as a closed set in an $(n+1)$-dimensional AR $T(X)$ of the same weight. If X is completely metrizable, compact, separable, or separable locally compact, then so is $T(X)$, respectively.*

Proof. When X is finite, X can be embedded in the interval $\mathbf{I}$. Then, we may assume that X is infinite. By Theorem 5.3.4, we may also assume that X is a dense subspace of a complete metric space $\widetilde{X}$ with $\dim \widetilde{X} = n$, where $\widetilde{X} = X$ if X is completely metrizable. By Corollary 5.2.6, $\widetilde{X}$ is homeomorphic to the inverse limit $\varprojlim \mathbf{K}$ of an inverse sequence $\mathbf{K} = (|K_i|_{\mathrm{m}}, f_i)_{i \in \mathbb{N}}$ of metric polyhedra and PL maps such that $\operatorname{card} K_i \leq w(X)$, $\dim K_i \leq n$, and each $f_i : K_{i+1} \to \operatorname{Sd} K_i$ is simplicial. Note that $\dim K_i = n$ for sufficiently large $i \in \mathbb{N}$. Otherwise, $\dim \widetilde{X} = \dim \varprojlim \mathbf{K} < n$ by Theorem 5.3.2. We identify $\widetilde{X} = \varprojlim \mathbf{K}$.

To consider the simplicial mapping cylinder Z_{f_i} of each simplicial map $f_i : K_{i+1} \to \operatorname{Sd} K_i$, we regard K_{i+1} as an ordered simplicial complex. Let $T_1 = 0 * \operatorname{Sd} K_1$ be the simplicial cone over $\operatorname{Sd} K_1$. Moreover, we may assume that $|Z_{f_i}| \cap |Z_{f_j}| = \emptyset$ for $|i-j| > 1$ and $|T_1| \cap |Z_{f_i}| = \emptyset$ for $i > 1$. We define

[10] In fact, every separable completely metrizable space with dim $\leq n$ can be embedded in ν^n as a closed set (cf. Remark 15 on Theorem 5.10.3).

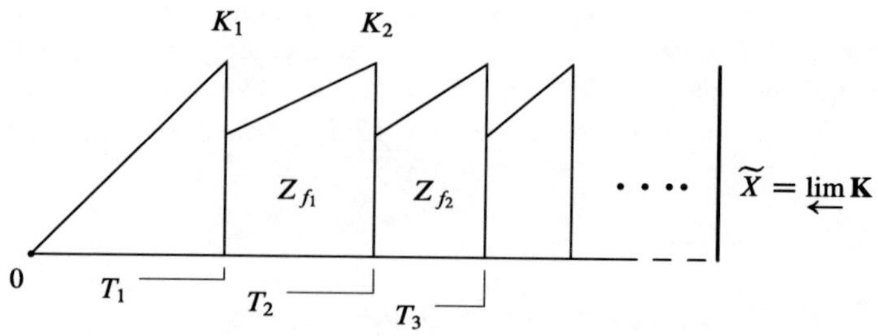

Fig. 6.21 The space $T(\widetilde{X})$

$$T_k = T_1 \cup \bigcup_{i=1}^{k-1} \mathrm{Sd}_{\mathrm{Sd}\,K_i} Z_{f_i} \quad \text{for } k > 1 \text{ and}$$

$$T = \bigcup_{i\in\mathbb{N}} T_i = T_1 \cup \bigcup_{i\in\mathbb{N}} \mathrm{Sd}_{\mathrm{Sd}\,K_i} Z_{f_i},$$

where $\mathrm{Sd}_{\mathrm{Sd}\,K_i} Z_{f_i}$ is the barycentric subdivision of Z_{f_i} relative to the subcomplex $\mathrm{Sd}\,K_i \subset Z_{f_i}$. Observe that T is a simplicial complex with $\operatorname{card} T^{(0)} \leq \aleph_0 w(X)$ and $T_1 \subset T_2 \subset \cdots$ are subcomplexes of T, where $\dim T = n+1$ because $\dim T_i = n+1$ for sufficiently large $i \in \mathbb{N}$ (Fig. 6.21). It should be noted that $\bigcup_{i\in\mathbb{N}} Z_{f_i}$, in general, is not a simplicial complex.

With respect to $\widetilde{X} \cap |T| = \emptyset$, we define $\gamma_i : |T| \cup \widetilde{X} \to |T_i|$, $i \in \mathbb{N}$, as follows:

$$\gamma_i(x) = \begin{cases} x & \text{if } x \in |T_i|, \\ \bar{c}_{f_i} \cdots \bar{c}_{f_j}(x) & \text{if } x \in |Z_{f_j}|,\ j \geq i, \\ p_i(x) & \text{if } x \in \widetilde{X}, \end{cases}$$

where $\bar{c}_{f_i} : Z_{f_i} \to \mathrm{Sd}\,K_i$ is the simplicial collapsing map and $p_i : \widetilde{X} = \varprojlim \mathbf{K} \to |K_i|_{\mathrm{m}}$ is the projection. Note that $\gamma_i\gamma_j = \gamma_i$ for $i \leq j$. Let $T(\widetilde{X})$ be the space $|T| \cup \widetilde{X}$ with the coarsest topology such that every $\gamma_i : |T| \cup \widetilde{X} \to |T_i|_{\mathrm{m}}$ is continuous, that is, the topology generated by sets $\gamma_i^{-1}(U)$, where $i \in \mathbb{N}$ and U is open in $|T_i|_{\mathrm{m}}$. It should be noted that if $U \subset |T_i| \setminus |K_i|$ then $\gamma_i^{-1}(U) = U$, which implies that $|T_i|_{\mathrm{m}}$ is a subspace of $T(\widetilde{X})$ and $|T|_{\mathrm{m}}$ is an open subspace of $T(\widetilde{X})$. Evidently, $|T|_{\mathrm{m}}$ is dense in $T(\widetilde{X})$.

For each $x \in \widetilde{X}$, $\{\gamma_i^{-1}(\mathrm{st}(p_i(x), K_i)) \mid i \in \mathbb{N}\}$ is a neighborhood basis at x in $T(\widetilde{X})$, where $\mathrm{st}(p_i(x), K_i) = |\mathrm{St}(c_{K_i}(p_i(x)), K_i)|$. Indeed, the open star $O_{T_{i+1}}(p_{i+1}(x))$ is open in $|T_{i+1}|_{\mathrm{m}}$ and

$$\begin{aligned} &\gamma_i(\gamma_{i+1}^{-1}(O_{T_{i+1}}(p_{i+1}(x)))) \subset \gamma_i(O_{T_{i+1}}(p_{i+1}(x))) \\ &\quad \subset \bar{c}_{f_i}(\mathrm{st}(p_{i+1}(x), Z_{f_i})) \subset \mathrm{st}(p_i(x), \mathrm{Sd}\,K_i) \subset \mathrm{st}(p_i(x), K_i), \end{aligned}$$

hence $x \in \gamma_{i+1}^{-1}(O_{T_{i+1}}(p_{i+1}(x))) \subset \gamma_i^{-1}(\mathrm{st}(p_i(x), K_i))$, which means that $\gamma_i^{-1}(\mathrm{st}(p_i(x), K_i))$ is a neighborhood of x in $T(\widetilde{X})$. Moreover, if U is an open set in $|T_i|_{\mathrm{m}}$ and $x \in \gamma_i^{-1}(U)$, then $U \cap |K_i|$ is an open neighborhood of $p_i(x) = \gamma_i(x)$ in $|K_i|_{\mathrm{m}}$. By Remark 16 at the end of Sect. 4.10,

$$\operatorname{mesh}_{\rho_{K_i}} f_{i,i+k}(K_{i+k}) \leq \operatorname{mesh}_{\rho_{K_i}} f_{i,i+k-1}(\operatorname{Sd} K_{i+k-1}) \leq 2 \cdot \left(\frac{n}{n+1}\right)^k,$$

where $f_{i,i+k} = f_i \cdots f_{i+k-1} : |K_{i+k}| \to |K_i|$.[11] Then, we can choose $k \in \mathbb{N}$ so that $f_{i,i+k}(\mathrm{st}(p_{i+k}(x), K_{i+k})) \subset U$, hence

$$\gamma_{i+k}^{-1}(\mathrm{st}(p_{i+k}(x), K_{i+k})) \subset \gamma_i^{-1}(f_{i,i+k}(\mathrm{st}(p_{i+k}(x), K_{i+k}))) \subset \gamma_i^{-1}(U).$$

Let $c_i : |T_{i+1}|_{\mathrm{m}} = |T_i| \cup |Z_{f_i}| \to |T_i|_{\mathrm{m}}$ be the strong deformation retraction extending the collapsing $\bar{c}_{f_i}$. Then, we have the following commutative diagram:

$$\begin{array}{ccccccc}
|T_1|_{\mathrm{m}} & \xleftarrow[\supset]{c_1} & |T_2|_{\mathrm{m}} & \xleftarrow[\supset]{c_2} & |T_3|_{\mathrm{m}} & \xleftarrow[\supset]{c_3} & \cdots \\
\cup & & \cup & & \cup & & \cdots \\
|K_1|_{\mathrm{m}} & \xleftarrow[f_1]{} & |K_2|_{\mathrm{m}} & \xleftarrow[f_2]{} & |K_3|_{\mathrm{m}} & \xleftarrow[f_3]{} & \cdots
\end{array}$$

We regard $\widetilde{X}$ as the inverse limit of the lower sequence. The space $T(\widetilde{X})$ can be regarded as the inverse limit of the upper sequence with $\gamma_i : T(\widetilde{X}) \to |T_i|_{\mathrm{m}}$ the projection (cf. Theorem 4.10.6).

As already noted, $|T|_{\mathrm{m}}$ is an open dense subspace of $T(\widetilde{X})$, hence $\widetilde{X}$ is a closed subspace of $T(\widetilde{X})$ and $w(T(\widetilde{X})) = w(|T|_{\mathrm{m}}) = w(X)$. Due to Theorem 4.5.9, each $|T_i|_{\mathrm{m}}$ is completely metrizable because T_i is finite-dimensional. Therefore, $T(\widetilde{X})$ is also completely metrizable by Proposition 4.10.9(2). Since $\dim |T_i|_{\mathrm{m}} = \dim T_i$ for each $i \in \mathbb{N}$ (Corollary 5.2.10), we have $\dim T(\widetilde{X}) = n + 1$ by Theorem 5.3.2 (or the Countable Sum Theorem 5.4.1).

For each $i \in \mathbb{N}$, let $\mathcal{Z}_i = \{c_i^{-1}(\sigma) \mid \sigma \in \operatorname{Sd} K_i\}$. Then,

$$\gamma_i = c_i \gamma_{i+1} \simeq_{\mathcal{Z}_i} \gamma_{i+1} \ \text{rel.}\ |T_i|$$

by the homotopy h^i extending $h^{f_i}(\gamma_{i+1} \times \mathrm{id}_{\mathbf{I}})|(T(\widetilde{X}) \setminus |T_i|) \times \mathbf{I}$. For each $x \in \widetilde{X}$ and $k \leq i$,

$$h^i(\gamma_k^{-1}(O_{K_k}(p_k(x))) \times \mathbf{I}) \subset \gamma_k^{-1}(\mathrm{st}(p_k(x), K_k)).$$

---

[11] In general, $f_{i,i+2} = f_i f_{i+1}$ is not simplicial with respect to K_{i+2} and $\operatorname{Sd}^2 K_i$ because $f_i(\hat{\tau})$ is not necessarily the barycenter of the simplex $f_i(\tau) \in \operatorname{Sd} K_i$ for $\tau \in K_{i+1}$.

Indeed, for each $y \in \gamma_k^{-1}(O_{K_k}(p_k(x)))$, choose $\sigma \in \operatorname{Sd} K_i$ so that $h^i(\{y\} \times \mathbf{I}) \subset c_i^{-1}(\sigma) \subset \gamma_i^{-1}(\sigma)$. Then, $h_0^i(y) = \gamma_i(y) \in c_i^{-1}(\sigma) \cap |K_i| = \sigma$. Choose a simplex $\tau \in K_k$ so that $\gamma_k(\sigma) \subset \tau$. Since $\gamma_k h_0^i(y) \in \tau$ and

$$\gamma_k h_0^i(y) = \gamma_k \gamma_i(y) = \gamma_k(y) \in O_{K_k}(p_k(x)),$$

it follows that $p_k(x) \in \tau$, i.e., $\tau \subset \operatorname{st}(p_k(x), K_k)$. Thus, we have

$$\gamma_k h^i(\{y\} \times \mathbf{I}) \subset \gamma_k \gamma_i^{-1}(\sigma) \subset \gamma_k(\sigma) \subset \tau \subset \operatorname{st}(p_k(x), K_k).$$

Connecting homotopies h^i, $i \in \mathbb{N}$, we can define a homotopy $h : T(\widetilde{X}) \times \mathbf{I} \to T(\widetilde{X})$ such that $h_0 = \mathrm{id}$, $h_{2^{-i+1}} = \gamma_i$ for each $i \in \mathbb{N}$, and $h(T(\widetilde{X}) \times (0, 1]) \subset |T|$. The continuity of h at a point $(x, 0) \in \widetilde{X} \times \{0\}$ follows from the fact that each $\gamma_k^{-1}(O_{K_k}(p_k(x)))$ is an open neighborhood of x in $T(\widetilde{X})$ and

$$\bigcup_{i \geq k} h^i(\gamma_k^{-1}(O_{K_k}(p_k(x))) \times \mathbf{I}) \subset \gamma_k^{-1}(\operatorname{st}(p_k(x), K_k)).$$

Hence, $|T|_{\mathrm{m}}$ is homotopy dense in $T(\widetilde{X})$, which implies $T(\widetilde{X})$ is an ANR (Corollary 6.6.7). Since $h_1(T(\widetilde{X})) = |0 * \operatorname{Sd} K_1|$ is contractible, so is $T(\widetilde{X})$, hence $T(\widetilde{X})$ is an AR (Corollary 6.2.9).

Except for the case where X is separable and locally compact, the desired space $T(X)$ is defined as the subspace $T(X) = |T| \cup X$ of $T(\widetilde{X})$. Indeed, X is closed in $T(X)$, $w(T(X)) = w(X)$, and $\dim T(X) = n + 1$. In the above, $h(T(X) \times \mathbf{I}) \subset T(X)$, so $T(X)$ is an AR for the same reason as $T(\widetilde{X})$.

When X is completely metrizable, since $\widetilde{X} = X$, $T(X) = T(\widetilde{X})$ is completely metrizable. When X is compact, since each K_i is finite due to Corollary 5.2.6, each $|T_i|_{\mathrm{m}}$ is compact, hence so is $T(X) = T(\widetilde{X})$ by Proposition 4.10.9(1). If X is separable then so is $\widetilde{X}$, hence each K_i is countable due to Corollary 5.2.6, which implies that T is also countable, so $T(X)$ is separable. If X is separable and locally compact, the above $T(X)$ is not locally compact at $0 \in K_0^{(0)}$ unless K_1 is finite. Thus, some modification is necessary for the additional statement to be valid.

(*The case that X is separable and locally compact*) In the above construction, each K_i is countable and locally finite and each f_i is proper by Proposition 5.2.6. If K_1 is disconnected, let $L_1, L_2, \ldots$ be the components of K_1 and take vertices $v_1 \in L_1^{(0)}, v_2 \in L_2^{(0)}, \ldots$. Adding new 1-simplexes $\langle v_1, v_2\rangle, \langle v_2, v_3\rangle, \ldots$ to K_1, we can assume that K_1 is connected. Moreover, instead of $K_0 = \{0\}$ and $T_1 = 0 * \operatorname{Sd} K_1$, let $K_0 = \omega \cup \{[i-1, i] \mid i \in \mathbb{N}\}$ be the natural triangulation of $\mathbb{R}_+ = [0, \infty)$ and $T_1 = Z_{f_0}$ the simplicial mapping cylinder for a proper simplicial map $f_0 : \operatorname{Sd} K_1 \to K_0$. Then, T would be countable and locally finite, hence $T(X)$ would be separable and locally compact.

Now, we will construct a proper simplicial map $f_0 : \operatorname{Sd} K_1 \to K_0$. For each $i \in \mathbb{N}$, let

$$V_i = \{v \in \operatorname{Sd} K_1^{(0)} \mid \langle v, u\rangle \in \operatorname{Sd} K_1 \text{ for some } u \in V_{i-1}\},$$

where $V_0 = \{v_0\}$, $v_0 \in \operatorname{Sd} K_1^{(0)}$. Then, $(\operatorname{Sd} K_1)^{(0)} = \bigcup_{i\in\omega} V_i$ because $\operatorname{Sd} K_1$ is connected. We can define a simplicial map $f_0 : \operatorname{Sd} K_1 \to K_0$ by

$$f_0(v) = \min\{i \in \omega \mid v \in V_i\} \text{ for each } v \in (\operatorname{Sd} K_1)^{(0)}.$$

Since $\operatorname{Sd} K_1$ is locally finite, each V_i is finite, hence the map $f_0 : |K_1|_{\mathrm{m}} = |\operatorname{Sd} K_1|_{\mathrm{m}} \to |K_0|_{\mathrm{m}} = \mathbb{R}_+$ is proper. Thus, each $|\widetilde{T}_i|_{\mathrm{m}}$ is locally compact and each c_i is also proper. Therefore, $T(X) = T(\widetilde{X})$ is locally compact by Proposition 4.10.9(3). Since $|K_0|_{\mathrm{m}} = \mathbb{R}_+$ is a deformation retract of $T(X)$, it follows that $T(X)$ is contractible, hence $T(X)$ is an AR. □

In Theorem 6.13.1 above, we cannot take $\dim T(X) = \dim X = n$, that is, an n-dimensional metrizable space cannot, in general, be embedded in an n-dimensional AR (nor an n-dimensional ANR). Neither the n-dimensional universal Nöbeling space ν^n nor the n-dimensional universal Menger compactum μ^n can be embedded in any n-dimensional ANR. Otherwise, they would be retracts of an n-dimensional ANR because they are n-dimensional AE(n)s by Theorems 6.11.2 and 6.12.5. But this is impossible because they are not ANRs by Theorem 6.11.2 (cf. 6.2.10(2)).

It should be noted that every separable n-dimensional metrizable space can be embedded into ν^n and μ^n, which are n-dimensional AE(n)s by Theorems 5.10.3 and 5.10.8. Without separability, we can obtain the same embedding theorem:

Theorem 6.13.2. *Every n-dimensional metrizable space X can be embedded in an n-dimensional metrizable AE(n) $S(X)$ with $w(S(X)) = w(X)$ as a closed set. If X is completely metrizable, compact, separable, or separable locally compact, then so is $S(X)$, respectively.*

To prove this theorem, we employ the following lemmas:

Lemma 6.13.3. *Let $f : K \to L$ be a simplicial map between simplicial complexes and X be a metrizable space with $\dim X \leq n-1$. Then, for each map $g : X \to |Z_f^{(n)}|_{\mathrm{m}}$, $g \simeq_{\mathcal{Z}} \bar{c}_f g = fg$ rel. $g^{-1}(|L|)$, where $\bar{c}_f : Z_f \to L$ is a simplicial collapsing and $\mathcal{Z} = \{\bar{c}_f^{-1}(\sigma) \mid \sigma \in L\}$.*

Proof. Let $f' = f|K^{(n-1)} : K^{(n-1)} \to L^{(n)}$. Then, observe

$$Z_f^{(n-1)} \subset Z_{f'} \subset Z_f^{(n)} \text{ and } |Z_{f'}| \setminus |K| = |Z_f^{(n)}| \setminus |K|.$$

By Theorem 5.2.9, we have a map $g' : X \to |Z_f^{(n-1)}|_{\mathrm{m}}$ such that $g'(x), g(x) \in \sigma_x$, where $\sigma_x = c_{Z_f}(g(x)) \in Z_f^{(n)}$ is the carrier of $g(x)$. Taking an Urysohn map $k : X \to \mathbf{I}$ with $k(g^{-1}(|L|)) = 0$ and $k(g^{-1}(|K|)) = 1$, we can define a homotopy $h : X \times \mathbf{I} \to |Z_f^{(n)}|_{\mathrm{m}}$ as follows:

$$h(x,t) = (1 - k(x)t)g(x) + k(x)tg'(x) \text{ for each } (x,t) \in X \times \mathbf{I}.$$

Then, $h_0 = g$, $h_t|g^{-1}(|L|) = g|g^{-1}(|L|)$ for each $t \in \mathbf{I}$, and $h_1(X) \subset |Z_{f'}|$. Hence, we have

$$g \simeq h_1 \simeq \bar{c}_{f'} h_1 = \bar{c}_f h_1 \simeq \bar{c}_f g \text{ rel. } g^{-1}(|L|),$$

where these homotopies are given by h_t, $\bar{h}_t^{f'} h_1$, and $\bar{c}_f h_{1-t}$, respectively. For each $x \in X \setminus g^{-1}(|L|)$,

$$h(\{x\} \times \mathbf{I}) \subset \sigma_x, \; \bar{c}_f h(\{x\} \times \mathbf{I}) \subset \bar{c}_f(\sigma_x) \text{ and}$$
$$\bar{h}^{f'}(\{h_1(x)\} \times \mathbf{I}) \subset \bar{c}_f^{-1}(\bar{c}_f(\sigma_x)).$$

This completes the proof. □

Lemma 6.13.4. *Let* $0 * K$ *be the simplicial cone over a simplicial complex* K. *Then, the* n*-skeleton* $(0 * K)^{(n)}$ *is* $(n-1)$*-connected.*

Proof. Regarding $0 * K$ as the simplicial mapping cylinder of the constant map of K, we can apply Lemma 6.13.3 to show that every map $g : \mathbf{S}^{n-1} \to (0 * K)^{(n)}$ is homotopic to the map of K to $\{0\}$, hence $g \simeq 0$. □

Proof of Theorem 6.13.2. Modifying the proof of Theorem 6.13.1, we will construct $S(X)$ as a closed subspace of $T(X)$. Let $S_1 = T_1^{(n)} = (0 * \operatorname{Sd} K_1)^{(n)}$,

$$S_k = S_1 \cup \bigcup_{i=1}^{k-1} \operatorname{Sd}_{\operatorname{Sd} K_i} Z_{f_i}^{(n)} \subset T_k^{(n)} \quad \text{for } k > 1 \text{ and}$$
$$S = \bigcup_{i \in \mathbb{N}} S_i = S_1 \cup \bigcup_{i \in \mathbb{N}} \operatorname{Sd}_{\operatorname{Sd} K_i} Z_{f_i}^{(n)} \subset T^{(n)}.$$

Since $Z_{f_i}^{(n)} = Z_{f_i|K_{i+1}^{(n-1)}} \cup K_{i+1}$, it follows that

$$\operatorname{Sd}_{\operatorname{Sd} K_i} Z_{f_i}^{(n)} = \operatorname{Sd}_{\operatorname{Sd} K_i} Z_{f_i|K_{i+1}^{(n-1)}} \cup \operatorname{Sd} K_{i+1},$$

which is not equal to $(\operatorname{Sd}_{\operatorname{Sd} K_i} Z_{f_i})^{(n)}$ if $\dim K_{i+1} = n$. Restricting the strong deformation retraction $c_k : |T_{k+1}|_{\mathrm{m}} \to |T_k|_{\mathrm{m}}$, we have the retraction $c'_k : |S_{k+1}|_{\mathrm{m}} \to |S_k|_{\mathrm{m}}$, which is no longer a strong deformation retraction.

Let $S(\widetilde{X}) = |S| \cup \widetilde{X}$ be the n-dimensional closed subspace of $T(\widetilde{X})$, which is regarded as the inverse limit of the middle sequence in the following commutative diagram:

$$
\begin{array}{ccccccc}
|T_1|_{\mathrm{m}} & \xleftarrow[\supset]{c_1} & |T_2|_{\mathrm{m}} & \xleftarrow[\supset]{c_2} & |T_3|_{\mathrm{m}} & \xleftarrow[\supset]{c_3} & \cdots \\
\cup & & \cup & & \cup & & \cdots \\
|S_1|_{\mathrm{m}} & \xleftarrow[\supset]{c'_1} & |S_2|_{\mathrm{m}} & \xleftarrow[\supset]{c'_2} & |S_3|_{\mathrm{m}} & \xleftarrow[\supset]{c'_3} & \cdots \\
\cup & & \cup & & \cup & & \cdots \\
|K_1|_{\mathrm{m}} & \xleftarrow[f_1]{} & |K_2|_{\mathrm{m}} & \xleftarrow[f_2]{} & |K_3|_{\mathrm{m}} & \xleftarrow[f_3]{} & \cdots
\end{array}
$$

Then, the projection $\gamma'_i : S(\widetilde{X}) \to |S_i|_{\mathrm{m}}$ is the restriction of the projection $\gamma_i : T(\widetilde{X}) \to |T_i|_{\mathrm{m}}$, where we no longer have that $\gamma'_i = c'_i\gamma'_i \simeq \gamma'_{i+1}$ rel. $|S_i|$.

The space $S(X)$ is the closed subspace $|S| \cup X = S(\widetilde{X}) \cap T(X)$ of $T(X)$, hence, if X is completely metrizable, compact or separable, then so is $S(X)$, respectively. When X is separable and locally compact, in the same way as $T(X)$, we can modify $S(X)$ to be separable and locally compact.

We will now apply Theorem 6.12.11 to prove that $S(X)$ is an AE(n). Since $S(X) \setminus X = |S|_{\mathrm{m}}$ is an ANE as a polyhedron with the metric topology, it suffices to show that $|S|_{\mathrm{m}}$ is $(n-1)$-homotopy dense in $S(X)$ and $(n-1)$-connected.

(*The* $(n-1)$-*homotopy denseness*) Let $g : Z \to S(X)$ be a map of a metrizable space Z with $\dim Z \leq n-1$. For each $i \in \mathbb{N}$, let $\mathcal{Z}_i = \{c_i'^{-1}(\sigma) \mid \sigma \in \operatorname{Sd} K_i\}$. Due to Lemma 6.13.3, we have

$$\gamma'_i g = c'_i\gamma'_{i+1}g \simeq_{\mathcal{Z}_i} \gamma'_{i+1}g \text{ rel. } g^{-1}(|S_i|)$$

by a $\mathcal{Z}_i$-homotopy $h^i : Z \times \mathbf{I} \to |S_{i+1}|_{\mathrm{m}} \subset S(X)$. Connecting these homotopies, we can define a homotopy $h : Z \times \mathbf{I} \to S(X)$ such that $h_0 = g$, $h_{2^{-i+1}} = \gamma'_i g$ for each $i \in \mathbb{N}$ and $h(Z \times (0,1]) \subset |S|$. The continuity of h at a point $(z,0) \in g^{-1}(X) \times \{0\} \subset Z \times \mathbf{I}$ is guaranteed by the following fact:

$$h^i(g^{-1}\gamma_k'^{-1}(O_{K_k}(p_k g(z))) \times \mathbf{I}) \subset \gamma_k'^{-1}(\operatorname{st}(p_k(g(z)), K_k)) \quad \text{for each } z \in g^{-1}(X) \text{ and } k \leq i.$$

This can be verified in a manner analogous to Theorem 6.13.1. Therefore, $|S|_{\mathrm{m}}$ is $(n-1)$-homotopy dense in $S(X)$.

(*The* $(n-1)$-*connectedness*) In the above, let $Z = \mathbf{S}^{n-1}$. Then, every map $g : \mathbf{S}^{n-1} \to S(X)$ is homotopic to a map $g' : \mathbf{S}^{n-1} \to |S_1|$. Since $S_1 = T_1^{(n)} = (0 * \operatorname{Sd} K_1)^{(n)}$ is $(n-1)$-connected by Lemma 6.13.4, we have $g \simeq g' \simeq 0$. This completes the proof. □

Notes for Chap. 6

For supplementary results and examples, refer to the following classical books of Borsuk and Hu. One can find interesting examples in Borsuk's book, Chap. VI. In Hu's book, ANEs are discussed not only for the class of metrizable spaces but also for a more general class.

- K. Borsuk, *Theory of Retracts*, Monog. Mat. **44** (Polish Sci. Publ., Warsaw, 1966).
- S.-T. Hu, *Theory of Retracts* (Wayne State Univ. Press, Detroit, 1965).

The theory has developed considerably since the publication of these books. Many important results have been gained and many problems have been solved. Some of them have been treated in this chapter while others have not. Many interesting examples can be seen in numerous existing articles. Especially, in 1994, Cauty constructed a metric linear space that is not an AR. This very important result will be proved in Sect. 7.12. For history of ANR theory (theory of retracts), refer to the article of Madešić [28].

Theorem 6.1.1 was established in [12]. The remark on 6.1.1 is taken from Michael [29]. Theorem 6.1.9(6) is due to Hanner [17].

Theorem 6.2.1 was established in [3], but the short proof presented here is due to Toruńczyk [35]. For another short proof, see [30]. Theorem 6.2.2 was the trick used in [23]. Theorem 6.2.3 was established in [18] and the proof presented here is due to Toruńczyk [35]. The first assertion of Theorem 6.2.4 was established in [10].

Theorem 6.3.2 was established in [13] and Theorem 6.3.4 was proved independently in [13] and [21]. For conditions that LEC spaces are ANRs, refer to [9]. The notion of ULC was introduced in [2], but we use the word "unified" instead of "uniformly." The example for Theorem 6.3.8 was constructed in [5] as a locally contractible compactum that is not an ANR, and it was shown in [13] that this is not LEC.

The first version of Theorem 6.4.1 is due to Borsuk [4].

Lemma 6.5.1 is proved by Dowker [11] in a more general setting (X is countably paracompact normal and A is a closed G_δ). Theorem 6.5.2 was established in [26] and Theorem 6.5.3 was essentially proved in successive stages by [37], and [17].

Theorems 6.6.1 and 6.6.2 are due to Lefschetz [27] and Hanner [17], respectively. The present proof of the implication (c) $\Rightarrow$ (a) in Theorem 6.6.2 follows Dowker's idea [11]. Theorem 6.7.5 was established in [25] (cf. [14]). The compact case of Theorem 6.7.8 can be obtained as a corollary of the result in [8].

Theorem 6.8.1 was established by Nguyen To Nhu [32] and the proof presented here is due to Sakai [33]. The concept of a uniform retract (or a uniform retraction) was introduced by Michael [31]; it is called a *regular retract* (or *retraction*) in [36]. Theorem 6.8.11 was proved by Michael [31] and Toruńczyk [36], independently. For characterizations of (finite-dimensional) uniform ARs and ANRs, refer to [31].

Theorem 6.9.1 was conjectured by Geoghegan [15] and proved by Cauty [7]. The proof presented here is due to Sakai [34].

The result of Sect. 6.10 was first proved in [19] for σ-compact spaces and then generalized in [16]. The property C was named in [20]. The definition of Sect. 6.10 is due to [1].

Theorem 6.13.1 was proved in [6]. The proof presented here is due to Kodama [24]. Theorem 6.13.2 was implicitly proved in [22].

References

1. D.F. Addis, J.H. Gresham, A class of infinite-dimensional spaces, Part I: Dimension theory and Alexandroff's problem. Fund. Math. **101**, 195–205 (1978)

2. F.D. Ancel, The role of countable dimensionality in the theory of cell-like relations. Trans. Am. Math. Soc. **287**, 1–40 (1985)
3. R. Arens, J. Eells, On embedding uniform and topological spaces. Pac. J. Math. **6**, 397–403 (1956)
4. K. Borsuk, Sur les prolongements des transformations continues. Fund. Math. **28**, 99–110 (1937)
5. K. Borsuk, Sur un espace compact localement contractile qui n'est pas un rétracte absolut de voisinage. Fund. Math. **35**, 175–180 (1948)
6. H.G. Bothe, Eine Einbettung m-dimensionaler Mengen in einen $(m + 1)$-dimensionalen absoluten Retrakt. Fund. Math. **52**, 209–224 (1963)
7. R. Cauty, Une caractérisation des rétractes absolus de voisinage. Fund. Math. **144**, 11–22 (1994)
8. Z. Čerin, Characterizing global properties in inverse limits. Pac. J. Math. **112**, 49–68 (1984)
9. D.W. Curtis, Some theorems and examples on local equiconnectedness and its specializations. Fund. Math. **72**, 101–113 (1971)
10. C.H. Dowker, An imbedding theorem for paracompact metric spaces. Duke Math. J. **14**, 639–645 (1947)
11. C.H. Dowker, Homotopy extension theorems. Proc. Lond. Math. Soc. (3) **6**, 100–116 (1956)
12. J. Dugundji, An extension of Tietze's theorem. Pac. J. Math. **1**, 353–367 (1951)
13. J. Dugundji, Locally equiconnected spaces and absolute neighborhood retracts. Fund. Math. **57**, 187–193 (1965)
14. J. Dydak, J. Segal, in *Shape Theory, An introduction*. Lecture Notes in Math. **688** (Springer, Berlin, 1978)
15. R. Geoghegan, Presented problems, in *Proceedings of International Conference on Geometric Topology*, Warsaw, 1978, ed. by K. Borsuk, A. Kirkor (Polish Sci. Publ., Warsaw, 1980), pp. 462–463
16. J.H. Gresham, A class of infinite-dimensional spaces, Part II: An extension theorem and theory of retracts. Fund. Math. **107**, 237–245 (1980)
17. O. Hanner, Some theorems on absolute neighborhood retracts. Arkiv Mat. **1**, 389–408 (1951)
18. F. Hausdorff, Erweiterung einer stetigen Abbildung. Fund. Math. **30**, 40–47 (1938)
19. W.E. Haver, Locally contractible spaces that are absolute neighborhood retracts. Proc. Am. Math. Soc. **40**, 280–284 (1973)
20. W.E. Haver, A covering property for metric spaces, in *Topology Conference, Proceedings of a Conference Held at Virginia Polytechnic Institute and State University*, Blacksburg, 1973, ed. by R.F. Dickman Jr., P. Fletcher. Lecture Notes in Math. **375** (Springer, Berlin, 1974), pp. 108–113
21. C.H. Himmelberg, Some theorems on equiconnected and locally equiconnected spaces. Trans. Am. Math. Soc. **115**, 43–53 (1965)
22. Y. Iwamoto, K. Sakai, Strong n-shape theory. Topology Appl. **122**, 253–267 (2002)
23. V.L. Klee, Some topological properties of convex sets. Trans. Am. Math. Soc. **78**, 30–45 (1955)
24. Y. Kodama, On embeddings of spaces into ANR and shapes. J. Math. Soc. Jpn. **27**, 533–544 (1975)
25. G. Kozlowski, Images of ANR's (unpublished manuscript)
26. A.H. Kruse, P.W. Liebnitz, An application of a family homotopy extension theorem to ANR spaces. Pac. J. Math. **16**, 331–336 (1966)
27. S. Lefschetz, *Topics in Topology*. Annals of Math. Studies **10** (Princeton University Press, Princeton, 1942)
28. S. Mardešić, Absolute neighborhood retracts and shape theory. Chap.9 in *History of Topology*, ed. by I.M. James (Elsevier Sci. B.V., Amsterdam), pp. 241–269, (1999)
29. E. Michael, Some extension theorems for continuous functions. Pac. J. Math. **3**, 789–806 (1953)
30. E. Michael, A short proof of Arens–Eells embedding theorem. Proc. Am. Math. Soc. **15**, 415–416 (1964)
31. E. Michael, Uniform AR's and ANR's. Compositio Math. **39**, 129–139 (1979)

32. Nguyen To Nhu, Investigating the ANR-property of metric spaces. Fund. Math. **124**, 243–254 (1984); Corrections. Fund. Math. **141**, 297 (1992)
33. K. Sakai, The completions of metric ANR's and homotopy dense subsets. J. Math. Soc. Jpn. **52**, 835–846 (2000)
34. K. Sakai, An alternative proof of Cauty-Geoghegan's characterization of ANRs. Topology Appl. **122**, 407–414 (2002)
35. H. Toruńczyk, A short proof of Hausdorff's theorem on extending metrics. Fund. Math. **77**, 191–193 (1972)
36. H. Toruńczyk, Absolute retracts as factors of normed linear spaces. Fund. Math. **86**, 53–67 (1974)
37. J.H.C. Whitehead, Note on a theorem due to Borsuk. Bull. Am. Math. Soc. **54**, 1125–1132 (1948)

Chapter 7
Cell-Like Maps and Related Topics

A compact set $A \neq \emptyset$ in X is said to be **cell-like** in X if A is contractible in every neighborhood of A in X. A compactum X is **cell-like** if X is cell-like in some *metrizable* space that contains X as a subspace. It will be seen that X is cell-like in every ANR that contains X as a subspace (Theorem 7.1.2). A **cell-like (CE) map** is a perfect (surjective) map $f : X \to Y$ such that each fiber $f^{-1}(y)$ is cell-like. The concept of cell-like maps is very important in Geometric Topology. It has been mainly developed in Shape Theory and Decomposition Theory. For infinite-dimensional manifolds (in particular Hilbert cube manifolds), this concept is one of the main tools.

In the first six sections of this chapter, we will discuss several fundamental properties of cell-like maps and related concepts. The remainder of the chapter will be devoted to some associated topics. In Sect. 7.10, we will construct an infinite dimensional compactum with finite cohomological dimension, which implies that there exists a cell-like map of a finite-dimensional compactum onto an infinite-dimensional compactum. In Sect. 7.12, we will use this example to construct a linear metric space that is not an AR.

This chapter is based on results in Chaps. 2–6. In Sect. 7.7 and 7.10, we will use some results from Algebraic Topology without proofs. Thus, these sections are not self-contained. Moreover, the construction of Sect. 7.12 requires an open cell-like map of a finite-dimensional compactum onto an infinite-dimensional compactum. Using Walsh's result on open maps, the cell-like map in Sect. 7.10 can be remade to be open. However, we will not give the proof of Walsh's result because it is beyond the scope of this book.

7.1 Trivial Shape and Related Properties

In this section, we introduce some properties related to cell-like compacta. First, note that every contractible compactum is cell-like but that the converse is not true. The $\sin(1/x)$-curve

K. Sakai, *Geometric Aspects of General Topology*, Springer Monographs in Mathematics, DOI 10.1007/978-4-431-54397-8_7, © Springer Japan 2013

$$\{0\} \times [-1, 1] \cup \left\{ (x, y) \in \mathbb{R}^2 \;\middle|\; y = \sin \frac{1}{x},\ 0 < x \leq 1 \right\}$$

is a typical example of a non-contractible cell-like compactum. The cell-likeness of this compactum comes from the following:

Proposition 7.1.1. *Let $X_1 \supset X_2 \supset \cdots$ be a nested sequence of contractible non-empty compacta. Then, $X = \bigcap_{i \in \mathbb{N}} X_i$ is cell-like.*

Proof. Every neighborhood U of X in X_1 contains some X_i. Since X_i is contractible, X is contractible in U. □

It is said that a non-empty closed set A in X has **Property UV^*** in X (simply A is **UV^*** in X) if each neighborhood U of A in X contains a neighborhood V of A that is contractible in U. A metrizable space X is said to be **UV^*** if X is UV^* in some ANR that contains X as a closed subspace.[1] It is said that X has **trivial shape** if any map from X to an arbitrary ANR is null-homotopic; equivalently, any map from X to an arbitrary polyhedron is null-homotopic (Corollary 6.6.5).

Theorem 7.1.2. *For a metrizable space X ($\neq \emptyset$), the following are equivalent:*

(a) *X has trivial shape;*
(b) *X is contractible in every ANR that contains X as a closed set;*
(c) *X is contractible in every neighborhood in some metrizable space that contains X as a closed set;*
(d) *X is UV^* in any ANR that contains X as a closed set;*
(e) *X is UV^*;*
(f) *X is UV^* in some metrizable space that contains X as a closed set.*

In particular, a compactum is cell-like if and only if it has trivial shape.

Proof. Consider the following diagram of implications. Every implication is trivial except for the two marked with asterisks.

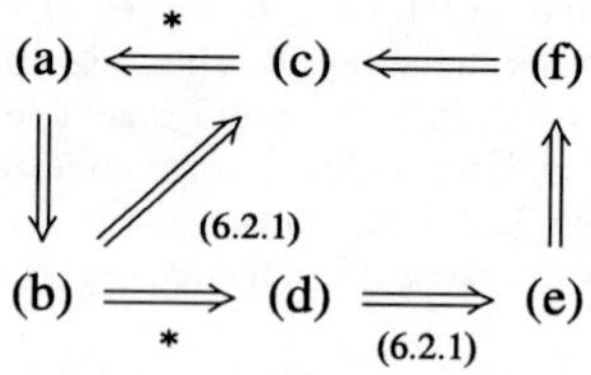

(c) ⇒ (a): Assume that X is closed in a metrizable space M and X is contractible in every neighborhood in M. For any ANR Y, every map $f : X \to Y$ extends over an open neighborhood U of X in M. Let $\tilde{f} : U \to Y$ be an extension of f. There exists a contraction $h : X \times \mathbf{I} \to U$. Then, $\tilde{f}h : X \times \mathbf{I} \to Y$ is a homotopy such that $\tilde{f}h_0 = f$ and $\tilde{f}h_1$ is constant.

[1] In other literature, **Property UV^*** is called **Property UV^∞**.

(b) ⇒ (d): Let Y be an ANR containing X as a closed subspace. For each open neighborhood U of X in Y, we have a contraction $h : X \times \mathbf{I} \to U$ by (b). Since U is an ANR, we have a neighborhood V of X in U and a homotopy $\tilde{h} : V \times \mathbf{I} \to U$ such that $\tilde{h}_0 = \mathrm{id}$ and $\tilde{h}_1$ is a constant map. Thus, X is UV^* in Y. □

Due to Theorem 7.1.2, a UV^* compactum is equal to a cell-like compactum. The term "cell-like" is used only for compacta but "UV^*" is used without compactness.

Note. When X is a compactum, the term "metrizable" in condition (c) can be replaced by the term "normal." Indeed, assume that X is contractible in every neighborhood in some normal space Y that contains X. We can regard X as a closed subspace of the Hilbert cube $\mathbf{Q} = [-1, 1]^{\mathbb{N}}$. Then, it suffices to show that X is contractible in every neighborhood U in $\mathbf{Q}$. By the coordinate-wise application of the Tietze Extension Theorem 2.2.2, we have a map $f : Y \to \mathbf{Q}$ with $f|X = \mathrm{id}$. Because $f^{-1}(U)$ is a neighborhood of X in Y, we have a contraction $h : X \times \mathbf{I} \to f^{-1}(U)$. Hence, $fh : X \times \mathbf{I} \to U$ is a contraction, that is, X is contractible in U.

Let $n \in \omega$. A closed set $A \subset X$ has **Property UV^n** in X (or simply A is **UV^n** in X) if each neighborhood U of A in X contains a neighborhood V of A such that, for each $0 \leq i \leq n$, every map $f : \mathbf{S}^i \to V$ is null-homotopic in U. We say that A has **Property UV^∞** in X (or A is **UV^∞** in X) if it is UV^n in X for every $n \in \mathbb{N}$. A metrizable space X is said to be **UV^n** (resp. **UV^∞**) if X is UV^n (resp. UV^∞) in some ANR that contains X as a closed subspace.[2]

Proposition 7.1.3. *A metrizable space X is UV^n (resp. UV^∞) if and only if X is UV^n (resp. UV^∞) in every ANR that contains X as a closed subspace.*

Proof. Since the UV^∞ case follows from the UV^n case and the "if" part is trivial, it suffices to show the "only if" part of the UV^n case.

Let Y_1 and Y_2 be ANRs containing X as a closed subspace. Then, we have maps $\varphi : W_1 \to Y_2$ and $\psi : W_2 \to Y_1$ such that $\varphi|X = \psi|X = \mathrm{id}$. Assume that X is UV^n in Y_1. For each open neighborhood U of X in Y_2, $\varphi^{-1}(U)$ is a neighborhood of X in Y_1, which contains a neighborhood V' of X such that for each $0 \leq i \leq n$, every map $g : \mathbf{S}^i \to V'$ is null-homotopic in $\varphi^{-1}(U)$. Because U is an ANR, there is a $\mathcal{U} \in \mathrm{cov}(U)$ that is an $\hbar$-refinement of $\{U\}$ (Corollary 6.3.5). We can easily find a neighborhood V of X in U such that $\varphi\psi|V$ is $\mathcal{U}$-close to id and $V \subset \psi^{-1}(V')$. Then, for each map $f : \mathbf{S}^i \to V$, the map $\psi f : \mathbf{S}^i \to V'$ is null-homotopic in $\varphi^{-1}(U)$, hence

$$f \simeq (\varphi\psi|V)f = \varphi\psi f \simeq 0 \text{ in } U.$$

Therefore, X is UV^n in Y_2. Thus, we have the "only if" part. □

Note that X is UV^n in X itself if and only if X is n-connected. Then, we have the following:

Corollary 7.1.4. *An ANR is UV^n if and only if it is n-connected. Consequently, a UV^∞ ANR is the same as an AR.* □

---

[2]Our **Property UV^∞** is weaker than in most other literature (cf. the previous footnote 1).

We show that Property UV^0 characterizes connectedness.

Proposition 7.1.5. *A metrizable space X is UV^0 if and only if X is connected.*

Proof. Regard X as a closed set in a normed linear space Y (Theorem 6.2.1). It is easy to show that if X is not connected then X is not UV^0 in Y, that is, the "only if" part holds. Each neighborhood U of X in Y has a cover $\mathcal{B}$ consisting of open balls. If X is connected, then, for each two $B, B' \in \mathcal{B}[X]$, there are $B_0, B_1, \ldots, B_n \in \mathcal{B}$ such that $B_0 = B_n = B$, and each B_i meets B_{i-1}, which implies that $\mathrm{st}(X, \mathcal{B})$ is path-connected. Thus, we have the "if" part. □

According to Theorem 7.1.2, Property UV^* characterizes the trivial shape. For a finite-dimensional space, we have the following:

Theorem 7.1.6. *An n-dimensional metrizable space X has trivial shape if and only if X is UV^n. Consequently, an n-dimensional compactum is cell-like if and only if it is UV^n.*

Proof. Since the "only if" part is trivial, it suffices to show the "if" part. Assuming that X is UV^n in an ANR Y, we will show that X is contractible in each open neighborhood U of X in Y. Choose open neighborhoods $U = V_{n+1} \supset \cdots \supset V_1 \supset V_0$ of X in Y so that every map $f : \mathbf{S}^i \to V_i$ is null-homotopic in V_{i+1}. Since V_0 is an ANR and $\dim X = n$, we can apply Theorem 6.12.7 and Corollary 6.3.5 to obtain a simplicial complex L with $\dim L \leq n$ and maps $f : X \to |L|$, $g : |L| \to V_0$ such that $gf \simeq \mathrm{id}_X$ in V_0. Let $K = v_0 * L$ be the simplicial cone over L with v_0 the cone vertex. Then, $\dim K \leq n + 1$. For each $0 \leq i \leq n + 1$, we denote $K_i = K^{(i)} \cup L$. Note that $K_0 = \{v_0\} \cup L$. Taking any point $h_0(v_0) \in V_0$, we can extend g to a map $h_0 : |K_0| \to V_0$. Suppose that g extends to a map $h_i : |K_i| \to V_i$. For each $(i + 1)$-simplex $\tau \in K \setminus L$, $h_i|\partial\tau : \partial\tau \to V_i$ extends to a map $h_\tau : \tau \to V_{i+1}$. We define a map $h_{i+1} : |K_{i+1}| \to V_{i+1}$ by $h_{i+1}|\tau = h_\tau$, which is an extension of g. By induction, we have a map $h = h_{n+1} : |K| \to V_{n+1} \subset U$ such that $h||L| = g$. Thus, $g \simeq 0$ in U, hence $\mathrm{id}_X \simeq gf \simeq 0$ in U, that is, X is contractible in U. By virtue of Theorem 7.1.2, X has trivial shape. □

Due to Corollary 4.10.11, every compactum X is the inverse limit of an inverse sequence (X_i, f_i) of compact polyhedra and PL maps. The following lemma is useful when treating the inverse limits:

Lemma 7.1.7. *Let X be the inverse limit of an inverse sequence $(X_i, f_i)_{i\in\mathbb{N}}$ of compacta with projections $p_i : X \to X_i$, $i \in \mathbb{N}$, and let P be a space with the homotopy type of a simplicial complex.*

(1) *For every map $g : X \to P$, there exist some $i_0 \in \mathbb{N}$ and maps $g_i : X_i \to P$, $i \geq i_0$, such that $g_i p_i \simeq g$.*
(2) *Let $g, g' : X_{i_0} \to P$ be maps. If $gp_{i_0} \simeq g' p_{i_0}$, then there is some $j_0 \geq i_0$ such that $gf_{i_0,j} \simeq g' f_{i_0,j}$ for every $j \geq j_0$, where $f_{i,j} = f_i \cdots f_{j-1}$ for $i < j$ and $f_{i,i} = \mathrm{id}_{X_i}$.*

Proof. Since P has the homotopy type of an ANR (Corollary 6.6.5), we may assume that P is an ANR. Recall $X = \varprojlim(X_i, f_i) \subset \prod_{i\in\mathbb{N}} X_i$ and $p_i = \mathrm{pr}_i|X$, $i \in \mathbb{N}$ (cf. Sect. 4.10). Then, $f_i p_{i+1} = p_i$. For each $i \in \mathbb{N}$, let $z_i \in X_i$ be fixed and define an embedding $\varphi_i : X_i \to \prod_{i\in\mathbb{N}} X_i$ by

$$\varphi_i(x) = (f_{1,i}(x), \ldots, f_{i-1,i}(x), x, z_{i+1}, z_{i+2}, \ldots).$$

Then, $\lim_{i\to\infty} \varphi_i p_i = \mathrm{id}$. For each $n \in \mathbb{N}$, let

$$X_n^* = \left\{x \in \prod_{i\in\mathbb{N}} X_i \;\middle|\; x(i) = f_{i,n}(x(n)) \text{ for each } i \le n\right\}.$$

Then, $\bigcap_{i\in\mathbb{N}} X_i^* = X$.

(1): Because P is an ANR, g extends to a map $\tilde{g} : U \to P$ over an open neighborhood U of X in $\prod_{i\in\mathbb{N}} X_i$. Then, $X_n^* \subset U$ for some $n \in \mathbb{N}$. For each $i \ge n$, since $\varphi_i(X_i) \subset X_i^* \subset X_n^*$, we have a map $g_i = \tilde{g}\varphi_i : X_i \to P$. Since $\lim_{i\to\infty} \varphi_i p_i = \mathrm{id}$, we have $\lim_{n\le i\to\infty} g_i p_i = g$. Hence, $g_i p_i \simeq g$ for sufficiently large $i \ge n$.

(2): Extending a homotopy from gp_i to $g' p_i$ over the following set

$$\Xi = (X \times \mathbf{I}) \cup (\prod_{i\in\mathbb{N}} X_i \times \{0, 1\}),$$

we have a map $h : \Xi \to P$ such that $h(x, 0) = g(x(i)) = g\mathrm{pr}_i(x)$ and $h(x, 1) = g'(x(i)) = g'\mathrm{pr}_i(x)$ for each $x \in \prod_{i\in\mathbb{N}} X_i$. Since P is an ANR, h extends to a map $\tilde{h} : U \times \mathbf{I} \to P$, where U is an open neighborhood of X in $\prod_{i\in\mathbb{N}} X_i$. Then, $X_{j_0}^* \subset U$ for some $j_0 \ge i_0$. For every $j \ge j_0$, $\varphi_j(X_j) \times \mathbf{I} \subset X_j^* \times \mathbf{I} \subset U$. Consequently, we have a homotopy $h' = \tilde{h}(\varphi_j \times \mathrm{id}_{\mathbf{I}}) : X_j \times \mathbf{I} \to P$. For each $x \in X_j$,

$$h'_0(x) = \tilde{h}(\varphi_j(x), 0) = h(\varphi_j(x), 0) = g\mathrm{pr}_i\varphi_j(x) = gf_{i,j}(x),$$

i.e., $h'_0 = gf_{i,j}$. Similarly, we have $h'_1 = g' f_{i,j}$. Therefore, $gf_{i,j} \simeq g' f_{i,j}$. □

Remark 1. As we saw in the proof above, in Lemma 7.1.7(1), when P is an ANR, there exist some $n \in \mathbb{N}$ and maps $g_i : X_i \to P$, $i \ge n$, such that $\lim_{n\le i\to\infty} g_i p_i = g$.

The cell-likeness of the inverse limits can be characterized as follows:

Theorem 7.1.8. *Let X be the inverse limit of an inverse sequence $(X_i, f_i)_{i\in\mathbb{N}}$ of compact ANRs and $p_i : X \to X_i$, $i \in \mathbb{N}$, be the projections. Then, the following are equivalent:*

(a) *X is cell-like;*
(b) *$p_i \simeq 0$ for every $i \in \mathbb{N}$;*
(c) *For each $i \in \mathbb{N}$, there is a $j \ge i$ such that $f_{i,j} \simeq 0$.*

Proof. The implication (b) ⇒ (c) follows from Lemma 7.1.7(2). Because $p_i = f_{i,j} p_j$, the implication (c) ⇒ (b) is trivial.

(a) ⇒ (b): By Theorem 7.1.2, X has trivial shape. Since every X_i is an ANR, the implication (a) ⇒ (b) is trivial.

(b) ⇒ (a): Let $g : X \to P$ be a map from X to an ANR P. By Lemma 7.1.7(1), there exist some $i \in \mathbb{N}$ and a map $g' : X_i \to P$ such that $g' p_i \simeq g$. Since $p_i \simeq 0$, it follows that $g \simeq 0$. Thus, X has trivial shape, hence it is cell-like by Theorem 7.1.2. □

Remark 2. The implications (c) ⇒ (b) ⇒ (a) hold without the assumption that each X_i is an ANR. This is a generalization of Proposition 7.1.1.

Concerning the UV^n properties, we have the following characterization:

Theorem 7.1.9. *Let X be the inverse limit of an inverse sequence $(X_i, f_i)_{i\in\mathbb{N}}$ of compact ANRs and $p_i : X \to X_i$, $i \in \mathbb{N}$, be the projections. Then, the following are equivalent:*

(a) *X is UV^n;*
(b) *For each $i \in \mathbb{N}$, there is a $j \geq i$ such that $f_{i,j}\alpha \simeq 0$ for every map $\alpha : \mathbf{S}^k \to X_j$, $0 \leq k \leq n$.*

Proof. We can regard X as a closed set in the Hilbert cube $\boldsymbol{Q}$.

(a) ⇒ (b): For each $i \in \mathbb{N}$, since X_i is an ANR, $p_i : X \to X_i$ extends to a map $\tilde{p}_i : U \to X_i$ over a neighborhood U of X in $\boldsymbol{Q}$. Then, U contains an open neighborhood V of X such that, for each $0 \leq i \leq n$, every map from $\mathbf{S}^i$ to V is null-homotopic in U. Because V is an ANR, applying Lemma 7.1.7(1) to the inclusion $X \subset V$, we can obtain some $j' \geq i$ and a map $g : X_{j'} \to V$ such that $gp_{j'} \simeq \mathrm{id}_X$ in V. Since $\tilde{p}_i g p_{j'} \simeq \tilde{p}_i|X = p_i = f_{i,j'}p_{j'}$, it follows from Lemma 7.1.7(2) that $\tilde{p}_i g f_{j',j} \simeq f_{i,j'} f_{j',j} = f_{i,j}$ for some $j \geq j'$. For $0 \leq k \leq n$ and each map $\alpha : \mathbf{S}^k \to X_j$, we have $gf_{j',j}\alpha \simeq 0$ in U, hence $f_{i,j}\alpha \simeq \tilde{p}_i g f_{j',j}\alpha \simeq 0$ in X_i.

(b) ⇒ (a): For each open neighborhood U of X in $\boldsymbol{Q}$, applying Lemma 7.1.7(1) to the inclusion $X \subset U$, we can obtain some $i \in \mathbb{N}$ and a map $g : X_i \to U$ such that $gp_i \simeq \mathrm{id}_X$ in U. Then, there is some $j \geq i$ such that $f_{i,j}\alpha \simeq 0$ for every map $\alpha : \mathbf{S}^n \to X_j$. Because X_j is an ANR, $p_j : X \to X_j$ extends to a map $\tilde{p}_j : V' \to X_j$ over a neighborhood V' of X in $\boldsymbol{Q}$. Since $gf_{i,j}\tilde{p}_j|X = gf_{i,j}p_j = gp_i \simeq \mathrm{id}_X$ in the ANR U, X has a neighborhood V in $\boldsymbol{Q}$ such that $V \subset V'$ and $gf_{i,j}\tilde{p}_j|V \simeq \mathrm{id}_V$ in U. For $0 \leq k \leq n$ and each map $\alpha : \mathbf{S}^k \to V$, we have $\alpha \simeq gf_{i,j}\tilde{p}_j\alpha \simeq 0$ in U. □

7.2 Soft Maps and the 0-Dimensional Selection Theorem

A map $f : X \to Y$ is said to be **soft** (***n*-soft**) provided that, for any metrizable space Z (with $\dim Z \leq n$) and any map $g : C \to X$ of a closed set C of Z, if fg extends to a map $h : Z \to Y$ then g extends to a map $\tilde{g} : Z \to X$ with $f\tilde{g} = h$.

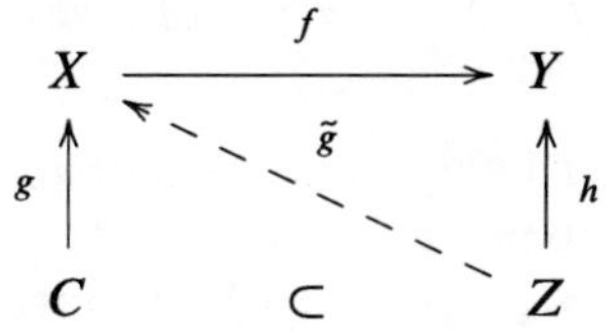

In the above, if Y is an AE or an ANE (an AE(n) or an ANE(n)) then so is X. Replacing the pair (Z, C) with a polyhedral pair[3] (with dim $\leq n$), we can introduce a **polyhedrally soft map** (a **polyhedrally *n*-soft map**). By the following proposition, every (n-)soft map is polyhedrally (n-)soft.

Proposition 7.2.1. *For a map $f : X \to Y$ to be polyhedrally soft (resp. polyhedrally n-soft), it is necessary and sufficient that, for any map $\alpha : \mathbf{S}^{i-1} \to X$, $i \in \mathbb{N}$ (resp. $i \leq n$), if $f\alpha$ extends to a map $\beta : \mathbf{B}^i \to Y$ then α extends to a map $\tilde{\alpha} : \mathbf{B}^i \to X$ with $f\tilde{\alpha} = \beta$.*

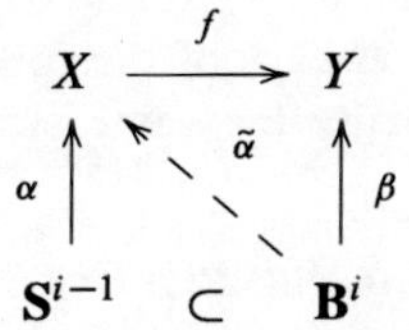

Consequently, a map $f : X \to Y$ is polyhedrally soft if and only if f is polyhedrally n-soft for every $n \in \omega$.

To show the sufficiency, let K be a simplicial complex with L a subcomplex of K, and let $g : |L| \to X$ and $h : |K| \to Y$ be maps with $fg = h||L|$. We inductively construct maps $g_i : |K^{(i)} \cup L| \to X$, $i \in \omega$ (resp. $i \leq n$), so that each g_i is an extension of g_{i-1} and $fg_i = h||K^{(i)} \cup L|$, where $g_{-1} = g$. This can be done by applying the condition to each simplex in $K^{(i)} \setminus (K^{(i-1)} \cup L)$. Then, the desired extension $\tilde{g} : |K| \to X$ of g can be defined by $\tilde{g}||K^{(i)}| = g_i||K^{(i)}|$ for each $i \in \omega$ (resp. $\tilde{g} = g_n$).

Evidently, every polyhedrally soft map is a weak homotopy equivalence and every polyhedrally n-soft map is an n-equivalence (cf. Sect. 4.13).

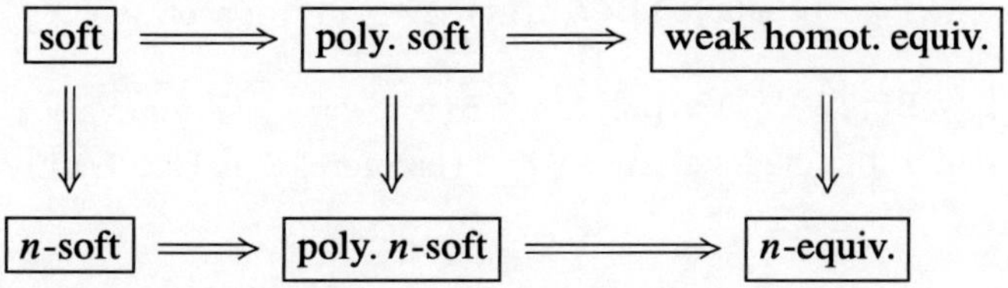

We have the following characterization of polyhedrally 0-soft maps:

[3]That is, $Z = |K|$ and $C = |L|$ for some simplicial complex K and a subcomplex $L \subset K$.

Proposition 7.2.2. *A map $f : X \to Y$ is polyhedrally 0-soft if and only if f is surjective.*

Proof. Because 0-dimensional polyhedra are discrete, it is trivial that a surjective map is polyhedrally 0-soft. To prove the converse, for each $y \in Y$, consider the pair $(\{y\}, \emptyset)$ in the definition of a polyhedrally 0-soft map. Then, we can find $x \in X$ such that $f(x) = y$. □

Proposition 7.2.3. *When Y is metrizable, every 0-soft map $f : X \to Y$ is a surjective open map.*

Proof. According to Proposition 7.2.2, f is surjective. Suppose that f is not open. Then, X has an open set U with $x \in U$ such that $f(x) \notin \operatorname{int} f(U)$. We have $y_n \in Y \setminus f(U)$, $n \in \mathbb{N}$, such that $f(x) = \lim_{n\to\infty} y_n$. Since $Z = \{f(x)\} \cup \{y_n \mid n \in \mathbb{N}\}$ is 0-dimensional, we have a map $g : Z \to X$ such that $g(f(x)) = x$ and $fg = \mathrm{id}_Z$, which causes a contradiction. □

If X is completely metrizable, the converse of Proposition 7.2.3 above is also true. This will be shown as a corollary of the following 0-Dimensional Selection Theorem, which can be proved by the same strategy as the Michael Selection Theorem 3.8.8.

Theorem 7.2.4 (0-Dimensional Selection Theorem). *Let X be a paracompact space with $\dim X = 0$ and $Y = (Y, d)$ be a metric space. Then, every lower semi-continuous (l.s.c.) closed-valued function $\varphi : X \to \mathrm{Cld}(Y)$ admits a selection if each $\varphi(x)$ is d-complete. In addition, if A is a closed set in X then each selection $f : A \to Y$ for $\varphi|A$ can extend to a selection $\tilde{f} : X \to Y$ for φ.*

Proof. We may assume that $\operatorname{diam} Y < 1$. By induction, we will construct maps $f_i : X \to Y$, $i \in \mathbb{N}$, such that

$$d(f_i(x), \varphi(x)) < 2^{-i} \text{ and } d(f_{i+1}(x), f_i(x)) < 2^{-i+2} \text{ for every } x \in X.$$

Assume that f_{n-1} has been obtained, where f_0 is any map. We define a closed-valued function $\psi : X \to \mathrm{Cld}(Y)$ as follows:

$$\psi(x) = \mathrm{cl}_Y\big(\varphi(x) \cap \mathrm{B}(f_{n-1}(x), 2^{-n+1})\big) \text{ for each } x \in X.$$

Let $W = \bigcup_{y\in Y} \mathrm{B}(y, 2^{-n+1}) \times \{y\}$. Since $\mathrm{B}(y, 2^{-n})^2 \subset W$ for every $y \in Y$, W is a neighborhood of the diagonal Δ_Y in Y^2. Therefore, ψ is l.s.c. by Lemmas 3.8.5 and 3.8.3.

For each $y \in Y$, let

$$V_y = \{x \in X \mid \psi(x) \cap \mathrm{B}(y, 2^{-n}) \neq \emptyset\}.$$

Since ψ is l.s.c., each V_y is open in X. By the 0-dimensionality, X has an open cover $\mathcal{U} \prec \{V_y \mid y \in Y\}$ with $\operatorname{ord}\mathcal{U} = 1$ (cf. Theorem 5.2.4). For each $U \in \mathcal{U}$, choose $y(U) \in Y$ so that $U \subset V_{y(U)}$. Then, we have a map $f_n : X \to Y$ such that $f_n(x) = y(U)$ for every $x \in U \in \mathcal{U}$. Observe that $d(f_n(x), \psi(x)) < 2^{-n}$ for every $x \in X$, which means that $d(f_n(x), \varphi(x)) < 2^{-n}$ and $d(f_n(x), f_{n-1}(x)) < 2^{-n+3}$.

The rest of the proof is similar to that of the Michael Selection Theorem 3.8.8. □

Now, as a corollary of Theorem 7.2.4, we can obtain the following characterization of 0-soft maps:

Corollary 7.2.5. *Let X is be a completely metrizable and Y be a metrizable space. Then, a map $f : X \to Y$ is 0-soft if and only if f is surjective and open.*

Proof. The "only if" part is Proposition 7.2.3. To see the "if" part, let Z be a 0-dimensional metrizable space and C a closed set in Z. Given maps $g : C \to X$ and $h : Z \to Y$ such that $fg = h|C$, note that $f^{-1}h : Z \to \mathrm{Cld}(X)$ is an l.s.c. closed-valued function (cf. Proposition 3.8.1) and g is a selection for $f^{-1}h|C$. We can apply Theorem 7.2.4 to extend g to a selection $\tilde{g} : Z \to X$. Then, $f\tilde{g} = h$. □

We have variety in the definition of softness. A map $f : X \to Y$ is said to be **homotopically soft (homotopically n-soft)** (resp. **approximately soft (approximately n-soft)**) provided that, for any metrizable space Z (with $\dim Z \leq n$), any map $g : C \to X$ of a closed set C in Z and each open cover $\mathcal{U}$ of Y, if fg extends to a map $h : Z \to Y$ then g extends $\tilde{g} : Z \to X$ such that $f\tilde{g} \simeq_{\mathcal{U}} h$ rel. C (resp. $f\tilde{g}$ is $\mathcal{U}$-close to h).

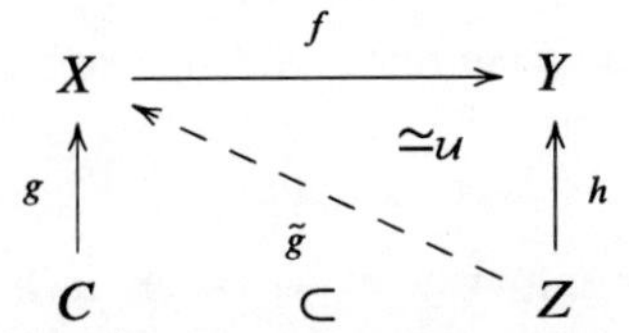

Replacing the pair (Z, C) with a pair of polyhedra (with $\dim \leq n$), we can introduce a **polyhedrally homotopically soft map** (a **polyhedrally homotopically n-soft map**) or a **polyhedrally approximately soft map** (a **polyhedrally approximately n-soft map**).

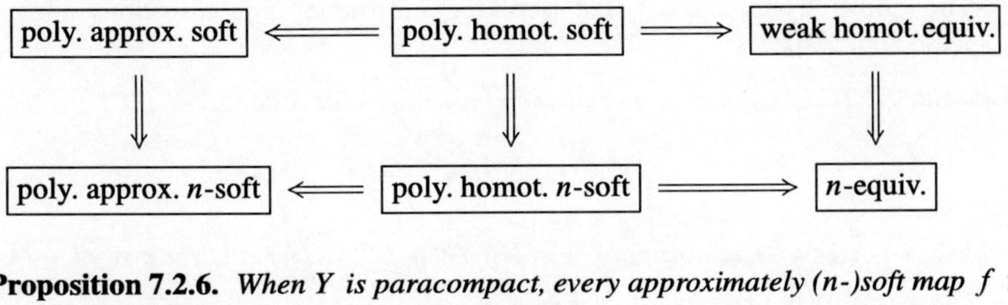

Proposition 7.2.6. *When Y is paracompact, every approximately $(n\text{-})$soft map $f : X \to Y$ is polyhedrally approximately $(n\text{-})$soft.*

Proof. Let K be a simplicial complex (with $\dim K \leq n$) and L be a subcomplex of K with a map $g : |L| \to X$ such that fg extends to a map $h : |K| \to Y$. For each open cover $\mathcal{U}$ of Y, we will show that g extends to a map $\tilde{g} : |K| \to X$ such that $f\tilde{g}$ is $\mathcal{U}$-close to h. Because Y is paracompact, $\mathcal{U}$ has an open star-refinement $\mathcal{V}$. Due to Whitehead's Theorem 4.7.11 on small subdivisions, replacing K with a subdivision, we may assume that $K \prec h^{-1}(\mathcal{V})$. Moreover, replacing K by $\operatorname{Sd}_L K$, we may also assume that L is full in K. We denote $\varphi = \mathrm{id}_{|K|} : |K| \to |K|_{\mathrm{m}}$. By Theorem 4.9.6, we have a map $\psi : |K|_{\mathrm{m}} \to |K|$ such that $\psi\varphi \simeq_K \mathrm{id}$ and $\varphi\psi \simeq_K \mathrm{id}$ by the straight-line homotopy. Since L is full in K, it follows that $\psi\varphi||L| \simeq_L \mathrm{id}$ and $\varphi\psi||L|_{\mathrm{m}} \simeq_L \mathrm{id}$ by the straight-line homotopy. Because f is approximately (n-)soft, the map $g\psi||L|$ extends a map $g' : |K|_{\mathrm{m}} \to X$ such that fg' is $\mathcal{V}$-close to $h\psi$.

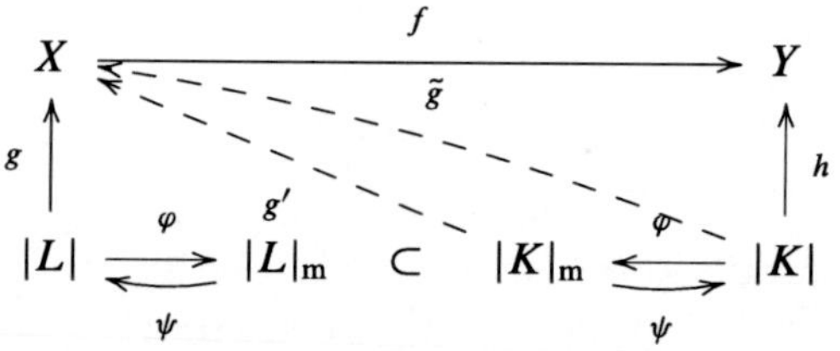

Since $K \prec h^{-1}(\mathcal{V})$, it follows that $L \prec g^{-1}f^{-1}(\mathcal{V})$, consequently

$$g'\varphi||L| = g\psi\varphi||L| \simeq_{f^{-1}(\mathcal{V})} g.$$

By the Homotopy Extension Theorem 4.3.3, g extends to a map $\tilde{g} : |K| \to X$ such that $\tilde{g} \simeq_{f^{-1}(\mathcal{V})} g'\varphi$, hence $f\tilde{g} \simeq_{\mathcal{V}} fg'\varphi$. Since $fg'\varphi$ is $\mathcal{V}$-close to $h\psi\varphi$ and $h\psi\varphi \simeq_{\mathcal{V}} h$, it follows that $f\tilde{g}$ is $\mathcal{U}$-close to h. □

Remark 3. In the above proof, if $f : X \to Y$ is homotopically (n-)soft, then $g\psi||L|$ can be extended to a map $g' : |K|_{\mathrm{m}} \to X$ such that $fg' \simeq_{\mathcal{V}} h\psi$ rel. $|L|_{\mathrm{m}}$, so $fg'\varphi \simeq_{\mathcal{V}} h\psi\varphi$ rel. $|L|$. But $f\tilde{g}||L| = h||L| \neq fg\psi\varphi||L| = fg'\varphi||L|$ and $h\psi\varphi||L| \neq h$ in general. Then, we cannot conclude that $f\tilde{g} \simeq_{\mathcal{U}} h$ rel. $|L|$. However, it is also true that every homotopically (n-)soft map $f : X \to Y$ is polyhedrally approximately (n-)soft when Y is paracompact. This will be shown in the next section.

The following lemma is easy but useful for constructing approximations of maps to paracompact spaces:

Lemma 7.2.7. *Given a sequence of open covers of Y as follows:*

$$\mathcal{U} \overset{*}{\succ} \mathcal{U}_1 \overset{*}{\succ} \mathcal{U}_2 \overset{*}{\succ} \cdots,$$

let $\mathcal{W}_{i+1} = \operatorname{st}(\mathcal{W}_i, \mathcal{U}_{i+1})$, where $\mathcal{W}_1 = \mathcal{U}_1$. Then, $\operatorname{st}(\mathcal{W}_i, \mathcal{U}_i) \prec \mathcal{U}$ for every $i \in \mathbb{N}$.

This lemma can be shown by induction because

$$\operatorname{st}(\mathcal{W}_i, \mathcal{U}_i) = \operatorname{st}(\operatorname{st}(\mathcal{W}_{i-1}, \mathcal{U}_i), \mathcal{U}_i) \prec \operatorname{st}(\mathcal{W}_{i-1}, \operatorname{st}\mathcal{U}_i) \prec \operatorname{st}(\mathcal{W}_{i-1}, \mathcal{U}_{i-1}).$$

Theorem 7.2.8. *Let Y be a paracompact space. A map $f : X \to Y$ is polyhedrally homotopically soft if and only if f is polyhedrally homotopically n-soft for every $n \in \omega$.*

Proof. The "only if" part is obvious. To prove the "if" part, let K be a simplicial complex with L a subcomplex of K, and let $g : |L| \to X$ and $h : |K| \to Y$ be maps with $fg = h||L|$. Since Y is paracompact, each open cover $\mathcal{U} \in \operatorname{cov}(Y)$ has open refinements as follows:

$$\mathcal{U} \overset{*}{\succ} \mathcal{U}_0 \overset{*}{\succ} \mathcal{U}_1 \overset{*}{\succ} \mathcal{U}_2 \overset{*}{\succ} \cdots .$$

For each $i \in \mathbb{N}$, let $\mathcal{W}_i = \operatorname{st}(\mathcal{W}_{i-1}, \mathcal{U}_i)$, where $\mathcal{W}_0 = \mathcal{U}_0$. By Lemma 7.2.7, $\operatorname{st}(\mathcal{W}_i, \mathcal{U}_i) \prec \mathcal{U}$ for every $i \in \mathbb{N}$.

We will inductively construct maps $g_i : |K^{(i)} \cup L| \to X$, $h_i : |K| \to Y$ and $\mathcal{U}_i$-homotopies $\varphi^{(i)} : |K| \times \mathbf{I} \to Y$, $i \in \mathbb{N}$, such that

$$g_i||K^{(i-1)} \cup L| = g_{i-1},\ fg_i = h_i||K^{(i)} \cup L|,\ \varphi_0^{(i)} = h_{i-1},\ \varphi_1^{(i)} = h_i,$$

$$\varphi_t^{(i)}||K^{(i-1)} \cup L| = h_{i-1}||K^{(i-1)} \cup L| = fg_{i-1} \text{ for every } t \in \mathbf{I},$$

where $g_{-1} = g$ and $h_{-1} = h$. Suppose that g_{n-1}, h_{n-1}, and $\varphi^{(n-1)}$ have been defined.

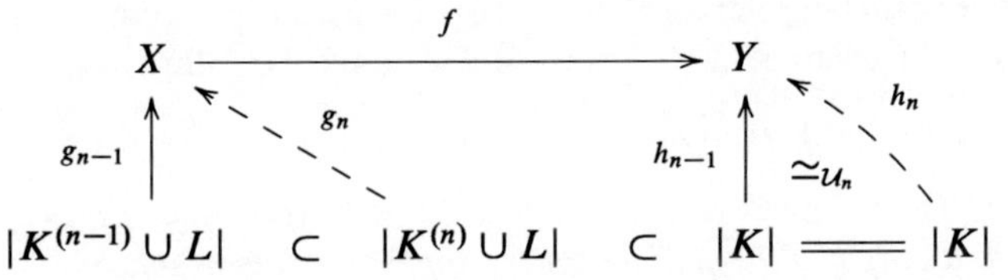

By the polyhedral homotopy n-softness of f, the map g_{n-1} extends to a map $g_n : |K^{(n)} \cup L| \to X$ such that

$$fg_n \simeq_{\mathcal{U}_n} h_{n-1}||K^{(n)} \cup L| \text{ rel. } |K^{(n-1)} \cup L|.$$

Using the Homotopy Extension Theorem 4.3.3, we have a $\mathcal{U}_n$-homotopy $\varphi^{(n)} : |K| \times \mathbf{I} \to Y$ such that

$$\varphi_0^{(n)} = h_{n-1},\ \varphi_1^{(n)}||K^{(n)} \cup L| = fg_n \text{ and}$$

$$\varphi_t^{(n)}||K^{(n-1)} \cup L| = h_{n-1}||K^{(n-1)} \cup L| \text{ for each } t \in \mathbf{I}.$$

The map h_n is defined by $h_n = \varphi_1^{(n)} : |K| \to Y$.

Now, we can extend g to the map $\tilde{g} : |K| \to X$ defined by $\tilde{g}||K^{(i)} \cup L| = g_i$ for each $i \in \omega$. Let $\varphi : |K| \times \mathbf{I} \to Y$ be a homotopy defined as follows: $\varphi_0 = f\tilde{g}$ and

$$\varphi(x, t) = \varphi^{(i)}(x, 2 - 2^{i+1}t) \text{ if } 2^{-i-1} \le t \le 2^{-i}, i \in \omega.$$

Then, $\varphi_1 = \varphi_0^{(0)} = h_{-1} = h$ and $\varphi_t||L| = h||L| = fg = f\tilde{g}||L|$ for every $t \in \mathbf{I}$. The continuity of φ at each point in $|K| \times \{0\}$ follows from the continuity of $\varphi|\sigma \times \mathbf{I}$ for each simplex $\sigma \in K$, where if $\dim \sigma = k$ then

$$\varphi_t|\sigma = h_k|\sigma = fg_k|\sigma = f\tilde{g}|\sigma = \varphi_0|\sigma \quad \text{for every } t \leq 2^{-k-1}.$$

For each $x \in |K|$ and $n \in \mathbb{N}$,

$$\begin{aligned}\varphi(\{x\} \times [2^{-n-1}, 1]) &= \varphi(\{x\} \times [2^{-n}, 1]) \cup \varphi^{(n)}(\{x\} \times \mathbf{I}) \\ &\subset \operatorname{st}(\varphi(\{x\} \times [2^{-n}, 1]), \mathcal{U}_n).\end{aligned}$$

Since $\varphi(\{x\} \times [2^{-1}, 1]) = \varphi^{(0)}(\{x\} \times \mathbf{I})$ is contained in some member of $\mathcal{U}_0 = \mathcal{W}_0$, it follows that $\varphi(\{x\} \times [2^{-n-1}, 1])$ is contained in some member of $\mathcal{W}_n$. In addition, note that $\varphi(\{x\} \times [0, 2^{-k-1}]) = \{f\tilde{g}(x)\}$ for sufficiently large $k \in \mathbb{N}$. Then, $\varphi(\{x\} \times \mathbf{I})$ is contained in some member of $\mathcal{U}$. Hence, φ is a $\mathcal{U}$-homotopy. Thus, we have $f\tilde{g} \simeq_{\mathcal{U}} h$ rel. L. □

Remark 4. If f is polyhedrally approximately n-soft for every $n \in \omega$, the above arguments are not valid to prove that f is polyhedrally approximately soft because we cannot apply the Homotopy Extension Theorem 4.3.3.

Every open cover $\mathcal{U}$ of an ANR (resp. a metrizable ANE$(n+1)$) Y has an $\hbar$-refinement (resp. h^n-refinement) $\mathcal{V} \in \operatorname{cov}(Y)$ by Proposition 6.3.1 and Theorem 6.3.4 (resp. Theorem 6.12.9). Therefore, we have the following:

Proposition 7.2.9. *Let Y be a metrizable space and $f : X \to Y$ be a map.*

(1) *When Y is LC^n, i.e., Y is an ANE$(n+1)$, f is (polyhedrally) homotopically n-soft if and only if f is (polyhedrally) approximately n-soft.*
(2) *When Y is LC$^\infty$, f is polyhedrally homotopically soft if and only if f is polyhedrally approximately soft.*
(3) *When Y is an ANR, f is homotopically soft if and only if f is approximately soft.* □

A polyhedrally approximately 0-soft map is simply a map with the dense image. We can state this formally as follows:

Proposition 7.2.10. *A map $f : X \to Y$ is polyhedrally approximately 0-soft if and only if $f(X)$ is dense in Y.*

Sketch of Proof. For the "if" part, note that 0-dimensional polyhedra are discrete. To see the "only if" part, for each $y \in Y$ and each open neighborhood V of y in Y, consider $(Z, C) = (\{y\}, \emptyset)$, $h = \mathrm{id}$, and $\mathcal{U} = \{Y \setminus \{y\}, V\}$ in the definition of a polyhedrally approximately 0-soft map.

7.3 Hereditary n-Equivalence and Local Connections

A map $f : X \to Y$ is called a **hereditary n-equivalence** if $f|f^{-1}(U) : f^{-1}(U) \to U$ is an n-equivalence for every open set U in Y, that is, it satisfies the following condition $(\pi)_i$ for each $i = 0, \ldots, n$:

$(\pi)_i$ For each map $\alpha : \mathbf{S}^{i-1} \to f^{-1}(U)$, if $f\alpha$ extends to a map $\beta : \mathbf{B}^i \to U$ then α extends to a map $\bar{\alpha} : \mathbf{B}^i \to f^{-1}(U)$ such that $f\bar{\alpha} \simeq \beta$ rel. $\mathbf{S}^{i-1}$,

where $\mathbf{B}^0 = \{0\}$ and $\mathbf{S}^{-1} = \emptyset$.

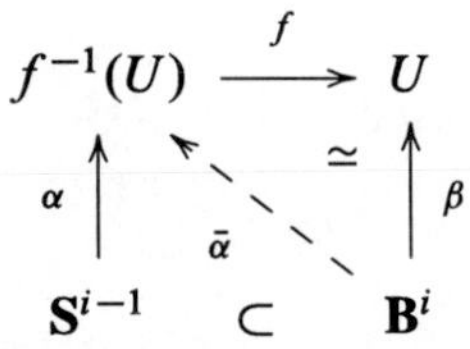

A map $f : X \to Y$ is a hereditary 0-equivalence if and only if every $y \in Y$ can be connected with an arbitrarily close point of $f(X)$ by a small path. A **hereditary weak homotopy equivalence** is a map $f : X \to Y$ such that $f|f^{-1}(U) : f^{-1}(U) \to U$ is a weak homotopy equivalence for every open set U in Y; equivalently, it is a hereditary n-equivalence for every $n \in \omega$. The following statements are easily proved:

- Every polyhedrally homotopically n-soft map is a hereditary n-equivalence.
- Every polyhedrally homotopically soft map is a hereditary weak homotopy equivalence.

Sketch of Proof. The second statement follows from the first statement. Let $f : X \to Y$ be a polyhedrally homotopically n-soft map and U be an open set in Y. For maps $\alpha : \mathbf{S}^{i-1} \to f^{-1}(U)$ and $\beta : \mathbf{B}^i \to U$ with $\beta|\mathbf{S}^{i-1} = f\alpha$, consider $\mathcal{U} = \{U, Y \setminus \beta(\mathbf{B}^i)\} \in \operatorname{cov}(Y)$.

If Y is paracompact, the converse statements are also true.

Theorem 7.3.1. *Let Y be a paracompact space. Then, a map $f : X \to Y$ is a hereditary n-equivalence if and only if f is polyhedrally homotopically n-soft.*

Proof. The "if" part has been shown previously. To prove the "only if" part, let (P, Q) be a pair of polyhedra with $\dim P \leq n$, and let $g : Q \to X$ and $h : P \to Y$ be maps with $fg = h|Q$. For each $\mathcal{U} \in \operatorname{cov}(Y)$, take open refinements as follows:

$$\mathcal{U} = \mathcal{U}_n \overset{*}{\succ} \mathcal{U}_{n-1} \overset{*}{\succ} \cdots \overset{*}{\succ} \mathcal{U}_0 \overset{*}{\succ} \mathcal{V}.$$

Then, P has a triangulation K such that Q is triangulated by a subcomplex of K and $K \prec h^{-1}(\mathcal{V})$ (Theorem 4.7.11). We will inductively define maps $g_i : Q \cup |K^{(i)}| \to X$ with homotopies $\varphi^{(i)} : (Q \cup |K^{(i)}|) \times \mathbf{I} \to Y$, $i = 0, \ldots, n$, such that

$$g_i|Q \cup |K^{(i-1)}| = g_{i-1},\ \varphi^{(i)}|(Q \cup |K^{(i-1)}|) \times \mathbf{I} = \varphi^{(i-1)},$$
$$\{\varphi^{(i)}(\sigma \times \mathbf{I}) \mid \sigma \in K^{(i)}\} \prec \mathcal{U}_i,\ \varphi_0^{(i)} = h|Q \cup |K^{(i)}| \text{ and } \varphi_1^{(i)} = fg_i,$$

where $g_{-1} = g$ and $\varphi^{(-1)} = h\mathrm{pr}_Q : Q \times \mathbf{I} \to Y$. Then, g can be extended to a map $g_n : P \to X$ such that $fg_n \simeq_{\mathcal{U}} h$ rel. Q.

Suppose that g_{i-1} and $\varphi^{(i-1)}$ have been defined. For each i-simplex $\sigma \in K$ with $\sigma \not\subset Q$, since $K \prec h^{-1}(\mathcal{V})$ and $\mathrm{st}(\mathcal{V}, \mathcal{U}_{i-1}) \prec \mathcal{U}_i$, we can choose $U_\sigma \in \mathcal{U}_i$ so that

$$\varphi^{(i-1)}((\sigma \times \{0\}) \cup (\partial\sigma \times \mathbf{I})) \subset \mathrm{st}(h(\sigma), \mathcal{U}_{i-1}) \subset U_\sigma.$$

Then, $fg_{i-1}(\partial\sigma) \subset U_\sigma$, i.e., $g_{i-1}(\partial\sigma) \subset f^{-1}(U_\sigma)$. Take a homeomorphism

$$\gamma_\sigma : \sigma \to (\sigma \times \{0\}) \cup (\partial\sigma \times \mathbf{I})$$

such that $\gamma_\sigma(x) = (x, 1)$ for each $x \in \partial\sigma$. Consider the map $h_\sigma = \varphi^{(i-1)}\gamma_\sigma : \sigma \to U$. Then, $h_\sigma|\partial\sigma = fg_{i-1}|\partial\sigma$. Because $f|f^{-1}(U_\sigma) : f^{-1}(U_\sigma) \to U_\sigma$ is an n-equivalence, $g_{i-1}|\partial\sigma$ extends to a map $g_\sigma : \sigma \to f^{-1}(U_\sigma)$ such that $fg_\sigma \simeq h_\sigma$ rel. $\partial\sigma$.

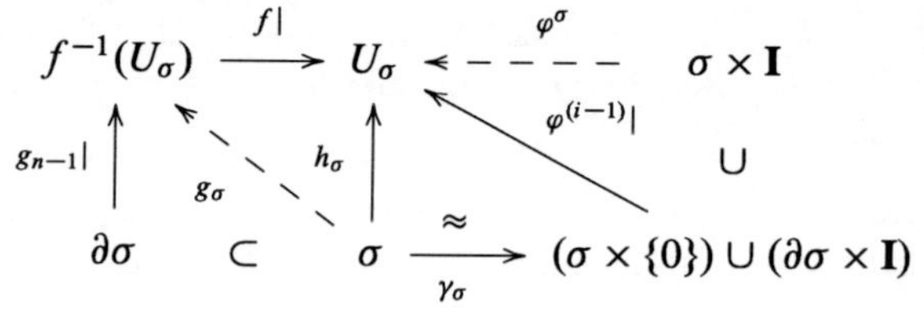

Using a homotopy from h_σ to fg_σ rel. $\partial\sigma$, we can easily construct a homotopy $\varphi^\sigma : \sigma \times \mathbf{I} \to U_\sigma$ such that $\varphi_1^\sigma = fg_\sigma$ and

$$\varphi^\sigma|(\sigma \times \{0\}) \cup (\partial\sigma \times \mathbf{I}) = h_\sigma \gamma_\sigma^{-1}.$$

The last condition means that $\varphi_0^\sigma = h|\sigma$ and $\varphi^\sigma|\partial\sigma \times \mathbf{I} = \varphi^{(i-1)}|\partial\sigma \times \mathbf{I}$. Summing up these g_σ and φ^σ, we can obtain the desired extensions g_i and $\varphi^{(i)}$ of g_{i-1} and $\varphi^{(i-1)}$, respectively. □

Combining Theorems 7.3.1 and 7.2.8, we have the following:

Corollary 7.3.2. *Let Y be a paracompact space. Then, a map $f : X \to Y$ is a hereditary weak homotopy equivalence if and only if f is a polyhedrally homotopically soft map.* □

As in the case of polyhedrally homotopically (n-)soft maps, it is easy to see that every homotopically n-soft map is a hereditary n-equivalence and every homotopically soft map is a hereditary weak homotopy equivalence. Then, by Corollary 7.3.2, we have the following:

Corollary 7.3.3. *When Y is paracompact, every homotopically (n-)soft map $f : X \to Y$ is a polyhedrally homotopically (n-)soft map.* □

A map $f : X \to Y$ is called a **local n-connection** (resp. a **strong local n-connection**) if $f(X)$ is dense in Y and every neighborhood U of each $y \in Y$ contains a neighborhood V of y such that, for each $0 \leq i < n$, every map $g : \mathbf{S}^i \to f^{-1}(V)$ is null-homotopic in $f^{-1}(U)$ and, for each map $g : \mathbf{S}^n \to f^{-1}(V)$, $fg \simeq 0$ in U (resp. $g \simeq 0$ in $f^{-1}(U)$, i.e., for each $0 \leq i \leq n$, every map $g : \mathbf{S}^i \to f^{-1}(V)$ is null-homotopic in $f^{-1}(U)$).

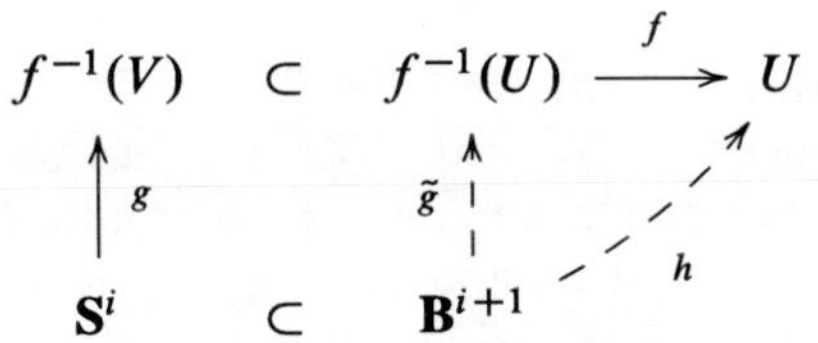

The following are direct consequences of the definitions:

- Every hereditary $(n+1)$-equivalence is a strong local n-connection.
- Every strong local n-connection is a local n-connection.
- Every local n-connection is a strong local $(n-1)$-connection.

Moreover, note that a map $f : X \to Y$ is a strong local n-connection if and only if $f(X)$ is dense in Y and each open cover $\mathcal{U}$ of Y has an open refinement $\mathcal{V}$ such that $f^{-1}(\mathcal{V})$ is a C^n-refinement of $f^{-1}(\mathcal{U})$. We call f a **local ∞-connection** if f is a (strong) local n-connection for every $n \in \omega$. A map $f : X \to Y$ is called a **local $*$-connection** if $f(X)$ is dense in Y and every neighborhood U of each $y \in Y$ contains a neighborhood V of y such that $f^{-1}(V)$ is contractible in $f^{-1}(U)$. If f is a closed map, each non-empty fiber $f^{-1}(y)$ has trivial shape. A perfect local $*$-connection is simply a cell-like map.

Theorem 7.3.4. *Let Y be a paracompact space. Then, for a map $f : X \to Y$, the following statements are equivalent:*

(a) *f is a strong local n-connection;*
(b) *f is polyhedrally approximately $(n+1)$-soft;*
(c) *f is a hereditary $(n+1)$-equivalence.*

Proof. The equivalence (b) ⇔ (c) is the statement of Theorem 7.3.1 and the implication (c) ⇒ (a) has been proved. Consequently, it remains to show the implication (a) ⇒ (b).

(a) ⇒ (b): Since f is a strong local n-connection, each $\mathcal{U} \in \operatorname{cov}(Y)$ has the following open refinements:

$$\mathcal{U} \overset{*}{\succ} \mathcal{V}_{n+1} \overset{*}{\succ} \mathcal{U}_{n+1} \succ \mathcal{V}_n \overset{*}{\succ} \cdots \overset{*}{\succ} \mathcal{U}_1 \succ \mathcal{V}_0 \overset{*}{\succ} \mathcal{U}_0$$

such that $f^{-1}(\mathcal{U}_{i+1}) \underset{C^i}{\succ} f^{-1}(\mathcal{V}_i)$. Then, it follows that

$$f^{-1}(\mathcal{V}_{n+1}) \overset{*}{\succ} f^{-1}(\mathcal{U}_{n+1}) \underset{C^n}{\succ} f^{-1}(\mathcal{V}_n) \overset{*}{\succ}$$
$$\cdots \overset{*}{\succ} f^{-1}(\mathcal{U}_1) \underset{C^0}{\succ} f^{-1}(\mathcal{V}_0) \overset{*}{\succ} f^{-1}(\mathcal{U}_0).$$

By the same arguments as in the proof of Lemma 6.12.2, we can prove that

$$f^{-1}(\mathcal{V}_0) \underset{L^{n+1}}{\prec} f^{-1}(\mathcal{V}_{n+1}).$$

Let (P, Q) be a pair of polyhedra with $\dim P \leq n+1$, and let $g : Q \to X$ and $h : P \to Y$ be maps such that $fg = h|Q$. Then, P has a triangulation K such that Q is triangulated by a subcomplex of K and $K \prec h^{-1}(\mathcal{U}_0)$ (Theorem 4.7.11). Since $f(X)$ is dense in Y, we can extend g to a map $g_0 : Q \cup |K^{(0)}| \to X$ such that fg_0 is $\mathcal{U}_0$-close to $h|Q \cup |K^{(0)}|$. Then, g_0 is a partial $f^{-1}(\mathcal{V}_0)$-realization of K, which extends to a full $f^{-1}(\mathcal{V}_{n+1})$-realization $\tilde{g} : P = |K| \to X$.

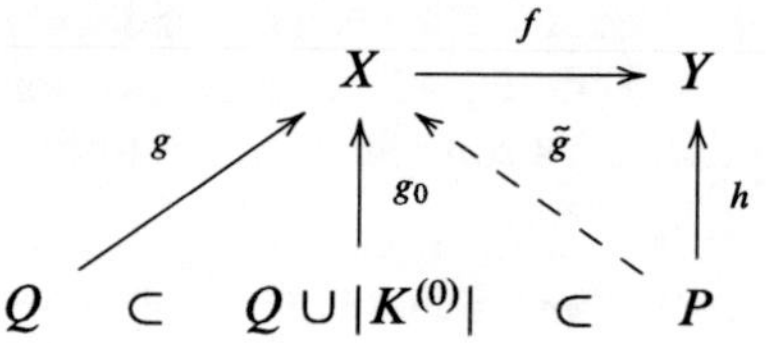

Each $x \in P$ is contained in some $\tau \in K$. Then, we have $V \in \mathcal{V}_{n+1}$ such that $\tilde{g}(\tau) \subset f^{-1}(V)$, i.e., $f\tilde{g}(\tau) \subset V$. On the other hand, $h(\tau)$ is contained in some $V_0 \in \mathcal{V}_0$. Take $v \in \tau^{(0)}$. Then, $f\tilde{g}(v) = fg_0(v)$ and $h(v)$ are contained in the same $U_0 \in \mathcal{U}_0$, hence $f\tilde{g}(x), h(x) \in \mathrm{st}(U_0, \mathcal{V}_{n+1})$. Consequently, $f\tilde{g}$ is $\mathcal{U}$-close to h. □

If Y is LC^n, we have the following proposition:

Proposition 7.3.5. *Let Y be LC^n and $f : X \to Y$ be a map.*

(1) *If f is polyhedrally approximately n-soft then f is a local n-connection.*
(2) *If f is polyhedrally approximately $(n+1)$-soft then f is a strong local n-connection.*

Proof. First, note that a polyhedrally approximately 0-soft map has the dense image by Proposition 7.2.10. Since Y is LC^0, every $y \in Y$ can be connected with an arbitrarily close point of $f(X)$ by a small path, which means that f is a local 0-connection.

Since Y is LC^n, every open neighborhood U of each $y \in Y$ contains a neighborhood V of y such that, for $0 \leq i \leq n$, every map $g : \mathbf{S}^i \to V$ extends to a map $\tilde{g} : \mathbf{B}^{i+1} \to U$. Now, let $0 \leq i \leq n$ and $g : \mathbf{S}^i \to f^{-1}(V)$ be a map. Then, fg extends to a map $h : \mathbf{B}^{i+1} \to U$.

$$\begin{array}{ccccc} f^{-1}(V) & \subset & f^{-1}(U) & \xrightarrow{f} & U \\ \uparrow{\scriptstyle g} & & & \nearrow{\scriptstyle h} & \\ \mathbf{S}^i & \subset & \mathbf{B}^{i+1} & & \end{array}$$

When f is polyhedrally approximately n-soft, if $0 \leq i \leq n-1$ then g extends to a map $\tilde{g} : \mathbf{B}^{i+1} \to X$ such that $f\tilde{g}$ is $\mathcal{U}$-close to h, where

$$\mathcal{U} = \{Y \setminus h(\mathbf{B}^{i+1}),\ U\} \in \operatorname{cov}(Y).$$

Then, $f\tilde{g}(\mathbf{B}^{i+1}) \subset U$, i.e., $\tilde{g}(\mathbf{B}^{i+1}) \subset f^{-1}(U)$. This means that f is a local n-connection. If f is polyhedrally approximately $(n+1)$-soft, this argument is valid for all $0 \leq i \leq n$, which shows that f is a strong local n-connection. □

Theorem 7.3.6. *Let Y be a metrizable space. Every local n-connection $f : X \to Y$ is polyhedrally homotopically n-soft.*

Proof. We may assume that $Y = (Y, d)$ is a metric space. In the case $n = 0$, it suffices to show that for each $y \in Y$ and for each $\varepsilon > 0$, there is an $x \in X$ such that $f(x)$ is connected with y by a path in the ε-neighborhood $\mathrm{B}(y, \varepsilon) \subset Y$. We can choose open neighborhoods $U_1 \supset U_2 \supset \cdots$ of y in Y such that, for every pair of points $x, x' \in f^{-1}(U_i)$, $f(x)$ and $f(x')$ are connected by a path in $\mathrm{B}(y, 2^{-i+1}\varepsilon)$. Since $f(X)$ is dense in Y, we have points $x_i \in f^{-1}(U_i)$. Taking paths $h_i : \mathbf{I} \to \mathrm{B}(y, 2^{-i+1}\varepsilon)$ with $h_i(0) = x_i$ and $h_i(1) = x_{i+1}$, we can define a path $h : \mathbf{I} \to \mathrm{B}(y, \varepsilon)$ as follows:

$$h(0) = y \ \text{ and } \ h(t) = h_i(2 - 2^i t) \ \text{ for } 2^{-i} \leq t \leq 2^{-i+1}.$$

Assuming the theorem is valid for $n-1$, we prove the theorem for n. Let (P, Q) be a pair of polyhedra with $\dim P = n$, and let $g : Q \to X$ and $h : P \to Y$ be maps such that $fg = h$. Each $\mathcal{U} \in \operatorname{cov}(Y)$ has refinements

$$\mathcal{U} \overset{*}{\succ} \mathcal{U}_1 \succ \mathcal{V}_1 \overset{*}{\succ} \mathcal{U}_2 \succ \mathcal{V}_2 \overset{*}{\succ} \cdots,$$

such that $\operatorname{mesh} \mathcal{U}_i < 2^{-i}$ and the following condition hold:

- Each $V \in \mathcal{V}_i$ is contained in some $U \in \mathcal{U}_i$ such that, for each map $\alpha : \mathbf{S}^n \to f^{-1}(V)$, $f\alpha \simeq 0$ in U, that is, $f\alpha$ extends to a map $\beta : \mathbf{B}^{n+1} \to U$.

For convenience, we denote $\mathcal{W}_i = \mathcal{U}_{i+2}$. Then,

$$\operatorname{st}^3 \mathcal{W}_i = \operatorname{st}(\mathcal{W}_i, \operatorname{st} \mathcal{W}_i) \prec \mathcal{V}_i \ \text{ and } \ \mathcal{W}_{i+1} \prec \mathcal{W}_i.$$

For each $i \in \mathbb{N}$, let K_i be a triangulation of P such that $K_i \prec h^{-1}(\mathcal{W}_i)$ and $K_{i+1} \lhd K_i$. Then, note that $|K_i^{(n-1)}| \subset |K_{i+1}^{(n-1)}|$.

By the inductive assumption, we have a map $g_i' : Q \cup |K_i^{(n-1)}| \to X$ such that $g_i'|Q = g$ and $fg_i' \simeq_{\mathcal{W}_i} h|Q \cup |K_i^{(n-1)}|$ rel. Q. Let

$$\psi^{(i)} : (Q \cup |K_i^{(n-1)}|) \times \mathbf{I} \to Y$$

be a $\mathcal{W}_i$-homotopy such that $\psi_0^{(i)} = h|Q \cup |K_i^{(n-1)}|$, $\psi_1^{(i)} = fg_i'$, and $\psi_t^{(i)}|Q = h|Q = fg$ for each $t \in \mathbf{I}$. Applying the Homotopy Extension Theorem 4.3.3, we can extend fg_i' to a map $h_i : P \to Y$ such that $h_i \simeq_{\mathcal{W}_i} h$ rel. Q. Since f is polyhedrally approximately n-soft by Theorem 7.3.4, we can extend g_i' to a map $g_i : P \to X$ such that fg_i is $\mathcal{W}_i$-close to h_i, hence fg_i is st $\mathcal{W}_i$-close to h. Let

$$\varphi_i : (P \times \{0,1\}) \cup (Q \times \mathbf{I}) \to X$$

be the map defined by $\varphi_i(x,t) = g(x)$ for each $(x,t) \in Q \times \mathbf{I}$, $\varphi_i(x,0) = g_i(x)$, and $\varphi_i(x,1) = g_{i+1}(x)$ for each $x \in P$. Then, $f\varphi_i$ extends to the map

$$\psi_i : (P \times \{0,1\}) \cup ((Q \cup |K_i^{(n-1)}|) \times \mathbf{I}) \to Y$$

defined as follows: $\psi_i|P \times \{0,1\} = f\varphi_i|P \times \{0,1\}$ and

$$\psi_i(x,t) = \begin{cases} \psi^{(i)}(x, 1-2t) & \text{if } t \leq \dfrac{1}{2}, \\ \psi^{(i+1)}(x, 2t-1) & \text{if } t \geq \dfrac{1}{2}. \end{cases}$$

Since f is polyhedrally approximately n-soft, we can extend φ_i to a map

$$\tilde{\varphi}_i : (P \times \{0,1\}) \cup ((Q \cup |K_i^{(n-1)}|) \times \mathbf{I}) \to X$$

such that $f\tilde{\varphi}_i$ is $\mathcal{W}_i$-close to ψ_i.

For each n-simplex $\sigma \in K_i$ with $\sigma \not\subset Q$, $h(\sigma)$ is contained in some $W \in \mathcal{W}_i$. Since fg_i is st $\mathcal{W}_i$-close to h, we have

$$fg_i(\sigma) \cup fg_{i+1}(\sigma) \subset \mathrm{st}(W, \mathrm{st}\,\mathcal{W}_i) \in \mathrm{st}^3\,\mathcal{W}_i \prec \mathcal{V}_i.$$

On the other hand, $\psi_i(\partial\sigma \times \mathbf{I}) \subset \mathrm{st}(W, \mathcal{W}_i)$ by the definition of ψ_i. Since $f\tilde{\varphi}_i$ is $\mathcal{W}_i$-close to ψ_i, it follows that

$$f\tilde{\varphi}_i(\partial\sigma \times \mathbf{I}) \subset \mathrm{st}(\mathrm{st}(W, \mathcal{W}_i), \mathcal{W}_i) \in \mathrm{st}^2\,\mathcal{W}_i \prec \mathcal{V}_i.$$

Consequently, there is some $V \in \mathcal{V}_i$ such that

$$f\tilde{\varphi}_i((\sigma \times \{0,1\}) \cup (\partial\sigma \times \mathbf{I})) = fg_i(\sigma) \cup fg_{i+1}(\sigma) \cup f\tilde{\varphi}_i(\partial\sigma \times \mathbf{I}) \subset V.$$

Then, we have $U \in \mathcal{U}_i$ and a map $\psi_\sigma : \sigma \times \mathbf{I} \to U$ such that

$$\psi_\sigma|(\sigma \times \{0,1\}) \cup (\partial\sigma \times \mathbf{I}) = f\tilde{\varphi}_i|(\sigma \times \{0,1\}) \cup (\partial\sigma \times \mathbf{I}).$$

Pasting these ψ_σ, we can obtain a $\mathcal{U}_i$-homotopy $\tilde{\psi}_i : P \times \mathbf{I} \to Y$, that is, $\tilde{\psi}_i|\sigma \times \mathbf{I} = \psi_\sigma|\sigma \times \mathbf{I}$ for each n-simplex $\sigma \in K_i$ with $\sigma \not\subset Q$. Observe that

$$\tilde{\psi}_i(x,t) = f\tilde{\varphi}_i(x,t) = f\varphi_i(x,t) = fg(x) \text{ for each } (x,t) \in Q \times \mathbf{I}.$$

Now, we define $\psi^* : P \times \mathbf{I} \to Y$ by $\psi^*_0 = h$ and

$$\psi^*(x,t) = \tilde{\psi}_i(x, 2 - 2^i t) \text{ if } 2^{-i} \le t \le 2^{-i+1}, i \in \mathbb{N}.$$

Clearly ψ^* is continuous at each point of $P \times (0,1]$. To verify the continuity of ψ^* at each point of $P \times \{0\}$, let $x \in P$ and $\varepsilon > 0$. By the continuity of h, we have a neighborhood U of x in P such that $\operatorname{diam} h(U) < \varepsilon/2$. Choose $k \in \mathbb{N}$ so that $2^{-k} < \varepsilon/2$. For each $x' \in U$ and $t \in (0, 2^{-k})$,

$$\begin{aligned} d(\psi^*(x',t), \psi^*(x,0)) &\le d(\psi^*(x',t), \psi^*(x',0)) + d(h(x'), h(x)) \\ &\le \operatorname{diam} \psi^*(\{x'\} \times [0, 2^{-k}]) + \operatorname{diam} U \\ &\le \sum_{i=k+1}^{\infty} \operatorname{diam} \tilde{\psi}_i(\{x'\} \times \mathbf{I}) + \varepsilon/2 \\ &< \sum_{i=k+1}^{\infty} \operatorname{mesh} \mathcal{U}_i + \varepsilon/2 < 2^{-k} + \varepsilon/2 < \varepsilon. \end{aligned}$$

Thus, ψ^* is continuous. Observe that ψ^* is a $\mathcal{U}$-homotopy and $\psi^*_t|Q = fg$ for every $t \in \mathbf{I}$. Then, $fg_1 = \psi^*_1 \simeq_{\mathcal{U}} h$ rel. Q. Therefore, f is polyhedrally homotopically n-soft. □

All the results can be summarized as follows:

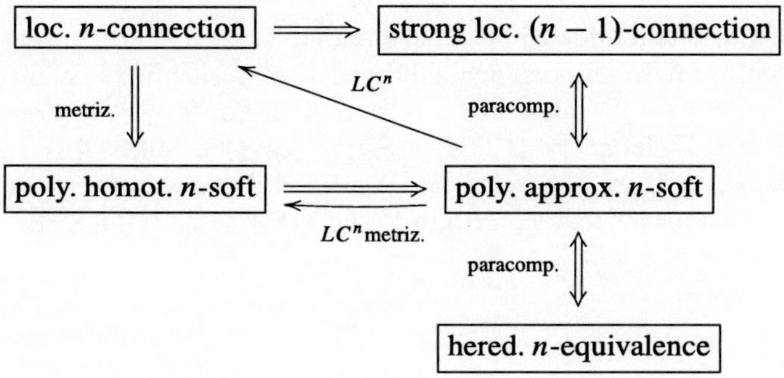

Corollary 7.3.7. *When Y be an LC^n metrizable space, the following five conditions for a map $f : X \to Y$ are equivalent:*

(a) *f is a local n-connection;*
(b) *f is a strong local $(n-1)$-connection ($f(X)$ is dense in Y when $n = 0$);*
(c) *f is polyhedrally approximately n-soft;*
(d) *f is polyhedrally homotopically n-soft;*
(e) *f is a hereditary n-equivalence.*

If Y is an LC^{n-1} paracompact space, conditions (b) and (c) are equivalent. If Y is an LC^n paracompact space, three conditions (a), (b), and (c) are equivalent, and conditions (d) and (e) are equivalent. □

Combining Proposition 7.2.8 with the above, we have

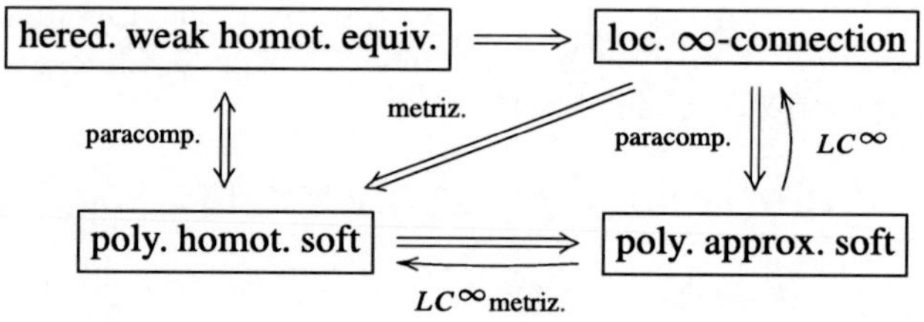

Corollary 7.3.8. *When Y is an LC^∞ metrizable space, the following four conditions for a map $f : X \to Y$ are equivalent:*

(a) *f is a local ∞-connection;*
(b) *f is polyhedrally approximately soft;*
(c) *f is polyhedrally homotopically soft;*
(d) *f is a hereditary weak homotopy equivalence.*

If Y is an LC^∞ paracompact space, conditions (a) *and* (b) *are equivalent and conditions* (c) *and* (d) *are equivalent.* □

Theorem 7.3.9. *Let $f : X \to Y$ be a local n-connection and Y be metrizable. Then, Y is LC^n.*

Proof. Let $y \in Y$ and U be an open neighborhood of y in Y. Because f is a local n-connection, we have an open neighborhood V of y such that $fg \simeq 0$ in U for each map $g : \mathbf{S}^i \to f^{-1}(V)$, where $0 \leq i \leq n$. For $0 \leq i \leq n$ and for each map $h : \mathbf{S}^i \to V$, let $\mathcal{U} = \{V, Y \setminus h(\mathbf{S}^i)\} \in \operatorname{cov}(Y)$. Since f is polyhedrally homotopically n-soft by Theorem 7.3.6, we have a map $g : \mathbf{S}^i \to X$ such that $fg \simeq_{\mathcal{U}} h$, which means that $fg \simeq h$ in V and $g(\mathbf{S}^i) \subset f^{-1}(V)$. Since $fg \simeq 0$ in U, it follows that $h \simeq 0$ in U. □

7.4 Fine Homotopy Equivalences Between ANRs

Recall that a map $f : X \to Y$ is a fine homotopy equivalence if f has a $\mathcal{U}$-homotopy inverse $g : Y \to X$ for each open cover $\mathcal{U}$ of Y, that is, $gf \simeq_{f^{-1}(\mathcal{U})} \mathrm{id}_X$ and $fg \simeq_{\mathcal{U}} \mathrm{id}_Y$.

Proposition 7.4.1. *Let $f : X \to Y$ be a fine homotopy equivalence and Y be regular.*

(1) *If Y is locally contractible then f is a local $*$-connection.*
(2) *If Y is LC^n then f is a strong local n-connection.*
(3) *If Y is LC^∞ then f is a local ∞-connection.*

Proof. Assertion (3) is a direct consequence of (2). The following is a proof of (1) (resp. (2)).

For each $y \in Y$ and each open neighborhood U of y in Y, choose open neighborhoods $V \subset W$ of y so that $\mathrm{cl}\, W \subset U$ and V is contractible in W (resp. $\alpha \simeq 0$ in W for every map $\alpha : \mathbf{S}^i \to V, 0 \le i \le n$) and let $\mathcal{U} = \{U, Y \setminus \mathrm{cl}\, W\} \in \mathrm{cov}(Y)$. Then, f has a $\mathcal{U}$-homotopy inverse $g : Y \to X$. Observe that $gf|f^{-1}(V) \simeq \mathrm{id}$ in $f^{-1}(U)$. On the other hand, $gf|f^{-1}(V) \simeq 0$ in $g(W)$ ($gf\alpha \simeq 0$ in $g(W)$ for every map $\alpha : \mathbf{S}^i \to f^{-1}(V), 0 \le i \le n$). Since $fg(W) \subset \mathrm{st}(W, \mathcal{U}) = U$, it follows that $g(W) \subset f^{-1}(U)$. Therefore, $f^{-1}(V)$ is contractible in $f^{-1}(U)$ ($\alpha \simeq 0$ in $f^{-1}(U)$ for every map $\alpha : \mathbf{S}^i \to f^{-1}(V), 0 \le i \le n$). □

Proposition 7.4.2. *Let $f : X \to Y$ be a fine homotopy equivalence and Y be paracompact.*

(1) *If X is an ANR, then f is approximately soft.*
(2) *If both X and Y are ANRs, then f is homotopically soft.*

Proof. Because (2) is a combination of (1) and Proposition 7.2.9(3), it suffices to show (1).

Let Z be a metrizable space and $g : C \to X$ be a map of a closed set C in Z such that fg extends to a map $h : Z \to Y$. For each $\mathcal{U} \in \mathrm{cov}(Y)$, let $\mathcal{V}$ be an open star-refinement of $\mathcal{U}$. We have a map $k : Y \to X$ such that $fk \simeq_{\mathcal{V}} \mathrm{id}_Y$ and $kf \simeq_{f^{-1}(\mathcal{V})} \mathrm{id}_X$. Then, $g \simeq_{f^{-1}(\mathcal{V})} kfg = kh|C$.

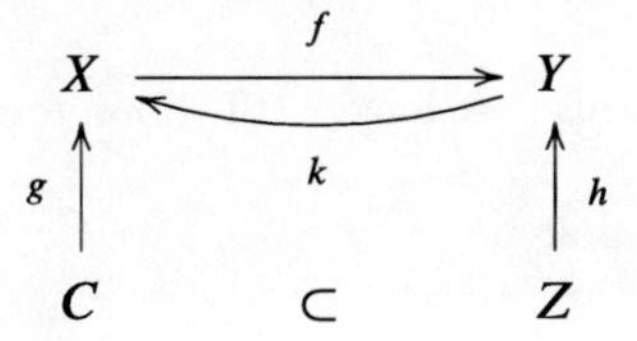

By the Homotopy Extension Theorem 6.4.1, g can be extended to a map $\tilde{g} : Z \to X$ that is $f^{-1}(\mathcal{V})$-homotopic to kh. Then, $f\tilde{g} \simeq_{\mathcal{V}} fkh \simeq_{\mathcal{V}} h$, so $f\tilde{g} \simeq_{\mathcal{U}} h$.[4] □

A fine homotopy equivalence between ANRs is characterized as follows:

Theorem 7.4.3. *For a map $f : X \to Y$ between ANRs, the following are equivalent:*

(a) *f is a fine homotopy equivalence;*
(b) *f is approximately (= homotopically) soft;*
(c) *f is polyhedrally approximately (= homotopically) soft;*
(d) *f is a hereditary weak homotopy equivalence;*
(e) *f is a local ∞-connection;*
(f) *f is a local $*$-connection.*

Proof. In the following diagram of implications, the equivalence among (c), (d), and (e) has been shown in Corollary 7.3.8. The implications (b) ⇒ (c) and (f) ⇒ (e) are trivial. The implications (a) ⇒ (b) and (a) ⇒ (f) have been shown in Propositions 7.4.2 and 7.4.1, respectively. Thus, it remains to show the implication (c) ⇒ (a).

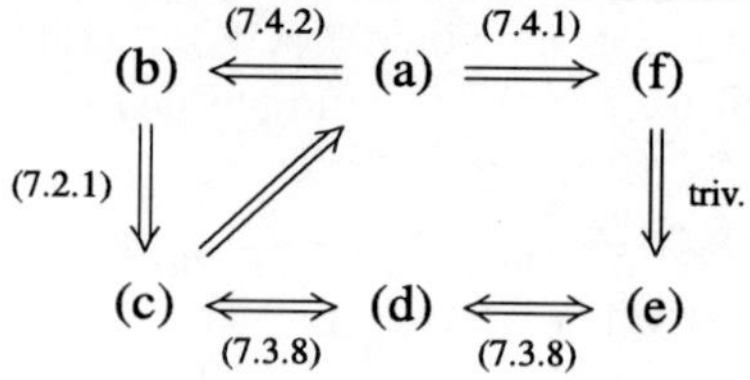

(c) ⇒ (a): For each $\mathcal{U} \in \operatorname{cov}(Y)$, let $\mathcal{V} \in \operatorname{cov}(Y)$ such that $\operatorname{st}^3 \mathcal{V} \prec \mathcal{U}$. Since Y is an ANR, we have a polyhedron P_Y with maps $\varphi_Y : Y \to P_Y$ and $\psi_Y : P_Y \to Y$ such that $\psi_Y \varphi_Y \simeq_{\mathcal{V}} \operatorname{id}_Y$, and apply (c) to obtain a map $g : P_Y \to X$ such that $fg \simeq_{\mathcal{V}} \psi_Y$.

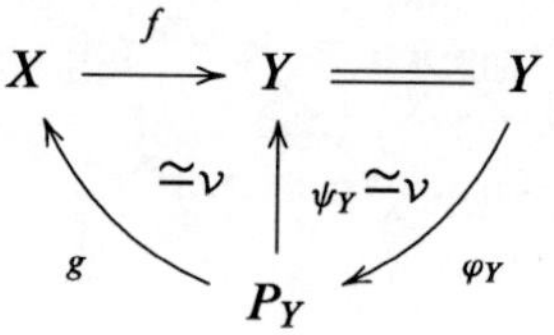

Since $fg\varphi_Y \simeq_{\mathcal{V}} \psi_Y \varphi_Y \simeq_{\mathcal{V}} \operatorname{id}_Y$, we have a st $\mathcal{V}$-homotopy $h : Y \times \mathbf{I} \to Y$ such that $h_0 = \operatorname{id}_Y$ and $h_1 = fg\varphi_Y$.

Choose $\mathcal{W} \in \operatorname{cov}(X)$ so that

$$\mathcal{W} \prec f^{-1}(\mathcal{V}) \text{ and } \mathcal{W} \prec (g\varphi_Y f)^{-1}(f^{-1}(\mathcal{V})).$$

---

[4]It is not shown that $f\tilde{g} \simeq_{\mathcal{U}} h$ rel. C.

Because X is an ANR, we have a polyhedron P_X with maps $\varphi_X : X \to P_X$ and $\psi_X : P_X \to X$ such that $\psi_X\varphi_X \simeq_{\mathcal{W}} \mathrm{id}_X$. We define a st $\mathcal{V}$-homotopy $\zeta : P_X \times \mathbf{I} \to Y$ and a map $\xi : P_X \times \{0,1\} \to X$ as follows:

$$\zeta(x,t) = h(f\psi_X(x),t);\ \xi(x,0) = \psi_X(x),\ \xi(x,1) = g\varphi_Y f\psi_X(x).$$

Then, $\zeta|P_X\times\{0,1\} = f\xi$. Indeed, $\zeta(x,0) = h_0 f\psi_X(x) = f\psi_X(x)$ and $\zeta(x,1) = h_1\psi_X(x) = fg\varphi_Y f\psi_X(x)$.

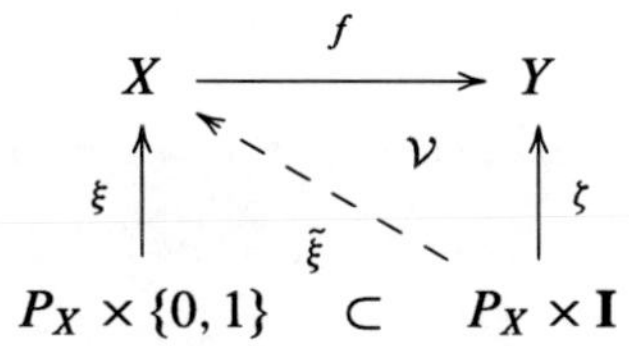

By (c), we can obtain an $f^{-1}(\mathrm{st}^2\,\mathcal{V})$-homotopy $\tilde{\xi} : P_X \times \mathbf{I} \to X$ that extends ξ. Let $\tilde{h} : X \times \mathbf{I} \to X$ be the $f^{-1}(\mathrm{st}^2\,\mathcal{V})$-homotopy defined by $\tilde{h}(x,t) = \tilde{\xi}(\varphi_X(x),t)$. Then,

$$\tilde{h}_0 = \tilde{\xi}_0\varphi_X = \psi_X\varphi_X \simeq_{\mathcal{W}} \mathrm{id}_X \text{ and}$$
$$\tilde{h}_1 = \tilde{\xi}_1\varphi_X = g\varphi_Y f\psi_X\varphi_X \simeq_{g\varphi_Y f(\mathcal{W})} g\varphi_Y f.$$

Since $\mathcal{W},\ g\varphi_Y f(\mathcal{W}) \prec f^{-1}(\mathcal{V})$ and $\mathrm{st}^3\,\mathcal{V} \prec \mathcal{U}$, it follows that $g\varphi_Y f \simeq_{f^{-1}(\mathcal{U})} \mathrm{id}_X$. □

Since every open set in an ANR is an ANR, Theorem 7.4.3 yields the following corollary:

Corollary 7.4.4. *Let $f : X \to Y$ be a fine homotopy equivalence between ANRs. Then, for every open set U in Y, $f|f^{-1}(U) : f^{-1}(U) \to U$ is also a fine homotopy equivalence.* □

Corollary 7.4.5. *Let $f : X \to Y$ be a fine homotopy equivalence between ANRs. If $A \subset Y$ is contractible in an open neighborhood U in Y, then $f^{-1}(A)$ is contractible in $f^{-1}(U)$.*

Proof. By Corollary 7.4.4, $f|f^{-1}(U) : f^{-1}(U) \to U$ has a homotopy inverse $g : U \to f^{-1}(U)$. Let $h : f^{-1}(U) \times \mathbf{I} \to f^{-1}(U)$ be a homotopy with $h_0 = \mathrm{id}$ and $h_1 = gf|f^{-1}(U)$. On the other hand, we have a contraction $k : A\times\mathbf{I} \to U$, that is, $k_0 = \mathrm{id}$ and k_1 is constant. Then, we can define a contraction $\varphi : f^{-1}(A) \times \mathbf{I} \to f^{-1}(U)$ as follows:

$$\varphi(x,t) = \begin{cases} h(x,2t) & \text{if } t \le 1/2, \\ gk(f(x),2t-1) & \text{if } t \ge 1/2. \end{cases}$$ □

Due to Proposition 6.7.1, a subset X of a metrizable space Y is homotopy dense in Y if and only if the inclusion $X \subset Y$ is a fine homotopy equivalence. Applying Theorem 7.4.3 to the inclusion $X \subset Y$, we have the following:

Corollary 7.4.6. *Let X and Y be ANRs such that X is a dense subset of Y. Then, the following are equivalent:*

(a) *X is homotopy dense in Y;*
(b) *For each open set U in Y, the inclusion $U \cap X \subset U$ is a weak homotopy equivalence, i.e., for each $n \in \mathbb{N}$, if a map $\alpha : \mathbf{S}^{n-1} \to U \cap X$ extends to a map $\beta : \mathbf{B}^n \to U$ then α extends to a map $\tilde{\alpha} : \mathbf{B}^n \to U \cap X$ such that $\tilde{\alpha} \simeq \beta$ in U rel. $\mathbf{S}^{n-1}$;*
(c) *Every neighborhood U of each y in Y contains a neighborhood V of y in Y such that every map $\alpha : \mathbf{S}^{n-1} \to V \cap X$ is null-homotopic in $U \cap X$ for each $n \in \mathbb{N}$.* □

Now, we prove the following theorem:

Theorem 7.4.7. *Let X and Y be ANRs and $f = \lim_{i\to\infty} f_i : X \to Y$ be the uniform limit of fine homotopy equivalences with respect to some $d \in \mathrm{Metr}(Y)$. Then, f is also a fine homotopy equivalence.*

Proof. According to Theorem 7.4.3, it suffices to show that f is a local ∞-connection. Since each $f_i(X)$ is dense in Y, it follows that $f(X)$ is dense. Every open neighborhood U of each y in Y contains an open neighborhood V of y such that V is contractible in U. For each $n \in \mathbb{N}$, let $\alpha : \mathbf{S}^n \to f^{-1}(V)$ be a map. Then, $f\alpha$ extends to a map $\beta : \mathbf{B}^{n+1} \to U$. Let

$$\delta = \mathrm{dist}(\beta(\mathbf{B}^{n+1}), Y \setminus U) > 0.$$

Since Y is an ANR, the open cover $\{\mathrm{B}(y, \delta/6) \mid y \in Y\}$ of Y has an h-refinement $\mathcal{V} \in \mathrm{cov}(Y)$ (Corollary 6.3.5). Because $f\alpha(\mathbf{S}^n)$ is compact, we have $\varepsilon > 0$ such that $\{\mathrm{B}(f\alpha(x), \varepsilon) \mid x \in \mathbf{S}^n\} \prec \mathcal{V}$. Indeed, find $x_i \in \mathbf{S}^n$ and $\varepsilon_i > 0$, $i = 1, \ldots, k$, such that

$$\{\mathrm{B}(f\alpha(x_i), 2\varepsilon_i) \mid i = 1, \ldots, k\} \prec \mathcal{V} \text{ and } f\alpha(\mathbf{S}^n) \subset \bigcup_{i=1}^{k} \mathrm{B}(f\alpha(x_i), \varepsilon_i).$$

Then, $\min\{\varepsilon_i \mid i = 1, \ldots, k\} > 0$ is the desired $\varepsilon > 0$. Choose $i \in \mathbb{N}$ so that f_i is ε-close to f, which implies that $f_i\alpha \simeq_{\delta/3} f\alpha = \beta|\mathbf{S}^n$. By the Homotopy Extension Theorem 6.4.1, $f_i\alpha$ extends to a map $\beta' : \mathbf{B}^{n+1} \to Y$ such that $\beta' \simeq_{\delta/3} \beta$. Since f_i is polyhedrally approximately soft by Theorem 7.4.3, α extends to a map $\tilde{\alpha} : \mathbf{B}^{n+1} \to X$ such that $d(f_i\tilde{\alpha}, \beta') < \delta/3$. Observe that

$$d(f\tilde{\alpha}, \beta) \leq d(f\tilde{\alpha}, f_i\tilde{\alpha}) + d(f_i\tilde{\alpha}, \beta') + d(\beta', \beta) < \delta.$$

Then, $f\tilde{\alpha}(\mathbf{B}^{n+1}) \subset U$, i.e., $\tilde{\alpha}(\mathbf{B}^{n+1}) \subset f^{-1}(U)$. □

7.5 Hereditary Shape Equivalences and UV^n Maps

Recall that $[X, Y]$ is the set of the homotopy classes of maps from X to Y. For any space P, a map $h : X \to Y$ induces a function $h^* : [Y, P] \to [X, P]$ defined by $h^*([f]) = [fh]$ for each map $f : Y \to P$.

$$\begin{array}{ccc} X & \xrightarrow{\quad h \quad} & Y \\ & {\scriptstyle g}\searrow \quad \simeq \quad \swarrow{\scriptstyle f} & \\ & P & \end{array}$$

Fact. *For every space X, Y, Z, and P, the following statements hold:*

(1) *The identity* $\mathrm{id}_X : X \to X$ *induces the identity* $\mathrm{id}_{[X,P]} : [X, P] \to [X, P]$, *i.e.,* $(\mathrm{id}_X)^* = \mathrm{id}_{[X,P]}$.
(2) *The composition $h_2h_1 : X \to Z$ of two maps $h_1 : X \to Y$ and $h_2 : Y \to Z$ induces the composition $h_1^*h_2^* : [Z, P] \to [X, P]$ of $h_2^* : [Z, P] \to [Y, P]$ and $h_1^* : [Y, P] \to [X, P]$, i.e., $(h_2h_1)^* = h_1^*h_2^*$.*

A map $h : X \to Y$ is said to be a **shape equivalence** if the function $h^* : [Y, P] \to [X, P]$ is a bijection for every ANR P, equivalently for every polyhedron P (Corollary 6.6.5). Observe the following:

(i) Every homotopy equivalence is a shape equivalence.
(ii) The composition of shape equivalences is also a shape equivalence.

Remark 5. We do not give the definition that X and Y have the same shape type, but it is not defined by the existence of a shape equivalence $h : X \to Y$. It should be noted that if there is a shape equivalence $h : X \to Y$ then X and Y have the same shape, but the converse does not hold.

The following is easy to prove:

Proposition 7.5.1. *The following are equivalent for a space $X \neq \emptyset$:*

(a) *X has trivial shape;*
(b) *The map of X to the singleton $\{0\}$ is a shape equivalence;*
(c) *For every $x \in X$, the inclusion $\{x\} \hookrightarrow X$ is a shape equivalence:*
(d) *There is some $x \in X$ such that the inclusion $\{x\} \hookrightarrow X$ is a shape equivalence.*

□

Corollary 7.5.2. *Let $h : X \to Y$ be a shape equivalence. Then, X has trivial shape if and only if Y has trivial shape. When X and Y are compacta, X is cell-like if and only if Y is cell-like.* □

A map $h : X \to Y$ is said to be a **hereditary shape equivalence** if $h|h^{-1}(A) : h^{-1}(A) \to A$ is a shape equivalence for any closed set A in Y. In this case, $h(X) = Y$ and each fiber $h^{-1}(y)$ has trivial shape. Then, every hereditary shape equivalence is cell-like if it is perfect.

Theorem 7.5.3. *Every closed fine homotopy equivalence $h : X \to Y$ between metrizable spaces is a hereditary shape equivalence.*

Proof. Let A be a closed set in Y and P be an ANR. Each map $f : h^{-1}(A) \to P$ extends to a map $\tilde{f} : U \to P$ from an open neighborhood U of $h^{-1}(A)$ in X. Because h is a closed map, there is an open neighborhood V of A in Y such that $h^{-1}(V) \subset U$. Let $\mathcal{V} = \{V, Y \setminus A\} \in \operatorname{cov}(Y)$. Since h is a fine homotopy equivalence, it has a $\mathcal{V}$-homotopy inverse $g : Y \to X$, i.e., $hg \simeq_{\mathcal{V}} \operatorname{id}$ and $gh \simeq_{h^{-1}(\mathcal{V})} \operatorname{id}$. Observe that $gh|h^{-1}(A) \simeq \operatorname{id}_{h^{-1}(A)}$ in $h^{-1}(V) \subset U$. Then, we have a map $\tilde{f}g|A : A \to P$ such that $\tilde{f}gh|h^{-1}(A) \simeq \tilde{f}|h^{-1}(A) = f$.

Now, let $f_0, f_1 : A \to P$ be maps such that $f_0h|h^{-1}(A) \simeq f_1h|h^{-1}(A)$. Since P is an ANR, A has an open neighborhood W in Y with maps $\tilde{f}_0, \tilde{f}_1 : W \to P$ that are extensions of f_0 and f_1, respectively. Then, $\tilde{f}_0h|h^{-1}(A) \simeq \tilde{f}_1h|h^{-1}(A)$. We can choose an open neighborhood U of $h^{-1}(A)$ in X so that $U \subset h^{-1}(W)$ and $\tilde{f}_0h|U \simeq \tilde{f}_1h|U$. As before, there is an open neighborhood V of A in Y such that $V \subset W$, $h^{-1}(V) \subset U$, and h has a $\mathcal{V}$-homotopy inverse $g : Y \to X$, where $\mathcal{V} = \{V, Y \setminus A\} \in \operatorname{cov}(Y)$. Since $hg|A \simeq \operatorname{id}_A$ in V and $g(A) \subset h^{-1}(V) \subset U$, it follows that

$$\begin{aligned} f_0 = \tilde{f}_0|A \simeq \tilde{f}_0hg|A &= (\tilde{f}_0h|U)(g|A) \\ &\simeq (\tilde{f}_1h|U)(g|A) = \tilde{f}_1hg|A \simeq \tilde{f}_1|A = f_1. \end{aligned}$$ □

A surjective map $f : X \to Y$ is called a ***UV** map** if each fiber $f^{-1}(y)$ is UV^*; equivalently, each fiber $f^{-1}(y)$ has trivial shape by Theorem 7.1.2.

- A perfect UV^* map is the same as a cell-like map.

For $n \in \omega \cup \{\infty,\}$, $f : X \to Y$ is a ***UV^n* map** if each fiber $f^{-1}(y)$ is UV^n. For a **closed** surjective map $f : X \to Y$, if X is an ANR, then the following equivalences hold:

$$\begin{aligned} f \text{ is } UV^n &\Leftrightarrow f \text{ is a strong local } n\text{-connection;} \\ f \text{ is } UV^\infty &\Leftrightarrow f \text{ is a local } \infty\text{-connection;} \\ f \text{ is } UV^* &\Leftrightarrow f \text{ is a local } *\text{-connection.} \end{aligned}$$

For closed surjective maps, we have the following version of Theorem 7.4.3:

Theorem 7.5.4. *For a closed surjective map $f : X \to Y$ between ANRs X and Y, the following are equivalent:*

(a) *f is a fine homotopy equivalence;*
(b) *f is a UV^∞ map (= a local ∞-connection);*
(c) *f is a UV^* map (= a local $*$-connection);*
(d) *f is a hereditary shape equivalence.*

If f is a perfect map, the following is also equivalent to the above:

(e) *f is a cell-like map.*

Proof. The equivalence among (a) through (c) has been obtained by Theorem 7.4.3. The implication (a) $\Rightarrow$ (d) has been shown in Theorem 7.5.3. Condition (d) implies that each fiber $f^{-1}(y)$ has trivial shape. Because X is an ANR, f is a closed UV^* map. Thus, we have (d) $\Rightarrow$ (c). □

Remark 6. Let X be the $\sin(1/x)$-curve given at the beginning of Sect. 7.1 and $f = \mathrm{pr}_1|X : X \to Y = \mathbf{I}$, where $\mathrm{pr}_1 : \mathbb{R}^2 \to \mathbb{R}$ is the projection onto the first factor. Then, f is cell-like (UV^*) but not a (fine) homotopy equivalence nor polyhedrally approximately 1-soft. Hence, in Theorem 7.5.4, it is essential that X is an ANR. Moreover, in general, a cell-like map is not a shape equivalence, hence it is not a hereditary shape equivalence. Such examples will be given in Theorems 7.7.5 and 7.7.8.

Corollary 7.5.5. *Let X, Y, and Z be ANRs. For each two cell-like maps $f : X \to Y$ and $g : Y \to Z$, the composition $gf : X \to Y$ is also a cell-like map.*

Proof. It is easy to show that the composition of hereditary shape equivalences is also a hereditary shape equivalence. Then, the result follows from Theorem 7.5.4. □

Remark 7. In general, the composition of cell-like maps is not cell-like. Such an example will be seen in Remark 11 in Sect. 7.7.

The following proposition will be used in Sect. 7.7.

Proposition 7.5.6. *For each map $f : A \to Y$ of a closed set A in a metrizable space X, let*

$$h : M_f \cup (X \times \{1\}) \to Y \cup_f X$$

be the map defined by $h|M_f = c_f$ and $h(x, 1) = x$ for each $x \in X \setminus A$, where c_f is the collapsing of the mapping cylinder. Then, h is a shape equivalence.

Proof. First, we extend f to the map $\tilde{f} : X \to Y \cup_f X$ by $\tilde{f}|X \setminus A = \mathrm{id}$, that is, $\tilde{f}$ is the restriction of the natural quotient map from $Y \oplus X$ onto $Y \cup_f X$. Then, $M_f \cup (X \times \{1\}) \subset M_{\tilde{f}}$ and $h = c_{\tilde{f}}|M_f \cup (X \times \{1\})$, where the collapsing $c_{\tilde{f}}$ is a homotopy equivalence, and is therefore a shape equivalence.

$$\begin{array}{ccc} M_f \cup (X \times \{1\}) & \subset & M_{\tilde{f}} \\ & {\scriptstyle h}\searrow \quad \swarrow{\scriptstyle c_{\tilde{f}}} & \\ & Y \cup_f X & \end{array}$$

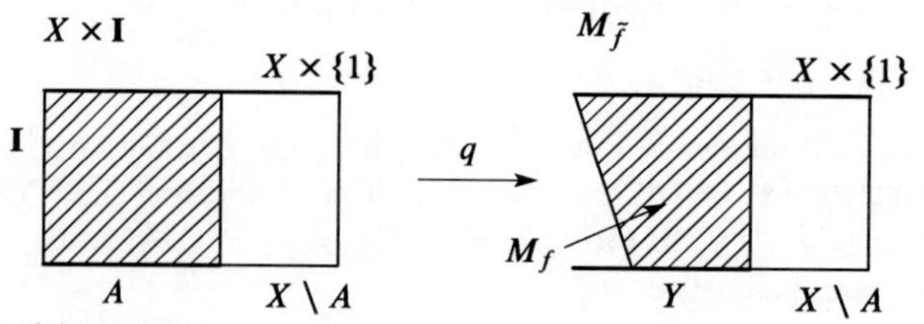

Fig. 7.1 $M_f \cup (X \times \{1\}) \subset M_{\tilde{f}}$

Since the composition of shape equivalences is a shape equivalence, it suffices to show that the inclusion $i : M_f \cup (X \times \{1\}) \hookrightarrow M_{\tilde{f}}$ is a shape equivalence, that is,

$$i^* : [M_{\tilde{f}}, P] \to [M_f \cup (X \times \{1\}),\ P]$$

is bijective for each ANR P. Identifying $X \setminus A = (X \setminus A) \times \{0\}$, we can regard the mapping cylinder $M_{\tilde{f}}$ as follows:

$$M_{\tilde{f}} = Y \cup_{f \circ \mathrm{pr}_A | A \times \{0\}} (X \times \mathbf{I}).$$

Let $q : Y \oplus (X \times \mathbf{I}) \to M_{\tilde{f}}$ be the natural quotient map. See Fig. 7.1.

To show that i^* is surjective, let $p : M_f \cup (X \times \{1\}) \to P$ be a map. Regarding $\varphi = pq|A \times \mathbf{I}$ as a homotopy such that φ_1 extends over X, we apply the Homotopy Extension Theorem 6.4.1 to obtain a map $\tilde{\varphi} : X \times \mathbf{I} \to P$ such that

$$\tilde{\varphi}|(A \times \mathbf{I}) \cup (X \times \{1\}) = pq|(A \times \mathbf{I}) \cup (X \times \{1\}).$$

Then, it follows that

$$\tilde{\varphi}|A \times \{0\} = pq|A \times \{0\} = (p|Y) f \mathrm{pr}_A|A \times \{0\}.$$

Thus, we have the map $\tilde{p} : M_{\tilde{f}} \to P$ such that $\tilde{p}|Y = p|Y$ and $\tilde{p}q|X \times \mathbf{I} = \tilde{\varphi}$, hence $\tilde{p}i = \tilde{p}|M_f \cup X \times \{1\} = p$.

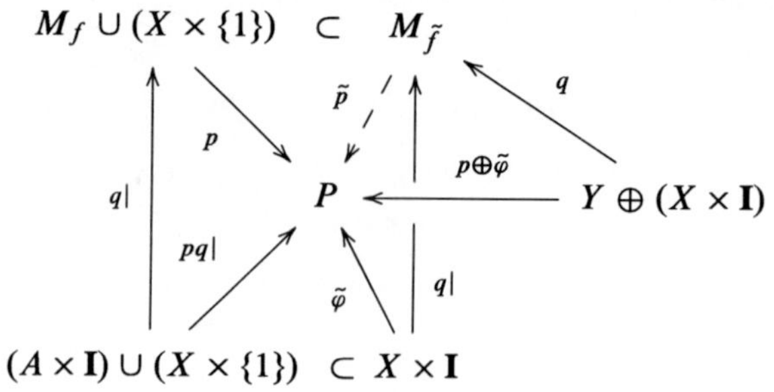

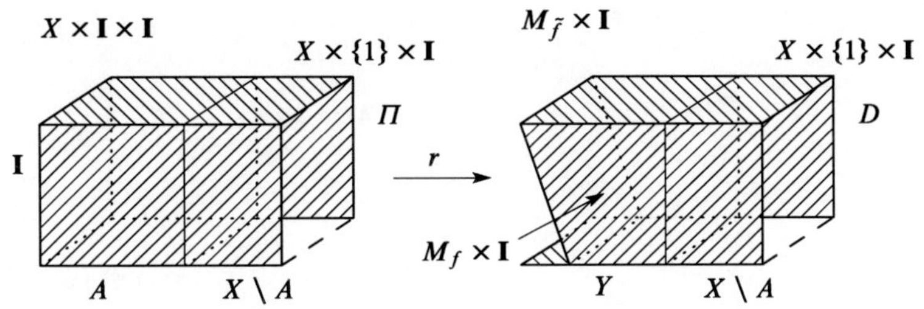

Fig. 7.2 $\big((M_f \cup (X \times \{1\})) \times \mathbf{I}\big) \cup \big(M_{\check{f}} \times \{0,1\}\big) \subset M_{\check{f}} \times \mathbf{I}$

To show that i^* is injective, let $p, p' : M_{\check{f}} \to P$ such that $pi \simeq p'i$. Pasting p, p' and a homotopy from pi to $p'i$, we define a map $\psi : D \to P$ on the set

$$D = \big((M_f \cup (X \times \{1\})) \times \mathbf{I}\big) \cup \big(M_{\check{f}} \times \{0,1\}\big),$$

which is regarded as the adjunction space $(Y \times \mathbf{I}) \cup_{\check{f}} \Pi$, where

$$\Pi = (A \times \mathbf{I} \times \mathbf{I}) \cup (X \times \{1\} \times \mathbf{I}) \cup (X \times \mathbf{I} \times \{0,1\}) \subset X \times \mathbf{I} \times \mathbf{I} \text{ and}$$
$$\check{f} = (f \circ \mathrm{pr}_A | A \times \{0\}) \times \mathrm{id}_{\mathbf{I}} : A \times \{0\} \times \mathbf{I} \to Y \times \mathbf{I}.$$

We make the following identification:

$$M_{\check{f}} \times \mathbf{I} = (Y \times \mathbf{I}) \cup_{\check{f}} (X \times \mathbf{I} \times \mathbf{I}),$$

where the natural quotient map is denoted by

$$r = q \times \mathrm{id}_{\mathbf{I}} : (Y \times \mathbf{I}) \oplus (X \times \mathbf{I} \times \mathbf{I}) = \big(Y \oplus (X \times \mathbf{I})\big) \times \mathbf{I} \to M_{\check{f}} \times \mathbf{I}.$$

Then, $r((Y \times \mathbf{I}) \oplus \Pi)$ is the domain of ψ. — Fig. 7.2.

Now, we regard the map $\psi r | (A \times \mathbf{I} \times \mathbf{I}) \cup (X \times \mathbf{I} \times \{0,1\})$ as a homotopy that realizes

$$\begin{aligned} &\psi r | (A \times \{0\} \times \mathbf{I}) \cup (X \times \{0\} \times \{0,1\}) \\ &\quad \simeq \psi r | (A \times \{1\} \times \mathbf{I}) \cup (X \times \{1\} \times \{0,1\}). \end{aligned}$$

Since $\psi r | (A \times \{1\} \times \mathbf{I}) \cup (X \times \{1\} \times \{0,1\})$ extends over $X \times \{1\} \times \mathbf{I}$, we apply the Homotopy Extension Theorem 6.4.1 to obtain a map $\varphi : X \times \mathbf{I} \times \mathbf{I} \to P$ such that $\varphi | \Pi = \psi r | \Pi$. Then, it follows that

$$\varphi | A \times \{0\} \times \mathbf{I} = \psi r | A \times \{0\} \times \mathbf{I} = (\psi | Y) \check{f}.$$

Therefore, we have the homotopy $\tilde{\psi} : M_{\tilde{f}} \times \mathbf{I} \to P$ such that $\tilde{\psi} r = \varphi$.

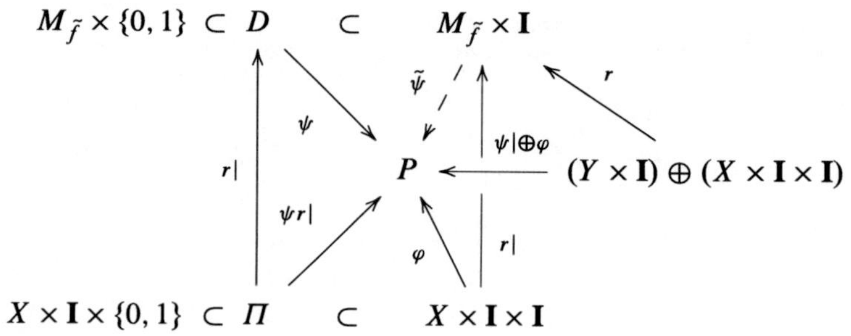

Since $\tilde{\psi}|M_{\tilde{f}} \times \{0,1\} = \psi|M_{\tilde{f}} \times \{0,1\}$, $\tilde{\psi}$ is a homotopy realizing $p \simeq p'$. □

7.6 The Near-Selection Theorem

Recall that X has Property C if, for any open covers $\mathcal{U}_n$ of X, $n \in \mathbb{N}$, X has an open cover $\mathcal{V} = \bigcup_{n\in\mathbb{N}} \mathcal{V}_n$ such that each $\mathcal{V}_n$ is pairwise disjoint and $\mathcal{V}_n \prec \mathcal{U}_n$ (see Sect. 6.10). Every countable-dimensional metrizable space has Property C (Proposition 6.10.4).

Theorem 7.6.1 (Haver's Near-Selection Theorem). *Let X be a σ-compact metrizable space with Property C, $Y = (Y,d)$ be a metric ANR, and $\varphi : X \to \mathrm{Comp}(Y)$ be a continuous set-valued function such that each $\varphi(x)$ is cell-like. For each map $\alpha : X \to (0,\infty)$, there exists a map $h : X \to Y$ such that $d(h(x),\varphi(x)) < \alpha(x)$ for every $x \in X$.*

Proof. We may assume that X is a metric space. Let $X = \bigcup_{n\in\mathbb{N}} X_n$, where $X_1 \subset X_2 \subset \cdots$ are compact. We can inductively choose $\delta_n > 0$ so that $\delta_n \le 2^{-n-3}$ and $\mathrm{N}(\varphi(x), 2\delta_n)$ is contractible in $\mathrm{N}(\varphi(x),\delta_{n-1})$ for each $x \in X_n$, where $\delta_0 = 2^{-3}$. Indeed, assume that $\delta_{n-1} > 0$ is chosen. Because $\varphi(x)$ is compact and UV^*, we have $\delta_x > 0$ such that $\mathrm{N}(\varphi(x),\delta_x)$ is contractible in $\mathrm{N}(\varphi(x),\delta_{n-1}/2)$, where $\delta_x \le \delta_{n-1}/2$. Since φ is continuous, each $x \in X_n$ has an open neighborhood V_x such that $d_H(\varphi(x),\varphi(x')) < \delta_x/2$ for any $x' \in V_x$. Observe that $x' \in V_x$ implies

$$\mathrm{N}(\varphi(x'),\delta_x/2) \subset \mathrm{N}(\varphi(x),\delta_x) \text{ and } \mathrm{N}(\varphi(x),\delta_{n-1}/2) \subset \mathrm{N}(\varphi(x'),\delta_{n-1}).$$

By the compactness of X_n, we have $x_1,\ldots,x_k \in X_n$ such that $X_n \subset \bigcup_{i=1}^k V_{x_i}$. Then, $\frac{1}{2}\min\{\delta_{x_i}/2 \mid i = 1,\ldots,k\} > 0$ is the desired $\delta_n > 0$.

By the continuity of φ and the compactness of X_n, we can take $\varepsilon_1 > \varepsilon_2 > \cdots > 0$ such that if $x \in X_n$ and $x' \in X$ with $d(x,x') < 2\varepsilon_n$ then $d_H(\varphi(x),\varphi(x')) < \delta_n$, that is, $\varphi(x) \subset \mathrm{N}(\varphi(x'),\delta_n)$ and $\varphi(x') \subset \mathrm{N}(\varphi(x),\delta_n)$.

We will construct collections $\mathcal{V}_1, \mathcal{V}_2, \ldots$ of open sets in X satisfying the following conditions:

(1) $\mathcal{V} = \bigcup_{i \in \mathbb{N}} \mathcal{V}_i \in \operatorname{cov}(X)$;
(2) Each $\mathcal{V}_i$ is pairwise disjoint;
(3) $\operatorname{mesh} \mathcal{V}_i < \varepsilon_i$;
(4) Each open set in $\mathcal{V}_i$ meets X_i;
(5) $\alpha(x) > 2^{-i-1}$ for $x \in \bigcup \mathcal{V}_i$;
(6) $\mathcal{V}_i \cap \mathcal{V}_j = \emptyset$ if $i \neq j$.

Let $k : \mathbb{N}^2 \to \mathbb{N}$ be a bijection such that $k(i,n) \geq i$ for every $(i,n) \in \mathbb{N}^2$ (e.g., $k(i,n) = 2^{n-1}(2i-1) \geq i$). Without loss of generality, we can assume that $\alpha(x) \leq 1$ for all $x \in X$. For each $i, j \in \mathbb{N}$, let

$$X_{i,j} = X_i \cap \alpha^{-1}([2^{-j}, 2^{-j+1}]).$$

Since each $X_{i,j}$ has Property C by Lemma 6.10.3, we have collections $\mathcal{V}_{j,k(i,n)}$, $n \in \mathbb{N}$, of open sets in

$$\alpha^{-1}\big((2^{-j} - 2^{-j-2}, 2^{-j+1} + 2^{-j-2})\big)$$

such that $\bigcup_{n \in \mathbb{N}} \mathcal{V}_{j,k(i,n)}$ covers $X_{i,j}$, each $\mathcal{V}_{j,k(i,n)}$ is pairwise disjoint, and

$$\operatorname{mesh} \mathcal{V}_{j,k(i,n)} < \varepsilon_{j+2k(i,n)-2}.$$

It can be assumed that the open sets of $\mathcal{V}_{j,k(i,n)}$ meet $X_{i,j}$, so they meet $X_{j+2k(i,n)-2}$ because $X_{i,j} \subset X_i \subset X_{j+2k(i,n)-2}$. Then, the following conditions hold:

(7) $\bigcup_{m \in \mathbb{N}} \mathcal{V}_{j,m}$ covers $\alpha^{-1}([2^{-j}, 2^{-j+1}])$;
(8) Each $\mathcal{V}_{j,m}$ is pairwise disjoint;
(9) $\operatorname{mesh} \mathcal{V}_{j,m} < \varepsilon_{j+2m-2}$;
(10) Every open set of $\mathcal{V}_{j,m}$ meets X_{j+2m-2};
(11) Every open set of $\mathcal{V}_{j,m}$ is contained in

$$\alpha^{-1}\big((2^{-j} - 2^{-j-2}, 2^{-j+1} + 2^{-j-2})\big).$$

We define $\mathcal{V}_i$ as follows:

$$\mathcal{V}_{2i-1} = \bigcup_{m=1}^{i} \mathcal{V}_{2i-2m+1,m} \quad \text{and} \quad \mathcal{V}_{2i} = \bigcup_{m=1}^{i} \mathcal{V}_{2i-2m+2,m}.$$

For $j, m \in \mathbb{N}$, let $n = j + 2m - 2$. If $n = 2i - 1$ then $j = n - 2m + 2 = 2i - 2m + 1$, and if $n = 2i$ then $j = n - 2m + 2 = 2i - 2m + 2$. Consequently, $\mathcal{V}_{j,m} \subset \mathcal{V}_n$. Therefore, (1) follows from (7); (2) follows from (8) and (11); (3) follows from (9); (4) follows from (10); (5) follows from (11). By replacing $\mathcal{V}_n$ with $\mathcal{V}_n \setminus \bigcup_{i<n} \mathcal{V}_i$, condition (6) is satisfied without failing the other conditions.

Let K be the nerve of $\mathcal{V} = \bigcup_{i\in\mathbb{N}} \mathcal{V}_i \in \operatorname{cov}(X)$ with $p : X \to |K|$ a canonical map. For each $V \in \mathcal{V}_i$, take $x_V \in V \cap X_i$ (cf. (4)) and $y_V \in \varphi(x_V)$. Then, we have a contraction

$$r^V : \mathrm{N}(\varphi(x_V), 2\delta_i) \times \mathbf{I} \to \mathrm{N}(\varphi(x_V), \delta_{i-1})$$

such that $r_1^V(\mathrm{N}(\varphi(x_V), 2\delta_i)) = \{y_V\}$ ($r_0^V = \mathrm{id}$). For each n-simplex $\sigma \in K$, let $V(\sigma) \in \mathcal{V}_{i(\sigma)}$ be the vertex of σ such that $i(\sigma) = \min\{i \in \mathbb{N} \mid \sigma^{(0)} \cap \mathcal{V}_i \neq \emptyset\}$ (cf. (2)), and let σ_0 be the $(n-1)$-face of σ with $V(\sigma) \notin \sigma_0$. For each $n \in \mathbb{N}$, let

$$K_n = \{\sigma \in K \mid \sigma^{(0)} \subset \textstyle\bigcup_{i=1}^n \mathcal{V}_i\}.$$

Then, $K = \bigcup_{n\in\mathbb{N}} K_n$.

We will inductively construct maps $g_n : |K_n| \to Y$, $n \in \mathbb{N}$, such that $g_n||K_{n-1}| = g_{n-1}$ and, for $\sigma \in K_n$, $t \in \mathbf{I}$, and $z \in \sigma_0$,

(∗) $g_n((1-t)z + tV(\sigma)) = r^{V(\sigma)}(g_n(z), t) \in \mathrm{N}(\varphi(x_{V(\sigma)}), \delta_{i(\sigma)-1})$.

Since $K_1 = K_1^{(0)} = \mathcal{V}_1$ by (2), we can first define $g_1 : |K_1| \to Y$ by $g_1(V) = y_V$ for each $V \in \mathcal{V}_1$. Suppose that $g_1, \ldots, g_{n-1}$ have been defined. For $j \leq n$, let

$$K_{n,j} = \{\sigma \in K \mid \sigma^{(0)} \subset \textstyle\bigcup_{i=j}^n \mathcal{V}_i\} \subset K_n.$$

Observe that $K_{n,n} = K_{n,n}^{(0)} = \mathcal{V}_n$ by (2), $K_{n,j} \subset K_{n,j-1}$, and $K_n = K_{n,1}$. We define $g_{n,n} : |K_{n,n}| \to Y$ by $g_{n,n}(V) = y_V$ for each $V \in \mathcal{V}_n$. Now, suppose that $g_{n,i} : |K_{n,i}| \to Y$, $i > j$, have been defined such that

(12) $g_{n,i}||K_{n,i+1}| = g_{n,i+1}$;
(13) $g_{n,i}||K_{n-1,i}| = g_{n-1}||K_{n-1,i}|$;
(14) $g_{n,i}(\sigma) \subset \mathrm{N}(\varphi(x_{V(\sigma)}), \delta_{i(\sigma)-1})$ for $\sigma \in K_{n,i}$;
(15) $g_{n,i}$ satisfies (∗) for $\sigma \in K_{n,i}$, $t \in \mathbf{I}$, and $z \in \sigma_0$.

For each $\sigma \in K_{n,j} \setminus K_{n,j+1}$, $V(\sigma) \in \mathcal{V}_j$ and $\sigma_0 \in K_{n,j+1}$. By the inductive hypothesis on $g_{n,j+1}$,

$$g_{n,j+1}(\sigma_0) \subset \mathrm{N}(\varphi(x_{V(\sigma_0)}), \delta_{i(\sigma_0)-1}) \subset \mathrm{N}(\varphi(x_{V(\sigma_0)}), \delta_j).$$

On the other hand, since $V(\sigma_0) \cap V(\sigma) \neq \emptyset$, it follows that

$$d(x_{V(\sigma_0)}, x_{V(\sigma)}) \leq \operatorname{diam} V(\sigma_0) + \operatorname{diam} V(\sigma) < \varepsilon_{i(\sigma_0)} + \varepsilon_j < 2\varepsilon_j.$$

Since $x_{V(\sigma)} \in X_j$, we have $\varphi(x_{V(\sigma_0)}) \subset \mathrm{N}(\varphi(x_{V(\sigma)}), \delta_j)$, hence $g_{n,j+1}(\sigma_0) \subset \mathrm{N}(\varphi(x_{V(\sigma)}), 2\delta_j)$. Then, $g_{n,j+1}$ can be extended to the map $g_{n,j} : |K_{n,j}| \to Y$ by

$$g_{n,j}((1-t)z + tV(\sigma)) = r^{V(\sigma)}(g_{n,j+1}(z), t) \in \mathrm{N}(\varphi(x_{V(\sigma)}), \delta_{j-1})$$
$$\text{for } \sigma \in K_{n,j} \setminus K_{n,j+1},\ t \in \mathbf{I} \text{ and } z \in \sigma_0,$$

where $j = i(\sigma)$. Then, $g_{n,j}$ satisfies (12), (14), and (15) by definition. Since g_{n-1} satisfies $(*)$, it follows that $g_{n,j}||K_{n-1,j}| = g_{n-1}||K_{n-1,j}|$, i.e., (13) is also satisfied. By downward induction on j, we can obtain the map $g_n = g_{n,1} : |K_n| = |K_{n,1}| \to Y$ that extends g_{n-1} and satisfies $(*)$.

Now, let $g : |K| \to Y$ be the map defined by $g||K_n| = g_n$. For each $x \in X$, let $\sigma \in K$ be the carrier of $p(x)$. Since p is a canonical map for the nerve K of $\mathcal{V}$, we have $x \in V(\sigma) \in \mathcal{V}_{i(\sigma)}$, hence $d(x, x_{V(\sigma)}) \le \operatorname{mesh} \mathcal{V}_{i(\sigma)} < \varepsilon_{i(\sigma)}$ by (3) and $\alpha(x) > 2^{-i(\sigma)-1}$ by (5). Since $x_{V(\sigma)} \in X_{i(\sigma)}$, it follows that

$$\begin{aligned} gp(x) &\in g(\sigma) \subset \mathrm{N}(\varphi(x_{V(\sigma)}), \delta_{i(\sigma)-1}) \\ &\subset \mathrm{N}(\mathrm{N}(\varphi(x), \delta_{i(\sigma)}), \delta_{i(\sigma)-1}) \subset \mathrm{N}(\varphi(x), 2\delta_{i(\sigma)-1}). \end{aligned}$$

Therefore, we have

$$d(gp(x), \varphi(x)) < 2\delta_{i(\sigma)-1} \le 2^{-i(\sigma)-1} < \alpha(x).$$

Thus, $gp : X \to Y$ is the desired map. □

7.7 The Suspensions and the Taylor Example

In this section, using a K-theory result of J.F. Adams, we will construct the Taylor example, that is, a compactum that is not cell-like but has the Hilbert cube $\boldsymbol{Q} = [-1, 1]^{\mathbb{N}}$ as its cell-like image. Moreover, using the Taylor example, we will construct a cell-like map of $\boldsymbol{Q}$ onto a compactum that is not an ANR. Here, we use the TORUŃCZYK CHARACTERIZATION OF THE HILBERT CUBE:[5]

Theorem 7.7.1 (TORUŃCZYK). *A space X is homeomorphic to the Hilbert cube $\boldsymbol{Q}$ if and only if X is a compact AR that has the disjoint cells property, that is, for each $n \in \mathbb{N}$, each pair of maps $f, g : \mathbf{I}^n \to X$ can be approximated by maps $f', g' : \mathbf{I}^n \to X$ with disjoint images, i.e., $f'(\mathbf{I}^n) \cap g'(\mathbf{I}^n) = \emptyset$.* □

The **suspension** ΣX of a space X is the following quotient space of $[-1, 1] \times X$:

$$\Sigma X = [-1, 1] \times X / \{\{-1\} \times X, \{1\} \times X\},$$

where $\{-1\} \times X$ and $\{1\} \times X$ are regarded as two distinct points. Let $q_X : [-1, 1] \times X \to \Sigma X$ be the quotient map. For each map $f : X \to Y$, there exists a unique map $\Sigma f : \Sigma X \to \Sigma Y$ such that $q_Y \circ (\mathrm{id} \times f) = (\Sigma f) \circ q_X$, that is, the following diagram is commutative:

[5] For this proof, refer to van Mill's book "Infinite-Dimensional Topology" mentioned in the Preface.

$$\begin{array}{ccc} [-1,1]\times X & \xrightarrow{\mathrm{id}\times f} & [-1,1]\times Y \\ {\scriptstyle q_X}\big\downarrow & & \big\downarrow{\scriptstyle q_Y} \\ \Sigma X & \xrightarrow[\Sigma f]{\qquad} & \Sigma Y. \end{array}$$

It should be noted that $\Sigma \mathrm{id}_X = \mathrm{id}_{\Sigma X}$ and $\Sigma g \Sigma f = \Sigma(gf)$ for each pair of maps $f : X \to Y$ and $g : Y \to Z$.

Observe that the suspension ΣX is the union of two cones

$$\mathbf{I} \times X/\{1\} \times X \;\text{ and }\; [-1,0] \times X/\{-1\} \times X.$$

Since the cone over a compact ANR is an AR by Corollary 6.5.5, the next proposition follows from 6.2.10(5)

Proposition 7.7.2. *For every compact ANR X, the suspension ΣX is an ANR.* □

Proposition 7.7.3. *For each cell-like compactum X, the suspension ΣX is also cell-like.*

Proof. Embed X into the Hilbert cube $\mathbf{Q}$. Then, it suffices to prove that ΣX is cell-like in $\Sigma \mathbf{Q}$. Let $q_{\mathbf{Q}} : [-1,1] \times \mathbf{Q} \to \Sigma \mathbf{Q}$ be the quotient map. For each neighborhood U of ΣX in $\Sigma \mathbf{Q}$, we can choose a neighborhood V of X in $\mathbf{Q}$ so that $q_{\mathbf{Q}}([-1,1] \times V) \subset U$. Since X is cell-like in $\mathbf{Q}$, there is a contraction $h : X \times \mathbf{I} \to V$, which induces the homotopy

$$\tilde{h} : \Sigma X \times \mathbf{I} \to q_{\mathbf{Q}}([-1,1] \times V) \subset U$$

such that $\tilde{h}_t = \Sigma h_t$ for each $t \in \mathbf{I}$. Then, $\tilde{h}_0 = \mathrm{id}$ and $\tilde{h}_1(\Sigma X) = h_1(X) \times [-1,1] \approx [-1,1]$, which implies that ΣX is contractible in U. Consequently, ΣX is cell-like in $\Sigma \mathbf{Q}$. □

Proposition 7.7.4. *Let $X = \varprojlim(X_i, f_i)$ be the inverse limit of an inverse sequence $(X_i, f_i)_{i\in\mathbb{N}}$ of compacta with $p_i : X \to X_i$, $i \in \mathbb{N}$, the projections of the inverse limit. Then, there exists the unique homeomorphism $h : \Sigma X \to \varprojlim(\Sigma X_i, \Sigma f_i)$ such that $q_i h = \Sigma p_i$ for each $i \in \mathbb{N}$, where each $q_i : \varprojlim(\Sigma X_i, \Sigma f_i) \to \Sigma X_i$ is the inverse limit projection.*

Proof. Let $q_X : [-1,1] \times X \to \Sigma X$ and $q_{X_i} : [-1,1] \times X_i \to \Sigma X_i$, $i \in \mathbb{N}$, be the quotient maps. For $(t,x) \in [-1,1] \times X$ and $i \in \mathbb{N}$,

$$(\Sigma f_i) q_{X_{i+1}}(t, x(i+1)) = q_{X_i}(t, f_i(x(i+1))) = q_{X_i}(t, x(i)).$$

Then, we can define a map $h : \Sigma X \to \varprojlim(\Sigma X_i, \Sigma f_i)$ as follows:

$$h q_X(t,x) = \big(q_{X_i}(t, x(i))\big)_{i\in\mathbb{N}} = \big((\Sigma p_i) q_X(t,x)\big)_{i\in\mathbb{N}},$$

which means that $q_i h = \Sigma p_i$ for each $i \in \mathbb{N}$. The uniqueness of the map h follows from the last condition.[6] We will show that h is bijective, which implies that h is a homeomorphism because ΣX is compact.

$$\begin{array}{ccccccc}
[-1,1]\times X & \xrightarrow{q_X} & \Sigma X & \xrightarrow{h} & \varprojlim(\Sigma X_i, \Sigma f_i) & \subset & \prod_{i\in\mathbb{N}} \Sigma X_i \\
\downarrow{\scriptstyle \mathrm{id}\times p_i} & & \downarrow{\scriptstyle \Sigma p_i} & \swarrow{\scriptstyle q_i} & & & \\
[-1,1]\times X_i & \xrightarrow[q_{X_i}]{} & \Sigma X_i & & & &
\end{array}$$

To prove that h is surjective, let $(q_{X_i}(t_i, x_i))_{i\in\mathbb{N}} \in \varprojlim(\Sigma X_i, \Sigma f_i)$. For each $i \in \mathbb{N}$,

$$q_{X_i}(t_i, x_i) = (\Sigma f_i) q_{X_{i+1}}(t_{i+1}, x_{i+1}) = q_{X_i}(t_{i+1}, f_i(x_{i+1})),$$

which implies that $t_1 = t_2 = \cdots = t$. If $t \neq \pm 1$ then $f_i(x_{i+1}) = x_i$ for every $i \in \mathbb{N}$. When $t = \pm 1$, we may assume that $f_i(x_{i+1}) = x_i$. Then, $x = (x_i)_{i\in\mathbb{N}} \in X$. It follows that $hq_X(t, x) = (q_{X_i}(t_i, x_i))_{i\in\mathbb{N}}$.

To show that h is injective, let $(t, x), (t', x') \in [-1, 1] \times X$ and assume that $hq_X(t, x) = hq_X(t', x')$. For each $i \in \mathbb{N}$,

$$q_{X_i}(t, p_i(x)) = q_i h q_X(t, x) = q_i h q_X(t', x') = q_{X_i}(t', p_i(x')).$$

Then, we have $t = t'$. If $t = t' = \pm 1$ then $q_X(t, x) = q_X(t', x')$. When $t = t' \neq \pm 1$, it follows that $p_i(x) = p_i(x')$ for every $i \in \mathbb{N}$, which means $x = x'$, hence $q_X(t, x) = q_X(t', x')$. □

For each $n \in \mathbb{N}$, the ***n*-fold suspension** $\Sigma^n X$ is inductively defined by $\Sigma^n X = \Sigma(\Sigma^{n-1} X)$, where $\Sigma^0 X = X$. For a map $f : X \to Y$, let

$$\Sigma^n f = \Sigma(\Sigma^{n-1} f) : \Sigma^n X = \Sigma(\Sigma^{n-1} X) \to \Sigma^n Y = \Sigma(\Sigma^{n-1} Y),$$

where $\Sigma^0 f = f$. The twofold suspension $\Sigma^2 X$ is also called the **double suspension**. Observe

$$\begin{aligned}
\Sigma^2 X &= [-1,1]^2 \times X / \big\{\{\pm 1\} \times [-1,1] \times X,\ \{(t, \pm 1)\} \times X \bigm| -1 < t < 1\big\} \\
&= \Sigma[-1,1] \times X / \big\{\{z\} \times X \bigm| z \in \partial\Sigma[-1,1]\big\},
\end{aligned}$$

where $\Sigma[-1,1] \approx \mathbf{B}^2$ and

$$\partial\Sigma[-1,1] = \big\{\{\pm 1\} \times [-1,1]\big\} \cup \big((-1,1) \times \{\pm 1\}\big) \approx \mathbf{S}^1.$$

[6] We can apply Theorem 4.10.6 to show the existence and the uniqueness of h.

Furthermore, regarding $[-1,1] = \Sigma\{0\}$, we can write $\Sigma[-1,1] = \Sigma^2\{0\}$. Then, we can also write

$$\Sigma^2 X = \Sigma^2\{0\} \times X \big/ \big\{\{z\} \times X \bigm| z \in \partial\Sigma^2\{0\}\big\}.$$

By induction, we can show that

$$\Sigma^n X = \Sigma^n\{0\} \times X \big/ \big\{\{z\} \times X \bigm| z \in \partial\Sigma^n\{0\}\big\},$$

where $\partial\Sigma^n\{0\}$ is the boundary $(n-1)$-sphere of $\Sigma^n\{0\} \approx \mathbf{B}^n$. Indeed, assume the above equality holds. Then, it follows that

$$\begin{aligned}
\Sigma^{n+1} X &= \Sigma\big(\Sigma^n X\big) \\
&= \Sigma\big(\Sigma^n\{0\} \times X \big/ \big\{\{z\} \times X \bigm| z \in \partial\Sigma^n\{0\}\big\}\big) \\
&= [-1,1] \times \Sigma^n\{0\} \times X \big/ \big\{\{\pm 1\} \times \Sigma^n\{0\} \times X, \\
&\qquad\qquad\qquad \{t\} \times \{z\} \times X \bigm| -1 < t < 1,\ z \in \partial\Sigma^n\{0\}\big\} \\
&= \Sigma^{n+1}\{0\} \times X \big/ \big\{\{z\} \times X \bigm| z \in \partial\Sigma^{n+1}\{0\}\big\},
\end{aligned}$$

where the last equality comes from the following facts:

$$\begin{aligned}
\Sigma^{n+1}\{0\} &= [-1,1] \times \Sigma^n\{0\} \big/ \big\{\{\pm 1\} \times \Sigma^n\{0\}\big\} \quad \text{and} \\
\partial\Sigma^{n+1}\{0\} &= \big\{\{\pm 1\} \times \Sigma^n\{0\}\big\} \cup \big((-1,1) \times \partial\Sigma^n\{0\}\big).
\end{aligned}$$

As we have just seen, the n-fold suspension $\Sigma^n X$ is the quotient space of $\Sigma^n\{0\} \times X$, where the quotient map is denoted by $q_X^n : \Sigma^n\{0\} \times X \to \Sigma^n X$. For each map $f : X \to Y$, the map $\Sigma^n f : \Sigma^n X \to \Sigma^n Y$ is induced by the map $\mathrm{id} \times f : \Sigma^n\{0\} \times X \to \Sigma^n\{0\} \times Y$, i.e., $\Sigma^n f$ is defined by the following commutative diagram:

$$\begin{array}{ccc}
\Sigma^n\{0\} \times X & \xrightarrow{\ \mathrm{id}\times f\ } & \Sigma^n\{0\} \times Y \\
{\scriptstyle q_X^n}\big\downarrow & & \big\downarrow{\scriptstyle q_Y^n} \\
\Sigma^n X & \xrightarrow[\ \Sigma^n f\ ]{} & \Sigma^n Y
\end{array}$$

Remark 8. Since $\Sigma^n\{0\} \approx \mathbf{I}^n$, we can regard

$$\Sigma^n X = \mathbf{I}^n \times X \big/ \big\{\{z\} \times X \bigm| z \in \partial\mathbf{I}^n\big\}.$$

In this case, $\Sigma^m\Sigma^n X \approx \Sigma^{m+n} X$ but $\Sigma^m\Sigma^n X \neq \Sigma^{m+n} X$ because

$$\Sigma^{m+n} X = \mathbf{I}^{m+n} \times X / \{\{z\} \times X \mid z \in \partial \mathbf{I}^{m+n}\} \text{ but}$$
$$\Sigma^m \Sigma^n X = \mathbf{I}^{m+n} \times X / \{\{y\} \times \mathbf{I}^n \times X,\ \{y'\} \times \{z\} \times X \mid y \in \partial \mathbf{I}^m,\ y' \in \mathbf{I}^m \setminus \partial \mathbf{I}^m,\ z \in \partial \mathbf{I}^n\}.$$

In his paper [1] (References in Notes for Chap. 7), Adams constructed a compact polyhedron A with a map $\alpha : \Sigma^r A \to A$ from the r-fold suspension of A onto A such that every composition

$$\alpha \circ \Sigma^r \alpha \circ \cdots \circ \Sigma^{(i-1)r} \alpha : \Sigma^{ir} A \to A$$

is essential, i.e., it is not null-homotopic. Using this map, we can construct a cell-like map that is not a shape equivalence, which is the subject of the following theorem:

Theorem 7.7.5 (Taylor). *There exists a cell-like map $f : X \to \boldsymbol{Q}$ of a compactum X onto the Hilbert cube $\boldsymbol{Q}$ such that X is not cell-like, that is, f is not a shape equivalence, where X is homeomorphic to the r-fold suspension $\Sigma^r X$.*

Proof. Let X be the inverse limit of the inverse sequence:

$$A \xleftarrow{\ \alpha\ } \Sigma^r A \xleftarrow{\ \Sigma^r \alpha\ } \Sigma^{2r} A \xleftarrow{\ \Sigma^{2r}\alpha\ } \cdots$$

where A is the compact polyhedron with the map $\alpha : \Sigma^r A \to A$ constructed by Adams. Let $p_i : X \to \Sigma^{(i-1)r} A$, $i \in \mathbb{N}$, be the inverse limit projections, where $\Sigma^0 A = A$. If X is cell-like, then $p_1 \simeq 0$ by Theorem 7.1.8. Note that every $\Sigma^{ir} A$ is an ANR by Proposition 7.7.2. Applying Lemma 7.1.7 (both (1) and (2)), we can find some $j \in \mathbb{N}$ such that

$$\alpha \circ \Sigma^r \alpha \circ \cdots \circ \Sigma^{(j-1)r} \alpha \simeq 0,$$

which is a contradiction. Thus, X is not cell-like. Moreover, by Proposition 7.7.4, $\Sigma^r X$ is homeomorphic to the inverse limit of

$$\Sigma^r A \xleftarrow{\ \Sigma^r \alpha\ } \Sigma^{2r} A \xleftarrow{\ \Sigma^{2r}\alpha\ } \Sigma^{3r} A \xleftarrow{\ \Sigma^{3r}\alpha\ } \cdots,$$

hence $\Sigma^r X \approx X$ (Corollary 4.10.4).

We regard $\Sigma^{ir} A$ as the quotient space of $\Sigma^{ir}\{0\} \times A$ with the quotient map $q_A^{ir} : \Sigma^{ir}\{0\} \times A \to \Sigma^{ir} A$ such that $q_A^{ir}(\{z\} \times A)$ is a singleton for each $z \in \partial \Sigma^{ir}\{0\}$. Let $f_i : \Sigma^{ir} A \to \Sigma^{ir}\{0\}$ be the map induced by the projection of $\Sigma^{ir}\{0\} \times A$ onto $\Sigma^{ir}\{0\}$. Because $\Sigma^{(i+1)r}\{0\} = \Sigma^{ir} \Sigma^r \{0\}$ is the quotient space of $\Sigma^{ir}\{0\} \times \Sigma^r\{0\}$, we have the map

$$\varphi_i : \Sigma^{(i+1)r}\{0\} = \Sigma^{ir} \Sigma^r \{0\} \to \Sigma^{ir}\{0\},$$

which is induced by the projection of $\Sigma^{ir}\{0\} \times \Sigma^{r}\{0\}$ onto $\Sigma^{ir}\{0\}$, that is, the map φ_i is defined by the following commutative diagram:

$$\begin{array}{ccc} & & \Sigma^{ir}\{0\} \times \Sigma^{r}\{0\} \\ & \swarrow^{\mathrm{pr}_{\Sigma^{ir}\{0\}}} & \downarrow q^{ir}_{\Sigma^{r}\{0\}} \\ \Sigma^{ir}\{0\} & \underset{\varphi_i}{\longleftarrow} & \Sigma^{ir}\Sigma^{r}\{0\} = \Sigma^{(i+1)r}\{0\} \end{array}$$

In the above, $\mathrm{pr}_{\Sigma^{ir}\{0\}}$ is a fine homotopy equivalence. Since $q^{ir}_{\Sigma^{r}\{0\}}$ is cell-like, it is also a fine homotopy equivalence. By Proposition 6.7.4, φ_i is a fine homotopy equivalence. On the other hand, we have the embedding $\psi_i : \Sigma^{ir}\{0\} \to \Sigma^{(i+1)r}\{0\}$ defined by $\psi_i(x) = q^{ir}_{\Sigma^{r}\{0\}}(x, 0)$. Therefore, $\varphi_i \psi_i = \mathrm{id}$.

By diagram chasing, we can see $f_i \Sigma^{ir}\alpha = \varphi_i f_{i+1}$. In the diagram below, the circumference and the upper right square commute. Moreover, $q^{(i+1)r}_{A}(q^{i}r_{\{0\}} \times \mathrm{id}) = q^{ir}_{\Sigma^{r}A}(\mathrm{id} \times q_1)$ is surjective. Hence, the botom rectangle also commutes.

$$\begin{array}{ccccc} & & \Sigma^{ir}\{0\} \times \Sigma^{r}\{0\} \times A & \xrightarrow{q^{ir}_{\Sigma^{r}\{0\}} \times \mathrm{id}} & \Sigma^{ir}\Sigma^{r}\{0\} \times A \\ & & \downarrow{\scriptstyle \mathrm{id} \times q^{r}_{A}} & & \| \\ & \Sigma^{ir}\{0\} \times A & \underset{\mathrm{id} \times \alpha}{\longleftarrow} & \Sigma^{ir}\{0\} \times \Sigma^{r}A & \Sigma^{(i+1)r}\{0\} \times A \\ & \downarrow{\scriptstyle q^{ir}_{A}} & & \downarrow{\scriptstyle q^{ir}_{\Sigma^{r}A}} & \downarrow{\scriptstyle q^{(i+1)r}_{A}} \\ \mathrm{pr}_{\Sigma^{ir}\{0\}} & \Sigma^{ir}A & \xleftarrow{\Sigma^{ir}\alpha} & \Sigma^{ir}\Sigma^{r}A = \!=\!= \Sigma^{(i+1)r}A & \mathrm{pr}_{\Sigma^{(i+1)r}\{0\}} \\ & \downarrow{\scriptstyle f_i} & & \downarrow{\scriptstyle f_{i+1}} & \\ & \Sigma^{ir}\{0\} & \underset{\varphi_i}{\longleftarrow} & \Sigma^{(i+1)r}\{0\} & \end{array}$$

Thus, we have the following commutative diagram:

$$\begin{array}{ccccccccc} A & \xleftarrow{\alpha} & \Sigma^{r}A & \xleftarrow{\Sigma^{r}\alpha} & \Sigma^{2r}A & \xleftarrow{\Sigma^{2r}\alpha} & \Sigma^{3r}A & \xleftarrow{\Sigma^{3r}\alpha} & \cdots \\ & & \downarrow{\scriptstyle f_1} & & \downarrow{\scriptstyle f_2} & & \downarrow{\scriptstyle f_3} & & \\ & & \Sigma^{r}\{0\} & \underset{\varphi_1}{\longleftarrow} & \Sigma^{2r}\{0\} & \underset{\varphi_2}{\longleftarrow} & \Sigma^{3r}\{0\} & \underset{\varphi_3}{\longleftarrow} & \cdots \end{array}$$

Let $Y = \varprojlim \Sigma^{ir}\{0\}$ be the inverse limit of the bottom sequence and $p_i : Y \to \Sigma^{ir}\{0\}$, $i \in \mathbb{N}$, the inverse limit projections. Then, we have the map $f = \varprojlim f_i : X \to Y$ defined by the maps f_i. It remains to show that $Y \approx \boldsymbol{Q}$ and $f : X \to Y$ is cell-like.

To show $Y \approx \boldsymbol{Q}$, we use the Toruńczyk characterization of $\boldsymbol{Q}$ (Theorem 7.7.1). As observed in the above, each bonding map φ_i is a fine homotopy equivalence, hence Y is a compact AR by Theorem 6.7.8. Recall that there are embeddings $\psi_i : \Sigma^{ir}\{0\} \to \Sigma^{(i+1)r}\{0\}$, $i \in \mathbb{N}$, such that $\varphi_i\psi_i = \mathrm{id}$. We can define an embedding $\bar{\psi}_i : \Sigma^{ir}\{0\} \to Y$, $i \in \mathbb{N}$, as follows:

$$\bar{\psi}_i(x) = (\varphi_{1,i}(x), \ldots, \varphi_{i-1,i}(x), x, \psi_{i,i+1}(x), \psi_{i,i+2}(x), \ldots),$$

where $\varphi_{j,i} = \varphi_j \cdots \varphi_{i-1}$ for $j < i$ and $\psi_{i,j} = \psi_{j-1} \cdots \psi_i$ for $j > i$. Then, $p_i\bar{\psi}_i = \mathrm{id}$ for each $i \in \mathbb{N}$ and $\lim_{i\to\infty} \bar{\psi}_i p_i = \mathrm{id}_Y$. Since $\Sigma^{ir}\{0\} \approx \mathbf{I}^{ir}$ for each $i \in \mathbb{N}$, we can apply Theorem 5.8.1 to see that Y has the disjoint cells property, that is, for each $n \in \mathbb{N}$ and $\mathcal{U} \in \mathrm{cov}(Y)$, each pair of maps $f, g : \mathbf{I}^n \to Y$ are $\mathcal{U}$-close to map $f', g' : \mathbf{I}^n \to Y$ with $f'(\mathbf{I}^n) \cap g'(\mathbf{I}^n) = \emptyset$. By the Toruńczyk characterization of the Hilbert cube, we have $Y \approx \boldsymbol{Q}$.

To prove that f is cell-like, let $y = (y_i)_{i\in\mathbb{N}} \in Y = \varprojlim \Sigma^{ir}\{0\}$. Consider the following commutative diagram:

$$\begin{array}{ccccccc}
\Sigma^r\{0\} \times A & \xleftarrow{\mathrm{id}\times\alpha q^r_A} & \Sigma^{2r}\{0\} \times A & \xleftarrow{\mathrm{id}\times\alpha q^r_A} & \Sigma^{3r}\{0\} \times A & \xleftarrow{\mathrm{id}\times\alpha q^r_A} & \cdots \\
\cup & & \cup & & \cup & & \\
\{y_1\} \times A & \longleftarrow & \{y_2\} \times A & \longleftarrow & \{y_3\} \times A & \longleftarrow & \cdots \\
\downarrow q_1 & & \downarrow q_2 & & \downarrow q_3 & & \\
f_1^{-1}(y_1) & \longleftarrow & f_2^{-1}(y_2) & \longleftarrow & f_3^{-1}(y_3) & \longleftarrow & \cdots \\
\cap & & \cap & & \cap & & \\
\Sigma^r A & \xleftarrow[\Sigma^r\alpha]{} & \Sigma^{2r} A & \xleftarrow[\Sigma^{2r}\alpha]{} & \Sigma^{3r} A & \xleftarrow[\Sigma^{3r}\alpha]{} & \cdots
\end{array}$$

The inverse limit of the third sequence is $f^{-1}(y)$. Recall that $X = \varprojlim \Sigma^{ir} A$ is a subspace of the product space $\prod_{i\in\mathbb{N}} \Sigma^{ir} A$. For each neighborhood U of $f^{-1}(y)$ in $\prod_{i\in\mathbb{N}} \Sigma^{ir} A$, we will show that $f^{-1}(y)$ is contractible in U, which implies that $f^{-1}(y)$ is cell-like by Theorem 7.1.2.

For each $j \in \mathbb{N}$, let

$$Y_j = \big\{x \in \textstyle\prod_{i\le j} \Sigma^{ir} A \bigm| x(j) \in f_j^{-1}(y_j), \\ \qquad \forall i < j,\ x(i) = \big(\Sigma^{ir}\alpha \circ \cdots \circ \Sigma^{(j-1)r}\alpha\big)(x(j))\big\}.$$

Then, $Y_j \times \prod_{i>j} \Sigma^{ir} A$ is compact and

$$f^{-1}(y) = \bigcap_{j\in\mathbb{N}} \left(Y_j \times \prod_{i>j} f_i^{-1}(y_i) \right) = \bigcap_{j\in\mathbb{N}} \left(Y_j \times \prod_{i>j} \Sigma^{ir} A \right) \subset U \subset \prod_{i\in\mathbb{N}} \Sigma^{ir} A.$$

Hence, we can find $j \in \mathbb{N}$ such that $Y_j \times \prod_{i>j} \Sigma^{ir} A \subset U$. For each $i > j$, $f_i^{-1}(y_i)$ is contractible in $\Sigma^{ir} A$. Indeed, recall

$$\Sigma^{ir} A = (\Sigma^{ir}\{0\} \times A) \big/ \big\{ \{z\} \times A \bigm| z \in \partial \Sigma^{ir}\{0\} \big\}.$$

Then, $f_i^{-1}(y_i)$ is a singleton if $y_i \in \partial\Sigma^{ir}\{0\}$. When $y_i \in \Sigma^{ir}\{0\} \setminus \partial\Sigma^{ir}\{0\}$, take a path $\gamma : \mathbf{I} \to \Sigma^{ir}\{0\}$ from $\gamma(0) = y_i$ to $\gamma(1) \in \partial\Sigma^{ir}\{0\}$ and define a homotopy $h : \{y_i\} \times A \times \mathbf{I} \to \Sigma^{ir}\{0\} \times A$ by $h(y_i, x, t) = (\gamma(t), x)$. Then, h induces a contraction $\tilde{h}$ of $f_i^{-1}(y_i)$ in $\Sigma^{ir} A$:

$$\begin{array}{ccc} \{y_i\} \times A \times \mathbf{I} & \xrightarrow{\;h\;} & \Sigma^{ir}\{0\} \times A \\ {\scriptstyle q_i \times \mathrm{id}} \downarrow & & \downarrow {\scriptstyle q_i} \\ f_i^{-1}(y_i) \times \mathbf{I} & \xrightarrow[\tilde{h}]{} & \Sigma^{ir} A \end{array}$$

It follows that $Y_{j+1} \times \prod_{i>j+1} f_i^{-1}(y_i)$ is contractible in $Y_j \times \prod_{i>j} \Sigma^{ir} A$, which implies that $f^{-1}(y)$ is contractible in U. Therefore, $f^{-1}(y)$ is cell-like. □

Remark 9. In the above proof, as mentioned in Remark 8, we can regard

$$\Sigma^n A = \mathbf{I}^n \times A \big/ \big\{ \{z\} \times A \bigm| z \in \partial \mathbf{I}^n \big\}.$$

In this case, recall that $\Sigma^{ir}\Sigma^r A \neq \Sigma^{(i+1)r} A$. For each $i \in \mathbb{N}$, the projection of $\mathbf{I}^{ir} \times A$ onto $\mathbf{I}^{ir}$ induces the map $f_i : \Sigma^{ir} A \to \mathbf{I}^{ir}$. In the literature explaining Taylor's example (as in Taylor's original paper [21], References in Notes for Chap. 7), the map $f : X \to \boldsymbol{Q}$ is constructed using the following diagram:

$$\begin{array}{ccccccccc} A & \xleftarrow{\;\alpha\;} & \Sigma^r A & \xleftarrow{\;\Sigma^r\alpha\;} & \Sigma^{2r} A & \xleftarrow{\;\Sigma^{2r}\alpha\;} & \Sigma^{3r} A & \xleftarrow{\;\Sigma^{3r}\alpha\;} & \cdots \\ & & {\scriptstyle f_1} \downarrow & & \downarrow {\scriptstyle f_2} & & \downarrow {\scriptstyle f_3} & & \\ & & \mathbf{I}^r & \xleftarrow[\mathrm{pr}]{} & \mathbf{I}^{2r} & \xleftarrow[\mathrm{pr}]{} & \mathbf{I}^{3r} & \xleftarrow[\mathrm{pr}]{} & \cdots \end{array}$$

However, the domain of the map $\Sigma^{ir}\alpha$ is $\Sigma^{ir}\Sigma^r A$, which is not $\Sigma^{(i+1)r}A$. If $\Sigma^{ir}\Sigma^r A$ is identified with $\Sigma^{(i+1)r}A$ by a homeomorphism then the above diagram is not commutative. Although $\Sigma^{ir}\Sigma^r A$ can be regarded as the quotient space of $\mathbf{I}^{(i+1)r} \times A$, the projection of $\mathbf{I}^{(i+1)r} \times A$ onto $\mathbf{I}^{(i+1)r}$ does not induce a map of $\Sigma^{ir}\Sigma^r A$ onto $\mathbf{I}^{(i+1)r}$.

Remark 10. We can also regard

$$\Sigma^n A = \mathbf{B}^n \times A / \{\{z\} \times A \mid z \in \mathbf{S}^{n-1}\}.$$

As above, $\Sigma^{(i+1)r}A \neq \Sigma^{ir}\Sigma^r A$. For each $i \in \mathbb{N}$, the projection of $\mathbf{B}^{ir} \times A$ onto $\mathbf{B}^{ir}$ induces the map $f_i : \Sigma^{ir}A \to \mathbf{B}^{ir}$. Then, we can construct a homeomorphism $\bar{\theta}_i : \Sigma^{ir}\Sigma^r A \to \Sigma^{ir}A$ so that the following diagram commutes:

$$\begin{array}{ccccc}
\Sigma^{ir}A & \xleftarrow{\Sigma^{ir}\alpha} & \Sigma^{ir}\Sigma^r A & \underset{\approx}{\xrightarrow{\bar{\theta}_i}} & \Sigma^{(i+1)r}A \\
{\scriptstyle f_i}\downarrow & & & & \downarrow{\scriptstyle f_{i+1}} \\
\mathbf{B}^{ir} & & \xleftarrow[\mathrm{pr}_{\mathbf{B}^{ir}}]{\qquad\qquad} & & \mathbf{B}^{(i+1)r}
\end{array}$$

In fact, we have a surjective map $\theta_i : \mathbf{B}^{ir} \times \mathbf{B}^r \to \mathbf{B}^{(i+1)r}$ defined by

$$\theta_i(y, z) = \left(y, \sqrt{1 - \|y\|^2}z\right) \text{ for each } (y, z) \in \mathbf{B}^{ir} \times \mathbf{B}^r.$$

Then, $\theta_i|(\mathbf{B}^{ir} \setminus \mathbf{S}^{ir-1}) \times \mathbf{B}^r$ is injective,

$$\theta_i^{-1}(\mathbf{S}^{(i+1)r-1}) = (\mathbf{S}^{ir-1} \times \mathbf{B}^r) \cup (\mathbf{B}^{ir} \times \mathbf{S}^{r-1}),$$

$$\theta_i^{-1}(\theta_i(y)) = \theta_i^{-1}(y, 0) = \{y\} \times \mathbf{B}^r \text{ for each } y \in \mathbf{S}^{ir-1}$$

and θ_i is $\mathbf{B}^{ir}$-preserving, that is, the following diagram commutes:

$$\begin{array}{ccc}
\mathbf{B}^{ir} \times \mathbf{B}^r & \xrightarrow{\theta_i} & \mathbf{B}^{(i+1)r} \\
& {\scriptstyle \mathrm{pr}_{\mathbf{B}^{ir}}}\searrow \quad \swarrow{\scriptstyle \mathrm{pr}_{\mathbf{B}^{ir}}|} & \\
& \mathbf{B}^{ir} &
\end{array}$$

Observe

$$\begin{aligned}
\Sigma^{ir}\Sigma^r A &= \mathbf{B}^{ir} \times \mathbf{B}^r \times A / \{\{y\} \times \mathbf{B}^r \times A,\ \{y'\} \times \{z\} \times A \mid \\
&\qquad\qquad y \in \mathbf{S}^{ir-1},\ y' \in \mathbf{B}^{ir} \setminus \mathbf{S}^{ir-1},\ z \in \mathbf{S}^{r-1}\} \\
&= \mathbf{B}^{ir} \times \mathbf{B}^r \times A / \{\theta_i^{-1}(z) \times A \mid z \in \mathbf{S}^{(i+1)r}\}.
\end{aligned}$$

Then, $\theta_i \times \mathrm{id}_A$ induces the desired homeomorphism $\bar{\theta}_i : \Sigma^{ir}\Sigma^r A \to \Sigma^{(i+1)r} A$. Indeed, let $q^n_X : \mathbf{B}^n \times X \to \Sigma^n X$ be the quotient map. Then, the following diagram commutes:

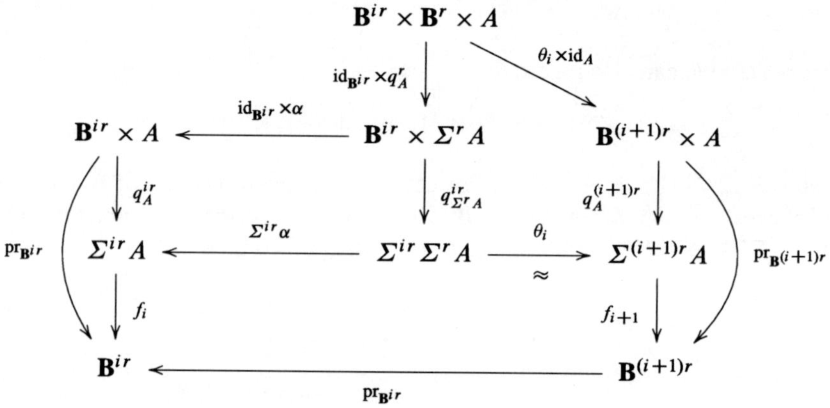

In this case, Y is defined as the inverse limit of the inverse sequence

$$\mathbf{B}^r \xleftarrow{\mathrm{pr}_{\mathbf{B}^r}} \mathbf{B}^{2r} \xleftarrow{\mathrm{pr}_{\mathbf{B}^{2r}}} \mathbf{B}^{3r} \xleftarrow{\mathrm{pr}_{\mathbf{B}^{3r}}} \cdots$$

However, it is not trivial that Y is homeomorphic to $\boldsymbol{Q}$.

Remark 11. Using the Taylor example, we can prove that the composition of cell-like maps is not a cell-like map in general. In fact, the composition $c \circ f : X \to \{0\}$ of the constant map $c : \boldsymbol{Q} \to \{0\}$ and Taylor's cell-like map $f : X \to \boldsymbol{Q}$ is not a cell-like map.

It should be noted that Taylor's cell-like map is not a shape equivalence by Corollary 7.5.2. Thus, in general, a cell-like map need not be a shape equivalence but we do have the following theorem:

Theorem 7.7.6. *Let $f : X \to Y$ be a cell-like map between metrizable spaces. If $f^{-1}(y)$ is a singleton except for finitely many $y \in Y$, then f is a hereditary shape equivalence.*

In this theorem, $Y \approx X/\{f^{-1}(y_1), \ldots, f^{-1}(y_k)\}$ for some finite $y_1, \ldots, y_k \in Y$. Since the composition of shape equivalences is a shape equivalence, this theorem can be easily reduced to the following special case:

Theorem 7.7.7. *For each cell-like compactum A in a metrizable space X, the quotient map $p : X \to X/A$ is a shape equivalence.*

Proof. In Proposition 7.5.6, when Y is a singleton, we can identify $Y \cup_f X = X/A$ and $M_f = A \times \mathbf{I}/A \times \{0\}$, where $M_f \cup (X \times \{1\})$ can be regarded as the following space:

$$Z = (A \times \mathbf{I}/A \times \{0\}) \cup (X \times \{1\}).$$

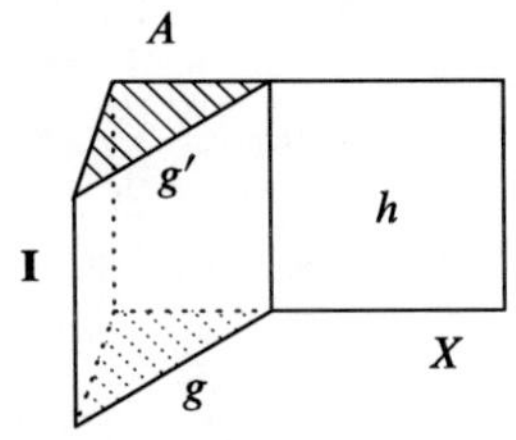

Fig. 7.3 Extending a homotopy $h : gi \simeq g'i$

Then, we have a shape equivalence $h : Z \to X/A$ such that $hi = q$, where $i(x) = (x, 1)$ for each $x \in X$. Thus, it suffices to show that $i : X \to Z$ is a shape equivalence, i.e., $i^* : [Z, P] \to [X, P]$ is bijective for each ANR P.

Let $f : X \to P$ be a map. According to Theorem 7.1.2, A has trivial shape, hence $f|A \simeq 0$. Then, we have a map $\tilde{f} : A \times \mathbf{I}/A \times \{0\} \to P$ such that $\tilde{f}i|A = f$, which extends to a map $g : Z \to P$ such that $gi = f$. Therefore, i^* is surjective.

To show that i^* is injective, let $g, g' : Z \to P$ be maps with a homotopy $h : X \times \mathbf{I} \to P$ such that $h_0 = gi$ and $h_1 = g'i$. Observe that

$$(A \times \{1\} \times \mathbf{I}) \cup ((A \times \mathbf{I}/A \times \{0\}) \times \{0, 1\}) \approx \Sigma A,$$

which is cell-like by Proposition 7.7.3, so it has trivial shape by Theorem 7.1.2. Regarding $(A \times \mathbf{I}/A \times \{0\}) \times \mathbf{I}$ as the cone over the above space, we know that $(A \times \mathbf{I}/A \times \{0\}) \times \mathbf{I}$ is contractible. Then, we can apply the Homotopy Extension Theorem 6.4.1 to obtain a map

$$h' : (A \times \mathbf{I}/A \times \{0\}) \times \mathbf{I} \to P$$

such that $h'(x, 1, t) = h(x, t)$ for each $x \in A$ and $t \in \mathbf{I}$, $h'(z, 0) = g(z)$ and $h'(z, 1) = g'(z)$ for each $z \in A \times \mathbf{I}/A \times \{0\}$. Then, h' extends to a homotopy $\tilde{h} : Z \times \mathbf{I} \to P$ such that $\tilde{h}i = h$, hence $\tilde{h}_0 = g$ and $\tilde{h}_1 = g'$. — Fig. 7.3. □

Using the Taylor example, we can also obtain the following theorem:

Theorem 7.7.8. *There exists a cell-like map $g : \boldsymbol{Q} \to Y$ of the Hilbert cube $\boldsymbol{Q}$ onto a compactum Y that is not cell-like, which means that g is not a shape equivalence.*

Proof. Let Q_0 and Q_1 be copies of $\boldsymbol{Q}$, $f : X \to Q_0$ be Taylor's cell-like map obtained in Theorem 7.7.5, and embed X into Q_1. We define $Y = Q_0 \cup_f Q_1$. The quotient map $g : Q_1 \to Y$ is cell-like. Indeed, for each $y \in Y$, if $y \in Q_1 \setminus X$ then $g^{-1}(y) = \{y\}$, and if $y \in Q_0$ then $g^{-1}(y) = f^{-1}(y)$ is cell-like.

Assume that Y is cell-like. Because we have a shape equivalence $h : M_f \cup (Q_1 \times \{1\}) \to Y$ by Proposition 7.5.6, it follows from Corollary 7.5.2 that $M_f \cup (Q_1 \times \{1\})$ is cell-like. On the other hand, the natural map

$$q : M_f \cup (Q_1 \times \{1\}) \to (M_f \cup (Q_1 \times \{1\}))/\{Q_0, Q_1 \times \{1\}\} \approx \Sigma X$$

is a shape equivalence by Theorem 7.7.6. Hence, ΣX is also cell-like. Using Proposition 7.7.3 inductively, it follows that $\Sigma^r X$ is cell-like. But $\Sigma^r X \approx X$ is not cell-like, which is a contradiction. Thus, Y is not cell-like. □

7.8 The Simplicial Eilenberg–MacLane Complexes

For each $n \in \mathbb{N}$, $\pi_n(\mathbf{S}^n) \neq \{0\}$ by the No Retraction Theorem 5.1.5. In fact, it is known that $\pi_n(\mathbf{S}^n)$ is an infinite cyclic group generated by $[\mathrm{id}_{\mathbf{S}^n}]$, that is, $\pi_n(\mathbf{S}^n) \cong \mathbb{Z}$. It follows from Theorem 5.2.3 that $\pi_m(\mathbf{S}^n) = \{0\}$ for $m < n$, and it is known that $\pi_m(\mathbf{S}^1) = \{0\}$ for any $m > 1$. However, for each $n > 1$, there is some $m > n$ such that $\pi_m(\mathbf{S}^n) \neq \{0\}$. For instance, $\pi_3(\mathbf{S}^2) \cong \mathbb{Z}$. For these facts, refer to any textbook on Homotopy Theory or Algebraic Topology.

There exists a space $K(\mathbb{Z}, n)$ such that

$$\pi_n(K(\mathbb{Z}, n)) \cong \mathbb{Z} \text{ and } \pi_m(K(\mathbb{Z}, n)) = \{0\} \text{ if } m \neq n.$$

Such a space $K(\mathbb{Z}, n)$ is called the **Eilenberg–MacLane space** (of type $(\mathbb{Z}, n)$), which is unique up to homotopy type. The unit circle $\mathbf{S}^1$ is $K(\mathbb{Z}, 1)$. It is easy to see the homotopical uniqueness of $K(\mathbb{Z}, 1)$, that is, if X is a path-connected space homotopy dominated by a simplicial complex such that $\pi_1(X) \cong \mathbb{Z}$ and $\pi_m(X) = \{0\}$ for any $m > 1$, then X has the homotopy type of $\mathbf{S}^1$. Indeed, let $f : \mathbf{S}^1 \to X$ be a map such that $[f] \in \pi_1(X)$ is a generator of $\pi_1(X) \cong \mathbb{Z}$. Then, f induces an isomorphism $f_\sharp : \pi_1(\mathbf{S}^1) \to \pi_1(X)$ because $f_\sharp([\mathrm{id}_{\mathbf{S}^1}]) = [f]$. For each $m > 1$, $f_\sharp : \pi_m(\mathbf{S}^1) \to \pi_m(X)$ is an isomorphism because $\pi_m(\mathbf{S}^1) = \pi_m(X) = \{0\}$. Therefore, f is a weak homotopy equivalence by Theorem 4.14.12, hence it is a homotopy equivalence by Corollary 4.13.10. In general, given an Abelian group G, there exists a space $K(G, n)$ such that $\pi_n(K(G, n)) \cong G$ and $\pi_m(K(G, n)) = \{0\}$ if $m \neq n$. The space $K(G, n)$ is also called the **Eilenberg–MacLane space** (of type (G, n)), which is unique up to homotopy type.[7]

For each $n > 1$, we will construct the space $K(\mathbb{Z}, n)$ as a *countable simplicial complex* such that $F(\partial \Delta^{(n+1)})$ is its subcomplex and $\partial \Delta^{(n+1)}$ is a mapping $(n+1)$-deformation retract of $|K(\mathbb{Z}, n)|$ for metrizable spaces. Here, a closed set A in a space X is a **mapping (n + 1)-deformation retract** of X for a class $\mathcal{C}$ of spaces if the following condition is satisfied:

(D_{n+1}) For any space $Z \in \mathcal{C}$ with $\dim Z \leq n + 1$ and each map $f : Z \to X$, there is a map $g : Z \to A$ such that $g|f^{-1}(A) = f|f^{-1}(A)$ and $f \simeq g$

---

[7] As usual, the Eilenberg–MacLane space $K(G, n)$ is constructed as a CW-complex, which has the homotopy type of a simplicial complex.

rel. $f^{-1}(A)$, that is, there exists a homotopy $h : Z \times \mathbf{I} \to X$ such that $h_0 = f, h_1 = g$, and $h_t|f^{-1}(A) = f|f^{-1}(A)$ for every $t \in \mathbf{I}$.

One should remark the following:

- If A is a mapping $(n+1)$-deformation retract of X for compact polyhedra then the inclusion $i : A \subset X$ is an $(n+1)$-equivalence.

In the above, when X is path-connected, the inclusion map $i : A \subset X$ induces the isomorphisms $i_\sharp : \pi_m(A) \to \pi_m(X)$, $m \leq n$, and the epimorphism $i_\sharp : \pi_{n+1}(A) \to \pi_{n+1}(X)$ (Theorem 4.14.12). Using this fact, we will prove the homotopical uniqueness of $K(\mathbb{Z}, n)$ (Theorem 7.8.6).

For polyhedra, we have the following:

Proposition 7.8.1. *Let K be a simplicial complex with L a subcomplex of K. If $|L|$ is a mapping $(n+1)$-deformation retract of $|K|$ for metrizable spaces and $L \subset K^{(n+1)}$, then $|L|$ is a retract of $|K^{(n+1)}|$.*

Proof. By Theorem 4.9.6 with Remark 14, $\varphi = \mathrm{id} : |K| \to |K|_{\mathrm{m}}$ is a homotopy equivalence that has a homotopy inverse $\psi : |K|_{\mathrm{m}} \to |K|$ such that the restriction $\psi||L|$ is a homotopy inverse of $\varphi||L| = \mathrm{id} : |L| \to |L|_{\mathrm{m}}$. Since $|L|$ is a mapping $(n+1)$-deformation retract of $|K|$ for metrizable spaces, there is a map $f : |K^{(n+1)}|_{\mathrm{m}} \to |L|$ such that $f||L| = \psi||L|$ and $f \simeq \psi$ rel. $|L|$. Thus, we have a map $f\varphi||K^{(n+1)}| : |K^{(n+1)}| \to |L|$. Because $f\varphi||L| = (\psi||L|)(\varphi||L|) \simeq \mathrm{id}$ in $|L|$, we can apply the Homotopy Extension Theorem 4.3.3 to obtain a retraction $r : |K^{(n+1)}| \to |L|$ homotopic to $f\varphi||K^{(n+1)}|$ in $|K|$. □

Before constructing the simplicial Eilenberg–MacLane complex, we will show the following:

Lemma 7.8.2. *Let (B_λ, S_λ), $\lambda \in \Lambda$, be pairwise disjoint copies of the pair $(\mathbf{B}^{n+2}, \mathbf{S}^{n+1})$ and $h : \bigoplus_{\lambda\in\Lambda} S_\lambda \to X$ be a map. Then, X is a mapping $(n+1)$-deformation retract of the adjunction space $X \cup_h \bigoplus_{\lambda\in\Lambda} B_\lambda$ for normal spaces.*

Proof. Let $f : Z \to X \cup_h \bigoplus_{\lambda\in\Lambda} B_\lambda$ be a map from a normal space Z with $\dim Z \leq n+1$. For each $\lambda \in \Lambda$, let $C_\lambda \subset B_\lambda$ such that $(B_\lambda, C_\lambda) \approx (\mathbf{B}^{n+2}, \frac{1}{2}\mathbf{B}^{n+2})$. Since $\dim f^{-1}(C_\lambda) \leq n+1$ and $\operatorname{bd} C_\lambda \approx \mathbf{S}^{n+1}$, we can apply Theorem 5.2.3 to obtain a map $f_\lambda : f^{-1}(C_\lambda) \to \operatorname{bd} C_\lambda$ such that $f_\lambda|f^{-1}(\operatorname{bd} C_\lambda) = f|f^{-1}(\operatorname{bd} C_\lambda)$. By Theorem 5.1.6(1), $f|f^{-1}(C_\lambda) \simeq f_\lambda$ rel. $f^{-1}(\operatorname{bd} C_\lambda)$ in C_λ. Because $\{f^{-1}(C_\lambda) \mid \lambda \in \Lambda\}$ is discrete in Z, we can define a map

$$f' : Z \to X \cup_h \bigoplus_{\lambda\in\Lambda}(B_\lambda \setminus \operatorname{int} C_\lambda)$$

by $f'|f^{-1}(C_\lambda) = f_\lambda$ for each $\lambda \in \Lambda$ and

$$f'|Z \setminus f^{-1}\big(\bigoplus_{\lambda\in\Lambda} \operatorname{int} C_\lambda\big) = f|Z \setminus f^{-1}\big(\bigoplus_{\lambda\in\Lambda} \operatorname{int} C_\lambda\big).$$

Then, $f' \simeq f$ rel. $f^{-1}(X)$. On the other hand, we have a retraction

$$r : X \cup_h \bigoplus_{\lambda \in \Lambda} (B_\lambda \setminus \operatorname{int} C_\lambda) \to X$$

such that $r \simeq \mathrm{id}$ rel. X. The map $g = rf' : Z \to X$ is the desired one because $g = rf' \simeq f' \simeq f$ rel. $f^{-1}(X)$. □

Proposition 7.8.3. *For every connected simplicial complex K and $n \in \mathbb{N}$, there exists a simplicial complex $\tilde{K}$ containing K as a subcomplex such that $\pi_{n+1}(|\tilde{K}|) = 0$ and $|K|$ is a mapping $(n+1)$-deformation retract of $|\tilde{K}|$ for normal spaces, hence $i_\sharp : \pi_m(|K|) \to \pi_m(|\tilde{K}|)$ is an isomorphism for $m \leq n$. Moreover, if K is countable then so is $\tilde{K}$.*

Proof. Here, we identify $\mathbf{S}^{n+1} = \partial \Delta^{n+2}$. By the Simplicial Approximation Theorem 4.7.14, we can write $\pi_{n+1}(|K|) = \{[h_\lambda] \mid \lambda \in \Lambda\}$,[8] where $h_\lambda : L_\lambda \to K$ is a simplicial map of $L_\lambda = \mathrm{Sd}^{n_\lambda} F(\partial \Delta^{n+2})$ for some $n_\lambda \in \mathbb{N}$. If K is countable, then Λ is also countable. For each $\lambda \in \Lambda$, we have $K_\lambda \lhd F(\Delta^{n+2})$ with $L_\lambda \subset K_\lambda$. For example, take a point $v \in \operatorname{rint} \Delta^{n+2}$ and let $K_\lambda = L_\lambda \cup \{v\} \cup \{\langle \sigma \cup \{v\} \rangle \mid \sigma \in L_\lambda\}$ (cf. Proof of Theorem 4.6.2). Then, $(|K_\lambda|, |L_\lambda|) \approx (\mathbf{B}^{n+2}, \mathbf{S}^{n+1})$. Regarding K_λ, $\lambda \in \Lambda$, as a pairwise disjoint collection, we have a simplicial complex $\bigoplus_{\lambda \in \Lambda} K_\lambda$ and its subcomplex $\bigoplus_{\lambda \in \Lambda} L_\lambda$. Let $h : \bigoplus_{\lambda \in \Lambda} L_\lambda \to K$ be the simplicial map defined by $h|L_\lambda = h_\lambda$ for each $\lambda \in \Lambda$. The desired simplicial complex $\tilde{K}$ can be defined as $\tilde{K} = Z(h) \cup \bigoplus_{\lambda \in \Lambda} K_\lambda$, where $Z(h)$ is the simplicial mapping cylinder of h. If K is countable, then so is $\tilde{K}$ because Λ is countable and each K_λ is finite. Since $|\tilde{K}|$ is homeomorphic to the adjunction space $|K| \cup_h \bigoplus_{\lambda \in \Lambda} |K_\lambda|$, it follows from Lemma 7.8.2 that $|K|$ is a mapping $(n+1)$-deformation retract of $|\tilde{K}|$. Hence, the inclusion $i : |K| \subset |\tilde{K}|$ is an $(n+1)$-equivalence. In particular, the inclusion i induces the epimorphism $i_\sharp : \pi_{n+1}(|K|) \to \pi_{n+1}(|\tilde{K}|)$. Since every $h_\lambda \simeq 0$ in $|\tilde{K}|$, it follows that $\pi_{n+1}(|\tilde{K}|) = i_\sharp(\pi_{n+1}(|K|)) = \{0\}$. □

Now, we will construct the **simplicial Eilenberg–MacLane complex** $K(\mathbb{Z}, n)$ for each $n \geq 2$. We apply Proposition 7.8.3 inductively to obtain a tower

$$F(\partial \Delta^{n+1}) = K_0 \subset K_1 \subset K_2 \subset \cdots$$

of countable simplicial complexes such that each $|K_{k-1}|$ is a mapping $(n+k)$-deformation retract of $|K_k|$ for normal spaces, $\pi_m(|K_k|) = \{0\}$ for $n \neq m \leq n+k$, and the inclusion $i_k : \partial \Delta^{n+1} \subset |K_k|$ induces the isomorphism $(i_k)_\sharp : \pi_n(\partial \Delta^{n+1}) \to \pi_n(|K_k|)$. Then, the countable simplicial complex $K(\mathbb{Z}, n) = \bigcup_{k \in \omega} K_k$ is the desired complex. Indeed, by virtue of Proposition 4.2.6, each map $f : \mathbf{S}^m \to |K(\mathbb{Z}, n)|$ has the image $f(\mathbf{S}^m)$ contained in some $|K_k|$ and if $f \simeq g$ in $|K(\mathbb{Z}, n)|$ for maps $f, g : \mathbf{S}^m \to |K(\mathbb{Z}, n)|$ then $f \simeq g$ in some $|K_k|$. So, it follows that

$$\pi_n(|K(\mathbb{Z}, n)|) \cong \pi_n(\partial \Delta^{n+1}) \cong \mathbb{Z} \text{ and } \pi_m(|K(\mathbb{Z}, n)|) = \{0\} \text{ if } m \neq n.$$

---

[8]It suffices to take a set $\{[h_\lambda] \mid \lambda \in \Lambda\}$ generating the group $\pi_{n+1}(|K|)$. Then, if $\pi_{n+1}(|K|)$ is finitely generated, we can take a finite set as Λ.

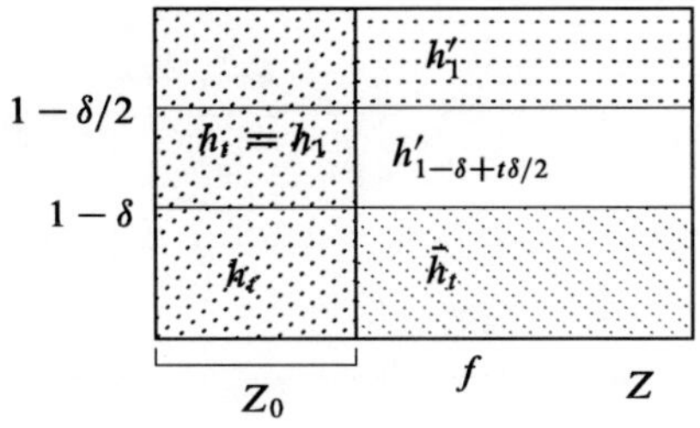

Fig. 7.4 Extending a homotopy h

By the same argument, it is easy to prove that $\partial\Delta^{n+1}$ is a mapping $(n+1)$-deformation retract of $|K(\mathbb{Z}, n)|$ for compact spaces.

To show that $\partial\Delta^{n+1}$ is a mapping $(n+1)$-deformation retract of $|K(\mathbb{Z}, n)|$ for metrizable spaces, we use the following lemma:

Lemma 7.8.4. *Suppose that A is a mapping $(n+1)$-deformation retract of an ANE X for metrizable spaces. Let $f : Z \to X$ be a map of a metrizable space Z with $\dim Z \le n+1$ and a homotopy $h : Z_0 \times \mathbf{I} \to X$ of a closed set Z_0 in Z with $\delta > 0$ such that $h_0 = f|Z_0$, $h_1(Z_0) \subset A$, $h_t = h_1$ for every $t \in [1-\delta, 1]$, and $h_t|Z_0 \cap f^{-1}(A) = f|Z_0 \cap f^{-1}(A)$ for every $t \in \mathbf{I}$. Then, h extends to a homotopy $\tilde{h} : Z \times \mathbf{I} \to X$ such that $\tilde{h}_0 = f$, $\tilde{h}_1(Z) \subset A$, $\tilde{h}_t = \tilde{h}_1$ for every $t \in [1-\delta/2, 1]$, and $\tilde{h}_t|f^{-1}(A) = f|f^{-1}(A)$ for every $t \in \mathbf{I}$.*

Proof. Because X is an ANE, h can be extended to a homotopy $\bar{h} : Z \times \mathbf{I} \to X$ such that $\bar{h}_0 = f$ and $\bar{h}_t|f^{-1}(A) = f|f^{-1}(A)$ for every $t \in \mathbf{I}$. Since A is a mapping $(n+1)$-deformation retract of X, we have a homotopy $h' : Z \times \mathbf{I} \to X$ such that

$$h'_0 = \bar{h}_{1-\delta},\ h'_1(Z) \subset A \ \text{ and } \ h'_t|\bar{h}_{1-\delta}^{-1}(A) = \bar{h}_{1-\delta}|\bar{h}_1^{-1}(A) \ \text{ for every } t \in \mathbf{I}.$$

The desired homotopy $\tilde{h} : Z \times \mathbf{I} \to X$ can be defined as follows:

$$\tilde{h}_t = \begin{cases} \bar{h}_t & \text{if } t \le 1-\delta, \\ h'_{2(t-1+\delta)/\delta} & \text{if } 1-\delta \le t \le 1-\delta/2, \\ h'_1 & \text{if } t \ge 1-\delta/2. \end{cases}$$

Indeed, $\tilde{h}_0 = \bar{h}_0 = f$ and $\tilde{h}_t = h'_1$ for every $t \in [1-\delta/2, 1]$. Since $Z_0 \cup f^{-1}(A) \subset \bar{h}_1^{-1}(A)$, it follows that $h'_t|Z_0 = \bar{h}_1|Z_0 = h_1$ and $h'_t|f^{-1}(A) = \bar{h}_1|f^{-1}(A) = f$ for every $t \in \mathbf{I}$. Then, $\tilde{h}|Z_0 \times \mathbf{I} = h$ and $\tilde{h}_t|f^{-1}(A) = f$ for every $t \in \mathbf{I}$. — Fig. 7.4. □

Proposition 7.8.5. *For each $n \ge 2$, $\partial\Delta^{n+1}$ is a mapping $(n+1)$-deformation retract of $|K(\mathbb{Z}, n)|$ for metrizable spaces.*

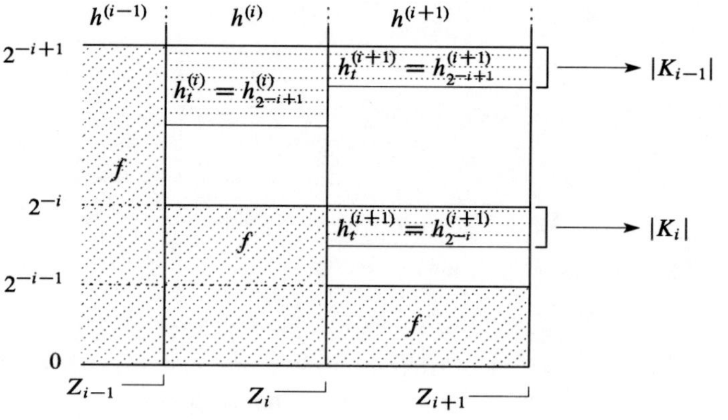

Fig. 7.5 Homotopies $h^{(i)}$

Proof. Observe that $|K(\mathbb{Z}, n)| = \varinjlim |K_k|$, where $(K_i)_{i\in\omega}$ is the tower defining $K(\mathbb{Z}, n)$. Then, each $|K_i|$ is a mapping $(n+1)$-deformation retract of K_{i+1} for normal spaces, hence for metrizable spaces. Let $f : Z \to |K(\mathbb{Z}, n)|$ be a map from a metrizable space Z with $\dim Z \leq n+1$. For each $i \in \omega$, let $Z_i = f^{-1}(|K_i|)$. Then, $Z_0 \subset Z_1 \subset \cdots$ and $Z = \bigcup_{i\in\omega} Z_i$. By induction, we define homotopies $h^{(i)} : Z_i \times \mathbf{I} \to |K_i|$, $i \in \omega$, such that

$$h^{(i)}|Z_{i-1} \times \mathbf{I} = h^{(i-1)},\ h_t^{(i)} = f|Z_i \ \text{ for } 0 \leq t \leq 2^{-i},$$

$$h_t^{(i)}|Z_i \cap f^{-1}(\partial\Delta^{n+1}) = f|Z_i \cap f^{-1}(\partial\Delta^{n+1}) \ \text{ for } t \in \mathbf{I},$$

$$h^{(i)}(Z_i \times [2^{-j}, 2^{-j+1}]) \subset |K_j|,\ h_{2^{-j}}^{(i)}(Z_i) \subset |K_j| \ \text{ for } j < i \text{ and}$$

$$h_t^{(i)} = h_{2^{-j}}^{(i)} \ \text{ for } 2^{-j} - 2^{-i-1} \leq t \leq 2^{-j},\ j < i.$$

The homotopy $h^{(0)} : Z_0 \times \mathbf{I} \to |K_0|$ is defined by $h_t^{(0)} = f|Z_0$ for $t \in \mathbf{I}$. Assume that $h^{(i)}$ has been defined. For each $j = 0, 1, \ldots, i+1$, let

$$L_j = \big(Z_i \times \mathbf{I}\big) \cup \big(Z_{i+1} \times [0, 2^{-j}]\big).$$

The homotopy $h^{(i)}$ can be extended over L_{i+1} by $(f|Z_{i+1}) \times \mathrm{id}$. Then, using Lemma 7.8.4 iteratively, we extend this over $L_i \subset \cdots \subset L_1 \subset L_0 = Z_{i+1} \times \mathbf{I}$ step-by-step. Thus, we can obtain the homotopy $h^{(i+1)} : Z_{i+1} \times \mathbf{I} \to |K_{i+1}|$ satisfying the conditions. — Fig. 7.5.

According to Theorem 2.8.6(2), each $z \in Z$ has a neighborhood V in Z such that $f(V) \subset |K_i|$ for some $i \in \omega$, which means that $V \subset Z_i$, so $z \in \operatorname{int} Z_i$. Therefore, $Z = \bigcup_{i\in\omega} \operatorname{int} Z_i$. Thus, we can define a homotopy $h : Z \times \mathbf{I} \to |K(\mathbb{Z}, n)|$ by $h|Z_i \times$

$\mathbf{I} = h^{(i)}$ for each $i \in \omega$. Then, $h_0 = f$, $h_1(Z) \subset \partial\Delta^{n+1}$ and $h_t|f^{-1}(\partial\Delta^{n+1}) = f|f^{-1}(\partial\Delta^{n+1})$ for every $t \in \mathbf{I}$. □

The homotopical uniqueness of the Eilenberg–MacLane spaces can be derived from the following theorem:

Theorem 7.8.6. *Let $n \in \mathbb{N}$ and suppose that X is a path-connected space homotopy dominated by a simplicial complex such that $\pi_n(X) \cong \mathbb{Z}$ and $\pi_m(X) = 0$ for every $m \neq n$. Then, X has the homotopy type of the simplicial Eilenberg–MacLane complex $K(\mathbb{Z}, n)$ constructed above.*

Proof. The case $n = 1$ has been seen already. For $n \geq 2$, let $(K_i)_{i\in\omega}$ be the tower in the definition of $K(\mathbb{Z}, n)$. Take a homeomorphism $\varphi : |K_0| \to \mathbf{S}^n$ and a map $\alpha : \mathbf{S}^n \to X$ such that $[\alpha]$ is a generator of $\pi_n(X) \cong \mathbb{Z}$. Let $f_0 = \alpha\varphi : |K_0| \to X$. To extend f_0 to a map $f : |K(\mathbb{Z}, n)| \to X$, we inductively construct maps $f_i : |K_i| \to X$, $i \in \mathbb{N}$, such that $f_i||K_{i-1}| = f_{i-1}$, and then f can be defined by $f||K_i| = f_i$. In fact, the construction of K_i from K_{i-1} is as follows: Taking a simplicial map $h : \bigoplus_{\lambda\in\Lambda} L_\lambda \to K_{i-1}$ such that each L_λ is a triangulation of $\mathbf{S}^{n+i}$ and $\{[h|L_\lambda] \mid \lambda \in \Lambda\} = \pi_{n+i}(|K_{i-1}|)$ (or $\{[h|L_\lambda] \mid \lambda \in \Lambda\}$ generates $\pi_{n+i}(|K_{i-1}|)$), we define $K_i = Z(h) \cup \bigoplus_{\lambda\in\Lambda} K_\lambda$, where K_λ is a triangulation of $\mathbf{B}^{n+i+1}$ with $L_\lambda \subset K_\lambda$. For each $\lambda \in \Lambda$, the map $f_{i-1}h|L_\lambda : \mathbf{S}^{n+i} \to X$ is null-homotopic because of $\pi_{n+i}(X) = \{0\}$. Hence, f_{i-1} extends over $Z(h|L_\lambda) \cup K_\lambda$. Thus, f_{i-1} extends to a map $f_i : |K_i| \to X$.

It remains to show that f is a homotopy equivalence. By virtue of Corollary 4.13.10, it suffices to verify that f is a weak homotopy equivalence. For $m \neq n$, $f_\sharp : \pi_m(|K(\mathbb{Z}, n)|) \to \pi_m(X)$ is an isomorphism because $\pi_m(|K(\mathbb{Z}, n)|) = \pi_m(X) = \{0\}$. Recall that the inclusion $|K_0| = \partial\Delta^{n+1} \subset |K(\mathbb{Z}, n)|$ is an $(n+1)$-equivalence, hence it induces the isomorphism from $\pi_n(|K_0|)$ onto $\pi_n(|K(\mathbb{Z}, n)|)$. On the other hand, $(f_0)_\sharp : \pi_n(|K_0|) \to \pi_n(X)$ is an isomorphism because it sends a generator of $\pi_n(|K_0|) \cong \mathbb{Z}$ to that of $\pi_n(X) \cong \mathbb{Z}$. Then, it follows that $f_\sharp : \pi_n(|K(\mathbb{Z}, n)|) \to \pi_n(X)$ is also an isomorphism. Consequently, f is a weak homotopy equivalence. □

For a pair (X, A) of spaces, a map $f : (Y, B) \to (Z, C)$ induces the map

$$f_* : \mathrm{C}((X, A), (Y, B)) \to \mathrm{C}((X, A), (Z, C)),$$

which is defined by $f_*(k) = f \circ k$ (cf. 1.1.3(1)).

Lemma 7.8.7. *Let $f, f' : (Y, B) \to (Z, C)$ with $f \simeq f'$ (as maps of pairs). For each pair (X, A) of spaces,*

$$f_* \simeq f'_* : \mathrm{C}((X, A), (Y, B)) \to \mathrm{C}((X, A), (Z, C)).$$

Proof. Let $h : (Y \times \mathbf{I}, B \times \mathbf{I}) \to (Z, C)$ be a homotopy from f to f'. We define

$$\tilde{h} : \mathrm{C}((X, A), (Y, B)) \times \mathbf{I} \to \mathrm{C}((X, A), (Z, C))$$

by $\tilde{h}(g,t) = h_t \circ g$. To see the continuity of $\tilde{h}$, for each compact set $K \subset X$ and each open set $U \subset Z$, let $(g,t) \in \tilde{h}^{-1}(\langle K; U\rangle)$. Since $h(g(K) \times \{t\}) = h_t(g(K)) \subset U$, we can choose open sets $V \subset Y$ and $W \subset \mathbf{I}$ so that $g(K) \times \{t\} \subset V \times W$ and $h(V \times W) \subset U$. Then, $g \in \langle K; V\rangle$ and $t \in W$. For each $g' \in \langle K; V\rangle$ and $s \in W$,

$$h_s(g'(K)) = h(g'(K) \times \{t'\}) \subset h(V \times W) \subset U,$$

hence $\tilde{h}(g',t') \in \langle K; U\rangle$. Thus, we have $\langle K; V\rangle \times W \subset \tilde{h}^{-1}(\langle K; U\rangle)$. Consequently, $\tilde{h}^{-1}(\langle K; U\rangle)$ is open in $\mathrm{C}(X,Y) \times \mathbf{I}$. □

Proposition 7.8.8. *For each pair (X,A) of spaces, $(Y,B) \simeq (Y',B')$ implies*

$$\mathrm{C}((X,A),(Y,B)) \simeq \mathrm{C}((X,A),(Y',B')).$$

Proof. Let $f : (Y,B) \to (Y',B')$ be a homotopy equivalence with g a homotopy inverse. Then, f and g induce the maps

$$f_* : \mathrm{C}((X,A),(Y,B)) \simeq \mathrm{C}((X,A),(Y',B')) \text{ and}$$
$$g_* : \mathrm{C}((X,A),(Y',B')) \simeq \mathrm{C}((X,A),(Y,B))$$

defined by $f_*(k) = f \circ k$ and $g_*(k) = g \circ k$, respectively. By virtue of Lemma 7.8.7, we have $g_* f_* = (gf)_* \simeq \mathrm{id}$ and $f_* g_* = (fg)_* \simeq \mathrm{id}$. □

Corollary 7.8.9. *For each simplicial complex K with $v_0 \in K^{(0)}$, the loop space $\Omega(|K|, x_0)$ has the homotopy type of a simplicial complex.*

Proof. By virtue of Proposition 7.8.8,

$$\Omega(|K|, x_0) = \mathrm{C}((\mathbf{I}, \partial\mathbf{I}), (|K|, x_0)) \simeq \mathrm{C}((\mathbf{I}, \partial\mathbf{I}), (|K|_{\mathrm{m}}, x_0)),$$

where the last space is an ANR by 1.1.3(5) and 6.1.9(9). Then, we have the result by Corollary 6.6.5. □

Recall that $\pi_n(\Omega(X,x_0), c_{x_0}) \cong \pi_{n+1}(X,x_0)$ for any pointed space (X,x_0) and $n \in \mathbb{N}$ (Theorem 4.14.4). Then, combining Theorem 7.8.6 with Corollary 7.8.9, we have the following corollary:

Corollary 7.8.10. *For each $n \in \mathbb{N}$, the loop space $\Omega(|K(\mathbb{Z}, n+1)|, v)$ has the homotopy type of $|K(\mathbb{Z}, n)|$, where $v \in |K(\mathbb{Z}, n+1)|$ is any point.* □

The **wedge sum** (or the **one-point union**) $X_1 \vee \cdots \vee X_k$ of pointed spaces $X_1 = (X_1, x_1), \ldots, X_k = (X_k, x_k)$ is defined as the quotient space

$$X_1 \vee \cdots \vee X_k = (X_1 \oplus \cdots \oplus X_k)/\{x_1, \ldots, x_n\},$$

which is also defined as the following subspace of the product space $\prod_{i=1}^{n} X_i$:

$$X_1 \vee \cdots \vee X_k = \{x \in \prod_{i=1}^{k} X_i \mid x(i) = x_i \text{ except for one } i\}.$$

When $(X_i, x_i) = (X, x_0)$ for every $i = 1, \ldots, k$, we write $\bigvee^k X$ instead of $X \vee \cdots \vee X$.

We use the following theorem but leave the proof to any textbook on Homotopy Theory or Algebraic Topology.

Theorem 7.8.11. *For every $n \geq 2$ and $k \in \mathbb{N}$, $\pi_i(\bigvee^k \mathbf{S}^n) = \{0\}$ for $i < n$ and $\pi_n(\bigvee^k \mathbf{S}^n) \cong \mathbb{Z}^k$.*[9]

Proposition 7.8.12. *For each $m > n$, $|F(\Delta^m)^{(n)}| \simeq \bigvee^k \mathbf{S}^n$, where $k = {}_mC_{n+1}$. Consequently, $\pi_i(|F(\Delta^m)^{(n)}|) = \{0\}$ for $i < n$ and $\pi_n(|F(\Delta^m)^{(n)}|) \cong \mathbb{Z}^k$.*

Proof. Since $P = |\mathrm{St}(\mathbf{e}_{m+1}, F(\Delta^m)^{(n)})|$ is a compact AR, the quotient space $|F(\Delta^m)^{(n)}|/P$ is a compact ANR by Theorem 6.5.3, where the quotient map $q : |F(\Delta^m)^{(n)}| \to |F(\Delta^m)^{(n)}|/P$ is a (fine) homotopy equivalence by Theorem 7.5.4. Thus, we have $|F(\Delta^m)^{(n)}| \simeq |F(\Delta^m)^{(n)}|/P$. The latter space is homeomorphic to $\bigvee^k \mathbf{S}^n$. □

In the construction of the simplicial Eilenberg–MacLane complex $K(\mathbb{Z}, n)$, replacing the starting complex $F(\partial\Delta^{n+1})$ with a connected simplicial complex L, we can obtain a connected simplicial complex $K(L, n)$ such that $\pi_i(|K(L, n)|) = \{0\}$ for $i > n$ and $|L|$ is a mapping $(n+1)$-deformation retract of $|K(L, n)|$ for metrizable spaces. Then, the inclusion $|L| \subset |K(L, n)|$ is an $(n+1)$-equivalence, hence $\pi_i(|K(L, n)|) \cong \pi_i(|L|)$ for $i \leq n$.

Now, for each $n \geq 2$ and $k \in \mathbb{N}$, let L be a triangulation of $\bigvee^k \mathbf{S}^n$. We can obtain the **simplicial Eilenberg–MacLane complex** $K(\mathbb{Z}^k, n)$ such that $|L|$ is a mapping $(n+1)$-deformation retract of $|K(\mathbb{Z}^k, n)|$ for metrizable spaces. It should be noted that $K(\mathbb{Z}^k, n)$ is also countable. In the same way as for Theorem 7.8.6, we can prove the homotopical uniqueness of $K(\mathbb{Z}^k, n)$. Indeed, let X be a path-connected space homotopy dominated by a simplicial complex such that $\pi_n(X) \cong \mathbb{Z}^k$ and $\pi_i(X) = \{0\}$ for every $i \neq n$. Using generators of $\pi_n(X)$, we define a map $\varphi : |L| \approx \bigvee^k \mathbf{S}^n \to X$. By the construction of $K(\mathbb{Z}^k, n)$, we can inductively extend φ to a map $f : |K(\mathbb{Z}^k, n)| \to X$. Then, f is a weak homotopy equivalence, so f is a homotopy equivalence by Corollary 4.13.10.

Since $\pi_n(|K(\mathbb{Z}, n)|^k) \cong \mathbb{Z}^k$ and $\pi_i(|K(\mathbb{Z}, n)|^k) = \{0\}$ for $i \neq n$ by Proposition 4.14.1, we have the following:

Proposition 7.8.13. *For each $n \geq 2$ and $k \in \mathbb{N}$, the product space $|K(\mathbb{Z}, n)|^k$ has the homotopy type of $|K(\mathbb{Z}^k, n)|$.* □

---

[9]Here, $\mathbb{Z}^k \cong \mathbb{Z} \oplus \cdots \oplus \mathbb{Z}$ (k many). When $n = 1$, this is not true but $\pi_1(\bigvee^k \mathbf{S}^1)$ is isomorphic to the free product $\mathbb{Z} * \cdots * \mathbb{Z}$ (k many) by Theorem of Seifert and Van Kampen.

In the rest of this section, we will construct the so-called simplicial Edwards–Walsh complex, which will be used in Sect. 7.10.

Proposition 7.8.14. *Let K be a simplicial complex and $n \geq 2$. There exist countable simplicial complexes K_σ, $\sigma \in K$, such that*

(1) $|K_\sigma| = \sigma$ *if* $\dim \sigma \leq n$;
(2) $K_{\sigma'} \subset K_\sigma$ *if* $\sigma' < \sigma$;
(3) $|K_\sigma| \cap |K_{\sigma'}| = |K_{\sigma \cap \sigma'}|$ *for each* $\sigma, \sigma' \in K$;
(4) $(|K_\sigma|, \partial\sigma) \approx (|K(\mathbb{Z}, n)|, \partial \Delta^{n+1})$ *for each* $(n+1)$*-simplex* $\sigma \in K$;
(5) $|K_\sigma| \simeq |K(\mathbb{Z}, n)|^{k(m)}$ *if* $\dim \sigma = m > n+1$ *(where* $k(m) = {}_m C_{n+1}$*)*;
(6) *If* $\dim \sigma \geq n+1$, *then* $|K_{\partial\sigma}|$ *is a mapping* $(n+1)$*-deformation retract of* $|K_\sigma|$ *for metrizable spaces, where* $K_{\partial\sigma} = \bigcup_{\sigma' < \sigma} K_{\sigma'}$.

Proof. For every $\sigma \in K^{(n)}$, let $K_\sigma = F(\sigma)$. Suppose that K_σ have been defined for all $\sigma \in K^{(m)}$. For each $(m+1)$-simplex $\sigma \in K$, let

$$L_{\sigma,i} = \bigcup \{K_{\sigma'} \mid \sigma' < \sigma,\ \dim \sigma' = i\},\ n < i \leq m.$$

Then, $F(\sigma)^{(n)} = L_{\sigma,n} \subset L_{\sigma,n+1} \subset \cdots \subset L_{\sigma,m} = K_{\partial\sigma}$ are simplicial complexes by (1), (2), and (3). Using (3) and (6), we can show that $|L_{\sigma,i}|$ is a mapping $(n+1)$-deformation retract of $|L_{\sigma,i+1}|$ for metrizable spaces. As observed in the above, we can construct a simplicial complex $K_\sigma = K(K_{\partial\sigma}, n)$ such that $\pi_i(|K_\sigma|) = \{0\}$ for $i > n$ and $|K_{\partial\sigma}|$ is a mapping $(n+1)$-deformation retract of $|K_\sigma|$ for metrizable spaces. Then, it follows that $\sigma^{(n)} = |L_{\sigma,n}|$ is a mapping $(n+1)$-deformation retract of $|K_\sigma|$ for metrizable spaces, which implies that the inclusion $\sigma^{(n)} \subset |K_\sigma|$ is an $(n+1)$-equivalence. Since $\pi_i(|K_\sigma|) \cong \pi_i(\sigma^{(n)})$ for $i \leq n$, we have $\pi_i(|K_\sigma|) = \{0\}$ for $i < n$ and $\pi_n(|K_\sigma|) \cong \mathbb{Z}^{k(m+1)}$ by Proposition 7.8.12. From the homotopy uniqueness and Proposition 7.8.13, it follows that

$$|K_\sigma| \simeq |K(\mathbb{Z}^{k(m+1)}, n)| \simeq |K(\mathbb{Z}, n)|^{k(m+1)}.$$

Thus, we have a simplicial complex K_σ satisfying (2), (5), and (6). Moreover, we can construct K_σ for every $(m+1)$-simplex $\sigma \in K$ so that $|K_\sigma| \cap |K_{\sigma'}| = |K_{\sigma \cap \sigma'}|$, that is, (3) is satisfied. By induction, we obtain the desired result. □

In Proposition 7.8.14, we have the following simplicial complex:

$$EW(K, n) = \bigcup_{\sigma \in K} K_\sigma = K^{(n)} \cup \bigcup_{\sigma \in K \setminus K^{(n)}} K_\sigma.$$

We call $EW(K, n)$ the **simplicial Edwards–Walsh complex** for K. Since each K_σ is countable, if K is countable then so is $EW(K, n)$. There exists a map $\varpi : |EW(K, n)| \to |K|$ such that $\varpi | |K^{(n)}| = \mathrm{id}$ and

$$\varpi^{-1}(\sigma) = |K_\sigma| \simeq |K(\mathbb{Z}, n)|^{k(\dim \sigma)} \text{ for each } \sigma \in K \setminus K^{(n)},$$

which is called an **associate map** of $EW(K,n)$. Indeed, we inductively define maps $\varpi_\sigma : |K_\sigma| \to \sigma, \sigma \in K$, such that $\varpi_\sigma^{-1}(\partial\sigma) = |K_{\partial\sigma}|$ and $\varpi_\sigma||K_{\sigma'}| = \varpi_{\sigma'}$ for each $\sigma' < \sigma$. Let $\varpi_\sigma = \mathrm{id}$ for every $\sigma \in K^{(n)}$ and assume that $\varpi_\sigma : |K_\sigma| \to \sigma$ have been defined for every $\sigma \in K^{(m)}$. For each $(m+1)$-simplex $\sigma \in K$, let $\varpi_{\partial\sigma} : |K_{\partial\sigma}| \to \partial\sigma$ be the map defined by $\varpi_{\partial\sigma}||K_{\sigma'}| = \varpi_{\sigma'}$ for each $\sigma' < \sigma$. Since $\varpi_{\partial\sigma} \simeq 0$ in σ, $\varpi_{\partial\sigma}$ can be extended to a map $\varpi_\sigma : |K_\sigma| \to \sigma$ by the Homotopy Extension Theorem 4.3.3, where we can modify ϖ_σ to satisfy the condition $\varpi_\sigma^{-1}(\partial\sigma) = |K_{\partial\sigma}|$ because $\operatorname{rint}\sigma$ is homotopy dense in σ and $|K_{\partial\sigma}|$ is a zero set in $|K_\sigma|$, i.e., $f^{-1}(0) = |K_{\partial\sigma}|$ for some map $f : |K_\sigma| \to \mathbf{I}$ (cf. Proposition 4.2.2).[10] Then, the map ϖ is defined by $\varpi||K_\sigma| = \varpi_\sigma$ for each $\sigma \in K$. We prove the following proposition regarding the map $\varpi : |EW(K,n)| \to |K|$:

Proposition 7.8.15. *For each simplicial complex K and $n \geq 2$, let $\varpi : |EW(K,n)| \to |K|$ be an associate map of the simplicial Edwards–Walsh complex $EW(K,n)$. Then, for each subcomplex L of K and a map $g : |L| \to |K(\mathbb{Z},n)|$, the composition $g\varpi|\varpi^{-1}(|L|)$ extends over $|EW(K,n)|$.*

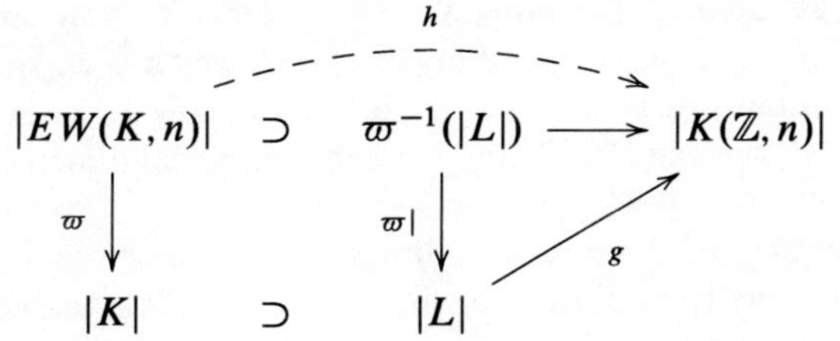

Proof. Because $|K(\mathbb{Z},n)|$ is $(n-1)$-connected, we can extend g over $|L| \cup |K^{(n)}|$ by skeleton-wise induction. Then, we can assume that $K^{(n)} \subset L$. For each $(n+1)$-simplex $\sigma \in K \setminus L$, $|K_{\partial\sigma}| = \partial\sigma$ and $(|K_\sigma|, \partial\sigma) \approx (|K(\mathbb{Z},n)|, \partial\Delta^{n+1})$, hence $\partial\sigma$ is a mapping $(n+1)$-deformation retract of $|K_\sigma|$ for metrizable spaces. By Proposition 7.8.1, $\partial\sigma$ is a retract of $|K_\sigma^{(n+1)}|$. Therefore, the composition $g\varpi|\varpi^{-1}(|L|)$ extends to a map

$$h' : \varpi^{-1}(|L|) \cup |EW(K,n)^{(n+1)}| \to |K(\mathbb{Z},n)|.$$

Since $\pi_m(|K(\mathbb{Z},n)|) = \{0\}$ for every $m > n$, we can extend h' over $|EW(K,n)|$ by skeleton-wise induction. □

[10]Since $\operatorname{Sd} L$ is a full subcomplex of $\operatorname{Sd} K_\sigma$, such a simplicial map f can be defined by $f((\operatorname{Sd} L)^{(0)}) = \{0\}$ and $f((\operatorname{Sd} K_\sigma)^{(0)} \setminus (\operatorname{Sd} L)^{(0)}) = \{1\}$.

7.9 Cohomological Dimension

In this section, using the Eilenberg–MacLane complexes defined in the previous section, we define the cohomological dimension and discuss its relationship with the (covering) dimension. We apply the cohomological dimension to prove that $\dim X \times P = \dim X + \dim P$ for every metrizable space X and every locally compact (or metric) polyhedron P (Theorem 7.9.7). This result was announced in Sect. 5.4.

In Theorem 5.2.3, the (covering) dimension is characterized as follows:

- For a normal space X, $\dim X \leq n$ if and only if, for each $m \geq n$ and each closed set A in X, every map $f : A \to \mathbf{S}^m$ extends over X.

Replacing $\mathbf{S}^m$ by $|K(\mathbb{Z}, m)|$, we define $\dim_{\mathbb{Z}} X \leq n$,[11] that is,

- $\dim_{\mathbb{Z}} X \leq n$ if and only if, for each $m \geq n$ and each closed set A in X, every map $f : A \to |K(\mathbb{Z}, m)|$ extends over X.

It is clear from this definition that $\dim_{\mathbb{Z}} X \leq n$ implies $\dim_{\mathbb{Z}} X \leq m$ for any $m > n$. Then, it can be defined that $\dim_{\mathbb{Z}} X = n$ if $\dim_{\mathbb{Z}} X \leq n$ and $\dim_{\mathbb{Z}} X \not< n$. Moreover, $\dim_{\mathbb{Z}} X = \infty$ means that $\dim_{\mathbb{Z}} X \not\leq n$ for any $n \in \mathbb{N}$. We call $\dim_{\mathbb{Z}} X$ the **cohomological dimension** of X.[12] Using the Eilenberg–MacLane complex $K(G, n)$ instead of $K(\mathbb{Z}, n)$, we can also define the cohomological dimension $\dim_G X$ with respect to G.

Due to Theorem 5.2.3, to assert $\dim X \leq n$, it suffices to examine the extensibility for maps from closed sets in X to the n-dimensional sphere $\mathbf{S}^n$, and it is not necessary to examine for $\mathbf{S}^m$ for $m > n$. For the cohomological dimension, we have the same situation, sated as follows:

Theorem 7.9.1. *For a metrizable space X, $\dim_{\mathbb{Z}} X \leq n$ if and only if X satisfies the following condition:*

(e)$_n$ *Every map $f : A \to |K(\mathbb{Z}, n)|$ of each closed set A in X extends over X.*

The "only if" part of Theorem 7.9.1 is obvious and the "if" part can be obtained by induction and the following lemma:

Lemma 7.9.2. *For a metrizable space X, the condition* (e)$_n$ *implies* (e)$_{n+1}$.

To prove this lemma, we use a path space. The path space $P(X, x_0)$ on a pointed space (X, x_0) is defined as follows:

$$P(X, x_0) = \mathrm{C}((\mathbf{I}, 0), (X, x_0)),$$

---

[11]In Chigogidze's book "Inverse Spectra," the notation $\dim_{\mathbb{Z}} X$ is used.

[12]The cohomological dimension $\dim_{\mathbb{Z}} X$ was originally defined as the maximum of the number $n \in \omega$ such that $\check{H}^n(X, A) \neq 0$ for every closed set A in X, where $\check{H}^n(X, A)$ is the n-th relative Čech cohomology group. This is why $\dim_{\mathbb{Z}}$ is called the cohomological dimension.

and it admits the compact-open topology. Let $e_1 : P(X, x_0) \to X$ be the map defined by $e_1(\alpha) = \alpha(1)$ (cf. 1.1.3(3)). Then, $e_1^{-1}(x_0)$ is simply the loop space $\Omega(X, x_0)$ (cf. Sect. 4.14).

A map $p : E \to B$ is called a **Hurewicz fibration** if it has the **homotopy lifting property** for an arbitrary space Z, that is, given a homotopy $h : Z \times \mathbf{I} \to B$ and a map $f : Z \to E$ with $pf = h_0$, there exists a homotopy $\tilde{h} : Z \times \mathbf{I} \to E$ such that $p\tilde{h} = h$ and $\tilde{h}_0 = f$.

Proposition 7.9.3. *For every pointed space (X, x_0), the following statements hold:*

(1) *The path space $P(X, x_0)$ is contractible;*
(2) *The map $e_1 : P(X, x_0) \to X$ is a Hurewicz fibration whose image is the path-component of X containing x_0;*
(3) *If $x \in X$ belongs to the same path-component as x_0 then $e_1^{-1}(x) \simeq e_1^{-1}(x_0) = \Omega(X, x_0)$;*
(4) *If A is a subset of the path-component of X containing x_0 and there is a contraction $h : A \times \mathbf{I} \to A$ such that $h_1(A) = \{a\}$ and $h_t(a) = a$ for every $t \in \mathbf{I}$, then $e_1^{-1}(A) \simeq \Omega(X, x_0)$.*

Proof. (1): We have the contraction $h : P(X, x_0) \times \mathbf{I} \to P(X, x_0)$ defined by $h(\alpha, t)(s) = \alpha((1 - t)s)$ for each $(\alpha, t) \in P(X, x_0) \times \mathbf{I}$ and $s \in \mathbf{I}$ (cf. 1.1.3(4)).

(2): It is trivial that the image of e_1 is the path-component of X containing x_0. To prove that $e_1 : P(X, x_0) \to X$ is a Hurewicz fibration, let $h : Z \times \mathbf{I} \to X$ be a homotopy and $f : Z \to P(X, x_0)$ a map with $e_1 f = h_0$, that is, $h_0(z) = e_1 f(z) = f(z)(1)$ for every $z \in Z$. We can define a map $h' : Z \times \mathbf{I} \times \mathbf{I} \to X$ as follows:

$$h'(z, t, s) = \begin{cases} f(z)((t + 1)s) & \text{if } 0 \le s \le (t + 1)^{-1}, \\ h(z, (t + 1)s - 1) & \text{if } (t + 1)^{-1} \le s \le 1. \end{cases}$$

Then, h' induces the map $\tilde{h} : Z \times \mathbf{I} \to \mathrm{C}(\mathbf{I}, X)$ defined by $\tilde{h}(z, t)(s) = h'(z, t, s)$. Observe that

$$\tilde{h}(z, t)(0) = f(z)(0) = x_0, \ \tilde{h}(z, 1)(s) = f(z)(s) \ \text{and}$$
$$e_1\tilde{h}(z, t) = \tilde{h}(z, t)(1) = h(z, t).$$

Thus, we have a homotopy $\tilde{h} : Z \times \mathbf{I} \to P(X, x_0)$ such that $\tilde{h}_0 = f$ and $e_1\tilde{h} = h$.

(3): Take any $\omega \in e_1^{-1}(x)$ and define maps $f : e_1^{-1}(x) \to e_1^{-1}(x_0)$ and $g : e_1^{-1}(x_0) \to e_1^{-1}(x)$ by $f(\alpha) = \alpha * \omega^{\leftarrow}$ and $g(\beta) = \beta * \omega$. It is easy to see that $gf \simeq \mathrm{id}$ and $gf \simeq \mathrm{id}$. Refer to Sect. 4.14

(4): Due to (3), it suffices to show that $e_1^{-1}(A) \simeq e_1^{-1}(a)$. By virtue of (2), we can obtain a homotopy $\tilde{h} : e_1^{-1}(A) \times \mathbf{I} \to e_1^{-1}(A)$ such that $\tilde{h}_0 = \mathrm{id}$ and $e_1\tilde{h} = h(e_1 \times \mathrm{id})$. Therefore,

$$\tilde{h}|e_1^{-1}(a) \times \mathbf{I} : e_1^{-1}(a) \times \mathbf{I} \to e_1^{-1}(a)$$

is a homotopy from id to $\tilde{h}_1|e_1^{-1}(a)$. Hence, the inclusion $e_1^{-1}(a) \subset e_1^{-1}(A)$ is a homotopy equivalence. □

The following can be proved by analogy to 6.1.9(9):

Lemma 7.9.4. *Let X_1, X_2 be closed sets in a compactum X such that $X_1 \cap X_2 = \emptyset$ or $X_1 \supset X_2$, and let Y be an ANE with $Y_1, Y_2 \subset Y$ that are also ANEs. Then, the space* $\mathrm{C}((X, X_1, X_2), (Y, Y_1, Y_2))$ *with the compact-open topology is an ANE.* □

Now, we will show Lemma 7.9.2.

Proof of Lemma 7.9.2. Let $f : A \to K(\mathbb{Z}, n+1)$ be a map from a closed set A in a metrizable space X. We will show that f extends over X. Consider the fibration

$$e_1 : E = P(|K(\mathbb{Z}, n+1)|, v_0) \to |K(\mathbb{Z}, n+1)|,$$

where $v_0 \in K(\mathbb{Z}, n+1)^{(0)}$. Then, E is an AE by Proposition 7.9.3(1) and 6.1.9(9) (or Lemma 7.9.4 above). For each subcomplex L of $K(\mathbb{Z}, n+1)$, $e_1^{-1}(|L|)$ is an ANE by Lemma 7.9.4 because

$$e_1^{-1}(|L|) = \mathrm{C}((\mathbf{I}, \{0\}, \{1\}), (|K(\mathbb{Z}, n+1)|, \{v_0\}, |L|)).$$

For each $i \in \omega$, we define

$$E_i = e_1^{-1}(|K(\mathbb{Z}, n+1)^{(i)}|) \subset E \quad\text{and}\quad A_i = f^{-1}(|K(\mathbb{Z}, n+1)^{(i)}|) \subset A.$$

Then, $E = \bigcup_{i \in \omega} E_i$ and $A = \bigcup_{i \in \omega} A_i$.

$$\begin{array}{ccccccc}
 & & e_1^{-1}(\sigma) & \subset & E_i & \subset & E = P(|K(\mathbb{Z}, n+1)|, v_0) \\
 & \overset{f_\sigma}{\nearrow} & & \overset{f_i}{\nearrow} & & & \downarrow e_1 \\
f^{-1}(\sigma) & \subset & A_i & \subset & A & \xrightarrow[f]{} & |K(\mathbb{Z}, n+1)|
\end{array}$$

We will inductively construct the maps $f_i : A_i \to E_i$, $i \in \omega$, such that $f_i|A_{i-1} = f_{i-1}$ and $f_i(f^{-1}(\sigma)) \subset e_1^{-1}(\sigma)$ for every $\sigma \in K(\mathbb{Z}, n+1)^{(i)}$. Let $f_0 : A_0 \to E_0$ be a map such that $e_1 f_0 = f|A_0$. For instance, it can be defined so that $f_0(f^{-1}(v))$ is a singleton for each vertex $v \in K(\mathbb{Z}, n+1)^{(0)}$. Assume that f_{i-1} has been constructed. For each i-simplex $\sigma \in K(\mathbb{Z}, n+1)$, $f_{i-1}(f^{-1}(\partial\sigma)) \subset e_1^{-1}(\partial\sigma)$. Due to Proposition 7.9.3(4) and Corollary 7.8.10,

$$e_1^{-1}(\sigma) \simeq \Omega(|K(\mathbb{Z}, n+1)|, v_0) \simeq |K(\mathbb{Z}, n)|.$$

Hence, we can apply the condition (e)$_n$ to extend $f_{i-1}|f^{-1}(\partial\sigma)$ over X. By restricting this extension, we have a map $f_\sigma : f^{-1}(\sigma) \to e_1^{-1}(\sigma)$ such that $f_\sigma|f^{-1}(\partial\sigma) = f_{i-1}|f^{-1}(\partial\sigma)$. Then, the desired map $f_i : A_i \to E$ can be defined by $f_i|f^{-1}(\sigma) = f_\sigma|f^{-1}(\sigma)$ for each i-simplex $\sigma \in K(\mathbb{Z}, n+1)$.

Now, we define $\tilde{f} : A \to E$ by $\tilde{f}|A_i = f_i$ for every $i \in \omega$. Because A is metrizable, each $x \in A$ has a neighborhood U_x in A such that $f(U_x) \subset |K(\mathbb{Z}, n+1)^{(k)}|$ for some $k \in \omega$ (cf. 4.2.16(5)). In other words, $U_x \subset A_k$, which means $\tilde{f}|U_x = f_k|U_x$. Therefore, $\tilde{f}$ is continuous. Moreover, for each $x \in A$, $f(x)$ is contained in an i-simplex $\sigma \in K(\mathbb{Z}, n+1)$. Then,

$$\tilde{f}(x) = f_i(x) \in f_i(f^{-1}(\sigma)) \subset e_1^{-1}(\sigma),$$

i.e., $e_1\tilde{f}(x) \in \sigma$. Thus, $e_1\tilde{f}$ is contiguous to f, which implies $e_1\tilde{f} \simeq f$. Since E is contractible by Proposition 7.9.3(1), the map $\tilde{f}$ extends over X, hence so does $e_1\tilde{f}$. Therefore, f also extends over X by the Homotopy Extension Theorem 6.4.1. □

Because $\mathbf{S}^1 \approx |K(\mathbb{Z}, 1)|$, the following is a direct consequence of Theorems 5.2.3 and 7.9.1:

- For every metrizable space X, $\dim_{\mathbb{Z}} X \leq 1$ if and only if $\dim X \leq 1$.

For finite-dimensional metrizable spaces, the cohomological dimension coincides with the (covering) dimension, which can be stated as follows:

Theorem 7.9.5. *For every metrizable space X, $\dim_{\mathbb{Z}} X \leq \dim X$. If X is finite-dimensional, then $\dim_{\mathbb{Z}} X = \dim X$.*

Proof. The case $\dim X = \infty$ is trivial. Then, it suffices to show the finite-dimensional case, i.e., $\dim X = n < \infty$ implies $\dim_{\mathbb{Z}} X = n$.

Let $f : A \to |K(\mathbb{Z}, n)|$ be a map from a closed set A in X. Since $\partial\Delta^{n+1}$ is a mapping $(n+1)$-deformation retract of $|K(\mathbb{Z}, n)|$ for metrizable spaces, f is homotopic to a map $g : A \to \partial\Delta^{n+1}$. Because of $\dim X = n$, g extends to a map $\tilde{g} : X \to \partial\Delta^{n+1} \approx \mathbf{S}^n$. Since $|K(\mathbb{Z}, n)|$ is an ANE, f also extends over X by the Homotopy Extension Theorem 6.4.1. Hence, we have $\dim_{\mathbb{Z}} X \leq n$.

Now, assume that $\dim_{\mathbb{Z}} X \leq n-1$. Let A be a closed set in X. Then, every map $f : A \to \partial\Delta^n \approx \mathbf{S}^{n-1}$ extends to a map $\tilde{f} : X \to |K(\mathbb{Z}, n-1)|$. Since $\partial\Delta^n$ is a mapping n-deformation retract of $|K(\mathbb{Z}, n-1)|$ for metrizable spaces, we have a map $g : X \to \partial\Delta^n$ such that $g|A = \tilde{f}|A = f$, that is, g is an extension of f. This means that $\dim X \leq n-1$, which is a contradiction. Consequently, $\dim_{\mathbb{Z}} X = n$. □

Now, using the cohomological dimension, we can prove the following lemma:

Lemma 7.9.6. *For every metrizable space X, $\dim X \times \mathbf{I} = \dim X + 1$, where we mean $\infty + 1 = \infty$.*

Proof. We may assume that $\dim X = n < \infty$. Since $\dim X \times \mathbf{I} \leq n+1$ by the Product Theorem 5.4.9, it suffices to show that $\dim X \times \mathbf{I} \geq n+1$. Since $X \times \mathbf{I}$ contains a copy of $\mathbf{I}$, we have the case $n = 0$. Consequently, we may assume that

$n > 0$. Since $\dim X = \dim_{\mathbb{Z}} X = n$ by Theorem 7.9.5, we have a map $f : A \to |K(\mathbb{Z}, n-1)|$ of a closed set A in X that cannot extend over X. Take a point $x_0 \in A$ and let $v = f(x_0) \in |K(\mathbb{Z}, n-1)|$. Let $i : A \subset X$ be the inclusion. Then, the following map is not surjective:

$$i^* : \mathrm{C}((X, x_0), (|K(\mathbb{Z}, n-1)|, v)) \to \mathrm{C}((A, x_0), (|K(\mathbb{Z}, n-1)|, v)).$$

Since the loop space $\Omega(|K(\mathbb{Z}, n)|, v)$ has the homotopy type of $K(\mathbb{Z}, n-1)$ by Corollary 7.8.10, we can replace $|K(\mathbb{Z}, n-1)|$ by $\Omega(|K(\mathbb{Z}, n)|, v)$. According to Proposition 4.14.2, we have the following commutative diagram:

$$\begin{array}{ccc}
\mathrm{C}((X \times \mathbf{I}, H_X), (|K(\mathbb{Z}, n)|, v)) & \xrightarrow{\varphi} & \mathrm{C}((X, x_0), (\Omega(|K(\mathbb{Z}, n)|, v), c_v)) \\
{\scriptstyle (i \times \mathrm{id}_{\mathbf{I}})^*} \downarrow & & \downarrow {\scriptstyle i^*} \\
\mathrm{C}((A \times \mathbf{I}, H_A), (|K(\mathbb{Z}, n)|, v)) & \xrightarrow{\varphi} & \mathrm{C}((A, x_0), (\Omega(|K(\mathbb{Z}, n)|, v), c_v)),
\end{array}$$

where both φ are bijective and the right i^* is not surjective, hence the left $(i \times \mathrm{id}_{\mathbf{I}})^*$ is not surjective. Then, we have a map $g : A \times \mathbf{I} \to |K(\mathbb{Z}, n)|$ such that $g(H_A) = v$ and g cannot extend over $X \times \mathbf{I}$, which means that $\dim_{\mathbb{Z}} X \times \mathbf{I} > n$. Because $X \times \mathbf{I}$ is finite-dimensional, $\dim X \times \mathbf{I} = \dim_{\mathbb{Z}} X \times \mathbf{I} \geq n + 1$ by Theorem 7.9.5. □

Theorem 7.9.7. *Let P be a locally compact polyhedron or $P = |K|_{\mathrm{m}}$ be the polyhedron of an arbitrary simplicial complex K with the metric topology. For every metrizable space X,*

$$\dim X \times P = \dim X + \dim P.$$

Proof. Since $\dim X \times P \leq \dim X + \dim P$ by the Product Theorem 5.4.9, it suffices to show that $\dim X \times P \geq \dim X + \dim P$. If $\dim P = \infty$ then $\dim X \times P = \infty$ because P can be embedded into $X \times P$ as a closed set. Then, we may assume that $\dim P = n < \infty$. In this case, P contains an n-simplex σ, which is homeomorphic to some $\mathbf{I}^n$. Inductively applying Lemma 7.9.6, we have

$$\dim X \times \sigma = \dim X \times \mathbf{I}^n = \dim X + n.$$

Since $X \times \sigma$ is a closed set in $X \times P$, it follows that $\dim X \times P \geq m + n$. □

In the remainder of this section, we will prove the following theorem, which means that any cell-like map does not raise the cohomological dimension:

Theorem 7.9.8. *Let $f : X \to Y$ be a cell-like map between compacta. Then,* $\dim_{\mathbb{Z}} Y \leq \dim_{\mathbb{Z}} X$.

This is the direct consequence of the following theorem:

Theorem 7.9.9. *Let $f : X \to Y$ be a UV^n map between compacta. If* $\dim_{\mathbb{Z}} X \leq n$ *then* $\dim_{\mathbb{Z}} Y \leq n$.

To prove this theorem, we need the following Vietoris–Begle-type mapping theorem:

Theorem 7.9.10. *Let $f : X \to Y$ be a UV^{n-1} map between compacta and P an ANR (or a polyhedron) with $\pi_i(P) = \{0\}$ for $i \geq n$. Then, $f^* : [Y, P] \to [X, P]$ is a bijection.*

Note that it suffices to prove only the ANR case. To prove Theorem 7.9.10, we will prove the following lemma:

Lemma 7.9.11. *Let $f : X \to Y$ be a UV^{n-1} map between compacta and P an ANR with $\pi_i(P) = \{0\}$ for $i \geq n$. Suppose that A is a closed set in Y and $f|f^{-1}(A)$ is an embedding. Then, a map $\alpha : A \to P$ from A to an ANR P extends over Y if $\alpha f|f^{-1}(A)$ extends over X.*

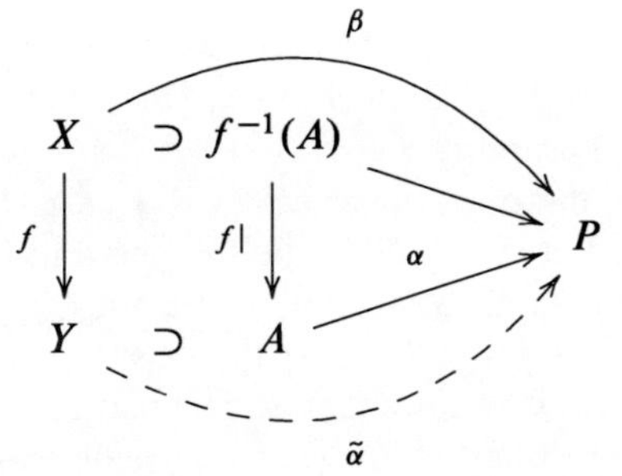

Proof. Let $\mathrm{pr}_1 : \boldsymbol{Q} = \mathbf{I}^{\mathbb{N}} \to \mathbf{I}$ be the projection of the Hilbert cube $\boldsymbol{Q}$ onto the first factor. Embed X and Y as closed sets in $\mathrm{pr}_1^{-1}(1) \approx \boldsymbol{Q}$ and let

$$Q_X = \boldsymbol{Q} \setminus (\mathrm{pr}_1^{-1}(1) \setminus X) \text{ and } Q_Y = \boldsymbol{Q} \setminus (\mathrm{pr}_1^{-1}(1) \setminus Y).$$

Since Q_X and Q_Y are homotopy dense in $\boldsymbol{Q}$, it is easy to extend f to a map $\tilde{f} : Q_X \to Q_Y$ such that $\tilde{f}(Q_X \setminus X) \subset Q_Y \setminus Y$, i.e., $\tilde{f}^{-1}(Y) = X$, which implies that $\tilde{f}^{-1}(y) = f^{-1}(y)$ for every $y \in Y$. The maps α and β extend to maps $\tilde{\alpha} : R \to P$ and $\tilde{\beta} : M \to P$, where M and R are open neighborhoods of X and A in Q_X and Q_Y, respectively. Because $\tilde{f}$ is closed and $\tilde{f}^{-1}(Y) = X \subset M$, we have an open neighborhood N of Y in Q_Y such that $\tilde{f}^{-1}(N) \subset M$.

Take open covers of P as follows:

$$\{P\} \underset{h}{\succ} \mathcal{W}_0 \overset{*}{\succ} \mathcal{W}_1 \underset{L}{\succ} \mathcal{W}_2 \overset{*}{\succ} \mathcal{W}_3.$$

Replacing R by a smaller one, we can assume that $\tilde{\beta}|\tilde{f}^{-1}(R)$ and $\tilde{\alpha}\tilde{f}|\tilde{f}^{-1}(R)$ are $\mathcal{W}_3$-close because

$$\tilde{\beta}|\tilde{f}^{-1}(A) = \beta|f^{-1}(A) = \alpha f|f^{-1}(A) = \tilde{\alpha}\tilde{f}|\tilde{f}^{-1}(A).$$

Choose an open cover $\mathcal{V}_0$ of Y in Q_Y so that

$$\mathrm{st}(A, \mathcal{V}_0) \subset R \text{ and } \mathcal{V}_0[A] \prec \tilde{\alpha}^{-1}(\mathcal{W}_3).$$

Let $\mathcal{U}_0$ be an open cover of Y in N with $\mathrm{st}\,\mathcal{U}_0 \prec \mathcal{V}_0$. We can inductively choose open covers $\mathcal{U}_i, \mathcal{V}_i$ of Y in N, $i = 1, \ldots, n$, as follows:

$$\tilde{f}^{-1}(\mathcal{V}_i) \underset{C^{n-1}}{\prec} \tilde{f}^{-1}(\mathcal{U}_{i-1}) \text{ and } \mathrm{st}\,\mathcal{U}_i \prec \mathcal{V}_i.$$

Indeed, assume that $\mathcal{V}_{i-1}$ has been obtained and take an open star-refinement $\mathcal{U}_{i-1}$ of $\mathcal{V}_{i-1}$. Each $y \in Y$ is contained in some $U_y \in \mathcal{U}_{i-1}$. Then, $f^{-1}(y) = \tilde{f}^{-1}(y) \subset \tilde{f}^{-1}(U_y)$. Since $f^{-1}(y)$ is UV^n, $f^{-1}(y)$ has an open neighborhood V'_y in M such that any map from $\mathbf{S}^j$ to V'_y extends to a map from $\mathbf{B}^{j+1}$ to $\tilde{f}^{-1}(U_y)$ for each $j \leq n-1$. Because $\tilde{f}$ is closed, y has an open neighborhood V_y in N such that $\tilde{f}^{-1}(V_y) \subset V'_y\ (\subset \tilde{f}^{-1}(U_y))$. Therefore, $\mathcal{V}_i = \{V_y \mid y \in Y\}$ is an open cover of Y in N that has the desired property.

Since Y is compact, the open cover $\{U \cap Y \mid U \in \mathcal{U}_n[Y]\}$ of Y has a finite subcover $\mathcal{U} \in \mathrm{cov}(Y)$. Let $K = N(\mathcal{U})$ be the nerve of $\mathcal{U}$ with $\varphi : Y \to |K|$ a canonical map. For each vertex $U \in K^{(0)} = \mathcal{U}$, choose a point $\psi_0(U) \in f^{-1}(U)$, where $\psi_0(U) \in f^{-1}(U \cap A)$ for $U \in \mathcal{U}[A]$. Thus, we have a map $\psi_0 : |K^{(0)}| \to X \subset M$. Since $\mathrm{st}\,\mathcal{U} \prec \mathcal{V}_n$, it follows that ψ_0 is a partial $\tilde{f}^{-1}(\mathcal{V}_n)$-realization of K. Assume that ψ_0 extends to a partial $\tilde{f}^{-1}(\mathcal{V}_{n-i+1})$-realization $\psi_{i-1} : |K^{(i-1)}| \to M$. For each i-simplex $\sigma \in K$, there is some $V_\sigma \in \mathcal{V}_{n-i+1}$ such that $\psi_{i-1}(\partial\sigma) \subset \tilde{f}^{-1}(V_\sigma)$. Then, we have some $U_\sigma \in \mathcal{U}_{n-i}$ such that $V_\sigma \subset U_\sigma$ and $\psi_{i-1}|\partial\sigma$ extends to a map $\psi_\sigma : \sigma \to \tilde{f}^{-1}(U_\sigma)$. Now, we can extend ψ_{i-1} to the map $\psi_i : |K^{(i)}| \to M$ by $\psi_i|\sigma = \psi_\sigma$ for each i-simplex $\sigma \in K$. For every $\sigma \in K$, there is $V_\sigma \in \mathcal{V}_{n-i+1}$ such that $\psi_{i-1}(\sigma^{(i-1)}) \subset \tilde{f}^{-1}(V_\sigma)$, hence

$$\begin{aligned}\psi_i(\sigma^{(i)}) &\subset \mathrm{st}(\tilde{f}^{-1}(V_\sigma), \tilde{f}^{-1}(\mathcal{U}_{n-i})) \\ &\in \mathrm{st}(\tilde{f}^{-1}(\mathcal{V}_{n-i+1}), \tilde{f}^{-1}(\mathcal{U}_{n-i})) \\ &\prec \mathrm{st}\,\tilde{f}^{-1}(\mathcal{U}_{n-i}) \prec \tilde{f}^{-1}(\mathcal{V}_{n-i}).\end{aligned}$$

Thus, ψ_i is a partial $\tilde{f}^{-1}(\mathcal{V}_{n-i})$-realization of K. By induction, we can obtain a partial $\tilde{f}^{-1}(\mathcal{V}_0)$-realization $\psi_n : K^{(n)} \to M$.

The nerve $L = N(\mathcal{U}[A])$ of $\mathcal{U}[A]$ is a subcomplex of K. For each $\sigma \in L$, we have $V \in \mathcal{V}_0$ such that $\psi_n(\sigma^{(n)}) \subset \tilde{f}^{-1}(V)$, where $V \cap A \neq \emptyset$ because $\psi_n(\sigma^{(0)}) \subset \tilde{f}^{-1}(V \cap A)$. Then, $\tilde{\alpha}(V)$ is contained in some $W \in \mathcal{W}_3$, hence

$$\tilde{\alpha}\tilde{f}\psi_n(\sigma^{(n)}) \subset \tilde{\alpha}(V) \subset W.$$

Since $\tilde{\beta}|\tilde{f}^{-1}(R)$ and $\tilde{\alpha}\tilde{f}|\tilde{f}^{-1}(R)$ are $\mathcal{W}_3$-close and $V \subset R$, it follows that

$$\tilde{\beta}\psi_n(\sigma^{(n)}) \subset \mathrm{st}(W, \mathcal{W}_3) \in \mathrm{st}\,\mathcal{W}_3 \prec \mathcal{W}_2.$$

Therefore, $\tilde{\beta}\psi_n||L^{(n)}|$ is a partial $\mathcal{W}_2$-realization of L, which extends to a full $\mathcal{W}_1$-realization $\gamma : |L| \to P$.

For each $a \in A$, $\varphi(a)$ is contained in some $\sigma \in L$, where $\gamma(\sigma)$ is contained in some $W \in \mathcal{W}_1$. Since $\sigma^{(0)} \subset \mathcal{U}[a]$, it follows that

$$\begin{aligned}\gamma(\sigma^{(0)}) &= \tilde{\beta}\psi_n(\sigma^{(0)}) = \beta\psi_0(\sigma^{(0)})\\ &\in \beta f^{-1}(\mathrm{st}(a,\mathcal{U}) \cap A) \subset \alpha(\mathrm{st}(a,\mathcal{U}) \cap A)\\ &\subset \mathrm{st}(\alpha(a), \tilde{\alpha}(\mathcal{U})) \subset \mathrm{st}(\alpha(a), \mathcal{W}_3),\end{aligned}$$

hence $W \cap \mathrm{st}(\alpha(a), \mathcal{W}_3) \neq \emptyset$. Then, α and $\gamma\varphi|A$ are $\mathcal{W}_0$-close, which implies $\alpha \simeq \gamma\varphi|A$.

On the other hand, we can define a map $\gamma_n : |K^{(n)}| \cup |L| \to P$ by $\gamma_n||K^{(n)}| = \tilde{\beta}\psi_n$ and $\gamma_n||L| = \gamma$. Since $\pi_i(P) = \{0\}$ for every $i \geq n$, by skeleton-wise induction, we can construct maps $\gamma_i : |K^{(i)}| \cup |L| \to P$, $i \geq n$, such that $\gamma_{i+1}||K^{(i)}| \cup |L| = \gamma_i$. Because $\dim K = m < \infty$ because K is finite, we have a map $\gamma_m\varphi : Y \to P$. As we saw in the above, $\gamma_m\varphi|A = \gamma\varphi|A \simeq \alpha$. By the Homotopy Extension Theorem 6.4.1, α can be extended over Y. □

Proof of Theorem 7.9.10. First, we show that f^* is surjective. Consider the mapping cylinder M_f, where we identify $X = X \times \{1\} \subset M_f$. Let $q : X \times \mathbf{I} \to M_f$ be the restriction of the quotient map. Then, q is a UV^{n-1} map and $q^{-1}(X) = X \times \{1\} \subset X \times \mathbf{I}$. For every map $\alpha : X \to P$, the map $\alpha\mathrm{pr}_X : X \times \mathbf{I} \to P$ is an extension of $\alpha q|q^{-1}(X)$.

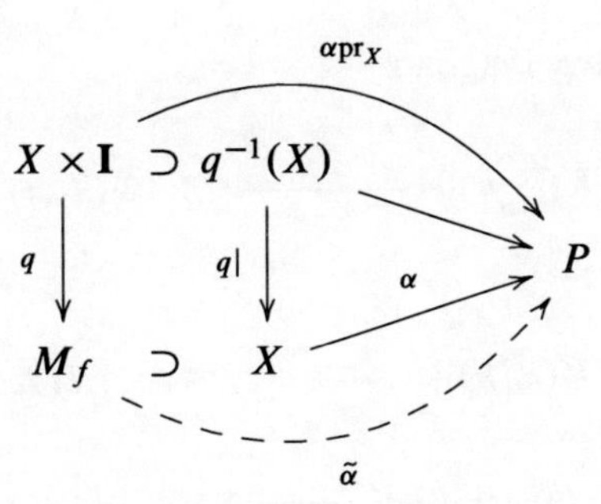

Applying Lemma 7.9.11, we have a map $\tilde{\alpha} : M_f \to P$. Since the collapsing $c_f : M_f \to Y$ is a strong deformation retraction, it follows that

$$(\tilde{\alpha}|Y)f = (\tilde{\alpha}|Y)c_f|X \simeq \tilde{\alpha}|X = \alpha.$$

Thus, we have $f^*[\tilde{\alpha}|Y] = [\alpha]$.

Next, we show that f^* is injective. The double mapping cylinder DM_f is defined as the adjunction space

$$DM_f = (Y \times \partial\mathbf{I}) \cup_{f \times \mathrm{id}_{\partial\mathbf{I}}} (X \times \mathbf{I}).$$

The map $f \times \mathrm{id}_{\mathbf{I}} : X \times \mathbf{I} \to Y \times \mathbf{I}$ induces the map $\bar{f} : DM_f \to Y \times \mathbf{I}$, which is a UV^{n-1} map and $\bar{f}|Y \times \partial\mathbf{I} = \mathrm{id}$. Let $\alpha_0, \alpha_1 : Y \to P$ be maps with $f^*[\alpha_0] = f^*[\alpha_1]$, i.e., $\alpha_0 f \simeq \alpha_1 f$. Then, a homotopy $h : X \times \mathbf{I} \to P$ from $\alpha_0 f$ to $\alpha_1 f$ induces the map $\bar{h} : DM_f \to P$.

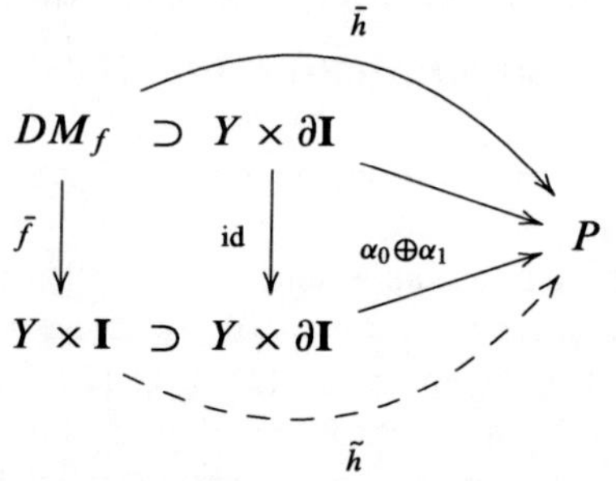

Applying Lemma 7.9.11, we have a map $\tilde{h} : Y \times \mathbf{I} \to P$ such that $\tilde{h}|Y \times \partial\mathbf{I} = \bar{h}|Y \times \partial\mathbf{I}$, which means that $\tilde{h}$ is a homotopy from α_0 to α_1. Consequently, it follows that $[\alpha_0] = [\alpha_1]$. □

Now, we prove Theorem 7.9.9.

Proof of Theorem 7.9.9. Let $f : X \to Y$ be a UV^n map with $\dim_{\mathbb{Z}} X \leq n$. For each closed set A in Y, we denote $f_A = f|f^{-1}(A) : f^{-1}(A) \to A$. Then, f_A is also a UV^n map. Let $i : A \subset Y$ and $j : f^{-1}(A) \subset X$ be the inclusions. We have the following commutative diagram:

$$\begin{array}{ccc}
[Y, |K(\mathbb{Z}, n)|] & \xrightarrow{\ i^*\ } & [A, |K(\mathbb{Z}, n)|] \\
{\scriptstyle f^*}\downarrow & & \downarrow{\scriptstyle f_A^*} \\
[X, |K(\mathbb{Z}, n)|] & \xrightarrow[\ j^*\ ]{} & [f^{-1}(A), |K(\mathbb{Z}, n)|].
\end{array}$$

According to Theorem 7.9.10, both vertical f^* and f_A^* are bijective. Because $\dim_{\mathbb{Z}} X \leq n$, j^* is surjective, consequently so is i^*. This means that $\dim_{\mathbb{Z}} Y \leq n$. □

As a corollary of Theorem 7.9.10, we have the following:

Corollary 7.9.12. *Let $f : X \to Y$ be a cell-like map between compacta and P an ANR (or a polyhedron) with $\pi_i(P) = \{0\}$ except for finitely many $i \in \mathbb{N}$. Then, $f^* : [Y, P] \to [X, P]$ is a bijection.* □

Remark 12. As we saw in the previous section, a cell-like map $f : X \to Y$ is not a shape equivalence even if X or Y is a compact AR. Therefore, the homotopy condition on P is essential in Corollary 7.9.12.

7.10 Alexandroff's Problem and the CE Problem

In Sect. 7.9, we have shown that $\dim_{\mathbb{Z}} X \leq \dim X$ for every compact space X and $\dim_{\mathbb{Z}} X = \dim X$ if X is finite-dimensional (Theorem 7.9.5). Then, it is natural to ask the following question:

- Does there exist an infinite-dimensional compactum with finite cohomological dimension $\dim_{\mathbb{Z}}$?

This problem is called **Alexandroff's Problem**. Recall that it has been shown in Sect. 7.9 that any cell-like map between compacta does not raise the cohomological dimension (Theorem 7.9.8). So, it is also natural to ask the following question:

- Do cell-like maps of compacta raise the dimension?

This is called the **CE Problem**. These two problems are equivalent. In fact, this section has the purpose of establishing the following theorem:

Theorem 7.10.1. *For every compactum X, the following are equivalent:*

(a) $\dim_{\mathbb{Z}} X \leq n$;
(b) *There exists a cell-like map $f : Y \to X$ for some compactum Y with $\dim Y \leq n$.*

In the next section, we will give an affirmative answer to Alexandroff's Problem, so the CE Problem is also positively answered.

To prove Theorem 7.10.1, we first establish a criterion to estimate the cohomological dimension of the inverse limit via the bonding maps. For each map $f : X \to Y$ from a compact space X to a metric space $Y = (Y, d)$ and $n, m \in \mathbb{N}$, we define $\alpha_n(f)$ and $\alpha_n^m(f)$ as follows:

$$\alpha_n(f) = \inf\{d(f, g) : g \in \mathrm{C}(X, Y),\ \dim g(X) \leq n\};$$

$$\alpha_n^m(f) = \sup\{\alpha_n(f|A) : A \in \mathrm{Cld}(X),\ \dim A \leq m\}.$$

Theorem 7.10.2. *Let $X = \varprojlim(X_i, f_i)$ be the inverse limit of an inverse sequence of compact metric polyhedra. Then, the following conditions are equivalent:*

(a) $\dim_{\mathbb{Z}} X \leq n$;
(b) $\lim_{j\to\infty} \alpha_n^{n+1}(f_{i,j}) = 0$ *for each* $i \in \mathbb{N}$,

where $f_{i,j} = f_i \cdots f_{j-1} : X_j \to X_i$ *for* $i < j$.

Proof. (b) ⇒ (a): Let A be a closed set in X and $g : A \to |K(\mathbb{Z}, n)|$ be a map. For each $i \in \mathbb{N}$, let $A_i = p_i(A)$, where $p_i : X \to X_i$ is the inverse limit projection. Then, $A = \varprojlim(A_i, f_i|A_{i+1})$ and each $p_i|A$ is the inverse limit projection. By Lemma 7.1.7, we have some $i \in \mathbb{N}$ and a map $g_i : A_i \to |K(\mathbb{Z}, n)|$ such that $g_i p_i|A \simeq g$. Since $|K(\mathbb{Z}, n)|$ is an ANE, g_i extends to a map $\tilde{g}_i : U \to |K(\mathbb{Z}, n)|$ over a neighborhood U of A_i in X_i. Let $\varepsilon = \frac{1}{4}\operatorname{dist}(A_i, X_i \setminus U) > 0$. Because X_i is a compact ANR, we can find $0 < \delta < \varepsilon$ such that every pair of δ-close maps to X_i are ε-homotopic in X_i. By (b), we can choose $j > i$ so that $\alpha_n^{n+1}(f_{i,j}) < \delta$. Let K be a triangulation of X_j with mesh $f_{i,j}(K) < \varepsilon$ (Corollary 4.7.7) and let L be the subcomplex of K with $|L| = \operatorname{st}(A_j, K)$. Then, we have a map $h' : |K^{(n+1)}| \to X_i$ such that $\dim h'(|K^{(n+1)}|) \le n$ and $d(h', f_{i,j}||K^{(n+1)}|) < \delta$, hence $h' \simeq_\varepsilon f_{i,j}||K^{(n+1)}|$. By the Homotopy Extension Theorem 6.4.1, h' extends to a map $h : |K| = X_j \to X_i$ such that $h \simeq_\varepsilon f_{i,j}$. Then, mesh $h(K) < 3\varepsilon$, hence $h(|L|) \subset U$ and $h||L| \simeq f_{i,j}||L|$ in U. Thus, we have the map $\tilde{g}_i h||L| : |L| \to |K(\mathbb{Z}, n)|$ and $\tilde{g}_i h||L| \simeq \tilde{g}_i f_{i,j}||L|$.

Recall that $\partial\Delta^{n+1}$ is a mapping $(n+1)$-deformation retract of $|K(\mathbb{Z}, n)|$ for metrizable spaces. Since $\dim h(|L^{(n+1)}|) \le n$ and $h(|L^{(n+1)}|) \subset U$, we have a map $g' : h(|L^{(n+1)}|) \to \partial\Delta^{n+1} \approx \mathbf{S}^n$ such that $g'|D = \tilde{g}_i|D$ and $g' \simeq \tilde{g}_i|h(|L^{(n+1)}|)$ rel. D, where

$$D = \tilde{g}_i^{-1}(\partial\Delta^{n+1}) \cap h(|L^{(n+1)}|).$$

Because $\dim h(|K^{(n+1)}|) \le n$, the map g' extends to a map $\tilde{g}' : h(|K^{(n+1)}|) \to \partial\Delta^{n+1} \approx \mathbf{S}^n$. Then, $\tilde{g}'h||L^{(n+1)}| = g'h||L^{(n+1)}| \simeq \tilde{g}_i h||L^{(n+1)}|$. By the Homotopy Extension Theorem 6.4.1, we can obtain a map $g'' : |L| \to |K(\mathbb{Z}, n)|$ such that $g''||L^{(n+1)}| = \tilde{g}'h||L^{(n+1)}|$ and $g'' \simeq \tilde{g}_i h||L|$. Thus, we have a map $g''' : |K^{(n+1)}| \cup |L| \to |K(\mathbb{Z}, n)|$ defined by $g'''||K^{(n+1)}| = \tilde{g}'h$ and $g'''||L| = g''$. On the other hand, for every $m > n$, every map from $\mathbf{S}^m$ to $|K(\mathbb{Z}, n)|$ extends over $\mathbf{B}^{m+1}$ because $\pi_m(|K(\mathbb{Z}, n)|) = 0$. By skeleton-wise induction, we can extend g''' to a map $g^* : |K| = X_j \to |K(\mathbb{Z}, n)|$. Then, it follows that

$$g^* p_j|A = g'' p_j|A \simeq \tilde{g}_i h p_j|A \simeq \tilde{g}_i f_{i,j} p_j|A = \tilde{g}_i p_i|A \simeq g,$$

hence g can be extended over X by the Homotopy Extension Theorem 6.4.1.

(a) ⇒ (b): For each $j \in \mathbb{N}$ and $\varepsilon > 0$, we have to find $k_0 \ge j$ so that $\alpha_n^{n+1}(f_{j,k}) < \varepsilon$ for every $k \ge k_0$. We assume that $\dim X_j > n$ because the other case is obvious. Let K be a triangulation of X_j with mesh $K < \varepsilon/2$ (Corollary 4.7.7). Let $EW(K, n)$ be the simplicial Edwards–Walsh complex for K. Recall that $K^{(n)} \subset EW(K, n) = \bigcup_{\sigma \in K} K_\sigma$, where K_σ, $\sigma \in K$, are simplicial complexes obtained by Proposition 7.8.14. Suppose that the map $p_j|p_j^{-1}(|K^{(n)}|) : p_j^{-1}(|K^{(n)}|) \to |EW(K, n)|$ extends to a map $g_m : p_j^{-1}(|K^{(m)}|) \to |EW(K, n)|$ for $m \ge n$ such that $g_m(p_j^{-1}(\sigma)) \subset |K_\sigma|$ for each $\sigma \in K^{(m)}$. For each $(m+1)$-simplex $\sigma \in K$, observe that

$$g_m(p_j^{-1}(\partial\sigma)) \subset \bigcup_{\sigma' < \sigma} |K_{\sigma'}| \subset |K_\sigma| \simeq |K(\mathbb{Z}, n)|^k \quad \text{for some } k \in \mathbb{N}.$$

Since $\dim_{\mathbb{Z}} p_j^{-1}(\sigma) \leq \dim_{\mathbb{Z}} X \leq n$, $g_m|p_j^{-1}(\partial\sigma)$ extends to a map $g_\sigma : p_j^{-1}(\sigma) \to |K_\sigma|$. By virtue of Proposition 7.8.14(3), the map g_{m+1} can be defined by $g_{m+1}|p_j^{-1}(\sigma) = g_\sigma$ for every $(m+1)$-simplex $\sigma \in K$.

Now, we have obtained the map $g_{\dim K} : X \to |EW(K,n)|$, which extends to a map $g : U \to |EW(K,n)|$ over a neighborhood U of X in $\prod_{j\in\mathbb{N}} X_j$. Observe that

$$g(\mathrm{pr}_j^{-1}(\sigma) \cap X) = g(p_j^{-1}(\sigma)) \subset |K_\sigma| \quad \text{for each } \sigma \in K.$$

Then, X has a neighborhood $V \subset U$ in $\prod_{i\in\mathbb{N}} X_i$ such that

$$g(\mathrm{pr}_j^{-1}(\sigma) \cap V) \subset \operatorname{int}\operatorname{st}(|K_\sigma|, EW(K,n)) \quad \text{for each } \sigma \in K.$$

We use the same notation X_i^* as in the proof of Lemma 7.1.7. Because X is compact, we can choose $k_0 > j$ so that

$$X_{k_0}^* = \left\{x \in \textstyle\prod_{i\in\mathbb{N}} X_i \;\middle|\; x(i) = f_{i,k}(x(k)) \text{ for } i < k\right\} \subset V.$$

Let $z \in \prod_{i\in\mathbb{N}} X_i$ be fixed and define an embedding $\varphi_i : X_i \to X_i^*$ by

$$\varphi_i(x) = (f_{1,i}(x), \ldots, f_{i-1,i}(x), x, z(i+1), z(i+2), \ldots).$$

Then, for every $k \geq k_0$, the following statement holds:

(*) If $x \in X_k$, $f_{j,k}(x) \in \sigma \in K$, and $g\varphi_k(x) \in |K_{\sigma'}|$, $\sigma' \in K$, then $\sigma \cap \sigma' \neq \emptyset$.

Indeed, since $f_{j,k} = \mathrm{pr}_j \varphi_k$, we have $\varphi_k(x) \in \mathrm{pr}_j^{-1}(\sigma) \cap V$, hence $g(\varphi_k(x)) \in \operatorname{int}\operatorname{st}(|K_\sigma|, EW(K,n))$. It follows that $|K_\sigma| \cap |K_{\sigma'}| \neq \emptyset$, which implies that $\sigma \cap \sigma' \neq \emptyset$ by Proposition 7.8.14(3).

Let A be a closed set in X_k with $\dim A \leq n+1$ and consider the restriction $f_{j,k}|A : A \to X_j = |K|$. Using Proposition 7.8.14(6), we can obtain a map $\psi : A \to |K^{(n)}|$ such that each $\psi(x)$ is contained in $\sigma^{(n)}$ for some $\sigma \in K$ with $g\varphi_k(x) \in |K_\sigma|$ and

$$\psi|(g\varphi_k)^{-1}(|K^{(n)}|) \cap A = g\varphi_j|(g\varphi_k)^{-1}(|K^{(n)}|) \cap A,$$

where the carrier of $f_{j,k}(x)$ in K meets σ by (*). Then, $d(f_{j,k}(x), \psi(x)) \leq 2 \operatorname{mesh} K < \varepsilon$. This means that $\alpha_n^{n+1}(f_{j,k}) < \varepsilon$. □

Let $X = (X, d_X)$ and (Y, d_Y) be metric spaces. A map $f : X \to Y$ is said to be **non-expansive** if $d_Y(f(x), f(x')) \leq d_X(x, x')$ for every $x, x' \in X$. An inverse sequence $(X_i, f_i)_{i\in\mathbb{N}}$ of metric spaces $X_i = (X_i, d_i)$ is said to be **non-expansive** if every bonding map f_i is non-expansive. For every inverse sequence $(X_i, f_i)_{i\in\mathbb{N}}$ of metrizable spaces, we have $d_i \in \operatorname{Metr}(X_i)$, $i \in \mathbb{N}$, such that $(X_i, f_i)_{i\in\mathbb{N}}$ is non-expansive. Indeed, for each $i \in \mathbb{N}$, let $d_i' \in \operatorname{Metr}(X_i)$. The desired metric d_i for X_i can be defined as follows:

$$d_i(x, y) = \sum_{j \le i} d'_j(f_{j,i}(x), f_{j,i}(y)),$$

where $f_{j,i} = f_j \cdots f_{i-1} : X_i \to X_j$ and $f_{i,i} = \mathrm{id}$.

Lemma 7.10.3. *Let X and Y be the inverse limits of inverse sequences $(X_i, f_i)_{i\in\mathbb{N}}$ and $(Y_i, g_i)_{i\in\mathbb{N}}$ with $p_i : X \to X_i$ and $q_i : Y \to Y_i$ the projections, respectively. Suppose that each Y_i has an admissible complete metric d_i such that $(Y_i, g_i)_{i\in\mathbb{N}}$ is non-expansive. Then, given maps $h_i : X_i \to Y_i$, $i \in \mathbb{N}$, such that $\sum_{i\in\mathbb{N}} d_i(h_i f_i, g_i h_{i+1}) < \infty$, there exists a map $h : X \to Y$ such that $d_i(h_i p_i, q_i h) \le \sum_{j\ge i} d_j(h_j f_j, g_j h_{j+1})$.*

Proof. Since $g_{i,j}$ is non-expansive for $i < j$, it follows that

$$\begin{aligned} d_i(g_{i,j}h_j p_j, g_{i,j+1}h_{j+1}p_{j+1}) &\le d_j(h_j p_j, g_j h_{j+1}p_{j+1}) \\ &= d_j(h_j f_j p_{j+1}, g_j h_{j+1}p_{j+1}) \le d_j(h_j f_j, g_j h_{j+1}). \end{aligned}$$

Then, for each $i < j < k$,

$$\begin{aligned} d_i(g_{i,j}h_j p_j, g_{i,k}h_k p_k) &\le d_i(g_{i,j}h_j p_j, g_{i,j+1}h_{j+1}p_{j+1}) \\ &\qquad + \cdots + d_i(g_{i,k-1}h_{k-1}p_{k-1}, g_{i,k}h_k p_k) \\ &\le d_j(h_j f_j, g_j h_{j+1}) + \cdots + d_{k-1}(h_{k-1}f_{k-1}, g_{k-1}h_k). \end{aligned}$$

Now, for each $\varepsilon > 0$, choose $k \in \mathbb{N}$ so that $\sum_{j\ge k} d_j(h_j f_j, g_j h_{j+1}) < \varepsilon$. Then, for each $j' > j \ge k$,

$$\begin{aligned} d_i(g_{i,j}h_j p_j, g_{i,j'}h_{j'}p_{j'}) &\le d_j(h_j f_j, g_j h_{j+1}) \\ &\quad + \cdots + d_{j'-1}(h_{j'-1}f_{j'-1}, g_{j'-1}h_{j'}) < \varepsilon. \end{aligned}$$

Therefore, $(g_{i,j}h_j p_j)_{j\ge i}$ is a Cauchy sequence in $\mathrm{C}(X, Y_i)$ with the sup-metric. Since $Y_i = (Y_i, d_i)$ is complete, so is $\mathrm{C}(X, Y_i)$, hence $(g_{i,j}h_j p_j)_{j\ge i}$ converges to a map $h'_i : X \to Y_i$. It should be noted that $g_{i-1}h'_i = h'_{i-1}$ for each $i \in \mathbb{N}$. We can define a map $h : X \to Y$ by $h(x) = (h'_i(x))_{i\in\mathbb{N}}$. For each $i \in \mathbb{N}$ and $\varepsilon' > 0$, choose $j > i$ so that $d_i(g_{i,j}h_j p_j, h'_i) < \varepsilon'$. Then,

$$\begin{aligned} d_i(h_i p_i, q_i h) &= d_i(h_i p_i, h'_i) \\ &\le d_i(h_i p_i, g_i h_{i+1}p_{i+1}) + d_i(g_{i,i+1}h_{i+1}p_{i+1}, g_{i,i+2}h_{i+2}p_{i+2}) \\ &\qquad + \cdots + d_i(g_{i,j-1}h_{j-1}p_{j-1}, g_{i,j}h_j p_j) + d_i(g_{i,j}h_j p_j, h'_i) \\ &\le d_i(h_i f_i, g_i h_{i+1}) + d_{i+1}(h_{i+1}f_{i+1}, g_{i+1}h_{i+2}) \\ &\qquad + \cdots + d_i(g_{i,j-1}h_{j-1}f_{j-1}, g_{i,j}h_j) + \varepsilon'. \end{aligned}$$

Therefore, $d_i(h_i p_i, q_i h) \leq \sum_{j\geq i} d_j(h_j f_j, g_j h_{j+1})$ for each $i \in \mathbb{N}$. This completes the proof. □

Now, we can prove Theorem 7.10.1.

Proof of Theorem 7.10.1. The implication (b) ⇒ (a) is obtained by combining Theorem 7.9.5 and Theorem 7.9.8.

(a) ⇒ (b): By Corollary 4.10.11, we may assume that $X = \varprojlim(X_i, f_i)$, where each X_i is a compact polyhedron. By $p_i : X \to X_i$, we denote the inverse limit projection. As observed in the above, we can give an admissible metric d_i for each X_i such that every $f_i : X_{i+1} \to X_i$ is non-expansive.

By the uniformly local contractibility of compact metric polyhedra and Theorem 7.10.2, we can inductively take two sequences $1 = \varepsilon_0 > \varepsilon_1 > \varepsilon_2 > \cdots > 0$ and $1 = k(1) < k(2) < \cdots \in \mathbb{N}$ so that

(1) $2\varepsilon_i < \varepsilon_{i-1}$ and the $2\varepsilon_i$-neighborhood $\mathrm{B}(x, 2\varepsilon_i)$ of each point $x \in X_{k(i)}$ is contractible in the ε_{i-1}-neighborhood $\mathrm{B}(x, \varepsilon_{i-1})$;
(2) $\alpha_n^{n+1}(f_{k(i),k(i+1)}) < \varepsilon_i/3$.

Then, (1) implies that $\varepsilon_i < 2^{-(i-j)}\varepsilon_j$ for every $i > j$. Hence, $\sum_{i>j} \varepsilon_i < \varepsilon_j$.

Triangulate each $X_{k(i)}$ by a simplicial complex K_i with mesh $K_i < \varepsilon_i/3$. By (2), we have a map $g'_i : |K_{i+1}^{(n+1)}| \to |K_i|$ such that

$$d_{k(i)}(g'_i, f_{k(i),k(i+1)}||K_{i+1}^{(n+1)}|) < \varepsilon_i/3 \text{ and } \dim g'_i(|K_{i+1}^{(n+1)}|) \leq n.$$

Pushing the image $g'_i(|K_{i+1}^{(n+1)}|)$ into $|K_i^{(n)}|$ by Theorem 5.2.9, we can obtain a map $g''_i : |K_{i+1}^{(n+1)}| \to |K_i^{(n)}|$ with $d_{k(i)}(g''_i, g'_i) \leq \operatorname{mesh} K_i < \varepsilon_i/3$. For each $i \in \mathbb{N}$, let $Y_i = |K_i^{(n)}|$ and $g_i = g''_i|Y_{i+1} : Y_{i+1} \to Y_i$. Thus, we have an inverse sequence $(Y_i, g_i)_{i\in\mathbb{N}}$ with $Y = \varprojlim(Y_i, g_i) \neq \emptyset$ (cf. 4.10.9(1)), where the inverse limit projection is denoted by $q_i : Y \to Y_i$. According to Theorem 5.3.2, we have $\dim Y \leq n$. We can regard $X = \varprojlim(X_{k(i)}, f_{k(i),k(i+1)})$ (Corollary 4.10.4). Because $d_{k(i)}(g_i, f_{k(i),k(i+1)}|Y_{i+1}) < 2\varepsilon_i/3$, we can apply Lemma 7.10.3 to the inclusions $h_i : Y_i \subset X_{k(i)}$, $i \in \mathbb{N}$, to obtain a map $h : Y \to X$ such that

$$d_{k(i)}(p_{k(i)}h, q_i) \leq \sum_{j\geq i} 2\varepsilon_j/3 < 2\varepsilon_{i-1}/3 \text{ for each } i \in \mathbb{N}.$$

We will now show that h is cell-like. For each point $x \in X$, we write $x_i = p_{k(i)}(x) \in X_{k(i)}$ for each $i \in \mathbb{N}$. Recall that $f_{k(i),k(i+1)}$ is non-expansive. For each $y \in Y_{i+1}$, we have

$$\begin{aligned} d_{k(i)}(x_i, g_i(y)) &\leq d_{k(i)}(f_{k(i),k(i+1)}(x_{i+1}), f_{k(i),k(i+1)}(y)) \\ &\qquad + d_{k(i)}(f_{k(i),k(i+1)}(y), g_i(y)) \\ &< d_{k(i+1)}(x_{i+1}, y) + 2\varepsilon_i/3. \end{aligned}$$

Then, it follows that

$$g_i(\overline{\mathrm{B}}(x_{i+1},\varepsilon_i)\cap Y_{i+1})\subset \mathrm{B}(x_i,\varepsilon_i+2\varepsilon_i/3)\cap Y_i\subset \overline{\mathrm{B}}(x_i,\varepsilon_{i-1})\cap Y_i.$$

Since $Y_i = |K_i^{(n)}|$ and $\operatorname{mesh} K_i < \varepsilon_i/3$, we have $Q_i = \overline{\mathrm{B}}(x_i,\varepsilon_{i-1})\cap Y_i \neq \emptyset$. Thus, we have an inverse sequence $(Q_i, g_i|Q_{i+1})_{i\in\mathbb{N}}$. The inverse limit $Q = \varprojlim(Q_i, g_i|Q_{i+1}) \neq \emptyset$ is a subspace of Y and $q_i|Q$, $i\in\mathbb{N}$, are the projections (cf. Propositions 4.10.8(2) and 4.10.9(1)).

We will show that $h^{-1}(x) = Q$. For each $y \in Q$,

$$\begin{aligned} d_{k(i)}(p_{k(i)}h(y),x_i) &\le d_{k(i)}(p_{k(i)}h(y),q_i(y)) + d_{k(i)}(q_i(y),x_i) \\ &< 2\varepsilon_{i-1}/3+\varepsilon_{i-1} < 2\varepsilon_{i-1}. \end{aligned}$$

Then, $\lim_{i\to\infty} d_{k(i)}(p_{k(i)}h(y),x_i) = 0$, which implies that $h(y) = x$, that is, $y \in h^{-1}(x)$. Conversely, if $y \in Y\setminus Q$ then $q_i(y)\notin Q_i$ for some $i\in\mathbb{N}$, which means that $d_{k(i)}(q_i(y),x_i) > \varepsilon_{i-1}$. Since $d_{k(i)}(p_{k(i)}h(y),q_i(y)) < 2\varepsilon_{i-1}/3$, it follows that $p_{k(i)}h(y) \neq x_i = p_{k(i)}(x)$, hence $h(y)\neq x$, i.e., $y\in Y\setminus h^{-1}(x)$.

It remains to prove that Q is cell-like. By Remark 2 of Theorem 7.1.8, for each $i\in\mathbb{N}$, it suffices to find $j > i$ so that $g_{i,j}|Q_j \simeq 0$. Observe that

$$\begin{aligned} g_{i+1,i+3}(Q_{i+3}) &= g_{i+1}g_{i+2}(\overline{\mathrm{B}}(x_{i+3},\varepsilon_{i+2})\cap Y_{i+3}) \\ &\subset g_{i+1}(\overline{\mathrm{B}}(x_{i+2},\varepsilon_{i+1})\cap Y_{i+2}) \\ &\subset \mathrm{B}(x_{i+1},2\varepsilon_{i+1})\cap Y_{i+1}, \end{aligned}$$

which implies that $g_{i+1,i+3}(Q_{i+3})$ is contractible in $\overline{\mathrm{B}}(x_{i+1},\varepsilon_i)$ by (1). Since $\dim g_{i+1,i+3}(Q_{i+3})\times \mathbf{I} \le n+1$, pushing this contraction into $|K_{i+1}^{(n+1)}|$ by Theorem 5.2.9, we can conclude that $g_{i+1,i+3}(Q_{i+3})$ is contractible in $\overline{\mathrm{B}}(x_{i+1},\varepsilon_i+\varepsilon_{i+1}/3)\cap|K_i^{(n+1)}|$. On the other hand, similar to g_i,

$$d_{k(i)}(x_i,g_i''(y)) < d_{k(i+1)}(x_{i+1},y) + 2\varepsilon_i/3 \quad \text{for each } y\in|K_i^{(n+1)}|.$$

Then, it follows that

$$\begin{aligned} g_i''\big(\overline{\mathrm{B}}(x_{i+1},\varepsilon_i+\varepsilon_{i+1}/3)\cap|K_i^{(n+1)}|\big) &\subset \overline{\mathrm{B}}(x_i,\varepsilon_i+\varepsilon_{i+1}/3+2\varepsilon_i/3)\cap Y_i \\ &\subset \overline{\mathrm{B}}(x_i,\varepsilon_{i-1})\cap Y_i = Q_i. \end{aligned}$$

Thus, $g_{i,i+3}(Q_{i+3})$ is contractible in Q_i. This completes the proof. □

In the rest of this section, we will give an affirmative answer to Alexandroff's Problem:

Theorem 7.10.4 (DRANISHNIKOV; DYDAK–WALSH). *There exists an infinite-dimensional compactum X with finite cohomological dimension* ($\dim_{\mathbb{Z}} X = 2$).

Remark 13. In Theorem 7.10.4, $\dim_{\mathbb{Z}} X = 3$ for the first example constructed by Dranishnikov, but $\dim_{\mathbb{Z}} X = 2$ due to Dydak–Walsh's construction, where one should recall that $\dim_{\mathbb{Z}} X \leq 1$ if and only if $\dim X \leq 1$.

Combining Theorem 7.10.4 with Theorem 7.10.1, we have the following positive answer to the CE Problem:

Corollary 7.10.5. *There exists a cell-like map $f : X \to Y$ from a 2-dimensional compactum X onto an infinite-dimensional compactum Y.* □

A map $f : X \to Y$ is said to be **monotone** if $f^{-1}(y)$ is connected for every $y \in Y$. The following profound theorem on approximations by open maps was proved by J.J. Walsh. It is beyond the scope of this book to present the proof; refer instead to [24] (References in Notes for Chap. 7).

Theorem 7.10.6 (J.J. WALSH). *Let M be a compact connected (topological) n-manifold (possibly with boundary), where $n \geq 3$. Every monotone map $f : M \to Y$ of M onto a metrizable space Y can be approximated by open maps $g : M \to Y$ such that $g^{-1}(y)$ and $f^{-1}(y)$ have the same shape type for each $y \in Y$.*[13] *In particular, every cell-like map $f : M \to Y$ can be approximated by a cell-like open map.*

Let $f : X \to Y$ be a cell-like map from a 2-dimensional compactum X onto an infinite-dimensional compactum Y as in Corollary 7.10.5. Embed the compactum X into $\mathbf{S}^5$ (or $\mathbf{I}^5$), and consider the adjunction space $Y \cup_f \mathbf{S}^5$ (or $Y \cup_f \mathbf{I}^5$). Restricting the quotient map, we can obtain a cell-like map $g : \mathbf{S}^5 \to Y \cup_f \mathbf{S}^5$ (or $g : \mathbf{I}^5 \to Y \cup_f \mathbf{I}^5$). Because $\mathbf{S}^5$ (or $\mathbf{I}^5$) is a compact connected 5-manifold (with boundary), we can apply Walsh's Theorem 7.10.6 to obtain the following corollary:

Corollary 7.10.7. *There exists a cell-like open map from $\mathbf{S}^5$ (or $\mathbf{I}^5$) onto an infinite-dimensional compactum.* □

To prove Theorem 7.10.4, we introduce the **cohomological dimension** of a map $f : X \to Y$. We define $\dim_{\mathbb{Z}} f \leq n$ provided that, for each map $g : A \to K(\mathbb{Z}, n)$ from a closed set A in Y, $gf|f^{-1}(A)$ extends over X.

---

[13] More generally, this theorem is valid for a quasi-open (or quasi-monotone) map, where f is said to be **quasi-open** or **quasi-monotone** if $f(U) = V$ for each open set V in Y and each component U of $f^{-1}(V)$.

$$
\begin{array}{ccccc}
 & & \overset{h}{\dashrightarrow} & & \\
X & \supset & f^{-1}(A) & \longrightarrow & |K(\mathbb{Z}, n)| \\
{\scriptstyle f}\big\downarrow & & {\scriptstyle f|}\big\downarrow & \nearrow_{g} & \\
Y & \supset & A & &
\end{array}
$$

We define $\dim_{\mathbb{Z}} f = n$ if $\dim_{\mathbb{Z}} f \leq n$ and $\dim_{\mathbb{Z}} f \not\leq n$. Then, it is obvious that $\dim_{\mathbb{Z}} \mathrm{id}_X = \dim_{\mathbb{Z}} X$. When $Y = |K|$ is the polyhedron of a simplicial complex K, we define the **cohomological dimension with respect to** K as follows: $\dim_{\mathbb{Z}}(f, K) \leq n$ provided that, for each subcomplex $L \subset K$ and each map $g : |L| \to |K(\mathbb{Z}, n)|$, the composition $gf|f^{-1}(|L|)$ extends over X. Using this terminology, Proposition 7.8.15 can be reformulated as follows:

- For each simplicial complex K and $n \geq 2$, let $\varpi : |EW(K, n)| \to |K|$ be an associate map of the simplicial Edwards–Walsh complex $EW(K, n)$. Then, $\dim_{\mathbb{Z}}(\varpi, K) \leq n$.

The following is trivial by definition:

Lemma 7.10.8. *Let $f : X \to |K|$ be a map with $\dim_{\mathbb{Z}}(f, K) \leq n$. Then, $\dim_{\mathbb{Z}}(f|Y, K) \leq n$ for every $Y \subset X$.* □

Lemma 7.10.9. *Let X be the inverse limit of an inverse sequence $(|K_i|, f_i)_{i \in \mathbb{N}}$ of compact metric polyhedra. If $\lim_{j\to\infty} \operatorname{mesh} f_{i,j}(K_j) = 0$ and $\dim_{\mathbb{Z}}(f_i, K_i) \leq n$ for each $i \in \mathbb{N}$ then $\dim_{\mathbb{Z}} X \leq n$.*

Proof. Let $g : A \to |K(\mathbb{Z}, n)|$ be a map from a closed set A in X. For each $i \in \mathbb{N}$, let $A_i = p_i(A)$, where $p_i : X \to |K_i|$ is the inverse limit projection. Then, $A = \varprojlim(A_i, f_i|A_{i+1})$ and each $p_i|A$ is the inverse limit projection. By Lemma 7.1.7, we have some $i \in \mathbb{N}$ and a map $g_i : A_i \to |K(\mathbb{Z}, n)|$ such that $g_i p_i|A \simeq g$. Then, g_i extends to a map $\tilde{g}_i : U \to |K(\mathbb{Z}, n)|$ over a neighborhood U of A_i in $|K_i|$. Since $\lim_{j\to\infty} \operatorname{mesh} f_{i,j}(K_j) = 0$, we can find $j \geq i$ so that $f_{i,j}(\operatorname{st}(A_j, K_j)) = \operatorname{st}(A_i, f_{i,j}(K_j)) \subset U$, where $\operatorname{st}(A_j, K_j) = |L|$ for some subcomplex L of K_j. Observe that

$$A_{j+1} = p_{j+1}(A) \subset f_j^{-1}(A_j) \subset f_j^{-1}(\operatorname{st}(A_j, K_j)) = f_j^{-1}(|L|) \subset |K_{j+1}|.$$

Because $\dim_{\mathbb{Z}}(f_j, K_j) \leq n$, there is a map $g_{j+1} : |K_{j+1}| \to |K(\mathbb{Z}, n)|$ such that

$$g_{j+1}|f_j^{-1}(\operatorname{st}(A_j, K_j)) = \tilde{g}_i f_{i,j+1}|f_j^{-1}(\operatorname{st}(A_j, K_j)).$$

Thus, we have a map $g_{j+1} p_{j+1} : X \to |K(\mathbb{Z}, n)|$. Then, it follows that

$$g_{j+1} p_{j+1}|A = \tilde{g}_i f_{i,j+1} p_{j+1}|A = g_i p_i|A \simeq g.$$

Therefore, g extends over X by the Homotopy Extension Theorem 6.4.1. □

For a pointed space (Y, y_0), let $\Omega(Y, y_0)$ be the loop space, that is,

$$\Omega(Y, y_0) = \mathbf{C}((\mathbf{I}, \partial\mathbf{I}), (Y, y_0)) = \mathbf{C}((\mathbf{S}^1, \mathbf{e}_1), (Y, y_0)),$$

where the constant loop c_{y_0} is the base point of $\Omega(Y, y_0)$. For each $i > 1$, $\Omega^i(Y, y_0) = \Omega(\Omega^i(Y, y_0), c_*)$, where c_* is the constant loop at the base point $* \in \Omega^{i-1}(Y, y_0)$ and $\Omega^0(Y, y_0)$ means $Y = (Y, y_0)$ itself. Here, omitting the base point, we simply write $\Omega^i Y$ instead of $\Omega^i(Y, y_0)$. To prove Theorem 7.10.4, we need three more lemmas whose proofs are not given because more algebraic preliminaries would be necessary. For their proofs, refer to the paper by Dydak and Walsh [8] (References in Notes for Chap. 7).

Lemma 7.10.10. [14] *Let (X, x_0), (Y, y_0) be pointed polyhedra, K be a countable simplicial complex, and $p : X \to |K|$ be a map with the following properties:*

(1) *$p^{-1}(|L|)$ is a subpolyhedron of X for every subcomplex L of K;*
(2) *There is some $i_0 \geq 2$ such that $[p^{-1}(\sigma), \Omega^i Y] = \{0\}$ for every $\sigma \in K$ and $i \geq i_0$.*

Then, p induces the isomorphisms

$$p^* : [(K, p(x_0)), (\Omega^i Y, *)] \to [(X, x_0), (\Omega^i Y, *)],\ i \geq i_0.$$

Let $EW(K, n)$ be the simplicial Edwards–Walsh complex for a simplicial complex K and $n \geq 2$. It should be noted that if K is countable then so is $EW(K, n)$. Then, an associate map $\varpi : |EW(K, n)| \to |K|$ has property (1) of Lemma 7.10.10 because $\varpi^{-1}(|L|)$ is a subpolyhedron of $|EW(K, n)|$ for every subcomplex L of K. Recall that $\varpi^{-1}(\sigma) = \sigma$ for each $\sigma \in K^{(n)}$ and $\varpi^{-1}(\sigma) \simeq |K(\mathbb{Z}, n)|^{k(\dim\sigma)}$ for each $\sigma \in K \setminus K^{(n)}$.

Lemma 7.10.11. *For each $n \geq 2$ and $k \geq 1$, $[|K(\mathbb{Z}^k, n)|, \Omega^i \mathbf{S}^m] = \{0\}$ if m is odd and $i \geq m$ or if m is even and $i \geq 2m - 1$.*

In Lemma 7.10.11, when $n = 2$ and $m = 3$, $[|K(\mathbb{Z}^k, 2)|, \Omega^i \mathbf{S}^3] = \{0\}$ for each $k \geq 1$ and $i \geq 3$. Then, an associate map $\varpi : |EW(K, 2)| \to |K|$ satisfies condition (2) of Lemma 7.10.10, where $Y = \mathbf{S}^3$, $n = 2$, and $i_0 = 3$. Consequently, ϖ induces isomorphisms

$$\varpi^* : [|K|, \Omega^3\mathbf{S}^3] \to [|EW(K, 2)|, \Omega^3\mathbf{S}^3].$$

---

[14] This lemma is formulated in a more general setting in [8] and called the COMBINATORIAL VIETORIS–BEGLE THEOREM, where X is a k-space (i.e., compactly generated) and property (1) is that the inclusion $p^{-1}(|L|) \subset X$ is a cofibration for every subcomplex L of K. Due to the Homotopy Extension Theorem 4.3.3, for an arbitrary simplicial complex K and any subcomplex L of K, the inclusion $|L| \subset |K|$ is a cofibration.

Therefore, for a map $f : |K| \to \Omega^3\mathbf{S}^3$,

$$f \not\simeq 0 \Leftrightarrow f\varpi \not\simeq 0.$$

Lemma 7.10.12. *Let $K = \bigcup_{i\in\mathbb{N}} K_i$ be a countable simplicial complex, where $K_1 \subset K_2 \subset \cdots$ are finite subcomplexes of K. Let (Y, y_0) be a pointed polyhedron and $i_0 \geq 3$ such that $\pi_i(Y, y_0)$ is finite for each $i \geq i_0$. For a map $f : |K| \to \Omega^i(Y, y_0)$, $i \geq i_0 - 1$, if $f||K_j| \simeq 0$ for every $j \in \mathbb{N}$ then $f \simeq 0$.*

It is known that $\pi_i(\mathbf{S}^{2n-1})$ is finite for every $i > 2n - 1$.[15] In particular, $\pi_i(\Omega^3\mathbf{S}^3) \cong \pi_{i+3}(\mathbf{S}^3)$ is finite for every $i \in \mathbb{N}$. Moreover, $\pi_m(\Omega^3\mathbf{S}^3) \cong \pi_{m+3}(\mathbf{S}^3) \neq \{0\}$ for some $m \geq 3$.[16] Then, we have a map $f : \mathbf{S}^m \to \Omega^3\mathbf{S}^3$ with $f \not\simeq 0$ for some $m \geq 3$.

Theorem 7.10.13 (Dydak–Walsh). *For a map $f : \mathbf{S}^m \to \Omega^3\mathbf{S}^3$ with $f \not\simeq 0$ $(m \geq 3)$, there exists a map $g : X \to \mathbf{S}^m$ of a compactum X with* $\dim_{\mathbb{Z}} X \leq 2$ *such that $g \not\simeq 0$.*

Proof. Let $K_1 = F(\partial\Delta^{m+1})$ and identify $|K_1| = \mathbf{S}^m$. Thus, we can regard $f : |K_1| \to \Omega^3\mathbf{S}^3$. We will inductively construct maps $g_i : |K_{i+1}| \to |K_i|$, $i \in \mathbb{N}$, so that $(|K_i|, g_i)_{i\in\mathbb{N}}$ satisfies the conditions of Lemma 7.10.9 for $n = 2$, and the compositions $fg_{1,i} = fg_1 \cdots g_{i-1} : |K_i| \to \Omega^3\mathbf{S}^3$ are not null-homotopic (i.e., $fg_1 \cdots g_i \not\simeq 0$). Then, $X = \varprojlim(|K_i|, g_i)$ and the inverse limit projection $p_1 : X \to |K_1|$ are the desired compactum and map. Indeed, $\dim_{\mathbb{Z}} X \leq 2$ by virtue of Lemma 7.10.9. Moreover, $fp_1 \not\simeq 0$ by Lemma 7.1.7(2), which implies that $p_1 \not\simeq 0$.

Let $EW(K_1, 2)$ be the simplicial Edwards–Walsh complex for K_1. Then, as observed above, an associate map $\varpi_1 : |EW(K_1, 2)| \to |K_1|$ satisfies conditions (1) and (2) of Lemma 7.10.10, hence it follows that $f\varpi_1 \not\simeq 0$. By Lemma 7.10.12, $f\varpi_1||K_2| \not\simeq 0$ for some finite subcomplex K_2 of $EW(K_1, 2)$. Let $g_1 = \varpi_1||K_2|$ and replace K_2 by a subdivision such that $\operatorname{mesh} g_1(K_2) < 2^{-1}$.

Next, let $\varpi_2 : |EW(K_2, 2)| \to |K_2|$ be an associate map of the simplicial Edwards–Walsh complex $EW(K_2, 2)$ for K_2. By the analogy of g_1, we can apply Lemma 7.10.10 to have $(fg_1)\varpi_2 \not\simeq 0$. By Lemma 7.10.12, $(fg_1)\varpi_2||K_3| \not\simeq 0$ for some finite subcomplex K_3 of $EW(K_2, 2)$. Let $g_2 = \varpi_2||K_3|$ and replace K_3 by a subdivision such that $\operatorname{mesh} g_1g_2(K_3) < 2^{-2}$ and $\operatorname{mesh} g_2(K_3) < 2^{-2}$.

[15]For example, see Theorem 7.1 in Chap. XI of Hu's book "Homotopy Theory." More generally, $\pi_i(\mathbf{S}^n)$ is finite for every $i > n$ except for $\pi_{4k-1}(\mathbf{S}^{2k})$ (cf. Hatcher's book "Algebraic Topology," p.339).

[16]In fact, $\pi_6(\mathbf{S}^3) \cong \mathbb{Z}_{12}$ by Theorem 16.1 in Chap. XI of Hu's book "Homotopy Theory."

$$\begin{array}{ccccccccc}
 & & & & |EW(K_1,2)| & & & & |EW(K_2,2)| \\
 & & & {}^{\varpi_1}\swarrow & \cup & & {}^{\varpi_2}\swarrow & & \cup \\
\Omega^3\mathbf{S}^3 & \xleftarrow[f]{} & |K_1| & \xleftarrow[g_1]{} & |K_2| & \xleftarrow[g_2]{} & & & |K_3| \xleftarrow[g_3]{} \cdots
\end{array}$$

Inductively, let $\varpi_i : |EW(K_i, 2)| \to |K_i|$ be an associate map of the simplicial Edwards–Walsh complex $EW(K_i, 2)$ for K_i. Applying Lemma 7.10.10 we have $(fg_1 \cdots g_{i-1})\varpi_i \not\simeq 0$. By Lemma 7.10.12, $(fg_1 \cdots g_{i-1})\varpi_i||K_{i+1}| \not\simeq 0$ for some finite subcomplex K_{i+1} of $EW(K_i, 2)$. Let $g_i = \varpi_i||K_{i+1}|$ and replace K_{i+1} by a subdivision such that mesh $g_j \cdots g_i(K_{i+1}) < 2^{-i}$ for every $j \leq i$. This completes the proof. □

Now, we can prove Theorem 7.10.4.

Proof of Theorem 7.10.4. An example X in Theorem 7.10.13 is an infinite-dimensional compactum with $\dim_{\mathbb{Z}} X \leq 2$. Since X has a map $g : X \to \mathbf{S}^m$ with $g \not\simeq 0$, we have $\dim X \geq m \geq 3$ by Proposition 5.2.8, which implies that X is infinite-dimensional by Theorem 7.9.5. Recall that $\dim_{\mathbb{Z}} X \leq 1$ if and only if $\dim X \leq 1$. Then, it follows that $\dim_{\mathbb{Z}} X = 2$. □

7.11 Free Topological Linear Spaces Over Compacta

Recall that the free topological linear space over a space X is a topological linear space $L(X)$ that contains X as a subspace and has the following extension property:

(LE) For an arbitrary topological linear space F, every map $f : X \to F$ of X uniquely extends to a linear map[17] $\tilde{f} : L(X) \to F$.

If the free topological linear space $L(X)$ exists then it is uniquely determined up to linear homeomorphism. For every Tychonoff space X, there exists the free topological linear space $L(X)$ over X (Theorem 3.9.2), which is regular (Lemma 3.9.1(2)). In addition, X is a Hamel basis for $L(X)$ (Lemma 3.9.1(1)).

In this section, to study topological and geometrical structures of $L(X)$, we will reconstruct $L(X)$ for a compactum X. Because $L(X)$ is a linear space with X a Hamel basis, it can be algebraically identified with the following linear space:

$$\mathbb{R}^X_f = \{\xi \in \mathbb{R}^X \mid \xi(x) = 0 \text{ except for finitely many } x \in X\},$$

[17] That is, a continuous linear function.

where each $x \in X$ is identified with $\delta_x \in \mathbb{R}^X_f$ defined by

$$\delta_x(y) = \begin{cases} 1 & \text{if } y = x, \\ 0 & \text{if } y \neq x. \end{cases}$$

We define $\gamma : \bigoplus_{n\in\mathbb{N}}(X^n \times \mathbb{R}^n) \to L(X)$ as follows:

$$\gamma(x, \lambda) = \sum_{i=1}^{n} \lambda(i)x(i) \text{ for } (x, \lambda) \in X^n \times \mathbb{R}^n.$$

We equip $L(X)$ with the topology such that γ is a quotient map. Identifying $X = X \times \{1\} \subset X \times \mathbb{R}$, $\gamma|X$ is the inclusion of X into $L(X)$, which is continuous. For an arbitrary topological linear space F, every map $f : X \to F$ uniquely extends to a linear map $\tilde{f} : L(X) \to F$ because X is a Hamel basis of $L(X)$. Since

$$\tilde{f}\gamma(x, \lambda) = \sum_{i=1}^{n} \lambda(i) f(x(i)) \text{ for } (x, \lambda) \in X^n \times \mathbb{R}^n,$$

it follows that $\tilde{f}\gamma : \bigoplus_{n\in\mathbb{N}}(X^n \times \mathbb{R}^n) \to F$ is continuous, which means that $\tilde{f}$ is continuous. Consequently, $L(X)$ has the property (LE). Then, it follows that the inclusion of X into $L(X)$ is an embedding.

Indeed, for each open set U in X and $x \in U$, let $f : X \to \mathbf{I}$ be a map with $f(x) = 0$ and $f(X \setminus U) = 1$, where it suffices to assume that X is Tychonoff. By (LE), f extends to a linear map $\tilde{f} : L(X) \to \mathbb{R}$. Then, $x \in \tilde{f}^{-1}((-\frac{1}{2}, \frac{1}{2})) \cap X \subset U$. This means that U is open with respect to the topology of X inherited from $L(X)$ (cf. Remark before Theorem 3.9.2).

Each $z \in L(X) \setminus \{0\}$ can be uniquely represented as follows:

$$z = \sum_{i=1}^{n} \lambda_i x_i, \ x_i \in X, \ \lambda_i \in \mathbb{R} \setminus \{0\},$$

where $x_i \neq x_j$ if $i \neq j$. This is called the **irreducible representation** of z. In this case, we denote $\mathrm{supp}(z) = \{x_1, \ldots, x_n\}$, which is called the support of z. For convenience, let $\mathrm{supp}(0) = \emptyset$.

To prove the Hausdorffness of $L(X)$, for each $z \neq z' \in L(X)$, it suffices to find a map $f : L(X) \to \mathbb{R}$ such that $f(z) \neq f(z')$. When $\mathrm{supp}(z) \neq \mathrm{supp}(z')$, we may assume that there is $x_0 \in \mathrm{supp}(z) \setminus \mathrm{supp}(z')$. By (LE), we have a linear map $f : L(X) \to \mathbb{R}$ such that

$$f(x_0) = 1 \text{ and } f((\mathrm{supp}(z) \cup \mathrm{supp}(z')) \setminus \{x_0\}) = 0.$$

Then, $f(z) \neq 0 = f(z')$. When $\operatorname{supp}(z) = \operatorname{supp}(z')$, we have the following irreducible representations:

$$z = \sum_{i=1}^{n} \lambda_i x_i,\ z' = \sum_{i=1}^{n} \lambda'_i x_i,\ x_i \in X,\ \lambda_i, \lambda'_i \in \mathbb{R} \setminus \{0\}.$$

Then, we may assume that $\lambda_0 \neq \lambda'_0$. By (LE), there exists a linear map $f : L(X) \to \mathbb{R}$ such that $f(x_0) = 1$ and $f(\operatorname{supp}(z) \setminus \{x_0\}) = 0$. Then, $f(z) = \lambda_0 \neq \lambda'_0 = f(z')$.

To prove that $L(X)$ is the free topological linear space over X, it remains to show the continuity of addition and scalar multiplication. For each $n \in \mathbb{N}$, we denote $\gamma_n = \gamma | X^n \times \mathbb{R}^n$ and we define

$$\begin{aligned} L_n(X) &= \{z \in L(X) \mid \operatorname{card} \operatorname{supp}(z) \leq n\} \\ &= \gamma_n(X^n \times \mathbb{R}^n) \subset L(X). \end{aligned}$$

For $r > 0$, let $L_n(X, r) = \gamma_n(X^n \times r\lozenge^n) \subset L_n(X)$, where

$$\lozenge^n = \{\lambda \in \mathbb{R}^n \mid \textstyle\sum_{i=1}^n |\lambda(i)| \leq 1\}.$$

Then, all $L_n(X)$ and $L_n(X, r)$ are closed in $L(X)$, hence the restrictions $\gamma | X^n \times \mathbb{R}^n$ and $\gamma | X^n \times r\lozenge^n$ are quotient maps.

Indeed, for each $z \in L(X) \setminus L_n(X)$, let $\operatorname{supp}(z) = \{x_1, \ldots, x_m\}$, where $m = \operatorname{card} \operatorname{supp}(z) > n$. By (LE), we have a linear map $f : L(X) \to \mathbb{R}^m$ such that $f(x_i) = \mathbf{e}_i$ for each $i = 1, \ldots, m$. Then, $f^{-1}(\mathbb{R}^m \setminus \{0\})$ is an open neighborhood of z in $L(X)$ that does not meet $L_n(X)$. Therefore, $L_n(X)$ is closed in $L(X)$.

For each $z \in L(X) \setminus L_n(X, r)$, we have the irreducible representation:

$$z = \sum_{i=1}^{n} \lambda_i x_i,\ x_i \in X,\ \lambda_i \in \mathbb{R} \setminus \{0\}.$$

By (LE), we have a linear map $f : L(X) \to \mathbb{R}^n$ such that $f(x_i) = \mathbf{e}_i$ for each $i = 1, \ldots, n$. Let

$$U = \{\lambda \in \mathbb{R}^n \mid \textstyle\sum_{i=1}^n |\lambda(i)| > r\}.$$

Then, $f^{-1}(U)$ is an open neighborhood of z in $L(X)$ that misses $L_n(X, r)$. Therefore, $L_n(X, r)$ is closed in $L(X)$.

It should be noted that if X is compact metrizable then so is $L_n(X, r)$ for each $n \in \mathbb{N}$ and $r > 0$ because $\gamma | X^n \times r\lozenge^n : X^n \times r\lozenge^n \to L_n(X, r)$ is a perfect map (see 2.4.5(1)).

Now, we consider the tower $L_1(X, 1) \subset L_2(X, 2) \subset \cdots$, where $L(X) = \bigcup_{n \in \mathbb{N}} L_n(X, n)$. Then, we can show the following:

Lemma 7.11.1. $L(X) = \varinjlim L_n(X, n)$.

Proof. Let $A \subset L(X)$ and assume that $A \cap L_n(X, n)$ is closed in $L_n(X, n)$ for every $n \in \mathbb{N}$. Then, each $\gamma^{-1}(A) \cap (X^n \times n\lozenge^n)$ is closed in $X^n \times \mathbb{R}^n$. Let $x_0 \in X$ be fixed. For each $n < m \in \mathbb{N}$, we define the embedding $h_{n,m} : X^n \times \mathbb{R}^n \to X^m \times \mathbb{R}^m$ by

$$h_{n,m}(x_1,\ldots,x_n;\lambda_1,\ldots,\lambda_n) = (x_1,\ldots,x_n,x_0,\ldots,x_0;\lambda_1,\ldots,\lambda_n,0,\ldots,0).$$

Because $\gamma h_{n,m} = \gamma|X^n \times \mathbb{R}^n$ and $h_{n,m}^{-1}(X^m \times m\Diamond^m) = X^n \times m\Diamond^n$, we have

$$h_{n,m}^{-1}(\gamma^{-1}(A) \cap (X^m \times m\Diamond^m)) = \gamma^{-1}(A) \cap (X^n \times m\Diamond^n).$$

Hence, $\gamma^{-1}(A) \cap (X^n \times m\Diamond^n)$ is closed in $X^n \times \mathbb{R}^n$ for each $m > n$. Since $X^n \times \mathbb{R}^n$ has the weak topology with respect to $\{X^n \times m\Diamond^n \mid m > n\}$, it follows that $\gamma^{-1}(A) \cap (X^n \times \mathbb{R}^n)$ is closed in $X^n \times \mathbb{R}^n$. Then, $\gamma^{-1}(A)$ is closed in $\bigoplus_{n\in\mathbb{N}}(X^n \times \mathbb{R}^n)$, which means that A is closed in $L(X)$. □

Remark 14. For a compactum X, as mentioned before Lemma 7.11.1, each $L_n(X,n)$ is also compact metrizable. Thus, $L(X)$ is the direct limit of a tower of compacta. Therefore, $L(X)$ is perfectly normal and paracompact by Corollary 2.8.8, hence it is hereditarily normal (= completely normal) (Theorem 2.2.7).

Recall that $\varinjlim X_n \times \varinjlim Y_n = \varinjlim(X_n \times Y_n)$ if each X_n and Y_n are locally compact (Proposition 2.8.4). Using this fact, we can prove the following:

Lemma 7.11.2. *$L(X)$ is a topological linear space.*

Proof. Since $L(X) = \varinjlim L_n(X,n)$ by Lemma 7.11.1 and each $L_n(X,n)$ is compact, it follows that $L(X) \times L(X) = \varinjlim(L_n(X,n) \times L_n(X,n))$ and

$$\gamma_n \times \gamma_n : (X^n \times n\Diamond^n) \times (X^n \times n\Diamond^n) \to L_n(X,n) \times L_n(X,n)$$

is a quotient map. Then, the addition $\mathfrak{a}$ is continuous according to the following commutative diagram:

$$\begin{array}{ccc} (X^n \times n\Diamond^n) \times (X^n \times n\Diamond^n) & \xrightarrow{\bar{\mathfrak{a}}} & X^{2n} \times 2n\Diamond^{2n} \\ {\scriptstyle \gamma_n\times\gamma_n}\downarrow & & \downarrow{\scriptstyle \gamma_{2n}} \\ L_n(X,n) \times L_n(X,n) & \xrightarrow[\mathfrak{a}]{} & L_{2n}(X,2n), \end{array}$$

where $\bar{\mathfrak{a}}((x,\lambda),(x',\lambda')) = ((x,x'),(\lambda,\lambda'))$.

Since $\mathbb{R} = \varinjlim[-n,n]$, we have $L(X) \times \mathbb{R} = \varinjlim(L_n(X,n) \times [-n,n])$ and

$$\gamma_n \times \mathrm{id} : (X^n \times n\Diamond^n) \times [-n,n] \to L_n(X,n) \times [-n,n]$$

is a quotient map. The scalar multiplication $\mathfrak{m}$ is continuous because of the following commutative diagram:

$$\begin{array}{ccc}
(X^n \times n\Diamond^n) \times [-n,n] & \xrightarrow{\bar{\mathfrak{m}}} & X^{n^2} \times n^2\Diamond^{n^2} \\
\Big\downarrow \gamma_n \times \mathrm{id} & & \Big\downarrow \gamma_{n^2} \\
L_n(X,n) \times [-n,n] & \xrightarrow[\mathfrak{m}]{} & L_{n^2}(X,n^2),
\end{array}$$

where $\bar{\mathfrak{m}}((x,\lambda),t) = (x,\ldots,x,t\lambda,0,\ldots,0)$. □

Consequently, we have arrived at the following theorem:

Theorem 7.11.3. *For every compact space X, $L(X)$ is the free topological linear space over X.* □

Let X and Y be compact spaces. For each map $f : X \to Y$, we have a unique continuous linear map $f_\natural : L(X) \to L(Y)$ which is an extension of f. This is functorial, that is, $(gf)_\natural = g_\natural f_\natural$ for any maps $f : X \to Y$ and $g : Y \to Z$, and $\mathrm{id}_{L(X)} = (\mathrm{id}_X)_\natural$. Thus, we have a covariant functor from the category of compact spaces into the category of topological linear spaces.

Next, we will look into the geometrical structure of the free topological linear space $L(X)$.

Lemma 7.11.4. *For each $n < m \in \mathbb{N}$ and $r > 0$,*

$$L_n(X) \cap L_m(X,r) = L_n(X,r).$$

Proof. Clearly, $L_n(X,r) \subset L_n(X) \cap L_m(X,r)$. For $z \in L_n(X) \cap L_m(X,r)$, we can write $z = \sum_{i=1}^{p} t_i x_i$, where $p \le n$, $t_i \ne 0$, and $x_i \ne x_j$ for $i \ne j$. On the other hand, because $z \in L_m(X,r)$, there is $(y,\lambda) \in X^m \times r\Diamond^m$ such that $z = \sum_{j=1}^{m} \lambda(j)y(j)$. Then, we have

$$\sum_{i=1}^{p} |t_i| = \sum_{i=1}^{p} \Bigg| \sum_{y(j)=x_i} \lambda(j) \Bigg| \le \sum_{i=1}^{p} \sum_{y(j)=x_i} |\lambda(j)| \le \sum_{j=1}^{m} |\lambda(j)| \le r.$$

We define $(y',\lambda') \in X^n \times r\Diamond^n$ as follows:

$$y'(i) = \begin{cases} x_i & \text{if } i \le p, \\ x_1 & \text{if } i > p, \end{cases} \quad \text{and} \quad \lambda'(i) = \begin{cases} t_i & \text{if } i \le p, \\ 0 & \text{if } i > p. \end{cases}$$

Then, it follows that

$$z = \sum_{i=1}^{p} t_i x_i = \sum_{i=1}^{n} \lambda'(i) y'(i) = \gamma_n(y',\lambda') \in L_n(X,r). \quad \square$$

Proposition 7.11.5. *Each $L_n(X)$ is closed in $L(X)$, which is the direct limit of the tower $L_n(X,1) \subset L_n(X,2) \subset \cdots$, i.e., $L_n(X) = \varinjlim L_n(X,m)$.*

Proof. For each $m \geq n \in \mathbb{N}$, $L_n(X) \cap L_m(X,m) = L_n(X,m)$ is closed in $L(X)$ by compactness. Thus, $L_n(X)$ is closed in $L(X)$ by Lemma 7.11.1, hence it follows that $L_n(X) = \varinjlim L_n(X,m)$ (cf. Remark for Proposition 2.8.1). □

Now, for each $n \in \mathbb{N}$, let

$$S_n(X) = L_n(X) \setminus L_{n-1}(X) = \{z \in L(X) \mid \operatorname{card}\operatorname{supp}(z) = n\},$$

where $L_0(X) = \{0\}$. Then, each $S_n(X)$ is open in $L_n(X)$ by Proposition 7.11.5. Let $V_1, \ldots, V_n$ be n many pairwise disjoint open sets in X. Observe that $U = \gamma_n(\prod_{i=1}^n V_i \times (\mathbb{R} \setminus \{0\})^n)$ is an open set in $S_n(X)$ and

$$\gamma_n^{-1}(U) = \bigoplus_{\theta \in \mathfrak{S}_n} \prod_{i=1}^{n} V_{\theta(i)} \times (\mathbb{R} \setminus \{0\})^n,$$

where $\mathfrak{S}_n$ is the n-th symmetric group. For each $\theta \in \mathfrak{S}_n$, the restriction $\gamma_n | \prod_{i=1}^n V_{\theta(i)} \times (\mathbb{R} \setminus \{0\})^n$ is a homeomorphism onto U as it is a bijective quotient map.

In general, a map $p : \tilde{X} \to X$ is called a **covering projection** if each $x \in X$ has an open neighborhood U **evenly covered** by p, that is, there are disjoint open sets U_λ, $\lambda \in \Lambda$, such that $p^{-1}(U) = \bigcup_{\lambda \in \Lambda} U_\lambda$ and each $p|U_\lambda$ is a homeomorphism onto U. Therefore, we have obtained the following proposition:

Proposition 7.11.6. *For each $n \in \mathbb{N}$, $\gamma_n|\gamma_n^{-1}(S_n(X))$ is a covering projection over $S_n(X)$ such that* card $\gamma_n^{-1}(x) = n!$ *for each $x \in S_n(X)$. In particular, each point of $S_n(X)$ has an open neighborhood that is homeomorphic to an open set in $X^n \times \mathbb{R}^n$.* □

As a corollary, we have the following:

Corollary 7.11.7. *If X is compact metrizable, then each $S_n(X)$ is locally compact, metrizable, and σ-compact.*

Proof. It follows from Proposition 7.11.6 that $S_n(X)$ is locally compact and locally metrizable. Since a locally metrizable paracompact space is metrizable (2.6.7(4)), $S_n(X)$ is metrizable (this also follows from the fact that the perfect image of a metrizable space is metrizable (2.4.5(1))). Because of the perfect normality of $L_n(X)$, the open set $S_n(X)$ is F_σ in $L_n(X)$. The σ-compactness of $S_n(X)$ follows from that of $L_n(X)$. □

By Corollary 5.4.4 and Hanner's Theorem 6.2.10(4), we also have the following corollaries:

Corollary 7.11.8. *For a finite-dimensional compactum X, each $S_n(X)$ is finite-dimensional, hence so is $L_n(X)$. Consequently, $L(X)$ is a countable union of finite-dimensional compact sets, hence $L(X)$ is strongly countable-dimensional.* □

As for the following, recall that a locally ANR paracompact space is an ANR (6.2.10(4)).

Corollary 7.11.9. *For a compact ANR X, each $S_n(X)$ is an ANR.* □

The following lemma will be used in the next section.

Lemma 7.11.10. *For each $n \in \mathbb{N}$ and $0 < r < s$, the following statements hold:*

(i) *$\sum_{i=1}^n t_i x_i \in S_n(X) \cap L_n(X, r)$ implies $\sum_{i=1}^n |t_i| \leq r$;*
(ii) *$S_n(X) \cap L_n(X, s)$ is a neighborhood of $S_n(X) \cap L_n(X, r)$ in $S_n(X)$.*

Proof. (i) This is trivial (cf. the proof of Lemma 7.11.4).

(ii) For each $z = \sum_{i=1}^n t_i x_i \in S_n(X) \cap L_n(X, r)$, we have disjoint open sets $U_1, \ldots, U_n$ in X such that $x_i \in U_i$. Because $0 < \sum_{i=1}^n |t_i| \leq r < s$ by (i), we can choose $0 < a_i < t_i < b_i$ so that $0 < \sum_{i=1}^n |t_i'| < s$ if $a_i < t_i' < b_i$. Then,

$$U_z = \gamma_n\big(\textstyle\prod_{i=1}^n U_i \times \prod_{i=1}^n (a_i, b_i)\big) \subset S_n(X) \cap L_n(X, s)$$

is an open neighborhood of z in $S_n(X)$, which implies the desired result. □

7.12 A Non-AR Metric Linear Space

By the Dugundji Extension Theorem 6.1.1, every locally convex topological linear space is an AE. In this section, we show that the local convexity is essential. This involves the following theorem:

Theorem 7.12.1. *There exists a σ-compact metric linear space that is not an AE, hence it is not an AR.*

To construct such a space, we use the free topological linear space $L(X)$ over a compactum X. Note that $L(X)$ is the direct limit of compacta according to the remark after Lemma 7.11.1. Then, $L(X)$ is perfectly normal and paracompact, hence it is also hereditarily normal (= completely normal) (Remark 14). Let $\mathfrak{M}_X$ be the set of all continuous metrics d on $L(X)$ such that $(L(X), d)$ is a metric linear space. Conventionally, open sets, neighborhoods, closures, etc. in $(L(X), d)$ are called d-open sets, d-neighborhoods, d-closures, etc. In addition, continuous maps with respect to d are said to be d-continuous.

Lemma 7.12.2. *Each neighborhood U of $\mathbf{0} \in L(X)$ is a d-neighborhood of $\mathbf{0}$ for some $d \in \mathfrak{M}_X$. In particular, $\mathfrak{M}_X \neq \emptyset$.*

Proof. Since $L(X)$ is a perfectly normal topological linear space (Remark 14), $L(X)$ has open sets $U_1 \supset U_2 \supset \cdots$ such that

$$U_1 \subset U,\ U_{i+1} + U_{i+1} \subset U_i,\ [-1,1]U_{i+1} \subset U_i \text{ and } \{\mathbf{0}\} = \bigcap_{i \in \mathbb{N}} U_i.$$

For each $x \in L(X)$, because each U_i is a neighborhood of $\mathbf{0} = 0x \in L(X)$, we can find $s > 0$ such that $sx \in U_i$, i.e., $x \in s^{-1}U_i$. By Proposition 3.4.1, $\{U_i \mid i \in \mathbb{N}\}$ is a neighborhood basis at $\mathbf{0}$ in some topology of $L(X)$, which makes $L(X)$ a topological linear space. By Theorem 3.6.1, this topology is metrizable and coarser than the original topology. Moreover, U is a neighborhood of $\mathbf{0}$ in this topology because $U_1 \subset U$. Thus, we have the desired result. □

Lemma 7.12.3. *For a compactum X, the following statements hold:*

(i) *Given $d_i \in \mathfrak{M}_X$, $i \in \mathbb{N}$, there exists some $d \in \mathfrak{M}_X$ such that* $\mathrm{id}_{L(X)} : (L(X), d) \to (L(X), d_i)$ *is continuous for every $i \in \mathbb{N}$;*
(ii) *Each open set U in $L(X)$ is d-open for some $d \in \mathfrak{M}_X$;*
(iii) *For an arbitrary metric space Y and $A \subset L(X)$, each map $f : A \to Y$ is d-continuous for some $d \in \mathfrak{M}_X$;*
(iv) *For each metrizable subset $A \subset L(X)$ and $d_0 \in \mathfrak{M}_X$, there is $d \in \mathfrak{M}_X$ such that $d \geq d_0$ and $d|A^2 \in$* Metr(A).

Proof. (i) Note that the product space $F = \prod_{i \in \mathbb{N}} (L(X), d_i)$ is a metrizable topological linear space. Let $\eth : L(X) \to F$ be the diagonal injection (i.e., $\eth(x)(i) = x$ for each $i \in \mathbb{N}$). Then, $\eth$ is continuous and linear. The desired metric can be defined by $\eth$ and an admissible metric for F. (For example, $d(z, z') = \sup_{i \in \mathbb{N}} \min\{i^{-1},\ d_i(z, z')\}$ or $d(z, z') = \sum_{i \in \mathbb{N}} \min\{2^{-i},\ d_i(z, z')\}$.)

(ii) For each $x \in U$, $U - x$ is a neighborhood of $\mathbf{0}$ in $L(X)$. By Lemma 7.12.2, $U - x$ is a d_x-neighborhood of $\mathbf{0}$ for some $d_x \in \mathfrak{M}_X$. Choose $r_x > 0$ so that $\mathrm{B}_{d_x}(\mathbf{0}, r_x) \subset U - x$ and let $V_x = \mathrm{B}_{d_x}(\mathbf{0}, r_x) + x \subset U$. Note that V_x is also open in $L(X)$. On the other hand, we have a tower $D_1 \subset D_2 \subset \cdots$ of compacta with $L(X) = \varinjlim D_i$. Then, $U = \bigcup_{i \in \mathbb{N}} (U \cap D_i)$ and each $U \cap D_i$ is separable metrizable. It is easy to prove that U is covered by a countable subcollection of $\{V_x \mid x \in U\}$, say $U = \bigcup_{i \in \mathbb{N}} V_{x_i}$. By (i), we have $d \in \mathfrak{M}_X$ such that $\mathrm{id}_{L(X)} : (L(X), d) \to (L(X), d_{x_i})$ is continuous for every $i \in \mathbb{N}$. Since each V_{x_i} is d-open, $U = \bigcup_{i \in \mathbb{N}} V_{x_i}$ is also d-open.

(iii) For each $x \in A$ and $i \in \mathbb{N}$, choose an open neighborhood $V_i(x)$ of x in $L(X)$ so that $\operatorname{diam} f(V_i(x) \cap A) < 1/i$. Then, $V_i(x)$ is d_i^x-open for some $d_i^x \in \mathfrak{M}_X$ by (ii). Here, we write $L(X) = \varinjlim D_i$ as in the proof of (ii). Because $A = \bigcup_{n \in \mathbb{N}} (A \cap D_n)$ and each $A \cap D_n$ is separable metrizable, we can obtain $\{x_j \mid j \in \mathbb{N}\} \subset A$ such that $A \subset \bigcup_{j \in \mathbb{N}} V_i(x_j)$ for every $i \in \mathbb{N}$. By (i), we have $d \in \mathfrak{M}_X$ such that $\mathrm{id}_{L(X)} : (L(X), d) \to (L(X), d_i^{x_j})$ is continuous, hence every $V_i(x_j)$ is d-open. For each $x \in A$ and $i \in \mathbb{N}$, choose $j \in \mathbb{N}$ so that $x \in V_i(x_j)$. Since $V_i(x_j)$ is a d-open neighborhood of x and $\operatorname{diam} f(V_i(x_j) \cap A) < 1/i$, it follows that f is d-continuous.

(iv) Take $d' \in \mathrm{Metr}(A)$ and apply (iii) to obtain $d'' \in \mathfrak{M}_X$ such that $\mathrm{id}_A : A \to (A, d')$ is d''-continuous. Note that $\mathrm{id}_A : A = (A, d') \to (A, d''|A^2)$ is also continuous, hence it is a homeomorphism. Therefore, $d''|A^2 \in \mathrm{Metr}(A)$. We define a metric on $L(X)$ as follows:

$$d(z, z') = \max\{d''(z, z'),\ d_0(z, z')\}.$$

Then, it is easy to see that $d \in \mathfrak{M}_X$ is the desired metric. □

Lemma 7.12.4. *Let E and F be metrizable topological linear spaces and $h : E \to F$ a continuous linear surjection. If E and F are ARs, then for each open set U in F, $h|h^{-1}(U) : h^{-1}(U) \to U$ is a fine homotopy equivalence.*

Proof. For every neighborhood V of each $y \in F$, choose an open neighborhood W of $\mathbf{0} \in F$ so that $tW \subset W$ for $t \in \mathbf{I}$ and $W + y \subset V$. Then,

$$h^{-1}(W + y) = h^{-1}(W) + x \quad \text{for each } x \in h^{-1}(y).$$

Since $th^{-1}(W) \subset h^{-1}(W)$ for $t \in \mathbf{I}$, it follows that $h^{-1}(W)$ is contractible. Thus, $h^{-1}(W + y)$ is a contractible open neighborhood of $h^{-1}(y)$ in $h^{-1}(V)$. Hence, h is a local $*$-connection. It follows from Theorem 7.4.3 that h is a fine homotopy equivalence. Due to Corollary 7.4.4, for each open set U in F, $h|h^{-1}(U) : h^{-1}(U) \to U$ is also a fine homotopy equivalence. □

Since there exists a cell-like open map from a finite-dimensional compactum onto an infinite-dimensional compactum (Corollary 7.10.7), Theorem 7.12.1 can be derived from the following theorem:

Theorem 7.12.5. *Let X be an infinite-dimensional compactum with a cell-like open map $f : Y \to X$ of a finite-dimensional compact ANR Y. Then, the free topological linear space $L(X)$ over X has a continuous metric d such that $(L(X), d)$ is a metric linear space but is not an AR.*

Proof. Let $\dim Y \leq m$. Because X is infinite-dimensional, we have a closed set A in X and a map $g : A \to \mathbf{S}^m$ that cannot extend over X (Theorem 5.2.3). Since $\mathbf{S}^m$ is an ANE for normal spaces (Theorem 5.1.6(2)), A has an open neighborhood W in $L(X)$ such that $\mathbf{0} \notin \mathrm{cl}\, W$ and g extends to a map $\bar{g} : \mathrm{cl}\, W \to \mathbf{S}^m$. Observe that

$$\bar{g} f_{\natural}|Y \cap f_{\natural}^{-1}(\mathrm{cl}\, W) = \bar{g} f|Y \cap f_{\natural}^{-1}(\mathrm{cl}\, W).$$

Later, using Haver's Near-Selection Theorem 7.6.1 and the fact that $\dim Y \leq m$ and f is a cell-like open map, we will prove the following claim:

Claim. There exists an open neighborhood U of $X \cup \mathrm{cl}\, W$ in $L(X)$ and the map $\bar{g} f_{\natural}$ extends to a map $h : f_{\natural}^{-1}(U) \to \mathbf{S}^m$.

By virtue of Lemma 7.12.3, we have $d \in \mathfrak{M}_X$ such that U and W are d-open and $\bar{g}$ is d-continuous. Again, by Lemma 7.12.3, we have $d' \in \mathfrak{M}_Y$ such that

$$f_\natural : (L(Y), d') \to (L(X), d) \text{ and } h : (f_\natural^{-1}(U), d') \to \mathbf{S}^m$$

are continuous. Thus, we have the following commutative diagram of metric spaces and maps:

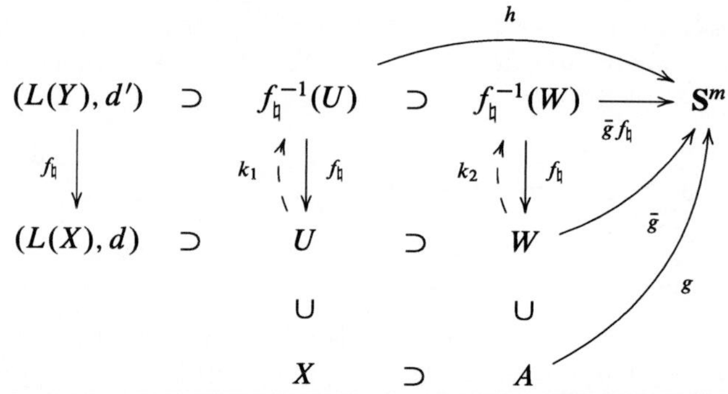

Since $L(Y)$ is a countable union of finite-dimensional compact sets by Corollary 7.11.8, it follows that $(L(Y), d')$ is (strongly) countable-dimensional. Therefore, $(L(Y), d')$ is an AR by Corollaries 6.10.1 and 6.2.9.

Suppose that $(L(X), d)$ is an AR. Then, $f_\natural | f_\natural^{-1}(U)$ and $f_\natural | f_\natural^{-1}(W)$ are homotopy equivalences by virtue of Lemma 7.12.4. Let $k_1 : U \to f_\natural^{-1}(U)$ and $k_2 : W \to f_\natural^{-1}(W)$ be their homotopy inverses. Observe that

$$hk_1|W \simeq hk_1 f_\natural k_2 \simeq hk_2 = \bar{g} f_\natural k_2 \simeq \bar{g}|W.$$

Thus, we have $hk_1|A \simeq g$. By the Homotopy Extension Theorem 6.4.1, g extends over U. This contradicts the fact that g cannot extend over X. Consequently, $(L(X), d)$ is not an AR. □

To prove the Claim, we need the following lemma, which easily follows from Lemma 7.12.3:

Lemma 7.12.6. *There exists some* $d_Y \in \mathfrak{M}_Y$ *such that* $f_\natural^{-1}(\operatorname{cl} W)$ *and every* $f_\natural^{-1}(L_n(X))$ *are* d_Y*-closed and* $\bar{g} f_\natural | f_\natural^{-1}(\operatorname{cl} W) : f_\natural^{-1}(\operatorname{cl} W) \to \mathbf{S}^m$ *is* d_Y*-continuous.* □

Recall that $\operatorname{Comp}(Z)$ is the space of all non-empty compact sets in a space Z with the Vietoris topology. When $Z = (Z, d)$ is a metric space, the Vietoris topology of $\operatorname{Comp}(Z)$ is induced by the Hausdorff metric d_H. Then, by Theorem 5.12.5(3), if Z is compact metrizable then so is $\operatorname{Comp}(Z)$. One should also recall that $S_n(X) = L_n(X) \setminus L_{n-1}(X)$ is metrizable by Corollary 7.11.7.

Lemma 7.12.7. *For each* $z \in S_n(X)$, $\varphi_n(z) = f_{\natural}^{-1}(z) \cap S_n(Y)$ *is a cell-like compactum, and the set-valued function* $\varphi_n : S_n(X) \to \operatorname{Comp}(S_n(Y))$ *is continuous.*

Proof. We write $z = \sum_{i=1}^{n} t_i x_i$ by the irreducible representation. Since $f^{-1}(x_i) \cap f^{-1}(x_j) = \emptyset$ for $i \neq j$, it follows that

$$\varphi_n(z) = \sum_{i=1}^{n} t_i f^{-1}(x_i) = \left\{ \sum_{i=1}^{n} t_i y_i \mid y_i \in f^{-1}(x_i) \right\} \approx \prod_{i=1}^{n} f^{-1}(x_i),$$

where each $f^{-1}(x_i)$ is a cell-like compactum. Then, $\varphi_n(z)$ is also a cell-like compactum.

To show that φ_n is upper semi-continuous (u.s.c.), let U be an open set in $S_n(Y)$ such that $\varphi_n(z) = \sum_{i=1}^{n} t_i f^{-1}(x_i) \subset U$. Choose disjoint open sets $V_1, \dots, V_n$ in Y and $\delta > 0$ so that

$$f^{-1}(x_i) \subset V_i,\ \delta < \min\left\{ |t_i| \mid i = 1, \dots, n \right\} \text{ and } \sum_{i=1}^{n} (t_i - \delta, t_i + \delta) V_i \subset U.$$

Because f is a closed map, each x_i has an open neighborhood W_i in X such that $f^{-1}(W_i) \subset V_i$. Let

$$W = \prod_{i=1}^{n} W_i \times \prod_{i=1}^{n} (t_i - \delta, t_i + \delta) \subset X^n \times \mathbb{R}^n.$$

Then, $\gamma_n(W)$ is an open neighborhood of z in $S_n(X)$. For each $(x, \lambda) \in W$,

$$\varphi_n(\gamma_n(x, \lambda)) = \sum_{i=1}^{n} \lambda(i) f^{-1}(x(i)) \subset \sum_{i=1}^{n} (t_i - \delta, t_i + \delta) V_i \subset U.$$

Therefore, φ_n is u.s.c.

Next, to prove that φ_n is l.s.c., let U be an open set in $S_n(Y)$ that meets $\varphi_n(z) = \sum_{i=1}^{n} t_i f^{-1}(x_i)$. Choose $y_i \in f^{-1}(x_i)$, $i = 1, \dots, n$, so that $\sum_{i=1}^{n} t_i y_i \in U$. Since $y_i \neq y_j$ for $i \neq j$, we have disjoint open sets $V_1, \dots, V_n$ in Y and $\delta > 0$ such that

$$y_i \in V_i,\ \delta < \min\left\{ |t_i| \mid i = 1, \dots, n \right\} \text{ and } \sum_{i=1}^{n} (t_i - \delta, t_i + \delta) V_i \subset U.$$

Because f is an open map, each $f(V_i)$ is an open neighborhood of $x_i = f(y_i)$ in X. Now, let

$$W = \prod_{i=1}^{n} f(V_i) \times \prod_{i=1}^{n} (t_i - \delta, t_i + \delta) \subset X^n \times \mathbb{R}^n.$$

Then, $\gamma_n(W)$ is an open neighborhood of z in $S_n(X)$. For each $(x, \lambda) \in W$, $\varphi_n(\gamma_n(x, \lambda)) \cap U \neq \emptyset$. Indeed, choose $y_i' \in V_i$, $i = 1, \ldots, n$, so that $f(y_i') = x(i)$. Then, it follows that

$$\sum_{i=1}^{n} \lambda(i) y_i' \in \sum_{i=1}^{n} \lambda(i) f^{-1}(x(i)) = \varphi_n(\gamma_n(x, \lambda)) \quad \text{and}$$

$$\sum_{i=1}^{n} \lambda(i) y_i' \in \sum_{i=1}^{n} (t_i - \delta, t_i + \delta) V_i \subset U.$$

Therefore, φ_n is l.s.c. □

Now, we will prove the Claim.

Proof of Claim. In the following, we will inductively construct closed sets $V_n \subset V_n' \subset L_n(X)$ and maps $h_n : f_\natural^{-1}(V_n' \cup \operatorname{cl} W) \to \mathbf{S}^m$, $n \in \mathbb{N}$, so as to satisfy the conditions below:

(1) $L_n(X) \cap (X \cup \operatorname{cl} W) \subset \operatorname{int}_n V_n \subset V_n \subset \operatorname{int}_n V_n'$;
(2) $L_n(X) \cap V_{n+1} = V_n$ and $L_n(X) \cap \operatorname{int}_{n+1} V_{n+1} = \operatorname{int}_n V_n$;
(3) $h_n | f_\natural^{-1}(\operatorname{cl} W) = \bar{g} f_\natural | f_\natural^{-1}(\operatorname{cl} W)$;
(4) $h_{n+1} | f_\natural^{-1}(V_n' \cup \operatorname{cl} W) = h_n$,

where we denote $\operatorname{int}_n A = \operatorname{int}_{L_n(X)} A$ for $A \subset L_n(X)$. Then, $V = \bigcup_{n \in \mathbb{N}} V_n$ is a closed neighborhood of $X \cup \operatorname{cl} W$ in $L(X) = \varinjlim L_n(X)$ by (1) and (2). Since $L(Y) = \varinjlim L_n(Y)$ and $f_\natural^{-1}(L_n(X)) \supset L_n(Y)$, it follows from Proposition 2.8.2 and (2) that

$$f_\natural^{-1}(V) = \varinjlim \left(f_\natural^{-1}(V) \cap L_n(Y)\right) = \varinjlim \left(f_\natural^{-1}(V_n) \cap L_n(Y)\right).$$

Then, by (4), we have the map $h_\infty : f_\natural^{-1}(V) \to \mathbf{S}^m$ such that $h_\infty | f_\natural^{-1}(V_n) = h_n | f_\natural^{-1}(V_n)$, hence $h_\infty | f_\natural^{-1}(\operatorname{cl} W) = \bar{g} f_\natural | f_\natural^{-1}(\operatorname{cl} W)$ by (3). Thus, we have $U = \operatorname{int} V$ and $h = h_\infty | f_\natural^{-1}(U)$, which are needed in the Claim.

The first step. Because $\dim Y \leq m$, the map $\bar{g} f | Y \cap f_\natural^{-1}(\operatorname{cl} W)$ extends over Y (Theorem 5.2.3), hence the map $\bar{g} f$ extends to a map $h_0 : Y \cup f_\natural^{-1}(\operatorname{cl} W) \to \mathbf{S}^m$. Since $\mathbf{S}^m$ is an ANE for normal spaces (Theorem 5.1.6(2)), h_0 extends to a map $\bar{h}_0 : G_0 \to \mathbf{S}^m$ from an open neighborhood G_0 of $Y \cup f_\natural^{-1}(\operatorname{cl} W)$ in $L(Y)$. Consider the following diagram:

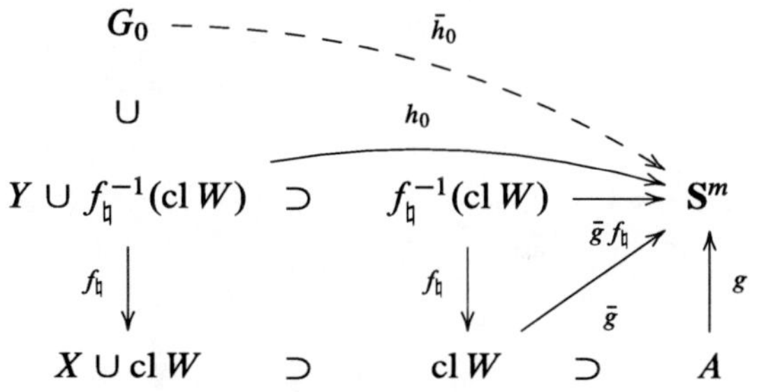

Recall that $\varphi_1 : S_1(X) \to \mathrm{Comp}(S_1(Y))$ is the continuous set-valued function defined by $\varphi_1(z) = f_\natural^{-1}(z) \cap S_1(Y)$ (Lemma 7.12.7). Then, we have the following open set in $S_1(X)$:

$$M_1 = \{z \in S_1(X) \mid \varphi_1(z) \subset G_0\},$$

which is open in $L_1(X)$ because $S_1(X)$ is open in $L_1(X)$. Observe that

$$\varphi_1(x) = f^{-1}(x) \subset Y \subset G_0 \quad \text{for } x \in X \text{ and}$$

$$\varphi_1(z) \subset f_\natural^{-1}(z) \subset f_\natural^{-1}(\mathrm{cl}\, W) \subset G_0 \quad \text{for } z \in L_1(X) \cap \mathrm{cl}\, W,$$

hence $L_1(X) \cap (X \cup \mathrm{cl}\, W) \subset M_1$. Therefore, we can find closed sets $V_1 \subset V_1' \subset L_1(X)$ such that

$$L_1(X) \cap (X \cup \mathrm{cl}\, W) \subset \mathrm{int}_1\, V_1 \subset V_1 \subset \mathrm{int}_1\, V_1' \subset V_1' \subset M_1.$$

Since $f_\natural$ is continuous, we have the continuous set-valued function

$$\tilde{\varphi}_1 = \varphi_1 f_\natural | f_\natural^{-1}(M_1) : f_\natural^{-1}(M_1) \to \mathrm{Comp}(S_1(Y) \cap G_0).$$

For each $y \in f_\natural^{-1}(M_1)$, $\tilde{\varphi}_1(y) = \varphi_1(f_\natural(y))$ is a cell-like compactum by Lemma 7.12.7. Since $S_1(Y)$ is finite-dimensional, σ-compact, and metrizable (Corollaries 7.11.7 and 7.11.8), so is the F_σ-set $f_\natural^{-1}(M_1 \setminus \mathrm{cl}\, W) = f_\natural^{-1}(M_1) \setminus f_\natural^{-1}(\mathrm{cl}\, W)$ in $S_1(Y)$. Moreover, $S_1(Y) \cap G_0$ is an ANR because so is $S_1(Y)$ (Corollary 7.11.9). By virtue of Lemma 7.12.3(iv), we have $d_1 \in \mathfrak{M}_Y$ such that $d_1 \geq d_Y$ and $d_1 | S_1(Y)^2 \in \mathrm{Metr}(S_1(Y))$, where d_Y is the metric obtained by Lemma 7.12.6. Note that $f_\natural^{-1}(\mathrm{cl}\, W)$ is d_1-closed and $\bar{g} f_\natural | f_\natural^{-1}(\mathrm{cl}\, W)$ is d_1-continuous. We can apply Haver's Near-Selection Theorem 7.6.1 to obtain a map

$$\xi_1 : f_\natural^{-1}(M_1 \setminus \mathrm{cl}\, W) \to S_1(Y) \cap G_0$$

such that $d_1(\xi_1(y), \tilde{\varphi}_1(y)) < d_1(y, f_\natural^{-1}(\mathrm{cl}\, W))$ for every $y \in f_\natural^{-1}(M_1 \setminus \mathrm{cl}\, W)$.

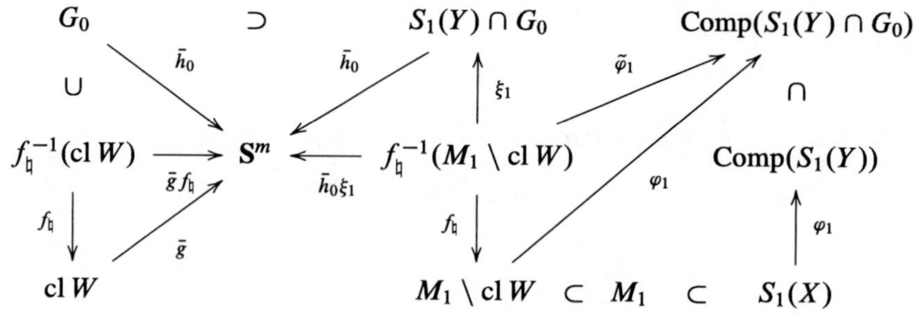

Then, we can define $h_1 : f_\natural^{-1}(V_1' \cup \mathrm{cl}\, W) \to \mathbf{S}^m$ as follows:

$$h_1(y) = \begin{cases} \bar{g} f_\natural(y) & \text{if } y \in f_\natural^{-1}(\mathrm{cl}\, W), \\ \bar{h}_0 \xi_1(y) & \text{if } y \in f_\natural^{-1}(V_1' \setminus \mathrm{cl}\, W). \end{cases}$$

We will verify the continuity of h_1. Since $f_\natural^{-1}(V_1' \setminus \mathrm{cl}\, W)$ is an open set in $f_\natural^{-1}(V_1' \cup \mathrm{cl}\, W)$, we have to show the continuity of h_1 at each $y \in f_\natural^{-1}(\mathrm{cl}\, W)$. To this end, we will show that h_1 is d_1-continuous at y because $\mathrm{id} : L(Y) \to (L(Y), d_1)$ is continuous. Since $h_1 | f_\natural^{-1}(\mathrm{cl}\, W) = \bar{g} f_\natural | f_\natural^{-1}(\mathrm{cl}\, W)$ is d_1-continuous, it suffices to prove that

$$\lim_{i \to \infty} d_1(y_i, y) = 0,\ y_i \in f_\natural^{-1}(V_1' \setminus \mathrm{cl}\, W) \Rightarrow \lim_{i \to \infty} \bar{h}_0 \xi_1(y_i) = \bar{g} f_\natural(y).$$

Since $\tilde{\varphi}_1 : f_\natural^{-1}(M_1) \to \mathrm{Comp}(S_1(Y) \cap G_0)$ is u.s.c. and

$$d_1(\xi_1(y_i), \tilde{\varphi}_1(y_i)) < d_1(y_i, f_\natural^{-1}(\mathrm{cl}\, W)) \leq d_1(y_i, y),$$

it follows that $\{\xi_1(y_i) \mid i \in \mathbb{N}\} \cup \tilde{\varphi}_1(y)$ is compact, hence $(\xi_1(y_i))_{i \in \mathbb{N}}$ has a convergent subsequence $(\xi_1(y_{i_j}))_{j \in \mathbb{N}}$. Since $d_1(\xi_1(y_{i_j}), \tilde{\varphi}_1(y)) \to 0$, $(\xi_1(y_{i_j}))_{j \in \mathbb{N}}$ converges to some

$$z \in \tilde{\varphi}_1(y) = \varphi_1(f_\natural(y)) = f_\natural^{-1}(f_\natural(y)) \cap S_1(Y) \subset f_\natural^{-1}(\mathrm{cl}\, W).$$

Then, $(\bar{h}_0 \xi_1(y_{i_j}))_{j \in \mathbb{N}}$ converges to $\bar{h}_0(z) = \bar{g} f_\natural(z) = \bar{g} f_\natural(y)$. By the same argument, any subsequence of $(\bar{h}_0 \xi_1(y_i))_{i \in \mathbb{N}}$ has a subsequence converging to $\bar{g} f_\natural(y)$. This means that $\lim_{i \to \infty} \bar{h}_0 \xi_1(y_i) = \bar{g} f_\natural(y)$.

The inductive step. Assume that V_{n-1}, V_{n-1}', and h_{n-1} have been constructed. Because $\mathbf{S}^m$ is an ANE for normal spaces, h_{n-1} extends to a map $\bar{h}_{n-1} : G_{n-1} \to \mathbf{S}^m$ from an open neighborhood G_{n-1} of $f_\natural^{-1}(V_{n-1}' \cup \mathrm{cl}\, W)$ in $L(Y)$. Recall that $\varphi_n : S_n(X) \to \mathrm{Comp}(S_n(Y))$ is the continuous set-valued function defined by $\varphi_n(z) = f_\natural^{-1}(z) \cap S_n(Y)$ (Lemma 7.12.7). Let

$$M_n = \{z \in S_n(X) \mid \varphi_n(z) \subset G_{n-1}\} \text{ and } N_n = M_n \cup \mathrm{int}_{n-1} V'_{n-1}.$$

Then, M_n is open in $S_n(X)$, hence in $L_n(X)$.

We will show that N_n is open in $L_n(X)$. Suppose that N_n is not open in $L_n(X)$. Then, $N_n \cap L_n(X,m)$ is not open in $L_n(X,m)$ for some $m \in \mathbb{N}$ because $L_n(X) = \varinjlim L_n(X,m)$ (Proposition 7.11.5). Since X is compact metrizable, $L_n(X,m) = \gamma_n(X^n \times m\Diamond^n)$ is also compact metrizable. Then, there is some $z \in N_n \cap L_n(X,m)$ that is the limit of $z_i \in L_n(X,m)\setminus N_n$. Since M_n is open in $L_n(X)$, it follows that $z \in \mathrm{int}_{n-1} V'_{n-1} \subset L_{n-1}(X)$. Then, $L_{n-1}(X)$ contains only finitely many z_i. Otherwise, infinitely many z_i would be contained in $\mathrm{int}_{n-1} V'_{n-1} \subset N_n$. Consequently, we may assume that $z_i \in S_n(X) = L_n(X) \setminus L_{n-1}(X)$ for every $i \in \mathbb{N}$. Since $z_i \notin M_n$, we have $y_i \in \varphi_n(z_i) \setminus G_{n-1}$. Recall that $\varphi_n(z_i) = f_\natural^{-1}(z_i) \cap S_n(Y)$. Then,

$$f_\natural(y_i) = z_i \in L_n(X,m) \cap S_n(X) = \gamma_n(X^n \times m\Diamond^n) \cap S_n(X).$$

Since $y_i \in S_n(Y)$, it follows that $y_i \in L_n(Y,m) = \gamma_n(Y^n \times m\Diamond^n)$. Because $L_n(Y,m)$ is compact, we may assume that $\lim_{i\to\infty} y_i = y \in L_n(Y,m)$. Then,

$$f_\natural(y) = \lim_{i\to\infty} f_\natural(y_i) = \lim_{i\to\infty} z_i = z,$$

hence $y \in f_\natural^{-1}(z) \subset f_\natural^{-1}(V'_{n-1}) \subset G_{n-1}$. Therefore, $y_i \in G_{n-1}$ for sufficiently large $i \in \mathbb{N}$, which is a contradiction. Thus, N_n is open in $L_n(X)$.

For each $z \in S_n(X) \cap \mathrm{cl}\, W$, we have

$$\varphi_n(z) \subset f_\natural^{-1}(z) \subset f_\natural^{-1}(\mathrm{cl}\, W) \subset G_{n-1}.$$

Hence, $S_n(X) \cap \mathrm{cl}\, W \subset M_n$. Then, it follows that

$$\begin{aligned} L_n(X) \cap (X \cup \mathrm{cl}\, W) &= (S_n(X) \cap \mathrm{cl}\, W) \cup \big(L_{n-1}(X) \cap (X \cup \mathrm{cl}\, W)\big) \\ &\subset M_n \cup \mathrm{int}_{n-1} V'_{n-1} = N_n. \end{aligned}$$

Since $L(X)$ is hereditarily normal (= completely normal) (Remark 14), so is $L_n(X)$, hence, we can find an open set $\tilde{V}$ in $L_n(X)$ such that

(5) $\big(L_n(X) \cap (X \cup \mathrm{cl}\, W)\big) \cup \mathrm{int}_{n-1} V_{n-1} \subset \tilde{V}$ and
(6) $\mathrm{cl}\, \tilde{V} \cap \big((L_{n-1}(X) \setminus V_{n-1}) \cup (L_n(X) \setminus N_n)\big) = \emptyset$.

Let $V_n = V_{n-1} \cup \mathrm{cl}\, \tilde{V} \subset N_n$. Then, it follows from (6) that

$$L_{n-1}(X) \cap V_n \subset V_{n-1} \subset L_{n-1}(X) \cap V_n,$$

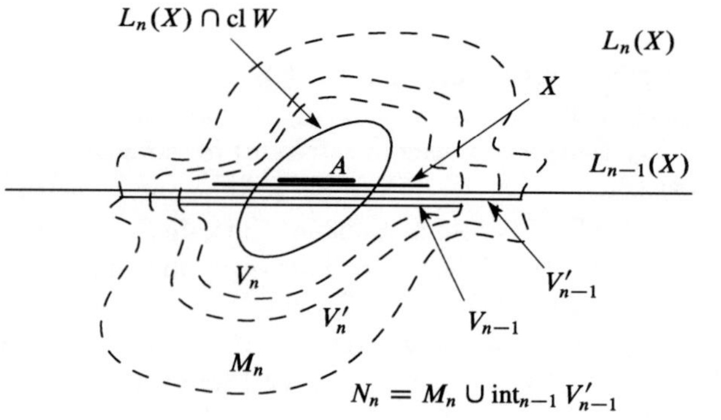

Fig. 7.6 M_n, N_n, and V_n

hence $L_{n-1}(X) \cap V_n = V_{n-1}$. Moreover, by (5),

$$\begin{aligned}\mathrm{int}_{n-1} V_{n-1} \subset L_{n-1}(X) \cap \tilde{V} &\subset L_{n-1}(X) \cap \mathrm{int}_n V_n \\ &\subset \mathrm{int}_{n-1}(L_{n-1}(X) \cap V_n) = \mathrm{int}_{n-1} V_{n-1}.\end{aligned}$$

Thus, we have $L_{n-1}(X) \cap \mathrm{int}_n V_n = \mathrm{int}_{n-1} V_{n-1}$. Choose a closed set $V'_n \subset L_n(X)$ so that $V_n \subset \mathrm{int}_n V'_n \subset V'_n \subset N_n$. Thus, we have obtained V_n and V'_n satisfying conditions (1) and (2) (Fig. 7.6).

Because $f_\natural$ is continuous, we have the continuous set-valued function

$$\tilde{\varphi}_n = \varphi_n f_\natural | f_\natural^{-1}(M_n) : f_\natural^{-1}(M_n) \to \mathrm{Comp}(S_n(Y) \cap G_{n-1}),$$

where $S_n(Y) \cap G_{n-1}$ is an ANR because so is $S_n(Y)$ (Corollary 7.11.9). Since $S_n(Y)$ is finite-dimensional σ-compact and metrizable (Corollaries 7.11.7 and 7.11.8), so is the F_σ-set $f_\natural^{-1}(M_n \backslash \mathrm{cl}\, W) = f_\natural^{-1}(M_n) \backslash f_\natural^{-1}(\mathrm{cl}\, W)$ in $S_n(Y)$. By the analogy of d_1, we apply Lemma 7.12.3(iv) to obtain $d_n \in \mathfrak{M}_Y$ such that $d_n \geq d_Y$ and $d_n | S_n(Y)^2 \in \mathrm{Metr}(S_n(Y))$. Then, $f_\natural^{-1}(L_{n-1}(X) \cup \mathrm{cl}\, W) = f_\natural^{-1}(L_{n-1}(X)) \cup f_\natural^{-1}(\mathrm{cl}\, W)$ is d_n-closed and $\bar{g} f_\natural | f_\natural^{-1}(\mathrm{cl}\, W)$ is d_n-continuous.

Due to Lemma 7.11.10(ii), $S_n(Y) \cap L_n(Y, m+1)$ is a neighborhood of $S_n(Y) \cap L_n(Y, m)$ in $S_n(Y)$ for each $m \in \mathbb{N}$. Then, we have an l.s.c. function $\zeta_n : S_n(Y) \to (0, \infty)$ defined by

$$\zeta_n(y) = \tfrac{1}{2} d_n(y, S_n(Y) \setminus L_n(Y, m+1))$$
$$\text{if } y \in S_n(Y) \cap L_n(Y, m) \setminus L_n(Y, m-1),\ m \in \mathbb{N}.$$

Indeed, assume that $y \in S_n(Y) \cap L_n(Y, m) \setminus L_n(Y, m-1)$ and $\zeta_n(y) > t$, i.e., $\frac{1}{2} d_n(y, S_n(Y) \setminus L_n(Y, m+1)) > t$. Then, the following V_y is a neighborhood of y in $S_n(Y)$:

$$V_y = \{y' \in S_n(Y) \mid \tfrac{1}{2}d_n(y', S_n(Y) \setminus L_n(Y, m+1)) > t\} \\ \cap L_n(Y, m+1) \setminus L_n(Y, m-1).$$

For each $y' \in V_y$, if $y' \in L_n(Y, m)$ then

$$\zeta_n(y') = \tfrac{1}{2}d_n(y, S_n(Y) \setminus L_n(Y, m+1)) > t,$$

and if $y' \notin L_n(Y, m)$ then

$$\zeta_n(y') = \tfrac{1}{2}d_n(y, S_n(Y) \setminus L_n(Y, m+2)) \\ \geq \tfrac{1}{2}d_n(y, S_n(Y) \setminus L_n(Y, m+1)) > t.$$

Applying Theorem 2.7.6, we have a map $\eta : S_n(Y) \to (0, 1)$ such that $0 < \eta_n(y) < \zeta_n(y)$. Then, it follows that

$$\eta_n(y) < \tfrac{1}{2}d_n(y, S_n(Y) \setminus L_n(Y, m+1)) \\ \text{for each } y \in S_n(Y) \cap L_n(Y, m) \text{ and } m \in \mathbb{N}.$$

Because each $\tilde{\varphi}_n(y)$ is compact, we can define $\bar{\eta}_n : f_\natural^{-1}(M_n \setminus \operatorname{cl} W) \to (0, 1)$ by $\bar{\eta}_n(y) = \min \eta_n\tilde{\varphi}_n(y)$. To prove the continuity of $\bar{\eta}_n$, observe that the following set-valued function is continuous:

$$f_\natural^{-1}(M_n \setminus \operatorname{cl} W) \ni y \mapsto \eta_n\tilde{\varphi}_n(y) \in \operatorname{Comp}((0, 1)).$$

For each $y \in f_\natural^{-1}(M_n \setminus \operatorname{cl} W)$ and $\varepsilon > 0$, we have $\delta > 0$ such that

$$d_n(y, y') < \delta \Rightarrow d_H(\eta_n\tilde{\varphi}_n(y), \eta_n\tilde{\varphi}_n(y')) < \varepsilon,$$

where d_H is the Hausdorff metric on $\operatorname{Comp}((0, 1))$ induced by the metric $d(t, t') = |t - t'|$. When $d_H(\eta_n\tilde{\varphi}_n(y), \eta_n\tilde{\varphi}_n(y')) < \varepsilon$, we can find $t \in \eta_n\tilde{\varphi}_n(y)$ and $t' \in \eta_n\tilde{\varphi}_n(y')$ such that $|t - \bar{\eta}_n(y')| < \varepsilon$ and $|t' - \bar{\eta}_n(y)| < \varepsilon$. Then, $\bar{\eta}_n(y) \leq t < \bar{\eta}_n(y') + \varepsilon$ and $\bar{\eta}_n(y') \leq t' < \bar{\eta}_n(y) + \varepsilon$, which means that $|\bar{\eta}_n(y) - \bar{\eta}_n(y')| < \varepsilon$.

As in the first step, we apply Haver's Near Selection Theorem 7.6.1 to obtain a map

$$\xi_n : f_\natural^{-1}(M_n \setminus \operatorname{cl} W) \to S_n(Y) \cap G_{n-1}$$

satisfying

$$d_n(\xi_n(y), \tilde{\varphi}_n(y)) < \min\{d_n(y, f_\natural^{-1}(L_{n-1}(X) \cup \operatorname{cl} W)),\ \bar{\eta}_n(y)\}.$$

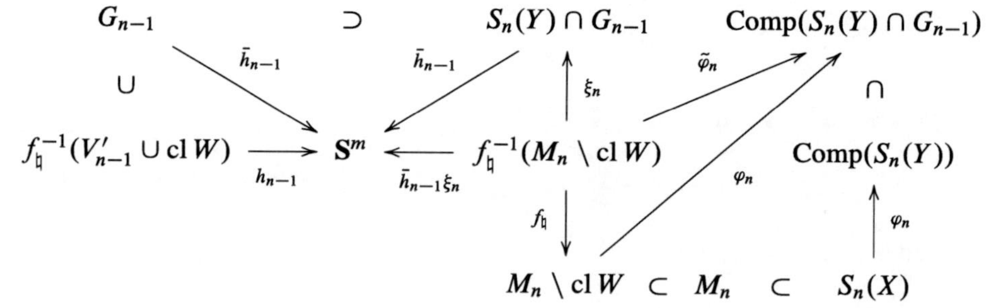

Here, it should be remarked that the distance between $\bar{h}_{n-1}\xi_n(y)$ and $\bar{h}_{n-1}(y)$ converges to $\mathbf{0}$ as y tends to $f_\natural^{-1}(\mathrm{cl}\, W)$ but it does not always decrease as y tends to $f_\natural^{-1}(V'_{n-1} \setminus \mathrm{cl}\, W)$.

We will show that the line segment from $\bar{h}_{n-1}\xi_n(y)$ to $\bar{h}_{n-1}(y)$ is contained in G_{n-1} if y is close to $f_\natural^{-1}(V'_{n-1} \setminus \mathrm{cl}\, W)$. To see this, let

$$R_n = \{y \in f_\natural^{-1}(M_n \setminus \mathrm{cl}\, W) \mid \langle y, \xi_n(y)\rangle \subset G_{n-1}\} \quad \text{and}$$
$$T_n = R_n \cup f_\natural^{-1}(\mathrm{int}_{n-1} V'_{n-1} \cup \mathrm{cl}\, W).$$

Note that $f_\natural^{-1}(M_n \setminus \mathrm{cl}\, W)$ is open in $f_\natural^{-1}(L_n(X) \cup \mathrm{cl}\, W)$. Since G_{n-1} is open in $L(Y)$ and ξ_n is continuous, it follows that R_n is open in $f_\natural^{-1}(L_n(X) \setminus \mathrm{cl}\, W)$, hence so in $f_\natural^{-1}(L_n(X) \cup \mathrm{cl}\, W)$.

We now show that T_n is open in $f_\natural^{-1}(L_n(X) \cup \mathrm{cl}\, W)$. Assume that T_n is not open in $f_\natural^{-1}(L_n(X) \cup \mathrm{cl}\, W)$. Because $f_\natural^{-1}(L_n(X) \cup \mathrm{cl}\, W)$ is closed in $L(Y) = \varinjlim L_n(Y, m)$ (Proposition 7.11.5), we have

$$f_\natural^{-1}(L_n(X) \cup \mathrm{cl}\, W) = \varinjlim \big(f_\natural^{-1}(L_n(X) \cup \mathrm{cl}\, W) \cap L_n(Y, m)\big).$$

Then, $T_n \cap L_n(Y, m)$ is not open in $f_\natural^{-1}(L_n(X) \cup \mathrm{cl}\, W) \cap L_n(Y, m)$ for some $m \in \mathbb{N}$, hence we have $z \in T_n \cap L_n(Y, m)$ which is the limit of

$$z_i \in f_\natural^{-1}(L_n(X) \cup \mathrm{cl}\, W) \cap L_n(Y, m) \setminus T_n,\ i \in \mathbb{N}.$$

Since R_n is open in $f_\natural^{-1}(L_n(X) \cup \mathrm{cl}\, W)$ and $z_i \notin R_n$, $i \in \mathbb{N}$, it follows that $z \in f_\natural^{-1}(\mathrm{int}_{n-1} V'_{n-1} \cup \mathrm{cl}\, W)$. Note that N_n is open in $L_n(X)$, $L_n(X) \cap \mathrm{cl}\, W \subset N_n$, and $\mathrm{int}_{n-1} V'_{n-1} \subset N_n$, hence $N_n \cup \mathrm{cl}\, W$ is a neighborhood of $f_\natural(z)$ in $L_n(X) \cup \mathrm{cl}\, W$. Therefore, we may assume that $z_i \in f_\natural^{-1}(N_n \cup \mathrm{cl}\, W)$ for every $i \in \mathbb{N}$. Then, $z_i \in f_\natural^{-1}(M_n \setminus \mathrm{cl}\, W)$ because $z_i \notin f_\natural^{-1}(\mathrm{int}_{n-1} V'_{n-1} \cup \mathrm{cl}\, W)$. Since $z_i \notin R_n$, we have $\langle z_i, \xi_n(z_i)\rangle \not\subset G_{n-1}$.

On the other hand, since $z_i \in L_n(Y, m)$, it follows that $f_\natural(z_i) \in L_n(X, m)$. Note that $f_\natural(z_i) \in M_n \subset S_n(X)$. Then, $f_\natural(z_i) \in S_n(X) \cap L_n(X, m)$. For $z' = \sum_{j=1}^n t_j y_j \in S_n(Y)$, by Lemma 7.11.10(i),

$$f_\natural(z') = \sum_{j=1}^{n} t_j f(y_j) \in S_n(X) \cap L_n(X, m) \Rightarrow z' \in L_n(Y, m).$$

Thus, it follows that

$$\tilde{\varphi}_n(z_i) = \varphi_n(f_\natural(z_i)) = f_\natural^{-1}(f_\natural(z_i)) \cap S_n(Y) \subset S_n(Y) \cap L_n(Y, m),$$

hence $\xi_n(z_i) \in L_n(Y, m+1)$ because

$$\begin{aligned} d_n(\xi_n(z_i), \tilde{\varphi}_n(z_i)) < \bar{\eta}_n(z_i) &= \min \eta_n \tilde{\varphi}_n(z_i) \\ &= \tfrac{1}{2} \min \left\{ d_n(y, S_n(Y) \setminus L_n(Y, m+1)) \;\middle|\; y \in \tilde{\varphi}_n(z_i) \right\}. \end{aligned}$$

Since $L_n(Y, m+1)$ is compact metrizable, so is $\mathrm{Comp}(L_n(Y, m+1))$ (Theorem 5.12.5(3)). Then, we may assume that $\xi_n(z_i)$ converges to some $a \in L_n(Y, m+1)$ and also $\tilde{\varphi}_n(z_i)$ converges to some $K \in \mathrm{Comp}(L_n(Y, m+1))$. Since $f_\natural(\tilde{\varphi}_n(z_i)) = f_\natural(z_i)$ converges to $f_\natural(z)$, it follows that $K \subset f_\natural^{-1}(f_\natural(z))$. Observe that

$$\begin{aligned} \lim_{i \to \infty} d_n(\xi_n(z_i), \tilde{\varphi}_n(z_i)) &\leq \lim_{i \to \infty} d_n\big(z_i, f_\natural^{-1}(L_{n-1}(X) \cup \mathrm{cl}\, W)\big) \\ &\leq \lim_{i \to \infty} d_n(z_i, z) = 0. \end{aligned}$$

Hence, $\lim_{i \to \infty} d_n(a, \tilde{\varphi}_n(z_i)) = 0$. Since $L_n(Y, m+1)$ is compact, d_n is admissible on $L_n(Y, m+1)$, hence $a \in K \subset f_\natural^{-1}(f_\natural(z))$. By the linearity of $f_\natural$, $f_\natural^{-1}(f_\natural(z))$ is a flat, which is convex. Then, it follows that

$$\langle z, a \rangle \subset f_\natural^{-1}(f_\natural(z)) \subset f_\natural^{-1}(V'_{n-1} \cup \mathrm{cl}\, W) \subset G_{n-1}.$$

Since $\langle z_i, \xi_n(z_i) \rangle$ converges to $\langle z, a \rangle$ in $\mathrm{Comp}(L_n(Y, m+1))$, $\langle z_i, \xi_n(z_i) \rangle \subset G_{n-1}$ for sufficiently large $i \in \mathbb{N}$, which is a contradiction. Consequently, T_n is open in $f_\natural^{-1}(L_n(X) \cup \mathrm{cl}\, W)$.

Now, choose an open set B in $f_\natural^{-1}(L_n(X) \cup \mathrm{cl}\, W)$ so that

$$f_\natural^{-1}(V'_{n-1} \cup \mathrm{cl}\, W) \subset B \subset \mathrm{cl}\, B \subset T_n$$

and let $\beta : f_\natural^{-1}(L_n(X) \cup \mathrm{cl}\, W) \to \mathbf{I}$ be a Urysohn map with

$$\beta(\mathrm{cl}\, B) = 0 \text{ and } \beta(f_\natural^{-1}(L_n(X) \cup \mathrm{cl}\, W) \setminus T_n) = 1.$$

From the definition of T_n, it follows that

$$(1-\beta(y))y + \beta(y)\xi_n(y) \in G_{n-1} \quad \text{for } y \in f_\natural^{-1}(M_n \setminus \operatorname{cl} W).$$

On the other hand,

$$\begin{aligned} f_\natural^{-1}(V_n' \cup \operatorname{cl} W) \setminus B \subset f_\natural^{-1}(N_n \cup \operatorname{cl} W) \setminus f_\natural^{-1}(V_{n-1}' \cup \operatorname{cl} W) \\ = f_\natural^{-1}(M_n \setminus \operatorname{cl} W). \end{aligned}$$

Then, we can define a map $h_n : f_\natural^{-1}(V_n' \cup \operatorname{cl} W) \to \mathbf{S}^m$ by

$$h_n(y) = \begin{cases} \bar{h}_{n-1}(y) & \text{if } y \in \operatorname{cl} B, \\ \bar{h}_{n-1}((1-\beta(y))y + \beta(y)\xi_n(y)) & \text{if } y \notin B. \end{cases}$$

Because $\bar{h}_{n-1}$ is an extension of h_{n-1}, the map h_n satisfies conditions (3) and (4). This completes the proof. □

Notes for Chap. 7

As mentioned at the beginning of this chapter, the concept of cell-like maps is profoundly related to Shape Theory and Decomposition Theory. For insight into these theories, refer to the following textbooks:

- S. Mardešić and J. Segal, *Shape Theory*, North-Holland Math. Library **26** (Elsevier Sci. B.V., Amsterdam, 1982)
- A. Chigogidze, *Inverse Spectra*, North-Holland Math. Library **53** (Elsevier Sci. B.V., Amsterdam, 1996)
- R.J. Davarmann, *Decomposition of Manifolds*, Pure and Appl. Math. **124** (Academic Press, Inc., Orlando, 1986)

Shape Theory was founded by K. Borsuk in 1968. As a textbook of Shape Theory, his own book gives a good introduction:

- K. Borsuk, *Theory of Shape*, Monog. Mat. **59** (Polish Sci. Publ., Warsaw, 1975)

To study cell-like maps, a background in Algebraic Topology is required. There exist many textbooks on Algebraic Topology. Among them, we recommend the following:

- E. Spanier, *Algebraic Topology* (McGraw-Hill, New York, 1966)
- S.-T. Hu, *Homotopy Theory* (Academic Press, Inc., New York, 1959)
- A. Hatcher, *Algebraic Topology* (Cambridge Univ. Press, Cambridge, 2002)

The paper [13] gives a good survey of cell-like maps up to the mid-1970s. A compactum A in an n-manifold M is called **cellular** if A has an arbitrarily small neighborhood in M that is homeomorphic to $\mathbf{B}^n$. In other words, A can be written as $A = \bigcap_{i\in\mathbb{N}} B_i$, where $B_i \approx \mathbf{B}^n$ and $B_{i+1} \subset \operatorname{int} B_i$. Every cellular compactum is cell-like but the converse does not hold. The Whitehead continuum is an example of non-cellular cell-like compacta (see Daverman's book, pp. 68–69).

The concept of the (n-)soft map was introduced by Shchepin in [19] and generalized to the (polyhedrally) approximately (n-)soft map in [20]. These are discussed in Chigogidze's book above. The 0-Dimensional Selection Theorem 7.2.4 was established by Michael [14, 15]. In [10], Kozlowski introduced the concept of the local n-connection and proved Theorem 7.3.6.

The Toruńczyk characterization of the Hilbert cube $\boldsymbol{Q}$ (Theorem 7.7.1) was established in [23]. This characterization (more generally, the Toruńczyk characterization of $\boldsymbol{Q}$-manifolds) is the main theme of van Mill's book "Infinite-Dimensional Topology" mentioned in the Preface. The author's second book "Topology of Infinite-Dimensional Manifolds" can also be referred to. The result of Adams used in Sect. 7.7 appeared in [1]. It also follows from the work of Toda [22]. The Taylor example in Theorem 7.7.5 was constructed in [21] (cf. [18]). The example of Theorem 7.7.8 is due to Keesling [9]. Using the Taylor example, many counter-examples are constructed in ANR Theory. For example, in [16], van Mill constructed a map $f : \boldsymbol{Q} \to Y$ of the Hilbert cube $\boldsymbol{Q}$ onto a non-AR compactum Y such that $f^{-1}(y) \approx \boldsymbol{Q}$ for every $y \in Y$, and in [17] he also constructed a separable metrizable space X such that, for each compact set A in an arbitrary metrizable space Y, every map $f : A \to X$ extends over Y but X is not an ANR.

Usually, the Eilenberg–MacLane complexes are constructed as CW-complexes, but we constructed them as simplicial complexes in Sect. 7.8.

Theorem 7.9.1 was originally proved by Cohen [4] but the proof presented here is based on an idea of S. Ferry, which appeared in Appendix A in [25]. The proof of Theore 7.9.12 (7.9.10) appeared in [25], and is based on Kozlowski's technique in [11]. Theorem 7.9.5 was established by Alexandroff [2]. It was shown by Edwards that Alexandroff's Problem is equivalent to the CE Problem. Theorems 7.10.1 and 7.10.2 are due to Edwards (see [25]). The existence of dimension-raising cell-like maps was first shown by A. Dranishnikov [5, 6]. Lemma 7.10.10 and Theorem 7.10.13 were established by Dydak and Walsh [8]. In [12], Kozlowski and Walsh proved that the image of a cell-like map of a compact 3-dimensional manifold is always finite-dimensional. It remains unsolved whether there exists a cell-like map of $\mathbf{I}^4$ onto an infinite-dimensional compactum. There are good surveys on Cohomological Dimension Theory [7]. Walsh's Theorem 7.10.6 on approximations by open maps was proved in [24].

The example of Theorem 7.12.1 was constructed in [3].

References

1. J.F. Adams, On the groups $J(X)$, IV. Topology **5**, 21–71 (1966); errata. Topology **7**, 331 (1968)
2. P. Alexandroff, On the dimension of normal spaces. Proc. Roy. Soc. Lond. Ser. A **189**, 11–39 (1947)
3. R. Cauty, Un espace métrique linéaire qui n'est pas un rétracte absolu. Fund. Math. **146**, 85–99 (1994)
4. H. Cohen, A cohomological definition of dimension for locally compact Hausdorff spaces. Duke Math. J. **21**, 209–224 (1954); A correction. Duke Math. J. **25**, 601 (1958)
5. A.N. Dranishnikov, Homological dimension theory (Russian). Uspekhi Mat. Nauk **43**(4), 11–55, 255 (1988); English transl. in Russ. Math. Surv. **43**, 11–63 (1988)
6. A.N. Dranishnikov, On a problem of P.S. Aleksandrov (Russian). Mat. Sb. (N.S.) **135**(177)(4), 551–557, 560 (1988); English transl. in Math. USSR-Sb. **63**, 539–545 (1989)
7. J. Dydak, Cohomological dimension theory, in *Handbook of Geometric Topology*, ed. by R.J. Daverman, R.B. Sher (Elsevier, Amsterdam, 2002), pp. 423–470
8. J. Dydak, J.J. Walsh, Infinite-dimensional compacta having cohomological dimension two: an application of the Sullivan conjecture. Topology **32**, 93–104 (1993)
9. J.E. Keesling, A non-movable trivial-shape decomposition of the Hilbert cube. Pull. Acad. Polon. Sci. Sér. Sci. Math. Astronom. Phys. **23**, 997–998 (1975)

10. G. Kozlowski, Factorization of certain maps up to homotopy. Proc. Am. Math. Soc. **21**, 88–92 (1969)
11. G. Kozlowski, Images of ANR's (unpublished manuscript)
12. G. Kozlowski, J. Walsh, Cell-like mappings on 3-manifolds. Topology **22**, 147–153 (1983)
13. R.C. Lacher, Cell-like mappings and their generalizations. Bull. Am. Math. Soc. **83**, 495–552 (1977)
14. E. Michael, Continuous selections, II. Ann. Math. **64**, 562–580 (1956)
15. E. Michael, Selected selection theorems. Am. Math. Mon. **63**, 233–238 (1956)
16. J. van Mill, A counterexample in ANR theory. Topology Appl. **12**, 315–320 (1981)
17. J. van Mill, Another counterexample in ANR theory. Proc. Am. Math. Soc. **97**, 136–138 (1986)
18. K. Sakai, Correcting Taylor's cell-like map. Glasnik Mat. Ser. III **46**, 483–487 (2011)
19. E.V. Shchepin, Topology of limit spaces of uncountable inverse spectra (Russian). Uspekhi Mat. Nauk **31**(5), 191–226 (1976); English transl. in Russ. Math. Surv. **31**(5), 155–191 (1976)
20. E.V. Shchepin, Soft maps of manifolds (Russian). Uspekhi Mat. Nauk **39**(5), 209–224 (1984); English transl. in Russ. Math. Surv. **39**(5), 251–270 (1984)
21. J.L. Taylor, A counterexample in shape theory. Bull. Am. Math. Soc. **81**, 629–632 (1975)
22. H. Toda, On unstable homotopy of spheres and classical groups. Proc. Nat. Acad. Sci. **46**, 1102–1105 (1960)
23. H. Toruńczyk, On CE-images of the Hilbert cube and characterization of Q-manifolds. Fund. Math. **106**, 31–40 (1980)
24. J.J. Walsh, Isotoping mappings to open mappings. Trans. Am. Math. Soc. **250**, 121–145 (1979)
25. J.J. Walsh, Dimension, cohomological dimension, and cell-like mappings, in *Shape Theory and Geometric Topology, Proc.* (Dubrovnik, 1981), ed. by S. Mardešić, J. Segal. Lecture Notes in Math. **870** (Springer, Berlin, 1981), pp. 105–118

Index

K. Sakai, *Geometric Aspects of General Topology*, Springer Monographs in Mathematics, DOI 10.1007/978-4-431-54397-8, © Springer Japan 2013